A Comparison Process for Mouse Pairs

This book proves some important new theorems in the theory of canonical inner models for large cardinal hypotheses, a topic of central importance in modern set theory. In particular, the author "completes" the theory of *Fine Structure and Iteration Trees* (FSIT) by proving a comparison theorem for mouse pairs parallel to the FSIT comparison theorem for pure extender mice, and then using the underlying comparison process to develop a fine structure theory for strategy mice.

Great effort has been taken to make the book accessible to non-experts so that it may also serve as an introduction to the higher reaches of inner model theory. It contains a good deal of background material, some of it unpublished folklore, and includes many references to the literature to guide further reading. An introductory essay serves to place the new results in their broader context.

This is a landmark work in inner model theory that should be in every set theorist's library.

JOHN R. STEEL is Professor of Mathematics at the University of California, Berkeley. He is a recipient of the Carol Karp Prize of the Association for Symbolic Logic, the Hausdorff Medal of the European Set Theory Society, and the Humboldt Prize.

LECTURE NOTES IN LOGIC

A Publication of The Association for Symbolic Logic

This series serves researchers, teachers, and students in the field of symbolic logic, broadly interpreted. The aim of the series is to bring publications to the logic community with the least possible delay and to provide rapid dissemination of the latest research. Scientific quality is the overriding criterion by which submissions are evaluated.

More information, including a list of the books in the series, can be found at http://aslonline.org/books/lecture-notes-in-logic/

LECTURE NOTES IN LOGIC 51

A Comparison Process for Mouse Pairs

JOHN R. STEEL

University of California, Berkeley

ASSOCIATION FOR SYMBOLIC LOGIC

CAMBRIDGE
UNIVERSITY PRESS

CAMBRIDGE
UNIVERSITY PRESS

Shaftesbury Road, Cambridge CB2 8EA, United Kingdom

One Liberty Plaza, 20th Floor, New York, NY 10006, USA

477 Williamstown Road, Port Melbourne, VIC 3207, Australia

314–321, 3rd Floor, Plot 3, Splendor Forum, Jasola District Centre,
New Delhi – 110025, India

103 Penang Road, #05–06/07, Visioncrest Commercial, Singapore 238467

Cambridge University Press is part of Cambridge University Press & Assessment,
a department of the University of Cambridge.

We share the University's mission to contribute to society through the pursuit of
education, learning and research at the highest international levels of excellence.

www.cambridge.org
Information on this title: www.cambridge.org/9781108840682

DOI: 10.1017/9781108886840

Association for Symbolic Logic
Richard A. Shore, Publisher
Department of Mathematics, Cornell University, Ithaca, NY 14853
http://aslonline.org

First published 2023

A catalogue record for this publication is available from the British Library.

ISBN 978-1-108-84068-2 Hardback

For Colleen, David, Bobby,
and my parents

CONTENTS

PREFACE

This book began life as a long research article titled *Normalizing iteration trees and comparing iteration strategies*. I found the main ideas behind the comparison process that motivates it in Spring 2015, and circulated a handwritten manuscript shortly afterward. I circulated a preliminary form of the present book in April 2016, and have revised and expanded it many times since then, as various significant gaps and errors showed up. The last major revisions took place in 2020-2021.

Beyond making the book correct, one of my goals has been to make it accessible. I have been encouraged here by the fact that the new definitions and results are actually quite elementary. They rest on the theory of *Fine structure and iteration trees* (FSIT), and can be seen as completing that theory in a certain way. The comparison theorem for pure extender mice that is at the heart of FSIT is deficient, in that how two mice compare depends on which iteration strategies are chosen to compare them. Here we remedy that defect, by developing a method for comparing the strategies. The result is a comparison theorem for *mouse pairs* parallel to the FSIT comparison theorem for pure extender mice. We then use the comparison process underlying that theorem to develop a fine structure theory for *strategy mice* parallel to the fine structure theory for pure extender mice of FSIT.

There are points at which descriptive set theory under determinacy hypotheses becomes relevant. At these points, it would help to have read the later sections of [70]. However, I have included enough material that the reader familiar with FSIT but shaky on determinacy should be able to follow the exposition. Our work here is motivated by the problem of analyzing ordinal definability in models of Axiom of Determinacy, but the prerequisite for following most of it is just inner model theory at the level of FSIT.

Inner model theory is sufficiently mature that many of its basic definitions and much of its notation have been streamlined and standardized over time. I have tried to contribute to that process, by adopting common definitions and notations where possible, and streamlining where I saw a chance to do so. Chapters 2 and 3 review and summarize the standard material on which the book rests, while Chapter 4 makes some revisions to it that are needed later.

The papers [77], [80], [78], [81], [64], [73], and [17], written in 2016-2018, have extended the work described here in various directions. The most important of the remaining open problems is whether, assuming determinacy, there actually are mouse pairs at every appropriate level of logical complexity. This is one instance of the fundamental iterability problem of inner model theory, still open after all these years. I suspect that it is the most accessible instance of that problem.

Acknowledgements. Special thanks to Xianghui Shi, who converted my handwritten manuscript to a LaTeX document in summer 2015, and thereby helped me over the "getting started" barrier. I worked on the first draft in Fall 2015 as a Simons Foundation fellow at the Isaac Newton Institute for Mathematical Sciences in the programme 'Mathematical, Foundational and Computational Aspects of the Higher Infinite'. My thanks to the institute and its staff for their support. Thanks also to the participants in the annual summer workshops in inner model theory, who helped me clarify the material here (and inspired the drawing below). Special thanks to Benjamin Siskind for his help in identifying and correcting some nontrivial errors. Finally, thanks to Nam Trang for his input at various stages, including help with tikzpicture, to Thomas Piecha for his help with formatting, to the referees for their input, and to Paul Larson for his work as editor.

My larger intellectual debts are too numerous to list carefully. The generation before mine opened up the beautiful world of large cardinals, determinacy, and inner models, and many people have contributed to its development since then. Ronald Jensen, Tony Martin, and Bill Mitchell have had a special influence on my own work. In the immediate environment of this book, the work of Grigor Sargsyan and Hugh Woodin on the analysis of HOD in models of the Axiom of Determinacy plays an important role. My thanks to all these people.

A mouse with a strategy
Gregor Schindler, 2019

Chapter 1

INTRODUCTION

In this book we shall develop a general comparison process for iteration strategies, and show how the process can be used to analyze ordinal definability in models of the Axiom of Determinacy. In this introduction, we look at the context and motivation for the technical results to come.

We begin with a broad overview of *inner model theory*, the subject to which this book belongs. Eventually we reach an outline of the ideas and results that are new here. The journey is organized so that the technical background needed to follow along increases as we proceed.

1.1. Large cardinals and the consistency strength hierarchy

Strong axioms of infinity, or as they are more often called, large cardinal hypotheses, play a central role in set theory. There are at least two reasons.

First, large cardinal hypotheses can be used to decide in a natural way many questions which cannot be decided on the basis of ZFC (the commonly accepted system of axioms for set theory, and hence all of mathematics). Many such questions come from *descriptive set theory*, the theory of simply definable sets of real numbers. For example, the hypothesis that there are infinitely many Woodin cardinals yields a systematic and detailed theory of the *projective* sets of reals, those that are definable in the language of second order arithmetic from real parameters. ZFC by itself yields such a theory at only the simplest levels of second order definability.

Second, large cardinal hypotheses provide a way of organizing and surveying all possible natural extensions of ZFC. This is due to the following remarkable phenomenon: for any natural extension T of ZFC which set theorists have studied, there seems to be an extension S of ZFC axiomatized by large cardinal hypotheses such that the consistency of T is provably (in ZFC) equivalent to that of S. The consistency strengths of the large cardinal hypotheses are linearly ordered, and usually easy to compare. Thus all natural extensions of ZFC seem to fall into a hierarchy linearly ordered by consistency strength, and calibrated by the large cardinal hypotheses.[1]

[1] Let $\mathrm{con}(T)$ be some natural formalization of the assertion that T is consistent. The consistency strength order is given by: $S \leq_{\mathrm{con}} T$ iff ZFC proves $\mathrm{con}(T) \rightarrow \mathrm{con}(S)$.

These two aspects of large cardinal hypotheses are connected, in that the consistency strength order on natural theories corresponds to the inclusion order on the set of their "sufficiently absolute" consequences. For example, if S and T are natural theories extending ZFC, and S has consistency strength less than or equal to that of T, then the arithmetic consequences of S are included in those of T. If in addition, S and T have consistency strength at least that of "there are infinitely many Woodin cardinals", then the consequences of S in the language of second order arithmetic are included in those of T. This pattern persists at still higher consistency strengths, with still more logically complicated consequences about reals and sets of reals being brought into a uniform order. This beautiful and suggestive phenomenon has a practical dimension as well: one way to develop the absolute consequences of a strong theory T is to compute a consistency strength lower bound S for T in terms of large cardinal hypotheses, and then work in the theory S. For one of many examples, the Proper Forcing Axiom (PFA) yields a canonical inner model with infinitely many Woodin cardinals that is correct for statements in the language of second order arithmetic, and therefore PFA implies all consequences of the existence of infinitely many Woodin cardinals that can be stated in the language of second order arithmetic.

One can think of the consistency strength of a theory as the degree to which it is committed to the existence of the higher infinite. Large cardinal hypotheses make their commitments explicitly: they simply say outright that the infinities in question exist. It is therefore usually easy to compare their consistency strengths. Other natural theories often have their commitments to the existence of the infinite well hidden. Nevertheless, set theorists have developed methods whereby these commitments can be brought to the surface, and compared. These methods have revealed the remarkable phenomenon described in the last paragraph, that natural theories appear to be wellordered by the degrees to which they are committed to the infinite, and that this degree of commitment corresponds exactly to the power of the theory to decide questions about concrete objects, like natural numbers, real numbers, or sets of real numbers.

We should emphasize that the paragraphs above describe a general pattern of existing theorems. There are many examples of natural theories whose consistency strengths have not yet been computed, and perhaps they, or some natural theory yet to be found, will provide counterexamples to the pattern described above. The pervasiveness of the pattern where we know how to compare consistency strengths is evidence that this will not happen.[2] The two methods whereby set theorists compare consistency strengths, forcing and inner model theory, seem to lead inevitably to the pattern. In particular, the wellorder of natural consistency strengths seems to correspond to the inclusion order on canonical minimal inner

[2]The pattern extends to weak subtheories of ZFC as well. This book is concerned only with theories having very strong commitments to infinity, and so we shall ignore subtheories of ZFC, but the linearity of the consistency strengths below that of ZFC is evidence of linearity higher up.

models for large cardinal hypotheses. Forcing and inner model theory seem sufficiently general to compare all natural consistency strengths, but at the moment, this is just informed speculation. So one reasonable approach to understanding the general pattern of consistency strengths is to develop our comparison methods further. In particular, inner model theory is in great need of further development, as there are quite important consistency strengths that it does not yet reach.

1.2. Inner model theory

The inner model program attempts to associate to each large cardinal hypothesis H a canonical minimal universe of sets M_H (an *inner model*) in which H is true. The stronger H is, the larger M_H will be; that is, $G \leq_{\text{con}} H$ if and only if $M_G \subseteq M_H$. Some of our deepest understanding of large cardinal hypotheses comes from the inner model program.

The inner models we have so far constructed have an internal structure which admits a systematic, detailed analysis, a *fine structure theory* of the sort pioneered by Ronald Jensen around 1970 ([18]). Thus being able to construct M_H gives us a very good idea as to what a universe satisfying H might look like. Inner model theory thereby provides evidence of the consistency of the large cardinal hypotheses to which it applies. (The author believes that this will some day include all the large cardinal hypotheses currently studied.) Since forcing seems to reduce any consistency question to the consistency question for some large cardinal hypothesis, it is important to have evidence that the large cardinal hypotheses themselves are consistent! No evidence is more convincing than an inner model theory for the hypothesis in question.

The smallest of the canonical inner models is the universe L of constructible sets, isolated by Kurt Gödel ([14]) in his 1937 proof that CH is consistent with ZFC. It was not until the mid 1960s that J. Silver and K. Kunen ([62], [27]) developed the theory of a canonical inner model going properly beyond L, by constructing M_H for $H =$ "there is a measurable cardinal".[3] Since then, progressively larger M_H for progressively stronger H have been constructed and studied in detail. (See for example [5], [31], and [32].) At the moment, we have a good theory of canonical inner models satisfying "there is a Woodin cardinal", and even slightly stronger hypotheses. (See [30], [35], and [66], for example.) One of the most important open problems in set theory is to extend this theory significantly further, with perhaps the most well-known target being models satisfying "there is a supercompact cardinal".

[3] ZFC is of course too weak, consistency-wise, to prove that there is such a model. Silver and Kunen worked in the theory ZFC + "there is a measurable cardinal". In the mid 1970s, Dodd and Jensen developed general methods for constructing the canonical inner model with a measurable under a wide assortment of hypotheses. See [5], [6], and [7].

Inner model theory is a crucial tool in calibrating consistency strengths: in order to prove that $H \leq_{\text{con}} T$, where H is a large cardinal hypothesis, one generally constructs a canonical inner model of H inside an arbitrary model of T. Because we do not have a full inner model theory very far past Woodin cardinals, we lack the means to prove many well-known conjectures of the form $H \leq_{\text{con}} T$, where H is significantly stronger than "there is a Woodin cardinal". Broadly speaking, there are great defects in our understanding of the consistency strength hierarchy beyond Woodin cardinals.

Inner model theory is also a crucial tool in developing the consequences for real numbers of large cardinal hypotheses. Indeed, the basics of inner model theory for Woodin cardinals were discovered in 1985-86 by D. A. Martin and the author, at roughly the same time they discovered their proof of Projective Determinacy, or PD. (Martin, Moschovakis, and others had shown in the 1960s and 70s that PD decides in a natural way all the classical questions about projective sets left undecided by ZFC alone.) This simultaneous discovery was not an accident, as the fundamental new tool in both contexts was the same: *iteration trees*, and the *iteration strategies* which produce them. Since then, progress in inner model theory has given us a deeper understanding of pure descriptive set theory, and the means to solve some old problems in that field.

The fundamental open problem of inner model theory is to extend the theory to models satisfying stronger large cardinal hypotheses. "There is a supercompact cardinal" is an old and still quite challenging target. One very well known test question here is whether (ZFC + "there is a supercompact cardinal") $\leq_{\text{con}}$ ZFC + PFA. The answer is almost certainly yes, and the proof almost certainly involves an inner model theory that is firing on all cylinders.[4] That kind of inner model theory we have now only at the level of many Woodin cardinals, but significant parts of the theory do exist already at much higher levels.[5]

1.3. Mice and iteration strategies

The canonical inner models we seek are often called *mice*. There are two principal varieties, the pure extender mice and the strategy mice.[6]

[4]A parallel, and still older, question is whether (ZFC + "there is a supercompact cardinal") $\leq_{\text{con}}$ ZFC + "there is a strongly compact cardinal".

[5]J. Baumgartner showed in the early 1980s that ZFC + PFA $\leq_{\text{con}}$ ZFC + "there is a supercompact cardinal". Supercompacts are far beyond Woodin cardinals, in the sense that there are many interesting consistency strengths strictly between the two, and in the sense that constructing canonical inner models for supercompacts presents significant new difficulties. Many set theoretic principles have been shown consistent relative to the existence of (sometimes many) supercompact cardinals, so inner-model-theoretic evidence of their consistency would be valuable.

[6]Strategy mice are sometimes called *hod mice*, because of their role in analyzing the hereditarily ordinal definable sets in models of the Axiom of Determinacy.

A pure extender premouse is a model of the form $L_\alpha[\vec{E}]$ where $\vec{E}$ is a coherent sequence of extenders. Here an extender is a system of ultrafilters coding an elementary embedding, and coherence means roughly that the extenders appear in order of strength, without leaving gaps. These notions were introduced by Mitchell in the 1970s[7], and they have been a foundation for work in inner model theory since then.

In this book, we shall assume that our premice have no long extenders on their coherent sequences.[8] Such premice can model superstrong, and even subcompact, cardinals. They cannot model κ^+-supercompactness. Long extenders lead to an additional set of difficulties.

An *iteration strategy* is a winning strategy for player II in the iteration game. For any premouse M, the iteration game on M is a two player game of length $\omega_1 + 1$.[9] In this game, the players construct a tree of models such that each successive node on the tree is obtained by an ultrapower of a model that already exists in the tree. I is the player that describes how to construct this ultrapower. He chooses an extender E from the sequence of the last model N constructed so far, then chooses another model P in the tree and takes the ultrapower of P by E. If the ultrapower is ill-founded then player I wins; otherwise the resulting ultrapower is the next node on the tree. Player II moves at limit stages λ by choosing a branch of the tree that has been visited cofinally often below λ, and is such that the direct limit of the embeddings along the branch is well-founded. If he fails to do so, he loses. If II manages to stay in the category of wellfounded models through all $\omega_1 + 1$ moves, then he wins. A winning strategy for II in this game is called an *iteration strategy* for M, and M is said to be *iterable* just in case there is an iteration strategy for it. Iterable pure extender premice are called *pure extender mice*.

Pure extender mice are canonical objects; for example, any real number belonging to such a mouse is ordinal definable. Let us say that a premouse M is *pointwise definable* if every element of M is definable over M. For any axiomatizable theory T, the minimal mouse satisfying T is pointwise definable. The canonicity of pure extender mice is due to their iterability, which, via the fundamental *Comparison Lemma*, implies that the pointwise definable pure extender mice are wellordered by inclusion. This is the *mouse order* on pointwise definable pure extender mice. The consistency strength of T is determined by the minimal mouse M having a generic extension satisfying T, and thus the consistency strength order on natural T is mirrored in the mouse order. However, in the case of the mouse order, we have *proved* that we have a wellorder; what we cannot yet do is tie natural T at high consistency strengths to it. As we climb the mouse order, the mice become

---

[7]See [31] and [32].

[8]An extender is short if all its component ultrafilters concentrate on the critical point. Otherwise, it is long.

[9]Iteration games of other lengths are also important, but this length is crucial, so we shall focus on it.

correct (reflect what is true in the full universe of sets) at higher and higher levels of logical complexity.

Iteration strategies for pointwise definable pure extender mice are also canonical objects; for example, a pointwise definable mouse has exactly one iteration strategy.[10] The existence of iteration strategies is at the heart of the fundamental problem of inner model theory, and for a pointwise definable M, to prove the existence of an iteration strategy is to define it. In practice, it seems necessary to give a definition in the simplest possible logical form. As we go higher in the mouse order, the logical complexity of iteration strategies must increase, in a way that keeps pace with the correctness of the mice they identify.

Our most powerful, all-purpose method for constructing iteration strategies is the *core model induction method.* Because iteration strategies must act on trees of length ω_1, they are not coded by sets of reals. Nevertheless, the fragment of the iteration strategy for a countable mouse that acts on countable iteration trees *is* coded by a set of reals. If this set happens to be absolutely definable (that is, Universally Baire) then the strategy can be extended to act on uncountable iteration trees in a unique way. There is no other way known to construct iteration strategies acting on uncountable trees. Thus, having an absolutely definable iteration strategy for countable trees is tantamount to having a full iteration strategy. The key idea in the core model induction is to use the concepts of descriptive set theory, under determinacy hypotheses, to identify a next relevant level of correctness and definability for sets of reals, a target level at which the next iteration strategy should be definable.

Absolute definability leads to determinacy. Thus at reasonably closed limit steps in a core model induction, one has a model M of $\mathsf{AD} + V = L(P(\mathbb{R}))$ that contains the restrictions to countable trees of the iteration strategies already constructed. Understanding the structure of HOD^M is important for going further.

1.4. HOD in models of determinacy

HOD is the class of all hereditarily ordinal definable sets. It is a model of ZFC[11], but beyond that, ZFC does not decide its basic theory, and the same is true of ZFC augmented by any of the known large cardinal hypotheses. The problem is that the definitions one has allowed are not sufficiently absolute. In contrast, the theory of HOD in determinacy models is well-determined, not subject to the vagaries of forcing.[12]

[10]This follows from Theorem 4.11 of [70], and the fact that any iteration strategy for a pointwise definable M has the Weak Dodd-Jensen property with respect to all enumerations of M.

[11]See [36].

[12]We mean here determinacy models of the form $M = L(\Gamma, \mathbb{R})$, where Γ is a proper initial segment of the universally Baire sets. If there are arbitrarily large Woodin cardinals, then for any sentence φ,

The study of HOD in models of AD has a long history. The reader should see [72] for a survey of this history. HOD was studied by purely descriptive set theoretic methods in the late 70s and 80s, and partial results on basic questions such as whether HOD $\models$ GCH were obtained then. It was known then that inner model theory, if only one could develop it in sufficient generality, would be relevant to characterizing the reals in HOD. It was known that HOD^M is close to M in various ways; for example, if $M \models \mathsf{AD}^+ + V = L(P(\mathbb{R}))$[13], then M can be realized as a symmetric forcing extension of HOD^M, so that the first order theory of M is part of the first order theory of its HOD.[14]

Just how relevant inner model theory is to the study of HOD in models of AD became clear in 1994, when the author showed that if there are ω Woodin cardinals with a measurable above them all, then $\mathrm{HOD}^{L(\mathbb{R})}$ up to $\theta^{L(\mathbb{R})}$ is a pure extender mouse.[15](See [65].) Shortly afterward, this result was improved by Hugh Woodin, who reduced its hypothesis to $\mathsf{AD}^{L(\mathbb{R})}$, and identified the full $\mathrm{HOD}^{L(\mathbb{R})}$ as a model of the form $L[M,\Sigma]$, where M is a pure extender premouse, and Σ is a partial iteration strategy for M. $\mathrm{HOD}^{L(\mathbb{R})}$ is thus a new type of mouse, sometimes called a *strategy mouse*, sometimes called a *hod mouse*. See [82] for an account of this work.

Since the mid-1990s, there has been a great deal of work devoted to extending these results to models of determinacy beyond $L(\mathbb{R})$. Woodin analyzed HOD in models of AD^+ below the minimal model of $\mathsf{AD}_{\mathbb{R}}$ fine structurally, and Sargsyan extended the analysis further, first to determinacy models below $\mathsf{AD}_{\mathbb{R}} +$ "θ is regular" (see [42] and [43]), and more recently, to models of still stronger forms of determinacy.[16] Part of the motivation for this work is that it seems to be essential in the core model induction: in general, the next iteration strategy seems to be a strategy for a hod mouse, not for a pure extender mouse. This idea comes from work of Woodin and Ketchersid around 2000. (See [25] and [52].)

1.5. Least branch hod pairs

The strategy mice used in the work just described have the form $M = L[\vec{E},\Sigma]$, where $\vec{E}$ is a coherent sequence of extenders, and Σ is an iteration strategy for M. The strategy information is fed into the model M slowly, in a way that is dictated

---

whether φ is true in all such HOD^M is absolute under set forcing. This follows easily from Woodin's theorem on the generic absoluteness of $(\Sigma^2_1)^{\mathrm{uB}}$ statements. See [69, Theorem 5.1].

[13] AD^+ is a technical strengthening of AD. It is not known whether AD $\Longrightarrow \mathsf{AD}^+$, but in every model of AD constructed so far, AD^+ also holds. In particular, the models of AD that are relevant in the core model induction satisfy AD^+.

[14] This is a theorem of Woodin from the early 1980s. Cf. [72].

[15] In a determinacy context, θ denotes the least ordinal that is not the surjective image of the reals.

[16] See [44]. Part of this work was done in collaboration with the author; see [75], [79], and [76]. The determinacy principles dealt with here are all weaker than a Woodin limit of Woodin cardinals.

in part by the determinacy model whose HOD is being analyzed. One says that the hierarchy of M is *rigidly layered*, or *extender biased*. The object (M,Σ) is called a rigidly layered (extender biased) *hod pair*.

Perhaps the main motivation for the extender biased hierarchy is that it makes it possible to prove a comparison theorem. There is no inner model theory without such a theorem. Comparing strategy mice necessarily involves comparing iteration strategies, and comparing iteration strategies is significantly more difficult than comparing extender sequences. Rigid layering lets one avoid the difficulties inherent in the general strategy comparison problem, while proving comparison for a class of strategy mice adequate to analyze HOD in the minimal model of $\mathsf{AD}_{\mathbb{R}}+$"$\theta$ is regular", and somewhat beyond. The key is that in this region, HOD does not have cardinals that are strong past a Woodin cardinal.

Unfortunately, rigid layering does not seem to help in comparing strategy mice that have cardinals that are strong past a Woodin. Moreover, it has serious costs. The definition of "hod premouse" becomes very complicated, and indeed it is not clear how to extend the definition of rigidly layered hod pairs much past that given in [44]. The definition of "rigidly layered hod premouse" is not uniform, in that the extent of extender bias depends on the determinacy model whose HOD is being analyzed. Fine structure, and in particular condensation, become more awkward. For example, it is not true in general that the pointwise definable hull of a level of M is a level of M. (The problem is that the hull will not generally be sufficiently extender biased.)

The more naive notion of hod premouse would abandon extender bias, and simply add the least missing piece of strategy information at essentially every stage. This was originally suggested by Woodin.[17] The focus of this book is a general comparison theorem for iteration strategies that makes it possible to use this approach, at least in the realm of short extenders. The resulting premice are called *least branch premice* (lpm's), and the pairs (M,Σ) are called *least branch hod pairs* (lbr hod pairs). Combining results of this book and [73], one has

THEOREM 1.5.1 ([73]). *Assume* $\mathsf{AD}^+ +$ *"there is an* (ω_1,ω_1) *iteration strategy for a pure extender premouse with a long extender on its sequence"; then*

(1) for any $\Gamma\subseteq P(\mathbb{R})$ *such that* $L(\Gamma,\mathbb{R})\models\mathsf{AD}_{\mathbb{R}}+$ *"there is no* (ω_1,ω_1) *iteration strategy for a pure extender premouse with a long extender on its sequence",* $\mathrm{HOD}^{L(\Gamma,\mathbb{R})}$ *is a least branch premouse, and*

(2) there is a $\Gamma\subseteq P(\mathbb{R})$ *such that* $L(\Gamma,\mathbb{R})\models\mathsf{AD}_{\mathbb{R}}+$ *"there is no* (ω_1,ω_1) *iteration strategy for a pure extender premouse with a long extender on its sequence", and* $\mathrm{HOD}^{L(\Gamma,\mathbb{R})}\models$ *"there is a subcompact cardinal".*

---

[17] There are some fine-structural problems with the precise method for inserting strategy information originally suggested by Woodin. The method for strategy insertion that is correct in detail is due to Schlutzenberg and Trang. Cf. [61].

Of course, one would like to remove the mouse existence hypothesis of Theorem 1.5.1, and prove its conclusion under AD^+ alone. Finding a way to do this is one manifestation of the long standing iterability problem we have discussed above. Although we do not yet know how to do this, the theorem does make it highly likely that in models of $\mathsf{AD}_{\mathbb{R}}$ that have not reached an iteration strategy for a pure extender premouse with a long extender, HOD is a least branch premouse. It also makes it very likely that there are such HOD's with subcompact cardinals. Subcompactness is one of the strongest large cardinal properties that can be represented with short extenders.[18]

Although we shall not prove Theorem 1.5.1 here, we shall prove an approximation to it that makes the same points. That approximation is Theorem 11.3.13 below.

Least branch premice have a fine structure much closer to that of pure extender models than that of rigidly layered hod premice. In this book we develop the basics, including the solidity and universality of standard parameters, and a form of condensation. In [81], the author and N. Trang have proved a sharper condensation theorem, whose pure extender version was used heavily in the Schimmerling-Zeman work ([49]) on $\square$ in pure extender mice. It seems likely that the rest of the Schimmerling-Zeman work extends as well.

Thus least branch hod pairs give us a good theory of HOD in the short extender realm, provided there are enough such pairs.[19] Below, we formulate a conjecture that we call *Hod Pair Capturing*, or HPC, that makes precise the statement that there are enough least branch hod pairs. HPC is the main open problem in the theory to which this book contributes.

1.6. Comparison and the mouse pair order

Let us first say more about the nature of least branch hod pairs (M,Σ). There are four requirements on Σ in the definition: strong hull condensation, quasi-normalizing well, internal lift consistency, and pushforward consistency. We shall describe these requirements informally, omitting some of the fine points, and give the full definitions later.

Recall that an iteration tree on a premouse M is *normal* iff the extenders $E_\alpha^{\mathcal{W}}$ used in $\mathcal{W}$ have lengths increasing with α, and each $E_\alpha^{\mathcal{W}}$ is applied to the longest possible initial segment of the earliest possible model in $\mathcal{W}$. For technical reasons we need to consider a slight weakening of the length-increasing requirement; we call the resulting trees *quasi-normal*. Our iteration strategies will act on finite *stacks* of quasi-normal trees, that is, sequences $s = \langle \mathcal{T}_0, \ldots, \mathcal{T}_n \rangle$ such that for all

[18] Until now, there was no very strong evidence that the HOD of a determinacy model could satisfy that there are cardinals that are strong past a Woodin cardinal.

[19] At least in the case that the background determinacy model satisfies $\mathsf{AD}_{\mathbb{R}} + V = L(P(\mathbb{R}))$. Some form of extender bias may be appropriate in other cases.

$k \leq n-1$, $\mathcal{T}_{k+1}$ is a quasi-normal tree on some initial segment of the last model in $\mathcal{T}_k$. We write $M_\infty(s)$ for the last model of $\mathcal{T}_n$, if there is one.

DEFINITION 1.6.1. Let Σ be an iteration strategy for a premouse P.

(1) (Tail strategy) If s is a stack by Σ and $Q \trianglelefteq M_\infty(s)$, then $\Sigma_{s,Q}$ is the strategy for Q given by: $\Sigma_{s,Q}(t) = \Sigma(s^\frown\langle Q,t\rangle)$.[20]
(2) (Pullback strategy) If $\pi\colon N \to P$ is elementary, then Σ^π is the strategy for N given by: $\Sigma^\pi(s) = \Sigma(\pi s)$, where πs is the lift of s by π to a stack on P.

In (2), elementarity must be understood fine structurally; our convention is that every premouse P has a degree of soundness attached to it, and elementarity means elementarity at that quantifier level.

Perhaps the most important regularity property of iteration strategies is *strong hull condensation.* To define it we need the notion of a *tree embedding* $\Phi\colon \mathcal{T} \to \mathcal{U}$, where $\mathcal{T}$ and $\mathcal{U}$ are normal trees on the same M. The idea of course is that Φ should preserve a certain amount of the iteration tree structure, but some care is needed in spelling out exactly how much. Φ is determined by a map $u\colon \mathrm{lh}(\mathcal{T}) \to \mathrm{lh}(\mathcal{U})$ and maps $\pi_\alpha\colon \mathcal{M}_\alpha^\mathcal{T} \to \mathcal{M}_{u(\alpha)}^\mathcal{U}$ having various properties. See Section 6.4.

DEFINITION 1.6.2. Let Σ be an iteration strategy for a premouse M; then Σ has *strong hull condensation* iff whenever s is a stack of normal trees by Σ and $N \trianglelefteq \mathcal{M}_\infty(s)$, and $\mathcal{U}$ is a normal tree on N by $\Sigma_{s,N}$, and $\Phi\colon \mathcal{T} \to \mathcal{U}$ is a tree embedding, with associated maps $\pi_\alpha\colon \mathcal{M}_\alpha^\mathcal{T} \to \mathcal{M}_{u(\alpha)}^\mathcal{U}$, then

(a) $\mathcal{T}$ is by $\Sigma_{s,N}$, and
(b) for all $\alpha < \mathrm{lh}(\mathcal{T})$, $\Sigma_{s^\frown\langle N,\mathcal{T}\restriction\alpha+1\rangle} = (\Sigma_{s^\frown\langle N,\mathcal{U}\restriction u(\alpha)+1\rangle})^{\pi_\alpha}$.

Strong hull condensation is a stronger version of the hull condensation property isolated by Sargsyan in [42].

The second important property is quasi-normalizing well. Given a stack $\langle\mathcal{T},\mathcal{U}\rangle$ on M with last model N such that $\mathcal{T}$ and $\mathcal{U}$ are normal, shuffling the extenders of $\mathcal{U}$ into $\mathcal{T}$ in a minimal way produces a normal tree $\mathcal{W} = W(\mathcal{T},\mathcal{U})$. If $\mathcal{U}$ has a last model R, we get a nearly elementary map $\pi\colon N \to R$. We call $W(\mathcal{T},\mathcal{U})$ the *embedding normalization* of $\langle\mathcal{T},\mathcal{U}\rangle$. The idea is simple, but there are many technical details.[21] It proves useful to consider a slightly less minimal shuffling $V(\mathcal{T},\mathcal{U})$ that we call the *quasi-normalization* of $\langle\mathcal{T},\mathcal{U}\rangle$. Even if $\mathcal{T}$ and $\mathcal{U}$ are normal, $V(\mathcal{T},\mathcal{U})$ may not be length-increasing, but it is nearly so. The reader should see Chapter 6 for full definitions.

DEFINITION 1.6.3. Let Σ be an iteration strategy for a premouse M. We say that Σ *quasi-normalizes well* iff whenever s is an stack on M by Σ, and $\langle\mathcal{T},\mathcal{U}\rangle$ is a stack of length 2 by Σ_s such that $\mathcal{T}$ and $\mathcal{U}$ are normal trees having last models, then

[20] For premice Q and R, $Q \trianglelefteq R$ iff the hierarchy of Q is an initial segment of that of R.

[21] Much of the general theory of normalization was developed independently by Schlutzenberg. See [57]. See also [21] and [63].

(a) $V(\mathcal{T},\mathcal{U})$ is by Σ_s, and
(b) letting $\mathcal{V} = V(\mathcal{T},\mathcal{U})$ and $\pi\colon \mathcal{M}^{\mathcal{U}}_\infty \to \mathcal{M}^{\mathcal{V}}_\infty$ be the map generated by quasi-normalization, we have that $\Sigma_{s^\frown\langle\mathcal{T},\mathcal{U}\rangle} = (\Sigma_{s^\frown\mathcal{V}})^\pi$.

The final basic regularity property of iteration strategies for pure extender premice is *internal lift consistency*. Suppose that s is a stack by Σ and $P \trianglelefteq Q \trianglelefteq M_\infty(s)$. Stacks t on P can be lifted to stacks t^+ on Q in a natural way. We say that Σ is internally lift consistent iff for all such s, P, and Q, $\Sigma_{s,P}(t) = \Sigma_{s,Q}(t^+)$. See Section 5.4.

For pairs (M,Σ) such that M is a strategy mouse, we require also that the internal strategy predicate of M be consistent with Σ. More generally, letting $\dot\Sigma$ be the predicate symbol used to record strategy information, we say that (M,Σ) is *pushforward consistent* iff whenever s is a stack by Σ and $N \trianglelefteq M_\infty(s)$, then $\dot\Sigma^N \subseteq \Sigma_{s,N}$.

If M is a pure extender premouse, and Σ is a strategy for M that has strong hull condensation, quasi-normalizes well, and is internally lift consistent, then we call (M,Σ) a *pure extender pair.* If M is a least branch premouse, and Σ is a strategy for M that has strong hull condensation, quasi-normalizes well, is internally lift consistent and pushforward consistent, then we call (M,Σ) a *least branch hod pair.* A pair of one of the two types is a *mouse pair.*

If (M,Σ) is a mouse pair, and s is a stack by Σ with last model N, then we call (N,Σ_s) an *iterate of* (M,Σ). If the branch M-to-N of s does not drop, we call it a non-dropping iterate. In that case, we have an iteration map $i_s\colon M \to N$. Let us write

$$(M,\Sigma) \trianglelefteq (R,\Lambda) \text{ iff } M \trianglelefteq R \text{ and } \Sigma = \Lambda_M.$$

We have no hope of showing anything about mouse pairs (M,Σ) unless we assume absolute definabilty for the iteration strategy. Here we assume Σ has scope HC, i.e., that M is countable and Σ is defined on finite stacks of countable trees, and we assume that we are in a model of AD^+.[22] The following is the main new result of the book.

THEOREM 1.6.4 (Comparison Lemma). *Assume* AD^+*, and let* (P,Σ) *and* (Q,Ψ) *be strongly stable*[23] *mouse pairs with scope* HC *of the same kind; then there are iterates* (R,Λ) *of* (P,Σ) *and* (S,Ω) *of* (Q,Ψ)*, obtained by normal trees* $\mathcal{T}$ *and* $\mathcal{U}$*, such that either*

(1) $(R,\Lambda) \trianglelefteq (S,\Omega)$ *and* P*-to-*R *does not drop, or*
(2) $(S,\Omega) \trianglelefteq (R,\Lambda)$ *and* Q*-to-*S *does not drop.*

Even for pure extender pairs, this theorem is new, because of the agreement between tail strategies it requires. In fact, it is no easier to prove the theorem for

[22] One could require that they be defined on countable stacks.

[23] Strong stability is a mild fine structural requirement. One can avoid it by slightly complicating the notion of *iterate*. See Definition 4.4.5 and Theorem 4.6.12.

pure extender pairs than it is to prove it for least branch hod pairs. The proof in both cases is the same, and it makes use of the properties of the iteration strategies we have isolated in the definition of mouse pair.

Working in the category of mouse pairs enables us to state a general Dodd-Jensen lemma. Let us say $\pi\colon (P,\Sigma) \to (Q,\Psi)$ is elementary iff π is elementary from P to Q, and $\Sigma = \Psi^\pi$. We shall show that an elementary submodel of a mouse pair is a mouse pair, and that the iteration maps associated to non-dropping iterations of a mouse pair are elementary.[24]

THEOREM 1.6.5 (Dodd-Jensen lemma). *Let (P,Σ) be a mouse pair, and (Q,Ψ) be an iterate of (P,Σ) via the stack s. Suppose $\pi\colon (P,\Sigma) \to (Q,\Psi)$ is elementary; then s does not drop, and for all ordinals $\eta \in P$, $i_s(\eta) \leq \pi(\eta)$.*

The proof is just the usual Dodd-Jensen proof; the point is just that the language of mouse pairs enables us to formulate the theorem in its proper generality. There is no need to restrict to mice with unique iteration strategies, as is usually done.

Similarly, we can define the mouse order in its proper generality, without restricting to mice with unique iteration strategies. If (P,Σ) and (Q,Ψ) are pairs of the same type, then $(P,\Sigma) \leq^* (Q,\Psi)$ iff (P,Σ) can be elementarily embedded into an iterate of (Q,Ψ). The Comparison and Dodd-Jensen theorems imply that $\leq^*$ is a prewellorder on each type.

1.7. Hod pair capturing

Least branch hod pairs can be used to analyze HOD in models of AD^+, provided that there are enough such pairs.

DEFINITION 1.7.1 (AD^+). (a) *Hod Pair Capturing* (HPC) is the assertion: for every Suslin-co-Suslin set A, there is a least branch hod pair (P,Σ) such that A is definable from parameters over $(\mathrm{HC},\in,\Sigma)$.

(b) *$L[E]$ capturing* (LEC) is the assertion: for every Suslin-co-Suslin set A, there is a pure extender pair (P,Σ) such that A is definable from parameters over $(\mathrm{HC},\in,\Sigma)$.

An equivalent (under AD^+) formulation would be that the sets of reals coding strategies of the type in question, under some natural map of the reals onto HC, are Wadge cofinal in the Suslin-co-Suslin sets of reals. The restriction to Suslin-co-Suslin sets A is necessary, for AD^+ implies that if (P,Σ) is a pair of one of the two types, then the codeset of Σ is Suslin and co-Suslin. This is the main result of [73], where it is also shown that the Suslin representation constructed is of optimal logical complexity.

---

[24]Neither is obvious. That iteration maps are elementary is a property of the iteration strategy known as pullback consistency. It follows from strong hull condensation.

Remark 1.7.2. HPC is a cousin of Sargsyan's *Generation of Full Pointclasses.* See [42] and [43, §6.1.].

Assuming AD^+, LEC is equivalent to the well known Mouse Capturing: for reals x and y, x is ordinal definable from y iff x is in a pure extender mouse over y. This equivalence is shown in [68]. (See especially Theorem 16.6.) We show in Theorem 10.4.3 below that under AD^+, LEC implies HPC. We do not know whether HPC implies LEC.

Granted $\mathsf{AD}_{\mathbb{R}}$ and HPC, we have enough hod pairs to analyze HOD.

THEOREM 1.7.3 ([73]). *Assume* $\mathsf{AD}_{\mathbb{R}}$ *and* HPC; *then* $V_\theta \cap \mathrm{HOD}$ *is the universe of a least branch premouse.*

Some techniques developed in [64] and [73] are needed to prove the theorem, so we shall not prove it here.

The natural conjecture is that LEC and HPC hold in all models of AD^+ that have not reached an iteration strategy for a premouse with a long extender. Because our capturing mice have only short extenders on their sequences, LEC and HPC cannot hold in larger models of AD^+.

DEFINITION 1.7.4. NLE ("No long extenders") is the assertion: there is no countable, $\omega_1 + 1$-iterable pure extender premouse M such that there is a long extender on the M-sequence.

CONJECTURE 1.7.5. *Assume* AD^+ *and* NLE; *then* LEC.

CONJECTURE 1.7.6. *Assume* AD^+ *and* NLE; *then* HPC.

As we remarked above, Conjecture 1.7.5 implies Conjecture 1.7.6. Conjecture 1.7.5 is equivalent to a slight strengthening of the usual Mouse Set Conjecture MSC. (The hypothesis of MSC is that there is no iteration strategy for a pure extender premouse with a superstrong, which is slightly stronger than NLE.) MSC has been a central target for inner model theorists for a long time.

1.8. Constructing mouse pairs

The basic source for mouse pairs is a background construction. In the simplest case, such a construction $\mathbb{C}$ builds pairs $(M_{\nu,k}, \Omega_{\nu,k})$ inductively, putting extenders on the $M_{\nu,k}$-sequence that are restrictions of nice extenders in V. The iteration strategy $\Omega_{\nu,k}$ is induced by an iteration strategy for V, and if we are constructing strategy premice, the relevant information about $\Omega_{\nu,k}$ is inserted into $M_{\nu,k}$ at the appropriate points. $M_{\nu,k+1}$ is the core of $M_{\nu,k}$. The construction breaks down if the standard parameter of $M_{\nu,k}$ behaves poorly, so that there is no core.

There is of course more to say here, and we shall do so later in the book. For now, let us note that the background universe for such a construction should be a model of ZFC that has lots of extenders, and yet knows how to iterate itself. In the AD^+ context, the following theorem of Woodin applies.[25]

THEOREM 1.8.1 (Woodin). *Assume* AD^+, *and let* Γ *be a good pointclass such that all sets in* Γ *are Suslin and co-Suslin; then for any real* x *there is a coarse* Γ*-Woodin pair* (N,Σ) *such that* $x \in N$.

Here, roughly speaking, N is a countable transitive model of ZFC with a Woodin cardinal and a term for a universal Γ set, and Σ is an iteration strategy for N that moves this term correctly, and is such that $\Sigma \cap N$ is definable over N. See Definition 7.2.3.

The following is essentially Theorem 10.4.1 to follow. It too is one of the main new results of the book.

THEOREM 1.8.2. *Assume* AD^+, *and let* (N,Σ) *be a coarse* Γ*-Woodin pair. Let* $\mathbb{C}$ *be a least branch construction in* N*; then* $\mathbb{C}$ *does not break down. Moreover, each of its levels* $(M^{\mathbb{C}}_{\nu,k}, \Omega^{\mathbb{C}}_{\nu,k})$ *is a least branch hod pair in* N*, and extends canonically to a least branch hod pair in* V.

Background constructions of the sort described in this theorem have an important role to play in our comparison process. Assume AD^+, and let (M,Ω) and (N,Σ) be mouse pairs of the same type. We compare (M,Ω) with (N,Σ) by putting M and N into a common Γ-Woodin universe N^*, where Σ and Ω are in $\Gamma \cap \check{\Gamma}$. We then iterate (M,Σ) and (N,Ω) into levels of a full background construction (of the appropriate type) of N^*. Here are some definitions encapsulating the method.

DEFINITION 1.8.3. Let (M,Σ) and (N,Ω) be mouse pairs of the same type; then

(a) (M,Σ) *iterates past* (N,Ω) iff there is a λ-separated iteration tree $\mathcal{T}$ by Σ on M whose last pair is (N,Ω).
(b) (M,Σ) *iterates to* (N,Ω) iff there is a λ-separated $\mathcal{T}$ as in (a) such that the branch M-to-N of $\mathcal{T}$ does not drop.
(c) (M,Σ) *iterates strictly past* (N,Ω) iff it iterates past (N,Ω), but not to (N,Ω).

λ-separation is a small strengthening of normality that is defined in Definition 4.4.8. One reason that it is important is that if $\mathcal{T}$ is λ-separated and $\mathcal{U}$ is a normal tree on $\mathcal{M}^{\mathcal{T}}_\infty$, then $W(\mathcal{T},\mathcal{U}) = V(\mathcal{T},\mathcal{U})$. That is, embedding normalization coincides with quasi-normalization in this case.

DEFINITION 1.8.4 (AD^+). Let (P,Σ) be a mouse pair; then $(*)(P,\Sigma)$ is the following assertion: Let (N,Ψ) be any coarse Γ-Woodin pair such that $P \in \mathrm{HC}^{N^*}$, and $\Sigma \in \Gamma \cap \check{\Gamma}$. Let $\mathbb{C}$ be a background construction done in N^* of the appropriate type, and let (R,Φ) be a level of $\mathbb{C}$. Suppose that (P,Σ) iterates strictly past all levels of $\mathbb{C}$ that are strictly earlier than (R,Φ); then (P,Σ) iterates past (R,Φ).

[25]See [26], and [71, Lemma 3.13].

If (M,Ω) is a mouse pair, and N is an initial segment of M, then we write Ω_N for the iteration strategy for trees on N that is induced by Ω. We can unpack the conclusion of Definition 1.8.4 as follows: suppose the comparison of P with R has produced a normal tree $\mathcal{T}$ on P with last model Q, with $\mathcal{T}$ by Σ, and S is an initial segment of both Q and R; then $\Sigma_{\mathcal{T},S} = \Phi_S$. Thus the least disagreement between Q and R is an extender disagreement. Moreover, if E on Q and F on R are the extenders involved in it, then $F = \varnothing$.

We shall show (cf. Theorems 8.4.3 and 9.5.6 below)

THEOREM 1.8.5. *Assume* AD^+*; then* $(*)(P,\Sigma)$ *holds, for all strongly stable mouse pairs* (P,Σ).

This theorem lets us compare two (or more) mouse pairs of the same type indirectly, by comparing them to the levels of an appropriate construction, done in a Γ-Woodin model, where both strategies are in $\Gamma \cap \check{\Gamma}$. One can show using the Woodinness that $\mathbb{C}$ reaches non-dropping iterates of both pairs[26]. This gives us a stage (M,Ω) of $\mathbb{C}$ such that one of the pairs iterates to it, while the other iterates past it.

1.9. The comparison argument

In what follows, we shall give fairly complete proofs of the theorems above. The book is long, partly because we wanted to make it accessible, and partly because we shall be forced to revise the basic definitions of [35] and [86] in various ways, so there is a limit to what we can simply quote. In addition, the need to compare strategies adds a layer of complexity to the proofs of the main fine structural theorems about strategy mice. Nevertheless, the main new ideas behind the strategy-comparison process itself are reasonably simple. We describe them now.

The first step is to focus on proving $(*)(P,\Sigma)$. That is, rather than directly comparing two strategies, we iterate them both into a common background construction and its strategy. In the comparison-of-mice context, this method goes back to Kunen ([27]), and was further developed by Mitchell, Baldwin ([2] and [3]), and the author.[27] Woodin and Sargsyan had used the method for strategy comparison in the hod mouse context. All these comparisons could be replaced by direct comparisons of the two mice or strategies involved, but in the general case of comparison of strategies, there are serious advantages to the indirect approach. There is no need to decide what to do if one encounters a strategy disagreement, because one is proving that that never happens. The comparison process is just the usual one of comparing least extender disagreements. Instead of the dual problems of designing a process and proving it terminates, one has a given process, and

[26]See Theorem 8.1.4.

[27]In unpublished 1985 notes titled "Large cardinals and Δ^1_3 wellorders".

knows why it should terminate: no strategy disagreements show up. The problem is just to show this. These advantages led the author to focus, since 2009, on trying to prove $(*)(P,\Sigma)$.

The main new idea that makes this possible is motivated by Sargsyan's proof in [42] that if Σ has branch condensation, then $(*)(P,\Sigma)$ holds.[28] Branch condensation is too strong to hold once P has extenders overlapping Woodin cardinals; we cannot conclude that $\Sigma(\mathcal{T}) = b$ from having merely realized $\mathcal{M}_b^{\mathcal{T}}$ into a Σ-iterate of P. We need some kind of realization of the entire phalanx $\Phi(\mathcal{T}^\frown b)$ in order to conclude that $\Sigma(\mathcal{T}) = b$. This leads to a weakening of branch condensation that one might call "phalanx condensation", in which one asks for a family of branch-condensation-like realizations having some natural agreement with one another. Phalanx condensation is still strong enough to imply $(*)(P,\Sigma)$, and might well be true in general for background-induced strategies. Unfortunately, Sargsyan's construction of strategies with branch condensation does not seem to yield phalanx condensation in the more general case. For one thing, it involves comparison arguments, and in the general case, this looks like a vicious circle. It was during one of the author's many attempts to break into this circle that he realized that certain properties related to phalanx condensation, namely normalizing well and strong hull condensation, could be obtained directly for background-induced strategies, and that these properties suffice for $(*)(P,\Sigma)$.

Let us explain this last part briefly. Suppose that we are in the context of Theorem 1.8.5. We have a premouse P with iteration strategy Σ that normalizes well and has strong hull condensation. We have N a premouse occuring in the fully backgrounded construction of N^*, where $P \in \mathrm{HC}^{N^*}$ and N^* captures Σ. We compare P with N by iterating away the least extender disagreement. It has been known since 1985 that only P will move. We must prove that no strategy disagreement shows up.

Suppose we have produced a λ-separated iteration tree $\mathcal{T}$ on P with last model Q, that $Q|\alpha = N|\alpha$, and that $\mathcal{U}$ is a normal tree on $R = Q|\alpha = N|\alpha$ of limit length played by both $\Sigma_{\mathcal{T},R}$ (the tail of Σ) and Ω_R, where Ω is the N^*-induced strategy for N. We wish to show that $\Sigma_{\mathcal{T},R}(\mathcal{U}) = \Omega_R(\mathcal{U})$. Because Σ is internally lift consistent, we can reduce to the case that $Q = R$.

Let $b = \Omega_R(\mathcal{U})$. We must see $b = \Sigma_{\mathcal{T},R}(\mathcal{U})$, that is, that $b = \Sigma(\langle \mathcal{T},\mathcal{U}\rangle)$. Since $\mathcal{T}$ is λ-separated, embedding normalization coincides with quasi-normalization. Let us consider

$$\mathcal{W}_c = W(\mathcal{T},\mathcal{U}^\frown c) = V(\mathcal{T},\mathcal{U}^\frown c)$$

for arbitrary cofinal branches c of $\mathcal{U}$. We shall see:

[28] Roughly, an iteration strategy Σ for M has *branch condensation* iff whenever $\mathcal{T}$ is an iteration tree of limit length by Σ, b is a cofinal branch of $\mathcal{T}$ with associated iteration map $i_b \colon M \to \mathcal{M}_b^{\mathcal{T}}$, $\pi \colon M \to N$ is an iteration map by Σ, and there is a $k \colon \mathcal{M}_b^{\mathcal{T}} \to N$ such that $\pi = k \circ i_b$, then $\Sigma(\mathcal{T}) = b$. See [42] for more detail.

(1) $\Sigma_{\mathcal{T},R}(\mathcal{U}) = c$ iff $\mathcal{W}_c$ is by Σ. The $\Longrightarrow$ direction follows at once from the fact that Σ quasi-normalizes well, and the $\Longleftarrow$ direction is proved in Section 6.6.
(2) Letting $i_b^* : N^* \to N_b^*$ come from lifting $i_b^{\mathcal{U}}$ to N^* via the iteration-strategy construction of [35], there is a tree embedding of $\mathcal{W}_b$ into $i_b^*(\mathcal{T})$. This is the key step in the proof. It is carried out in Chapter 8.
(3) $i_b^*(\Sigma) \subseteq \Sigma$ because Σ was captured by N^*, so $i_b^*(\mathcal{T})$ is by Σ.
(4) Thus $\mathcal{W}_b$ is by Σ, because Σ has strong hull condensation.
(5) So by (1), $\Sigma_{\mathcal{T},R}(\mathcal{U}) = b$.

Here is a diagram of the situation:

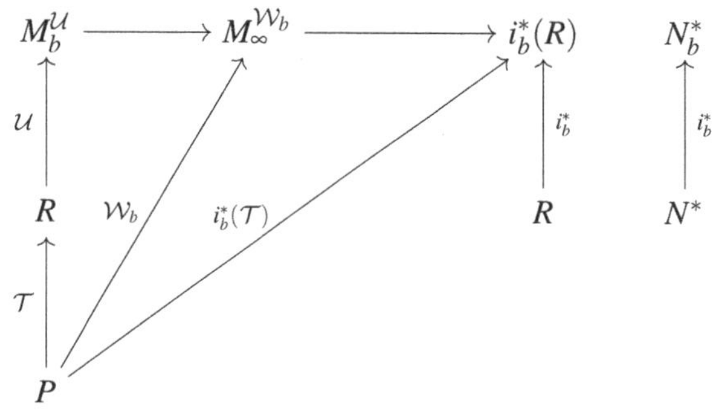

FIGURE 1.9.1. Proof of $(*)(P,\Sigma)$. $\mathcal{W}_b$ is a pseudo-hull of $i_b^*(\mathcal{T})$.

1.10. Plan of the book

Chapters 2 and 3 collect and organize some standard definitions and results from inner model theory. The book is aimed at people who have already encountered this material, via [35], [70], or [86] for example, but these chapters will serve as a bridge to the rest of the book.

In Section 3.6 we explain why this standard theory is not completely adequate to the problem of comparing iteration strategies. Roughly speaking, the problem is that the induced iteration strategies for the levels of a background construction are not connected sufficiently well to the iteration strategy for the background universe. Sections 3.7 and 3.8 analyze one of the two sources of this shortfall, and Chapter 4 removes both of them. This involves revising the notions of *premouse* and *iteration tree* slightly, and re-proving the standard fine structural results in the new setting.

Chapter 5 shows that the new definitions lead to background-induced iteration strategies that are better behaved in several ways. Chapters 6 and 7 push further in this direction, leading ultimately to Theorem 7.6.2, which says that pure extender background constructions, done in an appropriately iterable background universe, produce pure extender pairs.

In Chapter 8 we prove the main comparison theorem for pure extender pairs, Theorem 8.4.3. We shall adapt the proof of Theorem 8.4.3 to least branch hod pairs and to phalanx comparisons in Chapters 9 and 10, but the main steps all show up in this simpler situation, so we have begun with it. When we use the proof again in Chapters 9 and 10, we shall condense long stretches by pointing to the proof of Theorem 8.4.3.

Chapters 9 and 10 use the strategy-comparison process to develop the theory of least branch hod pairs. Chapter 11 uses this theory to analyze HOD in certain models of $\mathsf{AD}_{\mathbb{R}}$, and concludes with a discussion of further results that have been proved by the methods we develop here.

Chapter 2

PRELIMINARIES

Inner model theory deals with canonical objects, but inner model theorists have presented them in various ways. The conventions we use here are, for the most part, fairly common. For basic fine structural notions such as projecta, cores, standard parameters, fine ultrapowers, and degrees of elementarity, we shall stay close to Mitchell and Steel [35] and the paper [54] by Schindler and Zeman. We shall use Jensen indexing for the sequences of extenders from which premice are constructed; see for example Zeman's book [86]. In Chapter 4 we shall modify the notion of premouse slightly, by enlarging the standard parameters and associated cores. Until we get to that point, our notion of premouse is just the standard one determined by the conventions of [35], [54], and [86].[1]

Most of our terminology to do with iteration trees and iteration strategies traces back to Martin and Steel [30] and Mitchell and Steel [35], and is by now pretty standard. We do need to consider carefully iteration strategies defined on a wider class of iteration trees than is common, and so there is some less familiar terminology defined in Sections 2.6 and 2.7.

2.1. Extenders and ultrapowers

Our notation for extenders is standard.

DEFINITION 2.1.1. Let M be transitive and rudimentarily closed; then $E = \langle E_a \mid a \in [\theta]^{<\omega} \rangle$ is a (κ, θ)-extender over M with spaces $\langle \mu_a \mid a \in [\theta]^{<\omega} \rangle$ if and only if

(1) Each E_a is an (M, κ)-complete ultrafilter over $P([\mu_a]^{|a|}) \cap M$, with μ_a being the least μ such that $[\mu]^{|a|} \in E_a$.
(2) (Compatibility) For $a \subseteq b$ and $X \in M$, $X \in E_a \iff X^{ab} \in E_b$.
(3) (Uniformity) $\mu_{\{\kappa\}} = \kappa$.
(4) (Normality) If $f \in M$ and $f(u) < \max(u)$ for E_a a.e. u, then there is a $\beta < \max(a)$ such that for $E_{a \cup \{\beta\}}$ a.e. u, $f^{a, a\cup\{\beta\}}(u) = u^{\{\beta\}, a\cup\{\beta\}}$.

[1] The notion of premouse in [86], and its related fine structure, originate in Jensen's manuscripts [19] and [20].

The unexplained notation here can be found in [54, §8]. We shall often identify E with the binary relation $(a,X) \in E$ iff $X \in E_a$. One can also identify it with the other section-function of this binary relation, which is essentially the function $X \mapsto i_E^M(X) \cap \theta$. We call θ the *length* of E, and write $\theta = \mathrm{lh}(E)$. The *space of* E is

$$\mathrm{sp}(E) = \sup\{\mu_a \mid a \in [\mathrm{lh}(E)]^{<\omega}\}.$$

The *domain of* E is the family of sets it measures, that is, $\mathrm{dom}(E) = \{Y \mid \exists (a,X) \in E(Y = X \vee Y = [\mu_a]^{|a|} - X)\}$. If M is a premouse of some kind, we also write $M|\eta = \mathrm{dom}(E)$, where η is least such that $\forall (a,X) \in E(X \in M|\eta)$. By acceptability, $\eta = \sup(\{\mu_a^{+,M} \mid a \in [\theta]^{<\omega}\})$. We shall further abuse notation by writing $\eta = \mathrm{dom}(E)$ when M is determined by context.

The critical point of a (κ,θ)-extender is κ, and we use either $\mathrm{crit}(E)$ or κ_E to denote it. Given an extender E over M, we form the Σ_0 ultrapower

$$\mathrm{Ult}_0(M,E) = \{[a,f]_E^M \mid a \in [\mathrm{lh}(E)]^{<\omega} \text{ and } f \in M\},$$

as in [54, Theorem 8.4]. Our M will always be rudimentarily closed and satisfy the Axiom of Choice, so we have Łoś's Theorem for Σ_0 formulae, and the canonical embedding

$$i_E^M : M \to \mathrm{Ult}_0(M,E)$$

is cofinal and Σ_0 elementary, and hence Σ_1 elementary. By (1) and (3), $\kappa_E = \mathrm{crit}(i_E^M)$. By normality, $a = [a,\mathrm{id}]_E^M$, so $\mathrm{lh}(E)$ is included in the (always transitivized) wellfounded part of $\mathrm{Ult}_0(M,E)$. More generally,

$$[a,f]_E^M = i_E^M(f)(a).$$

If $X \subseteq \mathrm{lh}(E)$, then $E \restriction X = \{(a,Y) \in E \mid a \subseteq X\}$. $E \restriction X$ has the properties of an extender, except possibly normality, so we can form $\mathrm{Ult}_0(M, E \restriction X)$, and there is a natural factor embedding $\tau\colon \mathrm{Ult}_0(M,E\restriction X) \to \mathrm{Ult}_0(M,E)$ given by

$$\tau([a,f]_{E\restriction X}^M) = [a,f]_E^M.$$

In the case that $X = \nu > \kappa_E$ is an ordinal, $E \restriction \nu$ is an extender, and $\tau \restriction \nu$ is the identity. We say ν is a *generator of* E iff ν is the critical point of τ, that is, $\nu \neq [a,f]_E^M$ whenever $f \in M$ and $a \subseteq \nu$. Let

$$\nu(E) = \sup(\{\nu + 1 \mid \nu \text{ is a generator of } E\}).$$

So $\nu(E) \leq \mathrm{lh}(E)$, and E is equivalent to $E \restriction \nu(E)$, in that the two produce the same ultrapower.

We write

$$\lambda(E) = \lambda_E = i_E^M(\kappa_E).$$

Note that although E may be an extender over more than one M, $\mathrm{sp}(E)$, κ_E, $\mathrm{lh}(E)$, $\mathrm{dom}(E)$, $\nu(E)$, and $\lambda(E)$ depend only on E itself. If N is another transitive, rudimentarily closed set, and $P(\mu_a) \cap N = P(\mu_a) \cap M$ for all $a \in [\mathrm{lh}(E)]^{<\omega}$, then E is also an extender over N; moreover i_E^M agrees with i_E^N on $\mathrm{dom}(E)$. However, i_E^M

and i_E^N may disagree beyond that. We say E is *short* iff $\nu(E) \leq \lambda(E)$. It is easy to see that E is short if $\text{lh}(E) \leq \sup(i_E^M``((\kappa_E^+)^M))$. If E is short, then all its interesting measures concentrate on the critical point. When E is short, i_E^M is continuous at $\kappa^{+,M}$, and if M is a premouse, then $\text{dom}(E) = M|\kappa_E^{+,M}$. In this book, we shall deal almost exclusively with short extenders.

If we start with $j\colon M \to N$ with critical point κ, and an ordinal ν such that $\kappa < \nu \leq o(N)$, then for $a \in [\nu]^{<\omega}$ we let μ_a be the least μ such that $a \subseteq j(\mu)$, and for $X \subseteq [\mu_a]^{|a|}$ in M, we put

$$(a,X) \in E_j \iff a \in j(X).$$

E_j is an extender over M, called the (κ,ν)*-extender derived from* j. We have the diagram

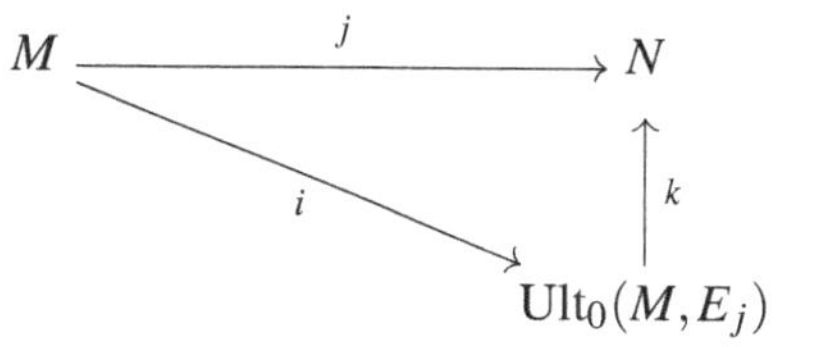

where $i = i_{E_j}^M$, and

$$k(i(f)(a)) = j(f)(a).$$

$k \restriction \nu$ is the identity. If E is an extender over M, then E is derived from i_E^M.

The *Jensen completion* of a short extender E over some M is the $(\kappa_E, i_E^M(\kappa_E^{+,M}))$ extender derived from i_E^M. E and its Jensen completion E^* are equivalent, in that $\nu(E) = \nu(E^*)$, and $E = E^* \restriction \text{lh}(E)$.

2.2. Pure extender premice

Our main results apply to premice of various kinds, both strategy premice and pure extender premice, with λ-indexing or ms-indexing for their extender sequences.[2,3] The comparison theorem for iteration strategies that is our first main goal holds in all these contexts. But the proof of this theorem requires a detailed fine structural analysis, and the particulars of the fine structure become important at certain points. We shall first prove the comparison theorem in the case of iteration strategies for what we shall call *pfs premice*. These are a variant on pure extender

[2] Why "mouse"? Like "quark", it is short and easily remembered. It has a fine tradition, going back to a discoverer of the concept. The longer, colorless "extender model" does have its place, but "mouse" is more flexible and distinctive.

[3] People sometimes speak of "strategic" mice or extender models, but this seems wrong to us. A strategic X (bomber, position, move, etc.) is an X that is incorporated in some strategy, not an X that has a strategy incorporated in it. A mouse that incorporates a strategy is a strategy mouse, just as a pie that incorporates apples is an apple pie.

premice with λ-indexing, the difference being that the soundness requirement has been relaxed. They are formally defined in Chapter 4. Until we get to that chapter, we shall deal primarily with the standard λ-indexed pure extender premice, as defined in [86].[4]

The reader should see [1, Definition 2.4] for further details on the following definition. A *potential Jensen premouse* is an acceptable J-structure

$$M = \langle J_\alpha^{\vec{E}}, \in, \vec{E}, \gamma, F \rangle$$

with various properties. $o(M) = \mathrm{OR} \cap M = \omega\alpha$. The language $\mathcal{L}_0$ of M has $\in$, predicate symbols $\dot{E}$ and $\dot{F}$, and a constant symbol $\dot{\gamma}$. We call $\mathcal{L}_0$ the *language of (pure extender) premice*.

If M is a potential Jensen premouse, then $\dot{E}^M$ is a sequence of extenders, and either $\dot{F}^M$ is empty (i.e., M is *passive*), or $\dot{F}^M$ codes a new extender being added to our model by M. The main requirements are

(1) (λ-indexing) If $F = \dot{F}^M$ is nonempty (i.e., M is *active*), then $M \models \mathrm{crit}(F)^+$ exists, and for $\mu = \mathrm{crit}(F)^{+M}$, $o(M) = i_F^M(\mu) = \mathrm{lh}(F)$. $\dot{F}^M$ is just the graph of $i_F^M \restriction (M|\mu)$.
(2) (Coherence) $i_F^M(\dot{E}^M) \restriction o(M) + 1 = \dot{E}^{M\frown}\langle \emptyset \rangle$.
(3) (Initial segment condition, J-ISC) If G is a whole proper initial segment of F, then the Jensen completion of G must appear in $\dot{E}^M$. If there is a largest whole proper initial segment, then $\dot{\gamma}^M$ is the index of its Jensen completion in $\dot{E}^M$. Otherwise, $\dot{\gamma}^M = 0$.
(4) If N is a proper initial segment of M, then N is a potential Jensen premouse.

Here an initial segment $G = F \restriction \eta$ of F is *whole* iff $\eta = \lambda_G$.

Since potential Jensen premice are acceptable J-structures, the basic fine structural notions apply to them. We recall some of them in the next section. We then define a *Jensen premouse* as a potential Jensen premouse all of whose proper initial segments are sound.

Figure 2.2.1 illustrates a common situation, one that occurs at successor steps in an iteration tree, for example.

There is a significant strengthening of the Jensen initial segment condition (3) above. If M is an active premouse, then we set

$$\nu(M) = \max(\nu(\dot{F}^M), \mathrm{crit}(\dot{F}^M)^{+,M}).$$

$\dot{F}^M \restriction \nu(M)$ is equivalent to $\dot{F}^M$, and so it is not in M. But

DEFINITION 2.2.1. Let M be an active premouse with last extender F; then M satisfies the ms-ISC (or is *ms-solid*) iff for any $\eta < \nu(M)$, $F \restriction \eta \in M$.

Clearly the ms-ISC implies the weakening of J-ISC in which we only demand that the whole proper initial segments of $\dot{F}^M$ belong to M. But for iterable M, this then implies the full J-ISC. (See [53].)

[4] The essential equivalence of λ-indexing with ms-indexing has been carefully demonstrated by Fuchs in [11] and [10].

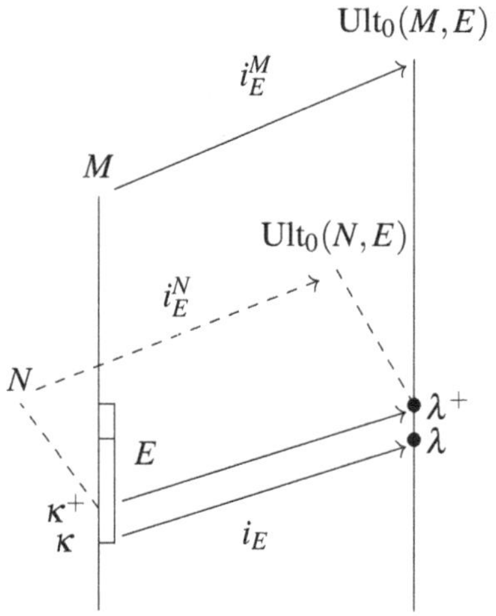

FIGURE 2.2.1. E is on the coherent sequence of M, $\kappa = \mathrm{crit}(E)$, and $\lambda = \lambda(E)$. $P(\kappa)^M = P(\kappa)^N = \mathrm{dom}(E)$, so $\mathrm{Ult}_0(M,E)$ and $\mathrm{Ult}_0(N,E)$ make sense. The ultrapowers agree with M below $\mathrm{lh}(E)$, and with each other below $\mathrm{lh}(E)+1$.

THEOREM 2.2.2 (ms-ISC). *Let M be an active premouse with last extender F, and suppose M is 1-sound and $(1, \omega, \omega_1 + 1)$-iterable; then M is ms-solid.*

This is essentially the initial segment condition of [35], but stated for Jensen premice. [35] goes on to say that the trivial completion of $F \restriction \eta$ is either on the M-sequence, or an ultrapower away. This is correct unless $F \restriction \eta$ is type Z. If $F \restriction \eta$ is type Z, then it is the extender of $F \restriction \xi$-then-U, where ξ is its largest generator, and U is an ultrafilter on ξ, and we still get $F \restriction \eta \in M$. (See [53]. Theorem 2.7 of [53] is essentially Theorem 2.2.2 above.)

If M is active, we let its *initial segment ordinal* be

$$\iota(M) = \sup(\{\eta + 1 \mid \dot{F}^M \restriction \eta \in M\}).$$

So M is ms-solid iff $\iota(M) = \nu(M)$. Theorem 2.2.2 becomes false when its soundness hypothesis is removed, since if $N = \mathrm{Ult}_0(M,E)$ where $\nu(M) \leq \mathrm{crit}(E) < \lambda_F$, then $\iota(N) = \iota(M) = \nu(M)$, but $\mathrm{crit}(E)$ is a generator of $i_E^M(F)$.

The proof of Theorem 2.2.2 requires a comparison argument based on iterability, and so in the context of this book, it is closer to the end of the development than to the beginning. In the theory of [35], the strong form of ms-solidity is an axiom on premice from the beginning, but comparison arguments are needed to show that the premice one constructs satisfy it.[5] One might similarly make ms-solidity an axiom on Jensen premice from the beginning, and so have it available earlier in the

[5] These are the results of [35, §9, §10] concerning bicephali and pseudo-premice.

game, so to speak. This would simplify a few things, but it is not standard, and we shall not do it here.[6]

We shall not use ms-premice, so henceforth we shall refer to potential Jensen premice as potential premice, or later, when we need to distinguish them from strategy premice, as *potential pure extender premice*.

2.3. Projecta and cores

Fine structure theory relies on a careful analysis of the condensation properties of mice; that is, of the extent to which Skolem hulls of a mouse M collapse to initial segments of M. Jensen's theory of projecta, standard parameters, and cores is the foundation for this analysis.

2.3.1. Sound premice and their reducts. Let M be an acceptable J-structure.[7] We define the projecta $\rho_i(M)$, standard parameters $p_i(M)$, and reducts ("Σ_i mastercodes") $M^i = M^{i,p_i(M)}$ by induction. At the same time we define k-solidity and k-soundness for M. We start with $A^0 = \emptyset$, and

$$M^0 = (M, A^0), \qquad \rho_0(M) = o(M), \qquad p_0(M) = \emptyset.$$

M is automatically 0-sound and 0-solid. The successor step is

$$\begin{aligned}\rho_{i+1}(M) &= \text{least } \alpha \text{ s.t. } \exists A \subseteq \alpha (A \text{ is boldface } \Sigma_1^{M^i} \text{ and } A \notin M^i), \\ &= \rho_1(M^i)\end{aligned}$$

and

$$p_{i+1}(M) = p_i(M) \cup r_{i+1},$$

where r_{i+1} is the lexicographically least descending sequence of ordinals from which a new subset of $\rho_1(M^i)$ can be Σ_1 defined over M^i. ($r_{i+1} = \emptyset$ is possible.) We then set

$$M^{i+1} = (M||\rho_{i+1}(M), A^{i+1}),$$

where

$$\begin{aligned}A^{i+1} &= \mathrm{Th}_1^{M^i}(M^i||\rho_{i+1} \cup r_{i+1}) \\ &= \{\langle \varphi, x\rangle \mid \varphi(u,v) \text{ is } \Sigma_1 \wedge M^i \models \varphi[x, r_{i+1}]\}.\end{aligned}$$

We say that M is $i+1$-*solid* iff r_{i+1} is solid and universal, and so is its image in the collapse of $\mathrm{Hull}_1^{M^i}(\rho_{i+1}(M) \cup r_{i+1})$.[8] In general, we do not care about M^k for $k \geq i+1$ if M is not $i+1$-solid; fine structure has broken down. One of our main

[6] Our definition of pfs premice in Chapter 4 does include a small fragment of ms-solidity. See Definition 4.1.2. *Dodd solidity* is a useful and still stronger form of the initial segment condition that iterable premice satisfy. See for example [53].

[7] See [54].

[8] See Definitions 2.3.10 and 2.3.11.

tasks in any construction of premice will be to show that the premice we produce are k-solid for all k. We say that M is $i+1$-*sound* iff M is i-sound, $i+1$-solid, and and $M^i = \mathrm{Hull}_1^{M^i}(\rho_{i+1}(M) \cup r_{i+1})$. In general, we will not care about M^{i+2} unless M is $i+1$-sound.

Notice that $o(M^i) = \rho_i(M)$, and $r_{i+1} \subseteq [\rho_{i+1}(M), \rho_i(M))$. We may sometimes identify A^i with a subset of ρ_i. We let

$$r_i(M) = p_i(M) \cap [\rho_i, \rho_{i-1}),$$
$$p_i(M) = p(M) \cup [\rho_i, o(M)) = \bigcup_{k \le i} r_i.$$

This completes our inductive definition of the M^i for $i < \omega$. If M is i-solid and i-sound for all $i < \omega$, then we say that M is ω-sound. In this case, we let $\rho_\omega(M)$ be the eventual value of $\rho_i(M)$ as $i \to \omega$. We define the reduct M^ω by

$$M^\omega = (M || \rho_\omega(M), A^i)_{k \le i},$$

where k is least such that $\rho_k(M) = \rho_\omega(M)$.

DEFINITION 2.3.1. A *Jensen premouse* is a pair $M = (\hat{M}, k)$ such that $k \le \omega$ and

(1) $\hat{M}$ is a k-sound potential premouse, and
(2) every proper initial segment of $\hat{M}$ is an ω-sound potential premouse.

We write $k = k(M)$.

We shall drop the qualifier "Jensen" until we start considering another sort of premouse in Chapter 4.

What we are calling a premouse is just a λ-indexed premouse in the usual sense, paired with a degree of soundness that it has. We usually abuse notation by identifying M with $\hat{M}$.[9]

Abusing notation this way, if M is a premouse, then we set $o(M) = \mathrm{ORD} \cap M$, so that $o(M) = \omega\alpha$ for $M = (J_\alpha^A, \dots)$. (The [54] convention differs slightly here.) We write $\hat{o}(M)$ for α itself. The index of M is

$$l(M) = \langle \hat{o}(M), k(M) \rangle.$$

If $\langle \nu, l \rangle \le_{\mathrm{lex}} l(M)$, then $M|\langle \nu, l \rangle$ is the initial segment N of M with index $l(N) = \langle \nu, l \rangle$. (So $\dot{E}^N = \dot{E}^M \cap N$, and when $\nu < \hat{o}(M)$, $\dot{F}^N = \dot{E}^M_{\omega\nu}$.) If $\nu \le \hat{o}(M)$, then we write $M|\nu$ for $M|\langle \nu, 0 \rangle$. We write $M||\nu$, or sometimes $M|\langle \nu, -1 \rangle$, for the structure that agrees with $M|\nu$ except possibly on the interpretation of $\dot{F}$, and satisfies $\dot{F}^{M|\nu} = \emptyset$. By convention, $k(M||\nu) = 0$.[10]

[9]The convention that each premouse has a distinguished degree of soundness is due to Itay Neeman. It is quite useful for simplifying statements about premice, while retaining precision.

[10]Many authors, for example [86], reverse the meanings of $M|\nu$ and $M||\nu$. We find it more logical to let $M||\nu$ stand for cutting M twice, first to $M|\nu$, and then again by throwing away the top extender.

Remark 2.3.2. $\langle \hat{M}, \omega \rangle$ is a premouse iff for all $k < \omega$, $\langle \hat{M}, k \rangle$ is a premouse. In contexts in which we care about $k(M)$, the important case is $k(M) < \omega$. If $k(M) = \omega$, one can usually just replace M with $N = (\text{rud}(\hat{M} \cup \{\hat{M}\}), \emptyset)$. For example, an ultrapower of M using M-definable functions is equivalent to an ultrapower of N using functions belonging to N. Some of the general statements about premice we make below may need small adjustments when $k(M) = \omega$.

We occasionally want to raise or lower a soundness degree.

DEFINITION 2.3.3. Let $M = (\hat{M}, k)$ be a premouse;

(a) for $i \leq k$, $M \downarrow i = (\hat{M}, i)$,
(b) $M^- = (\hat{M}, k \dot{-} 1)$,
(c) $M^+ = (\hat{M}, k+1)$.

Of course, M^+ is a premouse iff M is sound.[11]

DEFINITION 2.3.4. If P and Q are Jensen premice, then

(i) $P \trianglelefteq_0 Q$ iff there are μ and l such that $P = Q|\langle \mu, l \rangle$.
(ii) $P \triangleleft_0 Q$ iff $P \trianglelefteq_0 Q$ and $P \neq Q$.
(iii) $P \trianglelefteq Q$ iff there are μ and $l \geq 0$ such that $P = Q|\langle \mu, l \rangle$.
(iv) $P \triangleleft Q$ iff $P \trianglelefteq Q$ and $P \neq Q$.

The difference between the two initial segment notions lies in whether we regard $Q|\langle \mu, -1 \rangle$ as an initial segment of Q. In other words, if $Q|\mu$ is active, then $Q||\mu \trianglelefteq_0 Q$, but $Q||\mu \not\trianglelefteq Q$. Both notions have a role. If $P \trianglelefteq Q$ we say that P is an *initial segment* of Q, and if $P \triangleleft Q$ we say it is a proper initial segment. If $P \trianglelefteq_0 Q$ we say that P is a *weak initial segment* of Q, and if $P \triangleleft_0 Q$, it is a proper weak initial segment.

Note that if $\hat{P} = \hat{Q}$ but $k(P) < k(Q)$, then $P \triangleleft_0 Q$.

If $M = (\hat{M}, k)$ is a premouse, then its extender sequence is $\dot{E}^M = \dot{E}^{\hat{M}}$ together with a last, or top, extender $\dot{F}^M = \dot{F}^{\hat{M}}$. We speak of $\rho_i(M), M^i$, etc., instead of $\rho_i(\hat{M}), \hat{M}^i$, etc. Our soundness hypothesis means that for $n \leq k$, there is a natural surjection of M^n onto M.

DEFINITION 2.3.5. (a) If Q is an amenable J-structure, then h^1_Q is its canonical Σ_1 Skolem function.[12]

(b) If M is a premouse and $0 \leq n \leq k(M)$, then we define $d^n_M : M^n \xrightarrow{\text{onto}} M$ by

$$d^0_M = \text{id},$$
$$d^{n+1}_M(\langle \varphi, x \rangle) = d^n_M(h^1_{M^n}(\varphi, \langle x, r_{n+1}(M) \rangle)).$$

[11] If $k(M) = \omega$, then $M^+ = M^- = M$.

[12] For $\varphi(u, v) = \exists w \theta(u, v, w)$ where θ is Σ_0, $h_Q(\varphi, a) = b$ iff there is c such that $\langle b, c \rangle$ is $<_Q$ least such that $Q \models \theta[a, b, c]$.

(c) Let M be a premouse, $n \le k(M)$, and R be a relation on M. Let R^n be the relation on M^n given by

$$R^n(x_1,\dots,x_k) \iff R(d^n(x_1),\dots,d^n(x_k)).$$

Then R is $r\Sigma_{n+1}$ iff R^n is $\Sigma_1^{M^n}$.

(d) A function is $r\Sigma_n^M$ just in case its graph in $r\Sigma_n^M$.

The soundness requirement on premice is that when $k \le k(M)$, then $M = \text{ran}(d^k)$.[13] One can think of d_M^k as giving us a system of names for the elements of M, the names being parametrized by ordinals $< \rho_k(M)$ and involving a fixed name for $p_k(M)$.[14] The predicate A^k of M^k tells us the $r\Sigma_k$ truths of M about the objects named, which is somewhat more than just the Σ_k truths.

$r\Sigma_{n+1}^M$ versus $\Sigma_1^{M^n}$. It is possible to characterize the definability levels $r\Sigma_{i+1}^M$, without directly referring to the coding structures M^i. This is what is done in [35]. That has the advantage that we are primarily interested in premice, not codes for them. Also, the definition of $r\Sigma_{i+1}$ in [35] makes sense even when M is not i-sound.[15] On the other hand, certain things stand out better when we use the M^i. For example, $\Sigma_0^{M^i}$ has nice closure properties, and the prewellordering property for $\Sigma_1^{M^i}$ is often useful. A good compromise in many fine structural arguments is to focus on the case $i = 0$, where M is its own coding structure, and $r\Sigma_1 = \Sigma_1$. Usually, if this case works out, then the case $i > 0$ will work out too.

Jensen introduced the reducts M^i, and hence implicitly the definability levels $r\Sigma_{i+1}$ in the sound case, in order to prove Σ_{i+1} uniformization for $M \trianglelefteq L$. (See [18].) But it has turned out that the $r\Sigma_i$ stratification of definability is more useful for inner model theory than the usual Σ_i stratification. One wants level 2 formulae to be allowed a name for p_1, for example. In order to get a feel for what can be said in an $r\Sigma_n$ way, let us look more closely at $r\Sigma_2$.[16]

PROPOSITION 2.3.6. *Let M be a premouse and $k(M) \ge 1$; then*

(1) $d^1 = h \restriction M^1$, for some h that is Σ_1^M in p_1. Thus $d^1 \cap M^1$ is $\Sigma_0^{M^1}$.

(2) For $R \subseteq M^1$, R is $\Sigma_1^{M^1}$ iff R is $r\Sigma_2$; more generally

(3) let $S \subseteq M^1 \times M$, and let $R(x,y)$ iff $S(x,d^1(y))$; then R is $\Sigma_1^{M^1}$ iff S is $r\Sigma_2$.

(4) The predicates $x = p_1^M$, $x \ne p_1^M$, $x \in M^1$, and $x < \rho_1^M$ are each $r\Sigma_2$.

(5) Every $\Sigma_2^M(p_1^M)$ relation is $r\Sigma_2^M$.

---

[13] Clearly d_M^k"$M^k = d_M^k$"$\rho_k(M)$ when $k \ge 1$.

[14] M^k has a name for r_k, and a name for a name for r_{k-1}, and so on. Putting them together, it has a name for p_k.

[15] But the definition of $r\Sigma_{i+1}^M$ in [35] would not be very natural unless one were trying to capture $\Sigma_1^{M^i}$. Moreover, [35] only proves a few things about the unsound case.

[16] Jensen's Σ^* theory provides a treatment of the higher levels of definability over premice that is somewhat more general. See [83].

(6) The predicate $R(x,\alpha,q) \iff (\alpha < \rho_1^M \wedge x = Th_1^M(\alpha \cup \{p_1^M, q\}))$ *is* $r\Sigma_2$.

(7) d_M^1 *is an* $r\Sigma_2$ *function.*

Moreover, these statements are all true uniformly in M.

PROOF. For (1): $d^1(x) = y$ iff $h^1(x, p_1) = y$ iff $\langle \varphi, \langle x,y\rangle\rangle \in A^1$, where $\varphi(u,v)$ is a Σ_1 formula expressing $h^1((u)_0, v) = (u)_1$.

For (2): Let $R \subseteq M^1$, and let $S \subseteq M^1$ be its coded version, i.e., $S(x)$ iff $R(d^1(x))$. We must see that R is $\Sigma_1^{M^1}$ iff S is $\Sigma_1^{M^1}$. But $S(x)$ iff $\exists y \in M^1(d^1(x) = y \wedge R(y))$, and $R(x)$ iff $\exists y \in M^1(d^1(y) = x \wedge S(y))$, so this is true.

For (3): This is a calculation like that in (2). We omit it.

For (4): Clearly $x < \rho_1$ and $x \in M^1$ are $\Sigma_1^{M^1}$, so they are $r\Sigma_2$ by (2). To see that $x = p_1$ is $r\Sigma_2$, we must see that $d^1(x) = p_1$ is $\Sigma_1^{M^1}$. But $d^1(x) = p_1$ iff $\langle \varphi, x\rangle \in A^1$, where $\varphi(u,v)$ expresses $h^1(u, p_1) = v$.

(5) follows easily from (6). For (6): Let $S(x,\alpha,z)$ iff $R(x,\alpha,d^1(z))$. By (3), it is enough to show that S is $\Sigma_1^{M^1}$. But it is easy to see that $\mathrm{Th}_1^M(\alpha \cup \{d^1(z)\})$ can be reduced to $\mathrm{Th}_1^M(\alpha \cup \{z, p_1\})$ via a simple Σ_0 function, and the latter can be computed easily from $A^1 \cap M||\beta$ whenever $x, \alpha, z \in M||\beta$. So $S(x,\alpha,z)$ iff there is a $\beta < o(M^1)$ such that $A^1 \cap \beta$ certifies $S(x,\alpha,z)$. We are done if we show that the function $\beta \mapsto A^1 \cap M||\beta$ is $\Sigma_1^{M^1}$. (Note M^1 is closed under this function!) That is clear.

For (7), let $S(x,y)$ iff $d^1(x) = y$. By (3), it is enough to show the relation $R(x,z)$ iff $S(x, d^1(z))$ is $\Sigma_1^{M^1}$. But $R(x,z)$ iff $\langle \varphi, \langle x,z\rangle\rangle \in A^1$, where $\varphi(u,v)$ is a Σ_1 formula expressing $h^1((u)_0, v) = h^1((u)_1, v)$. ⊣

The predicates $x < \rho_1$ and $x = p_1$ are not (lightface) Σ_2^M in general. The predicate R identifying Σ_1^M theories in (6) is Π_2^M, but not Σ_2^M in general. So $r\Sigma_2$ goes strictly beyond Σ_2.[17]

Part (6) of Proposition 2.3.6 yields a normal form for $r\Sigma_2$ predicates: a predicate $R(x)$ is $r\Sigma_2^M$ iff there is a Σ_1^M predicate P such that for all x,

$$R(x) \iff \exists\alpha < \rho_1^M \exists q[P(x, \mathrm{Th}_1^M(\alpha \cup \{q, p_1^M\}))].$$

See [35], where this is essentially taken as the definition of $r\Sigma_2$.[18]

One can analyze $r\Sigma_n^M$ for $n > 2$ is a similar way. For $n \le k(M)$, the $\Sigma_1^{M^n}$ relations on M^n decode into the $r\Sigma_2^{M^{n-1}}$ relations on M^{n-1}, using $d^{n,n-1}\colon M^n \xrightarrow{\text{onto}} M^{n-1}$. Those decode into the $r\Sigma_3$ relations on M^{n-2}, using $d^{n-1,n-2}$, and so on. Eventually we have decoded the $\Sigma_1^{M^n}$ relations on M^n into the $r\Sigma_{n+1}^M$ relations on $M^0 = M$. At each decoding, an analog of Proposition 2.3.6 applies. This leads to:

LEMMA 2.3.7. *Let M be a premouse and* $1 \le n \le k(M)$*; then*

---

[17] One can show that the predicate $x = \rho_1^M$ is not $r\Sigma_2^M$ in general, because it is not always preserved by Σ_1 ultrapowers.

[18] In [35] the formulae in $\mathrm{Th}_1(\alpha \cup \{q\})$ are allowed to have Skolem terms in them. But one arrives at the same class $r\Sigma_2$, as explained in the appendix to §2 of [35].

(1) $d^n = h\restriction M^n$, for some function h that is $r\Sigma_n$ in $p_n(M)$. Thus $d^n \cap M^n$ is $\Sigma_0^{M^n}$.
(2) For $R \subseteq M^n$, R is $\Sigma_1^{M^n}$ iff R is $r\Sigma_{n+1}$; more generally
(3) let $S \subseteq M^n \times M$, and let $R(x,y)$ iff $S(x,d^n(y))$; then R is $\Sigma_1^{M^n}$ iff S is $r\Sigma_{n+1}$.
(4) The predicates $x = p_n^M$, $x \neq p_n^M$, $x \in M^n$, and $x < \rho_n^M$ are each $r\Sigma_{n+1}$.
(5) Every $\Sigma_{n+1}^M(p_n^M)$ relation is $r\Sigma_{n+1}^M$.
(6) The predicate $R(x,\alpha,q) \iff (\alpha < \rho_n^M \wedge x = Th_n^M(\alpha \cup \{p_n^M, q\}))$ is $r\Sigma_{n+1}$.
(7) d_M^n is an $r\Sigma_{n+1}^M$ function.

Moreover, these statements are all true uniformly in M.

Part (6) yields a normal form for $r\Sigma_{n+1}^M$ if we interpret Th_n^M as referring to the $r\Sigma_n$ theory in M. This is how $r\Sigma_{n+1}^M$ is defined in [35].[19]

LEMMA 2.3.8. *Let M be a premouse, and $0 \leq n \leq k(M)$; then*

(1) Every Boolean combination of $r\Sigma_n$ relations is $r\Sigma_{n+1}$.
(2) The class of $r\Sigma_{n+1}^M$ relations is closed under $\wedge, \vee, \exists x$, and substitution of partial $r\Sigma_{n+1}$ functions.
(3) Every $r\Sigma_{n+1}$ relation can be uniformized by a $r\Sigma_{n+1}$ function.

PROOF. (1) and (2) are easy. For (3), let us first uniformize the $r\Sigma_{n+1}$ relation $R(y,x)$ iff $d^n(x) = y$. (We omit M from the notation for readability.) Put

$$e^n(y) = x \iff (d^n(x) = y \wedge \forall w <_M x(d^n(w) \neq d^n(x))).$$

The first conjunct on the right is $r\Sigma_{n+1}$. The second conjunct is $\Sigma_1^{M^n}$, because one only needs $A^n \cap M||\beta$ where $x \in M||\beta$ to determine its truth. Thus the second conjunct is $r\Sigma_{n+1}$, so e^n is a $r\Sigma_{n+1}$ function.

Now let $R(x,y)$ be $r\Sigma_{n+1}$, and $S(u,v)$ iff $R(d^n(u), d^n(v))$, so that S is $\Sigma_1^{M^n}$. Let h be a $\Sigma_1^{M^1}$ function that uniformizes S. h is $r\Sigma_{n+1}$ by Lemma 2.3.7(2). Clearly

$$g = d^n \circ h \circ e^n$$

uniformizes R, and is $r\Sigma_{n+1}$. ⊣

We can use the name-finding function e^n to produce an $r\Sigma_{n+1}$ Skolem function. If $\varphi(u,v)$ is a Σ_1 formula in the language of M^n, let $\varphi^*(u,v)$ be the natural Σ_1 formula expressing "$\exists x \exists y(d^n(x) = u \wedge d^n(y) = v \wedge \varphi(x,y))$". The $r\Sigma_{n+1}$ relations are naturally indexed by the Σ_1 formulae of the form φ^*. We set

$$h_M^{n+1}(\varphi^*, x) = d^n(h_{M^n}^1(\varphi^*, e^n(x))).$$

DEFINITION 2.3.9. Let M be a premouse and $0 \leq n \leq k(M)$; then

(1) h_M^{n+1} is the *canonical $r\Sigma_{n+1}^M$ Skolem function* for M.
(2) For $X \subseteq M$, $\mathrm{Hull}_{n+1}^M(X) = \{h^{n+1}(\varphi, s) \mid s \in X^{<\omega} \wedge \varphi \in V_\omega\}$.
(3) $\mathrm{cHull}_{n+1}^M(X)$ is the transitive collapse of $\mathrm{Hull}_{n+1}^M(X)$.

---

[19] Lemma 2.3.7 holds trivially when $n = 0$, except possibly (6). If $\hat{o}(M)$ is a limit ordinal, then (6) holds.

It is clear that $\mathrm{Hull}^M_{n+1}(X)$ is closed under (lightface) $r\Sigma_{n+1}$ functions, and in particular, $p^M_n \in \mathrm{Hull}^M_{n+1}(X)$, and $\mathrm{Hull}^M_{n+1}(X)$ is closed under the coding and decoding functions e^n and d^n. Also[20,21]

$$\mathrm{Hull}^M_{n+1}(X) \cap M^n \prec_{\Sigma_1} M^n,$$

and under a natural notion of $r\Sigma_n$ formulae,

$$\mathrm{Hull}^M_{n+1}(X) \prec_{r\Sigma_{n+1}} M.$$

Solidity, universality, and cores.

DEFINITION 2.3.10. Let M be a premouse, $k \leq k(M)$, $\alpha < \rho_k(M)$ and $r \in [\rho_k(M)]^{<\omega}$; then

$$W^{\alpha,r}_M = \mathrm{cHull}^M_{k+1}(\alpha \cup r \cup p_k(M)).$$

If $\alpha \in p_{k+1}(M)$ and $r = p_{k+1}(M) - (\alpha+1)$, then we call $W^{\alpha,r}_M$ the *standard solidity witness* for α. We say $p_{k+1}(M)$ is *solid* iff all its standard solidity witnesses belong to M.[22]

DEFINITION 2.3.11. Let M be a premouse and $k \leq k(M)$. We say that r is $k+1$-*universal* over M if for $\rho = \rho_{k+1}(M)$ and $W = W^{\rho,r}_M$,

(a) $M|\rho^{+,M} = W|\rho^{+,W}$, and
(b) for any $A \subseteq \rho$, A is boldface $r\Sigma^M_{k+1}$ iff A is boldface $r\Sigma^W_{k+1}$.

This strengthens the notion of universality employed in [35] a bit. The strengthening will be useful later. The proof in [35] that premice produced in a background construction have universal parameters shows that the parameters are universal in this stronger sense.

The soundness required to qualify as a premouse is that for all $\langle \alpha, k\rangle \leq l(M)$, $p_k(M|\langle \alpha, l\rangle)$ is solid and k-universal over $M|\langle \alpha, k\rangle$. In particular, M must be $k(M)$-sound. It need not be $k(M)+1$-sound.

DEFINITION 2.3.12. Let M be a premouse; then

(a) $\rho^-(M) = \rho_{k(M)}(M)$,
(b) $\rho(M) = \rho_{k(M)+1}(M)$,
(c) $p(M) = p_{k(M)+1}(M)$, and
(d) $h_M = h^{k(M)+1}_M$.

We call $\rho(M), p(M)$, and h_M *the* projectum, parameter, and Skolem function of M.

[20] We are diverging here from the terminology of [54]. Their h^{n+1}_M is essentially our denotation function d^{n+1}_M. They call it the canonical $r\Sigma^M_{n+1}$ Skolem function, but it is not actually a Skolem function. Our h^{n+1}_M is the Skolem function, so the divergent terminology seems justified.

[21] The notation cHull for the transitive collapse of a Skolem hull is due to Schlutzenberg.

[22] Here we assume $k(M) < \omega$, as we often do without mention. Cf. Remark 2.3.2.

Let M be a premouse. We define

$$\mathfrak{C}(M) = \mathfrak{C}_{k(M)+1}(M) = \mathrm{cHull}^M_{k(M)+1}(\rho(M) \cup p(M)),$$

considered as an $\mathcal{L}_0$-structure. Let $\pi\colon \mathfrak{C}(M) \to M$ be the anticollapse, and $t = \pi^{-1}(p(M))$. We say that M is $k+1$-*solid*, or M *has a core*, iff $p_{k+1}(M)$ is $k+1$-universal over M, and t is $k+1$-solid over $\mathfrak{C}(M)$. This implies that t is $k+1$-universal over $\mathfrak{C}(M)$, that $p_{k+1}(M)$ is $k+1$-solid over M, and that $t = p_{k+1}(\mathfrak{C}(M))$. If M is $k(M)+1$-solid, then we call $\mathfrak{C}(M)$ the *core of* M, and the associated π is the *anticore map*. If M is $k(M)+1$-solid, then setting

$$k(\mathfrak{C}(M)) = k(M) + 1,$$

$\mathfrak{C}(M)$ is a premouse. We say

$$M \text{ is sound iff } M = \mathrm{Hull}^M_{k(M)+1}(\rho(M) \cup p(M)).$$

Equivalently, M is sound iff $\mathfrak{C}(M)$ exists, and $M = \mathfrak{C}(M)^-$.

We may occasionally say that M is $k+1$-solid for some $k > k(M)$. This just means that M^{k+1} exists, that is, that the process of starting with M and iteratively taking cores, setting $\mathfrak{C}_{k(M)}(M) = M$ and $\mathfrak{C}_{i+1}(M) = \mathfrak{C}(\mathfrak{C}_i(M))$, does not break down by reaching some non-solid $\mathfrak{C}_i(M)$ with $i \leq k$. M^{k+1} is the reduct which codes $\mathfrak{C}_{k+1}(M)$. We say that M is $k+1$ *sound* if M is $k+1$ solid, and $M = \mathfrak{C}_{k+1}(M)$. (If we ignore the distinguished soundness degrees, that is.)

For the notion of *generalized solidity witness*, see [54]. Roughly speaking, a generalized solidity witness for $\alpha \in p_1(M)$ is a transitive structure whose theory includes $\mathrm{Th}^M_1(\alpha \cup p_1(M) - (\alpha+1))$. If W is a generalized witness, then $\mathrm{Th}^M_1(\alpha \cup p_1(M) - (\alpha+1))$ is an initial segment of $\mathrm{Th}^W_1(\alpha \cup p_1(W) - (\alpha+1))$ in the natural prewellordering of Σ^M_1, so we can recover the standard witness from any generalized witness. Generalized witnesses are important because being a generalized witness for an $\alpha \in p_k(M)$ is an $r\Pi_k$ condition, hence preserved by $r\Sigma_k$ embeddings. Such embeddings may not preserve being a standard witness.[23]

Remark 2.3.13. We have defined cores here as they are defined in [54]. In [35] they are defined in slightly different fashion. First, [35] works directly with the $\mathfrak{C}_{k+1}(M)$, rather than with the reducts which code them. The translations indicated above show that is not a real difference; see [35], page 40. Second, if $k \geq 1$, then [35] puts the standard solidity witnesses for $p_k(M)$ into the hull collapsing to $\mathfrak{C}_{k+1}(M)$, and if $k \geq 2$, it also puts $\rho_{k-1}(M)$ into this hull if $\rho_{k-1}(M) < o(M)$. The definition from [54] used above does not do this directly. We are grateful to Schindler and Zeman for pointing out that nevertheless these objects do get into the cores as defined in [54], and therefore the two definitions of $\mathfrak{C}_{k+1}(M)$ are

---

[23] [35] defines the solidity witnesses to be theories: instead of $W^{\alpha,r}_M$, the witness is $\mathrm{Th}^M_{k+1}(\alpha \cup r)$. This does not quite work in all contexts, because one needs that the wellfoundedness of the $\in$ relation coded into this theory is preserved under embeddings mapping M into wellfounded models. Without knowing $W^{\alpha,r}_M \in M$, this is not clear. The corrected definition is due to Jensen.

equivalent. (For example, let $k = 2$ and let M be 1-sound, with $\alpha \in p_1(M)$. Let $r = p_1(M) \setminus (\alpha + 1)$. Let $\pi\colon \mathfrak{C}_2(M) \to M$ be the anticore map, and $\pi(\beta) = \alpha$ and $\pi(s) = r$. The relation "W is a generalized solidity witness for α, r" is Π_1 over M. (It is important to add *generalized* here. Being a standard witness is only Π_2.) Since π is Σ_2 elementary, there is a generalized solidity witness for β, s over $\mathfrak{C}_2(M)$ in $C_2(M)$. But any generalized witness generates the standard one (see [54, Lemma 7.4]), so the standard solidity witness U for β, s is in $\mathfrak{C}_2(M)$. Being the standard witness is Π_2, so $\pi(U)$ is the standard witness for α, r, and this witness is in $\mathrm{ran}(\pi)$, as desired. A similar calculation shows that being equal to ρ_1 can be expressed by a Π_3 formula, the Π_3 clause being "for all $\alpha < \rho_1$ there is a generalized witness for $\mathrm{Th}_1(\alpha \cup p_1)$". But π is $r\Sigma_3$ elementary, so $\pi(\rho_1(\mathfrak{C}_2(M))) = \rho_1(M)$.)

Remark 2.3.14. The pfs premice defined in Chapter 4 will differ from Jensen premice in that *all* the $\rho_i(M)$ for $i \leq k+1$ are added as points to the hull collapsing to the counterpart of $\mathfrak{C}_{k+1}(M)$.

Extension of embeddings. The extension-of-embeddings lemmas relate reducts to the structures they code. The downward extension of embeddings lemma tells us that if S is amenable and $\pi\colon S \to N^n$ is Σ_0, then there is a (unique) M such that $S = M^n$. The upward extension lemma tells us that if $\pi\colon M^n \to S$ is Σ_1 and preserves the wellfoundedness of certain relations (the important one being $\in^M$ as it is described in the predicate of M^n), then there is a unique N such that $S = N^n$.[24] In both cases, there is a unique $\hat{\pi}\colon M \to N$ extending π, given by $\hat{\pi}(d^n_M(x)) = d^n_N(\pi(x))$ for all $x \in M^n$.

Some care is needed in applying these lemmas. In the downward case, it is possible that $\pi\colon S \to N$ is Σ_0, N is a premouse, and S is not. The problem is that $\dot{F}^S$ may not measure all sets in S.[25] But if π is Σ_1 elementary, then S is a premouse, and this will pretty much always be the case in this book.

The importance of solidity shows up in our statement of the upward extension lemma. In order to conclude that N is a premouse and $\hat{\pi}$ is $r\Sigma_{n+1}$-elementary, we need to use the solidity of $p_n(M)$. When we decode the name for p_n that is part of S, we do indeed get a parameter $r = \hat{\pi}(p_n(M))$ that generates all of N modulo $o(S)$, simply by construction. $S = N^{n,r}$. So $\rho_n(N) \leq o(S)$, and $o(S) \leq \rho_n(N)$ can be shown using the amenability of S. Thus $p_n(N) \leq r$ in the parameter order.[26] But it is the fact that being a generalized solidity witness is preserved that lets us conclude that r is solid over N, and hence $p_n(N) = r$. It is easy to see no $t <_{\mathrm{lex}} r$ can generate r modulo $o(S)$, so if $r \neq p_n(N)$, then N is not n-sound, hence not a premouse.

---

[24]See Lemmas 5.10 and 5.11 of [54].

[25]One important context in which this happens is in the proof that iterable premice satisfy Jensen's $\square$ principle. See [49]. Premouse-like structures with a partial top extender are called *protomice*, and they are responsible for many of the difficulties overcome in [49].

[26]Regarding parameters as finite descending sequences of ordinals, this is the lexicographic order, so we often write $p \leq_{\mathrm{lex}} q$ for it.

Schindler and Zeman prove a more abstract upward extension lemma in [54]. They first define $M^{k,q}$ for arbitrary q, and then set $M^k = M^{k,p_k(M)}$. Their upward extension lemma then just asserts the existence of a premouse N such that $S = N^{n,\pi(p_n(M))}$. The more abstract lemma is useful in practice, because it separates the elementary facts to do with coding and decoding from the much less elementary, premouse-specific question as to whether π preserves the standard parameter.

We shall discuss the extension of embeddings lemmas in more detail in Section 4.1.

2.4. Elementarity of maps

Given premice M and N, $n = k(M) = k(N) < \omega$, and

$$\pi \colon M^n \to N^n$$

a Σ_0 elementary embedding on their n-th reducts, then by decoding the reducts we get a unique

$$\hat{\pi} \colon M \to N$$

that is Σ_n elementary and is such that $\pi \subseteq \hat{\pi}$. If π is Σ_1 elementary, then $\hat{\pi}$ is Σ_{n+1} elementary. The decoding is done iteratively, and yields that for $k < n$, $\hat{\pi} \colon M^k \to N^k$ is Σ_{n-k} or Σ_{n-k+1}, respectively. $\hat{\pi}$ is called the *n-completion* of π.[27] See Lemmas 5.8 and 5.9 of [54]. These lemmas record additional elementarity properties of $\hat{\pi}$, namely *$r\Sigma_{n+1}$-elementarity* if π is Σ_1, and *weak $r\Sigma_{n+1}$-elementarity* if π is only Σ_0. (See [54, §5.12].) Such maps are *cardinal preserving* in that $M \models$ "γ is a cardinal" iff $N \models$ "$\pi(\gamma)$ is a cardinal", except possibly the weakly $r\Sigma_0$ maps. In this case, we shall always just add cardinal preservation as an additional hypothesis. This leads us to:

DEFINITION 2.4.1. Let M and N be Jensen premice such that $k(M) = k(N)$, and $\pi \colon M \to N$; then letting $n = k(M)$,

(a) *π is weakly elementary* iff π is the n-completion of $\pi \restriction M^n$, and $\pi \restriction M^n \colon M^n \to N^n$ is Σ_0 and cardinal preserving.
(b) *π is elementary* iff π is weakly elementary, and $\pi \restriction M^n \colon M^n \to N^n$ is Σ_1.
(c) π is *cofinal* iff $\sup \pi``\rho_n(M) = \rho_n(N)$.

We should note that the reduct M^n has a name for $p_n(M)$ built into its language[28], so a weakly elementary $\pi \colon M \to N$ must by definition preserve $p_k(M)$ for $k \leq k(M)$. Weakly elementary maps are Σ_k elementary maps that preserve the p_k for $k \leq k(M)$, although this is not quite all there is to the concept.

[27]If $k(M) = k(N) = \omega$, then a Σ_0 elementary $\pi \colon M^\omega \to N^\omega$ determines a Σ_ω elementary $\hat{\pi} \colon M \to N$. As usual, our discussion is focused on the case $n < \omega$.

[28]Directly, only a name for $r_n(M)$, but indirectly a name for $p_n(M)$.

Formally, Definition 2.4.1 only applies when M and N have the same distinguished soundness degree. However, it is easy to see that if $\pi\colon M \to N$ is weakly elementary and $i < k(M)$, then $\pi\colon M \downarrow i \to N \downarrow i$ is elementary (and more). One could be pedantic and associate a degree to π itself, but we will not do that. The one caution is that π could be cofinal as a map from M to N, but not as a map from $M \downarrow i$ to $N \downarrow i$, where $i < k(M)$. That is quite common, in fact.

The elementary maps correspond to those which are near n-embeddings in the sense of [47]. The cofinal elementary maps correspond to the n-embeddings of [35]. When $n \geq 1$, the weakly elementary embeddings correspond to those that are n-apt in the sense of [47], $\Sigma_0^{(n)}$ in the sense of [86], or n-lifting in the sense of [56]. There are many other levels of elementarity isolated in these references. The elementarity notions that will come up in this book are defined in this and the next section in the context of Jensen premice. We adapt them to pfs premice in Section 4.3. There are essentially two: elementarity, and what we shall call *near elementarity*.[29]

Here are two basic sources of cofinal elementary maps.

LEMMA 2.4.2. *Let M be a solid premouse, $N = \mathfrak{C}(M)^-$, and let $\pi\colon N \to M$ be the anticore map; then letting $k = k(M) = k(N)$,*

(a) π is cofinal and elementary,
(b) if $\rho_{k+1}(M) = \rho_k(M)$, then $\pi = id$, and
(c) if $\pi \neq id$, then $\rho_{k+1}(N) \leq \operatorname{crit}(\pi) < \rho_k(N)$.

PROOF. We start with (a). By construction, π is the completion of $\hat{\pi} = \pi \restriction N^k$, which is Σ_1 elementary as a map from N^k to M^k. So π is elementary. Clearly $\operatorname{crit}(\pi) \geq \rho_{k+1}(M) = \rho_{k+1}(N)$. If $\rho_{k+1}(M) = \rho_k(M)$, then $N^k = M^k$ and since M and N are k-sound, $\hat{\pi}$ is the identity, hence cofinal. If $\rho_{k+1}(M) < \rho_k(M)$ and $\sup(\operatorname{ran}(\hat{\pi})) < \alpha < \rho_k(M)$, then $\operatorname{Th}_1^{M^k}(\rho(M) \cup r) = \operatorname{Th}_1^{M^k||\alpha}(\rho(M) \cup r)$, where $p(M) = p_k(M)^\frown r$ and $M^k||\alpha = (M||\alpha, A \cap \alpha)$. But $M^k||\alpha \in M^k$, so the new $\Sigma_1^{M^k}$ subset of $\rho(M)$ was not new after all. Thus again π is cofinal, and we have proved (a).

We have already observed that (b) holds. For (c): if $\operatorname{crit}(\pi) \geq \rho_k(N)$, then since π is cofinal and elementary, $N^k = M^k$. But as we noted, this implies $M = N$ and $\pi = \text{id}$. ⊣

The second basic source is fine structural ultrapowers. If M is a premouse with $n = k(M)$, and E is a short extender over M with $\kappa_E < \rho_n(M)$ and

$$P(\kappa_E)^M \subseteq \operatorname{dom}(E),$$

[29] We shall not use the notion of *weak n-embedding* defined in [35]. In the end, that notion is not very natural, and in a number of places it does not do the work that the authors of [35] thought that it did. In particular, there are problems with how it was used in the Shift Lemma, the copying construction, and the Weak Dodd-Jensen Lemma. These problems are discussed in [56], and a variety of ways to repair the earlier proofs are given.

then we set

$$\begin{aligned}\mathrm{Ult}(M,E) &= \mathrm{Ult}_n(M,E)\\ &= \text{decoding of } \mathrm{Ult}_0(M^n,E)\\ i_E^M &= \text{completion of canonical } i\colon M^n \to \mathrm{Ult}(M^n,E).\end{aligned}$$

By convention, $k(\mathrm{Ult}(M,E)) = k(M)$.

LEMMA 2.4.3. *Let M be a premouse and E be a short extender over M such that* $\mathrm{crit}(E) < \rho_{k(M)}(M)$ *and* $\mathrm{Ult}(M,E)$ *is wellfounded; then* $\mathrm{Ult}(M,E)$ *is a premouse, and the canonical embedding* $i_E^M : M \to \mathrm{Ult}(M,E)$ *is cofinal and elementary.*

PROOF. We assume here that $\mathrm{Ult}(M,E)$ is wellfounded, but one could make sense of these statements even if it is not. Let $n = k(M)$. It is easy to see from Łoś's Theorem that the canonical embedding $i\colon M^n \to \mathrm{Ult}_0(M^n,E)$ is Σ_1 and cofinal. If $n = 0$ and M is active, with $\kappa = \mathrm{crit}(\dot{F}^M)$, then also $i``\kappa^{+,M}$ is cofinal in $i(\kappa^{+,M})$. This implies that i preserves rQ sentences, and hence $\mathrm{Ult}(M,E)$ is a premouse.[30]

Let $S = \mathrm{Ult}_0(M^n,E)$ and $N = \mathrm{Ult}(M,E)$. We must see that $S = N^n$. The abstract part of the upward extension lemma tells us that $S = N^{n,\pi(p)}$, where $p = p_n(M)$. (Cf. [54, §4].) So we just need to see that $\pi(p) = p_n(N)$. But this follows from the fact that M is n sound, as we saw above: $\rho_n(N) \leq o(S)$ because $S = N^{n,\pi(p)}$, and $o(S) \leq \rho_n(N)$ can be shown using the amenability of S. So $p_n(N) \leq_{\mathrm{lex}} \pi(p)$ in the parameter order. But the solidity witnesses for p are moved by π to generalized solidity witnesses for $\pi(p)$, so $\pi(p)$ is solid, and hence $\pi(p) \leq_{\mathrm{lex}} p_n(N)$.

Thus $i_E^M : M \to \mathrm{Ult}(M,E)$ is the n-completion of i, and hence it is cofinal and elementary. ⊣

The coding and decoding involved in the definition of $\mathrm{Ult}(M,E)$ can obscure its properties. It is sometimes better to think of $\mathrm{Ult}(M,E)$ as an ordinary ultrapower formed using $r\Sigma_{k(M)}$ functions, as in [35]. More precisely, letting $n = k(M)$ and $\kappa = \mathrm{crit}(E)$,

$$\mathrm{Ult}(M,E) = \{[a,f]_E^M \mid a \in [\lambda]^{<\omega} \wedge \mathrm{dom}(f) = [\kappa]^{|a|} \wedge f \text{ is boldface } r\Sigma_n\}.$$

Of course, the coding into M^n is still present in the $r\Sigma_n$ definition of f, but it can sometimes help to think of $\mathrm{Ult}(M,E)$ this way. We do indeed get the same ultrapower. Each function $g \in M^n$ used in $\mathrm{Ult}_0(M^n,E)$ corresponds to the function

$$g^*(u) = d^n(g(u))),$$

which is $r\Sigma_n$ in the parameters g and $p_n(M)$. The decoding of $[a,g]_E^{M^n}$ is $[a,g^*]_E^M$. Conversely, if f is a boldface $r\Sigma_n$ function on $[\kappa]^{|a|}$, then $f = g^*$ for some $g \in M^n$.

DEFINITION 2.4.4. Let M be a premouse and $n \leq k(M)$, then

(1) $\mathrm{sk}_n = \{\langle n, \tau(u,v)\rangle \mid \tau(u,v) \text{ is } \Sigma_1 \text{ in the language of reducts}\}$.
(2) For $u,q \in M^n$ and $\tau \in \mathrm{sk}_n$, $f_{\tau,q}^M(u) = d_M^n(\langle \tau, \langle q,u\rangle\rangle)$.

[30] See [35, §2].

The $f^M_{\tau,q}$ parametrize the partial boldface $r\Sigma_n$ functions with domain contained in M^n.[31] One can think of τ as an $r\Sigma_n$ definition of f from q. One only needs parameters $q \in M^n$ because M is n-sound. Clearly $f^M_{\tau,q} = g^*$, where $g(u) = \langle \tau, \langle q, u \rangle \rangle$, so our two versions of $\mathrm{Ult}(M,E)$ are isomorphic, and we have

$$\mathrm{Ult}(M,E) = \{[a, f^M_{\tau,q}]^M_E \mid \tau \in \mathrm{sk}_n \wedge q \in M^n \wedge \mathrm{dom}(f^M_{\tau,q}) = [\kappa]^{|a|}\}.$$

One has the usual Łoś's Theorem for $r\Sigma_n$ formulae, so the ultrapower map is $r\Sigma_{n+1}$-elementary. Letting $i = i^M_E$, we get

$$[a, f^M_{\tau,q}]^M_E = i(f)(a) = f^{\mathrm{Ult}(M,E)}_{\tau,i(q)}(a).$$

Let us return to the general setting. Here are some basic facts regarding preservation of parameters and projecta. In stating them, we adopt the convention that if $\pi\colon M \to N$, where M and N are premice, then $\pi(o(M)) = o(N)$.

PROPOSITION 2.4.5. *Let M and N be Jensen premice with $n = k(M) = k(N)$, and $\pi\colon M \to N$ be weakly elementary; then*

(1) π is Σ_n elementary,
(2) $\pi(p_k(M)) = p_k(N)$ for all $k \le n$, and
(3) (a) $\sup \pi``\rho_k(M) \le \rho_k(N)$ for all $k \le n$,
(b) $\pi(\rho_k(M)) = \rho_k(N)$ for $k < n-1$, and
(c) $\rho_{n-1}(N) \le \pi(\rho_{n-1}(M))$.
(4) For any $\alpha < \rho_n(M)$ and $r \in M$, $\pi(Th^M_n(\alpha \cup \{r\}) = Th^N_n(\pi(\alpha) \cup \{\pi(r)\})$.

PROOF. (1) and (2) are part of the extension of embeddings lemmas. (3)(a) is also implicit there, since $\pi \restriction M^k\colon M^k \to N^k$ is a stage in completing $\pi \restriction M^n$.

(3)(c) is true by convention if $\rho_{n-1}(M) = o(M)$. For $k \le n$, let

$$\varphi_k(u,v) = \forall x \exists \alpha < u(x = h^k(\alpha, v)).$$

Here h^k is the canonical $r\Sigma_k$ Skolem function.[32] φ_k is the natural $r\Pi_{k+1}$ formula expressing the quoted one. $M \models \varphi_k[\rho_k(M), p_k(M)]$, so if $k \le n-1$ then $N \models \varphi_k[\pi(\rho_k(M)), p_k(N)]$, and therefore $\rho_k(N) \le \pi(\rho_k(M))$. This gives us (3)(c).

For (3)(b), note that "$x = \mathrm{Th}_k(y)$" can be expressed by a boolean combination of Π_k formulae. So "$v \le \rho_k$" can be expressed by a Π_{k+2} formula, namely

$$\psi_k(v) = \forall \alpha < v \exists x(x = \mathrm{Th}_k(\alpha \cup p_k)),$$

and "$\rho_k = \mathrm{OR}$" can be expressed by a Π_{k+2} sentence, namely $\theta_k =$ "$\forall \alpha \exists x(x = \mathrm{Th}_k(\alpha \cup p_k))$". If $k < n-1$ then π preserves φ_k, ψ_k, and θ. This yields (3)(b).

(4) follows from the fact that "$x = \mathrm{Th}_n(y)$" can be expressed by a Boolean combination of Π_n formulae. Or we can just note that as a predicate on the reduct M^n, $x = \mathrm{Th}_n(\alpha \cup p_n(M))$ is Σ_0, so preserved by π, and take α large enough that $r = h^n_M(\beta, p_n(M))$ for some $\beta < \alpha$. ⊣

[31] Our definition of $f^M_{\tau,q}$ here is slightly different than that in [35], but not in any important way.

[32] One could also use the decoding function d^k at this point.

Note that we do not necessarily have that $\pi(\rho_{n-1}(M)) = \rho_{n-1}(N)$, or $\rho_n(N) \leq \pi(\rho_n(M))$, or that π is $r\Sigma_{n+1}$-elementary on a set cofinal in $\rho_n(M)$. These are the additional requirements from [35] on weak n-embeddings. We do need to make use of the first two of these requirements later, so we make a definition.

DEFINITION 2.4.6. Let $\pi\colon M \to N$ be weakly elementary, and $k = k(M)$; then

(a) π *respects projecta* iff $\pi(\rho_j(M)) = \rho_j(N)$ for all $j < k$,
(b) π is *almost exact* iff $\rho_k(N) \leq \pi(\rho_k(M))$, and
(c) π is *exact* iff $\rho_k(N) = \pi(\rho_k(M))$.

We use here the convention that $\pi(o(M)) = o(N)$. Thus if $k(M) = 0$, then π respects projecta and is exact, and if $k(M) = 1$, then π respects projecta. So exactness only comes into play when $k(M) \geq 1$, and respecting projecta only comes into play when $k(M) \geq 2$. By Proposition 2.4.5, preservation of $\rho_{k(M)-1}$ is the only issue in (a). However, so far as we can see, weakly elementary maps may not preserve $\rho_{k(M)-1}$.

Elementary maps respect projecta and are almost exact, by the calculations we just did.[33]

PROPOSITION 2.4.7. *If $\pi\colon M \to N$ is elementary, then π respects projecta and is almost exact.*

PROOF. Let $n = k(M)$. We have $\rho_n(N) \leq \pi(\rho_n(M))$ because "$\rho_n \leq v$" is expressed by the Π_{n+1} formula $\varphi_n(v, p_n)$ displayed above. We get that $\pi(\rho_{n-1}(M)) = \rho_{n-1}(N)$ from the fact that π preserves $\varphi_{n-1}, \psi_{n-1}$, and θ_{n-1}. ⊣

As we shall see in the next section, the ultrapower map $\pi\colon M \to \mathrm{Ult}(M,E)$ may be discontinuous at $\rho^-(M)$. In that case, it is a cofinal, elementary map that is not exact.

In a similar vein,

PROPOSITION 2.4.8. *Let $\pi\colon M \to N$ be weakly elementary, $n = k(M)$, and suppose that if $n \geq 2$, then either $\rho_{n-1}(N) = o(N)$ or $\rho_{n-1}(N) \in \mathrm{ran}(\pi)$. Then π respects projecta.*

PROOF. If $\rho_{n-1}(N) = o(N)$, then $N \models \theta_{n-1}$. But θ_{n-1} is Π_{n+1} and π is Σ_n-elementary, so $M \models \theta_{n-1}$, so $\rho_{n-1}(M) = o(M)$. Similarly, if $\rho_{n-1}(N) = \pi(\mu)$, we get that $M \models (\varphi_{n-1} \wedge \psi_{n-1})[\mu]$, so $\mu = \rho_{n-1}(M)$. ⊣

The lifting maps that occur in the construction of background-induced iteration strategies are weakly elementary and respect projecta, but may not be elementary or almost exact. One can see why by considering the following simple examples.

Example 2.4.9. Let M be a premouse and let E be an extender on the M-sequence. Suppose that $E \restriction \lambda_E = (E^* \restriction \lambda_E) \cap M$, where E^* is an extender over V. Let

$$\sigma\colon \mathrm{Ult}(M,E) \to i^V_{E^*}(M)$$

[33] In fact, elementary maps are weak n-embeddings in the sense of [35].

be the natural map, given by completing

$$\sigma([a,f]_E^{M^k}) = [a,f]_{E^*}^V,$$

where $k = k(M)$, $a \subseteq \lambda_E$ is finite, and $f \in M^k$. Suppose $k = 0$, so $M^k = M$. It is clear that σ is Σ_0-elementary, but it may not be Σ_1-elementary. For example, let $\kappa = \text{crit}(E)$, and suppose $o(M) = \eta + \kappa$ for some η; then $o(\text{Ult}(M,E)) = i_E^M(\eta) + \kappa$, but $o(i_{E^*}^V(M)) = i_{E^*}^V(\eta) + i_{E^*}^V(\kappa)$. We have $\sigma(i_E^M(\eta)) = i_{E^*}^V(\eta)$ and $\sigma(\kappa) = \kappa$. Thus if $\varphi(u,v)$ is the Σ_1 formula "$u+v$ exists", then $\text{Ult}(M,E) \models \neg\varphi[i_E^M(\eta),\kappa]$ but $i_{E^*}^V(M) \models \varphi[\sigma(i_E^M(\eta)),\sigma(\kappa)]$.

One can construct similar failures of elementarity with $k(M) > 0$.

Example 2.4.10. Let M, E, E^*, and σ be as in Example 2.4.9, but suppose now that $k(M) = 1$, $\text{crit}(E) < \rho_1(M)$, and $\rho_1(M)$ has Σ_1 cofinality $\text{crit}(E)$ in M. By Lemma 2.5.6, i_E^M is discontinuous at $\rho_1(M)$, so

$$\rho_1(Ult(M,E)) = \sup i_E``\rho_1(M) < i_E(\rho_1(M)).$$

On the other hand, $\sigma(i_E(\rho_1(M)) = i_{E^*}(\rho_1(M)) = \rho_1(i_{E^*}(M))$. It follows that σ is not almost exact.

It is easy to see that the lifting maps σ in these examples do respect projecta. That is true in general of the lifting maps that occur in the construction of background-induced iteration strategies.

Note that if $\pi\colon M \to N$ is weakly elementary, and $k = k(M) = k(N)$, then π moves generalized solidity witnesses for $p_k(M)$ to generalized solidity witnesses for $p_k(N)$. For example, being a generalized witness for $p_1(M)$ is a Π_1 fact, so preserved by Σ_1 embeddings. If π is elementary, then it will move the standard solidity witnesses for $p_k(M)$ to the standard solidity witnesses for $p_k(N)$.

The preservation results above were confined to $\rho_k(M)$ and $p_k(M)$ for $k \leq k(M)$. We also need to consider what happens to $\rho_{k+1}(M)$ and $p_{k+1}(M)$. Here is the main fact concerning preservation by ultrapower maps.

DEFINITION 2.4.11. An extender E is *close to M* iff

(1) $\text{dom}(E) = \text{dom}(F)$, for some F on the sequence of M, and
(2) for all finite $a \subseteq \lambda(E)$,
   (a) E_a is Σ_1^M in parameters, and
   (b) for all $\alpha < \kappa_E^{+,M}$, $E_a \cap M|\alpha \in M$.

We say that E is *very close to M* iff E is close to M, and for all finite $a \subseteq \lambda(E)$, $E_a \in M$.

We have added item (1) to the standard definition of closeness from [35]. In Lemma 4.5.3 we show that all extenders used in a normal iteration tree are close to the models to which they are applied in this slightly stronger sense. We show also that they are often very close.

LEMMA 2.4.12. *Let M be a premouse and let E be an extender that is close to M such that $\rho(M) \leq \text{crit}(E)$, and let $N = \text{Ult}(M, E)$; then*

(a) for $A \subseteq \rho(M)$, A is $\Sigma_1^{M^k}$ iff A is $\Sigma_1^{N^k}$,
(b) $\rho(M) = \rho(N)$, and
(c) if $p(M)$ is solid, then $i_E^M(p(M)) = p(N)$, and $p(N)$ is solid.

PROOF. We assume $k(M) = 0$ for simplicity. Let $\rho = \rho(M)$ and $i = i_E^M$. If $A \subseteq \rho(M)$ and A is Σ_1^N in the parameter $[a, f]_E^M$, then by Łoś's Theorem, A is Σ_1^M in f and t, where t is such that E_a is Σ_1^M in t. This is because

$$\xi \in A \text{ iff } M \models \exists g \exists X \in E_a \forall u \in X \theta[\xi, f(u), g(u)],$$

where $\exists v\theta$ is the Σ_1 formula defining A from $[a, f]$ in N. The other direction in (a) is immediate.

This implies that $\rho \leq \rho(N)$. On the other hand, the amenability clause (2)(b) in closeness implies that $P(\rho)^M = P(\rho)^N$. So if $A \subseteq \rho$ is Σ_1^N but not in M, then A is Σ_1^N but not in N. So $\rho = \rho(N)$.

Toward (c), let $p = p(M)$ and $q = i(p)$. Let φ be a Σ_1 formula such that $\{\alpha \mid M \models \varphi[\alpha, p]\} \cap \rho \notin M$. Then $\{\alpha \mid N \models \varphi[\alpha, q]\} \cap \rho \notin N$, since the two sets are the same below ρ, and $P(\rho)^M = P(\rho)^N$. It follows that $p(N) \leq_{\text{lex}} q$. But if $\alpha \in p$, then there is a generalized solidity witness W for α such that W in M. Being a generalized witness is Π_1, so $i(W)$ is a generalized witness for $i(\alpha)$ in q. It is easy to see that the existence of these witnesses implies that $q \leq_{\text{lex}} p(N)$. So $q = p(N)$; moreover, $p(N)$ has solidity witnesses. ⊣

The solidity of $p(M)$ was crucial in showing that it is preserved this way by ultrapowers. This is perhaps the main reason one must prove that $p(M)$ is solid in order to get a reasonable theory going.

We shall resume our discussion of levels of elementarity at the end of the next section, after we have introduced a stronger notion of respect for projecta.

2.5. $r\Sigma_k$ cofinality and near elementarity

It is natural to ask whether there are ordinals that measure definable cofinalities in the same way that projecta measure definable cardinalities. There are, and we shall make use of them later.[34]

DEFINITION 2.5.1. Let M be a premouse and $k \leq k(M)$.

(a) For any $\gamma \leq o(M)$, $\text{cof}_k^M(\gamma)$ is the least η such that there is a partial boldface $r\Sigma_k^M$ function f such that $f``\eta$ is cofinal in γ.
(b) γ is *$r\Sigma_k$-singular* in M iff $\text{cof}_k^M(\gamma) < \gamma$, and $r\Sigma_k$-regular otherwise.
(c) $\eta_k^M = \text{cof}_k(\rho_k(M))^M$.

[34] Many of the mildly new lemmas in this section were proved independently by Farmer Schlutzenberg. See [59].

We say that an f as in (a) is a *witness that* $\text{cof}_k(\gamma) = \eta$.

We allow $\gamma = o(M)$ here. $\text{cof}_0(o(M)) = o(M)$, and $\text{cof}_0(\gamma)$ is just the usual cofinality of γ with respect to functions $f \in M$ if $\gamma < o(M)$. When $k = 1$, the $r\Sigma_1$ functions are just the Σ_1 functions, and this is a good special case to keep in mind.[35]

Clearly if $\gamma < \rho_k(M)$, where $k \leq k(M)$, then $\text{cof}_k^M(\gamma) = \text{cof}_0^M(\gamma)$. Equally clearly, $\text{cof}_k^M(\gamma) \leq \rho_k(M)$ for all γ, but this is somewhat misleading, because one cannot always take the witnessing function to be order preserving.

DEFINITION 2.5.2. Let f witness that $\text{cof}_k(\gamma) = \eta$. We say that f is *nice* iff $f\colon \eta \to \gamma$, and f is total and strictly order preserving,

PROPOSITION 2.5.3. *If* $\text{cof}_k^M(\gamma) < \rho_k(M)$*, then there is a nice witness to this fact. Moreover,* $\text{cof}_k^M(\gamma)$ *is the unique* $\eta < \rho_k(M)$ *such that* η *is* Σ_0*-regular in* M*, and there is a total, strictly order preserving,* $r\Sigma_k$ *function* $f\colon \eta \to \gamma$ *with range cofinal in* γ.

PROOF. Let f witness that $\text{cof}_k(\gamma) = \eta$. Let R be the prewellorder of η induced by f:

$$\alpha R \beta \text{ iff } f(\alpha) < f(\beta).$$

Because $\eta < \rho_k^M$, $R \in M$. No $X \in M$ such that $|X|^M < \eta$ can be R-cofinal, by the minimality of η. We define $h\colon \eta \to \eta$ by induction:

$$h(\gamma) = \text{least } \xi \text{ such that } \forall \beta < \gamma (\beta R \xi \wedge h(\beta) R \xi).$$

$h(\gamma)$ is defined because otherwise, $X = \gamma \cup h``\gamma$ is $R$-cofinal. It is clear that $h``\eta$ is R-cofinal, and $h \in M$, and $\alpha < \beta$ implies $h(\alpha) R h(\beta)$. So setting

$$g(\gamma) = f \circ h(\gamma),$$

g is a nice witness that $\text{cof}_k^M(\gamma) < \rho_k(M)$.

To see the uniqueness assertion, suppose that (f_0, η_0) and (f_1, η_1) were two such pairs. We define $h\colon \eta_0 \to \eta_1$ by $h(\alpha) =$ least β such that $f_0(\alpha) < f_1(\beta)$. Since we are below $\rho_k(M)$, $h \in M$. Clearly h is non-decreasing and $\text{ran}(h)$ is cofinal in η_1. Since η_1 is 0-regular, $\eta_1 \leq \eta_0$. Symmetrically, $\eta_0 \leq \eta_1$, so $\eta_0 = \eta_1$. ⊣

One can arrange that the nice witness to $\text{cof}_k^M(\rho_k(M)) = \eta_k^M$ is continuous at limit ordinals.

LEMMA 2.5.4. *Let* M *be a premouse and* $k = k(M)$*. Suppose* $\eta_k^M < \rho_k(M)$*; then there is a nice witness* f *to this fact such that* f *is continuous at limit ordinals.*

PROOF. Since $\eta_k^M < o(M)$, $k > 0$. Let

$$\eta = \eta_k^M = \eta_1^{M^{k-1}},$$

$$\rho = \rho_k(M) = \rho_1(M^{k-1}).$$

[35]We may occasionally use η_k^M, ρ_k^M, and p_k^M interchangeably with $\eta_k(M), \rho_k(M)$, and $p_k(M)$.

Let f be a nice witness that $\mathrm{cof}_1^{M^{k-1}}(\rho) = \eta$, say

$$f(\gamma) = \xi \text{ iff } M^{k-1} \models \varphi[\xi, \gamma, q]$$

where φ is Σ_1 and $q \in M^{k-1}$.

If $M \models$ "ρ is singular", then we can take $f \in M$ and the lemma is obvious, so assume ρ is regular in M. For $\theta < \rho_{k-1}(M)$, let $M^{k-1}||\theta = (M||\theta, A^M_{k-1} \cap M||\theta)$ and

$$f_\theta(\gamma) = \xi \text{ iff } M^{k-1}||\theta \models \varphi[\xi, \gamma, q].$$

Each f_θ is in M^{k-1}, and the function $\theta \mapsto f_\theta$ is Σ_1 over M^{k-1}. For any $\alpha < \eta$ there is a $\theta < \rho_{k-1}(M)$ such that $\alpha \subseteq \mathrm{dom}(f_\theta)$, since otherwise the function sending $\beta < \alpha$ to the least θ such that $\beta \in \mathrm{dom}(f_\theta)$ witnesses that $\mathrm{cof}_k^M(\rho) < \eta$. For $\alpha < \eta$, let

$$g(\alpha) = \text{least } \theta \text{ such that } \alpha \subseteq \mathrm{dom}(f_\theta),$$

and

$$h(\alpha) = \sup(\{f_{g(\alpha)}(\xi) \mid \xi < \alpha\}).$$

$h(\alpha) < \rho$ because $f_{g(\alpha)} \in M^{k-1}$. Clearly h is a continuous nice witness that $\mathrm{cof}_k^M(\rho) = \eta$. ⊣

Remark 2.5.5. Here is an example that shows the hypothesis $\mathrm{cof}_k(\gamma) < \rho_k$ in Proposition 2.5.3 is needed. Let M be a premouse such that $k(M) = 1$, $M \models \mathsf{KP}$, and such that for $\rho = \rho_1(M)$, $M \models$ "ρ^+ exists". (M could be the first initial segment of L satisfying KP plus "ω_1 exists". Then $\rho = \omega$, and $p_1(M) = \{\omega_1^M\}$.) Let $\gamma = \rho^{+,M}$. By our definition, $\mathrm{cof}_1^M(\gamma) \leq \rho$, but there can be no partial order preserving witness. This is because if f is order preserving, $\gamma - \mathrm{ran}(f)$ is also Σ_1, so $\mathrm{ran}(f) \in M$ by Δ_1-comprehension. But $\mathrm{cof}_0(\gamma)^M = \gamma$, so $\mathrm{ran}(f) \notin M$.

In this book, we shall only need to deal with $r\Sigma_k$-singularity when the cofinalities are $< \rho_k$, and hence nicely witnessed. The following lemma explains where it comes up.

LEMMA 2.5.6. *Let M be a premouse, $k = k(M)$, and let E be an extender over M such that* $\mathrm{crit}(E) < \rho_k(M)$. *Suppose that* $\mathrm{Ult}(M,E)$ *is wellfounded; then for any $\gamma \leq o(M)$ the following are equivalent:*

(a) $\mathrm{cof}_k^M(\gamma) = \mathrm{crit}(E)$,
(b) i_E^M is discontinuous at γ.

PROOF. Let $\kappa = \mathrm{crit}(E)$. We use the representation of $\mathrm{Ult}(M,E)$ in terms of equivalence classes $[a, f]_E^M$, where f is a total boldface $r\Sigma_k^M$ function with domain $[\kappa]^{|a|}$.

Suppose first that $\mathrm{cof}_k^M(\gamma) = \kappa$, and let f be a nice witness to this fact. By Łoś's Theorem, $\sup i_E``\gamma \leq [\{\kappa\}, f]_E^M < i_E^M(\gamma)$, so i_E^M is discontinuous at γ.

Conversely, suppose i^M_E is discontinuous at γ, and let

$$\sup i^M_E``\gamma \leq [a,f]^M_E < i^M_E(\gamma).$$

We may assume $\text{ran}(f) \subseteq \gamma$. By Łoś's Theorem, $\text{ran}(f)$ is cofinal in γ, so $\text{cof}_k(\gamma) \leq \kappa$. Suppose toward contradiction that $\text{cof}_k(\gamma) = \eta < \kappa$, and let g be a nice witness to this fact. For $u \in [\kappa]^{|a|}$ and $\alpha < \eta$, let

$$H(u,\alpha) \text{ iff } g(\alpha) < f(u).$$

H is $r\Sigma_k$ and bounded in $M||\rho^M_k$, so $H \in M$. Letting $H_\alpha = \{u \mid H(u,\alpha)\}$, we have $H_\alpha \in E_a$ for all $\alpha < \eta$, so we can find $u \in \bigcap_{\alpha<\eta} H_\alpha$. But then $f(u) \geq \gamma$, contradiction. ⊣

We allowed $\gamma = o(M)$ in Lemma 2.5.6. Here our understanding is that

$$i^M_E(o(M)) = o(\text{Ult}(M,E)).$$

Clearly $\text{cof}^M_0(o(M)) = o(M)$, so the Lemma says that i^M_E is continuous at $o(M)$ when $k(M) = 0$, which is of course true. If $k(M) > 0$, then it is possible that i^M_E is discontinuous at $o(M)$.

Lemma 2.5.6 provides us with an example of a cofinal, elementary $\pi\colon M \to N$ such that for $k = k(M)$, $\pi(\rho_k(M)) \neq \rho_k(N)$. Starting with a premouse $Q \models \mathsf{ZFC}+$ "κ is measurable", let $M = Q|\eta + 1$, where η is the κth cardinal of M above κ. It is easy to see that $\eta = \rho_1(M)$ and $\kappa = \text{cof}^M_1(\eta)$. Letting $\pi = i^M_E$ where $E \in M$ and $\text{crit}(E) = \kappa$, we have that $\pi``\rho_1(M) = \rho_1^{\text{Ult}(M,E)} < \pi(\rho_1(M))$. By taking a Σ_2 hull of M, we can arrange that $\rho_2(M) = \omega$, and π is an anticore map.

In the examples of the last paragraph, $\text{cof}_0(\rho_1(M)) = \kappa$. It is somewhat harder to construct an example of such a discontinuity when $\rho_1(M)$ is regular in M. That situation will be cause us some trouble in Section 3.2, so we digress here to show that it can indeed occur.

Recall that $\eta^M_k = \text{cof}^M_k(\rho_k(M))$. One can construct an initial segment M of L such that $\eta^M_1 < \rho_1(M) = \text{cof}^M_0(\rho_1(M))$. In order to have measurable cardinals or nontrivial anticore maps one must go beyond L, or course, but the same construction works if one starts with measurable cardinals.

PROPOSITION 2.5.7. *There is an $M \trianglelefteq L$ such that $\eta^M_1 < \rho_1(M) = \text{cof}^M_0(\rho_1(M))$.*

PROOF SKETCH. Work in L. Let $\mu < \kappa$, with μ regular and κ inaccessible. Let $N = J_{\kappa+\mu}$, $X = \text{Hull}^N_1(\{\kappa\})$ and $Y = \text{Hull}^N_1(\gamma \cup \{\kappa\})$, where $\gamma = X \cap \kappa$. Let M be the transitive collapse of Y, and $\pi\colon M \to N$ the uncollapse. Then $\rho_1(M) = \gamma = \text{crit}(\pi)$ and $\pi(\gamma) = \kappa$. γ is 0-regular in M. It is 1-singular with cofinality μ because if we let $\gamma_\alpha = \sup(\text{Hull}^{J_{\kappa+\alpha}}_1(\{\kappa\})) \cap \kappa$, then the function $f(\alpha) = \gamma_\alpha$ witnesses $\text{cof}^M_1(\gamma) = \mu$. ⊣

In the troublesome situation later on, both $\rho_1(M)$ and η^M_1 are measurable in M.

PROPOSITION 2.5.8. *There is a premouse M such that $k(M) = 2$, $\rho_2(M) = \omega$, $\rho_2(M) < \eta^M_1 < \rho_1(M) < o(M)$, and η^M_1 and $\rho_1(M)$ are measurable in M.*

PROOF. The same construction works, starting with μ and κ measurable in N, and $o(N) = \text{lh}(D) + \mu$, where D has critical point κ. (Now $X = \text{Hull}_1^N(\{D\})$.) ⊣

Proposition 2.5.8 gives us an anticore map $\pi\colon M \to \text{Ult}_1(M,E)$ that is discontinuous at ρ_1^M, while $\rho_1(M)$ is measurable in M. We discuss such anticore maps further in Section 3.8.

We shall need some facts about preservation of $\text{cof}_k(\gamma)$ under maps with some degree of elementarity. For elementary π we have

LEMMA 2.5.9. *Let $\pi\colon M \to N$ be elementary, $k \leq k(M)$, and let f be a nice witness that $\text{cof}_k^M(\gamma) = \eta$, where $\eta < \rho_k(M)$; then $\pi(f)$ is a nice witness that $\text{cof}_k^N(\pi(\gamma)) = \pi(\eta)$.*

PROOF. Let f be a nice witness that $\text{cof}_k^M(\gamma) = \eta$. If $k = 0$, then $f \in M$ and $\pi(f)$ is a nice witness that $\text{cof}_k^M(\pi(\gamma)) = \pi(\eta)$. Suppose now $k > 0$; we can still make sense of $\pi(f)$ by moving a definition of f. For example, suppose $k = 1$ and $f = \varphi^M$ where φ is Σ_1; then the fact that f is strictly order preserving with $\text{ran}(f) \subseteq \gamma$ is Π_1, and the fact that it is total and has range cofinal in γ is Π_2. Since $k(M) \geq 1$, π is Σ_2 elementary. Thus φ^N witnesses $\text{cof}_1^N(\pi(\gamma)) = \pi(\eta)$.

In general, we can fix $q \in M^k$ such that for all $\alpha < \eta$,

$$f(\alpha) = h_M^k(\alpha, q, p_k^M).$$

Let also $r \in M^k$ be such that

$$\gamma = h_M^k(r, p_k^M).$$

The fact that f is total, strictly order preserving, and maps into γ is a Σ_0 fact about $\text{Th}_k^M(\eta \cup \{q, r, p_k^M\})$. This theory is coded into the reduct $M^k||\xi$, where $\xi < \rho_k(M)$ is large enough.[36] The fact that $\text{ran}(f)$ is cofinal in γ is Π_1 over M^k.[37]

So letting

$$g(\alpha) = h_N^k(\alpha, \pi(q), p_k^N)$$

for $\alpha < \pi(\eta)$, we see that g is a nice witness that $\text{cof}_k^N(\pi(\gamma)) = \pi(\eta)$. ⊣

Lemma 2.5.9 leaves open whether the $r\Sigma_k$-regularity of ρ_k^M is preserved. Here we need a little more elementarity for π. If we assume that π is cofinal, as ultrapower and anticore maps are, then we can say more.

LEMMA 2.5.10. *Let $\pi\colon M \to N$ be cofinal and elementary, and $k = k(M) > 0$; then*

(1) $\rho_k(M)$ is $r\Sigma_k$-regular in M iff $\rho_k(N)$ is $r\Sigma_k$-regular in N.
(2) If π is continuous at $\rho_k(M)$, then $\pi(\rho_k(M)) = \rho_k(N)$ and $\pi(\eta_k(M)) = \eta_k^N$.
(3) If π is continuous at $\rho_k(M)$, then π is continuous at η_k^M.

---

36 Totality reduces to a Σ_0 fact about $M||\xi$ because the outer universal quantifier is bounded by η.

37 The Π_1 fact is that for all $\beta < o(M^k)$ and $s \in M||\beta$, if "$h^k(s, p_k)$ is defined" is in $A_M^k \cap M||\beta$, then "$\exists\alpha(h^k(s, p_k) < h^k(\alpha, q, p_k))$" is in $A_M^k \cap M||\beta$.

PROOF. For (1): Suppose f is a nice witness that $\mathrm{cof}_k^M(\rho_k(M)) = \eta$, where $\eta < \rho_k(M)$. Letting $f = f_{\tau,q}^M$ and $\pi(f) = f_{\tau,\pi(q)}^N$, we showed in the proof of Lemma 2.5.9 that $\pi(f)$ is total and order preserving on $\pi(\eta)$. It is clear that $\pi(f(\alpha)) = \pi(f)(\pi(\alpha))$, and from this we get that $\pi(f) \restriction \sup \pi``\eta$ is total and order preserving, with range cofinal in $\sup \pi``\rho_k^M = \rho_k^N$. Thus ρ_k^N is $r\Sigma_k$ singular in N.

Conversely, suppose $f_{\tau,q}^N$ with domain η is a nice witness that ρ_k^N is $r\Sigma_k$-singular in N. Let $\beta < \rho_k(M)$ be large enough that $\eta, q \in \pi(M||\beta)$. By the elementarity of π and the fact that $\sup \pi``\rho_k(M) = \rho_k(N)$, we see that $\{f_{\tau,r}^M(\xi) \mid \xi, r \in M||\beta\}$ is cofinal in $\rho_k(M)$. Thus $\rho_k(M)$ is $r\Sigma_k$-singular in M.

For (2): Since π is cofinal, $\rho_k(N) = \sup \pi``\rho_k(M) = \pi(\rho_k(N))$. If $\rho_k(M)$ is $r\Sigma_k$-regular in M, then by (1), $\pi(\eta_k^M) = \eta_k^N$. If $\eta_k^M < \rho_k(M)$, then $\pi(\eta_k^M) = \eta_k^N$ by Lemma 2.5.9.

For (3): We have $\pi(\rho_k(M)) = \rho_k(N)$. So if $\eta_k^M < \rho_k(M)$, (3) follows from Lemma 2.5.9, and if $\eta_k^M = \rho_k(M)$, it is trivial. ⊣

Remark 2.5.11. We do not know whether the converse to (3) in Lemma 2.5.10 is true in general. We showed in Lemma 2.5.6 that it holds for ultrapower maps. We shall show in Section 3.2 that it holds for anticore maps, granted that N is iterable.

Finally, if we assume elementarity one level up, the situation simplifies.

LEMMA 2.5.12. *Let $\pi\colon M \to N$ be elementary, and $1 \le k < k(M)$; then*

(1) $\pi(\rho_k(M)) = \rho_k(N)$,

(2) $\pi(\eta_k^M) = \eta_k^N$, and

(3) $\rho_{k+1}(M) \le \eta_k^M$ iff $\rho_{k+1}(N) \le \eta_k^N$.

PROOF. π respects projecta by Lemma 2.4.7, so we have (1). If $\eta_k^M < \rho_k(M)$, we get (2) from Lemma 2.5.9. Suppose then that $\rho_k(M)$ is $r\Sigma_k$-regular in M. We show that this fact is "Π_2 over M^k", hence preserved by π.

Let us assume first that $\rho_k(M) < \rho_{k-1}(M)$. Let $\langle \tau, q\rangle \in M^k$ be a name for $\rho_k(M)$, in the sense that $h_{M^{k-1}}^1(\tau, \langle q, p_k^M\rangle) = \rho_k(M)$. For $\beta < \rho_k(M)$, we have that

$$B_\beta = \{h_{M^{k-1}}^1(\tau, \langle r, p_k^M\rangle) \mid r \in M||\beta\}$$

is bounded in $\rho_k(M)$, because $\mathrm{cof}_k^M(\rho_k(M)) > \beta$. For $\gamma < \rho_k(M)$, there is a natural Σ_1 formula $\theta_{\beta,\gamma}(u,v)$ in the language of M^{k-1} such that

$$\begin{aligned} B_\beta \cap \rho_k^M \subseteq \gamma &\iff M^{k-1} \models \neg\theta_{\beta,\gamma}[\langle \beta,\gamma,q\rangle, p_k^M] \\ &\iff \langle \theta_{\beta,\gamma}, \langle \beta,\gamma,q\rangle\rangle \notin A_M^k \cap M||(\gamma+1). \end{aligned}$$

Moreover the map $\langle \beta,\gamma\rangle \mapsto \theta_{\beta,\gamma}$ is Σ_1 over M^k. We have that

$$M^k \models \forall\beta\exists\gamma\langle \theta_{\beta,\gamma}, \langle \beta,\gamma,q\rangle\rangle \notin A_M^k \cap M||(\gamma+1).$$

Since the right hand side is Π_2 over M^k, and $k < k(M)$, it passes to N^k. But $\langle \tau, \pi(q)\rangle$ is a name in N^{k-1} for $\pi(\rho_k(M))$, and $\pi(\rho_k(M)) = \rho_k(N)$. Thus the fact that $N^k \models \forall\beta\exists\gamma\langle\theta_{\beta,\gamma},\langle\beta,\gamma,\pi(q)\rangle\rangle \notin A^k_N \cap N||(\gamma+1)$ implies that $\rho_k(N)$ is $r\Sigma_k$-regular in N.

If $\rho_k(M) = \rho_{k-1}(M)$, the same proof works, but we no longer need the M^{k-1}-name for $\rho_k(M)$. From the point of view of M^{k-1}, it is just the class of ordinals. Thus we have (2).

(3) follows easily from (1) and (2) if $\eta^M_k = \rho_k(M)$, so assume $\eta^M_k < \rho_k(M)$. Because π is elementary and $k < k(M)$,

$$\sup \pi``\rho_{k+1}(M) \leq \rho_{k+1}(N) \leq \pi(\rho_{k+1}(M)).$$

So if $\rho_{k+1}(M) \leq \eta^M_k$ then $\rho_{k+1}(N) \leq \pi(\rho_{k+1}(M)) \leq \pi(\eta_k(M)) = \eta^N_k$, while if $\eta^M < \rho_{k+1}(M)$, then $\eta^N_k = \pi(\eta^M_k) < \sup \pi``\rho_{k+1}(M) \leq \rho_{k+1}(N)$. This proves (3). ⊣

Lemmas 2.5.9 and 2.5.12 mark a level of elementarity enjoyed by the lifting maps in the standard conversion systems of Chapter 3.

DEFINITION 2.5.13. Let $\pi\colon M \to N$, where M and N are premice with $k(M) = k(N)$; then π *strongly respects projecta* iff for all $k < k(M)$,

(1) $\pi(\rho_k(M)) = \rho_k(N)$,
(2) $\pi(\eta^M_k) = \eta^N_k$, and
(3) $\rho_{k+1}(M) \leq \eta_k(M)$ iff $\rho_{k+1}(N) \leq \eta^N_k$.

DEFINITION 2.5.14. Let $\pi\colon M \to N$, where M and N are premice with $k(M) = k(N)$; then π is *nearly elementary* iff

(1) π is weakly elementary and strongly respects projecta, and
(2) for $k = k(M)$, if f is a nice witness that $\mathrm{cof}^M_k(\rho_k(M)) = \eta$, where $\eta < \rho_k(M)$, then $\pi(f)$ is a nice witness that $\mathrm{cof}^N_k(\pi(\rho_k(M))) = \pi(\eta)$.

Remark 2.5.15. If $\pi\colon M \to N$ is weakly elementary and π preserves at least one nice witness that $\mathrm{cof}^M_k(\gamma) = \eta$, where $k = k(M)$ and $\eta < \rho_k(M)$, then π preserves all nice witnesses that $\mathrm{cof}^M_k(\gamma) = \eta$. For let f and g be such witnesses, and suppose that $\pi(f)$ is a nice witness that $\mathrm{cof}^N_k(\pi(\gamma)) = \pi(\eta)$. Since π is weakly elementary, $\pi(g)$ is a total, order preserving $r\Sigma^N_k$ function with domain $\pi(\eta)$. But the fact that $\mathrm{ran}(g)$ is cofinal in $\mathrm{ran}(f)$ is coded into $\mathrm{Th}_k(\eta \cup \{q\})$, where f and g are $r\Sigma^M_k$ in $q \in M^k$. So this fact is Σ_0 over M^k, hence passes to N^k. This implies $\mathrm{ran}(\pi(g))$ is cofinal in $\mathrm{ran}(\pi(f))$, so $\pi(g)$ is a nice witness that $\mathrm{cof}^N_k(\pi(\gamma)) = \pi(\eta)$.

Lemmas 2.5.9 and 2.5.12 imply that every elementary map is nearly elementary.[38] Examples 2.4.9 and 2.4.10 in the last section show that the converse is not true. These examples are typical of how nearly elementary maps that are not elementary arise. They come from factor embeddings from one ultrapower into another ultrapower that has been formed using a larger class of functions. The lifting maps that occur in the construction of background induced iteration

[38] In fact Lemma 2.5.9 implies that (2) holds for all γ, not just $\gamma = \rho_k(M)$. But it is not clear that the lift maps of Section 3.5 preserve Σ_k cofinality in this stronger sense.

strategies arise this way, and they are always nearly elementary, but generally not almost exact. Indeed, for such lift maps $\pi(\rho^-(M))$ can be strictly above, equal to, or strictly below $\rho^-(N)$.

Remark 2.5.16. Definition 2.5.14 records more information about the lifting maps of a standard conversion system than is customary. This additional information is useful if one wants to construct iteration strategies for Jensen premice that normalize well, using the standard background constructions of Jensen premice. One can see how it plays a role in that in Sections 3.7 and 3.8. But from Chapter 4 onward, we shall shift to a slightly different sort of premouse and background construction, because the construction of iteration strategies that normalize well is more natural in that context. Near elementarity is defined in that context in Section 4.3.

We do not have an example of a weakly elementary map that is not nearly elementary, but would guess that there must be one. Such maps do not play a role in this book, in any case. The levels of elementarity that are most important in what follows are: cofinal and elementary, elementary, nearly elementary.

Each of the classes of maps above (cofinal elementary, elementary, nearly elementary, and weakly elementary) are closed under composition.

Copying and the Shift Lemma. The copying construction propagates each level of elementarity. For example, let us look at one step of copying, as codified in the Shift Lemma.

Let us introduce some notation from Zeman's book [86].[39]

DEFINITION 2.5.17. Let P and Q be acceptable J-structures, and E and F be extenders over P and Q respectively; then we say that $\langle\pi,\varphi\rangle$ *embeds* $\langle P,E\rangle$ *into* $\langle Q,F\rangle$, and write

$$\langle\pi,\varphi\rangle\colon \langle P,E\rangle \to \langle Q,F\rangle,$$

iff $P\subseteq \mathrm{dom}(\pi)$, $\pi\restriction P\colon P\to Q$ is Σ_0 elementary, $\varphi\colon \lambda_E\to\lambda_F$ is order preserving, and for all $a\in[\lambda_E]^{<\omega}$ and X,

$$X\in E_a \iff \pi(X)\in F_{\varphi``a}.$$

We say $\langle\pi,\varphi\rangle$ Σ_1-*embeds* $\langle P,E\rangle$ into $\langle Q,F\rangle$, and write

$$\langle\pi,\varphi\rangle\colon \langle P,E\rangle \stackrel{*}{\to} \langle Q,F\rangle,$$

iff in addition π is Σ_1 elementary, E and F are close to P and Q respectively, and for all $a\in[\lambda_E]^{<\omega}$ there is a $q\in P$ and a Σ_1 formula θ such that

$$E_a=\{X\mid P\models\theta[X,q]\}$$

and

$$F_{\varphi``a}=\{X\mid Q\models\theta[X,\pi(q)]\}.$$

[39] See Sections 2.5 and 3.4.

DEFINITION 2.5.18. Let $\langle \pi, \varphi \rangle\colon \langle P, E\rangle \to \langle Q, F\rangle$; then we define

$$\sigma\colon \mathrm{Ult}_0(P,E) \to \mathrm{Ult}_0(Q,F)$$

by

$$\sigma([a,f]^P_E) = [\varphi``a, \pi(f)]^Q_F$$

and call σ the *copy map* associated to π, φ, P, Q, E, and F.

It is easy to see that σ is well defined.

LEMMA 2.5.19 (Shift Lemma 1). *Suppose that $\langle \pi, \varphi \rangle\colon \langle P, E\rangle \to \langle Q, F\rangle$, and let $\sigma\colon \mathrm{Ult}_0(P,E) \to \mathrm{Ult}_0(Q,F)$ be the associated copy map; then*

(1) σ is Σ_0 elementary,
(2) $\sigma \restriction \mathrm{lh}(E) = \varphi \restriction \mathrm{lh}(E)$,
(3) $\sigma \circ i^P_E = i^Q_F \circ \pi$, and
(4) σ is cofinal iff π is cofinal.

Moreover, if $\langle \pi, \varphi \rangle\colon \langle P, E\rangle \xrightarrow{*} \langle Q, F\rangle$, then σ is Σ_1 elementary.

PROOF. See [86, Lemmas 2.5.6 and 3.4.5]. To see the "moreover" part, notice that if $\theta(v)$ is Σ_1 and $\mathrm{Ult}(Q,F) \models \theta[[\varphi``a, \pi(f)]]$, then there is an $\alpha < o(Q)$ and a $Z \in F_{\varphi``a}$ such that for all $u \in Z$, $Q||\alpha \models \theta[f(u)]$. This fact can be pulled back to P, f, and E_a because $\langle \pi, \varphi\rangle$ is a Σ_1 embedding. ⊣

In practice, we often start with maps π and φ acting on premice or their reducts, as in the following immediate corollary.

COROLLARY 2.5.20 (Shift Lemma 2). *Let M, N, R, and S be premice. Let $\varphi\colon M \to N$ be Σ_0 elementary, E be an extender on the sequence of M, and $F = \varphi(E)$. Suppose $\pi\colon R \to S$ is nearly elementary, and let $k = k(R)$. Suppose also that*

(a) $R||\mathrm{dom}(E) = M||\mathrm{dom}(E)$, and $\mathrm{crit}(E) < \rho_k(R)$,
(b) $N||\mathrm{dom}(F) = S||\mathrm{dom}(F)$, and
(c) $\pi \restriction \mathrm{dom}(E) + 1 = \sigma \restriction \mathrm{dom}(E) + 1$.

Let σ_0 be the copy map associated to $\pi \restriction R^k\colon R^k \to S^k$; then σ_0 has a unique completion

$$\sigma\colon \mathrm{Ult}(R,E) \to \mathrm{Ult}(S,F),$$

and

(1) σ is nearly elementary,
(2) $\sigma \restriction \mathrm{lh}(E) = \varphi \restriction \mathrm{lh}(E)$,
(3) $\sigma \circ i^R_E = i^S_F \circ \pi$, and
(4) σ is cofinal iff π is cofinal.

Moreover, if $\langle \pi, \varphi \rangle\colon \langle R^k, E\rangle \xrightarrow{*} \langle S^k, F\rangle$, then σ is elementary.

Remark 2.5.21. Internal ultrapowers are the special case in which $M = R, N = S$, and $\pi = \varphi$. In this case, if π is elementary, then $\langle \pi, \varphi \rangle : \langle R^k, E \rangle \xrightarrow{*} \langle S^k, F \rangle$ holds simply because $\pi(E) = F$, so the copy map σ must be elementary.

We shall say that *the Shift Lemma applies to* (φ, π, E) iff the hypotheses of Corollary 2.5.20 hold. (Here M, N, R and S must be understood from context.) If we regard the ultrapowers as having been formed using definable functions, rather than coding into reducts, then the formula

$$\sigma([a, f]_E^P) = [\varphi(a), \pi(f)]_{\varphi(E)}^Q$$

holds, provided we let $\pi(f_{\tau,q}^P) = f_{\tau,\pi(q)}^Q$.

There are further results on the elementarity of copy maps in [47, Lemma 1.3], [86, §9.2], and [56].

Elementarity in various contexts. Here is a summary of some natural contexts in which these levels of elementarity play a role.

(i) The natural map from the core of M to M is elementary and cofinal.

(ii) Fine ultrapower maps, and more generally, the maps $\hat{i}^{\mathcal{T}}_{\alpha,\beta}$ along branches of an iteration tree, are elementary and cofinal.

(iii) If $\pi : M \to N$ is nearly elementary, and $\mathcal{T}$ is a semi-normal tree on M[40], then $\pi\mathcal{T}$ is semi-normal, and the copy maps $\pi_\alpha : \mathcal{M}_\alpha^{\mathcal{T}} \to \mathcal{M}_\alpha^{\pi\mathcal{T}}$ are nearly elementary. But it is possible that $\pi(\rho^-(M)) < \rho^-(N)$, which in turn means that $\mathcal{T}$ may be normal while $\pi\mathcal{T}$ is not. (See Remark 2.7.8 below.)

(iv) The Dodd-Jensen and Weak Dodd-Jensen lemmas hold in the category of nearly elementary maps.

(v) If π, M, N, and $\mathcal{T}$ are as in (iii), and in addition π is elementary, then all the π_α are elementary. Hence by Lemma 2.4.7 they are almost exact, so if $\mathcal{T}$ is normal, then $\pi\mathcal{T}$ is normal.

(vi) If π, M, N, and $\mathcal{T}$ are as in (iii), and π is cofinal and elementary, and $[0, \alpha)_T$ does not drop in model or degree, then the copy map π_α is cofinal and elementary. The no-dropping hypothesis is necessary here.

(vii) The maps $t_\tau^{\nu,\gamma}$ occuring in an embedding normalization are elementary. The maps σ_γ are nearly elementary, but may not be elementary or almost exact, so far as we can see. See Chapter 6.

(viii) The lifting maps that occur in a conversion system are nearly elementary. They are not in general elementary or almost exact. (See Section 3.4.)

2.6. Iteration trees on premice

Our notation and terminology regarding iteration trees is essentially that of [70].

[40]See Definition 2.6.4.

In an iteration tree $\mathcal{T}$ on a premouse M, we repeatedly form fine-structural ultrapowers. The α-th model of $\mathcal{T}$ is $\mathcal{M}_\alpha^{\mathcal{T}}$; the base model is $M = \mathcal{M}_0^{\mathcal{T}}$. $E_\alpha^{\mathcal{T}}$ is the *exit extender* taken from the sequence of $\mathcal{M}_\alpha^{\mathcal{T}}$ and used to form

$$\mathcal{M}_{\alpha+1}^{\mathcal{T}} = \mathrm{Ult}(\mathcal{M}_{\alpha+1}^{*,\mathcal{T}}, E_\alpha^{\mathcal{T}}),$$

where

$$\mathcal{M}_{\alpha+1}^{*,\mathcal{T}} = \mathcal{M}_\beta^{\mathcal{T}} | \langle \xi, k \rangle$$

for some $\beta = T\text{-pred}(\alpha+1)$, and some $\langle \xi, k \rangle \leq l(\mathcal{M}_\beta^{\mathcal{T}})$ such that $\mathrm{crit}(E_\alpha^{\mathcal{T}}) < \rho_k(\mathcal{M}_\beta^{\mathcal{T}} | \xi)$. We put $\alpha+1 \in D^{\mathcal{T}}$ iff $\mathcal{M}_{\alpha+1}^{*,\mathcal{T}} \lhd \mathcal{M}_\beta^{\mathcal{T}}$ iff $l(\mathcal{M}_{\alpha+1}^{*,\mathcal{T}}) < l(\mathcal{M}_\beta^{\mathcal{T}})$, and we say $\mathcal{T}$ drops at $\alpha+1$ in this case. So unlike [70], drops in degree yield elements of $D^{\mathcal{T}}$ too. If $\alpha \leq_T \beta$ and $(\alpha, \beta]_T \cap D^{\mathcal{T}} = \emptyset$, then the canonical embedding

$$i_{\alpha,\beta}^{\mathcal{T}} \colon \mathcal{M}_\alpha^{\mathcal{T}} \to \mathcal{M}_\beta^{\mathcal{T}}$$

is cofinal and elementary; that is, it is an n-embedding, where $n = k(\mathcal{M}_\alpha^{\mathcal{T}}) = k(\mathcal{M}_\beta^{\mathcal{T}})$.

Remark 2.6.1. All extenders in $\mathcal{T}$ are close to the models to which they are applied, so if $\mathrm{crit}(i_{\alpha,\beta}^{\mathcal{T}}) \geq \rho(\mathcal{M}_\alpha^{\mathcal{T}})$, then $\rho(\mathcal{M}_\alpha^{\mathcal{T}}) = \rho(\mathcal{M}_\beta^{\mathcal{T}})$ and $i_{\alpha,\beta}^{\mathcal{T}}(p(\mathcal{M}_\alpha^{\mathcal{T}})) = p(\mathcal{M}_\beta^{\mathcal{T}})$. In this case ("dropping to a mouse"), the core of $\mathcal{M}_\beta^{\mathcal{T}}$ is the pointwise image of the core of $\mathcal{M}_\alpha^{\mathcal{T}}$. In particular, if $\mathcal{M}_\alpha^{\mathcal{T}}$ is sound, then $i_{\alpha,\beta}^{\mathcal{T}}$ is the anticore embedding from $\mathfrak{C}(\mathcal{M}_\beta^{\mathcal{T}})$ to $\mathcal{M}_\beta^{\mathcal{T}}$. Thus the last mouse we dropped to, and the branch embedding acting on it, can be recovered from the final model along that branch. This is important in the comparison proof.

We shall also have a use for the natural partial embeddings that exist along branches that have dropped.

DEFINITION 2.6.2. Let $\mathcal{U}$ be an iteration tree, and $\alpha <_U \beta$. Then $\hat{\imath}_{\alpha,\beta}^{\mathcal{U}}$ is the natural map from a (perhaps proper!) weak initial segment of $\mathcal{M}_\alpha^{\mathcal{U}}$ into $\mathcal{M}_\beta^{\mathcal{U}}$. More precisely, letting

$$i_{\beta+1}^{*,\mathcal{U}} \colon \mathcal{M}_{\beta+1}^{*,\mathcal{U}} \to \mathrm{Ult}(\mathcal{M}_{\beta+1}^{*,\mathcal{U}}, E_\beta^{\mathcal{U}})$$

be the canonical embedding,

$$\hat{\imath}_{\alpha,\beta+1}^{\mathcal{U}} = i_{\beta+1}^{*\mathcal{U}} \circ \hat{\imath}_{\alpha,\gamma}^{\mathcal{U}}$$

if $\alpha \leq_U \gamma = U\text{-pred}(\beta+1)$, and

$$\hat{\imath}_{\alpha,\beta}^{\mathcal{U}}(x) = i_{\xi,\beta}^{\mathcal{U}}(\hat{\imath}_{\alpha,\xi}(x))$$

if β is a limit ordinal, and ξ is past the last drop in $[0,\beta)_U$.

It might have been more natural to have originally defined $i_{\alpha,\beta}^{\mathcal{U}}$ the way we just defined $\hat{\imath}_{\alpha,\beta}^{\mathcal{U}}$, but it is too late for that now. The difference between "$\hat{\imath}$" and "i" is barely visible anyway.

As we have defined it, the domain D of $i^{\mathcal{U}}_{\alpha,\beta}$ is a set, not a premouse. Clearly D is closed downward in the order of constructibility of $\mathcal{M}^{\mathcal{U}}_\alpha$, and thus there is a unique weak initial segment

$$\mathcal{D} \trianglelefteq_0 \mathcal{M}^{\mathcal{U}}_\alpha$$

such that D is the universe of $\mathcal{D}$, $k(\mathcal{D}) = 0$, and

$$\hat{\imath}_{\alpha,\beta} : \mathcal{D} \to \mathcal{M}^{\mathcal{U}}_\beta$$

is weakly elementary. Often it is possible to take $k(\mathcal{D}) > 0$. On the other hand, it could happen that D is the universe of $\mathcal{M}^{\mathcal{U}}_\alpha|\nu$, but $i^{\mathcal{U}}_{\alpha,\beta}$ is only weakly elementary on $\mathcal{M}^{\mathcal{U}}_\alpha||\nu$, and not on $\mathcal{M}^{\mathcal{U}}_\alpha|\nu$.

If $\mathcal{T}$ is an iteration tree, then $\mathrm{lh}(\mathcal{T})$ is the domain of its tree order, that is, $\mathrm{lh}(\mathcal{T}) = \{\alpha \mid \mathcal{M}^{\mathcal{T}}_\alpha \text{ exists}\}$. So if $\mathrm{lh}(\mathcal{T}) = \alpha + 1$, then $\mathcal{M}^{\mathcal{T}}_\alpha$ exists, but $E^{\mathcal{T}}_\alpha$ does not. $\mathcal{T} \restriction \beta$ is the initial segment $\mathcal{U}$ of $\mathcal{T}$ such that $\mathrm{lh}(\mathcal{U}) = \beta$. So $\mathcal{M}^{\mathcal{T}\restriction\alpha+1}_\alpha$ exists, but there is no exit extender $E^{\mathcal{T}\restriction\alpha+1}_\alpha$.

Remark 2.6.3. We allow iteration trees of length 1. Such a degenerate tree has no extenders, and thus consists of only its base model. This convention plays some role in the definitions of tree embeddings and strong hull condensation.

We do not need all the freedom to choose exit extenders and the models they are applied to that would make sense. The following definition discards some of it.

DEFINITION 2.6.4. Let M be a premouse, then a *semi-normal iteration tree* on M is a system

$$\mathcal{T} = \langle T, \langle (E_\alpha, M^*_{\alpha+1}) \mid \alpha + 1 < \mathrm{lh}(\mathcal{T}) \rangle \rangle$$

such that there are M_α for $\alpha < \mathrm{lh}(\mathcal{T})$ and D satisfying:

(1) $M_0 = M$, and T is a tree order;
(2) if $\alpha < \beta < \mathrm{lh}(\mathcal{T}) - 1$, then $\lambda(E_\alpha) \leq \lambda(E_\beta)$;
(3) if $\alpha + 1 < \mathrm{lh}(\mathcal{T})$, then E_α is on the M_α-sequence, and letting β be least such that either $\beta = \alpha$, or $\mathrm{crit}(E_\alpha) < \lambda(E_\beta)$,
  (a) T-pred$(\alpha+1) = \beta$, $M_\beta|\mathrm{lh}(E_\beta) \trianglelefteq M^*_{\alpha+1} \trianglelefteq M_\beta$, $P(\mathrm{crit}(E_\alpha)) \cap M^*_{\alpha+1} \subseteq \mathrm{dom}(E_\alpha)$, and $\mathrm{crit}(E_\alpha) < \rho_k(M^*_{\alpha+1})$, where $k = k(M^*_{\alpha+1})$,
  (b) $M_{\alpha+1} = \mathrm{Ult}(M^*_{\alpha+1}, E_\alpha)$,
  (c) $\alpha + 1 \in D$ iff $M^*_{\alpha+1} \neq M_\beta$;
(4) if $\lambda < \mathrm{lh}(\mathcal{T})$ is a limit ordinal, then $D \cap [0,\lambda)_T$ is finite, and M_λ is the direct limit of the M_α for $\alpha <_T \lambda$ under the $\hat{\imath}^{\mathcal{T}}_{\alpha,\eta}$; moreover $\lambda \notin D$.

We write $E_\alpha = E^{\mathcal{T}}_\alpha$, $D = D^{\mathcal{T}}$, and so on. For the tree order T we also write $\leq_T$. $[\alpha,\beta)_T = \{\gamma \mid \alpha \leq_T \gamma <_T \beta\}$, and so on. If $\mathrm{lh}(\mathcal{T}) = \alpha + 1$, then we call $[0,\alpha]_T$ the *main branch* of $\mathcal{T}$.

The agreement of models in a semi-normal tree is given by

LEMMA 2.6.5. *Let $\mathcal{T}$ be a semi-normal iteration tree, with models M_α and extenders E_α. Let $\alpha < \beta < \mathrm{lh}(\mathcal{T})$;*

(a) $M_\alpha|\lambda(E_\alpha)+1 = M_\beta|\lambda(E_\alpha)+1$, and $\lambda(E_\alpha) < \rho^-(M_\beta)$, and

(b) if $\alpha+1 \le \beta$ and $\mathrm{lh}(E_\alpha) < \mathrm{lh}(E_{\alpha+1})$, then $\mathrm{lh}(E_\alpha) < \lambda(E_{\alpha+1})$, $M_\alpha|\mathrm{lh}(E_\alpha) = M_\beta|\mathrm{lh}(E_\alpha)$, and $\mathrm{lh}(E_\alpha) \le \rho^-(M_\beta)$.

We omit the routine proof. The main point for (b) is that $\mathrm{lh}(E_\alpha)$ is a successor cardinal in $M_{\alpha+1}$, and $\lambda(E_{\alpha+1})$ is a limit cardinal in $M_{\alpha+1}|\mathrm{lh}(E_{\alpha+1})$. Part (a) shows that our requirement in Definition 2.6.4(3) on T-pred$(\beta+1)$ does not restrict the extenders we can take to be as $E_\beta^{\mathcal{T}}$. Note also that Definition 2.6.5(b) implies that the decreasing lengths in $\mathcal{T}$ must occur in finite intervals of the form $[\alpha, \alpha+n]$, ending when $\mathrm{lh}(E_{\alpha+n}) < \lambda(E_{\alpha+n+1})$, after which all lengths are $> \lambda(E_{\alpha+n+1})$. Thus semi-normal trees are very nearly length-increasing.

Clause (3)(a) of Definition 2.6.4 implies that if F is used after E along the same branch of $\mathcal{T}$, then $\lambda(E) \le \mathrm{crit}(F)$. One sometimes says that $\mathcal{T}$ is *non-overlapping*, or that (Jensen) generators are not moved. Clause (3)(b) says that $M^*_{\alpha+1}$ is somewhere between the shortest and longest initial segments of M_β to which we can apply the full E_α.

DEFINITION 2.6.6. Let $\mathcal{T}$ be a semi-normal iteration tree on a Jensen premouse; then for any $\beta < \mathrm{lh}(\mathcal{T})$,

$$\begin{aligned}\lambda_\beta^{\mathcal{T}} &= \sup\{\lambda_F \mid \exists\eta(\eta+1 \le_T \beta \wedge F = E_\eta^{\mathcal{T}})\}\\ &= \sup\{\lambda_F \mid \exists\eta(\eta+1 \le \beta \wedge F = E_\eta^{\mathcal{T}})\}.\end{aligned}$$

The two characterizations of $\lambda_\beta^{\mathcal{T}}$ are equivalent because we have demanded that semi-normal trees be λ-nondecreasing. $\lambda_\beta^{\mathcal{T}}$ is the sup of the "Jensen generators" of extenders used to produce $\mathcal{M}_\beta^{\mathcal{T}}$. We call them generators because

LEMMA 2.6.7. *Let $\mathcal{T}$ be semi-normal, $\alpha <_T \gamma+1 \le_T \beta$, T-pred$(\gamma+1) = \alpha$, and $D^{\mathcal{T}} \cap (\gamma+1, \beta)_T = \emptyset$; then for $k = k(\mathcal{M}^{*,\mathcal{T}}_{\gamma+1})$, $(\mathcal{M}_\beta^{\mathcal{T}})^k = \{\hat{\imath}_{\alpha,\beta}(f)(a) \mid f \in (\mathcal{M}^{*,\mathcal{T}}_{\gamma+1})^k \wedge a \in [\lambda_\beta^{\mathcal{T}}]^{<\omega}\}$.*

The proof is a simple induction on β.

Remark 2.6.8. The non-overlapping requirement is important because it implies that if the branch $[\alpha,\beta)_T$ does not drop, then from the branch embedding $i^{\mathcal{T}}_{\alpha,\beta}$ we can recover the sequence of models and extenders used along $[\alpha,\beta)_T$. For example, let E be the first extender used, that is, $E = E_\gamma$ where $\gamma+1 <_T \beta$ and T-pred$(\gamma+1) = \alpha$. Because generators are not moved, E is an initial segment of the extender G of $i^{\mathcal{T}}_{\alpha,\beta}$. That is, for $x \subseteq \mathrm{crit}(E)$ in M_α and $a \subseteq \lambda_E$ finite,

$$\begin{aligned}x \in E_a &\iff a \in i_E(x)\\ &\iff a \in i_{\gamma+1,\beta} \circ i_E(x)\end{aligned}$$

$$\iff a \in i_{\alpha,\beta}(x)$$
$$\iff x \in G_a.$$

By the Jensen initial segment condition, E is then the first whole initial segment of G that is not on the sequence of M_β. Having now recovered E, we can recover the factor embedding $i^{\mathcal{T}}_{\gamma+1,\beta}$, its extender G_1, and then the next extender used in $[\alpha,\beta]_T$, and so on.

If $\mathcal{T}$ is semi-normal, then T-pred$(\beta+1)$ is the largest α such that $\lambda^{\mathcal{T}}_\alpha \leq \text{crit}(E^{\mathcal{T}}_\beta)$. Another useful characterization is the following. Let θ be $\text{crit}(E^{\mathcal{T}}_\beta)^+$, as computed in $\mathcal{M}^{\mathcal{T}}_\beta | \text{lh}(E^{\mathcal{T}}_\beta)$. Then

$$T\text{-pred}(\beta+1) = \text{least } \alpha \text{ such that } \mathcal{M}^{\mathcal{T}}_\alpha|\theta = \mathcal{M}^{\mathcal{T}}_\beta|\theta.$$

Note here that θ is passive in $\mathcal{M}^{\mathcal{T}}_\beta$, so for α as on the right, θ is passive in $\mathcal{M}^{\mathcal{T}}_\alpha$. The formula may fail if we replace the $|$ by $||$, for when $\lambda_{E^{\mathcal{T}}_\alpha} = \text{crit}(E^{\mathcal{T}}_\beta)$, T-pred$(\beta+1)$ is $\alpha+1$, not α.

For the most part, we are interested in normal iteration trees.

DEFINITION 2.6.9. Let $\mathcal{T}$ be a semi-normal iteration tree; then

(1) $\mathcal{T}$ is *quasi-normal* iff whenever $\alpha+1 < \text{lh}(\mathcal{T})$, and $\beta = T\text{-pred}(\alpha+1)$, then $\mathcal{M}^{*,\mathcal{T}}_{\alpha+1} = \mathcal{M}^{\mathcal{T}}_\beta|\langle\eta,k\rangle$, where $\langle\eta,k\rangle \leq l(\mathcal{M}^{\mathcal{T}}_\beta)$ is largest so that $\text{crit}(E^{\mathcal{T}}_\alpha) < \rho_k(\mathcal{M}^{\mathcal{T}}_\beta|\eta)$.
(2) $\mathcal{T}$ is *length-increasing* iff whenever $\alpha < \beta < \text{lh}(\mathcal{T}) - 1$, then $\text{lh}(E^{\mathcal{T}}_\alpha) < \text{lh}(E^{\mathcal{T}}_\beta)$, and
(3) $\mathcal{T}$ is *normal* iff $\mathcal{T}$ is quasi-normal and length-increasing.

Another way of putting (1) is: $M^*_{\alpha+1}$ is the longest $\mathcal{P} \trianglelefteq \mathcal{M}^{\mathcal{T}}_\beta$ such that $\text{Ult}(\mathcal{P}, E_\alpha)$ makes sense. So we sometimes say that $\mathcal{T}$ is *maximal* if (1) holds. If $\mathcal{T}$ is a quasi-normal tree on M, then we can identify it with $\langle T, \langle E_\alpha \mid \alpha+1 < \text{lh}(\mathcal{T})\rangle\rangle$, the $M^{*,\mathcal{T}}_\xi$ being determined by maximality. We shall do this. Of course, M is relevant too, but often it will be understood from context. $\mathcal{T}$ can be a tree on more than one M.

If the maximality clause (1) fails, then we say that $\mathcal{T}$ has a *gratuitous drop* at $\alpha+1$. The possibility of such drops does not cause significant problems for the theorems we shall prove below, but it does further complicate the notation, along with a number of fine structural arguments. For that reason, we shall avoid non-maximal trees. There is some care needed in order to do that; see for example Remark 2.7.8. But in the end, the semi-normal iteration trees we deal with seriously will all be quasi-normal.

If $\mathcal{T}$ is normal, then for $\alpha < \beta$, $\mathcal{M}^{\mathcal{T}}_\alpha$ agrees with $\mathcal{M}^{\mathcal{T}}_\beta$ below $\text{lh}(E^{\mathcal{T}}_\alpha)$. They disagree at $\text{lh}(E^{\mathcal{T}}_\alpha)$. If $\mathcal{T}$ is merely quasi-normal, then only agreement up to $\lambda(E^{\mathcal{T}}_\alpha)$ is guaranteed. If $\mathcal{T}$ is normal, then $\alpha < \beta$ implies $\text{lh}(E_\alpha) < \lambda(E_\beta)$.

We do need to consider quasi-normal trees that are not normal, but only in limited circumstances.

Figure 2.6.1 shows how the agreement of models in a normal iteration tree is propagated when the tree is augmented by one new extender. (Figures like this were first drawn by Itay Neeman.)

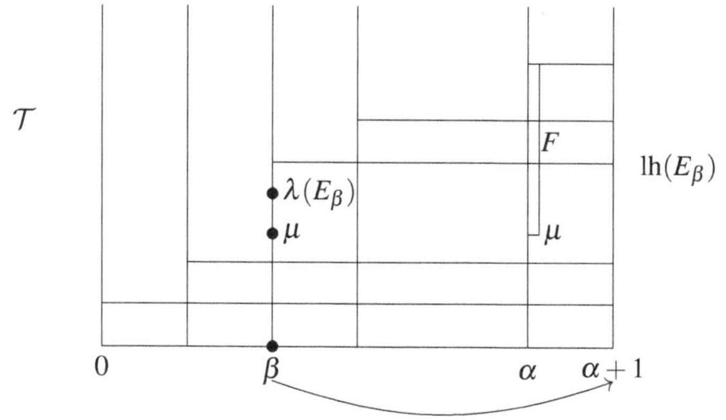

FIGURE 2.6.1. A normal tree $\mathcal{T}$, extended normally by F. The vertical lines represent the models, and the horizontal ones represent their levels of agreement. $\mathrm{crit}(F) = \mu$, and β is least such that $\mu < \lambda(E^{\mathcal{T}}_\beta)$. The arrow at the bottom represents the ultrapower embedding generated by F.

If one replaces the condition

$$\mathrm{crit}(E^{\mathcal{T}}_\alpha) < \lambda(E^{\mathcal{T}}_\beta)$$

by the condition

$$\mathrm{crit}(E^{\mathcal{T}}_\alpha) < \nu(E^{\mathcal{T}}_\beta)$$

in the definition of (Jensen) normality, one obtains a definition of *ms-normality*. (This is called s-normality in [10, §5].) In an ms-normal tree, if E is used before F on a branch, then F cannot move the generators of E, but it may move ordinals between $\nu(E)$ and $\lambda(E)$. In fact, there are some advantages to working with ms-normal trees, even in the context of Jensen premice. One is that full background constructions of Jensen-normally iterable M seem to require superstrong extenders in V (but see [41]). On the other hand, one can show granted only a Woodin with a measurable above that there is a ms-normally iterable Jensen mouse with a Woodin cardinal, granted that there is in V a Woodin with a measurable above it. ([35] yields an ms-iterable ms-mouse with a Woodin, and [11] and [10] then translates it to an ms-normally iterable Jensen mouse with a Woodin.) Nevertheless,

Definition 2.6.9 is the more common notion of normality in the setting of Jensen premice, and we are using it in this book. When we get to Section 4.4 we shall bring ms-normality back into the picture, in a subsidiary role. We believe that there are elementary simulations of Jensen normal trees by ms-normal trees, and vice-versa, but we have not verified this carefully.

Remark 2.6.10. ms-normal iterations preserve ms-solidity. As we remarked earlier, Jensen normal iterations may not.

We also need stacks of iteration trees.

DEFINITION 2.6.11. Let M be a premouse; then s is an *semi-normal M-stack* iff $s = \langle\langle P_\alpha \mid \alpha < \beta\rangle, \langle \mathcal{T}_\alpha \mid \alpha + 1 < \beta\rangle\rangle$, where

(1) $P_0 = M$,
(2) if $\alpha + 1 < \beta$, then $\mathcal{T}_\alpha$ is a semi-normal tree on P_α having a last model N, and $P_{\alpha+1} = N|\langle \nu, k\rangle$ for some ν, k,
(3) if $\alpha < \beta$ and α is a limit ordinal, then letting N be the direct limit of the P_β for $\beta < \alpha$, $P_\alpha = N|\langle \nu, k\rangle$ for some ν and k.

In (3), the direct limit is under the obvious partial maps $\hat{\imath}^s_{\xi,\gamma}\colon P_\xi \to P_\gamma$, for $\xi < \gamma < \alpha$. We demand that for $\alpha < \beta$ a limit, there are only finitely many drops along the branches producing these maps, and that the direct limit is wellfounded.

In clauses (2) and (3) we allow $k = -1$, with the convention that $N|\langle \nu, -1\rangle = N||\nu$. We may identify s with $\langle \mathcal{T}_0, P_1, \mathcal{T}_2, P_2, \ldots .\rangle$, and we may identify P_α with the pair $\langle \nu_\alpha, k_\alpha\rangle$ that determines it as in (2) or (3). So for example, if $\mathcal{T}$ is a quasi-normal tree on M, $P = \mathcal{M}^{\mathcal{T}}_\infty|\langle \nu, k\rangle$ for some ν, k, and $\mathcal{U}$ is a semi-normal tree on P, then $s = \langle \mathcal{T}, P, \mathcal{U}\rangle = \langle \mathcal{T}, \langle \nu, k\rangle, \mathcal{U}\rangle$ is a semi-normal M-stack.

If each $\mathcal{T}_\alpha$ is quasi-normal, then we call s a *quasi-normal M-stack.* Similarly, a normal M-stack is one whose component trees are all normal. We need nothing more than quasi-normal M stacks in this book, but we do need to allow gratuitous drops at the beginning of each quasi-normal tree $\mathcal{T}_\alpha$. If s is a semi-normal M-stack such that there are no such gratuitous drops, then we say s is a *maximal M-stack.*

For notational reasons, we allow some or all of the semi-normal trees in our M-stack to be empty. The empty tree is normal. Given an M-stack s as above, we write $P_\alpha(s), \mathcal{T}_\alpha(s), \nu_\alpha(s)$, and $k_\alpha(s)$ for the associated objects. $\mathcal{M}_0(s) = M$, $\mathcal{M}_{\alpha+1}(s)$ is the last model of $\mathcal{T}_\alpha(s)$, and $\mathcal{M}_\alpha(s)$ is the direct limit of the $\mathcal{M}_\xi(s)$ for $\xi < \alpha$ if α is a limit ordinal. (So $P_\alpha(s) \trianglelefteq \mathcal{M}_\alpha(s)$.) We write $\mathcal{U}(s)$ for $\mathcal{T}_{\mathrm{dom}(s)-1}(s)$, the last semi-normal tree in s. $\mathcal{U}(s)$ could have no last model. We write $M_\infty(s)$ for the last model of $\mathcal{U}(s)$, if it has one. If s has limit length, we let $\mathcal{M}_\infty(s)$ be the direct limit of the $\mathcal{M}_\alpha(s)$ for $\alpha < \mathrm{lh}(s)$ sufficiently large, provided this limit exists and is wellfounded.

If s is a maximal M-stack, then we identify s with its sequence of trees $\mathcal{T}_i(s)$, the $P_i(s)$ being determined by maximality. If s is not maximal, we must specify the base models of the $\mathcal{T}_i(s)$ as well.

2.7. Iteration strategies

What qualifies a premouse as a mouse, comparable with others of its kind, is an iteration strategy.

Let M be a premouse. $G(M,\theta)$ is the game of length θ in which I and II cooperate to produce a normal iteration tree on M, with II picking branches at limit steps, and being obliged to stay in the category of wellfounded models. See [70], where the game is called $G_k(M,\theta)$, for $k = k(M)$. A *θ-iteration strategy* for M is a winning strategy for II in $G(M,\theta)$. M is θ-iterable iff there is a θ-iteration strategy for M.

If λ is a limit ordinal, then $G(M,\lambda,\theta)$ is the game in which the players play λ rounds, the α-th round being a play of $G(N,\theta)$, where N is an initial segment, chosen by I, of the direct limit along the branch produced by the prior rounds. I moves at successor stages, by playing an extender or starting a new round if he wishes.[41] If the current round lasts θ moves, then there are no further rounds, and the game is over.[42]

II picks branches at limit stages, and his obligation is just to insure all models are wellfounded, including the direct limit of the base models in the final stack of length λ. Thus s is a normal M-stack of length α whose component normal trees have length $< \theta$ iff s is a position in $G(M,\lambda,\theta)$ that represents α completed rounds and is not yet a loss for II. A *(λ,θ)-iteration strategy* for M is a winning strategy for II in $G(M,\lambda,\theta)$, and M is (λ,θ)-iterable iff there is such a strategy. See [70]. Clearly $G(M,1,\theta) = G(M,\theta)$.[43]

DEFINITION 2.7.1. Let M be a premouse; then M is *countably iterable* iff every countable elementary submodel of M is (ω_1,ω_1+1)-iterable.

Countable iterability is what one needs to prove that M is well-behaved in a fine structural sense; for example, that its standard parameter is solid and universal.

Clearly one can modify these standard iteration games so that their outputs can be merely quasi-normal trees, or stacks of them. For example, if M is a premouse, we let $G^{\text{qn}}(M,\theta)$ be the variant of $G(M,\theta)$ in which player II must pick cofinal wellfounded branches at limit steps as before, and given that $\mathcal{T}$ with $\text{lh}(\mathcal{T}) = \alpha+1$ is the play so far, I is allowed to pick E_α from the $\mathcal{M}_\alpha = \mathcal{M}_\alpha^{\mathcal{T}}$ sequence such that $\lambda(E_\xi) \le \lambda(E_\alpha)$ for all $\xi < \alpha$. (Here $\mathcal{M}_0 = M$.) As before, we set

$$\xi = T\text{-pred}(\alpha+1) = \text{least } \beta \text{ s.t. } \text{crit}(E_\alpha) < \lambda(E_\beta),$$

and letting $\langle \nu,k\rangle$ be least such that $\rho(\mathcal{M}_\xi^{\mathcal{T}}|\langle \nu,k\rangle) \le \text{crit}(E_\alpha)$, or $\langle \nu,k\rangle = l(\mathcal{M}_\xi)$,

$$\mathcal{M}_{\alpha+1} = \text{Ult}(\mathcal{M}_\xi|\langle \nu,k\rangle, E_\alpha).$$

---

[41] For notational reasons, we allow I to move immediately from round α to round $\alpha+1$, without playing any extenders.

[42] Thus if M is countable, a position in $G(M,\omega_1,\omega_1)$ is a member of HC, and a strategy for it is a subset of HC.

[43] Up to minor details in how they are presented.

We write $\mathcal{M}_\xi|\langle \nu,k\rangle = \mathcal{M}^{*,\mathcal{T}}_{\alpha+1}$. II plays at limit ordinals as before. A quasi-normal tree on M is just a position in some $G^{\mathrm{qn}}(M,\theta)$ in which II has not yet lost.

For λ a limit ordinal or $\lambda = 1$, we let $G^{\mathrm{qn}}(M,\lambda,\theta)$ be the variant of $G(M,\lambda,\theta)$ whose output now is a stack of quasi-normal trees on M, that is, an M-stack, of length λ. II wins iff all models reached are wellfounded, and if $\lambda > 1$, there are finitely many drops along the sequence of base models, and their direct limit is wellfounded. Player I decides when new rounds begin, and may drop gratuitously in the model produced by the prior rounds before starting the next one. We allow him to move to the next round without playing any extenders. With these conventions, s is an M-stack iff s is a position representing a sequence of rounds in some $G^{\mathrm{qn}}(M,\lambda,\theta)$. Up to details in presentation, $G^{\mathrm{qn}}(M,\theta)$ and $G^{\mathrm{qn}}(M,1,\theta)$ are the same game.

Clearly, one could generalize further.

DEFINITION 2.7.2. $G^{\mathrm{sn}}(M,\theta)$ and $G^{\mathrm{sn}}(M,\lambda,\theta)$ are the analogs of $G^{\mathrm{qn}}(M,\theta)$ and $G^{\mathrm{qn}}(M,\lambda,\theta)$ whose outputs are merely semi-normal trees, or stacks of them, respectively.

DEFINITION 2.7.3. Let M be a premouse; then a (λ,θ)*-iteration strategy for* M is a winning strategy for II in one of the games $G(M,\lambda,\theta)$, $G^{\mathrm{qn}}(M,\lambda,\theta)$, or $G^{\mathrm{sn}}(M,\lambda,\theta)$.

Definition 2.7.3 should be regarded as provisional, in that we shall introduce iteration strategies defined on wider classes of iteration trees in Chapter 4. Definition 2.7.3 covers the sorts of iteration strategies for premice that we shall encounter before we get to Chapter 4.

Tail strategies. Iterates of an iterable structure are iterable, via a *tail strategy*. In general, if G is any game, Ω is a strategy for G, and p is a position in G, then we get a tail strategy Ω_p acting on extensions of p by setting $\Omega_p(q) = \Omega(p^\frown q)$. In the case that $G = G^{\mathrm{qn}}(M,\lambda,\theta)$ and Ω is a strategy for II, we shall for now just consider tail strategies determined by positions in which I has just begun a new round, and declared a base model for the tree to be played in that round.

DEFINITION 2.7.4. Let Ω be a winning strategy for II in $G^{\mathrm{qn}}(M,\lambda,\theta)$, let s be an M-stack according to Ω with $\mathrm{lh}(s) < \lambda$, and let $N = M_\infty(s)|\langle \nu,k\rangle$ for some ν,k; then $\Omega_{s,N}$ is the strategy for $G^{\mathrm{qn}}(N,\lambda - \mathrm{lh}(s),\theta)$ given by:

$$\Omega_{s,N}(t) = \Omega(s^\frown\langle N\rangle^\frown t),$$

for all N-stacks t. We set $\Omega_s = \Omega_{s,M_\infty(s)}$. If $N \trianglelefteq M$, then

$$\Omega_N = \Omega_{\langle\emptyset\rangle,N}.$$

We are assuming here that the position $s^\frown\langle N\rangle$ does include the information that N is the base model for a new round. There are other tails of Ω one might consider. For example, if p represents a quasi-normal tree on M played by Ω as part of round 1, to which I has not declared an end, then we have a tail strategy Ω_p. In this case,

Ω_P would act on quasi-normal extensions of the phalanx of $\mathcal{T}$. It will be some time before we consider tail strategies of that sort, and we are not introducing any notation for them now.

When $N = M_\infty(s)|\langle \nu, k\rangle$, we may write $\Omega_{s,\langle \nu,k\rangle}$ for $\Omega_{s,N}$. We write Ω_N or $\Omega_{\langle \nu,k\rangle}$ for $\Omega_{\langle\emptyset\rangle,N}$. So if Ω is a strategy for $G^{\mathrm{qn}}(M,2,\theta)$, then Ω_M is just the strategy for $G^{\mathrm{qn}}(M,\theta)$ that is its "M-tail".

It is also useful to have a notation for a join of strategies:

DEFINITION 2.7.5. Let Ω be a winning strategy for II in one of the games $G(M,\lambda,\theta)$, $G^{\mathrm{qn}}(M,\lambda,\theta)$, or $G^{\mathrm{sn}}(M,\lambda,\theta)$, and let s be an M-stack by Ω; then $\Omega_{s,<\nu} = \langle \Omega_{s,\langle \eta,k\rangle} \mid \eta < \nu \wedge k \leq \omega\rangle$.

Note that in general, $\Omega_{s,<\nu}$ is strictly weaker than $\Omega_{s,\langle \nu,0\rangle}$.

Our definitions so far allow the tails of an iteration strategy to be inconsistent with the strategy itself; for example, one could have a strategy Ω for $G^{\mathrm{qn}}(M,\lambda,\theta)$ such that $\Omega \neq \Omega_{\emptyset,M}$.[44] One could have more subtle inconsistencies, for example, $N \trianglelefteq M$ and some normal $\mathcal{T}$ by both Ω_M and Ω_N such that $\Omega_M(\mathcal{T}) \neq \Omega_N(\mathcal{T})$. The iteration strategies that we shall construct in Chapter 3 do not have such internal inconsistencies, and one of our main tasks will be to spell that out precisely and prove it. For example,

DEFINITION 2.7.6. Let Ω be a winning strategy for II in $G^{\mathrm{qn}}(M,\lambda,\theta)$; then Ω is *positional* iff whenever s and t are M-stacks by Ω of length $< \lambda$, and $N \trianglelefteq M_\infty(s)$ and $N \trianglelefteq M_\infty(t)$, then $\Omega_{s,N} = \Omega_{t,N}$.

The iteration strategies we shall construct are positional, but it is beyond the scope of this book to show that. We shall instead use some approximations to positionality here. We shall discuss this further in Sections 3.5 and 3.6.

Pullback strategies. Given $\pi\colon M \to N$ weakly elementary, we can copy an M-stack s to an N-stack πs, until we reach an illfounded model on the πs side. Thus if Ω is an strategy for N, we have the pullback strategy Ω^π for M.

More precisely, let $\mathcal{T}$ be a semi-normal tree on M. We define a semi-normal tree $\pi\mathcal{T}$ on N with the same tree order as $\mathcal{T}$, together with weakly elementary copy maps

$$\pi_\alpha \colon M_\alpha \to N_\alpha,$$

where $M_\alpha = \mathcal{M}^{\mathcal{T}}_\alpha$ and $N_\alpha = \mathcal{M}^{\pi\mathcal{T}}_\alpha$. $\pi_0 = \pi$, and if T-pred$(\alpha+1) = \beta$, then we let $\mathcal{M}^{*,\pi\mathcal{T}}_{\alpha+1} = \pi_\beta(\mathcal{M}^{*,\mathcal{T}}_{\alpha+1})$. By induction, we have that $\pi_\gamma \restriction \lambda(E^{\mathcal{T}}_\gamma) = \pi_\xi \restriction \lambda(E^{\mathcal{T}}_\gamma)$ whenever $\gamma \leq \xi$, with the agreement being up to $\mathrm{lh}(E^{\mathcal{T}}_\gamma)$ if $\mathrm{lh}(E^{\mathcal{T}}_\gamma) < \mathrm{lh}(E^{\mathcal{T}}_{\gamma+1})$. It follows that the Shift Lemma applies to $(\pi_\beta \restriction \mathcal{M}^{*,\mathcal{T}}_{\alpha+1}, \pi_\alpha, E^{\mathcal{T}}_\alpha)$, so we can let

$$E^{\pi\mathcal{T}}_\alpha = \pi_\alpha(E^{\mathcal{T}}_\alpha),$$

44For $N \trianglelefteq M$, Ω_N is the part of Ω that acts on plays where I exits the first round without playing any extenders, then drops to N at the beginning of the second round. Even if $N = M$, the change of rounds could affect how Ω plays.

and

$$\pi_{\alpha+1} = \text{copy map associated to } (\pi_\beta \restriction \mathcal{M}^{*,\mathcal{T}}_{\alpha+1}, \pi_\alpha, E^{\mathcal{T}}_\alpha).$$

By induction, the copy maps commute with the branch embeddings, that is, $\pi_\xi \circ \hat{\imath}^{\mathcal{T}}_{\gamma,\xi} = \hat{\imath}^{\pi\mathcal{T}}_{\gamma,\xi} \circ \pi_\xi$ when $\gamma \leq_T \xi$. So at limit steps λ we have a unique π_λ that commutes with the branch embeddings of $\mathcal{T}$ and $\pi\mathcal{T}$ along $[0,\lambda)_T$. If $\pi\mathcal{T}$ ever reaches an illfounded model, we stop the construction.

To copy a stack, we just repeat this process. For example, given $\pi\colon M \to N$ and $\mathcal{T}$ semi-normal on M with length $\alpha+1$, and $P \trianglelefteq \mathcal{M}^{\mathcal{T}}_\alpha$, $\pi\langle \mathcal{T}, \langle P, \mathcal{U}\rangle\rangle = \langle \pi\mathcal{T}, \langle \pi_\alpha(P), (\pi_\alpha \restriction P)\mathcal{U}\rangle\rangle$. Continuing this way, we can define πs for any M-stack s, so long as we do not encounter illfounded models in πs.

The Shift Lemma leads to

LEMMA 2.7.7. *Let $\pi\colon M \to N$ be weakly elementary, and let $\mathcal{T}$ be semi-normal on M; then*

(1) $\pi\mathcal{T}$ is semi-normal, and if $\mathcal{T}$ is length-increasing, then so is $\pi\mathcal{T}$.

(2) The copy maps π_α are weakly elementary, and if π is nearly elementary, then so are the π_α. If π is elementary, then so are the π_α.

(3) If π is elementary and $\mathcal{T}$ is maximal, then $\pi\mathcal{T}$ is maximal.

We omit the proof. Most of it is straightforward, but the part devoted to propagating the elementarity properties of the copy maps requires some fine structural analysis. See Section 4.5 for a proof of the Lemma 2.7.7 in the context of pfs premice.

Remark 2.7.8. There is a caution here. It is possible that $\mathcal{T}$ is normal, but $\pi\mathcal{T}$ is not even quasi-normal, because it is not maximal. Lemma 2.7.7 implies that if π is elementary, then the π-copy of a (quasi) normal tree is (quasi) normal. But one might have a nearly elementary π for which maximality is not preserved. For example, we might have $k(M) = 1$, and E on the M-sequence such that $\rho_1(M) \leq \text{crit}(E)$, but $\pi(\text{crit}(E)) < \rho_1(N)$. If $\mathcal{T}$ starts normally with E, it will drop to M^-, that is, to M with its degree reduced by one, and form $\text{Ult}(M^-, E)$. Our copying process then requires $\pi\mathcal{T}$ to start by forming $\text{Ult}(N^-, \pi(E))$, which for $\pi\mathcal{T}$ is a gratuitous drop.[45]

Nevertheless, if $\mathcal{T}$ is semi-normal and π is nearly elementary, then $\pi\mathcal{T}$ is semi-normal. Moreover, $D^{\mathcal{T}} = D^{\pi\mathcal{T}}$. The drops occur at the same places, and to models of the same degree, it is just that $\pi\mathcal{T}$ may sometimes drop further than it has to. If $\mathcal{T}$ is length-increasing, then there is a natural normal tree on N into which $\pi\mathcal{T}$ embeds. It is defined in Definition 4.5.19.

[45]Schlutzenberg [56] calls a semi-normal tree $\mathcal{T}$ *model maximal* iff whenever $T\text{-pred}(\alpha+1) = \beta$, then $\mathcal{M}^{*,\mathcal{T}}_{\alpha+1} = \mathcal{M}^{\mathcal{T}}_\beta | \langle \nu, k\rangle$ where ν (but perhaps not k) is as large as possible. If $\mathcal{T}$ is model maximal and π is nearly elementary, then $\pi\mathcal{T}$ is model maximal.

DEFINITION 2.7.9 (Pullback strategies). If Ω is a strategy for N, and $\pi\colon M \to N$ is nearly elementary, then Ω^π is the pullback strategy for M, given by

$$\Omega^\pi(s) = \Omega(\pi s),$$

for all s such that $\pi s \in \mathrm{dom}(\Omega)$.

We have not specified here what sort of iteration strategy Ω is, so Definition 2.7.9 is really a family of definitions. If π is elementary, then Ω^π is a strategy of the same type as Ω. If π is only nearly elementary, then $\pi\mathcal{T}$ may fail to be maximal, even if $\mathcal{T}$ is maximal, so setting $\Omega^\pi(s) = \Omega(\pi s)$ will only make sense if Ω is defined on stacks of possibly non-maximal trees. See Section 4.5 and Definition 4.6.5.

Universally Baire iteration strategies. We shall often be working with a countable premouse M, and an iteration strategy Σ for M that is defined on countable trees of some sort, with AD^+ as our background assumption. We can then extend Σ so that it acts on trees of length ω_1, because under AD^+, ω_1 is measurable. Here is a simple proposition along these lines.

PROPOSITION 2.7.10. *Assume* AD, *and let* Σ *be an* ω_1*-iteration strategy for a countable premouse* M*; then* Σ *can be extended to an* $\omega_1 + 1$ *strategy for* M.

PROOF. Let $\mathcal{T}$ be a normal tree of length ω_1 on M that is played by Σ. It will suffice to show $\mathcal{T}$ has a cofinal, wellfounded branch. But let $j\colon V \to N$ with $\mathrm{crit}(j) = \omega_1$ witness the measurability of ω_1. The pair $\langle \mathcal{T}, M\rangle$ can be coded by a set of ordinals A, and Łoś's Theorem holds for ultrapowers of wellordered structures, so $j\colon L[A] \to L[j(A)]$ is elementary. It follows that $j(\mathcal{T})$ is an iteration tree on M, $\mathcal{T} = j(\mathcal{T}) \upharpoonright \omega_1$, and $\omega_1 < \mathrm{lh}(j(\mathcal{T}))$. But this implies that $[0,\omega_1)_{j(T)}$ is a cofinal, wellfounded branch of $\mathcal{T}$. ⊣

Although it is quite easy to prove, this proposition stands at a key junction in inner model theory. The direct proofs of iterability only produce branches for countable iteration trees, even in the realm of linear iterations. Yet $\omega_1 + 1$-iterability is the minimal useful kind of iterability; for example, it is the kind needed to compare countable premice. All known proofs of $\omega_1 + 1$-iterability involve at some point producing an ω_1-strategy Σ, and showing that Σ is sufficiently absolutely definable that one can extend it to an $\omega_1 + 1$ strategy. In the proposition above, the absolute definability of Σ is evidenced by its membership in a model of AD. In contexts where one's goal is more ambitious than analyzing HOD in models of AD, the absolute definability of Σ has to be more finely calibrated, and a model of some fragment of AD that contains Σ constructed along with Σ. This leads into the core model induction method, our most all-purpose method for constructing iteration strategies.

Proposition 2.7.10, simple as it is, is one important reason that inner model theory and descriptive set theory have become so entangled in recent years.

When calibrating definablity in terms of pointclasses, the standard procedure is to code elements of HC (e.g., premice) by reals, and subsets of HC (e.g., ω_1-iteration strategies) by sets of reals. Of course, any reasonable way of doing this is fine, but we may as well spell one out. For $x \in \mathbb{R} = \omega^\omega$, we say $\mathrm{Cd}(x)$ iff $E_x =_{\mathrm{def}} \{\langle n,m\rangle \mid x(2^n 3^m) = 0\}$ is a wellfounded, extensional relation on ω. If $\mathrm{Cd}(x)$, then

$$\pi_x\colon (\omega, E_x) \cong (M, \in)$$

is the transitive collapse map, and

$$\mathrm{set}(x) = M \text{ and } \mathrm{set}_0(x) = \pi_x(0).$$

So Cd is Π^1_1, and set_0 maps Cd onto HC. For $A \subseteq \mathrm{HC}$, we let

$$\mathrm{Code}(A) = \{x \in \mathbb{R} \mid \mathrm{Cd}(x) \wedge \mathrm{set}_0(x) \in A\}.$$

If Σ is an iteration strategy with scope HC for a countable M, and Γ is a pointclass, then we sometimes say "$\Sigma \in \Gamma$" when we mean $\mathrm{Code}(\Sigma) \in \Gamma$.

Recall that a set $A \subseteq \mathbb{R}$ is *κ-Universally Baire* (κ-uB) iff there are trees T and U on some $\omega \times Z$ such that $p[T] = \mathbb{R} \setminus p[U]$ holds in $V[g]$ whenever g is V-generic for a poset of size $< \kappa$, and $p[T] = A$ holds in V. We call such a pair (T,U) a *κ-uB code* of A.[46] If κ is a limit of Woodin cardinals, then the κ-uB sets are the same as the $< \kappa$-homogeneously Suslin sets; moreover, if A is κ-uB, as witnessed by the pair of trees (T,U), then the theory of $(\mathrm{HC}, \in, p[T])$ is absolute for forcing of size $< \kappa$ (cf. [69]). This enables us to extend ω_1-iteration strategies that are κ-uB to κ-iteration strategies. As is well known, the extension is independent of the partucular uB code chosen. In fact, with a little care, we do not need the Woodin cardinals to make it.

PROPOSITION 2.7.11. *Let $A \subseteq \mathrm{HC}$, and suppose (T,U) is a κ-uB code of* $\mathrm{Code}(A)$. *For $b \in H_\kappa$, put*

$$b \in B \text{ iff } \mathrm{Col}(\omega, <\kappa) \Vdash \exists x \in p[T](\mathrm{set}_0(x) = b).$$

Then $(\mathrm{HC}, \in, A) \prec_{\Sigma_1} (H_\kappa, \in, B)$.

PROOF SKETCH. Note that $p[T]$ and $p[U]$ remain invariant in $V^{\mathrm{Col}(\omega,<\kappa)}$, in that if $\mathrm{set}_0(x) = \mathrm{set}_0(y)$, then $x \in p[T]$ iff $y \in p[T]$, and similarly for U. Also, whether $x \in p[T]$ for any and all x such that $\mathrm{set}_0(x) = b$ is decided by the empty condition. Suppose $(H_\kappa, \in, B) \models \varphi[a]$, where φ is Σ_1 and $a \in \mathrm{HC}$. Let $\pi\colon N \to V_\theta$ with N countable and transitive, and $\pi(\langle \bar{T}, \bar{U}\rangle) = \langle T, U\rangle$. Let $\pi(M) = H_\kappa$ and $\pi(\bar{B}) = B$. We have $\pi(a) = a$, and $(M, \in, \bar{B}) \models \varphi[a]$. Using $\bar{T}$ and $\bar{U}$ and a simple absoluteness argument, we see that $\bar{B} = A \cap M$. So $(M, \in, A \cap M) \models \varphi[a]$. But φ is Σ_1, so $(\mathrm{HC}, \in, A) \models \varphi[a]$, as desired. ⊣

In order to apply the proposition to iteration strategies, we have to be careful about how we present them. Given an ω_1 strategy Σ, let A_Σ be the set of all pairs $(\mathcal{T}, \alpha)$ such that $\mathcal{T}$ is a tree of limit length by Σ, and $\alpha \in \Sigma(\mathcal{T})$.

---

[46]The concept was first isolated and studied for its own sake by Q. Feng, M. Magidor, and W. H. Woodin. See [9]. There are earlier related results due to K. Schilling and R. Vaught in [45].

COROLLARY 2.7.12. *Let Σ be an ω_1-iteration strategy for a countable premouse P, and suppose that* $\mathrm{Code}(A_\Sigma)$ *is κ-uB; then there is a κ-iteration strategy Ψ extending Σ.*

PROOF. Let $B \subseteq H_\kappa$ be such that $(\mathrm{HC}, \in, A_\Sigma) \prec_{\Sigma_1} (H_\kappa, \in, B)$. It is not hard to see that $B = A_\Psi$, where Ψ is the desired extension of Σ. ⊣

Clearly, the extension Ψ to H_κ is independent of the particular κ-uB code of A_Σ chosen. We call Ψ the *canonical κ-extension of* Σ. Abusing language somewhat, we may say that a κ-iteration strategy is κ-uB when it is the canonical κ-extension of an ω_1-strategy. The extension process works equally well for (λ, ω_1)-strategies.

The following simple fact about such strategies is useful.

PROPOSITION 2.7.13. *Let Σ be a κ-uB κ-iteration strategy for some countable P, and $j\colon V \to M$ with M transitive; then $j(\Sigma) \cap H_\kappa \subseteq \Sigma$.*

PROOF. Let (T, U) be a κ-uB code for $\mathrm{Code}(A_\Sigma)$. Suppose $\mathcal{T} \in H_\kappa$ is by both Σ and $j(\Sigma)$, and has limit length λ. If $\alpha < \lambda$, and $\alpha \in j(\Sigma)(\mathcal{T})$, then letting $\mathrm{set}_0(x) = \langle \mathcal{T}, \alpha \rangle$ with x in $V^{\mathrm{Col}(\omega, <\kappa)}$, we get $x \in p[j(T)]$. As usual, this implies $x \notin p[U]$, and hence $x \in p[T]$. Thus $\alpha \in \Sigma(\mathcal{T})$, as desired. ⊣

We shall show in Lemma 7.6.7 below that the conclusion $j(\Sigma) \cap H_\kappa \subseteq \Sigma$ also follows from strong hull condensation for Σ.

2.8. Comparison and genericity iterations

For the sake of completeness, we sketch a proof of the Comparison Theorem for pure extender mice. The reader can find full details in [70], and some related history in the introduction to [74].

THEOREM 2.8.1. *Let P and Q be premice of size $\leq \theta$, and suppose Σ and Ψ are $\theta^+ + 1$-iteration strategies for P and Q respectively; then there are normal trees $\mathcal{T}$ by Σ and $\mathcal{U}$ by Ψ of size θ, with last models R and S, such that either*

(a) $R \trianglelefteq S$, and P-to-R does not drop, or

(b) $S \trianglelefteq R$, and Q-to-S does not drop.

PROOF SKETCH. We build $\mathcal{T}$ and $\mathcal{U}$ inductively, by "iterating away the least disagreement" at successor steps, and using our iteration strategies at limit steps. At step α we have $\mathcal{T}_\alpha$ and $\mathcal{U}_\alpha$ with last models P_α and Q_α respectively. We begin with $P_0 = P$, $Q_0 = Q$, and $\mathcal{T}_0 = \mathcal{U}_0$ being the empty tree. At step $\alpha + 1$, let

$$\gamma = \text{least } \beta \text{ such that } P_\alpha | \beta \neq Q_\alpha | \beta.$$

If there is no such β, the comparison is complete. Otherwise, let

$$\mathcal{T}_{\alpha+1} = \mathcal{T}_\alpha{}^\frown \langle E_\gamma^{P_\alpha} \rangle \quad \text{and} \quad \mathcal{U}_{\alpha+1} = \mathcal{U}_\alpha{}^\frown \langle E_\gamma^{Q_\alpha} \rangle.$$

Here $\mathcal{S}^\frown \langle E \rangle$ stands for the unique normal extension of $\mathcal{S}$ whose last extender used is E, with the understanding that $\mathcal{S}^\frown \langle E \rangle = \mathcal{S}$ if $E = \emptyset$. At limit steps we let

$\mathcal{T}_\lambda$ be $\bigcup_{\alpha<\lambda}\mathcal{T}_\alpha$, extended by the branch $\Sigma(\bigcup_{\alpha<\lambda}\mathcal{T}_\alpha)$ if this tree has limit length. Similarly on the $\mathcal{U}$ side.

We claim that the comparison is complete at some stage $\alpha<\theta$. For suppose not, and let $\mathcal{T}=\mathcal{T}_{\theta^++1}$ and $\mathcal{U}=\mathcal{U}_{\theta^++1}$ be the normal trees of length θ^++1 that result. Let

$$N=\mathcal{M}^{\mathcal{T}}_{\theta^+}|\theta^+=\mathcal{M}^{\mathcal{U}}_{\theta^+}|\theta^+$$

be the common lined up part at stage θ^+. Let $\pi\colon H\to V_\xi$ be elementary, where ξ is large, everything relevant is in $\operatorname{ran}(\pi)$, H is transitive, and $\theta<\operatorname{crit}(\pi)<\theta^+$. Let $\alpha=\operatorname{crit}(\pi)$. We have $\pi(\langle P,Q\rangle)=\langle P,Q\rangle$ and $\pi(\alpha)=\theta^+$, and it is not hard to see that

$$\pi(\mathcal{T}\restriction\alpha+1)=\mathcal{T},$$
$$\pi(\mathcal{U}\restriction\alpha+1)=\mathcal{U},$$
$$\pi\restriction\mathcal{M}^{\mathcal{T}}_\alpha=i^{\mathcal{T}}_{\alpha,\theta^+},$$

and

$$\pi\restriction\mathcal{M}^{\mathcal{U}}_\alpha=i^{\mathcal{U}}_{\alpha,\theta^+}.$$

Also,

$$P(\alpha)^{\mathcal{M}^{\mathcal{T}}_\alpha}=P(\alpha)^{\mathcal{M}^{\mathcal{T}}_{\theta^+}}=P(\alpha)^{\mathcal{M}^{\mathcal{U}}_{\theta^+}}=P(\alpha)^{\mathcal{M}^{\mathcal{U}}_\alpha}.$$

Thus π, $i^{\mathcal{T}}_{\alpha,\theta^+}$, and $i^{\mathcal{U}}_{\alpha,\theta^+}$ all generate the same (α,θ^+)-extender; call it G. Let E be the first extender used in $\mathcal{T}$ along the branch $[\alpha,\theta^+]_T$, and F the first extender used in $\mathcal{U}$ along $[\alpha,\theta^+]_U$. As we observed in Remark 2.6.8, E is the first whole initial segment of G that is not on the N-sequence, and similarly for F. Thus $E=F$. Since $\operatorname{lh}(E)=\operatorname{lh}(F)$, they were used at the same stage in the comparison. But we were iterating away disagreements, so $E\neq F$, contradiction.

This gives us $\mathcal{T}=\mathcal{T}_\alpha$ and $\mathcal{U}=\mathcal{U}_\alpha$ with last models R and S such that $R\trianglelefteq S$ or $S\trianglelefteq R$. If $R\triangleleft S$, then R is sound, and therefore the branch P-to-R did not drop, so we have conclusion (a). Similarly, if $S\triangleleft R$ we get conclusion (b). Thus we may assume $R=S$. It is now enough to show that one of the two branches P-to-R and Q-to-S did not drop. Assume otherwise, and let

$$C=\mathfrak{C}(R)=\mathfrak{C}(S)$$

be the core, and π the anticore map. We have that C occurs on both branches, and that π is the iteration map of the branch C-to-R of $\mathcal{T}$, and the iteration map of the branch C-to-S of $\mathcal{U}$. But as in the termination proof, this means the first extenders used in these two branches are the same, a contradiction. ⊣

Notice that although the successful comparison only involves trees of size θ, we really did need θ^++1-iterability to show that it exists. In particular, to compare countable mice, we need ω_1+1-iterability.

COROLLARY 2.8.2. *Let M and N be countably iterable premice such that $\rho^-(M)=\rho^-(N)=\omega$; then either $M\trianglelefteq N$ or $N\trianglelefteq M$.*

PROOF. Let $\mathcal{T}$ on M with last model R and $\mathcal{U}$ on N with last model S be as in Theorem 2.8.1, and suppose without loss of generality that $R \trianglelefteq S$ and M-to-R does not drop. Since $\rho^-(M) = \omega$, it is impossible to take an ultrapower of M without dropping, so $\mathcal{T}$ is empty and $M = R$. It is enough to show that $\mathcal{U}$ is also empty. But otherwise, N-to-S must drop, and letting $C = \mathfrak{C}(S)$, the last drop is to C, and the anticore map $\pi \colon C \to S$ is the same as the branch embedding of $\mathcal{U}$.[47] We have

$$\rho^-(M) = \omega < \text{crit}(\pi) < \rho^-(C) \leq \rho^-(S),$$

so if $\hat{M} = \hat{S}$, then $k(S) < k(M)$, contrary to $M \trianglelefteq S$. Thus $\hat{M} \neq \hat{S}$, which implies that $M \in S$. But this means $M \triangleleft S|\omega_1^S$. Since $S|\omega_1^S = C|\omega_1^C$, we get that $M \triangleleft C$, so $M < \mathcal{M}_\gamma^{\mathcal{U}}$ for $\gamma = U$-pred$(\alpha+1)$, so the comparison was over before we reached S, contradiction. ⊣

COROLLARY 2.8.3. *Let M and N be countably iterable premice such that $\rho^-(M) = \rho^-(N) = \omega$ and $o(M) = o(N)$; then $M = N$. Thus M is ordinal definable from $o(M)$.*

COROLLARY 2.8.4. *Let M be a countably iterable premouse; and $x \in P(\omega) \cap M$; then x is ordinal definable.*

PROOF. Let α be least such that x is definable over $M|\alpha$; then $\rho_k(M|\alpha) = \omega$ for some k, so $M|\alpha$ is ordinal definable, so x is ordinal definable. ⊣

To what extent is the converse of this corollary true? Does every ordinal definable real belong to an iterable premouse? In practice, the reals in mice are not just ordinal definable, but ordinal definable in a generically absolute way, because the ω_1+1-iteration strategies for countable premice that we have constructed so far are canonical extensions of Universally Baire ω_1-strategies. So the more reasonable question is whether every real that is ordinal definable in some $L(\Gamma, \mathbb{R})$, where Γ is a proper Wadge initial segment of the Universally Baire sets, belongs to a mouse. This seems quite plausible, but we do not as of now have a definition of *mouse* sufficiently general to state a precise conjecture here. The mouse capturing conjectures stated in Section 1.7 are the best we can do.

The Comparison Theorem gives us upper bounds on the definability of mice and the reals that belong to them. We get lower bounds from the capturing and correctness properties of the mice. One of the main tools for proving mouse capturing and correctness is the extender algebra.

Genericity iterations. The reader should see [70, §7] for basic information on the extender algebra and genericity iterations. The paper [8] gives a much more extensive treatment.

[70] and other expositions of genericity iterations of premice use ms-indexing and ms-normal trees. There is a small subtlety involved in carrying out the arguments using Jensen-normal trees, as we are doing in this book. Jensen-normal

47 In other words, $\pi = i^{\mathcal{U}}_{\alpha+1,\beta} \circ i^{*,\mathcal{U}}_{\alpha+1}$, where $C = \mathcal{M}^{*,\mathcal{U}}_{\alpha+1}$ and $S = \mathcal{M}^{\mathcal{U}}_{\beta}$.

genericity iterations must be allowed to drop, unless our identities are generated by superstrong extenders. However, this dropping will not occur along the main branch, so it is harmless. We explain this briefly now.

Let M be a premouse, and $\mu < \delta$ cardinals of M. We let $\mathbb{B} = \mathbb{B}^M_{\mu,\delta}$ be the ω-generator *extender algebra* determined by the extenders on the $M|\delta$-sequence with critical point $> \mu$. $\mathbb{B}$ is the Lindenbaum algebra of a certain infinitary theory T in the propositional language $\mathcal{L}_{\delta,0}$ generated by the sentence symbols A_n, for $n < \omega$. For $x \subset \omega$, $x \models A_n$ iff $n \in x$, and then $x \models \varphi$ for φ an arbitrary sentence of $\mathcal{L}_0$ has the natural meaning. The axioms of T are those sentences of the form

$$\bigvee_{\alpha<\kappa} \varphi_\alpha \longleftrightarrow \bigvee_{\alpha<\lambda} i_E(\langle \varphi_\xi : \xi < \kappa \rangle) \restriction \lambda,$$

whenever E is on the $M|\delta$-sequence, $\mathrm{crit}(E) = \kappa > \mu$, and $i_E(\langle \varphi_\xi : \xi < \kappa \rangle) \restriction \lambda \in M|\eta$, for some cardinal η of M such that $\eta < \lambda_E$. Let us write $T = T(M|\delta, \mu)$.

The usual argument shows that if δ is Woodin in M, then $M \models$ "$\mathbb{B}$ is δ-c.c.". It is also clear that if M comes from a background construction in V, then every $x \in V$ satisfies all axioms of T. This is because if E generates an axiom as above, and E^* is its background extender, then $E \restriction \eta = E^* \restriction \eta \cap M$, for all M-cardinals η. It is important here that η is a cardinal of M, since otherwise the connection between E and E^* may be less direct.

THEOREM 2.8.5. *Let M be a countable premouse, and Σ an $\omega_1 + 1$-iteration strategy for M. Suppose $\mu < \delta$ and $M \models$ "δ is Woodin", and let $x \subseteq \omega$; then there is a countable, normal iteration tree $\mathcal{U}$ of length $\alpha + 1$ such that*

(a) $[0, \alpha]_U$ *does not drop,*
(b) $\mathrm{crit}(i^{\mathcal{U}}_{0,\alpha}) > \mu$, *and*
(c) *x is $i^{\mathcal{U}}_{0,\alpha}(\mathbb{B}^M_{\mu,\delta})$- generic over $\mathcal{M}^{\mathcal{U}}_\alpha$.*

PROOF. We form $\mathcal{U}$ by iterating away the least disagreement between the theory $T(\mathcal{M}^{\mathcal{T}}_\alpha | \sup \hat{\imath}^{\mathcal{T}}_{0,\alpha}$"$\delta, \mu)$ and the truth about x. More precisely, $E^{\mathcal{U}}_\alpha$ is the first extender on the sequence of $\mathcal{M}^{\mathcal{U}}_\alpha$ with critical point above μ that induces an axiom of $T(\mathcal{M}^{\mathcal{U}}_\alpha | \sup \hat{\imath}^{\mathcal{U}}_{0,\alpha}$"$\delta, \mu)$ not satisfied by x. The rest is determined by the rules of Jensen normal trees. Note the hat above the i in the formula! $[0, \alpha)_U$ may have dropped. $\hat{\imath}_{0,\alpha}(\mu) = \mu$, but it may happen that $\hat{\imath}_{0,\alpha}(\delta)$ is undefined.

Just as in the proof of the Comparison Theorem, the construction of $\mathcal{U}$ terminates at some countable stage with a last model $\mathcal{M}^{\mathcal{U}}_\alpha$ such that x satisfies all the axioms of $T(\mathcal{M}^{\mathcal{U}}_\alpha | \sup \hat{\imath}^{\mathcal{U}}_{0,\alpha}$"$\delta, \mu)$. We must see that in this case, $[0, \alpha)_U$ has not dropped. Suppose that it has, and let $\xi + 1 \leq_U \alpha$ be the site of the last drop, and U-pred$(\xi + 1) = \gamma$. Let $E = E^{\mathcal{U}}_\gamma$, and let

$$\psi = \bigvee_{\alpha<\kappa} \varphi_\alpha \longleftrightarrow \bigvee_{\alpha<\lambda} i_E(\langle \varphi_\nu : \nu < \kappa \rangle) \restriction \lambda$$

be the bad axiom induced by E, and η a cardinal of $\mathcal{M}^{\mathcal{U}}_\gamma$ such that $\psi \in \mathcal{M}^{\mathcal{U}}_\gamma | \eta$.

Since we dropped when applying $E^{\mathcal{U}}_{\xi}$, $\eta \le \mathrm{crit}(E^{\mathcal{U}}_{\xi})$, so $i^{\mathcal{U}}_{\gamma,\alpha} \upharpoonright \eta$ is the identity. But also, $\mathcal{M}^{\mathcal{U}}_{\gamma} | \mathrm{lh}(E) \trianglelefteq \mathcal{M}^{*}_{\xi+1}$, so $i^{\mathcal{U}}_{\gamma,\alpha}(E)$ exists. Clearly, $i^{\mathcal{U}}_{\gamma,\alpha}(E)$ still induces ψ as an axiom of $T(\mathcal{M}^{\mathcal{U}}_{\alpha} | \sup i^{\mathcal{U}}_{0,\alpha}``\delta, \mu)$. Since x does not satisfy ψ, the genericity iteration did not terminate at α, contradiction. $\dashv$

2.9. Coarse structure

One must consider also iteration trees on transitive models M that are not equipped with any distinguished fine structural hierarchy. In that case, we shall always assume $M \models \mathsf{ZFC}$, for simplicity. In general, V^M_α plays the role that $M|\alpha$ would in the fine structural case. All extenders are total on the models to which they are applied, and all embeddings are fully elementary in the $\in$-language. We shall sometimes call such M, and associated objects like iteration trees or embeddings acting on them, *coarse*, in order to distinguish them from their fine-structural cousins.

DEFINITION 2.9.1. Let E be an extender over V; then E is *nice* iff

(a) E is strictly short, that is, $\mathrm{lh}(E) < \lambda(E)$,
(b) for some ν, $\mathrm{lh}(E)$ is the least strongly inaccessible η such that $\nu < \eta$,
(c) $V_{\mathrm{lh}(E)} \subseteq \mathrm{Ult}(V, E)$.

Nice E can be used to background extenders in a Jensen premouse, even though $\mathrm{lh}(E) < \lambda(E)$. The requirement of (b) enables us to avoid a counterexample to UBH for stacks of normal trees due to Woodin. See Theorem 7.3.17 below.

DEFINITION 2.9.2. Let $\mathcal{T}$ be an iteration tree on a coarse M; then

(a) $\mathcal{T}$ is *nice* iff whenever $\alpha+1 < \mathrm{lh}(\mathcal{T})$, then $\mathcal{M}^{\mathcal{T}}_{\alpha} \models$ "$E^{\mathcal{T}}_{\alpha}$ is nice".
(b) $\mathcal{T}$ is *quasi-normal* iff
   (i) if $\alpha < \beta$ and $\beta + 1 < \mathrm{lh}(\mathcal{T})$, then $\mathrm{lh}(E^{\mathcal{T}}_{\alpha}) \le \mathrm{lh}(E^{\mathcal{T}}_{\beta})$, and
   (ii) if $\alpha+1 < \mathrm{lh}(\mathcal{T})$, then T-pred$(\alpha+1)$ is the least β such that $\mathrm{crit}(E^{\mathcal{T}}_{\alpha}) < \mathrm{lh}(E^{\mathcal{T}}_{\beta})$.
(c) $\mathcal{T}$ is *normal* iff $\mathcal{T}$ is quasi-normal, and if $\alpha < \beta$ and $\beta+1 < \mathrm{lh}(\mathcal{T})$, then $\mathrm{lh}(E^{\mathcal{T}}_{\alpha}) < \mathrm{lh}(E^{\mathcal{T}}_{\beta})$.

This definition of normality and quasi-normality is only appropriate for nice trees, but all our coarse iteration trees will be nice, so that is ok. It would be possible to allow gratuitous dropping, but we shall not do that. Nice iteration trees do not drop anywhere. Moreover, we shall often restrict the choice of extenders in $\mathcal{T}$ even further.

DEFINITION 2.9.3. Let $\mathcal{T}$ be an iteration tree on M, where $M \models \mathsf{ZFC}$ is transitive, and let $(M, \mathcal{F})$ be amenable; then

(a) $\mathcal{T}$ is an *$\mathcal{F}$-tree* iff whenever $\alpha+1 < \mathrm{lh}(\mathcal{T})$, then $E^{\mathcal{T}}_{\alpha} \in i^{\mathcal{T}}_{0,\alpha}(\mathcal{F})$.

(b) $\mathcal{T}$ is *above* κ iff $\mathcal{T}$ is an $\mathcal{F}$-tree, where $\mathcal{F} = \{E \mid \text{crit}(E) > \kappa\}$.
(c) $\mathcal{T}$ is *based on* V_δ^M iff $\mathcal{T}$ is an $\mathcal{F}$-tree, where $\mathcal{F} = V_\delta^M$.
(d) A *putative* $\mathcal{F}$*-tree* on M is a system having all the properties of an $\mathcal{F}$-tree on M, except that its last model may be illfounded.

We sometimes think of an $\mathcal{F}$ tree on M as a tree on the pair $(M, \mathcal{F})$, with models of the form $(\mathcal{M}_\alpha^\mathcal{T}, i_{0,\alpha}(\mathcal{F}))$.

In Definition 2.9.3 we are not assuming that $\mathcal{T}$ is quasi-normal. It may be a stack of quasi-normal trees, in which case we may call it an $\mathcal{F}$-stack, or a putative $\mathcal{F}$-stack. The non-quasi-normal iteration trees on coarse premice that we consider will always be stacks of quasi-normal trees. One could venture further into the wilds, but we shall not do that. However, we do need to work with coarse trees that are quasi-normal but not normal.

DEFINITION 2.9.4. Let $M \models \mathsf{ZFC}$ be transitive, and $(M, \mathcal{F})$ be amenable; then
(a) $G^{\text{qn}}(M, \eta, \theta, \mathcal{F})$ is the variant of $G^{\text{qn}}(M, \eta, \theta)$ in which player I must choose his exit extenders from the current image of $\mathcal{F}$, and
(b) an $(\eta, \theta, \mathcal{F})$-iteration strategy for M is a winning strategy for player II in $G^{\text{qn}}(M, \eta, \theta, \mathcal{F})$.

Player I is allowed to drop gratuitously at the beginning of a round of $G^{\text{qn}}(M, \eta, \theta, \mathcal{F})$. In this coarse case, we demand that if $(P, \mathcal{G})$ is the current last model, I can only drop to models of the form $(V_\alpha^P, \mathcal{G} \cap V_\alpha^P)$ such that α is an inaccessible cardinal in P.[48]

In general, we shall only make use of iteration strategies for coarse M that choose branches that, when allowed to act on the largest possible base model, become the unique cofinal wellfounded branch.

DEFINITION 2.9.5. Let M be transitive and $(M, \mathcal{F}) \models \mathsf{ZFC}(\dot{A})$, and let $\lambda, \theta \in$ OR; then
(a) M is *strongly uniquely* $(\lambda, \theta, \mathcal{F})$*-iterable* iff there is a $(\lambda, \theta, \mathcal{F})$-iteration strategy Σ for M such that whenever $\mathcal{T}$ is a tree by Σ of limit length, then $\Sigma(\mathcal{T})$ is the unique cofinal, wellfounded branch of $\mathcal{T}$.
(b) M is *strongly uniquely* $(\theta, \mathcal{F})$*-iterable for normal trees* iff M is strongly uniquely $(1, \theta, \mathcal{F})$-iterable.

We say that M is strongly uniquely (λ, θ)-iterable above κ, or for trees based on V_δ^M, iff M is strongly uniquely $(\lambda, \theta, \mathcal{F})$-iterable for the associated $\mathcal{F}$. Notice that strong unique iterability is more than just having a unique iteration strategy; that strategy must be to choose the unique cofinal, wellfounded branch.

It will sometimes help to restrict the sort of $\mathcal{F}$ we consider.

DEFINITION 2.9.6. $(w, \mathcal{F})$ is a *coherent pair* iff $\mathcal{F}$ is a set of nice extenders, and w is a wellorder of some V_δ such that $\mathcal{F} \subseteq V_\delta$, and for all $E \in \mathcal{F}$,

[48]The inaccessibility requirement just simplifies a few things.

(a) If $\eta < \mathrm{lh}(E)$ and $E \upharpoonright \eta$ is nice, then $E \upharpoonright \eta \in \mathcal{F}$,

moreover, letting $i_E : V \to M = \mathrm{Ult}(V,E)$ be the canonical embedding,

(b) $i_E(\mathcal{F}) \cap V^M_{\mathrm{lh}(E)+1} = \mathcal{F} \cap V^M_{\mathrm{lh}(E)+1}$, and

(c) $i_E(w) \cap V^M_{\mathrm{lh}(E)+1} = w \cap V^M_{\mathrm{lh}(E)+1}$.

We let $\delta(w)$ be the δ such that w wellorders V_δ.

DEFINITION 2.9.7. $(M, \in, w, \mathcal{F})$ is a *coarse extender premouse* iff M is a transitive model of ZFC, and $M \models (w, \mathcal{F})$ is a coherent pair.

Coarse extender premice will serve as the background universes in which we construct fine structural, pure extender premice. $\mathcal{F}$ will serve as the class of background extenders used to certify extenders that we add to the sequence of our evolving premouse.

DEFINITION 2.9.8. Let $\mathcal{F}$ be a set of extenders and $A \subseteq V_\delta$, and $\kappa < \delta$; then κ is *A-reflecting in δ via extenders in $\mathcal{F}$* iff $\forall \beta < \delta \exists E \in \mathcal{F}(\mathrm{crit}(E) = \kappa \wedge i^V_E(A) \cap V_\beta = A \cap V_\beta)$. We say δ is *Woodin via extenders in $\mathcal{F}$* (or *δ is $\mathcal{F}$-Woodin*) iff $\forall A \subseteq \delta \exists \kappa < \delta$ (κ is A reflecting in δ via extenders in $\mathcal{F}$).

PROPOSITION 2.9.9. *Let δ be Woodin, and w be a wellorder of V_δ. Let*

$$\mathcal{F} = \{E \in V_\delta \mid E \text{ is nice and } i^V_E(w) \cap V^{Ult(V,E)}_{\mathrm{lh}(E)+1} = w \cap V^{\mathrm{Ult}(V,E)}_{\mathrm{lh}(E)+1}\}.$$

Then $(w, \mathcal{F})$ is a coherent pair, and δ is Woodin via extenders in $\mathcal{F}$.

PROOF. Let $E \in \mathcal{F}$ and suppose $\eta < \mathrm{lh}(E)$ and $E \upharpoonright \eta$ is nice. Since η is the least inaccessible above some ν, the factor embedding $\pi\colon \mathrm{Ult}(V, E \upharpoonright \eta) \to \mathrm{Ult}(V,E)$ is such that $\mathrm{crit}(\pi) > \eta$. Thus $i_{E \upharpoonright \eta}(w) \cap V_{\eta+1} = i_E(w) \cap V_{\eta+1}$, so $E \upharpoonright \eta \in \mathcal{F}$.

Any $E \in \mathcal{F}$ satisfies part (c) in Definition 2.9.6 by definition. But also, if $G \in \mathrm{Ult}(V,E)$ and $\mathrm{lh}(G) \leq \mathrm{lh}(E)$, part (c) for E implies that $G \in i_E(\mathcal{F})$ iff $G \in \mathcal{F}$. Thus $(w, \mathcal{F})$ is a coherent pair.

To see that δ is $\mathcal{F}$-Woodin, fix $A \subseteq \delta$, and let κ be (A,w)-reflecting in δ. Standard arguments then show that κ is A-reflecting via extenders in $\mathcal{F}$. ⊣

DEFINITION 2.9.10. Let $(w, \mathcal{F})$ be a coherent pair and $\delta = \delta(w)$; then $(w, \mathcal{F})$ is *maximal* iff $\mathcal{F} = \{E \in V_\delta \mid E$ is nice and $i^V_E(w) \cap V^{Ult(V,E)}_{\mathrm{lh}(E)+1} = w \cap V^{\mathrm{Ult}(V,E)}_{\mathrm{lh}(E)+1}\}$.

We have not made it part of the definition of coherent pair that the extenders in $\mathcal{F}$ be linearly ordered by the Mitchell order, that is, that for all $E, F \in \mathcal{F}$ such that $E \neq F$, $E \in \mathrm{Ult}(V,F)$ or $F \in \mathrm{Ult}(V,E)$. This is because, as the proposition shows, one can obtain coherent pairs witnessing the Woodinness of a Woodin cardinal directly, without going into inner model theory. If one adds Mitchell linearity to the requirements, it is not clear how to do this, whatever large cardinal one starts with. It seems necessary to replace V by a canonical inner model M constructed from $\vec{F}$, and prove a comparison lemma (the *Bicephalus Lemma*) that guarantees that $F_\alpha \cap M$ is unique in some sense.

Perhaps the simplest way to obtain a Mitchell linear coarse premouse $(M, w, \mathcal{F})$ with a $\mathcal{F}$-Woodin cardinal is to start with a fine premouse M with a Woodin cardinal δ, and then let w be its canonical wellorder, and $\mathcal{F}$ consist of those extenders E such that $M \models$ "E is nice" and the Jensen completion of E is on the M-sequence.[49]

The Mitchell order is wellfounded, so if $\mathcal{F}$ is Mitchell linear, then we can enumerate it in increasing Mitchell order. The resulting $\vec{F}$ is coarsely coherent, in the following sense.

DEFINITION 2.9.11. A sequence $\vec{F} = \langle F_\alpha \mid \alpha < \mu \rangle$ is *coarsely coherent* iff each F_α is a nice extender over V, and

(1) if G is a nice initial segment of F_α, then $G = F_\beta$ for some $\beta < \alpha$,
(2) if $\beta < \alpha$, then $\text{lh}(F_\beta) \leq \text{lh}(F_\alpha)$, and
(3) $i \colon V \to \text{Ult}(V, F_\alpha)$ is the canonical embedding, and $\vec{E} = i(\vec{F})$, then $\langle E_\xi \mid \text{lh}(E_\xi) \leq \text{lh}(F_\alpha) \rangle = \langle F_\xi \mid \xi < \alpha \rangle$.

An iteration tree on a coarse extender premouse $(M, w, \mathcal{F})$, is just an $\mathcal{F}$-iteration tree on M. That is, all extenders used must be taken from $\mathcal{F}$ and its images. Similarly for $\mathcal{F}$-stacks of quasi-normal trees. So the trees in an $\mathcal{F}$-stack are nice. In the coarse case, iteration trees do not have any necessary drops, and we prohibit gratuitous dropping and decreasing lengths just to keep things simple. Thus all $\mathcal{F}$-stacks are maximal.

The following lemma shows one way Mitchell linearity is useful.

LEMMA 2.9.12. *Let $(M, w, \mathcal{F})$ be a Mitchell linear coarse premouse, and let Σ be an $\mathcal{F}$-iteration strategy for M; then for any N, there is at most one normal $\mathcal{F}$-iteration tree played according to Σ whose last model is N.*

PROOF. Let $\vec{F}$ be the coarsely coherent sequence associated to $\mathcal{F}$. Let $\mathcal{T}$ and $\mathcal{U}$ be distinct such trees. Because both are played by Σ and normal, there must be a β such that $\mathcal{T} \upharpoonright \beta + 1 = \mathcal{U} \upharpoonright \beta + 1$, but $G \neq H$, where $G = E^{\mathcal{T}}_\beta$ and $H = E^{\mathcal{U}}_\beta$. Both G and H are taken from $i(\vec{F})$, where $i = i^{\mathcal{T}}_{0,\beta} = i^{\mathcal{U}}_{0,\beta}$. Say G occurs before H in $i(\vec{F})$. Then $G \in N$ because $\mathcal{U}$ is strictly length increasing. But $G \notin N$ because $G \notin \mathcal{M}^{\mathcal{T}}_{\beta+1}$, and $\mathcal{T}$ is length non-decreasing. ⊣

Assuming AD^+, one can construct Mitchell linear coarse premice $(M, w, \mathcal{F})$ via the Γ-Woodin construction.[50] These M can have a Woodin cardinal δ, and yet be correct for predicates in some complicated pointclass Γ. We shall have that δ is countable in V, and $(M, w, \mathcal{F})$ is strongly uniquely (ω_1, ω_1)-iterable. The same construction also produces coarse strategy premice.[51] We say more about this in Section 7.2.

---

[49]By a theorem due to Schlutzenberg and the author, if M is countably iterable, then $\mathcal{F} = \{E \mid M \models$ "E is nice"$\}$. Schlutzenberg proved a much stronger result in this direction in [58].

[50]See [71, §3] and [68, §10].

[51]However, it does not directly produce coarse strategy mice with Woodin cardinals.

Chapter 3

BACKGROUND-INDUCED ITERATION STRATEGIES

We construct a mouse M by adding extenders to its coherent sequence, one by one. If we add E, then $M|\text{lh}(E)$ must be a premouse, and this imposes a fairly severe restriction on E. Nevertheless, no first-order requirement like premousehood can guarantee that we are building a standard structure, one that can be compared with others of its kind. We need to be building an *iterable* premouse. Moreover, it is not enough that $M|\text{lh}(E)$ be iterable, for we need the full M to be iterable, and when we add E, we do not know what M will be.

The standard way to solve these difficulties is to demand a *background certificate* E^* for E. What exactly one demands of E^* depends on the context. In this book we shall ask that E^* be a nice extender over V such that $E \subseteq E^*$. In contexts where one is trying to construct mice without assuming there are large cardinals at all, much more care is needed at this point, and the iterability proofs become more difficult.

In any of its forms, the background certificate demand conflicts with the demand that our mice be sound. The standard way to solve that difficulty is to "core down" at every step, replacing the current approximation to M by its core. There are highly nontrivial comparison arguments involved in showing that this core exists, and agrees sufficiently with M that the process of adding certified extenders and coring down converges to anything.[1] These arguments rely on the iterability of M.

The existence of full background extender certificates means that we can lift iteration trees on M to iteration trees on V, and thus use an iteration strategy Σ^* for V to induce an iteration strategy Σ for M. This of course does not solve the iterability problem for M, it just reduces it to the problem for V. But some such reduction, ideally using weaker background certificates, seems inevitable in any construction of iteration strategies for premice. M cannot see the iteration trees with respect to which it must be iterable, but V can see their lifts. Moreover, those lifts can be taken to be simple (for example, use only nice extenders) in ways that the trees on M being lifted are not.

[1]These are the solidity, universality, and condensation theorems of [35, §8]. The analogous results for pfs mice are proved in Section 4.10, and those for strategy mice are proved in Sections 9.6 and 10.3.

In this book, we shall be looking very carefully at how an iteration strategy Σ^* for V induces iteration strategies for the premice occurring in a full background extender construction. In our applications, V will satisfy "I am strongly uniquely $\vec{F}$-iterable", where $\vec{F}$ is the sequence of background extenders used in the construction, and Σ^* will be the corresponding $\vec{F}$-iteration strategy. In Section 3.1 we describe the well known construction of pure extender premice. In Sections 3.2 through 3.4 we describe the standard lifting procedure, and in Section 3.5 we define the iteration strategies induced by this procedure.

Section 3.6 describes two ways in which this framework can lead to ill-behaved iteration strategies. Section 3.8 analyzes the first of these two problems more closely.[2] It seems essential to our method for comparing iteration strategies that they not exhibit such behavior, and this will lead us in Chapter 4 to modify many of our basic definitions, including the definitions of premouse, background construction, iteration tree, and induced strategy. Thus the background-induced iteration strategies we describe in this chapter will not literally play any role in the rest of the book. Nevertheless, it seems best to introduce the standard notions first. They are not too far from the revised notions.

3.1. Full background extender constructions

We shall use much of the notation of [41] in this context. The reader might also look at [1], on which it relies, and at [35, §11].

DEFINITION 3.1.1. Let $(w, \mathcal{F})$ be a coherent pair. A *$(w, \mathcal{F})$-construction above κ* is a full background construction in which the background extenders are nice extenders in $\mathcal{F}$, have critical points $> \kappa$, have strictly increasing strengths, and are minimal (first in Mitchell order, then in w).

More precisely, such a construction $\mathbb{C}$ consists of w, $\mathcal{F}$, premice $M^{\mathbb{C}}_{\nu,k}$, with $k(M_{\nu,k}) = k$, and extenders $F^{\mathbb{C}}_{\nu}$ obtained as follows. (In the notation of [35], $M_{\nu,k} = \mathcal{C}_k(\mathcal{N}_\nu)$, and $F^{\mathbb{C}}_{\nu}$ is a choice of background extender for the last extender of $M_{\nu,0} = \mathcal{N}_\nu$.) We let $M_{0,0}$ be the passive premouse with universe V_ω. For any k, ν,

$$M_{\nu,k+1} = \text{core}(M_{\nu,k}) =_{\text{def}} \mathfrak{C}(M_{\nu,k}).$$

We stop the construction if this core does not exist, that is, if the standard parameter of $M_{\nu,k}$ is not solid and universal. Supposing that $\mathbb{C}$ does not stop, that is, $M_{\nu,k+1}$ is defined, we have that $M_{\nu,k+1}$ agrees with $M_{\nu,k}$ to their common value for ρ^+_{k+1}. (If there are no cardinals of $M_{\nu,k}$ strictly greater than $\rho(M_{\nu,k})$, then $M_{\nu,k+1} = (M_{\nu,k})^+$; that is, the two are equal, except the distinguished soundness degree is increased by one.)

[2]We call them the *resurrection consistency problem* and the *background coherence problem*.

For $k < \omega$ sufficiently large, $M_{\nu,k} = M_{\nu,k+1}$, except of course that its associated k has changed. That is, $\hat{M}_{\nu,k}$ is eventually constant as $k \to \omega$. We set

$$\hat{M}_{\nu,\omega} = \text{eventual value of } \hat{M}_{\nu,k} \text{ as } k \to \omega,$$

and

$$\hat{M}_{\nu+1,0} = \text{rud closure of } \hat{M}_{\nu,\omega} \cup \{\hat{M}_{\nu,\omega}\}, \\ \text{arranged as a passive premouse,}$$

and

$$M_{\nu+1,0} = (\hat{M}_{\nu+1,0}, \emptyset).$$

Finally, if ν is a limit, put

$$M^{<\nu} = \text{unique passive } P \text{ such that for all premice } N, \\ N \lhd P \text{ iff } N \lhd M_{\alpha,0} \text{ for all sufficiently large } \alpha < \nu.$$

One can use the agreement between mice and their cores to show that if ν is a cardinal, then $\nu \leq o(M^{<\nu})$. We explain further below.

There are two possibilities now: we may add a new extender to the sequence, or we may not.

Extender-active option. We may set

$$M_{\nu,0} = (M^{<\nu}, F),$$

where F is such that $(M^{<\nu}, F)$ is a Jensen premouse, and F has a certificate in the sense of Definition 3.1.2 below. The Bicephalus Lemma states that, under a natural iterability hypothesis, there is at most one certifiable F such that $(M^{<\nu}, F)$ is a premouse. Nevertheless, this unique F may have many certificates. We let $F_\nu^{\mathbb{C}}$ be the unique certificate for F specified below.

Extender-passive option. We may set

$$M_{\nu,0} = M^{<\nu}.$$

In this case, we let $F_\nu^{\mathbb{C}} = \emptyset$.

We say that $\mathbb{C}$ is *extender-active at* ν iff $F_\nu^{\mathbb{C}} \neq \emptyset$, and extender-passive at ν otherwise. We say $\mathbb{C}$ is *maximal* iff $\mathbb{C}$ is extender-active at ν whenever there is an F meeting the requirements of the extender-active option at ν.

The requirements on certificates are[3]

DEFINITION 3.1.2. A *background certificate* for F (relative to $\mathbb{C}$, at ν) is an extender F^* with the following properties.

(i) $F^* \in \mathcal{F}$,
(ii) $F^* \upharpoonright \lambda_F \cap M^{<\nu} = F \upharpoonright \lambda_F$, and
(iii) $\lambda_F < \mathrm{lh}(F^*)$, and $\forall \tau < \nu \; (\mathrm{lh}(F_\tau^{\mathbb{C}}) < \mathrm{lh}(F^*))$.

[3] See Definition 2.1 of [41].

Regarding (ii), notice that $i_{F^*}(\kappa_F) > \lambda_F$ because F^* is short, so $F^* \restriction (\lambda_F + 1) \cap M^{<\nu} \neq F \restriction (\lambda_F + 1)$. Also, F^* is nice, so $\{\mathrm{lh}(F_\tau^{\mathbb{C}}) \mid \tau < \nu\}$ is bounded in $\mathrm{lh}(F^*)$.

If $\mathbb{C}$ is extender-active at ν, then we then let $F_\nu^{\mathbb{C}}$ be the unique certificate for F such that

(*) $F_\nu^{\mathbb{C}}$ is a certificate for F, is minimal in the Mitchell order among all certificates for F, and is w-least among all Mitchell order minimal certificates for F.

Since $\mathcal{F}$ is closed under initial segment, it follows that $\mathrm{lh}(F_\nu^{\mathbb{C}})$ is the *least* strongly inaccessible η such that $\lambda_F < \eta$ and $\forall \tau < \nu (\mathrm{lh} F_\tau^{\mathbb{C}} < \eta)$.

DEFINITION 3.1.3. $\mathbb{C}$ is a *background construction* (for pure extender mice) if and only if $\mathbb{C} = \langle w, \mathcal{F}, \langle M_{\nu,k}, F_\nu \mid \langle \nu, k\rangle <_{\mathrm{lex}} \mathrm{lh}(\mathbb{C})\rangle\rangle$, where

(1) $(w, \mathcal{F})$ is a coherent pair, and
(2) $\langle M_{\nu,k}, F_\nu \mid \langle \nu, k\rangle <_{\mathrm{lex}} \mathrm{lh}(\mathbb{C})\rangle$ meets the requirements above.

We say that $\mathbb{C}$ is *maximal* iff it adds an extender whenever there is one that meets the requirements of the extender-active case. We write $w = w^{\mathbb{C}}$, $\mathcal{F} = \mathcal{F}^{\mathbb{C}}$, $M_{\nu,k} = M_{\nu,k}^{\mathbb{C}}$, and $F_\nu = F_\nu^{\mathbb{C}}$.

Remark 3.1.4. w is only used if there is more than one Mitchell minimal certificate for F. So if $\mathcal{F}$ is Mitchell linear, then w plays no role, and we call $\mathbb{C}$ an $\mathcal{F}$-construction.

If context permits, we may suppress mention of $(w, \mathcal{F})$, and call $\langle M_{\nu,k}^{\mathbb{C}}, F_\nu^{\mathbb{C}} \mid \langle \nu, k\rangle <_{\mathrm{lex}} \mathrm{lh}(\mathbb{C})\rangle$ a background construction. $\mathrm{lh}(\mathbb{C})$ is the length of $\mathbb{C}$, and $\vec{F}^{\mathbb{C}}$ is the sequence of all background certificates $F_\nu^{\mathbb{C}}$ actually used by $\mathbb{C}$.

Here is a simple lemma on the agreement of models in a construction. Recall that $\rho^-(M) = \rho_{k(M)}(M)$.

LEMMA 3.1.5. *Let $\mathbb{C}$ be a background construction, with levels $M_{\nu,k} = M_{\nu,k}^{\mathbb{C}}$.*

(a) Let $\langle \mu, l\rangle <_{\mathrm{lex}} \langle \nu, k\rangle < \mathrm{lh}(\mathbb{C})$, and suppose that whenever

$$\langle \mu, l\rangle \leq_{\mathrm{lex}} \langle \eta, j\rangle \leq_{\mathrm{lex}} \langle \nu, k\rangle,$$

then $\rho^-(M_{\mu,l}) \leq \rho^-(M_{\eta,j})$; then $M_{\mu,l} \lhd M_{\nu,k}$.

(b) Let $\gamma < o(M_{\nu,k})$ be a cardinal of $M_{\nu,k}$ such that $\gamma \leq \rho^-(M_{\nu,k})$, and suppose $P \trianglelefteq M_{\nu,k}$ is such that $\rho^-(P) = \gamma$; then

(i) there is a unique $\langle \mu, l\rangle \leq_{\mathrm{lex}} \langle \nu, k\rangle$ such that $P = M_{\mu,l}$, moreover

(ii) if $P = M_{\mu,l}$, then $\gamma \leq \rho^-(M_{\eta,j})$ whenever $\langle \mu, l\rangle \leq_{\mathrm{lex}} \langle \eta, j\rangle \leq_{\mathrm{lex}} \langle \nu, k\rangle$.

PROOF. For (a): We have $\hat{M}_{\mu,l+1} = \hat{M}_{\mu,l}$, so $M_{\mu,l} \lhd M_{\mu,l+1}$. The agreement of a mouse with its core then gives $M_{\mu,l} \lhd M_{\eta,j}$ by induction on $\langle \eta, j\rangle$.

For (b): Let $\langle \mu, l\rangle$ be least such that $P \trianglelefteq M_{\mu,l}$. We claim that $P = M_{\mu,l}$. Suppose that $P \lhd M_{\mu,l}$; then $l > 0$, say $l = n+1$. Since P is not an initial segment of $M_{\mu,n}$, $\rho(M_{\mu,n}) = \rho^-(M_{\mu,n+1}) < \gamma$. Since γ is a cardinal and $\leq \rho_k(M_{\nu,k})$, we cannot have $M_{\mu,n+1} \trianglelefteq M_{\nu,k}$. But then by part (a), we have a least $\langle \eta, j\rangle$ between $\langle \mu, n+1\rangle$ and $\langle \nu, k\rangle$ such that $\rho^-(M_{\eta,j}) < \rho^-(M_{\mu,n+1})$. So $M_{\eta,j}$ collapses $\rho^-(M_{\mu,n+1})$, which by (a) again gives us $\langle \eta_1, j_1 \rangle$ strictly between $\langle \eta, j \rangle$ and $\langle \nu, k \rangle$ such that $\rho^-(M_{\eta_1,j_1}) < \rho^-(M_{\eta,j})$. And so on. We get an infinite descending sequence of projecta realized between $\langle \mu, l \rangle$ and $\langle \nu, k \rangle$, contradiction.

This shows that $P = M_{\mu,l}$. The proof also showed (b)(ii) for $\langle \mu, l \rangle$. But this implies that $\langle \mu, l \rangle$ is unique, for otherwise we have $\langle \eta, j \rangle$ strictly between $\langle \mu, l \rangle$ and $\langle \nu, k \rangle$ such that $P = M_{\eta,j}$. We can then apply (a) to see that $M_{\mu,l} \lhd M_{\eta,j}$, that is, $P \lhd P$, contradiction. ⊣

LEMMA 3.1.6. *Let $\mathbb{C}$ be a background construction; then for any premouse N, there is at most one $\langle \nu, k \rangle$ such that $N = M^{\mathbb{C}}_{\nu,k}$.*

PROOF. Notice that $N = M_{\nu,k}$ implies that that $k = k(N)$. It is certainly possible that $\hat{M}_{\nu,k} = \hat{M}_{\nu,k+1}$.

If the lemma fails, we have $\nu < \mu$ and k such that $N = M_{\nu,k} = M_{\mu,k}$. Let

$$\rho = \inf\{\rho_j(M_{\eta,j}) \mid \langle \nu, k \rangle \leq_{\text{lex}} \langle \eta, j \rangle \leq_{\text{lex}} \langle \mu, k \rangle\}.$$

Since $\nu < \mu$, Lemma 3.1.5(a) yields $\gamma < o(M_{\nu,k})$ such that γ is a cardinal of $M_{\nu,k}$ and $\rho < \gamma \leq \rho_k(M_{\nu,k})$. But now let $\langle \eta, j \rangle$ be least such that $\rho_j(M_{\eta,j}) = \rho$ and $\langle \nu, k \rangle \leq_{\text{lex}} \langle \eta, j \rangle$. We have that $M_{\eta,j} \trianglelefteq M_{\mu,k}$. If $\eta < \mu$, then γ is no longer a cardinal in $M_{\mu,k}$, so $M_{\mu,k} \neq M_{\nu,k}$, contradiction. Thus $\eta = \mu$, and $j \leq k$. But then we have

$$\rho_j(M_{\mu,k}) = \rho < \rho_k(M_{\nu,k}) = \rho_k(M_{\mu,k}),$$

contradiction. ⊣

By the lemma, we may define

DEFINITION 3.1.7. Let $\mathbb{C}$ be a background construction; then

(a) $\text{lev}(\mathbb{C}) = \{M^{\mathbb{C}}_{\nu,k} \mid \langle \nu, k \rangle < \text{lh}(\mathbb{C})\}$, and

(b) for $P, Q \in \text{lev}(\mathbb{C})$, $P <_{\mathbb{C}} Q$ iff $\exists \nu, k, \mu, l (P = M_{\nu,k} \wedge Q = M_{\mu,l} \wedge \langle \nu, k \rangle <_{\text{lex}} \langle \mu, l \rangle)$.

Lemma 3.1.5 gives us a useful way to think about $\mathbb{C}$. Let M be a level of $\mathbb{C}$, and let P_α enumerate in $\lhd$-increasing order the $Q \lhd M$ such that $\rho^-(Q) \leq \rho^-(M)$ and $\rho^-(Q)$ is a cardinal of M. Lemma 3.1.5 implies that each P_α is a level of $\mathbb{C}$. If $\rho^-(M) = o(M)$ (for example, if $k(M) = 0$), then the set of P_α's is cofinal in $<_{\mathbb{C}}$ below M. The P_α and their limit points are what M itself can see of the construction below M. The levels of $\mathbb{C}$ between the P_α are a part of the connection between M and its background universe, and in general only visible in the background universe. What M can see are its cardinals, and the levels of $\mathbb{C}$ that added new subsets of the cardinals that are below $\rho^-(M)$.

Lemma 3.1.6 implies that any extender F can be the last extender of at most one $M^{\mathbb{C}}_{\nu,\mathbb{C}}$, so we may define

DEFINITION 3.1.8. Let $\mathbb{C}$ be a background construction, and suppose $M^{\mathbb{C}}_{\nu,0}$ is active, with last extender F; then $B^{\mathbb{C}}(F) = F^{\mathbb{C}}_{\nu}$.

The Mitchell minimality of our certificates has some simple consequences.

LEMMA 3.1.9. *Let* $\mathbb{C}$ *be a background construction,* $M^{\mathbb{C}}_{\nu,0} = (M^{<\nu}, F)$*, and* $F^* = F^{\mathbb{C}}_{\nu}$*; then*

(a) $\mathrm{lh}(F^*)$ *is the least strongly inaccessible* η *such that* $\lambda_F < \eta$ *and* $\forall \tau < \nu(\mathrm{lh}(F^{\mathbb{C}}_{\tau}) < \eta)$*,*
(b) $\{\mathrm{lh}(F^{\mathbb{C}}_{\tau}) \mid \tau < \nu\}$ *is bounded in* $\mathrm{lh}(F^*)$*, and*
(c) $i^V_{F^*}(M^{<\nu}) \models \lambda_F$ *is not measurable.*

PROOF. Let $M = M^{<\nu}, F$, and F^* be as in the hypothesis. Let η be the least strongly inaccessible η such that $\lambda_F < \eta$ and $\forall \tau < \nu(\mathrm{lh}(F^{\mathbb{C}}_{\tau}) < \eta)$. F^* is nice, so the lengths are bounded in η. Clearly $F^* \restriction \eta$ is also a certificate for F, and $F^* \restriction \eta \in \mathcal{F}$ because $(w, \mathcal{F})$ is a coherent pair. By Mitchell minimality, $F^* = F^* \restriction \eta$, as desired in (a). This also proves (b).

For (c), suppose toward contradiction that $i_{F^*}(M) \models \lambda_F$ is measurable ; then in $i_{F^*}(V)$ we have a background E^* for the order zero total measure on λ_F of $i_{F^*}(M)$. By the agreement lemma 3.1.5, $\mathrm{crit}(E^*) = \lambda_F$. Let $\eta = \mathrm{lh}(F^*)$. We have $V_\eta \subseteq i_{F^*}(V)$, so η is still the least inaccessible above λ_F in $i_{F^*}(V)$. So $V_\eta \subseteq \mathrm{Ult}(V, E^*)$ holds in $i_{F^*}(V)$, and hence in V. $E^* \in i_{F^*}(\mathcal{F})$, so $E^* \restriction \eta \in i_{F^*}(\mathcal{F})$, so $E^* \restriction \eta \in \mathcal{F}$ by coherence. But then let

$$G^* = i^V_{E^* \restriction \eta}(F^*) \restriction \eta.$$

$G^* \in \mathcal{F}$ by the coherence of $\mathcal{F}$. It is easy to see that G^* still backgrounds F, and satisfies (i)-(iii), so that it is a certificate for F. However,

$$G^* = i^V_{E^* \restriction \eta}(F^* \restriction \lambda_F) \restriction \eta,$$

so $G^* \in \mathrm{Ult}(V, F^*)$. This contradicts the Mitchell minimality of F^*. ⊣

Part (c) of this lemma will be important in Chapter 4.

There is a natural coherence lemma for maximal w-constructions. Its hypotheses include the uniqueness of certified extenders that is the conclusion of the Bicephalus Lemma.

DEFINITION 3.1.10. Let $\mathbb{C} = \langle w, \mathcal{F}, \langle (M_{\tau,k}, F_\tau) \mid \langle \tau, k \rangle < \mathrm{lh}(\mathbb{C}) \rangle \rangle$ be a background construction, and $\langle \gamma, 0 \rangle \leq \mathrm{lh}(\mathbb{C})$; then

$$\mathbb{C} \restriction \gamma = \langle w \cap V_\eta, \mathcal{F} \cap V_\eta, \langle (M_{\tau,k}, F_\tau) \mid \tau < \gamma \wedge k < \omega \rangle \rangle,$$

where

$$\eta = \eta^{\mathbb{C}}_\gamma = \sup(\{\mathrm{lh}(F^{\mathbb{C}}_\tau) + 1 \mid \tau < \gamma\}).$$

We call $(M^{<\gamma}, \emptyset)$ the *last model* of $\mathbb{C} \restriction \gamma$.

If $\langle \gamma, 0 \rangle < \mathrm{lh}(\mathbb{C})$, then the last model of $\mathbb{C} \restriction \gamma$ is just $M^{\mathbb{C}}_{\gamma,0} || o(M^{\mathbb{C}}_{\gamma,0})$.

LEMMA 3.1.11. *Let* $\mathbb{C}$ *be a maximal background construction above* κ*. Suppose* $M^{\mathbb{C}}_{\nu,0} = (M^{<\nu}, F)$ *where* $F \neq \emptyset$*, and let* $F^* = F^{\mathbb{C}}_\nu$ *and* $\mathbb{D} = i_{F^*}(\mathbb{C})$*; then*

(1) $\mathbb{D}\restriction\nu = \mathbb{C}\restriction\nu$,
(2) $M^{\mathbb{D}}_{\nu,0} \neq M^{\mathbb{C}}_{\nu,0}$; *moreover, if F is the unique $\mathcal{F}^{\mathbb{C}}$-certifiable G such that $(M^{<\nu},G)$ is a premouse, then* $M^{\mathbb{D}}_{\nu,0} = (M^{<\nu},\emptyset)$,
(3) $(M^{<\nu},\emptyset) \lhd_0 i_{F^*}(M^{<\nu})$, *and*
(4) *if* $\xi < \nu$, *and* $\mathbb{C}\restriction\xi$ *has last model* N *such that* $o(N) < \text{crit}(F^*)$, *then* $\mathbb{C}\restriction\xi \in V_{\text{crit}(F^*)}$.

PROOF. Let η be the least strongly inaccessible such that $\lambda_F < \eta$ and $\forall\tau < \nu(\text{lh}\,F^{\mathbb{C}}_\tau < \eta)$, so that $\eta = \text{lh}(F^*)$ by Lemma 3.1.9. $\mathbb{C}\restriction\nu$ uses only extenders in $\mathcal{F}\cap V_\eta$ as backgrounds, moreover, it uses one whenever possible. Since $i_{F^*}(w)\cap V_\eta = w\cap V_\eta$, $i_{F^*}(\mathcal{F})\cap V_\eta = \mathcal{F}\cap V_\eta$, and $\mathbb{D}$ adds an extender whenever possible, $\mathbb{C}\restriction\nu = \mathbb{D}\restriction\nu$.

For (2), suppose $M^{\mathbb{D}}_{\nu,0} = (M^{<\nu},G)$. If $G\neq\emptyset$, then the certificate G^* for G in $\mathbb{D}$ satisfies $\text{lh}(G^*) = \eta$ by Lemma 3.1.9 in $\text{Ult}(V,F^*)$, so $G^*\in\mathcal{F}$ by coherence. Thus G^* is also a certificate for G in V, so if $F = G$ then F^* is not a Mitchell minimal certificate for F, contradiction. Thus $F\neq G$, and if $G\neq\emptyset$ then F is not the unique $\mathcal{F}^{\mathbb{C}}$-certifiable G such that $(M^{<\nu},G)$ is a premouse.

For (3): By (2), it is enough to show that there is no $P\in\text{lev}(\mathbb{D})$ such that $(M^{<\nu},\emptyset) <_{\mathbb{D}} P <_{\mathbb{D}} i_{F^*}(M^{<\nu})$ and $\rho(P) < \lambda_F$. Suppose there were, and let μ be the infimum of all such $\rho(P)$. Let $\gamma = \mu^{+,M^{<\nu}}$. Since λ_F is a limit cardinal in $M^{<\nu}$, $\gamma < \lambda_F$, and hence γ is a cardinal in $\text{Ult}(M^{<\nu},F)$ by coherence. On the other hand, γ is not a cardinal of $i_{F^*}(M^{<\nu})$ because some P as above collapsed it. But we have a factor embedding $\pi\colon \text{Ult}(M^{<\nu},F)\to i_{F^*}(M^{<\nu})$, with $\text{crit}(\pi) = \lambda_F$, so $\pi(\gamma) = \gamma$. This is a contradiction.

For (4): In $\text{Ult}(V,F^*)$, N is the last model of $i_{F^*}(\mathbb{C})\restriction\xi = \mathbb{C}\restriction\xi$, moreover, $\mathbb{C}\restriction\xi\in V_{\text{lh}(F^*)+2}$ by the way we chose F^*. Thus letting $\kappa = \text{crit}(F^*)$,

$$\text{Ult}(V,F^*) \models \text{“}N \text{ is the last model of some } \mathbb{C}\restriction\xi \text{ in } V_{i_{F^*}(\kappa)}\text{”}.$$

But $i_{F^*}(N) = N$, so pulling this back under i_{F^*} yields (4). ⊣

Remark 3.1.12. Once we have iteration strategies and comparison in place, we can strengthen Lemma 3.1.11. The extender uniqueness hypothesis of (2) is true, so $M^{\mathbb{D}}_{\nu\,0} = (M^{<\nu},\emptyset)$. In fact, $\text{lh}(F)$ is a cardinal in $i_{F^*}(M^{<\nu})$. See Section 4.9.

Remark 3.1.13. From one point of view, our background constructions are very slow to add extenders. There must be a nice background extender that coheres with the construction so far. Nevertheless, we shall see in Section 8.1 that they can capture much of the strength of their background universes.

In general, if F is an extender somewhere on the $M^{\mathbb{C}}_{\nu,k}$ sequence, then the certificate that justifies F is $B^{\mathbb{C}}(\sigma(F))$, where σ is a *resurrection map*. We obtain the resurrection maps by composing anticore maps, but there are some subtleties, which we discuss in the next section, where we give the full definition.

3.2. Resurrection maps

Associated to a construction $\mathbb{C}$ we have *resurrection maps* that act on initial segments N of some $M_{\nu,k}$, and trace them back via anticore maps to an origin as some $M_{\eta,l}$. The picture is somewhat complicated by the fact that a given N may have more than one origin. Our definitions have the effect that we always trace back to the earliest possible origin, which is the only reasonable thing to do, because there may be no last origin.

Our definitions are very close to those of [35]. Our notation is close to that of [1], with a few changes that will reduce the number of subscripts in various formulae.

Let $\mathbb{C}$ be a background construction, Q a model of $\mathbb{C}$ and $N \trianglelefteq Q$. We shall define $R = \mathrm{Res}_{\mathrm{Q}}[N]$ and $\sigma = \sigma_{\mathrm{Q}}[N]$. We shall have $R \leq_{\mathbb{C}} Q$, $k(R) = k(N)$, and $\sigma\colon N \to R$ is elementary.[4] We call $\mathrm{Res}_{\mathrm{Q}}[N]$ the *complete resurrection* of N from stage Q. For S such that $\mathrm{Res}_{\mathrm{Q}}[N] \leq_{\mathbb{C}} S \leq_{\mathbb{C}} Q$, we shall also define the partial resurrection $\mathrm{Res}_{\mathrm{Q,S}}[N]$ and its map $\sigma_{\mathrm{Q,S}}[N]$. The complete resurrection results from composing partial resurrections.

Any level of $\mathbb{C}$ is its own complete resurrection, so

$$\mathrm{Res}_{\mathrm{Q}}[Q] = Q, \text{ and } \sigma_{\mathrm{Q}}[Q] = \mathrm{id}.$$

The remainder of the definition is by induction on the place of Q in $<_{\mathbb{C}}$. We maintain inductively

(∗) If $R <_{\mathbb{C}} Q$ and $\rho^-(R) \leq \rho(S)$ for all S such that $R \leq_{\mathbb{C}} S <_{\mathbb{C}} Q$, then
  (i) $R \trianglelefteq Q$,
  and for all $N \trianglelefteq R$ and Y such that $\mathrm{Res}_R[N] \leq_{\mathbb{C}} Y \leq_{\mathbb{C}} R$,
  (ii) $\mathrm{Res}_{\mathrm{Q,Y}}[N] = \mathrm{Res}_{\mathrm{R,Y}}[N]$, and
  (iii) $\sigma_{\mathrm{Q,Y}}[N] = \sigma_{\mathrm{R,Y}}[N]$.

This enables us to resurrect from limit levels in an unambiguous way.

Suppose first that $Q = M_{\nu,k+1}$, and let

$$\pi\colon Q^- \to X$$

be the anticore map, where $X = M_{\nu,k}$. π is cofinal and elementary. Let

$$\mu = \rho(X)^{+,X} = \rho^-(Q)^{+,Q}.$$

$X|\mu = Q|\mu$ by solidity and universality.[5] We define $\mathrm{Res}_{\mathrm{Q,X}}[N]$ and $\sigma_{\mathrm{Q,X}}[N]$ for $N \triangleleft Q$ by

$$\mathrm{Res}_{\mathrm{Q,X}}[N] = \begin{cases} N & \text{for } N \triangleleft Q|\mu, \\ \pi(N) & \text{otherwise,} \end{cases}$$

[4] It may not be cofinal.

[5] $\mu = o(X)$ is possible.

and

$$\sigma_{\mathrm{Q,X}}[N] = \begin{cases} \mathrm{id} \restriction N & \text{for } N \lhd Q|\mu, \\ \pi \restriction N & \text{otherwise.} \end{cases}$$

We are allowing here $N \notin Q$, that is, we may have $N = (\hat{Q}, n)$ where $n \leq k$. In that case $\pi(N) = (\hat{X}, n)$.

Remark 3.2.1. If $\mathrm{crit}(\pi) \neq \rho(X)$, then $N = \pi(N)$ when $N \lhd Q|\mu$, so there is no real case split. It is possible that $\mathrm{crit}(\pi) = \rho(X)$, however. That leads to the resurrection consistency problem, and will ultimately force us to change the way we take cores.

If $N = Q^-$, then $\mathrm{Res}_{\mathrm{Q,X}}[N] = X$ is the complete resurrection of N from Q, and we write $\mathrm{Res}_{\mathrm{Q}}[N]$ for it, and $\sigma_{\mathrm{Q}}[N]$ for its map. (I.e. π.) More generally,

$$\mathrm{Res}_{\mathrm{Q}}[N] = \mathrm{Res}_{\mathrm{X}}[\mathrm{Res}_{\mathrm{Q,X}}[N]],$$
$$\sigma_{\mathrm{Q}}[N] = \sigma_{\mathrm{X}}[\mathrm{Res}_{\mathrm{Q,X}}[N]] \circ \sigma_{\mathrm{Q,X}}[N],$$

and for Y such that $\mathrm{Res}_{\mathrm{Q}}[N] \leq_{\mathbb{C}} Y \leq_{\mathbb{C}} X$,

$$\mathrm{Res}_{\mathrm{Q,Y}}[N] = \mathrm{Res}_{\mathrm{X,Y}}[\mathrm{Res}_{\mathrm{Q,X}}[N]], \text{ and}$$
$$\sigma_{\mathrm{Q,Y}}[N] = \sigma_{\mathrm{X,Y}}[\mathrm{Res}_{\mathrm{Q,X}}[N]] \circ \sigma_{\mathrm{Q,X}}[N].$$

It is easy to verify our induction hypothesis $(*)$. The key is that for $N \lhd Q|\mu$, we have set $\mathrm{Res}_{\mathrm{Q,X}}[N] = N$, so $\mathrm{Res}_{\mathrm{Q}}[N] = \mathrm{Res}_{\mathrm{X}}[N]$. In other words, we are following N backward under the earliest anticore maps that apply to it.

Now let us consider the limit case, that is, the case that $Q = M_{\nu,0}$ for some ν. Let $N \lhd Q$, and let $N \trianglelefteq R \lhd Q$ be such that

$$\rho^-(R) = \inf(\{\rho(S) \mid N \trianglelefteq S \lhd Q\}).$$

By Lemma 3.1.5, $R \trianglelefteq S$ for all S such that $R \leq_{\mathbb{C}} S <_{\mathbb{C}} Q$. We let

$$\mathrm{Res}_{\mathrm{Q}}[N] = \mathrm{Res}_{\mathrm{R}}[N],$$
$$\sigma_{\mathrm{Q}}[N] = \sigma_{\mathrm{R}}[N],$$

and for Y such $\mathrm{Res}_{\mathrm{Y}}[N] \leq_{\mathbb{C}} Y \leq_{\mathbb{C}} R$,

$$\mathrm{Res}_{\mathrm{Q,Y}}[N] = \mathrm{Res}_{\mathrm{R,Y}}[N],$$
$$\sigma_{\mathrm{Q,Y}}[N] = \sigma_{\mathrm{R,Y}}[N].$$

Finally, for S such that $R \leq_{\mathbb{C}} S <_{\mathbb{C}} Q$, we set

$$\mathrm{Res}_{\mathrm{Q,S}}[N] = N, \text{ and}$$
$$\sigma_{\mathrm{Q,S}}[N] = \mathrm{id}.$$

By $(*)$, the definitions of $\mathrm{Res}_{\mathrm{Q}}[N]$ and the rest are independent of the choice of R. It is easy to verify that $(*)$ continues to hold.

PROPOSITION 3.2.2. *(a)* $\mathrm{Res}_{\mathrm{Q}}[N]$ *is the* $<_{\mathbb{C}}$*-least* X *such that* $\mathrm{Res}_{\mathrm{Q,X}}[N]$ *is defined.*

(b) $k(N) = k(\mathrm{Res}_{\mathrm{Q,X}}[N])$, *and* $\sigma_{\mathrm{Q,X}}[N]$ *is elementary.*

(c) If $P \lhd N$, *then* $\mathrm{Res}_{\mathrm{Q}}[P] <_{\mathbb{C}} \mathrm{Res}_{\mathrm{Q}}[N]$.

(d) If $P \lhd N$ *and* $\mathrm{Res}_{\mathrm{Q,X}}[N]$ *is defined, then* $\mathrm{Res}_{\mathrm{Q,X}}[P] \lhd \mathrm{Res}_{\mathrm{Q,X}}[N]$.

(e) Suppose that $\mathrm{Res}_{\mathrm{Q}}[N] \leq_{\mathbb{C}} X \leq_{\mathbb{C}} Y \leq_{\mathbb{C}} Q$; *then*

(i) $\mathrm{Res}_{\mathrm{Q,X}}[N] = \mathrm{Res}_{\mathrm{Y,X}}[\mathrm{Res}_{\mathrm{Q,Y}}[N]]$, *and*

(ii) $\sigma_{\mathrm{Q,X}}[N] = \sigma_{\mathrm{Y,X}}[\mathrm{Res}_{\mathrm{Q,Y}}[N]] \circ \sigma_{\mathrm{Q,Y}}[N]$.

(f) Suppose $k(N) > 0$ *and* $\mathrm{Res}_{\mathrm{Q,X}}[N]$ *is defined; then* $\mathrm{Res}_{\mathrm{Q,X}}[N^-] = (\mathrm{Res}_{\mathrm{Q,X}}[N])^-$.

(g) If $\mathrm{Res}_{\mathrm{Q}}[N] = M_{\nu,k+1}$, *then* $\mathrm{Res}_{\mathrm{Q}}[N^-] = M_{\nu,k}$. *Moreover, if* $\pi \colon (M_{\nu,k+1})^- \to M_{\nu,k}$ *is the anticore map, then* $\pi \circ \sigma_{\mathrm{Q}}[N] = \sigma_{\mathrm{Q}}[N^-]$.

These are easy to prove by induction on the rank of Q in $<_{\mathbb{C}}$.

The resurrection map $\sigma_{\mathrm{Q}}[N]$ can be naturally factored using the *N-dropdown sequence of Q*. The dropdown sequence is the trace in Q of those corings between $\mathrm{Res}_{\mathrm{Q}}[N]$ and Q that contributed directly to replacing $\mathrm{Res}_{\mathrm{Q}}[N]$ by N.

DEFINITION 3.2.3. Let $N \lhd Q$. The *N-dropdown sequence of Q* is given by

(a) $A_0 = N$,

(b) A_{i+1} is the least $B \unlhd Q$ such that $A_i \lhd B$ and $\rho^-(B) < \rho^-(A_i)$.

We write $A_i = A_i(Q,N)$, and let $n(Q,N)$ be the largest i such that A_i is defined.

One place dropdown sequences show up is the following. Suppose $\mathcal{T}$ is a maximal iteration tree, $Q = \mathcal{M}_\beta^{\mathcal{T}}$, $N = Q|\mathrm{lh}(E_\beta^{\mathcal{T}})$, $\mathcal{T}$-pred$(\alpha+1) = \beta$; then $\mathcal{T}$ drops at $\alpha+1$ iff $\rho^-(A_i(Q,N)) \leq \mathrm{crit}(E_\alpha^{\mathcal{T}})$ for some i. In the case that $\mathcal{T}$ does drop, $E_\alpha^{\mathcal{T}}$ will be applied to $A_i(Q,N)^-$, where i is least such that $\rho^-(A_i(Q,N)) \leq \mathrm{crit}(E_\alpha^{\mathcal{T}})$.

Let Q be a level of a background construction $\mathbb{C}$. Let $N \lhd Q$, and let $\langle A_i \mid i \leq n \rangle$ be the N-dropdown sequence of Q. We can factor the partial resurrections $\mathrm{Res}_{\mathrm{Q}}[A_i]$ and $\sigma_{\mathrm{Q}}[A_i]$, starting with $i = n$ and working down to $i = 0$, where we reach a natural factoring of the complete resurrection of $A_0 = N$. This was done in [35, §11]. There are some complications, and we do not really need this analysis here, so we omit it. The revised resurrection maps of Chapter 4 factor in a simpler way, as we show in Lemma 4.7.13.

Ths case split in the definition of $\sigma_{\mathrm{Q,X}}[P]$ leads to the possibility of inconsistent resurrection maps.

DEFINITION 3.2.4. Let $\mathbb{C}$ be a background construction, $X <_{\mathbb{C}} Q$, and $P \lhd N \lhd Q$. We say that *the (Q,X) resurrections of P and N are consistent* iff

(a) $\mathrm{Res}_{\mathrm{Q,X}}[P] = \sigma_{\mathrm{Q,X}}[N](P)$, and

(b) $\sigma_{\mathrm{Q,X}}[P] = \sigma_{\mathrm{Q,X}}[N] \restriction P$.

We shall see in Section 3.6 why such inconsistencies are a problem for us, and in Section 3.7 that they actually do occur in the standard constructions we are describing now. We do get a limited form of resurrection consistency in the standard constructions.

LEMMA 3.2.5. *Let $\mathbb{C}$ be a background construction and $P \lhd N \lhd Q$. Suppose that* $\mathrm{Res}_{\mathrm{Q}}[N] \leq_{\mathbb{C}} Y \leq_{\mathbb{C}} Q$*; then*

(a) $\mathrm{Res}_{\mathrm{Q,Y}}[P] \trianglelefteq \sigma_{\mathrm{Q,Y}}[N](P)$*, and*

(b) $\sigma_{\mathrm{Q,Y}}[P] \restriction \rho = \sigma_{\mathrm{Q,Y}}[N] \restriction \rho$*, where* $\rho = \inf(\{\rho(S) \mid P \trianglelefteq S \lhd N\})$.

PROOF. By induction on Q. Let us just consider the successor step.[6] Let $Q = \mathfrak{C}(X)$. Let $\kappa = \rho(X)$ and $\mu = \kappa^{+,X}$, and let $\pi\colon Q^- \to X$ be the anticore map.

For (a): If $\mu \leq o(P)$, then $\mathrm{Res}_{\mathrm{Q,X}}[P] = \pi(P)$ and $\mathrm{Res}_{\mathrm{Q,X}}[N] = \pi(N)$, so we have (a) when $X = Y$. We get (a) for $Y <_{\mathbb{C}} X$ by induction. If $o(N) < \mu$, then $\mathrm{Res}_{\mathrm{Q,X}}[P] = P$, $\mathrm{Res}_{\mathrm{Q,X}}[N] = N$, so again we have (a) by induction. Finally, if $o(P) < \mu \leq o(N)$, then $P \trianglelefteq \pi(P)$ because X is solid, so

$$\mathrm{Res}_{\mathrm{Q,X}}[P] = P \trianglelefteq \pi(P) \trianglelefteq \pi(N) = \mathrm{Res}_{\mathrm{Q,X}}[N].$$

This and induction yield (a).

For (b): $N \lhd Q$, so $N \trianglelefteq Q^-$. Assume first $\mu \leq o(P)$, so that $\mathrm{Res}_{\mathrm{Q,X}}[P] = \pi(P)$, $\mathrm{Res}_{\mathrm{Q,X}}[N] = \pi(N)$, and $\sigma_{\mathrm{Q,X}}[P]$ and $\sigma_{\mathrm{Q,X}}[N]$ are restrictions of π, so we have (b) when $X = Y$. To apply our induction hypothesis and get (b) when $Y <_{\mathbb{C}} X$, we must show that

$$\sup \pi``\rho \leq \inf(\{\rho(S) \mid \pi(P) \trianglelefteq S \lhd \pi(N)\}).$$

This follows from the elementarity of π.[7]

Assume next $o(N) < \mu$. Then all the relevant one-step resurrections are the identity, so we can apply induction.

Finally, we have the case $o(P) < \mu \leq o(N)$. By the definition of ρ we have $\rho \leq \kappa \leq \mathrm{crit}(\pi)$, so $\sup \pi``\rho = \rho$. $\mathrm{Res}_{\mathrm{Q,X}}[P] = P$ and $\mathrm{Res}_{\mathrm{Q,X}}[N] = \pi(N)$. The elementarity of π guarantees that for all S such that $P \trianglelefteq S \trianglelefteq N$, $\rho^-(S) \geq \kappa$. We have then by induction that for all relevant $Y <_{\mathbb{C}} X$, $\sigma_{\mathrm{X,Y}}[P] \restriction \rho = \sigma_{\mathrm{X,Y}}[\pi(N)] \restriction \rho$, and this finishes the proof of (b). ⊣

We shall use part (b) of the lemma in Section 3.4.

3.3. A Shift Lemma for conversion stages

Let $\mathbb{C}$ be a background construction, Q be a level of $\mathbb{C}$, and let $\psi\colon M \to Q$ be sufficiently elementary. Given a quasi-normal tree $\mathcal{T}$ on M, we shall use ψ and the background extenders provided by $\mathbb{C}$ to lift $\mathcal{T}$ to a nice, normal iteration tree

[6]The limit step is trivial.

[7]If $N \in Q$ the statement is clearly preserved. Otherwise we have $N = Q|\langle \hat{o}(Q), n\rangle$ for some $n < k$, where $k = k(Q)$. The statement $\forall S \in Q(P \trianglelefteq S \trianglelefteq N) \Longrightarrow \nu \leq \rho^-(P))$ is Π_1, hence preserved. But $\rho_0(Q), \ldots, \rho_{k-2}(Q)$ are also preserved because π is elementary on Q^-, and this covers all the remaining cases except $S = N = Q^-$. In this case, $\rho_{k-1}(X) = \sup \pi``\rho_{k-1}(Q) \geq \sup \pi``\rho$, so it works out too.

$\mathcal{T}^*$ on V. The apparatus associated to such a lifting is called a *conversion system*, and we shall describe it in detail in the next section. Such systems are generated in an inductive process that produces at each stage α a lift map $\psi_\alpha\colon \mathcal{M}_\alpha^{\mathcal{T}} \to Q_\alpha$, where Q_α is a level of the construction $i_{0,\alpha}^{\mathcal{T}^*}(\mathbb{C})$ of $\mathcal{M}_\alpha^{\mathcal{T}^*}$. The information here constitutes something we shall call a *conversion stage*.

DEFINITION 3.3.1. $\langle M, \psi, Q, \mathbb{C}, R\rangle$ is a *conversion stage* iff

(1) R is a transitive model of ZFC, and $(R, \mathbb{C})$ is amenable,
(2) $(R, \mathbb{C}) \models$ "$\mathbb{C}$ is a maximal background construction", and
(3) $Q \in \mathrm{lev}(\mathbb{C})$, and $\psi\colon M \to Q$ is nearly elementary.

It is convenient here to allow $\mathrm{lh}(\mathbb{C}) = o(R)$. In practice, $(R, w, \mathbb{C})$ will satisfy the relativised form of ZFC, and most often, $\mathbb{C} \in R$. In that case, $(R, \in, w^{\mathbb{C}}, \mathcal{F}^{\mathbb{C}})$ is a coarse premouse, and $\mathbb{C}$ is its unique maximal $(w^{\mathbb{C}}, \mathcal{F}^{\mathbb{C}})$-construction.

Definition 3.3.1 includes the requirement that $(w, \mathbb{C})$ be maximal because it is needed in Lemma 3.1.11 on the coherence properties of constructions, and that lemma is useful. One could convert iteration trees using arbitrary constructions, but it is convenient to assume maximality. If we need to restrict the extenders added in $\mathbb{C}$, we do so by restricting $\mathcal{F}^{\mathbb{C}}$.

We now prove a relative of the Shift Lemma that captures some of what happens at the successor step in a conversion process.

LEMMA 3.3.2 (Shift Lemma for Conversion Systems). *Let $\langle M, \psi, Q, \mathbb{C}, R\rangle$ be a conversion stage. Let E be an extender over M such that E is close to M,* $\mathrm{crit}(E) < \rho^-(M)$, *and* $\psi(\mathrm{crit}(E)) < \rho^-(Q)$. *Let E^* be an extender over R, and* $\varphi\colon \mathrm{dom}(E) \cup \lambda(E) \to \mathrm{dom}(E^*) \cup \mathrm{lh}(E^*)$ *be such that*

(i) $(a, X) \in E$ *iff* $(\varphi(a), \varphi(X)) \in E^*$, *and*
(ii) $\varphi \upharpoonright \mathrm{dom}(E) = \psi \upharpoonright \mathrm{dom}(E)$.

Let $i = i_E^M$ and $i^* = i_{E^*}^R$ be the ultrapower embeddings, and assume that $\mathrm{Ult}(R, E^*)$ *is wellfounded. There is then a nearly elementary map* $\sigma\colon \mathrm{Ult}(M, E) \to i^*(Q)$ *given by*[8]

$$\sigma([a, f]_E^M) = [\varphi(a), \psi(f)]_{E^*}^R.$$

Moreover

(a) $\sigma \upharpoonright \lambda(E) = \varphi \upharpoonright \lambda(E)$,
(b) $\sigma \circ i = i^* \circ \psi$,
(c) $\langle \mathrm{Ult}(M, E), \sigma, i^*(Q), i^*(\mathbb{C}), \mathrm{Ult}(R, E^*)\rangle$ *is a conversion stage, and*
(d) for all $x \in Q$, $x \in \mathrm{ran}(\psi)$ iff $i^*(x) \in \mathrm{ran}(\sigma)$.

PROOF. The agreement between ψ and φ implies that σ is well defined and weakly elementary. Parts (a) and (b) are straightforward. Part (c) is clear, once we check that σ is nearly elementary.

[8]The definition of σ can be understood either by means of reducts, or by letting $\psi(f_{\tau,q}^M) = f_{\tau,\psi(q)}^Q$ for $\tau \in \mathrm{sk}_k$ and $q \in M$.

Let us show that σ strongly respects projecta. For let $M_1 = \mathrm{Ult}(M,E)$, $Q_1 = i^*(Q)$, and $n < k(M)$. Since i is elementary, it strongly respects projecta, so $i(\langle \eta_n^M, \rho_n(M)\rangle) = \langle \eta_n^{M_1}, \rho_n(M_1)\rangle$. i^* strongly respects projecta because it is Σ_ω elementary, and ψ strongly respects projecta by hypothesis. Thus

$$\begin{aligned}\psi_1(\langle \eta_n^{M_1}, \rho_n(M_1)\rangle) &= \psi_1 \circ i(\eta_n^M, \rho_n(M)\rangle) \\ &= i^* \circ \psi(\langle \eta_n^M, \rho_n(M)\rangle) \\ &= \langle \eta_n^{Q_1}, \rho_n(Q_1)\rangle.\end{aligned}$$

Also, $\rho_{n+1}(M_1) \leq \eta_n^{M_1}$ iff $\rho_{n+1}(M) \leq \eta_n^M$ iff $\rho_{n+1}(Q) \leq \eta_n^Q$ iff $\rho_{n+1}(Q_1) \leq \eta_n^{Q_1}$. Thus ψ_1 strongly respects projecta.

Let us check clause (2) in Definition 2.5.14, that σ preserves nice cofinality witnesses appropriately. Let $k = k(M)$, $\rho = \rho_k(M)$, $\eta = \eta_k(M)$, $\rho_1 = \rho_k(M_1)$, and $\eta_1 = \eta_k^{M_1}$. Let suppose first that i is continuous at ρ. It follows that i is continuous at η, and i maps ρ and η to ρ_1 and η_1. If $\eta = \rho$, then $\eta_1 = \rho_1$, so clause (2) in the definition of near elementarity is vacuously true. Assume then that $\eta < \rho$, and let f be a nice witness to the fact that $\mathrm{cof}_k^M(\rho) = \eta$. Because ψ is nearly elementary, $\psi(f)$ is a nice witness that $\mathrm{cof}_k^Q(\psi(\rho)) = \psi(\eta)$, so $i^* \circ \psi(f)$ is a nice witness that $\mathrm{cof}_k^{Q_1}(i^* \circ \psi(\rho)) = i^* \circ \psi(\eta)$. Since i is elementary, $i(f)$ is a nice witness that $\mathrm{cof}_k(\rho_1)) = \eta_1$. By commutativity, $\sigma(i(f))$ is therefore a nice witness that $\mathrm{cof}_k^{Q_1}(\sigma(\rho_1)) = \eta_1$. By Remark 2.5.15, σ preserves all nice witnesses that $\mathrm{cof}_k^{M_1}(\rho_1) = \eta_1$.

Suppose next that i is discontinuous at ρ. It follows that $\eta < \rho$, $\mathrm{crit}(E) = \eta$, and $\rho_1 = \sup i``\rho$. Letting f be a nice witness that $\mathrm{cof}_k^M(\rho) = \eta$, we get that $\eta_1 = \eta$ and $i(f) \restriction \eta$ is a nice witness that $\mathrm{cof}_k^{M_1}(\rho_1) = \eta_1$. By near elementarity, $\psi(f)$ is a nice witness that $\mathrm{cof}_k^Q(\psi(\rho)) = \psi(\eta)$. But $\mathrm{crit}(E^*) = \psi(\eta)$, so $i^* \circ \psi(f) \restriction \psi(\eta)$ is a nice witness that $\mathrm{cof}_k^{Q_1}(\gamma) = \psi(\eta)$, where $\gamma = \sup(\mathrm{ran}(i^* \circ \psi(f) \restriction \psi(\eta))$. Since $\sigma(i(f)) = i^*(\psi(f))$, the nice witness $i(f) \restriction \eta$ will be preserved by σ if $\sigma(\eta) = \psi(\eta)$ and $\sigma(\rho_1) = \gamma$. But

$$\sigma(\eta) = \sigma([\{\eta\}, \mathrm{id}]_E^M) = [\{\psi(\eta)\}, \mathrm{id}]_{E^*}^R = \psi(\eta),$$

and

$$\sigma(\rho_1) = \sigma(\sup(i(f)``\eta)) = \sup(\sigma \circ i(f)``\sigma(\eta)) = \sup(i^* \circ \psi(f)``\psi(\eta)) = \gamma.$$

Thus σ preserves some nice witness that $\mathrm{cof}_k^{M_1}(\rho_1) = \eta_1$, so by Remark 2.5.15, it preserves all such nice witnesses.

Thus σ is nearly elementary. For part (d) of the lemma: $x \in \mathrm{ran}(\psi)$ implies $i^*(x) \in \mathrm{ran}(\sigma)$ by commutativity. Suppose $i^*(x) \in \mathrm{ran}(\sigma)$, say $i^*(x) = \sigma(y)$, where $y = [a, f]_E^M)$. Fixing f, we can assume that a is $<_0$ minimal such that $y = [a, f]_E^M$, where $<_0$ is the parameter order, that is, the lexicographic order on descending sequences of ordinals. It is enough by commutativity to see that $y \in \mathrm{ran}(i)$, that is, that f is constant on a set $X \in E_a$.

Suppose not; then we get a set $X \in E_a$ such that f is 1-1 on X as follows: for $u \in [\text{crit}(E)]^{|a|}$, let

$$g(u) = \begin{cases} 0 & \text{if } \exists v <_0 u(f(v) = f(u)), \\ f(u) & \text{otherwise.} \end{cases}$$

It is important here that g remains an $r\Sigma_k$ function, of course. But $\text{Th}_k^M(\text{crit}(E) \cup q) \in M$, where $f = f^M_{\tau,q}$, and from this theory we can define g in an $r\Sigma_k$ way. Since we chose a to be $<_0$ minimal,

$$i(g)(a) = i(f)(a),$$

and so f agrees with g on a set $X \in E_a$, namely $X = \{u \mid g(u) \neq 0\}$. Clearly f is 1-1 on X.

But ψ is sufficiently elementary that $\psi(f)$ is 1-1 on $\psi(X)$; moreover $\psi(X) \in E^*_{\varphi(a)}$, and $i^*(x) = \sigma(y) = [\varphi(a), \psi(f)]^R_{E^*}$. It follows that $\psi(f)(u) = x$ for E_a a.e. u, so $\psi(f)$ is not 1-1, contradiction. ⊣

We pause to describe briefly how this lemma fits into the construction of conversion systems in the next section. Suppose that $\langle M, \psi, Q, \mathbb{C}, R\rangle$ is a conversion stage, and that we are at the first stage in our conversion process, so that M is the base model of the tree $\mathcal{T}$ we are converting. Let $E = E_0^{\mathcal{T}}$, so that E is on the M sequence. We have $\psi(E)$ on the Q sequence. We resurrect a background extender for $\psi(E)$ by setting

$$\varphi = \sigma_Q[Q|\text{lh}(\psi(E))] \circ \psi,$$

and

$$E^* = B^{\mathbb{C}}(\varphi(E)).$$

Let $k = k(M)$, and suppose $\text{crit}(E) < \rho_k(M)$. Since $\text{crit}(\psi(E)) < \rho_k(Q)$, we have $\text{dom}(\psi(E)) \leq \rho_k(Q)$, so since $k = k(Q)$, $\sigma_Q[Q|\text{lh}(\psi(E)]$ is the identity on $\text{dom}(\psi(E))$, and $\varphi \upharpoonright \text{dom}(E) = \psi \upharpoonright \text{dom}(E)$. Thus the hypotheses of Lemma 3.3.2 hold, and it produces $\langle \text{Ult}(M,E), \psi_1, i_{E^*}(Q), i_{E^*}(\mathbb{C}), \text{Ult}(R,E^*)\rangle$ as our next conversion stage.

Suppose next that $k = k(M) = k(Q)$ and $\rho_k(M) \leq \text{crit}(E) < \rho_{k-1}(M)$. In this case we let $Q_0 = \text{Res}_Q[Q^-]$ and $\psi_0 = \sigma_Q[Q^-] \circ \psi$. We get that $\langle M^-, \psi_0, Q_0, \mathbb{C}, R\rangle$ is a conversion stage, and that ψ_0 has the agreement with φ needed for Lemma 3.3.2. Applying that lemma yields $\langle \text{Ult}(M^-,E), \psi_1, i_{E^*}(Q_0), i_{E^*}(\mathbb{C}), \text{Ult}(R,E^*)\rangle$ as the next conversion stage. We have dropped at both levels, although the drop at the Q-level may be unnecessary.[9]

Of course it is possible that one needs to drop further than one degree when E is applied to an initial segment of M. It is also possible that M is not the base model

[9]If $\psi(\text{crit}(E)) < \rho_k(Q)$, we also have the option of not dropping at the Q level, thereby producing a different kind of conversion stage. Conversion systems that never drop at the Q level unless they must have some interest.

of $\mathcal{T}$, and that E comes from some model of $\mathcal{T}$ strictly after M. We deal with the general case in the next section.

3.4. Conversion systems

Let $\mathbb{C}$ be a maximal $(w,\mathcal{F})$-construction, and M be a level of $\mathbb{C}$. Given a quasi-normal tree $\mathcal{T}$ on M, we can use the background extenders provided by $\mathbb{C}$ to lift $\mathcal{T}$ to a nice, normal iteration tree $\mathcal{T}^*$ on V. More generally, we shall start with a conversion stage $c = \langle M, \psi, Q, \mathbb{C}, R\rangle$, and then lift a quasi-normal $\mathcal{T}$ on M to a nice, normal tree $\mathcal{T}^*$ on R, using ψ and the background extenders of $\mathbb{C}$. The apparatus associated to such a lifting is called a *conversion system.*

The particular conversion system we introduce here is essentially the same as the used in [35].[10] Still other conversion systems are possible.[11] We call the system we are about to define lift$(\mathcal{T},c)$, or lift$(\mathcal{T},M,\psi,Q,\mathbb{C},R)$ if we want to display the components of c. If the lifting process does not break down by producing an illfounded model, we shall have

$$\text{lift}(\mathcal{T},c) = \langle \mathcal{T}^*, \langle c_\alpha \mid \alpha < \text{lh}(\mathcal{T})\rangle\rangle,$$

where $c = c_0$, and the c_α are conversion stages. For $\alpha < \text{lh}(\mathcal{T})$, let

$$c_\alpha = \langle M_\alpha, \psi_\alpha, Q_\alpha, \mathbb{C}_\alpha, R_\alpha\rangle.$$

We shall maintain by induction

$(1)_\alpha$ (a) $\mathcal{T}^* \restriction \alpha+1$ is a nice, normal iteration tree on R with the same tree order as $\mathcal{T}$,

(b) for all $\nu \leq \alpha$, c_ν is a conversion stage, moreover, $M_\nu = \mathcal{M}_\nu^{\mathcal{T}}$, $R_\nu = \mathcal{M}_\nu^{\mathcal{T}^*}$, and $\mathbb{C}_\nu = i_{0,\nu}^{\mathcal{T}^*}(\mathbb{C})$.

The lifting maps commute appropriately with the embeddings of $\mathcal{T}$ and $\mathcal{T}^*$. Drops in model in $\mathcal{T}$ are mirrored by drops in the construction at the background level. Let $i_{\xi,\nu} = i_{\xi,\nu}^{\mathcal{T}}$ and $i^*_{\xi,\nu} = i_{\xi,\nu}^{\mathcal{T}^*}$.

$(2)_\alpha$ Let $\xi <_T \nu \leq \alpha$; then

(a) $Q_\nu \leq_{\mathbb{C}_\nu} i^*_{\xi,\nu}(Q_\xi)$,

(b) $(\xi,\nu]_T$ drops in model or degree iff $Q_\nu <_{\mathbb{C}_\nu} i^*_{\xi,\nu}(Q_\xi)$, and

(c) if $(\xi,\nu]_T$ does not drop in model or degree, then $Q_\nu = i^*_{\xi,\nu}(Q_\xi)$ and $\psi_\nu \circ i_{\xi,\nu} = i^*_{\xi,\nu} \circ \psi_\xi$.

Having defined lift$(\mathcal{T} \restriction \alpha+1, c)$, where $\alpha+1 < \text{lh}(\mathcal{T})$, we set

$$H_\alpha = \psi_\alpha(E_\alpha),$$
$$X_\alpha = Q_\alpha | \text{lh}(H_\alpha),$$

[10] See also Definition 2.2 of [41].

[11] One could convert arbitrary semi-normal trees in essentially the same way, but doing so adds some complications that we have decided to avoid.

$$G_\alpha = \sigma_{Q\alpha}[X_\alpha](H_\alpha),$$
$$Y_\alpha = \mathrm{Res}_{Q\alpha}[X_\alpha],$$
$$G^*_\alpha = B^{\mathbb{C}_\alpha}(G_\alpha).$$

Here $\sigma_{Q\alpha}$ is the resurrection map of $\mathbb{C}_\alpha$. H_α is the last extender of X_α, and its complete resurrection G_α is the last extender of Y_α. We then proceed as in [35], by setting

$$E^{\mathcal{T}^*}_\alpha = G^*_\alpha.$$

The ψ_α will agree with one another in a way that lets us keep the conversion going. The agreement involves maps $\mathrm{res}_\alpha \in R_\alpha$, defined when $\alpha + 1 < \mathrm{lh}(\mathcal{T})$, that connect the ordinals below $o(X_\alpha)$ to the background universe R_α. We set[12]

$$\mathrm{res}_\alpha = \sigma_{Q\alpha}[X_\alpha]^{\mathbb{C}_\alpha},$$

so that

$$\mathrm{res}_\alpha \colon X_\alpha \to Y_\alpha,$$

and $G_\alpha = \mathrm{res}_\alpha(H_\alpha) = \mathrm{res}_\alpha \circ \psi_\alpha(E_\alpha)$.

As with the other induction hypotheses, our agreement hypotheses at α concern $\mathcal{T} \restriction \alpha + 1$ and the objects which depend on it. In other words, they are hypotheses on $\mathrm{lift}(\mathcal{T} \restriction \alpha + 1, c)$, and objects which depend on E_α do not play a role in them. Notice that the ordinals associated to G_ν are ordered by

$$\lambda(G_\nu) < o(Y_\nu) = \mathrm{lh}(G_\nu) < \mathrm{lh}(G^*_\nu) < \lambda(G^*_\nu).$$

Let

$$\xi_\nu = \text{unique } \xi \text{ such that } Y_\nu = M^{\mathbb{C}_\nu}_{\xi,0}.$$

$(3)_\alpha$ If $\nu < \mu \le \alpha$, then
- (a) $Y_\nu || o(Y_\nu) = Q_\mu | o(Y_\nu)$,
- (b) $\mathrm{res}_\nu \circ \psi_\nu \restriction \lambda(E_\nu) = \psi_\mu \restriction \lambda(E_\nu)$,
- (c) $\lambda(G_\nu)$ and $\lambda(G^*_\nu)$ are cardinals of Q_μ, and $\lambda(G^*_\nu) \le \rho^-(Q_\mu)$,
- (d) $\lambda(G^*_\nu) \le \psi_\mu(\lambda(E_\nu))$, and
- (e) $\mathbb{C}_\nu \restriction \xi_\nu = \mathbb{C}_\mu \restriction \xi_\nu$, and $M^{\mathbb{C}_\mu}_{\xi_\nu,0}$ is passive.

By (d), $\mathrm{res}_\nu \circ \psi_\nu(\lambda(E_\nu)) < \psi_\mu(\lambda(E_\nu))$, so the agreement in (b) cannot be strengthened. Clause (c) is useful because it implies the resurrection map $\sigma_{Q_\mu}[X_\mu]$ of $\mathbb{C}_\mu$ has critical point $\ge \lambda(G^*_\nu)$. This means that replacing X_μ by Y_μ does not disturb the agreement with Y_ν we had already.

In addition to $(3)_\alpha$, we have the agreement of models implicit in the fact that $\mathcal{T} \restriction \alpha + 1$ is quasi-normal, and $\mathcal{T}^* \restriction \alpha + 1$ is normal. In particular, R_ν agrees below $\mathrm{lh}(G^*_\nu)$ with all R_μ for $\mu \ge \nu$.

[12]We shall often omit superscripts like $\mathbb{C}_\alpha$ in the displayed formula. The construction in which a resurrection is taking place is usually clear from context.

Notation: $(\dagger)_\alpha$ is the conjunction of $(1)_\alpha$ through $(3)_\alpha$.

Again, $(\dagger)_\alpha$ involves objects that are associated to lift$(\mathcal{T} \restriction \alpha+1, c)$.

CLAIM 1. *Assume $(\dagger)_\alpha$, and let $\nu < \mu \leq \alpha$; then*
(a) $\lambda(G^*_\nu) \leq \lambda(H_\mu)$, and $\text{res}_\mu \restriction \lambda(G^*_\nu) = id$,
(b) $Y_\nu || o(Y_\nu) = Y_\mu | o(Y_\nu)$,
(c) $\xi_\nu < \xi_\mu$.

PROOF. By $(3)_\alpha$(d),

$$\lambda(G^*_\nu) \leq \psi_\mu(\lambda(E_\nu)) \leq \psi_\mu(\lambda(E_\mu)) = \lambda(H_\mu).$$

Also, $\lambda(G^*_\nu) \leq \rho^-(Q_\mu)$ by $(3)_\alpha$, so $\text{res}_\mu = \sigma_{Q_\mu}[X_\mu]$ is the identity on $\lambda(G^*_\nu)$ by Lemma 3.2.5. Thus we have (a) of the claim.

For (b): we have $Y_\nu || o(Y_\nu) = Q_\mu | o(Y_\nu)$ by $(3)_\alpha$. But $o(Y_\nu) = \text{lh}(G_\nu)$, so we have just shown that $o(Y_\nu) < o(X_\mu)$, and res_μ is the identity on $o(Y_\nu)$. Hence $Y_\nu || o(Y_\nu) = Y_\mu | o(Y_\nu)$.

For (c): $\mathbb{C}_\nu \restriction \xi_\nu = \mathbb{C}_\mu \restriction \xi_\nu$ has last model $Y_\nu || o(Y_\nu)$. Since $\text{lh}(G_\nu) < \lambda(H_\mu) < \text{lh}(G_\mu)$, $\xi_\nu \neq \xi_\mu$. If $\xi_\mu < \xi_\nu$, then $Y_\mu || o(Y_\mu) <_{\mathbb{C}_\nu} Y_\nu || o(Y_\nu)$, and yet $Y_\mu || o(Y_\mu)$ is not an initial segment of $Y_\nu || o(Y_\nu)$. It follows that there is $\kappa < \lambda(G_\mu)$ such that κ is a cardinal in Y_μ but not in Y_ν. (Take $\kappa = \rho^{+,Y_\mu}$, where ρ is the smallest projectum associated to a stage of $\mathbb{C}_\nu$ between ξ_μ and ξ_ν. We must have $\rho < \lambda(G_\mu)$.) But $Y_\nu || o(Y_\nu) \trianglelefteq Y_\mu$ by (b), so this is impossible. ⊣

The step from α to $\alpha + 1$ in the conversion process goes as follows. Let

$$(E, H, X, Y, G, G^*) = (E_\alpha, H_\alpha, X_\alpha, Y_\alpha, G_\alpha, G^*_\alpha),$$

and let $\beta = T\text{-pred}(\alpha+1)$. We shall apply Lemma 3.3.2, the Shift Lemma for Conversions, with $\varphi = \text{res}_\alpha \circ \psi_\alpha$ as the embedding of E_α into G^*. If we are not forced to drop at the M level, then the conversion stage we move up by i_{G^*} is just c_β. Otherwise the conversion stage we move up from β to $\alpha+1$ is obtained from c_β via an appropriate resurrection inside $\mathbb{C}_\beta$.

Let us show that we obtain a normal extension of $\mathcal{T}^* \restriction \alpha+1$ by setting $\beta = T^*\text{-pred}(\alpha+1)$. Let

$$\kappa = \text{crit}(E),$$
$$\kappa^* = \text{crit}(G) = \text{res}_\alpha \circ \psi_\alpha(\kappa).$$

CLAIM 2. *(1) Suppose $\gamma < \alpha$; then*
(a) $\text{lh}(G^*_\gamma) < \lambda(G^*_\gamma) \leq \text{lh}(G^*)$, and
(b) $\kappa < \lambda(E_\gamma)$ iff $\kappa^* < \text{lh}(G^*_\gamma)$.
(2) $\mathcal{T}^* \restriction \alpha+2$ is normal.

PROOF. We have

$$\text{lh}(G^*_\gamma) < \lambda(G^*_\gamma) \leq \lambda(H) \leq \lambda(G) < \text{lh}(G^*),$$

by Claim 1.

For (1)(b), suppose first $\kappa < \lambda(E_\gamma)$; then $\text{res}_\gamma \circ \psi_\gamma(\kappa) < \lambda(G_\gamma)$, so $\psi_\alpha(\kappa) < \lambda(G_\gamma)$ by $(3)_\alpha$. But $\text{res}_\alpha \restriction \lambda(G^*_\gamma) = \text{id}$ by Claim 1. So $\kappa^* < \lambda(G) < \text{lh}(G^*)$, as desired. Suppose next $\lambda(E_\gamma) \le \kappa$; then $\psi_\alpha(\lambda(E_\gamma)) \le \psi_\alpha(\kappa)$, so

$$\lambda(G^*_\gamma) \le \psi_\alpha(\kappa) \le \text{res}_\alpha \circ \psi_\alpha(\kappa) = \kappa^*,$$

as desired.

Clearly (1) implies that $\mathcal{T}^* \restriction \alpha + 2$ is normal. ⊣

Remark 3.4.1. We have simply assumed here that $R_{\alpha+1}$ is wellfounded. This then implies that $M_{\alpha+1}$ and the $Q_{\alpha+1}$ we are about to define are wellfounded. If $R_{\alpha+1}$ is illfounded, we just stop the construction of $\text{lift}(\mathcal{T}, c)$.

$(\dagger)_\alpha$ also implies

CLAIM 3. *(a)* $\text{res}_\alpha \circ \psi_\alpha \restriction \lambda(E_\beta) = \text{res}_\beta \circ \psi_\beta \restriction \lambda(E_\beta)$.
(b) Suppose $\alpha + 1 \notin D^{\mathcal{T}}$; *then*
 (i) $\psi_\beta \restriction \text{dom}(E) + 1 = \psi_\alpha \restriction \text{dom}(E) + 1$, *and*
 (ii) res_β *and* res_α *are the identity on* $\psi_\beta(\text{dom}(E) + 1)$.

PROOF. For (a): this is clear if $\beta = \alpha$, so assume $\beta < \alpha$. Then $(3)_\alpha$ implies that ψ_α agrees with $\text{res}_\beta \circ \psi_\beta$ on $\lambda(E_\beta)$, and Claim 1(a) implies that res_α is the identity on $\text{res}_\beta \circ \psi_\beta(\lambda(E_\beta))$. This proves (a).

For (b): Note that $\text{dom}(E) < \lambda(E_\beta)$, so $\psi_\beta(\text{dom}(E)) < \lambda(H_\beta)$. Since we are not dropping in $\mathcal{T}$, E is total on M_β and $\text{dom}(E) \le \rho^-(M_\beta)$. Since ψ_β is nearly elementary, $\sup \psi_\beta``\rho^-(M_\beta) \le \rho^-(Q_\beta)$. Thus $\psi_\beta(\text{dom}(E)) \le \rho^-(Q_\beta)$ and $\psi_\beta(\text{dom}(E))$ is a cardinal initial segment of Q_β. It follows that

$$\psi_\beta(\text{dom}(E)) \le \inf(\{\rho(S) \mid X_\beta \trianglelefteq S \triangleleft Q_\beta\}).$$

This implies that $\sigma_{Q_\beta}[X_\beta] \restriction \psi_\beta(\text{dom}(E)) + 1 = \text{id}$.[13] If $\beta = \alpha$, we have (b)(ii). If $\beta < \alpha$, then res_α is the identity on $\lambda(G_\beta)$ by Claim 1, so res_α is the identity on $\text{res}_\beta \circ \psi_\beta(\text{dom}(E)) + 1 = \psi_\beta(\text{dom}(E)) + 1$. Thus we have (b)(ii) in either case. From this and (a) we get (b)(i). ⊣

We define $\psi_{\alpha+1}$ and $Q_{\alpha+1}$ by cases.

The non-dropping case: $\alpha + 1 \notin D^{\mathcal{T}}$.

We are in case (b) of Claim 3. So ψ_α agrees with ψ_β on $\text{dom}(E)$, $\psi_\beta(\text{dom}(E)) = \text{dom}(H)$, and res_α is the identity on $\text{dom}(H)$, so that $\text{dom}(H) = \text{dom}(G)$. This means we can apply Lemma 3.3.2, the Shift Lemma for Conversions, with its inputs being $\langle M_\beta, \psi_\beta, Q_\beta, \mathbb{C}_\beta, R_\beta \rangle$ and $\varphi = \text{res}_\alpha \circ \psi_\alpha$. That is, we set

$$Q_{\alpha+1} = i^*_{\beta,\alpha+1}(Q_\beta)$$

[13]See Lemma 3.2.5(b). It is possible that $\psi_\beta(\text{dom}(E)) = \rho^-(M_\beta)$, but in this case $\rho^-(M_\beta)$ is a successor cardinal in M_β, so it cannot be equal to $\text{crit}(\sigma_{Q_\beta}[X_\beta])$.

and

$$\psi_{\alpha+1}([a,f]_E^{M_\beta})) = [\mathrm{res}_\alpha \circ \psi_\alpha(a), \psi_\beta(f)]_{G^*}^{R_\beta}.$$

By Lemma 3.3.2, $\langle M_{\alpha+1}, \psi_{\alpha+1}, Q_{\alpha+1}, \mathbb{C}_{\alpha+1}, R_{\alpha+1}\rangle$ is a conversion stage.

One can factor $\psi_{\alpha+1}$ in a natural way. By the usual Shift Lemma we have a map

$$\sigma\colon \mathrm{Ult}(M_\beta, E) \to \mathrm{Ult}(Q_\beta, G)$$

given by

$$\sigma([a,f]_E^{M_\beta}) = [\mathrm{res}_\alpha \circ \psi_\alpha(a), \psi_\beta(f)]_G^{Q_\beta}.$$

We also have

$$\tau\colon \mathrm{Ult}(Q_\beta, G) \to i_{G^*}(Q_\beta)$$

given by

$$\tau([a,f]_G^{Q_\beta}) = [a,f]_{G^*}^{R_\beta}.$$

$\tau \restriction \lambda_G = \mathrm{id}$ and $\tau(\lambda_G) = \lambda_{G^*}$. (The restriction of τ to $\mathrm{lh}(G)$ is just the factor map that certified G by G^* in $\mathbb{C}_\alpha$.) Clearly $\psi_{\alpha+1} = \tau \circ \sigma$. We have the diagram

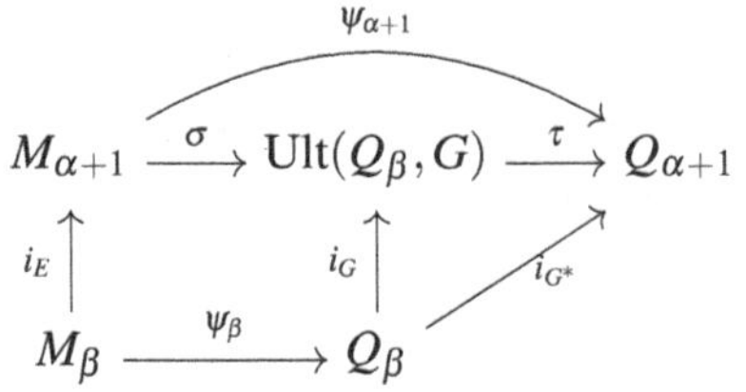

Let us check that our induction hypotheses continue to hold.

CLAIM 4. *In case 1, $(\dagger)_{\alpha+1}$ holds.*

PROOF. We have already verified (1) of $(\dagger)_{\alpha+1}$. The commutativity condition (2) is easy based on the diagram above.

Let us now check the agreement hypotheses (3). Clause (a): We must show that $Y_\alpha || o(Y_\alpha) = Q_{\alpha+1}|o(Y_\alpha)$. By Lemma 3.1.11, this is true if we replace $Q_{\alpha+1}$ with $i_{G^*}(Y_\alpha)$. But $Q_\beta|\kappa^* = Y_\beta \restriction \kappa^* = Q_\alpha \restriction \kappa^* = Y_\alpha \restriction \kappa^*$. $o(Y_\alpha) < i_{G^*}(\kappa^*)$, so

$$\begin{aligned} Y_\alpha || o(Y_\alpha) &= i_{G^*}(Y_\alpha)|o(Y_\alpha) \\ &= i_{G^*}(Q_\beta)|o(Y_\alpha) \\ &= Q_{\alpha+1}|o(Y_\alpha). \end{aligned}$$

By induction, we then get that $Y_\nu || o(Y_\nu = Q_{\alpha+1}|o(Y_\nu)$ for all $\nu < \alpha$.

Clause (b): $\psi_{\alpha+1}$ agrees with $\mathrm{res}_\alpha \circ \psi_\alpha$ on $\lambda(G_\alpha)$. If $\nu < \alpha$, then $(3)_\alpha$ implies that $\mathrm{res}_\nu \circ \psi_\nu$ agrees with ψ_α on $\lambda(G_\nu)$, and hence with $\mathrm{res}_\alpha \circ \psi_\alpha$ on $\lambda(G_\nu)$. Thus $\mathrm{res}_\nu \circ \psi_\nu$ agrees with $\psi_{\alpha+1}$ on $\lambda(G_\nu)$, as desired.

Clause (c): Let us just consider the case $\nu = \alpha$ and $\mu = \alpha + 1$, since the rest then follows easily by induction. We must see that $\lambda(G)$ is a limit cardinal in $Q_{\alpha+1}$. But $\lambda(G)$ is inaccessible in $\mathrm{Ult}(Q_\beta, G)$, and $\tau \restriction \lambda(G) = \mathrm{id}$, so this follows. $\lambda(G^*)$ is a cardinal of $R_{\alpha+1}$[14], so it is a cardinal of $Q_{\alpha+1}$. Finally, $\lambda(G^*) \leq \rho^-(Q_{\alpha+1})$ because $\mathrm{crit}(G^*) < \rho^-(Q_\beta)$ and $Q_{\alpha+1} = i_{G^*}(Q_\beta)$.

Clause (d): The new case is $\mu = \alpha + 1$. $\psi_{\alpha+1}(\lambda(G_\alpha)) = \lambda(G^*_\alpha)$ by the definition of $\psi_{\alpha+1}$. Now let $\nu < \alpha$. We have

$$\lambda(G^*_\nu) \leq \psi_\alpha(\lambda(E_\nu))$$

by induction, so if $\lambda(E_\nu) < \lambda(E_\alpha)$, then

$$\psi_{\alpha+1}(\lambda(E_\nu)) = \mathrm{res}_\alpha \circ \psi_\alpha(\lambda(E_\nu)) \geq \psi_\alpha(\lambda(E_\nu)) \geq \lambda(G^*_\nu),$$

as desired. If $\lambda(E_\nu) = \lambda(E_\alpha)$, then using Claim 2, part (1)(a),

$$\psi_{\alpha+1}(\lambda(E_\nu)) = \psi_{\alpha+1}(\lambda(E_\alpha)) = \lambda(G^*_\alpha) > \lambda(G^*_\nu),$$

which is again what we want.[15]

Clause (e): It is enough to show $\mathbb{C}_\alpha \restriction \xi_\alpha = \mathbb{C}_{\alpha+1} \restriction \xi_\alpha$, and $M^{\mathbb{C}_{\alpha+1}}_{\xi_\alpha,0}$ is passive, since the rest of (e) then follows from $(3)_\alpha$(e). But letting $\mathbb{D} = i^{R_\alpha}_{G^*}(\mathbb{C}_\alpha)$, $\mathbb{C}_\alpha \restriction \xi_\alpha = \mathbb{D} \restriction \xi_\alpha$ and $M^{\mathbb{D}}_{\xi_\alpha,0}$ is passive, by Lemma 3.1.11. Thus we are done if $\beta = \alpha$, so assume $\beta < \alpha$. This implies $\kappa^* = \mathrm{crit}(G^*) < \mathrm{lh}(G^*_\beta)$, so $\kappa^* < \xi_\beta$, so $\mathbb{C}_\beta \restriction \kappa^* = \mathbb{C}_\alpha \restriction \kappa^*$, so $\mathbb{C}_{\alpha+1} \restriction i_{G^*}(\kappa^*) = \mathbb{D} \restriction i_{G^*}(\kappa^*)$. But $\xi_\alpha < i_{G^*}(\kappa^*)$, so we are done. $\dashv$

The dropping case: $\alpha + 1 \in D^{\mathcal{T}}$.

Let $J = M^{*,\mathcal{T}}_{\alpha+1}$, so that $J \trianglelefteq M_\beta$ and

$$M_{\alpha+1} = \mathrm{Ult}(J, E),$$

and let

$$K = \psi_\beta(J).$$

Here if $J = M_\beta \downarrow n$, then we understand K to be $Q_\beta \downarrow n$. Since ψ_β is nearly elementary, $\psi_\beta \restriction J$ is elementary, and[16]

$$\mathrm{crit}(H) < \sup \psi_\beta``\rho^-(J) = \rho^-(K).$$

$\sigma_{Q_\beta}[K] \circ \psi_\beta$ is elementary, so

$$d = \langle J, \sigma_{Q_\beta}[K] \circ \psi_\beta, \mathrm{Res}_{Q_\beta}[K], \mathbb{C}_\beta, R_\beta \rangle$$

[14]Because $R_{\alpha+1}$ and $\mathrm{Ult}(R_\alpha, G^*)$ have the same $V_{\lambda(G^*)+1}$.

[15]This shows why (d) of $(3)_\alpha$ cannot be strengthened to $\nu < \mu \Longrightarrow \psi_\mu(\lambda(E_\nu)) = \lambda(G^*_\nu)$. This is only true when $\mathcal{T}$ is normal.

[16]It is possible that $J = M^-_\beta$, $K = Q^-_\beta$, and $\mathrm{crit}(H) < \rho(K) = \rho^-(Q_\beta)$. In that case, what we are about to do will constitute an unneccessary drop at the Q-level, one that a different conversion system might avoid.

is a conversion stage. This is the stage we shall move up to $c_{\alpha+1}$ via i_{G^*}. In order to do that we must see that $\mathrm{res}_\alpha \circ \psi_\alpha$ agrees with $\sigma_{Q_\beta}[K] \circ \psi_\beta$ on $\mathrm{dom}(E)$. But $\mathrm{res}_\alpha \circ \psi_\alpha$ agrees with $\mathrm{res}_\beta \circ \psi_\beta$ on $\mathrm{dom}(E)$, so it is enough to show

CLAIM 5. res_β *agrees with* $\sigma_{Q_\beta}[K]$ *on* $\mathrm{dom}(H)$.

PROOF. $\mathrm{res}_\beta = \sigma_{Q_\beta}[X_\beta]$, so by Lemma 3.2.5, it is enough to show that for all S such that $X_\beta \trianglelefteq S \triangleleft K$, $\mathrm{dom}(H) \leq \rho(S)$. Our definition of J guarantees that for all S such that $M_\beta | \mathrm{lh}(E_\beta) \trianglelefteq S \triangleleft J$, $\mathrm{dom}(E) \leq \rho(S)$. Since ψ_β is nearly elementary as a map on the whole of M_β, it preserves this fact. (In the worst case, $S = J^-$, so $\rho(S) = \rho^{--}(M_\beta)$, which is mapped by ψ_β to $\rho^{--}(Q_\beta) = \rho(K^-)$.) So indeed $\mathrm{dom}(H) \leq \rho(S)$ for all S such that $X_\beta \trianglelefteq S \triangleleft K$. ⊣

By the Shift Lemma for conversion stages, letting

$$\psi_{\alpha+1}([a,f]^J_E) = [\mathrm{res}_\alpha \circ \psi_\alpha(a), \sigma_{Q_\beta}[K] \circ \psi_\beta(f)]^{R_\beta}_{G^*},$$

and

$$Q_{\alpha+1} = i_{G^*}(\mathrm{Res}_{Q_\beta}[K]),$$

we get the next conversion stage $c_{\alpha+1}$. The induction hypotheses $(\dagger)_{\alpha+1}$ are easy to verify.

This completes the successor step in our inductive definition of $\mathrm{lift}(\mathcal{T}, c)$. Now suppose γ is a limit ordinal $< \mathrm{lh}(\mathcal{T})$. We define $\mathcal{T}^* \restriction \gamma+1$ by setting $[0,\gamma]_{T^*} = [0,\gamma]_T$. If this results in $\mathcal{M}^{\mathcal{T}^*}_\gamma$ being illfounded, then we stop the conversion. So suppose that $\mathcal{M}^{\mathcal{T}^*}_\gamma$ is wellfounded. Induction hypothesis (2) then tells us that $D^{\mathcal{T}} \cap [0,\gamma)_T$ is finite. Let $\alpha <_T \gamma$ be large enough that $D^{\mathcal{T}} \cap \gamma \subseteq \alpha$. By (2) we have $i^*_{\alpha,\xi}(Q_\alpha) = Q_\xi$ for all $\xi \in [\alpha,\gamma)_T$. We set

$$Q_\gamma = i^*_{\alpha,\gamma}(Q_\alpha),$$

and define $\psi_\gamma \colon M_\gamma \to Q_\gamma$ by letting

$$\psi_\gamma(i^{\mathcal{T}}_{\xi,\gamma}(x)) = i^*_{\xi,\gamma}(\psi_\xi(x))$$

for all $\xi \in [\alpha,\gamma)_T$. By (2), ψ_γ is well-defined. It is now easy to check that $(\dagger)_\gamma$ holds.

DEFINITION 3.4.2. Let $c = \langle M, \psi, Q, \mathbb{C}, R\rangle$ be a conversion stage, and let $\mathcal{T}$ be a quasi-normal iteration tree on M; then

(1) $\mathrm{lift}(\mathcal{T}, c) = \langle \mathcal{T}^*, \langle c_\alpha \mid \alpha < \mathrm{lh}(\mathcal{T})\rangle\rangle$ is the conversion system defined above. We write $\mathcal{T}^* = \mathrm{lift}(\mathcal{T}, c)_0$ for its tree component, and $\mathbb{C}_\xi = i^{\mathcal{T}^*}_{0,\xi}(\mathbb{C})$.

(2) $\mathrm{stg}(\mathcal{T}, c, \alpha) = c_\alpha = \langle \mathcal{M}^{\mathcal{T}}_\alpha, \psi_\alpha, Q_\alpha, \mathbb{C}_\alpha, \mathcal{M}^{\mathcal{T}^*}_\alpha\rangle$ is the conversion stage occurring at α in $\mathrm{lift}(\mathcal{T}, c)$.

(3) $\mathrm{res}_\xi(\mathcal{T}, c) = \mathrm{res}_\xi = \sigma_{Q_\xi}[Q_\xi | \mathrm{lh}(\psi_\xi(E^{\mathcal{T}}_\xi))]^{\mathbb{C}_\xi}$. We call res_ξ the ξ-th *generator map* associated to $\mathrm{lift}(\mathcal{T}, c)$.

We need only determine $\mathcal{T}^*$, and the ψ_α and Q_α, in order to determine lift$(\mathcal{T},c)$, so we may abuse notation by writing

$$\text{lift}(\mathcal{T},M,\psi,Q,\mathbb{C},R) = \langle \mathcal{T}^*, \langle Q_\alpha \mid \alpha < \text{lh}(\mathcal{T})\rangle, \langle \psi_\alpha \mid \alpha < \text{lh}(\mathcal{T})\rangle\rangle.$$

DEFINITION 3.4.3. In the special case of Definition 3.4.2 that $M = Q$ and $\psi = \text{id}$, we set

$$\text{lift}(\mathcal{T},M,\mathbb{C},R) = \text{lift}(\mathcal{T},M,\text{id},M,\mathbb{C},R).$$

We also let

$$\text{lift}(\mathcal{T},M,\mathbb{C}) = \text{lift}(\mathcal{T},M,\mathbb{C},V)$$

in the case that $R = V$ (the universe of all sets).

3.5. Induced iteration strategies

We are most interested in the case that the background universe is iterable. Suppose that $c = \langle M,\psi,Q,\mathbb{C},R\rangle$ is a conversion stage, and that Σ^* is a $(\theta,\vec{F}^{\mathbb{C}})$ iteration strategy for the background universe R; then Σ^* induces a strategy Σ for M as follows: for $\mathcal{T}$ quasi-normal on M,

$$\mathcal{T} \text{ is by } \Sigma \iff \text{lift}(\mathcal{T},c)_0 \text{ is by } \Sigma^*.$$

We write

$$\Sigma = \Omega(c,\Sigma^*)$$

for this induced strategy. When $M \in \text{lev}(\mathbb{C})$, we set

$$\Omega(\mathbb{C},M,R,\Sigma^*) = \Omega(\langle M,\text{id},M,\mathbb{C},R\rangle,\Sigma^*).$$

We write $\Omega(\mathbb{C},M,\Sigma^*)$ when R can be understood from context. We may occasionally use the notation lift$(\mathcal{T},c,\Sigma^*)$ for the largest initial segment of lift$(\mathcal{T},c)$ that is by Σ^*. So $\mathcal{T}$ is by $\Omega(\mathbb{C},M,\Sigma^*)$ iff lift$(\mathcal{T},c) =$ lift$(\mathcal{T},c,\Sigma^*)$. We have shown above that the lifted tree $\mathcal{T}^*$ is normal, even if $\mathcal{T}$ itself is only quasi-normal. So Σ^* need only be defined on nice, normal iteration trees, and in fact, only on those produced by the conversion process.

If Σ^* is defined on stacks of normal trees, of any length, then we can extend the lifting process and the induced strategy Σ for M so that it is defined on stacks of quasi-normal trees of the same length. For example, let

$$c = \langle M,\psi,Q,\mathbb{C},R\rangle$$

be a conversion stage, and Σ^* an $(\eta,\theta,\vec{F}^{\mathbb{C}})$ iteration strategy for R, where $\eta > 1$. Let $\Omega = \Omega(c,\Sigma^*)$, and $\mathcal{T}$ be a quasi-normal tree on M by Ω having last model $M^{\mathcal{T}}_\alpha$, and let $N \trianglelefteq M^{\mathcal{T}}_\alpha$. We get a tail strategy for quasi-normal trees on N as follows. Letting

$$\text{stg}(\mathcal{T},c,\alpha) = \langle M_\alpha,\psi_\alpha,Q_\alpha,\mathbb{C}_\alpha,R_\alpha\rangle,$$

we set

$$d = \langle N, \text{res}_{Q\alpha}[N] \circ \psi_\alpha, \text{Res}_{Q\alpha}[N], \mathbb{C}_\alpha, R_\alpha \rangle,$$

and define the tail strategy $\Omega_{\mathcal{T},N}$ on quasi-normal trees of length $< \theta$ by

$$\mathcal{U} \text{ is by } \Omega_{\mathcal{T},N} \iff \text{lift}(\mathcal{U}, d)_0 \text{ is by } \Sigma^*_{\mathcal{T}^*, R_\alpha},$$

where of course $\mathcal{T}^* = \text{lift}(\mathcal{T}, c)_0$. Clearly we can continue this process so as to define a tail strategy $\Omega_{\mathcal{T},N,\mathcal{U},P}$, for any P that is an initial segment of the last model of $\mathcal{U}$, and so on.

DEFINITION 3.5.1. Let $c = \langle M, \psi, Q, \mathbb{C}, R \rangle$ be a conversion stage, and let Σ^* be an $(\lambda, \theta, \vec{F}^{\mathbb{C}})$-iteration strategy for R; then $\Omega(c, \Sigma^*)$ is the (λ, θ)-iteration strategy induced by Σ^* as above.

$\Omega(c, \Sigma^*)$ acts on stacks of trees of the same sort that Σ^* acts on. Again, when $M \in \text{lev}(\mathbb{C})$ and R can be understood from context, we write

$$\Omega(\mathbb{C}, M, \Sigma^*) = \Omega(\langle M, \text{id}, M, \mathbb{C}, R \rangle, \Sigma^*)$$

In our case of interest, Σ^* chooses unique wellfounded branches. This implies that Σ^* has important internal consistency properties. We shall elaborate in Chapter 7, but the proofs that coarse strategies witnessing unique iterability have these properties are quite straightforward. What takes a lot of effort is showing that these properties of Σ^* pass to induced strategies of the form $\Omega(\mathbb{C}, M, \Sigma^*)$. We shall discuss the obstacles in the next section.

3.6. Internal consistency for iteration strategies

Suppose $c = \langle M, \psi, Q, \mathbb{C}, R \rangle$ is a conversion stage, and Σ^* is a (λ, θ) iteration strategy for R that chooses unique wellfounded branches. Uniqueness implies that Σ^* has various useful internal consistency properties, such as positionality, pullback consistency, strategy coherence, normalizing well, and strong hull condensation. We would like to show that these properties pass to the induced strategy $\Omega(c, \Sigma^*)$ for M, but unfortunately, in many cases the connection between Σ^* and $\Omega(c, \Sigma^*)$ is not sufficiently tight that one can do this directly.

To illustrate the problems, let us look at some special cases of positionality. Let $c = \langle M, \psi, Q, \mathbb{C}, R \rangle$ and $\Omega = \Omega(c, \Sigma^*)$. It is easy to see that $\Omega = \Omega_M$. For $\Omega_M = \Omega_{\langle \emptyset, M \rangle}$ is obtained by lifting the empty tree on M to the empty tree on R, with lift map $\text{id} : M \to M$, then resurrecting M to itself with resurrection map $\text{id} : M \to M$. Thus Ω_M is the pullback of $\Omega(c, \Sigma^*)$ under the identity map, so $\Omega = \Omega_M$.

But now suppose $P \lhd N \lhd M$. Must we have $(\Omega_N)_P = \Omega_P$? A little thought shows that this is not at all obvious. Suppose for example that $M = Q$ and $\psi = \text{id}$, so that $\Omega = \Omega(\mathbb{C}, M, \Sigma^*)$. At the background level, we are executing either one or two empty trees, so we are not leaving R and $\mathbb{C}$. Let $X = \mathrm{Res}_{\mathrm{M}}[N]$, $\Omega(X) = \Omega(\mathbb{C}, X, \Sigma^*)$, $Y = \mathrm{Res}_{\mathrm{M}}[P]$, and $\Omega(Y) = \Omega(\mathbb{C}, Y, \Sigma^*)$. By definition

$$\begin{aligned}\Omega_P &= \Omega(Y)^{\sigma_{\mathrm{M}}[P]}\\ &= (\Omega(X)_{\mathrm{Res}_{\mathrm{M,X}}[P]})^{\sigma_{\mathrm{M,X}}[P]},\end{aligned}$$

because $\sigma_{\mathrm{M}}[P] = \sigma_{\mathrm{X,Y}}[\mathrm{Res}_{\mathrm{M,X}}[P]] \circ \sigma_{\mathrm{M,X}}[P]$. On the other hand

$$\begin{aligned}(\Omega_N)_P &= (\Omega(X)^{\sigma_{\mathrm{M}}[N]})_P\\ &= (\Omega(X)_{\sigma_{\mathrm{M}}[N](P)})^{\sigma_{\mathrm{M}}[N]}.\end{aligned}$$

If the (M, X) resurrections of N and P are consistent, that is, $\sigma_{\mathrm{M}}[N](P) = \mathrm{Res}_{\mathrm{M,X}}[P]$ and $\sigma_{\mathrm{M}}[N] \restriction P = \sigma_{\mathrm{M,X}}[P]$, then we get $\Omega_P = (\Omega_N)_P$. Otherwise, there seems to be no reason they should be equal.

Here is a definition that rules out these simple pathologies.

DEFINITION 3.6.1. Let Ω be a (λ, θ)-iteration strategy for M; then Ω is *mildly positional* iff

(a) $\Omega = \Omega_M$, and
(b) whenever s is a stack by Ω and $P \trianglelefteq N \trianglelefteq M_\infty(s)$, then $(\Omega_{s,N})_P = \Omega_{s,P}$.

Mild positionality seems like something one would want near the beginning.

In addition to possible resurrection inconsistency, there is a second obstacle to a direct proof of positionality for background-induced strategies. The following simple example illustrates the problem. Let $M \in \mathrm{lev}(\mathbb{C})$, where $\mathbb{C}$ is a background construction in V, and let $\Omega = \Omega(\mathbb{C}, M, \Sigma^*)$. Let E be a total extender on the M-sequence, let $\langle E \rangle$ be the iteration tree on M whose only extender is E, and let $N \triangleleft M|\mathrm{lh}(E)$. By coherence, $N \triangleleft \mathrm{Ult}(M, E)$. Positionality would imply that $\Omega_{\langle E \rangle, N} = \Omega_N$. Can one show this directly?[17]

The answer is yes in one central case. Suppose that $o(N) < \lambda_E$, $E = \dot{F}^M$, and $k(M) = 0$, so that $E = E^* \cap ([\lambda_E]^{<\omega} \times M)$, where $E^* = B^{\mathbb{C}}(E)$ is the background extender. Let $\mathbb{D} = i_{E^*}(\mathbb{C})$. Letting $Q = M||o(M)$, we have by coherence at the background level that for some ν,

$$Q = M^{\mathbb{C}}_{\nu,-1} = M^{\mathbb{D}}_{\nu,0}$$

and $\mathbb{C} \restriction \nu = \mathbb{D} \restriction \nu$, $F^{\mathbb{C}}_\nu = E^*$, and V agrees with $\mathrm{Ult}(V, E^*)$ to $\mathrm{lh}(E^*)$. Thus we may set

$$\begin{aligned}X &= \mathrm{Res}_{\mathrm{Q}}[N]^{\mathbb{C}} = \mathrm{Res}_{\mathrm{Q}}[N]^{\mathbb{D}},\\ \sigma &= \sigma_{\mathrm{Q}}[N]^{\mathbb{C}} = \sigma_{\mathrm{Q}}[N]^{\mathbb{D}}.\end{aligned}$$

[17]That $\Omega_{\langle E \rangle, N} = \Omega_N$ is an instance of *strategy coherence*. Strategy coherence is a consequence of positionality which, unlike full positionality, is essential for a theory of strategy mice. That $\Omega_{\langle E \rangle, N} = \Omega_N$ is also an instance of *normalizing well*, and this too is essential to the theory we are developing.

Let $R = i_{E^*}(M)$ and $\pi\colon \mathrm{Ult}(M,E) \to R$ be the lift map, so that $\pi \restriction \lambda_E = \mathrm{id}$ by the simplifying assumptions on E that we have made. No level of $\mathbb{D}$ past ν projects to or below λ_E, so

$$X = \mathrm{Res}_{\mathrm{R}}[N]^{\mathbb{D}},$$
$$\sigma = \sigma_{\mathrm{R}}[N]^{\mathbb{D}}.$$

We then get

$$\begin{aligned}\Omega_{\langle E\rangle,N} &= \Omega(\mathbb{D}, \mathrm{Res}_{\mathrm{R}}[\pi(N)], i_{E^*}(\Sigma^*))^{\sigma_{\mathrm{R}}[\pi(N)]\circ\pi}\\ &= \Omega(\mathbb{D}, X, i_{E^*}(\Sigma^*))^{\sigma}\\ &= \Omega(\mathbb{C}, X, \Sigma^*)^{\sigma}\\ &= \Omega_N.\end{aligned}$$

Here we use that Σ^* and $i_{E^*}(\Sigma^*)$ agree on $V_{\mathrm{lh}(E^*)}$, because they choose unique wellfounded branches.

There are two problems with converting this argument into a general one. First, if $M \neq M|\,\mathrm{lh}(E)$, then we must connect E to E^* using the resurrection map $\sigma_{\mathrm{M}}[M|\,\mathrm{lh}(E)]$. This works out fine unless the $(M, \mathrm{Res}_{\mathrm{M}}[M|\,\mathrm{lh}(E)])$ resurrections of $M|\,\mathrm{lh}(E)$ and N are inconsistent. We then have a problem like that above.

There is a second problem that has nothing to do with dropping and resurrection. Namely, the central case assumed $o(N) < \lambda_E$, but we need this form of positionality in the case that $\lambda_E \leq o(N) < \mathrm{lh}(E)$ too. Under the assumptions of the central case, the lift map $\pi\colon \mathrm{Ult}(M,E) \to i_{E*}(M)$ has critical point λ_E, so if $\lambda_E \leq o(N)$, then $\pi(N) \neq N$. This means that the background extenders used in computing $\Omega(\mathbb{D}, i_{E^*}(M), i_{E^*}(\Sigma^*))_{\pi(N)}$ may be completely different from those involved in computing $\Omega(\mathbb{C}, M, \Sigma^*)_N$.

One might call the first problem the *resurrection consistency problem*, and the second one the *background coherence problem*. Both problems have to do with the behavior of background-induced iteration strategies on stacks of normal trees. We shall address them in the next chapter.

3.7. Measurable projecta

We shall solve the resurrection consistency problem by moving to a slightly different notion of premouse. The transition is fairly simple: the main new requirement is that premice have no measurable projecta. We shall describe the new premice in detail in the next chapter.

This transition is motivated by the root cause of resurrection inconsistency in the background constructions of standard premice, namely, the existence of anticore maps with critical point the current projectum. We shall see in this section that such anticore maps occur precisely when $\rho_{k+1}(M_{\nu,k+1})$ is measurable in $M_{\nu,k+1}$. In the next section we look at how $\rho_k(M_{\nu,k+1})$ might be measurable in $M_{\nu,k+1}$.

We shall use some of the ideas in these two sections when we get to the new premice, but not the results themselves. The main idea is present in the proof of the following theorem.

THEOREM 3.7.1. *Let $\mathbb{C}$ be a background construction with levels $M_{\eta,j}$. Suppose $\rho(M_{\nu,k}) < \rho^-(M_{\nu,k})$, and let $\pi\colon M^-_{\nu,k+1} \to M_{\nu,k}$ be the anticore map; then for $\rho = \rho(M_{\nu,k})$,*

(1) there is no $M_{\nu,k}$-total E on the $M_{\nu,k}$-sequence such that $\text{crit}(E) = \rho$, and
(2) the following are equivalent:
  (a) $\text{crit}(\pi) = \rho$,
  (b) there is an $M_{\nu,k+1}$-total E on the $M_{\nu,k+1}$-sequence such that $\text{crit}(E) = \rho$,
  (c) the $M_{\nu,k+1}$ sequence has a total order zero measure D on ρ, and there is a (unique) elementary $\sigma\colon \text{Ult}(M^-_{\nu,k+1}, D) \to M_{\nu,k}$ such that $\sigma(\rho) = \rho$ and $\pi = \sigma \circ i_D$.

PROOF. For (1), we use the amenable closure argument.[18] Let E be a total-on-$M_{\nu,k}$ extender from the $M_{\nu,k}$-sequence such that $\rho = \text{crit}(E)$. We may assume E has order zero. Let

$$\gamma = \rho^{+,M_{\nu,k}}$$

and

$$P = M_{\nu,k}|\text{lh}(E).$$

Suppose first that $P = M_{\nu,k}$, so that $k = k(P) = 0$. Let $E^* = B^{\mathbb{C}}(E)$ be the background extender, and $A \subseteq \rho$ the new Σ_1^P subset of ρ. We have that $A \cap \alpha \in P$ for all $\alpha < \rho$, but $A \cap \rho \notin P$. By coherence and the fact that $E \subseteq E^*$, we get

$$P = \text{Ult}(P, E)||\text{lh}(E) = i_{E^*}(P)||\text{lh}(E).$$

But $i_{E^*}(A)$ is amenable to $i_{E^*}(P)$ below $i_{E^*}(\rho)$, so

$$A = i_{E^*}(A) \cap \rho \in i_{E^*}(P).$$

The factor embedding from $\text{Ult}(P, E)$ to $i_{E^*}(P)$ has critical point λ_E, and so γ is a cardinal of $i_{E^*}(P)$, and thus $A \in i_{E^*}(P)||\gamma$. But then $A \in P$, contradiction.

So we may assume $P \lhd M_{\nu,k}$. We have that $\rho(P) \leq \gamma$ because E has order zero, and if $\rho(P) \leq \rho = \rho(M_{\nu,k})$, then $P = M_{\nu,j}$ for $j < k$ is impossible because $\rho_k(M_{\nu,k}) > \rho$, and $P \in M_{\nu,k}$ is impossible because γ is a cardinal of $M_{\nu,k}$, so we have a contradiction. Thus $\rho(P) = \gamma$.

But now we have $P^+ \trianglelefteq M_{\nu,k}$, and $\rho^-(P^+) = \gamma$. We can therefore apply Lemma 3.1.5 to find $\mu < \nu$ such that $P^+ = M_{\mu,1}$. That is, $P = M^-_{\mu,1}$, and $\rho(P) = \gamma = \rho(M_{\mu,0})$. Let $\pi\colon M^-_{\mu,1} \to M_{\mu,0}$ be the anticore map, and let $E^* = B^{\mathbb{C}}(\pi(E))$.

[18]This sort of argument was first discovered and exploited by Hugh Woodin in the fine structure theory of mice with long extenders. See [41].

Again, let $A \subset \rho$ be the new $\Sigma_{k+1}^{M_{\nu,k}}$ subset of ρ. We have $A \cap \alpha \in M_{\nu,k}$ for all $\alpha < \rho$ and $A \cap \rho \notin M_{\nu,k}$. However, $M_{\nu,k}|\gamma = P|\gamma = M_{\mu,0}|\gamma$, and γ is a cardinal in all these models. So $A \cap \alpha \in M_{\mu,0}$ for all $\alpha < \rho$ and $A \cap \rho \notin M_{\mu,0}$. But then we can apply the amenable closure argument above: $A = i_{F^*}(A) \cap \rho$, so $A \in i_{E^*}(M_{\mu,0})$ by the elementarity of i_{F^*}. But $M_{\mu,0}$ and $i_{E^*}(M_{\mu,0})$ agree to their common ρ^+, namely γ. Hence $A \in M$, contradiction.

For (2), clearly (c) implies (b), and (b) implies (a) by part (1). So we must see that (a) implies (c). For this, we just inspect the standard proof that $M_{\nu,k+1}$ is solid and universal.[19] Let $M = M_{\nu,k}$ and $H = M^-_{\nu,k+1}$. We compare the phalanx (M, H, ρ) with M. We can assume here that M is countable, and that our iteration strategy Σ for M has the Weak Dodd-Jensen property.[20] Σ induces a pullback strategy $\Sigma^{(\mathrm{id},\pi)}$ for (M, H, ρ). Let $\mathcal{T}$ and $\mathcal{U}$ be the resulting trees on (M, H, ρ) and M.

The standard proof shows that the final model on both sides is the same. Call it Q. We also get that Q is above H in $\mathcal{T}$, and the branches H-to-Q of $\mathcal{T}$ and M-to-Q of $\mathcal{U}$ do not drop. The branch embeddings $i\colon H \to Q$ and $j\colon M \to Q$ are such that $\mathrm{crit}(i) \geq \rho$ and $\mathrm{crit}(j) \geq \rho$.[21] We have also that $i(p(H)) = j(p(M)) = p(Q)$.

Since $i(p(H)) = j(p(M))$, and $i \restriction \rho = j \restriction \rho = \mathrm{id}$, we get

$$\pi = j^{-1} \circ i.$$

But $\mathrm{crit}(j) > \rho$ by part (1), and $\mathrm{crit}(\pi) = \rho$, so $\mathrm{crit}(i) = \rho$.

By (1) and $\mathrm{crit}(j) > \rho$, ρ is not the critical point of a total on Q extender from Q. It follows that the first extender used in i is the order zero total measure on ρ from the H-sequence. Call this D. We get the desired $\sigma\colon \mathrm{Ult}(H, D) \to M$ by setting $\sigma = j^{-1} \circ k$, where k is the branch tail of H-to-Q, that is, $i = k \circ i_D$. ⊣

Remark 3.7.2. The equivalence of (2)(a) and (2)(b) requires only that we are dealing with a mouse and its core. Part (1), and the equivalent (2)(c), rely on amenable closure. So these only work for the cores taken in a background construction. It is easy to produce a counterexample otherwise, by taking $M = \mathrm{Ult}(H, E)$ for E not of order zero.

DEFINITION 3.7.3. For M a premouse,

(a) M is *projectum-measurable* iff there is a total-on-M extender E on the M-sequence such that $\mathrm{crit}(E) = \rho^-(M)$.
(b) (M, D) is a *pfs violation* iff D is a total-on-M extender on the M-sequence, and $\mathrm{crit}(D) = \rho^-(M) < \rho^-(M^-)$.
(c) A pfs violation (M, D) has *order zero* iff D has order zero.

[19] See [35, §8] or [70, §5].

[20] See Lemma 4.6.10, and [39].

[21] For the reader who would like his memory jogged: We apply the Weak Dodd-Jensen property to the iteration maps in $\mathcal{U}$ and the (id, π)-lift of $\mathcal{T}$. This and the fact that we were iterating disagreements shows that $\mathcal{T}$ ends with Q above H, and that H-to-Q does not drop. The M side cannot end with P such that $Q \lhd P$ because otherwise the new subset of ρ would be in P, hence M. So $P = Q$, and the M-to-P branch of $\mathcal{U}$ does not drop. Its embedding j has $\mathrm{crit}(j) \geq \rho$ because $\rho = \rho(H) = \rho(Q) = \rho(P)$.

Clearly, N is projectum-measurable iff there is a (unique) order zero pfs violation of the form (N,D). We will sometimes say that (N,κ,D) is a pfs violation, if (N,D) is a pfs violation and $\kappa = \text{crit}(D) = \rho_{k(N)}(N)$.

COROLLARY 3.7.4. *Let $\mathbb{C}$ be a background construction, $X \in \text{lev}(\mathbb{C})$, $M = \mathcal{C}(X)$, and let $\pi\colon M^- \to X$ be the anticore map. Equivalent are*

(a) $\text{crit}(\pi) = \rho(X)$,

(b) M is projectum-measurable,

(c) there is a D such that (M,D) is an order zero pfs violation.

PROOF. This is immediate from Theorem 3.7.1. The only thing to check is that if $\text{crit}(\pi) = \rho(X) = \rho^-(M)$, then $\rho^-(M) < \rho^-(M^-)$. Say $k = k(X)$, so that $k+1 = k(M)$ and $k = k(M^-)$. We must see that $\rho_{k+1}(M) < \rho_k(M)$. But $\rho_{k+1}(M) = \rho_{k+1}(X) < \rho_k(X)$ because π is not the identity, and $\pi``\rho_k(M)$ is cofinal in $\rho_k(X)$ while $\pi \restriction \rho_{k+1}(M) = \text{id}$. Thus $\rho_{k+1}(M) < \rho_k(M)$. ⊣

We should note that although $\rho(M)$ is officially *the* projectum of M, it is $\rho^-(M)$ that is relevant for projectum-measurability.[22]

"pfs" stands for "projectum-free spaces". In one version of fine structure theory for mice with long extenders, it is important that no projectum be the space corresponding to a long generator. See for example [41]. It turns out that one can also avoid projecta being spaces, that is, critical points, in the short extender realm.

When we first add E, as the top extender of $M^{\mathbb{C}}_{\nu,0}$, we have $\kappa_E \neq \rho(M_{\nu,0})$ by amenable closure. If $\bar{E}$ is the image of E in $M_{\mu,l}$ under some corings that correspond to projecta $> \kappa_E$, then still $\kappa_{\bar{E}} = \kappa_E \neq \rho(M_{\nu,l})$. But we may reach a first stage where $\rho(M_{\mu,l}) < \kappa_E$, and it may be that κ_E collapses to $\rho(M_{\mu,l})$ when we core down in the standard way, making $\rho_{l+1}(M_{\nu,l+1})$ measurable in $M_{\nu,l+1}$.

Our revised background constructions will avoid this by putting $\rho_{l+1}(M_{\nu,l})$ as a point into the hull that collapses to $M_{\nu,l+1}$. A straightforward generalization of Theorem 3.7.1 then implies that $\rho_{l+1}(M_{\nu,l+1})$ is not measurable in $M_{\nu,l+1}$. Doing just this does not rule out measurable projecta in $M_{\nu,l+1}$, however, because $\rho_l(M_{\mu,l+1})$ might be measurable in $M_{\mu,l+1}$. This could happen if the anticore map from $M^-_{\mu,l+1}$ to $M_{\mu,l}$ does not preserve ρ_l, or equivalently, is discontinuous at $\rho_l(M_{\nu,l+1})$. In other words, there can be measurable projecta that do not trace back to the critical points of anticore maps, but rather to their discontinuities.

Our solution here will be to put $\rho_l(M_{\nu,l})$ into the hull collapsing to $M_{\nu,l+1}$ as well. Of course, our solutions involve relaxing the standard soundness requirements on premice, so we shall need to see that we still have a fine structure theory. This amounts to showing that the new parameters are preserved by the relevant maps. The proof of Theorem 3.7.1 will help with that, as will the ideas in the next section.

---

[22] If (M,D) is a pfs violation, then $k(M) \geq 1$, so our previous advice to focus on the case $k(M) = 0$ needs to be adjusted. The case $k(M) = 1$ seems to be representative of the general case.

3.8. Projecta with measurable cofinality

Let $\mathbb{C}$ be a background construction in the sense of Definition 3.1.3. We wish to characterize the measurable projecta of initial segments of levels of $\mathbb{C}$ that do not trace back to a projectum-critical anticore map. The source for them is an anticore map $\pi\colon M^{-}_{\nu,k+1} \to M_{\nu,k}$ that is discontinuous at $\rho_k(M_{\nu,k+1})$, and so we shall need some lemmas on the $r\Sigma_k$ cofinality of ρ_k.

Our notation for the $r\Sigma_k$ cofinality of $\rho_k(M)$ is η^M_k. Section 2.5 contains a general discussion of the notion, and some elementary preservation lemmas that we shall use here. We shall also need two further, less elementary results concerning the preservation of η^M_k under anticore maps.

THEOREM 3.8.1. *Let X be an $(\omega_1, \omega_1 + 1)$ iterable premouse, $Q = \mathcal{C}(X)$ be its core, and $\pi\colon Q^- \to X$ be the anticore map. Let $k = k(X)$, and suppose that π is discontinuous at $\rho_k(Q)$; then for $\mu = \eta^M_k$,*

(1) $\rho_{k+1}(Q) \leq \mu < \rho_k(Q)$, and
(2) $Q \models \mu$ is measurable.

PROOF. The case $k = 1$ is typical of the general one, so the reader could just assume $k = 1$ below. Since we wish to prove a first order property of X, we may assume X is countable, and by [39], we have an iteration strategy Σ for X with the Weak Dodd-Jensen property relative to some enumeration $\vec{e}$ of X. We assume that $\vec{e}$ begins with $p(X)$.

The key is that π can be derived from an iteration map. More precisely, let $\rho = \rho_{k+1}(X)$, and consider the comparison of (X, Q^-, ρ) with X, using Σ and the iteration strategy for (X, Q^-, ρ) that Σ induces.

The solidity/universality argument of [35] shows that the resulting iteration trees $\mathcal{T}$ on (X, Q^-, ρ) and $\mathcal{U}$ on X have a common last model R, that R is above Q^- in $\mathcal{T}$, and that neither Q^--to-R nor X-to-R drops. Let

$$i\colon Q^- \to R$$

and

$$j\colon X \to R$$

be the branch embeddings. $\mathrm{crit}(i) \geq \rho$ by the rules of $\mathcal{T}$, and $\mathrm{crit}(j) \geq \rho$ because $\rho = \rho(R)$. Using the Weak Dodd-Jensen property together with the solidity of the standard parameter, we get

$$i = j \circ \pi.$$

These claims are all proved in full in [35].

Now suppose π is discontinuous at $\rho_k(Q)$. It follows from the elementarity of the iteration maps i and j that

$$i(\rho_k(Q)) = j \circ \pi(\rho_k(Q)) > j(\rho_k(X)) \geq \rho_k(R) \geq \sup i``\rho_k(Q).$$

Thus i is discontinuous at $\rho_k(Q)$.

Let us look at the first discontinuity along the branch Q^--to-R in $\mathcal{T}$. We get Y on the branch and E such that $\mathrm{Ult}(Y,E)$ is also on the branch, with

$$i = l \circ i_E^Y \circ h,$$

where $h\colon Q^- \to Y$ is continuous at $\rho_k(Q)$, i_E^Y is discontinuous at $h(\rho_k(Q))$, and $l\colon \mathrm{Ult}(Y,E) \to R$ is the branch tail embedding. All models in Q^--to-R have degree k, and ultrapower embeddings are cofinal and elementary, so $\rho_k(Y) = \sup h``\rho_k(Q) = h(\rho_k(Q))$. Let $\mu = \mathrm{crit}(E)$. We must have $\mu < h(\rho_k(Q))$, since otherwise our branch from Q to R would have a drop. Since $\mathrm{Ult}(Y,E) = \mathrm{Ult}_k(Y,E)$ consists of $[a,f]_E^Y$ such that f is $r\Sigma_k$, i_E is discontinuous at γ iff $\mathrm{cof}_k^Y(\gamma) = \mu$. (See Lemma 2.5.6.) Thus $\eta_k^Y = \mu$. The normal measure $E_{\{\mu\}}$ is in Y, and $\rho_{k+1}^Y = \rho \le \mu$ because Y is above Q^- in $\mathcal{T}$. So we have what we want at Y, and we just need to pull this back under h.

Lemma 2.5.9 lets us do that. By induction on models Z in the branch Q^--to-Y, letting $l\colon Q^- \to Z$ be the branch embedding, we show that $l(\rho_k(Q)) = \rho_k(Z)$ and $l(\eta_k^Q) = \eta_k^Z$. The lemma takes us past successor steps, and limit steps are easy. It follows that $\mu \in \mathrm{ran}(h)$, and $h^{-1}(\mu) = \eta_k^Q$. Since $\rho_{k+1}(Q) \le \mathrm{crit}(h)$ and $\rho_{k+1}(Q) = \rho_{k+1}(Y) \le \mu$, we get $\rho_{k+1}(Q) \le h^{-1}(\mu)$. ⊣

The converse to the theorem does not hold in general. If you start with Q such that $k(Q) = 2$ and $\rho_1(Q)$ has measurable cofinality μ in Q, with $\rho_2(Q) \le \mu < \rho_1(Q)$, and then take $X = \mathrm{Ult}_1(Q^-, E)$ where $\rho_2(Q) \le \mathrm{crit}(E) < \rho_1(Q)$ and $\mathrm{crit}(E) \neq \mu$, then the anticore map $\pi = i_E$ is continuous at $\rho_1(Q)$.

But if $Q = M_{\nu,k+1}^{\mathbb{C}}$ and $X = M_{\nu,k}^{\mathbb{C}}$ and $\pi\colon Q^- \to X$ is the anticore map, then we do get the converse. The difference here is that the iteration from which π is derived has to hit all measurable cardinals in the interval $[\rho_{k+1}(Q), \rho_k(Q))$ along its main branch (many times). But we can prove the converse without going into that:

THEOREM 3.8.2. *Let $\mathbb{C}$ be a background construction, and let $Q = M_{\nu,k+1}^{\mathbb{C}}$, $X = M_{\nu,k}^{\mathbb{C}}$, and $\pi\colon Q^- \to X$ be the anticore map. The following are equivalent:*

(1) π is discontinuous at $\rho_k(Q)$,

(2) $\rho_{k+1}(Q) \le \eta_k^Q < \rho_k(Q)$ and $Q \models$ "η_k^Q is measurable".

PROOF. (1) implies (2) by Theorem 3.8.1. Assume (2). Since X is a level of $\mathbb{C}$ and $\pi(\eta_k^Q) < \rho_{k(X)}(X)$, $\pi(\eta_k^Q)$ is measurable in V. We must have $\rho_{k+1}(Q) < \pi(\eta_k^Q)$, since otherwise $\rho_{k+1}(Q) = \eta_k^Q = \pi(\eta_k^Q)$, and we can use the amenable closure argument to get a contradiction.[23] But $|\eta_k^Q| = |\rho_{k+1}(Q)|$ in V, so $\eta_k^Q < \pi(\eta_k^Q)$.

$\pi(\eta_k^Q)$ is regular in V and $|\eta_k^Q| = |\rho_{k+1}(Q)|$, so π is discontinuous at η_k^Q. By Lemma 2.5.10(3), π is discontinuous at $\rho_k(Q)$. ⊣

[23]Let $A \subseteq \rho_{k+1}(Q)$ be the new set. Let E^* background the resurrection of the order zero measure of X on $\pi(\eta_k^Q)$. We get $A \in i_{E^*}(X)$ because A was amenable to X, so $A \in X$ by coherence, contradiction.

We have shown in Theorem 3.7.1 that projectum-critical uncorings give rise to pfs violations, but Theorem 3.8.2 shows that not every pfs violation in an initial segment of some $Q \in \text{lev}(\mathbb{C})$ resurrects to a projectum-critical uncoring. In the situation described in (1) and (2) of Theorem 3.8.2, it could be that (Q^-, D) is a pfs violation, but $\pi(\rho^-(Q^-)) \neq \rho^-(X)$, so $(X, \pi(D))$ is not a pfs violation. This leads to the following definition.

DEFINITION 3.8.3. Let (M,D) be an order zero pfs violation, and $k = k(M)$. We say that (M,D) is *weak* iff

(a) $\rho_{k+1}(M) \leq \eta_k^M < \rho_k(M)$, and

(b) there is a total measure on the M-sequence with critical point η_k^M.

(M,D) is *strong* iff it is not weak.

One can show that the weak pfs violations are precisely those whose resurrection does not end with a projectum-critical uncoring.

By pursuing these ideas further, one can obtain well-behaved iteration strategies for the premice M reached in a background construction $\mathbb{C}$ done in a coarse premouse R that itself has a coarsely well-behaved iteration strategy. "Well-behaved" means having the internal consistency properties discussed in Section 3.6, as well as others we shall discuss later. But in order to do this, one must restrict the class of iteration trees on M to which its strategy applies, and one must change the way we have converted trees on M to trees on R. The changes involve keeping close track of which pfs violations in M and its iterates lift to strong pfs violations under the maps of our conversion system.

One could probably develop a theory of strategy comparison along these lines, but it becomes quite complicated. There is a better way to solve the resurrection consistency problem.

Chapter 4

MORE MICE AND ITERATION TREES

Let $\mathbb{C}$ be a background construction. Resurrection inconsistency arises in $\mathbb{C}$ when we have $X \in \text{lev}(\mathbb{C})$ with projectum $\rho = \rho(X)$, and for $Q = \mathcal{C}(X)$ and $\pi\colon Q^- \to X$ the anticore map, $\pi(\rho) \neq \rho$. For P such that $X|\rho \trianglelefteq P \triangleleft X|\rho^{+,X}$ we then have $\text{Res}_{Q,X}[P] = P$, whereas $\text{Res}_{Q,X}[N](P) = \pi(P) \neq P$ whenever $X|\rho^{+,X} \trianglelefteq N \trianglelefteq Q^-$. All resurrection inconsistencies in $\mathbb{C}$ trace back to situations of this kind, that is, to anticore maps whose critical point is the current projectum.

This suggests that we modify our constructions by always putting $\rho_{k+1}(M_{\nu,k})$ as a point into the hull collapsing to $M_{\nu,k+1}$. This would suffice to eliminate resurrection inconsistency, but it turns out that the definitions and basic facts are simpler if we also put $\rho_k(M_{\nu,k})$ into the hull as a point. The $\rho_i(M_{\nu,k})$ for $i < k$ get into the hull automatically, so the net effect is that our revised background constructions will set

$$M_{\nu,k+1} = \text{cHull}_{k+1}^{M_{\nu,k}}(\rho_{k+1} \cup \{p_i(M_{\nu,k}), \rho_i(M_{\nu,k}) \mid i \leq k+1\}).$$

The anticollapse map[1] now has critical point $> \rho$. For such revised constructions we can define resurrection maps that always follow the anticore maps, and hence are all consistent with one another.

We shall see that such constructions produce premice M such that no projectum of M is measurable in M. We call this property *projectum-free spaces*, and the new premice *pfs premice*.

Since we do not always core down "all the way", the new $M_{\nu,k+1}$ will not always be $k+1$-sound in the old sense. We must therefore isolate the sense in which the new $M_{\nu,k+1}$ is sound, and develop the basic fine structure theory around this notion. This is mostly a matter of adapting known techniques, but there are some new issues related to the preservation of projecta by iteration maps. Part of the difficulty is that the proofs must generalize to the context of strategy mice.

Many of the deeper theorems in fine structure are proved by combining phalanx comparisons with the Dodd-Jensen property. We saw two such proofs at the end of the last chapter, and we shall see several more as we develop the fine structure of

[1]Recall that cHull stands for the transitive collapse of the hull in question.

pfs premice at the end of this chapter. The method is very powerful.[2] Some of the less straightforward points in the fine structure of pfs premice have to do with their behavior in phalanx comparisons.

4.1. Mice with projectum-free spaces

We are now going to revise our background constructions so that $\rho_k(M_{\nu,k})$ and $\rho_{k+1}(M_{\nu,k})$ are always put into the hull collapsing to $M_{\nu,k+1}$. In this respect, the new constructions are similar to those in the long extender fine structure theory of [41], in which one sometimes does not core down "all the way".

We call the new constructions *PFS constructions*, and the new premice that they produce *pfs premice*. Here "pfs" stands for "projectum-free spaces".

At bottom, the new constructions and premice are the same as the old ones, it is just that the old ones are being presented differently. Theorems 3.7.1, 3.8.1, and 3.8.2 tell us how to translate between the two hierarchies. We shall need their main elements in the basic fine structure of the new premice, but we do not need to make the translation between hierarchies explicit. At various points we do need to look at structures obtained by taking one or two steps back toward the standard hierarchy.

What does change when we move to PFS constructions and premice is what extenders are available on the coherent sequence to be used in forming iteration trees, and how those iteration trees are converted to iteration trees on the background universe. One could instead stick to the old constructions and premice, while making changes in which iteration trees on them are allowed, and how those trees are converted to trees on the background universe, that reflect what would happen with their PFS equivalents. This seems to lead to a maze of special cases whose only motivation is a suppressed translation to the pfs hierarchy. So we have elected not to go that route.[3]

Projecta, cores, and premice. Let us describe in more detail the first order properties of pfs premice. Many of the elementary definitions, lemmas, and definability calculations in Chapter 2 go over to the new premice with little change, but there are some new elements.

A *potential pfs premouse* is a potential Jensen premouse in the sense of Section 2.2, but we want to impose a weak form ms-solidity from the beginning.

DEFINITION 4.1.1. Let M be a potential Jensen premouse, and E be an extender on the M-sequence. We say that E has the *weak* ms-ISC iff letting $\kappa = \mathrm{crit}(E)$, the Jensen completion of $E_{\{\kappa\}}$ is on the $M|\mathrm{lh}(E)$-sequence.

---

[2]The author first encountered this method in the unpublished paper [4] by Dodd. The method was more fully developed in [35] and [39].

[3]After developing it in some detail. The results of the last two sections in Chapter 3 are useful, but there is more to it.

Notice that E determines $M|\mathrm{lh}(E)$, so the terminology makes sense. If E is itself the Jensen completion of $E_{\{\kappa\}}$, then E has the weak ms-ISC.

DEFINITION 4.1.2. *M is a potential pfs premouse* iff M is a potential Jensen premouse in the sense of Section 2.2, and whenever E is an extender on the sequence $\dot{E}^M$, then E has the weak ms-ISC.

Notice that we do not require that $\dot{F}^M$ have the weak ms-ISC. This would be wrong, because if $E = \dot{F}^M$ is the Jensen completion of $E_{\{\kappa_E\}}$, and $i\colon M \to \mathrm{Ult}_0(M,G)$ is such that $\kappa_E < \mathrm{crit}(i) < \lambda_E$, then $i(E)$ does not have the weak ms-ISC. In a similar vein, if $\mathbb{C}$ is a background construction, then the last extender of $M^{\mathbb{C}}_{\nu,0}$ may not have the weak ms-ISC. We shall show below that the last extender of $M^{\mathbb{C}}_{\nu,1}$ does have the weak ms-ISC, and more generally, if M is 1-sound and iterable, then $\dot{F}^M$ has the weak ms-ISC.

Iterable Jensen premice satisfy the full ms-ISC, but one needs a comparison argument to show this. It is better for our exposition to record this fragment of it as an axiom before we get to comparison.[4]

DEFINITION 4.1.3. Let M be a potential pfs premouse; then

(a) κ is *measurable by the M-sequence* iff κ is the critical point of an M-total extender on the M-sequence, and
(b) the *order zero measure of M on κ* is the first M-total extender on the M sequence that has critical point κ.

As usual, the M-sequence includes $\dot{F}^M$. We may later slip into writing "$M \models \kappa$ is measurable" or "κ is measurable in M" when we mean that κ is measurable by the M-sequence. This is not so bad, because for iterable M the two are equivalent, by a comparison argument.[5] But in our current pre-comparison stage we need to make the distinction, and it is pretty much always measurability by the M-sequence that matters.

The weak ms-ISC justifies the terminology in (b).

PROPOSITION 4.1.4. *Let M be a potential pfs premouse, and E be on the M-sequence, with $\kappa = \kappa_E$.*

(1) The following are equivalent:
  (a) E is the order zero measure of M on κ,
  (b) E is total on M, and κ is not measurable by the $\mathrm{Ult}_0(M,E)$ sequence.

---

[4]We are aiming to show that PFS constructions produce structures $M_{\nu,k}$ with a complete fine structure theory. The proof is an induction on $\langle \nu, k \rangle$, and the function of the premouse axioms is to isolate enough about the first order theory $M_{\nu,k}$ that we can use this information, together with an iteration strategy for $M_{\nu,k}$, to develop the theory of $M_{\nu,k+1}$. We are therefore free to include in the premouse axioms anything we can prove is part the theory of $M_{\nu,k+1}$. Of course it is nice to have a small set of axioms.

[5]This is a theorem that traces back to Kunen and Mitchell. The modern, definitive form of it was proved by Farmer Schlutzenberg in [55].

(2) *If E is the order zero measure of M on κ, and $E \neq \dot{F}^M$, then E is the Jensen completion of $E_{\{\kappa\}}$.*

PROOF. For (1): Let E be the first M-total measure on the M-sequence, and suppose toward contradiction that F witnesses that κ is measurable by the sequence $N = \mathrm{Ult}_0(M,E)$. Since $\kappa \notin \mathrm{ran}(i_E)$, $F \neq \dot{F}^N$, and thus F has the weak ms-ISC. So letting G be the Jensen completion of F_κ, G is on the sequence of N. It follows that $\rho(N|\mathrm{lh}(G)) \leq \kappa^{+,N} = \kappa^{+,M}$. But $G \in N$, so $|\mathrm{lh}(G)| = \kappa^+$ in N. On the other hand, $\lambda(E)$ is an inaccessible cardinal in N, so $\mathrm{lh}(G) < \lambda(E)$. By coherence, G is on the sequence of M before E, contradiction. Thus (a) implies (b). (b) implies (a) follows easily from coherence.

For (2): Let G be the Jensen completion of $E_{\{\kappa\}}$. Since $E \neq \dot{F}^M$, G is on the sequence of $M|\mathrm{lh}(E)$. If $G \neq E$, then G is on the sequence of $\mathrm{Ult}_0(M,E)$, contrary to (1)(b). Thus $G = E$, as desired. ⊣

We shall use these simple facts about measures of order zero quite often as we develop the theory of pfs premice, and this is why we have built the weak ms-ISC into the definition of potential pfs premice.

PROPOSITION 4.1.5. *There are sentences θ and φ in the language of potential premice such that θ is Π_2, φ is Σ_2, and for any potential pfs premouse M,*

$$M \models \theta \vee \varphi \text{ iff } \dot{F}^M \text{ satisfies the weak ms-ISC.}$$

PROOF. θ says that $\dot{F}$ is the Jensen completion of $\dot{F}_{\{\kappa\}}$, for $\kappa = \mathrm{crit}(\dot{F})$. It has the form $\forall\alpha\exists g([\{\kappa\},g]_{\dot{F}} = \alpha)$, so it is Π_2. φ says that for some E on $\dot{E}^M$, $E_{\{\kappa\}} = \dot{F}_{\{\kappa\}}$. Because all $E \in \dot{E}^M$ have the weak ms-ISC, this is equivalent to the Jensen completion of $\dot{F}_{\{\kappa\}}$ being on $\dot{E}^M$.[6] Clearly, φ is Σ_2 in the language of potential premice. ⊣

Proposition 4.1.5 shows that the weak ms-ISC for $\dot{F}$ is preserved under Σ_1 ultrapowers and Σ_2 hulls. As we saw above, it is not preserved under Σ_0 ultrapowers. The requirement on potential premice is that all $E \in \dot{E}^M$ satisfy the weak ms-ISC, and the proposition yields a Π_1 sentence capturing it. So like the other clauses in potential premouse-hood, this one is preserved under Σ_0 ultrapowers and Σ_1 hulls.

One can define projecta and cores using either the $r\Sigma_n$ hierarchy, or iterated Σ_1 definability over coding structures. Both (closely related!) points of view are useful. Our official definition here will use iterated Σ_1 definability. Let us recall some terminology from [54]: for any acceptable J-structure (N,B)[7]

$$\rho_1(N,B) = \text{least } \alpha \text{ s.t. } \exists A \subset \alpha(A \in \Sigma_1^{(N,B)} \wedge A \notin N),$$

[6]One can think of φ as asserting that there is a *generalized* weak ms-solidity witness for $\dot{F}$.

[7]It is enough for us to consider the case that N is a potential premouse and B is amenable to N.

$$\begin{aligned} p_1(N,B) &= \text{first standard parameter of } (N,B) \\ &= \text{lex-least descending sequence of ordinals } r \text{ such that} \\ &\quad \exists A \subseteq \rho_1(N,B)(A \notin N \wedge A \text{ is } \Sigma_1^{(N,B)} \text{ definable from } r.), \\ h^1_{(N,B)} &= \text{canonical } \Sigma_1 \text{ Skolem function of } (N,B). \end{aligned}$$

We allow $\rho_1(N,B) = o(N)$ and $p_1(N,B) = \emptyset$. We define solidity and universality by

DEFINITION 4.1.6. Let $M = (N,B)$ be an acceptable J-structure, and

$$r \in [o(N)]^{<\omega};$$

then

$$W_M^{\alpha,r} = \mathrm{cHull}_1^M(\alpha \cup r \setminus (\alpha+1)).$$

We call $W_M^{\alpha,r}$ the *standard solidity witness* for r at α. We say r is *solid over M* iff all its standard solidity witnesses belong to M.

DEFINITION 4.1.7. Let $M = (N,B)$ be an acceptable J-structure, and

$$r \in [o(N)]^{<\omega}.$$

We say that r is *universal* over M if for $\rho = \rho_1(M)$ and $W = W_M^{\rho,r}$,

(a) $M|\rho^{+,M} = W|\rho^{+,W}$, and
(b) for any $A \subseteq \rho$, A is boldface Σ_1^M iff A is boldface Σ_1^W.

It is easy to see that there is at most one parameter $r \in [o(M)]^{<\omega}$ that is both solid and universal over M.[8]

Now let M be a potential pfs premouse; we define its projecta $\rho_k = \rho_k(M)$, its cores $\mathfrak{C}_k = \mathfrak{C}_k(M)$, its standard parameters $p_k = p_k(M)$, and a parameter $w_k = w_k(M)$ that codes objects associated to p_k and ρ_k.[9] Simultaneously we define k-solidity and k-soundness for M, along with the k-th *strong core* $\bar{\mathfrak{C}}_k = \bar{\mathfrak{C}}_k(M)$ of M. M is k-solid iff $\mathfrak{C}_k(M)$ exists and is well behaved in certain ways, and M is k-sound iff M is k-solid and $M = \mathfrak{C}_k(M)$. The strong core is like the one we took in ordinary premice.

As we go, we also define the reducts

$$M^k = (M||\rho_k, A^k),$$

and surjections $d^k \colon M||\rho_k \to \mathfrak{C}_k$ that decode M^k into $\mathfrak{C}_k$.

We start with

$$\rho_0 = o(M),$$

[8]Suppose r and s were distinct such parameters, and let α be largest in $r \triangle s$. Suppose $\alpha \in r$; then for $\rho = \rho_1(M)$, one can compute $W^{\rho,s}$ from $W^{\alpha,r}$, so $W^{\rho,s} \in M$, contrary to the universality of s.

[9]To be pedantic, one should at this point distinguish potential pfs premice from potential Jensen premice with a label of some sort, because $\mathfrak{C}_k(M)$ etc. have already been defined for potential Jensen premice. We shall just let context make the distinction.

$$\bar{\mathfrak{C}}_0 = \mathfrak{C}_0 = M,$$
$$p_0 = w_0 = \emptyset,$$
$$A^0 = \emptyset,$$

and we say that M is 0-sound and 0-solid.

Now let

$$\rho_1 = \rho_1(M^0) = \rho_1(M),$$
$$p_1 = p_1(M^0) = p_1(M),$$
$$\bar{\mathfrak{C}}_1 = \mathrm{cHull}_1^M(\rho_1 \cup \{p_1\}),$$

and

$$\mathfrak{C}_1 = \mathrm{cHull}_1^M(\rho_1 \cup \{p_1, \rho_1\}).$$

Let $\sigma : \bar{\mathfrak{C}}_1 \to M$ and $\pi : \mathfrak{C}_1 \to M$ be the anticollapse maps, and $\bar{p}_1 = \sigma^{-1}(p_1)$. We say that M^0 is *parameter solid* if p_1 is solid and universal over M and $\bar{p}_1$ is solid and universal over $\bar{\mathfrak{C}}_1$. We say that M^0 is *projectum solid* iff ρ_1 is not measurable by the M-sequence, and either

$$\mathfrak{C}_1 = \bar{\mathfrak{C}}_1,$$

or

$$\mathfrak{C}_1 = \mathrm{Ult}_0(\bar{\mathfrak{C}}_1, D)$$

where D is the order zero measure of $\bar{\mathfrak{C}}_1$ on ρ_1, and

$$\sigma = \pi \circ i_D.$$

We say that M is *weakly ms-solid* iff either M is passive, or the last extenders of $\mathfrak{C}_1$ and $\bar{\mathfrak{C}}_1$ satisfy the weak ms-ISC.

We say that M is *1-solid* iff M^0 is parameter solid, projectum solid, and weakly ms-solid. If M is not 1-solid, we stop our inductive definition. We say that M is 1-sound iff M is 1-solid and $M = \mathfrak{C}_1(M)$.

Let $\tau = \pi^{-1} \circ \sigma$, so that either τ is the identity, or $\tau = i_D$ for D the order zero measure of $\bar{\mathfrak{C}}_1$ on ρ_1. Using the elementarity of τ, we see that $\tau(\bar{p}_1) = p_1(\mathfrak{C}_1)$, and hence $p_1(\mathfrak{C}_1)$ is solid and universal over $\mathfrak{C}_1$.[10] Since $\pi(\rho_1) = \rho_1$, ρ_1 is not measurable by the $\mathfrak{C}_1$-sequence. Thus if M is 1-solid, then $\mathfrak{C}_1(M)$ is 1-sound.

Remark 4.1.8. We shall show that if M is reached in a PFS construction, then granted iterability, M is solid. The proof of projectum solidity is essentially the same as that of Theorem 3.7.1.

For any N, 2-solidity and $\mathfrak{C}_2(N)$ will be defined by looking at definability over $M = \mathfrak{C}_1(N)$, which is 1-sound. So let us assume now that M is 1-sound. For φ a Σ_1 formula and $x \in M||\rho_1$, let

$$d^1(<\varphi, x>) = h_M^1(\varphi, \langle x, p_1, \rho_1 \rangle).$$

[10] If b is a solidity witness for $\bar{p}_1$, then $\tau(b)$ is a generalized solidity witness for p_1. See Lemma 4.3.6.

d^1 is a partial map of $M||\rho_1$ onto M that is Σ_1^M in the parameters p_1 and ρ_1.[11] Let

$$w_1 = \langle \eta_1^M, \rho_1, p_1 \rangle,$$
$$A^1 = \{\langle \varphi, b \rangle \mid \varphi \text{ is } \Sigma_1 \wedge b \in M||\rho_1 \wedge M \models \varphi[b, w_1]\},$$

and

$$M^1 = (M||\rho_1, A^1).$$

M^1 is amenable, and codes the whole of M by soundness and the Σ_1 definability of d^1.[12] We let

$$\rho_2(M) = \rho_1(M^1),$$
$$p_2(M) = p_1(M^1),$$
$$\bar{\mathfrak{C}}_2(M) = \text{transitive collapse of } d^1 \circ h^1_{M^1}\text{``}(\rho_2 \cup \{p_2\}), \text{ and}$$
$$\mathfrak{C}_2(M) = \text{transitive collapse of } d^1 \circ h^1_{M^1}\text{``}(\rho_2 \cup \{p_2, \rho_2\}).$$

Let $\sigma\colon \bar{\mathfrak{C}}_2 \to M$ and $\pi\colon \mathfrak{C}_2 \to M$ be the anticollapse maps, and $\bar{p}_2 = \sigma^{-1}(p_2)$; then

(a) M^1 is *parameter solid* iff p_2 is solid and universal over M^1 and $\bar{p}_2$ is solid and universal over the reduct $(\bar{\mathfrak{C}}_2)^1$ of $\bar{\mathfrak{C}}_2$.

(b) M^1 is *projectum solid* iff ρ_2 is not measurable by the M-sequence, and either $\bar{\mathfrak{C}}_2 = \mathfrak{C}_2$, or $\mathfrak{C}_2 = \text{Ult}(\bar{\mathfrak{C}}_2, D)$ and $\sigma = \pi \circ i_D$, for i_D the order zero measure of $\bar{\mathfrak{C}}_2$ on ρ_2.

(c) M^1 is *stable* iff either $\eta_1^M < \rho_2$, or η_1^M is not measurable by the M-sequence.

We say that M is *2-solid* iff M^1 is parameter solid, projectum solid, and stable.[13] If M is not 2-solid then we stop the inductive definition, and otherwise we continue. We say that M is 2-sound iff M is 2-solid and $M = \mathfrak{C}_2(M)$.

We shall show in Lemma 4.3.6 below that $\mathfrak{C}_2(M)$ is 2-sound, and letting $\pi\colon \mathfrak{C}_2(M) \to M$ be the anticollapse map,

$$\pi(w_1(\mathfrak{C}_2(M))) = w_1(M) \quad \text{and} \quad \pi(p_2(\mathfrak{C}_2(M))) = p_2(M).$$

Remark 4.1.9. Even when $\bar{\mathfrak{C}}_1(M) = \mathfrak{C}_1(M)$, $\bar{\mathfrak{C}}_2$ is not the usual second core of M described in Section 2.3, because w_1 codes ρ_1 and η_1, and M^1 has a name for w_1. The usual second core need not contain η_1 or ρ_1. Including names for ρ_1 and η_1 also affects what is meant by $p_2(M)$. When $\bar{\mathfrak{C}}_1(M) = \mathfrak{C}_1(M)$, $\mathfrak{C}_2$ can be as many as two ultrapowers away from the usual second core H, by order zero measures on ρ_2 and η_1^H. See Section 3.8. The usual second core will appear in our proof of parameter solidity.

---

[11] Our definition of d^1 ignores the case $\rho_1 = o(M)$. In that case, ρ_1 should be dropped from the right hand side. We shall ignore similar special cases in some of the formulae below.

[12] [35] would include the solidity witnesses for p_1 in w_1, but this is redundant. See Remark 2.3.13.

[13] If M is active, then since it is 1-sound, $\dot{F}^M$ has the weak ms-ISC. This passes automatically to $\mathfrak{C}_2$ and $\bar{\mathfrak{C}}_2$ by Proposition 4.1.5, so we do not need to make it part of the definition of 2-solidity.

The general inductive step is the same. Suppose M is k-sound. For φ a Σ_1 formula and $x \in M||\rho_k$, let

$$d^k(<\varphi, x>) = d^{k-1} \circ h^1_{M^{k-1}}(\varphi, \langle x, p_k, \rho_k \rangle),$$

so that d^k is a partial map of $M||\rho_k$ onto M. Let

$$\begin{aligned} w_k &= \langle \eta_k^M, \rho_k, p_k \rangle, \\ A^k &= \{\langle \varphi, b \rangle \mid \varphi \text{ is } \Sigma_1 \wedge b \in M||\rho_k \wedge M^{k-1} \models \varphi[b, w_k]\}, \end{aligned}$$

and

$$M^k = (M||\rho_k, A^k).$$

Then[14]

$$\begin{aligned} \rho_{k+1} &= \rho_1(M^k), \\ p_{k+1} &= p_1(M^k), \\ \bar{\mathfrak{C}}_{k+1} &= \text{transitive collapse of } d^k \circ h^1_{M^k}``(\rho_{k+1} \cup p_{k+1}), \\ \bar{p}_{k+1} &= \sigma^{-1}(p_{k+1}), \end{aligned}$$

and

$$\mathfrak{C}_{k+1} = \text{transitive collapse of } d^k \circ h^1_{M^k}``(\rho_{k+1} \cup \{p_{k+1}, \rho_{k+1}\}).$$

Here $\sigma\colon \bar{\mathfrak{C}}_{k+1} \to M$ is the anticollapse map.[15] Let $\pi\colon \mathfrak{C}_{k+1} \to M$ be the anticollapse. Soundness and solidity at $k+1$ are defined by

DEFINITION 4.1.10. Let M be a k-sound potential pfs premouse; then

(a) M^k is *parameter solid* iff p_{k+1} and $\bar{p}_{k+1}$ are solid and universal over M^k and $(\bar{\mathfrak{C}}_{k+1})^k$ respectively,

(b) M^k is *projectum solid* iff ρ_{k+1} is not measurable by the M-sequence, and either
   (i) $\bar{\mathfrak{C}}_{k+1} = \mathfrak{C}_{k+1}$, or
   (ii) $\mathfrak{C}_{k+1} = \text{Ult}_k(\bar{\mathfrak{C}}_{k+1}, D)$, where D is the order zero measure of $\bar{\mathfrak{C}}_{k+1}$ on ρ_{k+1}, and $\sigma = \pi \circ i_D$, and

(c) M^k is *stable* iff either $\eta_k^M < \rho_{k+1}$, or η_k^M is not measurable by the M-sequence.

We say that M is $k+1$-*solid* iff M^0 is weakly ms-solid, and M^k is parameter solid, projectum solid, and stable. (Stability holds trivially if $k(M) = 0$.) We say that M is $k+1$-*sound* iff M is $k+1$-solid and $M = \mathfrak{C}_{k+1}(M)$. We say that M is ω-*sound* iff M is k-sound for all $k < \omega$.

---

[14]If $\rho_k = \rho_{k-1}$, then one should omit the constant symbol for ρ_k from A^k. Similarly, if $\eta_k^M = \rho_{k-1}$ then there is no constant for η_k^M in A^k.

[15]Notice that $w_k \in \text{ran}(\sigma)$.

By Lemma 4.3.6 below, if M is $k-1$-sound and k-solid, then $\mathfrak{C}_k(M)$ is k-sound, so the definitions above apply to it. We define

$$\mathfrak{C}_{k+1}(M) = \mathfrak{C}_{k+1}(\mathfrak{C}_k(M)),$$
$$\bar{\mathfrak{C}}_{k+1}(M) = \bar{\mathfrak{C}}_{k+1}(\mathfrak{C}_k(M)),$$

and so on. This lets us define $\mathfrak{C}_k(M)$ for all k, even if M is not 1-sound.[16]

DEFINITION 4.1.11. A *pfs premouse of type 1* is a pair $M = (\hat{M}, k)$ such that $\hat{M}$ is a potential pfs premouse, and

(a) $\hat{M}$ is k-sound,
(b) whenever P is an initial segment of $\hat{M}$ such that $o(P) < o(\hat{M})$, then P is an ω-sound potential pfs premouse.

We write $k(M) = k$, and say that $\hat{M}$ is the *bare premouse* associated to M, and identify $\hat{M}$ with M when context permits.

If M is an active premouse and $k(M) \geq 1$, then by Proposition 4.1.5, $\dot{F}^M$ has the weak ms-ISC.

All levels of the model we construct in Section 4.7 will be type 1 pfs premice. However, ultrapowers can produce a second, less important type, as we shall discuss in the next section.

The notations $M|\langle \nu, k\rangle$, $M|\nu$, $M||\nu$ for initial segments of ordinary premice apply to pfs premice as well. So does our notation for degree changes:

DEFINITION 4.1.12. Let $M = (\hat{M}, k)$ be a pfs premouse; then $M^+ = (\hat{M}, k+1)$, $M^- = (\hat{M}, k-1)$, and $M \downarrow n = (\hat{M}, n)$. (Here $\omega + 1 = \omega - 1 = \omega$, and $0 - 1 = 0$.)

Our k-free conventions also apply:

DEFINITION 4.1.13. Let M be a pfs premouse and $k = k(M)$; then

$$\rho(M) = \rho_{k+1}(M),$$
$$p(M) = p_{k+1}(M),$$
$$\mathfrak{C}(M) = \mathfrak{C}_{k+1}(M), \text{ and}$$
$$\bar{\mathfrak{C}}(M) = \bar{\mathfrak{C}}_{k+1}(M)$$

are *the* projectum, standard parameter, core, and strong core of M. $k(\mathfrak{C}(M)) = k(\bar{\mathfrak{C}}(M)) = k+1$. We let $\rho^-(M) = \rho_k(M)$. We say that M is *solid* iff M is $k+1$-solid, M is *sound* iff M is $k+1$-sound. Similarly, we say that M is *parameter solid* (respectively *projectum solid, stable*) iff M^k is parameter solid (respectively projectum solid, stable).

[16] It is important, however, that if M is not k-sound, then Σ_{k+1} definability over M itself plays no role in the definition of $\mathfrak{C}_{k+1}(M)$.

So M is solid iff M is stable, parameter solid, and projectum solid, and if $k(M) = 0$, then M is weakly ms-solid. Our definitions are such that $\bar{\mathfrak{C}}(M)$ and $\mathfrak{C}(M)$ may exist even though M is not solid. This is not our case of interest, but it is convenient when we are proving solidity.

If M is a pfs premouse, then its projecta are not measurable in M. In fact, we have

LEMMA 4.1.14. *Let M be a pfs premouse; then whenever E on the M sequence and is not total on M, then* $\text{crit}(E)$ *is not a cardinal of M.*

PROOF. Let E be an extender on the M sequence that is not total on M. We then have N such that $M|\text{lh}(E) \trianglelefteq N \triangleleft M||o(M)$ and $\rho(N) \leq \text{crit}(E)$. Since N^+ is a pfs premouse, $\rho(N) < \text{crit}(E)$. This implies that $\text{crit}(E)$ is not a cardinal of M, as desired. ⊣

We can go on to define the class of $r\Sigma_k$ relations by

DEFINITION 4.1.15. Let M be a pfs premouse, $k = k(M)$, and R be a relation on M. Let d^k be the decoding function defined above, and let R^k be the relation on M^k given by

$$R^k(x_1, \ldots, x_n) \iff R(d^k(x_1), \ldots, d^k(x_n)).$$

Then R is $r\Sigma^M_{k+1}$ iff R^k is $\Sigma^{M^k}_1$.

A^{k+1}_M is essentially the $r\Sigma^M_{k+1}$ theory of parameters in

$$\rho_{k+1}(M) \cup \{p_i(M), \rho_i(M), \eta_i(M) \mid i \leq k+1\}.$$

The class of $r\Sigma^M_{k+1}$ relations has various closure and structural properties that help to calculate definability over premice. These are laid out in Section 2.3 for ordinary premice, and those results all go over to the context of pfs premice. The main difference is that now $r\Sigma_{k+1}$ definitions are allowed to use names for η_k and ρ_k.

In particular, suppose $k = k(M)$; then d^k is $r\Sigma^M_k$ in the parameters p_k, ρ_k, and η_k. It has an inverse that is also $r\Sigma^M_k$ in these parameters, given by

$$e^k(y) = x \iff (d^k(x) = y \wedge \forall w <_M x(d^k(w) \neq d^k(x))).$$

We can use e^k to produce an $r\Sigma_{k+1}$ Skolem function. If $\varphi(u,v)$ is a Σ_1 formula in the language of M^k, let $\varphi^*(u,v)$ be the natural Σ_1 formula expressing "$\exists x \exists y(d^k(x) = d^k(u) \wedge d^k(y) = d^k(v) \wedge \varphi(x,y))$". The $r\Sigma^M_{k+1}$ relations are naturally indexed by the Σ_1 formulae of the form φ^*. We set

$$h^{k+1}_M(\varphi^*, x) = d^k(h^1_{M^k}(\varphi^*, e^k(x))),$$

and call h^{k+1}_M is the *canonical* $r\Sigma^M_{k+1}$ *Skolem function* for M. For $X \subseteq M$, the associated Skolem hull is

$$\text{Hull}^M_{k+1}(X) = \{h^{k+1}(\varphi, s) \mid s \in X^{<\omega} \wedge \varphi \in V_\omega\},$$

and

$$\text{cHull}^M_{k+1}(X) = \text{transitive collapse of Hull}^M_{k+1}(X).$$

$\text{Hull}^M_{k+1}(X)$ is closed under (lightface) $r\Sigma_{k+1}$ functions, and in particular, $p_k(M)$, $\rho_k(M)$, and η^M_k are all in $\text{Hull}^M_{k+1}(X)$, and $\text{Hull}^M_{k+1}(X)$ is closed under the coding and decoding functions e^k and d^k.

We can then characterize the core and strong core of M, where M has type 1 and $k = k(M)$, by

$$\bar{\mathfrak{C}}_{k+1}(M) = \text{cHull}^M_{k+1}(\rho_{k+1}(M) \cup p_{k+1}(M))$$

and

$$\mathfrak{C}_{k+1}(M) = \text{cHull}^M_{k+1}(\rho_{k+1}(M) \cup \{p_{k+1}(M), \rho_{k+1}(M)\}).$$

$w_k(M) = \langle \eta_k(M), \rho_k(M), p_k(M) \rangle$ belongs to both hulls.

4.2. Other soundness patterns

The levels of our model will all be pfs premice of type 1, but ultrapowers of limited elementarity can produce a second type. Here "type" refers to soundness pattern, not to the structure as a bare premouse. For the most part we can avoid bare premice with this unusual soundness pattern, but it smooths some definitions if we call them, when paired with their degree k, pfs premice of type 2.[17]

To see how type 2 premice arise, suppose M is a pfs premouse of type 1 and $k(M) = 1$. Suppose that E is an extender over M such that $\text{crit}(E) = \eta^M_1 < \rho_1(M)$, and let

$$Q = \text{Ult}_1(M, E).$$

We can produce Q by decoding $\text{Ult}_0(M^1, E)$, or more directly as the set of all $[a, f]^M_E$ such that f is Σ^M_1 in some parameter.[18] Then

$$\rho_1(Q) = \sup i_E``\rho_1(M) < i_E(\rho_1(M)).$$

For the moment, let us regard Q as a bare premouse. How should we define $k(Q)$? If $M \neq \bar{\mathfrak{C}}_1(M)$, that is, $\rho_1(M) \notin \text{Hull}^M_1(\rho_1(M) \cup \{p_1(M)\})$, then $i_E(\rho_1(M)) \notin \text{Hull}^Q_1(\rho_1(Q) \cup \{\rho_1(Q), p_1(Q)\})$, so Q is not 1-sound. If we set $k(Q) = 1$, then we do not get a pfs premouse of type 1.

Ultrapowers like this show up in the proofs of solidity and universality. In that context, we can replace Q by $\bar{Q} = \text{Ult}_1(\bar{\mathfrak{C}}_1(M), E)$, which becomes a pfs premouse of type 1 when we set $k(\bar{Q}) = 1$. So far as we know, this replacement can be done without loss whenever ultrapowers like Q appear. That is what we shall do in practice. Nevertheless, some definitions work better if we set $k(Q) = 1$, and call

[17]Type 2 premice can also be produced by Skolem hulls of limited elementarity, such as those that show up in the proof of $\square_\kappa$, or in the full normalization argument sketched in Claim 6.1.8.

[18]The discussion after Definition 2.4.4 explains why the two versions of $\text{Ult}_1(M, E)$ are isomorphic.

$(Q,1)$ a pfs premouse of type 2. Q is is almost 1-sound, in that it is Σ_1 generated by one additional point, and Σ_1 ultrapowers will preserve this.

DEFINITION 4.2.1. Let N be a pfs premouse of type 1; then

(a) N *strongly sound* iff N is sound and $N = \bar{\mathfrak{C}}(N)^-$,

(b) N has *type 1A* iff $k(N) = 0$, or $k(N) > 0$ and N^- is strongly sound. Otherwise N has *type 1B*.

Thus N is strongly sound iff it is its own strong core, up to the degree change. Letting $k = k(N)$, N has type 1A iff $\rho_k(N) = \rho_{k-1}(N)$ or $\rho_k(N) \in \mathrm{Hull}_1^{N^{k-1}}(\rho_k(N) \cup p_k(N))$.

DEFINITION 4.2.2. Let (M,A) be an acceptable J-structure. If

$$\mathrm{Hull}_1^{(M,A)}(\rho_1(M,A) \cup p_1(M,A)) = M,$$

then we set $\hat{\rho}_1(M,A)) = 0 = \hat{\eta}_1(M,A)$. Otherwise, let

$$\hat{\rho}_1(M,A) = \text{least } \xi \text{ such that } \xi \notin \mathrm{Hull}_1^{(M,A)}(\rho_1(M,A) \cup p_1(M,A)),$$

and

$$\hat{\eta}_1(M,A) = \mathrm{cof}_1^{(M,A)}(\hat{\rho}_1(M,A)).$$

DEFINITION 4.2.3. Let N be a pfs premouse of type 1; then

$$\hat{\rho}(N) = \hat{\rho}_1(N^{k(N)})$$

and

$$\hat{\eta}(N) = \hat{\eta}_1(N^{k(N)}).$$

If $k = k(N)$, then we shall also write $\hat{\rho}_{k+1}(N) = \hat{\rho}_1(N^k)$ and $\hat{\eta}_{k+1}(N) = \hat{\eta}_1(N^k)$.

DEFINITION 4.2.4. Let N be a pfs premouse of type 1 and $k(N) = k < \omega$; then N is *almost sound* iff

(a) N is solid,

(b) $N^k = \mathrm{Hull}_1^{N^k}(\rho_1(N^k) \cup \{p_1(N_0^k), \hat{\rho}_1(N^k)\})$,

(c) if $\rho_1(N^k) \leq \hat{\rho}_1(N^k)$, then letting

$$(H,B) = \mathrm{cHull}_1^{N^k}(\rho_1(N^k) \cup p_1(N_0^k)),$$

with anticollapse map $\pi\colon (H,B) \to N^k$, we have

$$N^k = \mathrm{Ult}((H,B),D),$$

where D is the order zero measure of H on $\hat{\rho}_1(N^k)$, and $\pi = i_D$, and

(d) if $\rho_1(N^k) < \hat{\rho}_1(N^k)$, then $\hat{\eta}_1(N^k) < \rho_1(N^k)$.

PROPOSITION 4.2.5. *Let N be a pfs premouse of type 1 and $k(N) = m < \omega$; then the following are equivalent:*

(1) N is sound,
(2) N is almost sound and $\hat{\rho}(N) \leq \rho(N)$.

PROOF. This is follows at once from the definitions. ⊣

DEFINITION 4.2.6. N is a *pfs premouse* iff N is a potential pfs premouse and $k(N) = 0$, or $k(N) > 0$ and N^- is an almost sound pfs premouse of type 1. We let $\rho^-(N) = \rho(N^-) = \rho_{k(N)}(N)$ and $\hat{\rho}^-(N) = \hat{\rho}(N^-) = \hat{\rho}_{k(N)}(N)$.

DEFINITION 4.2.7. Let M be a pfs premouse; then M has *type 2* iff M^- is not sound.

Let $k = k(N)$. Our definitions are such that N has type 1A iff $\hat{\rho}_k(N) = 0$, type 1B iff $\hat{\rho}_k = \rho_k(N)$, and type 2 iff $\hat{\rho}_k(N) > \rho_k(N)$. All proper initial segments of a pfs premouse have type 1. Clearly, if $k(N) = 0$ then N has type 1. Here are two more simple consequences of the definitions:

PROPOSITION 4.2.8. *Let N be a pfs premouse of type 2; then*

(a) $\bar{\mathfrak{C}}(N^-) = \mathfrak{C}(N^-)$, *and*
(b) $N = \mathrm{Ult}_{k-1}(\bar{\mathfrak{C}}(N^-), D)^+$, *where D is the order zero measure of* $\bar{\mathfrak{C}}(N^-)$ *on* $\hat{\rho}_k(N)$.

PROOF. Let $k = k(N)$. Since $\rho_k(N) < \hat{\rho}_k(N)$,

$$\rho_k(N) \in \mathrm{Hull}_1^{N^{k-1}}(\rho_k(N) \cup \{p_k(N)\})$$

by the definition of $\hat{\rho}_k(N)$. This implies that the core and strong core of N^- coincide, so we have (a).

Part (b) follows from clause (c) in Definition 4.2.4. ⊣

PROPOSITION 4.2.9. *Let N be a pfs premouse and* $k = k(N) > 0$. *Suppose that* $\rho_k(N) \leq \hat{\rho}_k(N)$, *and let* $A \subseteq \hat{\rho}_k(N)$ *be* $\Sigma_1^{N^{k-1}}$ *in parameters; then* $A = B \cap \hat{\rho}_k(N)$, *where B is* $\Sigma_1^{N^{k-1}}$ *in parameters from* $\rho_k(N) \cup p_k(N)$.

PROOF. By soundness, A is $\Sigma_1^{N^{k-1}}$ in parameters from $\rho_k(N) \cup \{\hat{\rho}_k(N), p_k(N)\}$. Since N is almost sound, $N^{k-1} = \mathrm{Ult}(R^{k-1}, D)$, where R^+ is the strong core of N, and $\mathrm{crit}(D) = \hat{\rho}_k(N)$. It is enough to show there is a B such that $B \cap \hat{\rho}_k(N) = A$ and B is $\Sigma_1^{N^{k-1}}$ in parameters from $\mathrm{ran}(i_D)$. But let

$$\xi \in A \iff N^{k-1} \models \exists v \theta[v, \xi, \hat{\rho}_k(N), p_k(N), \alpha]$$

where $\alpha < \rho_1(N^{k-1})$ and θ is Σ_0. By Łoś's Theorem,

$$\xi \in A \iff \exists g \exists X \in D \forall u \in X R^{k-1} \models \theta[g(u), \xi, u, p(R^{k-1})].$$

So for $\xi < \hat{\rho}_k(N)$,

$$\xi \in A \iff N^{k-1} \models \exists g \exists X \in i(D) \forall u \in X \theta[g(u), \xi, u, p(N^{k-1})].$$

Letting B be the set of ξ such that the right hand side holds, B is $\Sigma_1^{N^{k-1}}$ in parameters in $\{i(D)\} \cup \rho_k(N) \cup p_k(N)$, hence in parameters from $\operatorname{ran}(i_D)$, and $B \cap \hat{\rho}_k(N) = A$. ⊣

Let define another set of coding structures.

DEFINITION 4.2.10. Let M be a pfs premouse and $k = k(M) > 0$; then

$$\hat{w}_k(M) = \langle \hat{\eta}_k^M, \hat{\rho}_k(M), p_k(M) \rangle,$$

$$\hat{A}_M^k = \{ \langle \varphi, b \rangle \mid \varphi \text{ is } \Sigma_1 \wedge b \in M || \rho_k(M) \wedge M^{k-1} \models \varphi[b, \hat{w}_k(M)] \},$$

and

$$\hat{M}^k = (M || \rho_k, \hat{A}^k).$$

For $k \geq 1$, $\hat{M}^k$ is decoded by the function $\hat{d}^k = \hat{d}_M^k$, where for φ a Σ_1 formula and $x \in M || \rho_k(M)$,

$$\hat{d}^k(\langle \varphi, x \rangle) = d^{k-1} \circ h^1_{M^{k-1}}(\varphi, \langle x, p_k, \hat{\rho}_k(M) \rangle).$$

Thus $\hat{d}^k$ is a partial map of $M || \rho_k$ onto M.[19]

The notations $M | \langle \nu, k \rangle$, $M | \nu$, $M || \nu$ for initial segments of type 1 premice apply to type 2 premice as well. So does our notation for degree changes, our k-free notation for projecta, cores, and parameters, and so on.

4.3. Elementarity for premouse embeddings

Let us define elementarity and near elementarity for maps on pfs premice. As before, anticore and ultrapower maps are elementary, while the lifting maps in a conversion system may be only nearly elementary.

DEFINITION 4.3.1. Let M and N be pfs premice, and $k = k(M) = k(N)$. Let $\pi \colon \hat{M}^k \to \hat{N}^k$ be Σ_0 elementary, and let $\sigma \colon M \to N$ be given by

$$\sigma(\hat{d}_M^k(x)) = \hat{d}_N^k(\pi(x))$$

for all $x \in \hat{M}^k$; then we call σ the *completion* of π.

Notice that "$\hat{d}_M^k(x) = \hat{d}_M^k(y)$" is decided by the Σ_0 theory of $\hat{M}^k$, so the completion of π is well defined. Equivalently σ is the unique map extending π such that

$$\sigma(h_1^{M^{k-1}}(x, p_k(M), \hat{w}_k(M))) = h_1^{N^{k-1}}(\pi(x), p_k(N), \hat{w}_k(N))),$$

and for all $i < k$ such that $i > 0$,

$$\sigma(h_1^{M^{i-1}}(x, p_i(M), w_i(M))) = h_1^{N^{i-1}}(\pi(x), p_i(N), w_i(N)).$$

[19]If M is of type 1B, $\hat{M}^k = M^k$, so $\hat{d}_M^k = d_M^k$. If M is of type 1A, then $\hat{A}_M^k$ and A_M^k are simply interdefinable, but not equal. We used M^k in forming the core $\mathfrak{C}_{k+1}(M)$. We shall use $\hat{M}^k$ when keeping track of elementarity for the ultrapower $\operatorname{Ult}_k(M, E)$.

It is easy to see that $\sigma \restriction M^{k-1} : M^{k-1} \to N^{k-1}$ is Σ_1 elementary, and σ is also its completion.[20] The full σ is thus Σ_k elementary.

Suppose that $\sigma : M \to N$ is the completion of some $\pi : \hat{M}^k \to \hat{N}^k$. Clearly $\sigma(w_i(M)) = w_i(N)$ for all $i < k$, and $\sigma(\hat{w}_k(M)) = \hat{w}_k(N)$. Ultrapower maps that are discontinuous at $\rho_k(M)$ show that $\sigma(w_k(M)) \neq w_k(N)$ is possible. Some other simple observations:

(i) If M has type 1A, then M^- is strongly sound and $\hat{w}_k(M) = \langle 0, 0, p_k(M) \rangle$. This implies N has type 1A. In this case, σ may or may not preserve w_k.

(ii) If M has type 1B, that is, $\rho_k(M) = \hat{\rho}_k(M)$, then N has type 1B or type 2, both being possible. In this case $w_k(M) = \hat{w}_k(M)$, and N has type 1B iff $w_k(N) = \hat{w}_k(N)$ iff $\sigma(w_k(M)) = w_k(N)$.

DEFINITION 4.3.2. Let M and N be pfs premice such that $k = k(M) = k(N)$, and let $\pi : M \to N$; then

(a) π is *nearly elementary* iff $\pi \restriction \hat{M}^k$ is a Σ_0 elementary and cardinal preserving map from $\hat{M}^k$ to $\hat{N}^k$, and π is its completion.

(b) π is *elementary* iff π is nearly elementary, and $\pi \restriction \hat{M}^k$ is Σ_1 elementary as a map from $\hat{M}^k$ to $\hat{N}^k$.

(c) π is *cofinal* iff $\pi``\rho_k(M)$ is cofinal in $\rho_k(N)$.

(d) π is *almost exact* iff $\rho_k(N) \leq \pi(\rho_k(M))$.

(e) π is *exact* iff $w_k(N) = \pi(w_k(M))$.

Of course, elementarity is really a property of (π, M, N), not just π. π may be elementary as a map from M^- to N^-, but not as a map from M to N. When M and N are not clear from context, we shall specify them.

Clearly if π is either exact or cofinal, then π is almost exact. The proof of Lemma 2.4.7 shows that if π is elementary, then π is almost exact. Almost exact embeddings preserve type 2.

PROPOSITION 4.3.3. *Suppose M has type 2, and $\sigma : M \to N$ is nearly elementary and almost exact; then N has type 2.*

PROOF. Letting $k = k(M)$, we have $\rho_k(M) < \hat{\rho}_k(M)$. But then

$$\rho_k(N) \leq \sigma(\rho_k(N)) < \sigma(\hat{\rho}_k(M)) = \hat{\rho}_k(N),$$

as desired. ⊣

The lifting maps of a conversion system will be nearly elementary maps whose target models always have type 1, but whose domain models may have type 2.

Exactness requires that both η_k and ρ_k be preserved, but in practice, preservation of one implies preservation of the other. The elementarity hypothesis in the following lemma will come up very often as we proceed.

---

[20] M^{k-1} must have type 1, so $\hat{M}^{k-1}$ and M^{k-1} are simply interdefinable.

LEMMA 4.3.4. *Let* $\pi\colon M \to N$ *be the completion of* $\pi \restriction \hat{M}^k$. *Suppose that as a map from* $\hat{M}^k$ *to* $\hat{N}^k$, π *is either* Σ_2 *elementary, or cofinal and* Σ_1 *elementary; then equivalent are*

(a) $\pi(\rho_k(M)) = \rho_k(N)$,
(b) $\pi(\eta_k^M) = \eta_k^N$,
(c) π *is exact.*

PROOF. If $\pi \restriction \hat{M}^k$ is cofinal and Σ_1 elementary, then the proof is essentially identical to the proof of Lemma 2.5.10. (Notice that in this case π is continuous at $\rho_1(Q)$.) If π is Σ_2 elementary, then the proof of Lemma 2.5.12 applies. We shall not go through the definability calculations in those proofs again here. They do need our stronger elementarity hypothesis on $\pi \restriction \hat{M}^k$. ⊣

Given pfs premice M and N of degree k, and a Σ_0 elementary, cardinal preserving map $\pi\colon \hat{M}^k \to \hat{N}^k$, there is a unique nearly elementary $\sigma\colon M \to N$ that completes σ. But we may only be given π and one of the two premice, and want to construct the other. If we start with N, we are taking a hull to get M. If we start with M, we are taking an ultrapower to produce N. The upward and downward extension lemmas describe the basic facts about these constructions.

Downward extension and anticore maps. Let (P,B) be Σ_0-elementarily equivalent to $\hat{M}^k$, where M is a pfs premouse. Can we conclude that $(P,B) = \hat{N}^k$ for some pfs premouse N? The predicate of $\hat{M}^k$ codes M^{k-1} because $M^{k-1} = \mathrm{Hull}_1^{M^{k-1}}(\rho_k(M) \cup \{\hat{w}_k(M)\})$, so

$$M^{k-1} = \mathrm{Dec}(\hat{M}^k),$$

where Dec stands for a certain simple decoding procedure whose details can be found in [54].[21] $\hat{A}^k_M$ codes enough about this procedure that we can apply the procedure to (P,B), and let

$$(Q,C) = \mathrm{Dec}((P,B)).$$

B is a theory containing a name $\dot{w} = \langle \dot{\eta}, \dot{\rho}, \dot{p} \rangle$ that was interpreted in M^{k-1} as standing for $\hat{w}_k(M)$, as well as names for each ordinal $< o(P)$.[22] The decoding (Q,C) is determined by the fact that

$$Q = \mathrm{Hull}_1^{(Q,C)}(o(P) \cup \{\dot{w}^{(Q,C)}\}),$$

and by the fact that the Σ_1 theory in (Q,C) of parameters in $o(P) \cup \{\dot{w}^{(Q,C)}\}$ is B.

If $k = 1$, we are done decoding, and if $k > 1$, we can just decode again, because (Q,C) is Σ_1-elementarily equivalent to M^{k-1}. In the end we should get $(N,\emptyset)$ such that $k(N) = 0$ and $(P,B) = \hat{N}^k$.

There are two issues here. First, Q or the further decodings of (P,B) may be illfounded. Second, the parameter names occuring in B may not get correctly

---

[21] In the notation of [54], $\hat{M}^k = M^{k,q}$, where $q = \hat{w}_k(M)$.

[22] Literally speaking, the "names" $\dot{\eta}, \dot{\rho}, \dot{p}$ are variables that were assigned to the corresponding objects in M^{k-1}.

interpreted in Q by the decoding function. In the case of downward extension, Q is embedded into $M||\rho_{k-1}(M)$, so wellfoundedness is not a problem, but we must strengthen the Σ_0 elementarity requirement of [54, Lemma 3.1] in order to insure that $\dot{w}$ is correctly interpreted. We also require that M is of type 1.

LEMMA 4.3.5 (Downward extension of embeddings). *Let M be a pfs premouse of type 1, $k(M) = k < \omega$, and $\pi\colon (P,B) \to M^k$ be such that either*

(a) π is Σ_2 elementary, or

(b) π is cofinal and Σ_1 elementary;

then there is a unique type 1 pfs premouse N such that $k(N) = k$ and $N^k = (P,B)$, and a unique elementary and exact map $\sigma\colon N \to M$ extending π.

PROOF. If $k = 0$, then this is simply the assertion that P is a pfs premouse of degree zero. Our assumptions imply that π preserves rQ formulae, so this follows by the standard proof. See [86] or [35].

The rest of the proof is by induction on k. The case $k = 1$ is representative, so let us first assume that $k = 1$. Let $(Q,C) = \mathrm{Dec}((P,B))$, so that $C = \emptyset$ and Q is a pfs premouse of degree zero. Let $\sigma\colon Q \to M^-$ be the one step completion of π, given by

$$\sigma(h^1_Q(\alpha, \dot{w}^Q)) = h^1_{M^-}(\pi(\alpha), w_1(M^-)).$$

σ is at least Σ_2 elementary, since π is at least Σ_1 elementary. We must see that Q is 1-sound and $\dot{w}^Q = w_1(Q)$, or equivalently, $\sigma(w_1(Q)) = w_1(M)$. Q is weakly ms-solid by Proposition 4.1.5, so we are left with parameter solidity and projectum solidity.

CLAIM 1. $\dot{\rho}^Q = o(P) = \rho_1(Q)$.

PROOF. $Q = h^1_Q``(o(P) \cup \{\dot{w}^Q\})$, so $\rho_1(Q) \leq o(P)$. Since B is amenable to P, $o(P) \leq \rho_1(Q)$, so $o(P) = \rho_1(Q)$. On the other hand, it is a Π_2 fact about M^1 that $\dot{\rho}^{M^-} = o(M^1)$, so this passes to (P,B), and we get that $\dot{\rho}^Q = o(P)$. ⊣

Thus $\sigma(\rho_1(Q)) = \rho_1(M)$.

CLAIM 2. *$\dot{p}^Q = p_1(Q)$, and Q is parameter solid.*

PROOF. The standard parameter of the strong core $\bar{\mathfrak{C}}(M^-)$ is universal, so

$$P(\rho_1(M^-)) \cap M \subseteq h^1_{M^-}``(\rho_1(M^-) \cup \{p_1(M^-)\}).$$

This is a Π_2 fact in M about $\rho_1(M^-)$ and $p_1(M^-)$, so it goes down under σ, and

$$P(\rho_1(Q)) \cap Q \subseteq h^1_Q``(\rho_1(Q) \cup \{\dot{p}^Q\}).$$

It follows that $\mathrm{Th}^Q_1(\rho_1(Q) \cup \{\dot{p}^Q\}) \notin Q$, so $p_1(Q) \leq_{\mathrm{lex}} \dot{p}^Q$. On the other hand, $p_1(M)$ is solid, so we can use the argument in Remark 2.3.13 to show that $\dot{p}^Q$ is solid over Q. It follows that $\dot{p}^Q \leq_{\mathrm{lex}} p_1(Q)$, so $\dot{p}^Q = p_1(Q)$, and $p_1(Q)$ is solid and universal over Q.

For the remainder of parameter solidity, let $\tau\colon \bar{\mathfrak{C}}_1(Q) \to Q$ be the anticore map. ($\tau = \mathrm{id}$ is possible, but not the interesting case.) Let $r = \tau^{-1}(p_1(Q)) = \tau^{-1} \circ \sigma^{-1}(p_1(M))$. Again, the proof in Remark 2.3.13 shows that r has solidity witnesses in $\bar{\mathfrak{C}}_1(Q)$, and thus r is solid and universal over $\bar{\mathfrak{C}}_1(Q)^-$. This finishes the proof that Q is parameter solid. ⊣

Thus we have $Q = \mathfrak{C}_1(Q)^-$.

CLAIM 3. *$\bar{\mathfrak{C}}_1(Q) = \mathfrak{C}_1(Q)$ iff $\bar{\mathfrak{C}}_1(M) = \mathfrak{C}_1(M)$.*

PROOF. $\bar{\mathfrak{C}}_1(M) = \mathfrak{C}_1(M)$ iff the sentence $\theta =$ "$\exists \alpha < \dot{\rho}(h^1(\alpha, \dot{p}) = \dot{\rho})$" is in A^1_M. But $\theta \in A^1_M$ iff $\theta \in B$ iff $\bar{\mathfrak{C}}_1(Q) = \mathfrak{C}_1(Q)$. ⊣

CLAIM 4. *Q is projectum solid.*

PROOF. $\sigma(\rho_1(Q)) = \rho_1(M)$, so $\rho_1(Q)$ is not measurable in Q. We are done if $\bar{\mathfrak{C}}_1(Q) = \mathfrak{C}_1(Q)$, so assume not. Let

$$\tau\colon \bar{\mathfrak{C}}_1(Q)^- \to Q$$

be the anticore map, and

$$i_D\colon \bar{\mathfrak{C}}_1(M)^- \to \mathrm{Ult}_0(\bar{\mathfrak{C}}_1(M)^-, D) = M^-$$

be the anticore map at the M level, where $D \in \bar{\mathfrak{C}}_1(M)$ is the order zero measure on $\rho_1(M)$ witnessing that M^- is projectum solid. Let $j = i_D^{-1} \circ \sigma \circ \tau$. The appropriate diagram is

$$\begin{array}{ccc} S = \bar{\mathfrak{C}}_1(M) & \xrightarrow{\ i_D\ } & M \\ {\scriptstyle j}\uparrow & & \uparrow{\scriptstyle \sigma} \\ R = \bar{\mathfrak{C}}_1(Q) & \xrightarrow{\ \tau\ } & Q \end{array}$$

To see that j is well defined, note that

$$\begin{aligned} \sigma \circ \tau(h^1_R(\alpha, p_1(R)) &= h^1_M(\sigma \circ \tau(\alpha), p_1(M)) \\ &= i_D(h^1_S(\sigma \circ \tau(\alpha), p_1(S)) \end{aligned}$$

for all $\alpha < \rho_1(R)$, so $\mathrm{ran}(\sigma \circ \tau) \subseteq \mathrm{ran}(i_D)$. We claim that $j(\rho_1(R)) = \rho_1(S)$. This is because

$$\tau(\rho_1(R)) = \text{least } \xi \text{ in } \mathrm{Hull}^Q_1(\rho_1(Q) \cup \{p_1(Q)\}) \setminus \rho_1(Q),$$

so

$$\begin{aligned} \sigma \circ \tau(\rho_1(R)) &= \text{least } \xi \text{ in } \mathrm{Hull}^M_1(\rho_1(M) \cup \{p_1(M)\}) \setminus \rho_1(M) \\ &= i_D(\rho_1(S)). \end{aligned}$$

To see the step from line 1 to line 2, recall that the language of B has names for $\rho_1(Q)$ and $p_1(Q)$. Letting $\tau(\rho_1(R)) = h^1_Q(\alpha, p_1(Q))$, line 1 becomes a $\Pi_1^{(P,B)}$

assertion about $\alpha, \rho_1(Q)$, and $p_1(Q)$. Since $\pi\colon (P,B) \to M^1$ is Σ_1 elementary, this assertion holds in M^1 about $\sigma(\alpha), \rho_1(M)$, and $p_1(M)$. That yields line 2.

Since $j(\rho_1(R)) = \rho_1(S)$, we must have $j(F) = D$, for F the order zero measure of R on $\rho_1(R)$. The reader can easily check that $\tau = i_F^R$, so F witnesses projectum solidity for Q. ⊣

Q is trivially stable, that is 0-stable, so Q is 1-sound.

So far we have used only that π is Σ_1 elementary, but our next claim uses the stronger elementarity hypotheses.

CLAIM 5. $\dot{\eta}^Q = \eta_1^Q$.

PROOF. We have shown that $\sigma(\rho_1(Q)) = \rho_1(M)$. The claim then follows from Lemma 4.3.4. ⊣

Our claims imply that $Q^1 = (P,B)$. Taking $N = Q^+$, N is a pfs premouse of type 1, and $\sigma\colon N \to M$ is elementary and exact.

This finishes the case $k = 1$. The case $k = n+1$ where $n \geq 1$ is quite similar. Letting $(Q,C) = \mathrm{Dec}((P,B))$, we get $\psi\colon (Q,C) \to M^n$ by setting

$$\psi(h^1_{(Q,C)}(\alpha, \dot{w}^{(Q,C)})) = h^1_{M^n}(\pi(\alpha), w_{n+1}(M)).$$

ψ is Σ_2 elementary, so by induction we can complete it to a map $\sigma\colon N \to M$ that is elementary and exact at level n, and such that $N^n = (Q,C)$. We need to show that $N^{n+1} = (P,B)$, and for that, the key is that $w_{n+1}(N) = \dot{w}^{(Q,C)}$. But by definition, $w_{n+1}(N) = \langle \eta_1^{(Q,C)}, \rho_1^{(Q,C)}, p_1^{(Q,C_0)} \rangle$, where C_0 is C restricted to the sublanguage without names for $\rho_n(N)$ and η_n^N. Moreover, π is sufficiently elementary that the proof given in the case $k = 1$ shows that $\dot{w}^{(Q,C)} = \langle \eta_1^{(Q,C)}, \rho_1^{(Q,C)}, p_1^{(Q,C_0)} \rangle$. ⊣

Anticore maps are cofinal and Σ_1 elementary on the associated reducts, so we can apply part (2) of Downward Extension to them.

LEMMA 4.3.6. *Let R be a solid pfs premouse of type 1, $P = \bar{\mathfrak{C}}(R)$, and $M = \mathfrak{C}(R)$. Let $\sigma\colon P^- \to R$ and $\pi\colon M^- \to R$ be the anticore maps, and $\tau = \pi^{-1} \circ \sigma$; then*

(1) P^- and M^- are pfs premice of type 1, and σ, τ, and π are cofinal, elementary, and exact,
(2) M is a pfs premouse of type 1, and
(3) $\pi(p(M^-)) = p(R)$.

PROOF. Let $k = k(R)$. Let

$$\begin{aligned}
(Q,B) &= \mathrm{cHull}_1^{R^k}(\rho_1(R^k) \cup \{p_1(R^k)\}),\\
\sigma_0 &= \text{anticollapse map},\\
(N,C) &= \mathrm{cHull}_1^{R^k}(\rho_1(R^k) \cup \{\rho_1(R^k), p_1(R^k))\}),\\
\pi_0 &= \text{anticollapse map},\\
\tau_0 &= \pi_0^{-1} \circ \sigma_0.
\end{aligned}$$

σ_0 and π_0 are Σ_1 elementary, hence so is τ_0. σ_0 is cofinal because the new set $X \subseteq \rho_1(N^k)$ is Σ_1 over N^k, and since $X \notin N^k$, the minimal witnesses to facts of the form $\alpha \in X$ for $\alpha < \rho_1(N^k)$ must be cofinal in $o(N^k)$. But these minimal witnesses are all in $\operatorname{ran}(\sigma_0)$, so σ_0 is cofinal. It follows that π_0 and τ_0 are cofinal.

By the Downward Extension Lemma, (Q,B) can be decoded to a pfs premouse, and we have defined $\bar{\mathfrak{C}}(R)^-$ to be its decoding, and σ to be the completion of σ_0. Similarly, (N,C) decodes to a pfs premouse, $\mathfrak{C}(R)^-$ is this premouse, and π is the completion of π_0. This proves (1).

R is parameter solid, so $p_1(R^k)$ is solid and universal over R^k and $\sigma_0^{-1}(p_1(R^k))$ is solid and universal over (Q,B), and $\sigma_0^{-1}(p_1(R^k)) = p_1(Q,B)$. This easily implies that $\pi_0^{-1}(p_1(R^k)) = p_1(N,C)$, so we have (3).

Since R is solid, $\rho_{k+1}(R)$ is not sequence-measurable in R, and since

$$\pi(\rho_{k+1}(M)) = \rho_{k+1}(R),$$

$\rho_{k+1}(M)$ is not sequence-measurable in M. Similarly,

$$\pi(\eta_k^M) = \eta_k^R,$$

so if $\rho_{k+1}(M) \leq \eta_k^M$, then η_k^M is not sequence-measurable in M. Together with (3), this implies that M is $k+1$-sound, that is, $k(M)$-sound, so we have (2). ⊣

Upward extension and ultrapower maps. The paradigm for upward extension is the canonical embedding $\pi\colon \hat{M}^k \to (P,B) = \operatorname{Ult}_0(\hat{M}^k, E)$. Given that (P,B) decodes to a wellfounded bare premouse N, we want to show that (N,k) is a pfs premouse, $\hat{N}^k = (P,B)$, and π can be completed to an elementary $\sigma\colon M \to N$. As we saw above, σ could be discontinuous at $\rho_k(M)$, and hence not exact. Moreover, M could be of type 1 while (N,k) is of type 2.

The reader should keep this paradigm in mind, but we can state the lemma more abstractly. In the abstract version, we allow π to be any appropriately elementary embedding.[23]

LEMMA 4.3.7 (Upward extension of embeddings). *Let M be a pfs premouse, $k(M) = k < \omega$, and $\pi\colon \hat{M}^k \to (P,B)$ be Σ_0 elementary. Suppose that all decodings of (P,B) are wellfounded, and either*

(1) π is Σ_2 elementary, or

(2) π is cofinal and Σ_1 elementary.

Then there is a unique pfs premouse N such that $k(N) = k$ and $(P,B) = \hat{N}^k$, and a unique elementary $\sigma\colon M \to N$ such that $\pi \subseteq \sigma$.

PROOF. If $k = 0$, the lemma just asserts that P is a pfs premouse and π is elementary. This follows from the fact that rQ formulae go up under π. The rest of the proof is by induction on k. The case that $k = 1$ is representative of the general one, except for some points concerning η_{k-1}^M that we shall handle when we get to them.

[23] In other words, we replace E by a possibly long extender.

Suppose that $k = 1$, and let

$$(Q,C) = \mathrm{Dec}((P,B)),$$
$$k(Q) = 0.$$

By assumption, Q is wellfounded, and $C = \emptyset$ since the predicate of $\hat{M}^1$ codes the theory of $(M^-,\emptyset)$. Let $\sigma\colon M^- \to Q$ be the map

$$\sigma(h^1_{M^-}(\alpha,\hat{w}_1(M))) = h^1_Q(\pi(\alpha),\dot{w}^Q).$$

It is easy to see that σ is Σ_2 elementary. This implies that Q is a pfs premouse of degree zero, and if it is active, then its last extender has the weak ms-ISC. Moreover, $Q = \mathrm{Hull}^Q_1(o(P) \cup \{\dot{w}^Q\})$, so $\rho_1(Q) \le o(P)$. But if $\alpha < o(P)$ and $r \in Q$, say $r = h^1_Q(\gamma,\dot{w}^Q)$, then $B \cap \max(\alpha,\gamma)+1$ codes $\mathrm{Th}^Q_1(\alpha \cup \{r\})$. Since (P,B) is amenable, we get that $o(P) \le \rho_1(Q)$, and hence

$$o(P) = \rho_1(Q).$$

CLAIM 1. *$\rho_1(M) < o(M)$ iff $\rho_1(Q) < o(Q)$.*

PROOF. If $\rho_1(M) \le h^1_M(\alpha,\hat{w}_1(M))$, where $\alpha < \rho_1(M)$, $o(P) \le h^1_Q(\pi(\alpha),\dot{w}^Q)$ because π is Σ_1 elementary. If $\exists\beta < o(P)(o(P) \le h^1_Q(\beta,\dot{w}^Q))$, and π is Σ_2 elementary, then we can pull this $\Sigma_2^{(P,B)}$ fact back to $\hat{M}^k$. If π is Σ_1 and cofinal, we can find $\alpha < \rho_1(M)$ such that $o(P) \le h^1_Q(\beta,\dot{w}^Q)$ for some $\beta < \alpha$, and then use the fact that $\pi(\mathrm{Th}^M_1(\alpha \cup \{\hat{w}_1(M)\})) = \mathrm{Th}^Q_1(\pi(\alpha) \cup \{\dot{w}^Q\})$. ⊣

If $\rho_1(M) = o(M)$, then $P = Q$ and $\sigma = \pi$. We do need to see that $\sigma(\hat{w}_1(M)) = \hat{w}_1(Q)$ in this case (where $\sigma(o(M)) = o(Q)$ by convention). This can be shown using the proof of Lemma 2.5.10 when π is cofinal and Σ_1 elementary, and the proof of Lemma 2.5.12 when π is Σ_2 elementary. We omit further detail, and assume $\rho_1(M) < o(M)$ and $\rho_1(Q) < o(Q)$ henceforth.

If π is Σ_2 elementary, then σ is Σ_3 elementary, and this makes it easier to show that σ has the preservation properties we require.

CLAIM 2. *Suppose that π is Σ_2 elementary; then*

(1) $\rho_1(Q) = \sigma(\rho_1(M))$,
(2) $\hat{w}_1(Q) = \sigma(\hat{w}_1(M))$,
(3) *Q is almost sound,*
(4) *σ is elementary and exact as a map from M to Q^+, moreover M and Q^+ have the same type.*

PROOF. Let $\rho_1(M) = h^1_M(\alpha,\hat{w}_1(M))$, where $\alpha < \rho_1(M)$. This fact can be expressed as

$$\hat{M}^1 \models \theta[\alpha],$$

where θ is a Π_2 formula.[24] Thus

$$(P,B) \models \theta[\pi(\alpha)],$$

so $h^1_Q(\pi(\alpha), \dot{w}^Q) = o(P) = \rho_1(Q)$. Since $\sigma(\rho_1(M)) = h^1_Q(\pi(\alpha), \dot{w}^Q)$, we have (1).

For (2), let us show first that $\sigma(p_1(M)) = p_1(Q)$. Let $r = \sigma(p_1(M))$. Being a solidity witness for $p_1(M)$ is Π_2 over M, so preserved by σ. Thus it suffices to show that $\mathrm{Th}_1(\rho_1(Q) \cup r) \notin Q$. But for $\rho = \rho_1(M)$,

$$M \models \forall A \subseteq \rho \exists \alpha < \rho (A = h^1(\alpha, \hat{w}_1(M)) \cap \rho).$$

The formula on the right is Π_2, so it holds of $\rho_1(Q)$ and $\dot{w}^Q$ in Q. Thus $\mathrm{Th}_1(\rho_1(Q) \cup r) \notin Q$, so $\sigma(p_1(M)) = p_1(Q)$.

Let us show that $\sigma(\hat{\rho}_1(M)) = \hat{\rho}_1(Q)$. Suppose first that M has type 1A, that is, $M = \mathrm{Hull}^M_1(\rho_1(M) \cup p_1(M))$. This is a Π_2 fact, so $Q = \mathrm{Hull}^Q_1(\rho_1(Q) \cup p_1(Q))$, so $\hat{\rho}_1(Q) = 0 = \sigma(\hat{\rho}_1(M))$. Suppose next that M does not have type 1A, so that $\hat{\rho}_1(M)$ is the least ordinal in M that is not in $\mathrm{Hull}^M_1(\rho_1(M) \cup p_1(M))$. Then

$$\hat{\rho}_1(M) = \text{unique } \eta \text{ such that } M \models \theta[\eta, \rho_1(M), p_1(M)],$$

where θ is the natural Π_2 formula, and hence

$$\begin{aligned} \sigma(\hat{\rho}_1(M)) &= \text{unique } \eta \text{ such that } Q \models \theta[\eta, \rho_1(Q), p_1(Q)] \\ &= \hat{\rho}_1(Q). \end{aligned}$$

Finally, $\sigma(\hat{\eta}_1(M)) = \sigma(\mathrm{cof}^M_1(\hat{\rho}_1(M))) = \mathrm{cof}^Q_1(\hat{\rho}_1(Q)) = \hat{\eta}_1(Q)$ by the calculation in the proof of Lemma 2.5.12. This finishes the proof of (2).

For (3), we show first that Q is solid, that is, 1-solid. We showed above that Q is parameter solid. Stability is trivial, since $\eta^Q_0 = o(Q)$.[25] Let us check projectum solidity. Since $\rho_1(M)$ is not sequence-measurable in M, $\rho_1(Q)$ is not sequence-measurable in Q. If M has type 1A or type 2, then then the same is true of Q, and there is nothing more to check in projectum solidity. So assume M has type 1B. Since M is projectum solid, $M^- = \mathrm{Ult}(R,D)$, where $R^+ = \bar{\mathfrak{C}}_1(M)$ is the strong core of M^-, and D is the order zero measure of R on $\rho_1(M)$. Let $S = \bar{\mathfrak{C}}_1(Q)^-$, and $\tau \colon S \to Q$ be the anticore map. We get the diagram

$$\begin{array}{ccc} S = \bar{\mathfrak{C}}_1(Q)^- & \xrightarrow{\ \tau\ } & Q \\ {\scriptstyle j}\uparrow & & \uparrow{\scriptstyle \sigma} \\ R = \bar{\mathfrak{C}}_1(M)^- & \xrightarrow{\ i_D\ } & M^- \end{array}$$

---

[24] θ has a Π_1 clause stating that $o(\hat{M}^1) \le h^1_M(\alpha, \hat{w}_1(M))$, and a Π_2 clause stating that $h^1_M(\alpha, \hat{w}_1(M)) \le o(\hat{M}^1)$.

[25] If $k > 1$, we use the fact that $\sigma(\rho_k(M)) = \rho_k(Q)$ and $\sigma(\eta^M_{k-1}) = \eta^Q_{k-1}$, together with the stability of M^-, to conclude that Q is stable.

Here $j = \tau^{-1} \circ \sigma \circ i_D$. This makes sense because

$$\begin{aligned}\tau(\rho_1(Q)) &= \text{least } \xi \text{ in } \mathrm{Hull}_1^Q(\rho_1(Q) \cup \{p_1(Q)\}) - \rho_1(Q)\\ &= \sigma(\text{least } \xi \text{ in } \mathrm{Hull}_1^M(\rho_1(M) \cup \{p_1(M)\}) - \rho_1(M))\\ &= \sigma \circ i_D(\rho_1(M)).\end{aligned}$$

Since also $\tau(p_1(S)) = \sigma \circ i_D(p_1(S))$, $\mathrm{ran}(\tau) \subseteq \mathrm{ran}(\sigma \circ i_D)$, so j is well defined and the diagram commutes. Our calculation also shows that $j(\rho_1(M)) = \rho_1(Q)$. It is now easy to see that

$$Q = \mathrm{Ult}(S, j(D))$$

and τ is the ultrapower embedding, as desired. This finishes the proof that Q is 1-solid.

$Q = \mathrm{Hull}_1^Q(\rho_1(Q) \cup \{\hat{w}_1(Q)\})$, as required by almost soundness. The remaining requirement is that if $\hat{\rho}_1(Q) \notin \mathrm{Hull}_1^Q(\rho_1(Q) \cup p_1(Q))$, then $Q = \mathrm{Ult}_0(R, U)$, where R is the transitive collapse of $\mathrm{Hull}_1^Q(\rho_1(Q) \cup p_1(Q))$, U is the order zero measure of R on $\hat{\rho}_1(Q)$, and i_U is the anticollapse map. But if $\hat{\rho}_1(Q) \notin \mathrm{Hull}_1^Q(\rho_1(Q) \cup p_1(Q))$ then $\hat{\rho}_1(M) \notin \mathrm{Hull}_1^M(\rho_1(M) \cup p_1(M))$. Letting D be the order zero measure of M on $\hat{\rho}_1(M)$ that we get from almost soundness for M^-, we can take $U = j(D)$ for the appropriate j, just as we did in the proof that Q is projectum solid.

This finishes the proof of (3). Our calculations have also established (4). ⊣

In view of Claim 2, we may assume that π is cofinal and Σ_1 elementary. We do so for the remainder of the proof.

The calculations below are perhaps better motivated if we regard Q as an ultrapower of M^- via the long extender of π, where the ultrapower is formed using functions that are $\Sigma_1^{M^-}$ in parameters. In short,

$$\begin{aligned}Q &= \mathrm{Ult}_1(M^-, E_\pi)\\ &= \{\pi(f)(a) \mid a \in [o(P)]^{<\omega} \wedge f \in \mathcal{F}\},\end{aligned}$$

where $\mathcal{F}$ consists of all functions f that are Σ_1^M in parameters and have domain $[\xi]^{<\omega}$ for some $\xi < o(P)$. (Since we are now assuming π is cofinal, each component measure $(E_\pi)_a$ concentrates on bounded subsets of $\rho_1(M)$.) $\pi(f)$ is interpreted by moving the Σ_1 definition of f, as usual. Our σ is the canonical embedding from M^- to $\mathrm{Ult}_1(M^-, E_\pi)$.

Let

$$\delta = \max(\rho_1(M), \hat{\rho}_1(M)).$$

$\delta = \hat{\rho}_1(M)$ unless M has type 1A.

CLAIM 3. *Let $A \subseteq \sigma(\delta)$ be such that $A \in Q$; then there is a B such that*

$$B \in \mathrm{Hull}_1^Q(\rho_1(Q) \cup \{\sigma(p_1(M))\})$$

such that $B \cap \sigma(\delta) = A$.

PROOF. Let $A = \pi(f)(a)$. We may assume that $\operatorname{ran}(f) \subseteq P(\delta)^M$, so f is itself essentially a subset of δ that is Σ_1^M in parameters.

By Lemma 4.2.9, there is a function g that is Σ_1^M in parameters from $\rho_1(M) \cup \{p_1(M)\}$ and such that $f(u) = g(u) \cap \delta$ for all $u \in \operatorname{dom}(f)$. Let $B = \pi(g)(a)$; then it is easy to see that B works. ⊣

CLAIM 4. $\sigma(p_1(M)) = p_1(Q)$.

PROOF. Let $p = p_1(M)$ and $r = \sigma(p)$. M has solidity witnesses for r, and being such a witness is a Π_2^M fact, so it is preserved by σ. Thus it is enough to show that $\mathrm{Th}_1^Q(\rho_1(Q) \cup \{r\}) \notin Q$. But $\rho_1(Q) \leq \sigma(\delta)$, so by Claim 3 every subset of $\rho_1(Q)$ in Q is coded into $\mathrm{Th}_1^Q(\rho_1(Q) \cup \{r\})$. It follows that $\mathrm{Th}_1^Q(\rho_1(Q) \cup \{r\}) \notin Q$. ⊣

We can now complete the case that M has type 1A.

CLAIM 5. *If M has type 1A, then Q is strongly sound, Q^+ has type 1A, and $\sigma\colon M \to Q^+$ is elementary.*

PROOF. Let us show that Q is solid. $p_1(Q)$ is solid by the proof of Claim 4, and it is easy to see that $Q = \bar{\mathfrak{C}}_1(Q)^-$, so Q is parameter solid. Q is trivially stable.[26] For projectum solidity, we must see that $\rho_1(Q)$ is not measurable in Q. If $\sigma(\rho_1(M)) = \rho_1(Q)$ this follows at once from projectum solidity for M, so assume that $\rho_1(Q) = \sup \sigma``\rho_1(M) < \sigma(\rho_1(M))$.

Letting $\sup \sigma``\rho_1(M) \leq \pi(g)(a) < \sigma(\rho_1(M))$, we see that $g``[\xi]^{|a|}$ is unbounded in $\rho_1(M)$ for some ξ. Thus $\eta_1^M < \rho_1(M)$. Let $\eta = \eta_1^M$, and let $f \in \mathcal{F}$ be a nice witness that $\operatorname{cof}_1^M(\rho_1(M)) = \eta$ such that f is continuous at limit ordinals. $f \restriction \xi \in M$ for all $\xi < \eta$, and the function

$$g(\xi) = f \restriction \xi$$

is in $\mathcal{F}$. Let

$$h = \pi(g)(\sup \pi``\eta),$$

noting here that π is discontinuous at η because it is discontinuous at $\rho_1(M)$. Then $h \in Q$, and it is easy to see that $\operatorname{ran}(h)$ is cofinal in $\rho_1(Q)$. Thus $\rho_1(Q)$ is Σ_0-singular in Q, and hence not measurable in Q.

This finishes the proof that Q is solid. Since $\bar{\mathfrak{C}}_1(Q)^- = Q$, Q is strongly sound. The rest of Claim 5 is clear. ⊣

Let us assume now that M has type 1B or type 2. Thus $\delta = \hat{\rho}_1(M)$.

CLAIM 6. $\sigma(\hat{w}_1(M)) = \hat{w}_1(Q)$.

PROOF. Let $\xi < \sigma(\hat{\rho}_1(M))$. We have $\xi = \pi(f)(a)$ for some $f \in \mathcal{F}$ such that $\operatorname{ran}(f) \subseteq \hat{\rho}_1(M)$. By Lemma 4.2.9, f is Σ_1^M in parameters from $\rho_1(M) \cup p_1(M)$, so $\pi(f)(a)$ is Σ_1^Q in parameters from $\rho_1(Q) \cup p_1(Q)$. Thus $\sigma(\hat{\rho}_1(M)) \leq \hat{\rho}_1(Q)$.

[26] Again, if $k > 1$ there is a little argument. We have $\rho_k(M) \leq \eta_{k-1}(M)$ iff $\rho_k(Q) \leq \eta_{k-1}(Q)$ because $\sigma(\eta_{k-1}^M) = \eta_{k-1}^Q$ and $\sup \sigma``\rho_k(M) = \rho_k(Q)$. So since M is stable, Q is stable.

On the other hand, $\hat{\rho}_1(M) \notin \mathrm{Hull}_1^M(\rho_1(M) \cup p_1(M))$, and this is a Π_1^M fact about $\hat{\rho}_1(M)$, so it is preserved by σ. Thus $\sigma(\hat{\rho}_1(M)) = \hat{\rho}_1(Q)$.

To see that σ preserves $\hat{\eta}$, assume first that $\hat{\eta}_1(M) < \hat{\rho}_1(M)$. Then $\hat{\eta}_1(M) < \rho_1(M)$ because M^- is almost sound. By Proposition 2.5.3 there is a nice witness f that $\mathrm{cof}_1^M(\hat{\rho}_1(M)) = \hat{\eta}_1(M)$, and by Lemma 2.5.9, $\sigma(f)$ is a nice witness that $\mathrm{cof}_1^Q(\hat{\rho}_1(Q)) = \sigma(\hat{\eta}_1(M))$, as desired.

Suppose next that $\hat{\eta}_1(M) = \hat{\rho}_1(M)$. Since M^- is almost sound, $\hat{\rho}_1(M) = \rho_1(M)$, so $\rho_1(M)$ is Σ_1 regular over M, and hence σ is continuous at $\rho_1(M)$. This implies that $\sigma(\rho_1(M)) = \rho_1(Q)$ and $\rho_1(Q)$ is Σ_1 regular over Q. (Cf. Lemma 2.5.10.)

We have already shown that σ preserves p_1. Thus $\sigma(\hat{w}_1(M)) = \hat{w}_1(Q)$. ⊣

CLAIM 7. *Q is almost sound.*

PROOF. We show first that Q is solid. We have already shown that $p_1(Q)$ is solid. Let $R = \bar{\mathfrak{C}}_1(M)^-$, and $M^- = \mathrm{Ult}(R, D)$ where D is the order zero measure of R on $\hat{\rho}_1(M)$. Let $S = \bar{\mathfrak{C}}_1(Q)^-$ and let $\tau\colon S \to Q^-$ be the anticollapse. Let $j = \tau^{-1} \circ \sigma \circ i_D$. We have the diagram from Claim 2:

$$\begin{array}{ccc} S = \bar{\mathfrak{C}}_1(Q) & \xrightarrow{\ \tau\ } & Q \\ \uparrow j & & \uparrow \sigma \\ R = \bar{\mathfrak{C}}_1(M) & \xrightarrow{\ i_D\ } & M \end{array}$$

$S = \mathrm{Ult}(R, E_\pi)$, and j is the ultrapower map. Since $\tau^{-1}(p_1(Q)) = j(p_1(R))$, $\tau^{-1}(p_1(Q))$ is solid and universal over S. Thus Q is parameter solid. Stability is trivial.[27]

One can check that $Q = \mathrm{Ult}(S, j(D))$ and $\tau = i_{j(D)}$. Thus if $\hat{\rho}_1(Q) = \rho_1(Q)$ then Q is projectum solid. If $\rho_1(Q) < \hat{\rho}_1(Q)$ then either $\sigma(\rho_1(M)) = \rho_1(Q)$ or $\rho_1(Q)$ is Σ_0 singular in Q, by the proof of Claim 5. In both cases, $\rho_1(Q)$ is not measurable in Q. Moreover, $\bar{\mathfrak{C}}_1(Q) = \mathfrak{C}_1(Q)$ if $\rho_1(Q) < \hat{\rho}_1(Q)$. Thus Q is projectum solid in this case too.

This finishes the proof that Q is solid. Clause (b) in the definition of almost soundness requires that $Q = \mathrm{Hull}_1^Q(\rho_1(Q) \cup \{p_1(Q), \hat{\rho}_1(Q)\})$, which is of course true. Clause (c) holds because $Q = \mathrm{Ult}(S, j(D))$ and $\tau = i_{j(D)}$. We proved clause (d) when we showed $\sigma(\hat{\eta}_1(M)) = \hat{\eta}_1(Q)$. ⊣

Thus Q^+ is a pfs premouse and $(P, B) = \hat{Q}^1$, so $\sigma\colon M \to Q^+$ is elementary.

This finishes the proof of the Upward Extension Lemma in the case $k = 1$. When $k > 1$, the proof yields a Σ_2 elementary map $\sigma_1 : M^{k-1} \to (Q, C)$, which then can be upwardly extended to $\sigma\colon M \downarrow 0 \to N$ by induction. $N \downarrow (k-1)$ is a pfs premouse and σ is elementary as a map from M^- to $N \downarrow (k-1)$ by induction. The proof above shows that $N \downarrow k$ is a pfs premouse and σ is elementary from M to $N \downarrow k$.

[27] If $k > 1$ we again use that $\rho_k(M) \le \eta_{k-1}(M)$ iff $\rho_k(Q) \le \eta_{k-1}(Q)$ and M^- is stable.

The only new points have to do with the preservation of η_{k-1}, and we have already described how to deal with them. ⊣

Concerning the exactness of the upward extension, we have

LEMMA 4.3.8. *Let $\sigma\colon M \to N$ be the completion of $\pi\colon \hat{M}^k \to \hat{N}^k$, where $k = k(M) = k(N)$; then*

(a) if π is Σ_2 elementary, then σ is exact, and

(b) if π is cofinal and Σ_1 elementary, then the following are equivalent:

(i) σ is not exact,

(ii) $\eta_k^M < \rho_k(M)$, and π is discontinuous at η_k^M.

The proof is implicit in the proof of Lemma 4.3.7, so we omit it.

Upward Extension concerns ultrapowers by possibly long extenders. We shall mostly apply it to ultrapowers by short extenders.

DEFINITION 4.3.9. Suppose that M is a pfs premouse, $k = k(M)$, and E is an extender over M such that $\mathrm{crit}(E) < \rho_k(M)$; then

$$\mathrm{Ult}(M,E) = \mathrm{Ult}_k(M,E)$$

is the full decoding of $\mathrm{Ult}_0(\hat{M}^k, E)$. Letting $\pi\colon \hat{M}^k \to \mathrm{Ult}_0(\hat{M}^k, E)$ be the canonical embedding, i_E^M is the completion of π. We call i_E^M the canonical embedding associated to $\mathrm{Ult}(M,E)$.

The canonical embedding π is cofinal and Σ_1 elementary as a map from $\hat{M}^k$ to $\mathrm{Ult}_0(\hat{M}^k, E)$, so it has a completion i_E^M. Moreover

COROLLARY 4.3.10. *Let $k = k(M)$, and $i\colon M \to \mathrm{Ult}(M,E)$ be the canonical embedding; then i is elementary, and i is exact iff $\mathrm{crit}(E) \neq \eta_k^M$.*

PROOF. M is elementary by Upward Extension. Since E is short, i is discontinuous at η_k^M iff $\mathrm{crit}(E) = \eta_k^M$. Thus we can apply Lemma 4.3.8. ⊣

We can also regard $\mathrm{Ult}_k(M,E)$ as the ultrapower of M formed using $r\Sigma_k^M$ functions. We discussed the equivalence between the two ways of looking at $\mathrm{Ult}_k(M,E)$ immediately after Definition 2.4.4.

Lemma 4.3.10 tells us that the fine structure related to $\rho_k(M)$ is preserved by k-ultrapowers with critical point $< \rho_k(M)$. We must also consider what happens to the fine structure related to $\rho_{k+1}(M)$ when we iterate between ρ_{k+1} and ρ_k. We only care about the level $k+1$ fine structure when M is of type 1 and stable. In this case, iterations between ρ_{k+1} and ρ_k will produce elementary, exact maps into further stable pfs premice of type 1. It is important here that the extenders being used are close to the models to which they are applied, a fact that we shall prove in Lemma 4.5.3.

LEMMA 4.3.11. *Let M be a stable type 1 pfs premouse, E be close to M, and $\rho(M) \leq \mathrm{crit}(E) < \rho^-(M)$. Let $N = \mathrm{Ult}(M,E)$ and $i = i_E^M$; then*

(a) N is a stable type 1 pfs premouse, and i is elementary and exact.

(b) $\rho(N) = \rho(M)$.
(c) *If* $k = k(M)$ *and* $A \subseteq \rho(M)$, *then* A *is boldface* $r\Sigma^M_{k+1}$ *iff* A *is boldface* $r\Sigma^N_{k+1}$.
(d) *If* M *is parameter solid, then* N *is parameter solid,* $i(p(M)) = p(N)$, *and* N *is not strongly sound. If in addition* $\mathrm{crit}(E) > \rho(M)$, *then* N *is not sound.*
(e) *If* M *is parameter solid and* N *is almost sound, then* M *is the strong core of* N, *and* E *is the order 0 measure of* M *on* $\hat{\rho}(N)$.
(f) *If* M *is projectum solid, then so is* N.

PROOF. Let $k = k(M)$. By closeness, $\mathrm{crit}(E)$ is sequence-measurable in M.[28] Since M is stable and $\rho(M) \leq \mathrm{crit}(E)$, $\mathrm{crit}(E) \neq \eta^M_k$. By Lemma 4.3.10, N is a pfs premouse and i is elementary and exact. Since M is stable and $i(\eta^M_k) = \eta^N_k$, N is stable.

The proof of Lemma 2.4.12 shows that $\rho(M) = \rho(N)$. Suppose $A \subseteq \rho(M)$ and A is $\Sigma_1^{N^k}$ in the parameter $[a, f] = [a, f]^{M^k}_E$. Let $\theta(u, v, w)$ be Σ_0 and such that

$$\alpha \in A \iff N^k \models \exists v \theta[\alpha, v, [a, f]].$$

Then by Łoś's Theorem,

$$\alpha \in A \iff \exists g \in M^k \exists X \in E_a(M^k \models \forall u \in X \theta[\alpha, g(u), f(u)]).$$

Since E is close to M, E is Σ_1^M in some parameter q, so the right hand side converts to an $r\Sigma^M_{k+1}$ definition of A from f and q.

For (d): Since i^M_E is exact, $i(w_k(M)) = w_k(N)$, so if M is parameter solid, then $i(p(M)) = p(N)$ and N is parameter solid by the proof of Lemma 4.3.10. N is not strongly sound because $\mathrm{crit}(E) \notin \mathrm{Hull}^N_{k+1}(\rho(N) \cup \{p(N)\})$. Finally, if $\mathrm{crit}(E) > \rho(M)$, then $\mathrm{crit}(E) \notin \mathrm{Hull}^N_{k+1}(\rho(N) \cup \{\rho(N), p(N)\})$, so N is not sound. This proves (d).

For (e), let $\nu = \hat{\rho}(N)$. N is not strongly sound, so since it is almost sound, $\rho(N) \leq \nu$, and $N = \mathrm{Ult}(R, D)$ where $R = \bar{\mathfrak{C}}(N)^-$ and D is the order 0 measure of R on ν, and i_D is the anticollapse map with range $\mathrm{Hull}^N_{k+1}(\rho(N) \cup \{p(N), w_k(N)\})$. In particular $\mathrm{ran}(i_D) \subseteq \mathrm{ran}(i_E)$, so we have the diagram

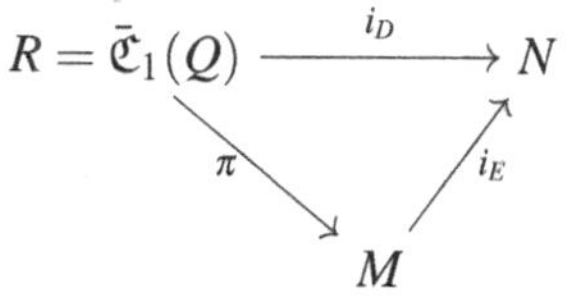

Here $\pi = i_D^{-1} \circ i_E$. Since i_D is the identity on ν, π and i_E are the identity on ν.

We claim that $\nu = \mathrm{crit}(E)$. For if $\nu < \mathrm{crit}(E)$, then $\nu \in \mathrm{ran}(i_E)$, and, since $N = \mathrm{Hull}^N_{k+1}(\rho(N) \cup \{p(N), \nu, w_k(N)\})$, we get $\mathrm{crit}(E) \in \mathrm{ran}(i_E)$, contradiction.

ν is not sequence-measurable in N, so since E is close to M, E must be the order 0 measure of M on ν. To finish the proof of (e) it is enough to show that M is strongly sound, for then $M = R$ and $E = D$.

[28]Here we use our slight strengthening of closeness as defined in [35].

If not, let $\mu = \hat{\rho}(M)$. Since $\pi \upharpoonright \nu = \text{id}$, $\nu \leq \mu$. Let φ, β, and γ be such that φ is a Σ_1 formula, $\beta < \rho(N)$, $\gamma < \rho_k(M)$, and

$$i_E(\mu) = \text{unique } \xi \text{ s.t. } N^k || i_E(\gamma) \models \varphi[\beta, p(N), \nu, \xi].$$

Since $\nu = [\{\nu\}, \text{id}]_E^{M^k}$, we have some $\alpha < \nu$ such that

$$\mu = \text{unique } \xi \text{ s.t. } M^k || \gamma \models \varphi[\beta, p(N), \alpha, \xi].$$

So $\mu \in \text{Hull}_{k+1}^M(\rho(M) \cup \{p(M), w_k(M)\})$, contradiction.

Finally, for part (f): if $\rho(M)$ is not measurable by the M-sequence, then $\text{crit}(E) > \rho(M)$ by closeness, so $\rho(N)$ is not measurable by the N-sequence. Moreover $\mathfrak{C}(M) = \mathfrak{C}(N)$ and $\bar{\mathfrak{C}}(M) = \bar{\mathfrak{C}}(N)$, so the rest of projectum solidity propagates to N as well. ⊣

We get at once

COROLLARY 4.3.12. *Let M be a solid pfs premouse, E be close to M, and $\rho(M) \leq \text{crit}(E) < \rho^-(M)$; then* $\text{Ult}(M,E)$ *is a solid pfs premouse, and* $\mathfrak{C}(M) = \mathfrak{C}(\text{Ult}(M,E))$.

Summary. Let M be a pfs premouse, and E an extender over M with $\text{crit}(E) < \rho^-(M)$. We have shown

(i) If M is solid and of type 1, then anticore map from $\mathfrak{C}(M)^-$ to M is elementary, cofinal, and exact.
(ii) The canonical embedding i_E^M is elementary and cofinal. It is exact iff $\text{crit}(E) \neq \eta_k^M$.
(iii) If $\rho(M) \leq \text{crit}(E)$, M is solid, and E is close to M, then i_E^M preserves the fine structure associated to $\rho(M)$. In particular, i_E^M is exact, and M and $\text{Ult}(M,E)$ have the same core and strong core.

4.4. Plus trees

It turns out that the good behavior of background-induced iteration strategies involves more than their action on stacks of normal, or even quasi-normal, iteration trees. We must consider their action on stacks of iteration trees whose constituents are what we shall call *plus trees*. We shall eventually show that our background-induced strategies are determined by how they act on single normal trees, but the proof involves a strategy comparison, and so this reduction is not available to us now. Plus trees come up in the comparison proof itself.

We need plus trees in order to deal with the "background coherence problem" for induced iteration strategies. Let us look more closely at that problem.

Suppose that $M = M_{\nu,k}^{\mathbb{C}}$ is reached in a background construction $\mathbb{C}$ in the sense of Chapter 3, and that $\Sigma = \Omega(\mathbb{C}, M, \Sigma^*)$ is the strategy for M determined by $\mathbb{C}$ and some strategy Σ^* for the background universe. Suppose that M is active, with last extender E, and that E^* is the background extender for E in $\mathbb{C}$. If we begin an iteration tree $\mathcal{T}$ on M by using E, then the lifted tree $\mathcal{T}^* = \text{lift}(\mathcal{T}, M, \mathbb{C})_0$ begins with E^*, and then the lifting process continues by using the natural factor map

$$\pi\colon \text{Ult}(M,E) \to i_{E^*}(M)$$

to lift the next extender. Our problems stem from the fact that π is not the identity on $\text{lh}(E)$, because

$$\lambda_E < \pi(\lambda_E).$$

Thus if F is indexed in M before E but after λ_E, then $\pi(F) \neq F$, and coherence at the premouse level is not mirrored by coherence at the background level.

As we saw in Chapter 3, this does not cause any problems in defining Σ. However, it does cause a serious problem if we want to reduce the good behavior of Σ on stacks of normal trees to the good behavior of Σ^* on such stacks. To see this in a simple case, let $\mathcal{T} = \langle E \rangle$, and let $\mathcal{U}$ be a normal tree on $M || \text{lh}(E)$ such that the stack $\langle \mathcal{T}, \mathcal{U} \rangle$ is by Σ. We would like to show that that $\mathcal{U}$ is by Σ. This is a simple instance of several different internal consistency properties of iteration strategies that we need in order to compare strategies.

Now $\Sigma_{\mathcal{T}}(\mathcal{U})$ is defined by looking at how $\pi\mathcal{U}$ is lifted via $i_{E^*}(\mathbb{C})$ in $\text{Ult}(V, E^*)$, and following $\Sigma^*_{\mathcal{T}^*}$ there. So we need to see that the lift $\mathcal{U}^*$ of $\mathcal{U}$ via $\mathbb{C}$ to a tree on V is by Σ^*, and what we know is that the lift $(\pi\mathcal{U})^*$ of $\pi\mathcal{U}$ using $i_{E^*}(\mathbb{C})$ to a tree on $\text{Ult}(V, E^*)$ is by the tail $\Sigma^*_{\mathcal{T}^*}$. But for F such that

$$\lambda(E) < \text{lh}(F) < \text{lh}(E),$$

there is no connection between the $\mathbb{C}$-background of F and the $i_{E^*}(\mathbb{C})$-background of $\pi(F)$, so $\mathcal{U}^*$ and $(\pi\mathcal{U})^*$ may have no connection.

The fact that the factor map $\pi\colon \text{Ult}(M,E) \to i_{E^*}(M)$ is not the identity at λ_E also leads to problems in other arguments that use the connection between Σ and Σ^* [29] One might think that ms-indexing would avoid these problems if we are working below superstrongs, but it does not. If $\nu(E)$ is a limit of generators of E and a generator of E^*, then $\nu(E) < \pi(\nu(E))$, so π is not the identity on the extenders that are ms-indexed before E.

There is one case where things work out. If we are working in ms-indexing, and E has a largest generator, then π is the identity on all extenders that are ms-indexed before E. This may seem like a very special case, but it turns out that we can always compare premice by iterating only by extenders having a largest generator, and we do not have to move to ms-indexing to do it.[30] We shall call the iteration trees involved here *λ-separated plus trees*.

[29] One cannot strengthen the background extender demand to $\lambda_E = \lambda_{E^*}$ in general, for then not all whole initial segments of E will be on the M-sequence.

[30] This applies to both Jensen premice and pfs premice. Our main interest now and henceforth is pfs premice.

DEFINITION 4.4.1. Let M be a pfs premouse, and E be an extender on the M-sequence; then

(1) E^+ is the extender with generators $\lambda_E \cup \{\lambda_E\}$ that represents $i_F^{\mathrm{Ult}(M,E)} \circ i_E^M$, where F is the order zero total measure on λ_E in $\mathrm{Ult}(M,E)$,
(2) $\hat{\lambda}(E^+) = \lambda_E$,
(3) $\mathrm{lh}(E^+) = \mathrm{lh}(E)$, and
(4) $o(E^+) = (\mathrm{lh}(E)^+)^{\mathrm{Ult}(M,E^+)}$.

$o(E^+)$ is where the order zero measure on λ_E of $\mathrm{Ult}(M,E)$ would be ms-indexed. It is not hard to see that $\mathrm{Ult}(M,E^+)$ agrees with the 0-ultrapower $\mathrm{Ult}_0(M||\mathrm{crit}(E)^{+,M}, E^+)$ past $o(E^+)$. It is easy to code E^+ as an amenable subset of $\mathrm{lh}(E)$ that is Σ_0 over $M|\mathrm{lh}(E)$, and of course, E is Σ_0 over $(M||\mathrm{lh}(E), E^+)$. So E and E^+ have the same information.

DEFINITION 4.4.2. G is *of plus type* iff $G = E^+$, for some extender E that is on the sequence of a pfs premouse M. In this case, we let $G^- = E$. The *extended M-sequence* consists of all extenders E such that either E or E^- is on the M-sequence. If E is on the extended M-sequence, then

$$\varepsilon(E) = \begin{cases} \mathrm{lh}(E) & \text{if } E \text{ is of plus type,} \\ \lambda(E) & \text{otherwise.} \end{cases}$$

We wish to consider iteration trees that are allowed to use extenders of the form E^+, where E is on the coherent sequence of the current model. To unify notation, if E is an extender on the sequence of some premouse, let us set

(i) $\hat{\lambda}(E) = \lambda(E) = \hat{\lambda}(E^+)$,
(ii) $E^- = E$, and
(iii) $o(E) = (\mathrm{lh}(E)^+)^{\mathrm{Ult}(M,E)} = o(E^+)$.

DEFINITION 4.4.3. Let M be a pfs premouse; then a *plus tree* on M is a system $\mathcal{T} = \langle T, \langle E_\alpha \mid \alpha+1 < \mathrm{lh}(\mathcal{T})\rangle\rangle$ such that there are M_α and $i_{\alpha,\beta}$ and D satisfying:

(1) $M_0 = M$, and T is a tree order;
(2) if $\alpha+1 < \mathrm{lh}(\mathcal{T})$, then E_α is on the extended M-sequence, and
 (a) if $\xi < \alpha$, then $\hat{\lambda}(E_\xi) \le \hat{\lambda}(E_\alpha)$, and
 (b) if $\xi < \alpha$ and E_ξ is of plus type, then $\mathrm{lh}(E_\xi) < \hat{\lambda}(E_\alpha)$;
(3) if $\alpha+1 < \mathrm{lh}(\mathcal{T})$, then letting β be least such that either $\beta = \alpha$, or $\mathrm{crit}(E_\alpha) < \hat{\lambda}(E_\beta)$,
 (a) T-pred$(\alpha+1) = \beta$,
 (b) $M_{\alpha+1} = \mathrm{Ult}(M^*_{\alpha+1}, E_\alpha)$, for $M^*_{\alpha+1}$ the shortest initial segment N of M_β such that $\rho(N) \le \mathrm{crit}(E_\alpha)$, if one exists, and $M^*_{\alpha+1} = M_\beta$ otherwise,
 (c) $\alpha+1 \in D$ iff $M^*_{\alpha+1} \ne M_\beta$
 (d) $\hat{\imath}_{\beta,\alpha+1} = i^*_{\alpha+1}\colon M^*_{\alpha+1} \to M_{\alpha+1}$ is the canonical embedding, and
(4) if $\lambda < \mathrm{lh}(\mathcal{T})$ is a limit ordinal, then $D \cap [0,\lambda)_T$ is finite, and M_λ is the direct limit of the M_α for $\alpha <_T \lambda$ under the $\hat{\imath}^{\mathcal{T}}_{\alpha,\eta}$; moreover $\lambda \notin D$.

It may seem that clause (3) of Definition 4.4.3 allows generators to move along branches of $\mathcal{T}$. The worry would be the case that $\beta = \xi + 1$, where $E_\xi = F^+$ for some F, so that $\hat{\lambda}(E_\xi) = \lambda_F$. But in this case, the only important generators of E_ξ are in $\lambda_F \cup \{\lambda_F\}$. Clause (3) requires that generators below $\lambda_F = \hat{\lambda}(E_\xi)$ are not moved. λ_F itself has no total measures in M_β, and hence in M_α. There are no partial extenders on the sequence of M_α with critical point λ_F because the proper initial segments of M_α are projectum solid. (See Lemma 4.1.14.) Thus E_α is not moving any important generators of E_ξ. It is quite possible that $\text{crit}(E_\alpha) < \lambda(E_\xi)$, however.

The analog of $\lambda_\beta^\mathcal{T}$ (the sup of the Jensen generators of $\mathcal{M}_\beta^\mathcal{T}$) in our current context is

DEFINITION 4.4.4. Let $\mathcal{T}$ be a plus tree on a pfs premouse; then for any $\beta < \text{lh}(\mathcal{T})$,

$$\begin{aligned}\varepsilon_\beta^\mathcal{T} &= \sup\{\varepsilon(F) \mid \exists\eta(\eta+1 \leq_T \beta \wedge F = E_\eta^\mathcal{T})\}\\ &= \sup\{\varepsilon(F) \mid \exists\eta(\eta+1 \leq \beta \wedge F = E_\eta^\mathcal{T})\}.\end{aligned}$$

The two characterizations of $\varepsilon_\beta^\mathcal{T}$ are equivalent because we have demanded that plus trees be ε-nondecreasing.[31] If the branch of $\mathcal{T}$ from α to β does not drop, then $\mathcal{M}_\beta^\mathcal{T}$ is generated from $\varepsilon_\beta^\mathcal{T} \cup \text{ran}(i_{\alpha,\beta}^\mathcal{T})$, as in Lemma 2.6.7.

We shall show in Lemma 4.5.3 that in any plus tree, all extenders used are close to the model to which they are applied.[32] For now, let us simply assume this.

The branch embeddings in a plus tree are elementary, but the pattern of soundness types can be complicated. Along non-dropping branches type can change and the branch embeddings may not be exact. If the type becomes type 2, then our first drop can be to an $M_{\alpha+1}^*$ that is only almost sound. At and after that drop, the premice along this branch are all type 1, and further drops are to sound premice.

Fortunately, we can avoid this complexity in practice by restricting ourselves to base models that are strongly stable, in the following sense.

DEFINITION 4.4.5. Let M be a pfs premouse and $k = k(M)$; then M is *strongly stable* iff there is no M-total extender E on the M-sequence such that $\text{crit}(E) = \eta_k^M$.

If $k(M) = 0$, then M is strongly stable, since $\eta_0^M = o(M)$. When the base model of a plus tree is strongly stable, then its branch embeddings are exact, and its models exhibit the familiar soundness pattern.

LEMMA 4.4.6. *Let M be a strongly stable pfs premouse of type 1, and let $\mathcal{T}$ be a plus tree on M; then*

(i) all $\mathcal{M}_\alpha^\mathcal{T}$ have type 1,

(ii) all branch embeddings are elementary and exact,

[31] Hence plus trees are by definition quasi-normal.

[32] In the sense of Definition 2.4.11.

(iii) whenever $\alpha+1 \in D^{\mathcal{T}}$, then $M^*_{\alpha+1}$ is sound, and
(iv) if $\alpha+1 \in D^{\mathcal{T}}$, $\alpha+1 \leq_T \beta$, and $D^{\mathcal{T}} \cap (\alpha+1,\beta]_T = \emptyset$, then
 (a) M_β is solid, $M^*_{\alpha+1} = \mathfrak{C}(M_\beta)^-$, and $i_{\alpha+1,\beta} \circ i^{*,\mathcal{T}}_{\alpha+1}$ is the anticore map, and
 (b) If $k = k(M_\beta)$ and $A \subseteq \rho(M_\beta)$, then A is boldface $r\Sigma_{k+1}$ over $M^*_{\alpha+1}$ iff A is boldface $r\Sigma_{k+1}$ over M_β.

PROOF SKETCH. Let $k = k(M)$ and $\eta = \eta_k^M$. If $[0,\gamma]_T \cap D^{\mathcal{T}} = \emptyset$, then (by induction on γ), $i_{0,\gamma}(\eta) = \eta_k^{M_\gamma}$ and $i_{0,\gamma}$ is exact and continuous at $\rho_k(M)$, and M_γ is strongly stable and of type 1. If $\gamma = \mathcal{T}\text{-pred}(\alpha+1)$ and $\alpha+1 \in D^{\mathcal{T}}$, then $M^*_{\alpha+1}$ is a proper initial segment of a type 1 premouse, hence sound. Moreover, since $\rho(M^*_{\alpha+1}) \leq \text{crit}(E_\alpha)$ and E_α is close to $M^*_{\alpha+1}$, Lemma 2.4.12 applies, and we get that $M_{\alpha+1}$ has type 1, and (iv) holds when $\beta = \alpha+1$. We continue this way by induction. ⊣

Remark 4.4.7. All initial segments of a premouse must be stable, but in general they will not all be strongly stable. M can be strongly stable while M^- is not. If $\eta^M_{k-1} < \rho_k$ and η^M_{k-1} is measurable in M, then M^- is not strongly stable, although M^- is stable (and in fact, solid).

Plus trees are maximal by definition, but not necessarily length increasing. We say *the plus case occurs at α* iff E_α is of plus type.

DEFINITION 4.4.8. Let $\mathcal{T}$ be a plus tree on M; then
(a) $\mathcal{T}$ is *normal* (or *length-increasing*) iff whenever $\alpha < \beta < \text{lh}(\mathcal{T})-1$, then $\text{lh}(E^{\mathcal{T}}_\alpha) < \text{lh}(E^{\mathcal{T}}_\beta)$,
(b) $\mathcal{T}$ is *λ-tight* iff for all $\alpha+1 < \text{lh}(\mathcal{T})$, $E^{\mathcal{T}}_\alpha$ is not of plus type, and
(c) $\mathcal{T}$ is *λ-separated* iff for all $\alpha+1 < \text{lh}(\mathcal{T})$, $E^{\mathcal{T}}_\alpha$ is of plus type.

A λ-tight iteration tree is just an ordinary quasi-normal iteration tree. One can re-organize any plus tree $\mathcal{T}$ as a λ-tight tree $\mathcal{U}$ in a fairly straightforward way. This is not important for us, but for the sake of completeness, here is the rough idea. $\mathcal{T}$ and $\mathcal{U}$ agree until we reach α such that $E^{\mathcal{T}}_\alpha = F^+$ for some F. At that point $\mathcal{U}$ uses F, and then the order zero measure D of $\text{Ult}(M_\alpha, F)$, in two steps. The last models are now the same again, that is, $\mathcal{M}^{\mathcal{T}}_{\alpha+1} = \mathcal{M}^{\mathcal{U}}_{\alpha+2}$. The difficulty arises if $E = E^{\mathcal{T}}_{\alpha+1}$ is such that $\lambda_F < \text{crit}(E) < \lambda_D$. In that case E is applied to $\mathcal{M}^{\mathcal{T}}_{\alpha+1}$ in $\mathcal{T}$, and to $\mathcal{M}^{\mathcal{U}}_{\alpha+1}$ in $\mathcal{U}$, by the rules of normality for the two types of tree. The relationship of last models is now that $\mathcal{M}^{\mathcal{T}}_{\alpha+2} = \text{Ult}(\mathcal{M}^{\mathcal{U}}_{\alpha+3}, i_E(D))$. In general, the simulation of $\mathcal{T}$ by $\mathcal{U}$ continues to make use of such correspondences.

This reduction of plus trees $\mathcal{T}$ to λ-tight trees $\mathcal{U}$ is of no use to us, however, because we want to study background-induced iteration strategies, and the conversions $\text{lift}(\mathcal{T}, c)$ and $\text{lift}(\mathcal{U}, c)$ dictated by a given conversion system could be completely unrelated. Thus $\mathcal{T}$ might be according to an induced strategy $\Omega(c, \Sigma^*)$ while $\mathcal{U}$ is not. It seems that one can only rule this out after having proved a comparison theorem for iteration strategies.

In fact, our initial results on the good behavior of background induced iteration strategies will apply at the other extreme, to their restrictions to λ-separated trees. Notice that λ-separated trees are normal, by (2)(b) of Definition 4.4.3. Every extender E used in a λ-separated tree has a largest generator, and we shall see that this helps with the background coherence problem. The results of Section 8.1 show that λ-separated trees are enough for comparison, and Theorem 5.5.2 shows that background induced iteration strategies are determined by their action on λ-separated trees. In fact, the main results of this book would not be affected if we simply restricted all iteration strategies to stacks of λ-separated trees. We shall not do that, however.

Remark 4.4.9. The example above shows that there can be distinct finite normal plus trees with the same last model. This cannot happen if both trees are λ-tight, or both trees are λ-separated.

The agreement of models in a plus tree is a bit awkward to state. It is easy to see that any plus tree $\mathcal{T}$ breaks up into disjoint maximal finite intervals in which the exit extenders have strictly decreasing length. That is, $\mathrm{lh}(\mathcal{T})$ can be partitioned into intervals $[\alpha, \alpha+n]$, where $0 \leq n < \omega$, such that

(i) for all $\beta < \alpha$, $\mathrm{lh}(E_\beta) < \mathrm{lh}(E_\alpha)$,

(ii) for all $i < n$, $E_{\alpha+i}$ is not of plus type, and

$$\hat{\lambda}(E_{\alpha+i}) \leq \hat{\lambda}(E_{\alpha+i+1}) < \mathrm{lh}(E_{\alpha+i+1}) < \mathrm{lh}(E_{\alpha+i}),$$

and

(iii) $\mathrm{lh}(E_{\alpha+n}) < \hat{\lambda}(E_{\alpha+n+1})$, or $\alpha+n+1 = \mathrm{lh}(\mathcal{T})$.

Of course $n = 0$ is possible. Part (iii) implies $\mathrm{lh}(E_{\alpha+n}) < \hat{\lambda}(E_\beta)$ for all $\beta > \alpha+n$. Part (ii) is justified by clause (2)(c) in Definition 4.4.3. We call $[\alpha, \alpha+n]$ a *maximal delay interval*, and we say that $\alpha+n$ *ends a delay interval*.

It may seem pointless to allow decreasing lengths, because given a maximal delay interval $[\alpha, \alpha+n]$, we could have just skipped using $E_\alpha, \dots, E_{\alpha+n-1}$, and taken $E_{\alpha+n}$ out of $\mathcal{M}^{\mathcal{T}}_\alpha$ to continue the iteration. Doing this everywhere would produce a normal iteration tree $\mathcal{S}$ with the same last model as $\mathcal{T}$, differing only in that the nontrivial delay intervals in $\mathcal{T}$ are eliminated. More precisely, let

$$i \colon \eta \to \{\xi \mid \xi \text{ begins a delay interval in } \mathcal{T}\}$$

be the increasing enumeration. Suppose we have defined $\mathcal{S} \upharpoonright \xi+1$, and

$$\mathcal{M}^{\mathcal{S}}_\xi = \mathcal{M}^{\mathcal{T}}_{i(\xi)}.$$

Let $i(\xi)+n$ end the delay interval in $\mathcal{T}$ that starts at $i(\xi)$, so that $i(\xi+1) = i(\xi)+n+1$. We set

$$E^{\mathcal{S}}_\xi = E^{\mathcal{T}}_{i(\xi)+n}.$$

Let $T\text{-pred}(i(\xi)+n+1) = \gamma$, and $\beta = i(\eta)$ begin the delay interval to which γ belongs, and $P \trianglelefteq \mathcal{M}^{\mathcal{T}}_\gamma$ be what $E_{i(\xi)+n}$ is applied to in $\mathcal{T}$. One can easily check that $P \trianglelefteq \mathcal{M}^{\mathcal{T}}_{\beta}$.[33] So we can let $\eta = \mathcal{S}\text{-pred}(\xi+1)$ and $\mathcal{M}^{\mathcal{S}}_{\xi+1} = \text{Ult}(P, E^{\mathcal{S}}_{\xi})$ and continue.

DEFINITION 4.4.10. Let $\mathcal{T}$ be a plus tree, and let $\mathcal{S}$ be the plus tree defined above, whose models are precisely those $\mathcal{M}^{\mathcal{T}}_{\alpha}$ such that α begins a delay interval in $\mathcal{T}$, and whose exit extenders are just those $E^{\mathcal{T}}_{\alpha+n}$ such that $\alpha+n$ ends a delay interval in $\mathcal{T}$. We call $\mathcal{S}$ the *normal companion* of $\mathcal{T}$, and write $\mathcal{S} = \mathcal{T}^{\text{nrm}}$.

The following lemma says the branches of $\mathcal{T}$ that do not drop infinitely often are in one-one correspondence with the branches of $\mathcal{T}^{\text{nrm}}$ that do not drop infinitely often. Branches corresponding this way have the same direct limit models and branch embeddings.

LEMMA 4.4.11. *Let $\mathcal{T}$ be a plus tree on a premouse M; then*

(a) if T-pred$(\gamma+1)$ does not begin a delay interval in $\mathcal{T}$; then $\gamma+1 \in D^{\mathcal{T}}$, and
(b) if $\mathcal{T}$ has limit length and b is a cofinal branch of $\mathcal{T}$ such that $D^{\mathcal{T}} \cap b$ is finite, then all sufficiently large $\eta \in b$ begin a delay interval.

We essentially gave the proof of (a) while defining $\mathcal{T}^{\text{nrm}}$, and (b) follows at once from (a).

So why bother with $\mathcal{T}$, why not just use $\mathcal{T}^{\text{nrm}}$? The answer is that we shall be considering trees by some iteration strategy Σ. It may happen that $\mathcal{T}$ is by Σ, but its normal companion is not. In the strategy-comparison proof, we have to live with the possibility that this happens when Σ is a background-induced strategy. We shall eventually show that background-induced strategies are not pathological in this way, but the proof involves a strategy comparison. Until we get to that point, we need to deal with plus trees that are not length-increasing.

Lemma 4.4.11 simplifies in the coarse case:

LEMMA 4.4.12. *Let $\mathcal{T}$ be a nice iteration tree on a transitive model M of* ZFC*; then for any $\gamma+1 < \text{lh}(\mathcal{T})$, T-pred$(\gamma+1)$ begins a delay interval in $\mathcal{T}$. Thus if b is a branch of $\mathcal{T}$ of limit length, then every $\eta \in b$ begins a delay interval.*

PROOF. Let $\alpha = T\text{-pred}(\gamma+1)$. If α does not begin a delay interval, then we have $\beta < \alpha$ such that $\text{lh}(E^{\mathcal{T}}_{\beta}) = \text{lh}(E^{\mathcal{T}}_{\alpha})$.[34] But $\mathcal{T}$ is non-overlapping, so then $T\text{-pred}(\gamma+1) \leq \beta$, contradiction. $\dashv$

So in the coarse case, at any limit ordinal λ, the cofinal branches of $\mathcal{T}$ and their direct limits are in one-one correspondence with the cofinal branches of $\mathcal{T}^{\text{nrm}}$. In the coarse case our focus is iteration strategies that pick unique cofinal well-founded branches, and for such strategies, $\mathcal{T}$ and $\mathcal{T}^{\text{nrm}}$ are completely equivalent. Nevertheless, it is a occasionally convenient for bookkeeping reasons to permit non-length-increasing iteration trees in the coarse case.

[33] If $\beta < \gamma$, then $\hat{\lambda}(E^{\mathcal{T}}_{\beta}) \leq \text{crit}(E^{\mathcal{S}}_{\xi}) < \hat{\lambda}(E^{\mathcal{T}}_{\gamma}) < \text{lh}(E^{\mathcal{T}}_{\beta})$, and $P \triangleleft \mathcal{M}^{\mathcal{T}}_{\beta} | \text{lh}(E^{\mathcal{T}}_{\beta})$.

[34] Lengths are nondecreasing in the coarse case.

It is not hard to show that if $\mathcal{T}$ is a plus tree of limit length λ on a pfs premouse M, and b is a cofinal branch of $\mathcal{T}$ that drops at most finitely often, then b corresponds to a branch of $\mathcal{T}^{\text{nrm}}$.

PROPOSITION 4.4.13. *Let $\mathcal{U}$ be a plus tree with models $M_\alpha = \mathcal{M}^{\mathcal{U}}_\alpha$ and extenders $E_\alpha = E^{\mathcal{U}}_\alpha$. Let $\alpha < \beta < \text{lh}(\mathcal{U})$, then*

(1) $M_\alpha || \hat{\lambda}(E_\alpha) = M_\beta || \hat{\lambda}(E_\alpha)$;
(2) if α ends a delay interval, then
  (a) $M_\alpha || \text{lh}(E_\alpha) = M_\beta || \text{lh}(E_\alpha)$, and
  (b) E^-_α is indexed at $\text{lh}(E_\alpha)$ on the M_α sequence, but $\text{lh}(E_\alpha)$ is a cardinal of M_β;
(3) if $o(E_\alpha) \leq \hat{\lambda}(E_{\alpha+1})$, then M_β agrees with $\text{Ult}(M_\alpha, E_\alpha)$ below $o(E_\alpha)$, and
(4) if $\text{lh}(E_\alpha) \leq \hat{\lambda}(E_{\alpha+1}) < o(E_\alpha)$, then $\text{lh}(E_\alpha) < \text{crit}(E_{\alpha+1})$, and $\text{lh}(E_\alpha)$ is a cutpoint of $M_{\alpha+1}$, and $\mathcal{U} = \mathcal{U} \upharpoonright (\alpha+1)^\frown \mathcal{W}$, where $\mathcal{W}$ is a tree above $\text{lh}(E_\alpha)$ on some level of $M_{\alpha+1}$ that projects to $\text{lh}(E_\alpha)$.

We omit the elementary proof. Note that the increased agreement described in (2)(a)(b) holds whenever $E^{\mathcal{U}}_\alpha$ is of plus type, by clause (2)(c) in the definition of plus trees.

For the most part, what we need from the proposition is

COROLLARY 4.4.14. *Let $\mathcal{U}$ be a normal plus tree, $M_\alpha = \mathcal{M}^{\mathcal{U}}_\alpha$, and $E_\alpha = E^{\mathcal{U}}_\alpha$; then for $\alpha < \beta < \text{lh}(\mathcal{U})$,*

(1) $M_\alpha || \text{lh}(E_\alpha) = M_\beta | \text{lh}(E_\alpha)$,
(2) $\text{lh}(E_\alpha)$ is a cardinal of M_β, so $M_\alpha | \text{lh}(E_\alpha) \neq M_\beta | \text{lh}(E_\alpha)$, and
(3) if $\alpha + 1 \leq_T \beta$, then $\text{lh}(E_\alpha) \leq \rho^-(M_\beta)$.

Part (3) is easy to prove by induction. It comes down to the fact that if $\text{Ult}(M,E)$ exists, then $\rho^-(\text{Ult}(M,E)) = \sup i_E``\rho^-(M)$.

4.5. Copy maps, lifted trees, and levels of elementarity

The Shift Lemma and copying construction work as they did with ordinary premice.[35] Let us adopt the definitions from Section 2.5 related to copy maps, starting with Definition 2.5.17, with the understanding that now we are talking about pfs premice.

In particular, suppose M, N, P, and Q are pfs premice, $\pi\colon P \to Q$ is nearly elementary, $\varphi\colon M \to N$ is Σ_0 elementary, and E is an extender on the M-sequence

[35] Sections 9.1 and 9.2 of [86] have a very careful treatment of copying, both for ordinary iteration trees, and for phalanx based iteration trees. There are a number of subtleties that come up, especially in the phalanx case.

such that $\text{crit}(E) < \rho^-(P)$. Let $F = \varphi(E)$, and suppose the agreement between P and M, Q and N, and π and φ is such that

$$\langle \pi, \varphi \rangle \colon (P,E) \to (Q,F),$$

as defined in Definition 2.5.17. The agreement guarantees that $\text{crit}(F) < \rho^-(Q)$. Letting $k = k(P) = k(Q)$, Lemma 2.5.19 gives us a Σ_0 elementary, cardinal preserving

$$\sigma_0 \colon \text{Ult}_0(\hat{P}^k, E) \to \text{Ult}(\hat{Q}^k, F).$$

Assuming that $S = \text{Ult}(Q,F)$ is wellfounded, we have that $\hat{S}^k = \text{Ult}_0(\hat{Q}^k, F)$, $R = \text{Ult}(P,E)$ is wellfounded, and $\hat{R}^k = \text{Ult}_0(\hat{P}^k, E)$. Thus there is a unique nearly elementary

$$\sigma \colon \text{Ult}(P,E) \to \text{Ult}(Q,F)$$

that completes σ_0.[36] We call σ the *copy map* associated to π, φ and E.

Stronger elementarity hypotheses on π lead to stronger elementarity conclusions regarding σ. In particular

(1) If π is cofinal, then σ_0 is cofinal, and hence σ is cofinal and elementary.
(2) If π is elementary and $\langle \pi, \varphi \rangle \colon (P,E) \overset{*}{\to} (Q,F)$, then σ_0 is Σ_1 elementary, so σ is elementary.
(3) If π is almost exact, then σ is almost exact.
(4) If π is exact and $\text{crit}(E) \neq \eta_k^P$, then σ is exact.

We can copy plus trees as we did ordinary ones. Given pfs premice M and N, $\pi \colon M \to N$ nearly elementary, and $\mathcal{T}$ a plus tree on M, we define an iteration tree $\pi\mathcal{T}$ on N with the same tree order as $\mathcal{T}$, together with nearly elementary copy maps

$$\pi_\alpha \colon M_\alpha \to N_\alpha,$$

where $M_\alpha = \mathcal{M}_\alpha^{\mathcal{T}}$ and $N_\alpha = \mathcal{M}_\alpha^{\pi\mathcal{T}}$. Let $E_\alpha = E_\alpha^{\mathcal{T}}$ and $F_\alpha = E_\alpha^{\pi\mathcal{T}}$. The system $\pi\mathcal{T}$ will have all the properties of a plus tree, except that it may not be maximal. We shall have by induction that the copy maps commute with the branch embeddings of $\mathcal{T}$ and $\pi\mathcal{T}$, and agree with one another, in that

(1) if $\beta \leq \alpha$, then $\pi_\alpha \upharpoonright \varepsilon(E_\beta) = \pi_\beta \upharpoonright \varepsilon(E_\beta)$ and $N_\alpha | \varepsilon(F_\beta) = N_\beta | \varepsilon(F_\beta)$, and
(2) if $\beta \leq_T \alpha$, then $\pi_\alpha \circ \hat{\imath}_{\beta,\alpha}^{\mathcal{T}} = \hat{\imath}_{\beta,\alpha}^{\mathcal{U}} \circ \pi_\beta$.

Set $\pi_0 = \pi$. The successor step is as follows: let $E = E_\alpha$, $\beta = T\text{-pred}(\alpha+1)$, and

$$F = \pi_\alpha(E),$$
$$P = \mathcal{M}_{\alpha+1}^{*,\mathcal{T}},$$
$$Q = \pi_\beta(P).$$

[36]We are given the pfs premice R and S here, so we do not need anything more than Σ_0 elementarity for σ_0 to complete it.

Here if $E = G^+$, where G is on the M_α sequence, then $\pi_\alpha(E) = \pi_\alpha(G)^+$, with the usual convention if G is the last extender predicate of M_β. Similarly, if $c(P) = o(M_\beta)$ then $Q = N_\beta \downarrow k(P)$. The agreement between π_α and π_β implies that β is least such that $\text{crit}(F) < \hat{\lambda}(F_\beta)$, and $\hat{\lambda}(F_\xi) \leq \hat{\lambda}(F)$ for all $\xi < \alpha$, with $\text{lh}(F_\xi) < \hat{\lambda}(F)$ if F_ξ has plus type. We also get

$$\langle \pi_\beta, \pi_\alpha \rangle \colon (P, E) \to (Q, F),$$

so the Shift Lemma applies. We let $F_\alpha = F$, which by the rules for semi-normal trees results in $\beta = \pi\mathcal{T}\text{-pred}(\alpha + 1)$ and

$$N_{\alpha+1} = \text{Ult}(Q, F).$$

We let

$$\pi_{\alpha+1} = \text{copy map associated to } (\pi_\beta \restriction P, \pi_\alpha, E).$$

One can easily check that our inductive hypotheses are preserved. At limit steps λ we use commutativity to obtain $\pi_\lambda : M_\lambda \to N_\lambda$. If we ever reach an illfounded N_α, the copying stops.

DEFINITION 4.5.1. Let M and N be pfs premice, and $\pi \colon M \to N$ be nearly elementary; then $\pi\mathcal{T}$ is the *copied tree* defined above, and the $\pi_\alpha : \mathcal{M}_\alpha^{\mathcal{T}} \to \mathcal{M}_\alpha^{\pi\mathcal{T}}$ are the *copy maps associated to $\mathcal{T}$ and π*.

The system $\pi\mathcal{T}$ can fail to be maximal in the following way. Adopting the notation above for the step from α to $\alpha + 1$, we might have $P = M_\beta^-$, so that $Q = N_\beta^-$, and yet $\text{crit}(F) < \rho(Q)$, so that F should be applied to N_β, not N_β^-, in a maximal tree. This cannot happen if π_β is almost exact, for then $\rho^-(M_\beta) \leq \text{crit}(E)$ implies $\rho^-(N_\beta) \leq \text{crit}(F)$. It cannot happen if $P \lhd M_\beta^-$, since then π_β is exact as a map from P^+ to Q^+.

We shall show in Proposition 4.5.17 that if the initial $\pi \colon M \to N$ is elementary, then all the copy maps $\pi_\alpha : M_\alpha \to N_\alpha$ are elementary, and hence almost exact. Thus in this case, $\pi\mathcal{T}$ is maximal, and hence a plus tree.[37] The key is to show by induction that, in the notation of the successor step above,

$$\langle \pi_\beta, \pi_\alpha \rangle \colon (P, E) \xrightarrow{*} (Q, F).$$

In other words, for every a, π_β moves some Σ_1^P definition of E_a to a Σ_1^Q definition of $F_{\pi_\alpha(a)}$. An easy calculation shows that this implies that $\pi_{\alpha+1}$ is elementary.[38]

DEFINITION 4.5.2. An extender E is *close to M* iff

(1) $\text{dom}(E) = \text{dom}(F)$, for some F on the sequence of M, and

(2) for all finite $a \subseteq \varepsilon(E)$,

---

[37]That elementarity is propagated by copying seems to have appeared first as Lemma 1.3 of [47]. See also Lemma 9.2.5 of [86], where a more detailed proof is given for Jensen indexed mice.

[38]See [86], Lemma 3.4.5.

(a) E_a is Σ_1^M in parameters, and
(b) for all $\alpha < \kappa_E^{+,M}$, $E_a \cap M|\alpha \in M$.

We say that E is *very close to* M iff E is close to M, and for all finite $a \subseteq \varepsilon(E)$, $E_a \in M$.

We have replaced $\lambda(E)$ in Definition 2.4.11 by $\varepsilon(E)$ to allow for the possibility that E has plus type; otherwise there is no change.

Clearly if E is on the M-sequence, then it is close to M. If $E = \dot{F}^M$ is the last extender of M, then it may fail to be very close to M. The following refinement of the Closeness Lemma[39] generalizes these simple facts. It says that if E is applied to M in $\mathcal{T}$, then E is close to M, and either E is very close to M or the situation has a special form similar to the case that $E = \dot{F}^M$.

LEMMA 4.5.3 (Closeness Lemma). *Let $\mathcal{T}$ be a plus tree, with models M_ξ and extenders E_ξ, and let $M_\xi^* = \mathcal{M}_\xi^{*,\mathcal{T}}$. Let T-pred$(\alpha+1) = \beta$ where $\beta < \alpha$, and set $E = E_\alpha$; then either*

(1) E is very close to M_α, or
(2) $\beta <_T \alpha$, and letting $\eta+1$ be least in $(\beta, \alpha]_T$,
  (i) $M_{\eta+1}^* \trianglelefteq M_{\alpha+1}^*$, $k(M_{\eta+1}^*) = 0$, $M_{\eta+1}^*$ is active,
  (ii) $D^{\mathcal{T}} \cap (\eta+1, \alpha]_T = \emptyset$,
  (iii) $E^- = i_{\eta+1,\alpha} \circ i_{\eta+1}^*(F)$, where F is the last extender of $M_{\eta+1}^*$, and $\rho(M_{\eta+1}^*) \leq \mathrm{dom}(F) < \mathrm{crit}(i_{\eta+1}^*)$, and
  (iv) if $M_{\eta+1}^* \in M_{\alpha+1}^*$, then E is very close to $M_{\alpha+1}^*$.

In case (1), E is very close to $M_{\alpha+1}^*$. In case (2), E is close to $M_{\eta+1}^*$, and hence to $M_{\alpha+1}^*$.

PROOF. The proof is by induction on α.

Let $\lambda = \hat{\lambda}(E_\beta)$. We have that $\mathrm{dom}(E) < \lambda$ and λ is an inaccessible cardinal in M_ξ for all $\xi > \beta$, moreover $M_\beta|\lambda \lhd M_{\alpha+1}^*$.

CLAIM 1. *If E is not very close to M_α, then E is close to $M_{\alpha+1}^*$, and (2) of the lemma holds.*

PROOF. Let us fix $a \subseteq \varepsilon(E)$ such that $E_a \notin M_\alpha$. It follows that M_α is active, with last extender $E^- = \dot{F}^{M_\alpha}$. Moreover $E_a \notin M_\beta|\lambda$, so $E_a \notin M_\xi$ for all $\xi > \beta$ because $\mathrm{dom}(E) < \lambda$ and λ is inaccessible in M_ξ when $\xi > \beta$.

Now let $\eta+1 \in [0, \alpha]_T$ be least such that $\beta < \eta+1$. We show first that there are no drops in model in $(\eta+1, \alpha]_T$. For suppose otherwise, and let $\gamma+1 \leq_T \alpha$ be largest such that, setting $\xi = T$-pred$(\gamma+1)$, we have $M_{\gamma+1}^* \in M_\xi$. Since E_a is boldface $\Sigma_1^{M_\alpha}$, E_a is boldface $\Sigma_1^{M_{\gamma+1}^*}$, by Lemma 2.4.12 and the fact that $\mathrm{dom}(E) < \hat{\lambda}(E_\eta) \leq \mathrm{crit}(i_{\gamma+1,\alpha} \circ i_{\gamma+1}^*)$. But then $E_a \in M_\xi$. Since $\xi \geq \eta+1 > \beta$, we have a contradiction.

[39] Lemma 6.1.5 of [35].

It follows that $M_{\eta+1}$ is active, and $E^- = i_{\eta+1,\alpha}(G)$, where $G = \dot{F}^{M_{\eta+1}}$. Moreover, $G = i^*_{\eta+1}(F)$, where $F = \dot{F}^{M^*_{\eta+1}}$ and $\rho(M^*_{\eta+1}) \leq \text{dom}(F)$. But then E^- is in $\text{ran}(\hat{\imath}_{\eta+1,\alpha} \circ i^*_{\eta+1})$, so since $\text{crit}(E) < \hat{\lambda}(E_\eta)$, we get

$$\text{crit}(E) < \text{crit}(E_\eta).$$

Thus $\beta = T\text{-pred}(\alpha+1) \leq T\text{-pred}(\eta+1)$. But $T\text{-pred}(\eta+1) \leq \beta$, by the definition of $\eta+1$. So $\beta = T\text{-pred}(\eta+1)$. Finally, $\text{dom}(E) < \text{crit}(E_\eta)$, so $E_a \notin M^*_{\eta+1}$, so $\rho_1(M^*_{\eta+1}) < \text{crit}(E_\eta)$, so $k(M^*_{\eta+1}) = 0$.

Since $\text{crit}(E) < \text{crit}(E_\eta)$, we have $M^*_{\eta+1} \trianglelefteq M^*_{\alpha+1}$. Thus $\eta+1$ satisfies requirements (2)(i)-(iii) of our lemma. Moreover, if $c \subseteq \varepsilon(E)$ is finite, then E_c is boldface Σ_1 over $M^*_{\eta+1}$. Since $\text{dom}(E) = \text{dom}(F)$ and F is on the $M^*_{\eta+1}$ sequence, E is close to $M^*_{\alpha+1}$. Moreover, if $M^*_{\eta+1} \in M^*_{\alpha+1}$, then E is very close to $M^*_{\alpha+1}$. This gives us (2)(iv). ⊣

CLAIM 2. *If E is very close to M_α, then E is very close to $M^*_{\alpha+1}$.*

PROOF. If $a \subseteq \varepsilon(E)$ is finite, then $E_a \in M_\alpha$, so $E_a \in M_\alpha|\lambda$, so $E_a \in M^*_{\alpha+1}$. Thus we need only see that $\text{dom}(E) = \text{dom}(F)$ for some F on the $M^*_{\alpha+1}$ sequence. But letting $\kappa = \text{crit}(E)$, we have $E_{\{\kappa\}} \in M_\alpha|\text{lh}(E)$, and therefore E does not have order zero. Let F be the Jensen completion of $E_{\{\kappa\}}$. The weak initial segment condition implies that F is on the sequence of M_α, hence on the sequence of $M_\alpha|\lambda$, hence on the sequence of $M^*_{\alpha+1}$. Since $\text{dom}(F) = \text{dom}(E)$, we are done. ⊣

Clearly the two claims yield the lemma. ⊣

Remark 4.5.4. One might guess that in general, if E is on the $M_{\beta+1}$ sequence and $\text{dom}(E) < \hat{\lambda}(E_\beta)$, then $\text{dom}(E) = \text{dom}(F)$ for some F on the M_β sequence. This is not true, however. One can construct a simple counterexample in which E has order zero, $\beta = 1$, and $T\text{-pred}(\beta+1) = 0$.

Remark 4.5.5. It seems possible that $T\text{-pred}(\alpha+1) = \alpha$, and E_α is very close to M_α but not to $M^*_{\alpha+1}$. But $M_\alpha|\text{lh}(E_\alpha) \trianglelefteq M^*_{\alpha+1}$ in this case, so E_α is close to $M^*_{\alpha+1}$ for a reason that copy and lift maps will preserve. See Remark 2.5.21.

We need to analyze alternative (2) of the Closeness Lemma a bit more. Let $\mathcal{T}$ be a plus tree and $\beta <_T \alpha$. Let $\beta <_T \gamma \leq_T \alpha$ and $T\text{-pred}(\gamma) = \beta$. We say that $\hat{\imath}^{\mathcal{T}}_{\beta,\alpha}$ *has a well-supported extender* iff

(i) $D^{\mathcal{T}} \cap (\gamma, \alpha]_T = \emptyset$, so that $\hat{\imath}^{\mathcal{T}}_{\beta,\alpha} = i^{\mathcal{T}}_{\gamma,\alpha} \circ i^{*,\mathcal{T}}_{\gamma}$, and

(ii) $\varepsilon^{\mathcal{T}}_\alpha \leq \sup(\text{ran}(\hat{\imath}^{\mathcal{T}}_{\beta,\alpha}))$.

Recall here that $\varepsilon^{\mathcal{T}}_\alpha = \sup(\{\varepsilon(G) \mid G \text{ is used in } [0,\alpha)_T\})$. If the extender of $\hat{\imath}^{\mathcal{T}}_{\beta,\alpha}$ is well supported, then its component measures concentrate on bounded subsets of $M^{*,\mathcal{T}}_\gamma$. For $c \subseteq \varepsilon$ finite, let

$$\mu_c = \text{least } \xi \text{ such that } c \subseteq \hat{\imath}^{\mathcal{T}}_{\beta,\alpha}(\xi).$$

Then the well supported extender of $\hat{\imath}^{\mathcal{T}}_{\beta,\alpha}$ is

$$(c,X) \in I^{\mathcal{T}}_{\beta,\alpha} \text{ iff } c \in [\varepsilon_\alpha]^{<\omega} \wedge X \subseteq [\mu_c]^{|c|} \wedge c \in \hat{\imath}_{\beta,\alpha}(X).$$

DEFINITION 4.5.6. Let $\mathcal{T}$ be a plus tree, and $I = I^{\mathcal{T}}_{\beta,\alpha}$; then I is *very close to* M iff for all finite $a \subseteq \varepsilon^{\mathcal{T}}_\alpha$, $I_a \in M$.

LEMMA 4.5.7. *Let $\mathcal{T}$ be a plus tree, $\beta <_T \alpha$, and suppose that $\hat{\imath}^{\mathcal{T}}_{\beta,\alpha}$ has a well supported branch extender I. Suppose that all extenders used in $(\beta,\alpha]_T$ are very close to the models to which they are applied; then I is very close to $M^{*,\mathcal{T}}_\gamma$, where γ is the least ordinal in $(\beta,\alpha]_T$.*

PROOF. Let us drop the superscript $\mathcal{T}$ when we can.

Let $\gamma \leq_T \alpha$ and $T\text{-pred}(\gamma) = \beta$. We shall show by induction on $\eta \leq_T \alpha$ that $I_{\beta,\eta}$ is very close to M^*_γ. The limit case is easy, so assume that

$$J = I_{\beta,\eta},$$

is very close to M^*_γ, and let $\xi + 1 \leq_T \alpha$, $T\text{-pred}(\xi+1) = \eta$, and let $c \subseteq \varepsilon^{\mathcal{T}}_{\xi+1} = \varepsilon(E_\xi)$ be finite. Let

$$I = I_{\beta,\xi+1}.$$

Since E_ξ is very close to M_η, we can assume that $c = a \cup b$, where $a \subseteq \varepsilon_\eta$ and f is a function in M^*_γ such that

$$\begin{aligned}(E_\xi)_b &= [a,f]^{M^*_\gamma}_J\\ &= i_{\gamma,\eta} \circ i^*_\gamma(f)(a).\end{aligned}$$

By induction, $J_a \in M^*_\gamma$. We can then compute I_c within M^*_γ as an iterated product. For $A \subseteq [\mu]^n$ and $u \in [\mu]^k$ where $k < n < \omega$, let $A_u = \{v \mid u \cup v \in A\}$; then for $A \subseteq [\mu_c]^{|c|}$,

$$\begin{aligned}A \in I_c &\text{ iff } a \cup b \in i_{\eta,\xi} \circ i_{\gamma,\eta} \circ i^*_\gamma(A)\\ &\text{ iff } i_{\gamma,\eta} \circ i^*_\gamma(A)_a \in i_{\gamma,\eta} \circ i^*_\gamma(f)(a)\\ &\text{ iff } i^{M^*_\gamma}_J(A)_a \in [a,f]^{M^*_\gamma}_J\\ &\text{ iff for } J_a \text{ a.e. } t,\ A_t \in f(t).\end{aligned}$$

Thus I_c in M^*_γ. ⊣

LEMMA 4.5.8. *Let $\mathcal{T}$ be a plus tree and $\beta = T\text{-pred}(\alpha+1) < \alpha$. Suppose that E_α is not very close to M_α. Let $\eta + 1 \leq_T \alpha$ be such that $\beta = T\text{-pred}(\eta+1)$; then*

(a) if $\beta <_T \gamma+1 \leq_T \alpha$, then E_γ is very close to both M_γ and $M^*_{\gamma+1}$, and

(b) the branch extender of $\hat{\imath}^{\mathcal{T}}_{\beta,\alpha}$ is well supported, and very close to $M^{*,\mathcal{T}}_{\eta+1}$.

PROOF. We begin with (a).

CLAIM 1. *E_η is very close to M_η.*

PROOF. Suppose not, and let $\kappa = \text{crit}(E_\alpha)$ and $\mu = \text{crit}(E_\eta)$, so that $\kappa < \mu < o(M^*_{\eta+1})$ and $\rho_1(M^*_{\eta+1}) < \mu$ by the Closeness Lemma. Let $F = \dot{F}^{M^*_{\eta+1}}$, so that $\kappa = \text{crit}(F)$.

Suppose first that $\eta = \beta$. Since $\rho_1(M^*_{\eta+1}) < \mu$, $\text{lh}(E_\beta) \leq o(M^*_{\eta+1})$. But $E_\beta \neq F$ because they have different critical points. Thus $E_\beta \in M^*_{\eta+1}$, so E_β is very close to $M^*_{\eta+1}$.

Suppose next that $\eta > \beta$, and suppose toward contradiction that E_η is not very close to M_η. Then clause (2) of the Closeness Lemma applies with η replacing α, so $\text{crit}(F) = \text{crit}(E_\eta)$. But $\kappa < \mu$, contradiction. ⊣

CLAIM 2. *Let $\beta <_T \xi = T\text{-pred}(\gamma+1)$ and $\gamma+1 \leq_T \alpha$; then E_γ is very close to M_γ and $M^*_{\gamma+1}$.*

PROOF. The proof is the same as that of Claim 1. Now $M_\xi = M^*_{\gamma+1}$ by (2)(ii) of Lemma 4.5.3. If $\xi = \gamma$, then E_γ cannot be $\dot{F}^{M_\gamma}$ because $\kappa < \text{crit}(E_\gamma)$. Thus E_γ is very close to M_γ, and hence to $M_\xi = M^*_{\gamma+1}$. So we may assume $\xi < \gamma$. By the Closeness Lemma, if E_γ is not very close to M_γ, then it has the same critical point as $\dot{F}^{M_\xi}$. But $\text{crit}(E_\gamma) > \kappa = \text{crit}(\dot{F}^{M_\xi})$.

Thus E_γ is very close to M_γ, and since $\xi < \gamma$, it is also very close to $M_\xi = M^*_{\gamma+1}$. ⊣

This proves (a).

Let G be the last extender of $M^*_{\eta+1}$; then all critical points along $(\beta, \alpha]_T$ are below the current image of λ_G, and $\hat{\imath}_{\beta,\alpha}$ is continuous at λ_G. Thus $\varepsilon^{\mathcal{T}}_\alpha \leq \sup \hat{\imath}_{\beta,\alpha}``\lambda_G$, so $I = I^{\mathcal{T}}_{\beta,\alpha}$ is well supported. Each I_c concentrates on $[\mu_c]^{|c|}$, where $\mu_c < \lambda_G$. I is very close to $M^*_{\eta+1}$ by the claims and Lemma 4.5.7. ⊣

Let us record the what we have shown about α such that $T\text{-pred}(\alpha+1) < \alpha$ and E_α is not very close to M_α in a definition.

DEFINITION 4.5.9. Let $\mathcal{T}$ be a plus tree, with models M_ξ and extenders E_ξ, and let $M^*_\xi = \mathcal{M}^{*,\mathcal{T}}_\xi$. We say α is *special in $\mathcal{T}$* iff letting $E = E_\alpha$ and $\beta = T\text{-pred}(\alpha+1)$,

(i) $\beta <_T \alpha$, and letting η be least in $(\beta, \alpha]_T$,
(ii) M^*_η is active, $k(M^*_\eta) = 0$, and letting $F = \dot{F}^{M^*_\eta}$, $\rho(M^*_\eta) \leq \text{dom}(F)$,
(iii) $D^{\mathcal{T}} \cap (\eta, \alpha]_T = \emptyset$, and for $i = i^{\mathcal{T}}_{\eta,\alpha} \circ i^{*,\mathcal{T}}_\eta$, $\text{dom}(F) < \text{crit}(i)$ and $E^- = i(F)$,
(iv) if $\eta \leq_T \gamma+1 \leq_T \alpha$, then E_γ is very close to M_γ and to $M^*_{\gamma+1}$, and
(v) $M^*_{\eta+1} \trianglelefteq M^*_{\alpha+1}$.

If α is special in $\mathcal{T}$, then M_α is active with last extender E_α. Moreover, letting η be as in the definition, the branch extender $I^{\mathcal{T}}_\eta$ is well supported and very close to $M^*_{\eta+1}$, by the proof of Lemma 4.5.8(b).

We have shown

THEOREM 4.5.10. *Let $\mathcal{T}$ be a plus tree and T-pred$(\alpha+1) < \alpha$; then either $E_\alpha^{\mathcal{T}}$ is very close to both $\mathcal{M}_\alpha^{\mathcal{T}}$ and $\mathcal{M}_{\alpha+1}^{*,\mathcal{T}}$, or α is special in $\mathcal{T}$.*

We do not know whether there can be special nodes α such that E_α is very close to M_α.

If α is special in $\mathcal{T}$, then $(E_\alpha)_c$ is $\Sigma_1^{M^*_{\eta+1}}$ in the parameter $(I_\alpha)_c$, where $\eta+1 \leq_T \alpha$ and T-pred$(\eta+1) = T$-pred$(\alpha+1)$. Let us record the Σ_1 definition.

DEFINITION 4.5.11. Let M be an active pfs premouse and $\bar{E} = \dot{F}^M$. Let $\kappa = \text{crit}(E)$ and $\lambda = \lambda(E)$. Let $M \models$ "I is a κ^+ complete ultrafilter on $[\nu]^n$", where $\kappa < \nu < \lambda$; then we define ultrafilters U_I and U_I^+ over $P([\kappa]^{<\omega}) \cap M$ by: for $X \subseteq [\kappa]^{<\omega}$ such that $X \in M$,

$$X \in U_I \text{ iff for } I \text{ a.e. } u,\, X \in \bar{E}_u$$

and

$$X \in U_I^+ \text{ iff for } I \text{ a.e. } u,\, X \in \bar{E}^+_{u\cup\{\lambda\}}.$$

We say that I is a *good code* of U_I and U_I^+ over M.

LEMMA 4.5.12. *(a) Let M be a pfs premouse and I be a good code of U over M; then U is Σ_1^M in the parameter I.*
(b) Let α be a special node of $\mathcal{T}$, $\beta = T$-pred$(\alpha+1)$, and $c \subseteq \varepsilon_\alpha^{\mathcal{T}}$ be finite; then $(I_{\beta,\alpha}^{\mathcal{T}})_c$ is a good code of $(E_\alpha)_c$ over $M_{\alpha+1}^{*,\mathcal{T}}$.

PROOF. For (a), we calculate that $X \in U_I$ iff

$$M \models \exists\xi\exists Z(Z = \dot{F}\cap M||\xi \wedge \exists Y \in I \forall u \in Y(u,X) \in Z).$$

For (b), let $I = I_{\beta,\alpha}^{\mathcal{T}}$, and suppose that X is in $(E_\alpha)_c$. Let $X \in M_{\alpha+1}^*||\xi$, where $\xi < \text{dom}(E) = \text{dom}(\bar{E})$, for $\bar{E}$ the last extender of $M^*_{\alpha+1}$. Let G be the fragment $\bar{E} \cap [\varepsilon(E)]^{<\omega} \times M^*_{\alpha+1}||\xi$. We have that $G \in M^*_{\alpha+1}$ and $(c,X) \in \hat{\imath}_{\beta,\alpha}(G)$. But $c = [c,\text{id}]_I^{M^*_{\alpha+1}}$, so pulling back under $\hat{\imath}_{\beta,\alpha}$, $(u,X) \in G$ for I_c a.e. u. Thus $X \in U_{I_c}$ or $X \in U_{I_c}^+$, depending on the type of E_α. The reverse inclusion follows from the fact that U and $(E_\alpha)_c$ are both ultrafilters. ⊣

The Σ_1 definitions given Lemma 4.5.12 will give us the uniformity of closeness in various copying and lifting constructions. When the extenders are very close to the models to which they are applied, then the uniformity is given by

DEFINITION 4.5.13. Let P and Q be pfs premice, and E and F be (possibly long) extenders. We say that $\langle\pi,\varphi\rangle\colon (P,E) \overset{**}{\to} (Q,F)$ if and only if

(1) $\langle\pi,\varphi\rangle\colon (P,E) \to (Q,F)$,
(2) E is very close to P, and
(3) for all finite $c \subseteq \varepsilon(E)$, $\pi(E_c) = F_{\varphi(c)}$.

In practice, F will be very close to Q, but the definition only requires that all $F_{\varphi(c)}$ for $c \subseteq \varepsilon(E)$ belong to Q. The long extenders to which we shall apply the definition will be well supported branch extenders.

LEMMA 4.5.14. *Suppose that $\langle \pi, \varphi \rangle : (P, E) \to (Q, F)$; then the following are equivalent*

(a) $\langle \pi, \varphi \rangle : (P, E) \overset{**}{\to} (Q, F)$,

(b) E is very close to P, and $\langle \pi, \varphi \rangle : (P, E) \overset{*}{\to} (Q, F)$.

PROOF. Clearly (a) implies (b). Assume (b) holds, and let $c \subseteq \varepsilon(E)$ be finite. Let $U = E_c$, and let $\theta(v_0, v_1)$ be a Σ_1 formula and $r \in P$ be such that $X \in E_c$ iff $P \models \theta[X, r]$ and $X \in F_{\varphi(c)}$ iff $Q \models \theta[X, \pi(r)]$. Then $P \models \forall X(\theta(X, r) \to X \in U)$. This is a Π_1 fact about U and r, and π is elementary, so $Q \models \forall X(\theta(X, \pi(r)) \to X \in \pi(U))$. It follows that $\pi(U) = F_{\varphi(c)}$, since the two are ultrafilters. ⊣

LEMMA 4.5.15. *Let M, N, P, and Q be premice. Let $\varphi : M \to N$ and $\pi : P \to Q$ be elementary, and let E be an extender such that E^- is on the sequence of M. Suppose that E is very close to M and P, and that the Shift Lemma applies to (π, φ, E); then $\langle \pi, \varphi \rangle : (P, E) \overset{**}{\to} (Q, \varphi(E))$.*

PROOF. Let $U = E_b$, where $b \subset \varepsilon(E)$ is finite. We have $\pi(U) = \varphi(U)$ by the agreement between π and φ. The fact about the parameters U and b that $U = E_b$ is expressible by a Π_1 formula $\theta(v_0, v_1)$ interpreted over the structure $M|\mathrm{lh}(E)$. Since φ is elementary, it preserves θ, and thus $\pi(U) = \varphi(U) = \varphi(E)_{\varphi(b)}$, as desired. ⊣

Notice that in Lemma 4.5.15 we needed the map φ on extenders to be elementary. The Shift Lemma itself only requires that φ be Σ_0 elementary.

We can extend the uniformity here by replacing φ by an appropriate embedding of one branch extender into another. Unfortunately, it takes longer to state the resulting lemma than it does to understand it. We shall apply it to copying below, and to other kinds of lifting later on.

LEMMA 4.5.16. *Let $\mathcal{T}$ and $\mathcal{U}$ be plus trees with models M_ξ and N_ξ respectively. Let $\alpha <_T \beta$, and $r : [\alpha, \beta]_T \to \mathrm{lh}(\mathcal{U})$ be such that*

(i) $\xi <_T \eta$ iff $r(\xi) <_U r(\eta)$, and

(ii) if $\xi, \eta \in \mathrm{dom}(r)$ and $\xi = T\text{-pred}(\eta)$, then $r(\xi) = U\text{-pred}(r(\eta))$.

Suppose that all extenders used in $[\alpha, \beta]_T$ are very close to the models to which they are applied, and that the branch extenders of $\hat{\imath}^{\mathcal{T}}_{\alpha,\beta}$ and $\hat{\imath}^{\mathcal{U}}_{r(\alpha),r(\beta)}$ are well supported. Let $M^* \trianglelefteq M_\alpha$ be the domain of $\hat{\imath}^{\mathcal{T}}_{\alpha,\beta}$ and $N^* \trianglelefteq N_{r(\alpha)}$ be the domain of $\hat{\imath}^{\mathcal{U}}_{r(\alpha),r(\beta)}$, and suppose we have maps π_ξ for $\xi \in [\alpha, \beta]_T$ such that

(i) $\pi_\alpha : M^* \to N^*$ is elementary,

(ii) for $\xi > \alpha$, $\pi_\xi : M_\xi \to N_{r(\xi)}$ is elementary,

(iii) if $\xi < \delta$ then $\pi_\delta \circ \hat{\imath}^{\mathcal{T}}_{\xi,\delta} = \hat{\imath}^{\mathcal{U}}_{r(\xi),r(\delta)} \circ \pi_\xi$, and

(iv) if $\xi, \eta \in \mathrm{dom}(r)$ *and* $\xi = T\text{-pred}(\eta)$*, then*

$$\langle \pi_\xi, \pi_\eta \rangle\colon (M_\xi, E^{\mathcal{T}}_{\eta-1}) \overset{**}{\to} (N_{r(\xi)}, E^{\mathcal{U}}_{r(\eta)-1}).$$

Then

$$\langle \pi_\alpha, \pi_\beta \rangle\colon (M^*, I^{\mathcal{T}}_{\alpha,\beta}) \overset{**}{\to} (N^*, I^{\mathcal{U}}_{r(\alpha),r(\beta)}).$$

PROOF SKETCH. Fixing α, we show this by induction on β. The limit step is trivial. At the successor step, we have $T\text{-pred}(\eta) = \xi$, and know that $I^{\mathcal{T}}_{\alpha,\xi}$ is appropriately embedded into $I^{\mathcal{U}}_{r(\alpha),r(\xi)}$. By (iv), we also have $E^{\mathcal{T}}_{\eta-1}$ appropriately embedded into $E^{\mathcal{U}}_{r(\eta)-1}$. Let $I = I^{\mathcal{T}}_{\alpha,\eta}$ and $J = I^{\mathcal{U}}_{r(\alpha),r(\beta)}$. Inspecting the proof of Lemma 4.5.7, we see that π_α moves the iterated product measure corresponding to I_c to the iterated product measure corresponding to $J_{\pi_\eta(c)}$, as desired. ⊣

Let us look at copying now.

LEMMA 4.5.17 (Copy Lemma). *Let M and N be pfs premice, $\pi\colon M \to N$ be nearly elementary, and $\mathcal{T}$ on M be a plus tree, and let $\pi\mathcal{T}$ be the copied tree, with associated copy maps π_α. Let $E_\alpha = E^{\mathcal{T}}_\alpha$; then for $\alpha < \beta$,*

(1) π_α is nearly elementary,

(2) $\pi_\alpha \restriction \varepsilon(E_\alpha) = \pi_\beta \restriction \varepsilon(E_\alpha)$, and

(3) if $\alpha <_T \beta$, then $\pi_\alpha \circ \hat{\imath}^{\mathcal{T}}_{\beta,\alpha} = \hat{\imath}^{\pi\mathcal{T}}_{\beta,\alpha} \circ \pi_\beta$.

Moreover, if π is elementary, then all the π_α are elementary, and $\pi\mathcal{T}$ is a plus tree.

PROOF. Let M_ξ and E_ξ be the models and extenders of $\mathcal{T}$, and N_ξ and F_ξ the models and extenders of $\pi\mathcal{T}$. Let $M^*_\xi = \mathcal{M}^{*,\mathcal{T}}_\xi$, $N^*_\xi = \mathcal{M}^{*,\pi\mathcal{T}}_\xi$, $i_{\delta,\gamma} = i^{\mathcal{T}}_{\delta,\gamma}$, and $j_{\delta,\gamma} = i^{\pi\mathcal{T}}_{\delta,\gamma}$.

Parts (1)-(3) are a routine induction. Letting $\beta = T\text{-pred}(\alpha+1)$, (2) implies that

$$\langle \pi_\beta, \pi_\alpha \rangle\colon (M^*_{\alpha+1}, E_\alpha) \to (N^*_{\alpha+1}, F_\alpha).$$

We then get $\pi_{\alpha+1}$ and (1)-(3) at $\alpha+1$ from Lemma 2.5.19.[40]

Suppose now that π is elementary. We shall show by induction that π_ξ is elementary. This is easy if ξ is a limit ordinal, so suppose $\xi = \alpha+1$. Let $E = E_\alpha$ and $F = F_\alpha$, and let $\beta = T\text{-pred}(\alpha+1)$ and $\lambda = \hat{\lambda}(E_\beta)$. The case $\beta = \alpha$ is straightforward, and covered by Remark 2.5.21, so let us assume $\beta < \alpha$.

Suppose first that E is very close to M_α, and hence to $M^*_{\alpha+1}$. Then by Lemma 4.5.15, $\langle \pi_\beta, \pi_\alpha \rangle\colon (M^*_{\alpha+1}, E) \overset{**}{\to} (N^*_{\alpha+1}, F)$, so $\pi_{\alpha+1}$ is elementary.

Suppose next that α is special in $\mathcal{T}$. We must then be in the situation described in Lemma 4.5.3(2). Here is a diagram

[40] Concerning (2), the agreement is actually on $\lambda(E_\alpha)$ if E_α does not have plus type, and on $\mathrm{lh}(E_\alpha)+1$ if it does. Moreover, π_α agrees with $\pi_{\alpha+1}$ on $\mathrm{lh}(E_\alpha)+1$ in any case, although it may agree less with later π_β if E_α is not of plus type.

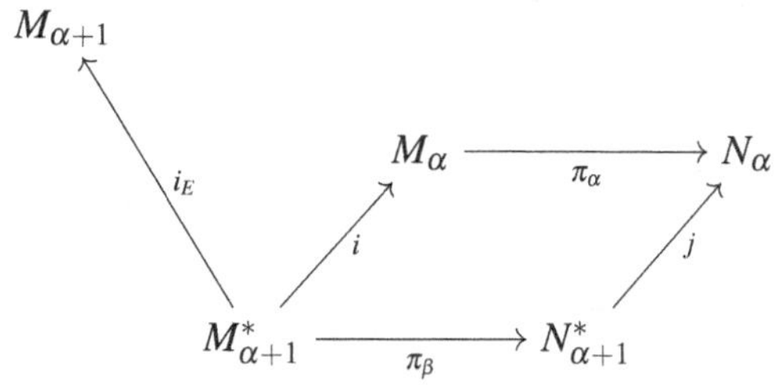

Here $i = \hat{\imath}^{\mathcal{T}}_{\beta,\alpha}$ and $j = \hat{\imath}^{\pi\mathcal{T}}_{\beta,\alpha}$. By Lemma 4.5.8, all extenders used in i are very close to the models from which they are taken, and to the models to which they are applied. Letting I and J be the well supported branch extenders associated to i and j, we have

$$\langle \pi_\beta, \pi_\alpha \rangle \colon (M^*_{\alpha+1}, I) \overset{**}{\to} (N^*_{\alpha+1}, J)$$

by Lemma 4.5.16. Let us show now that

$$\langle \pi_\beta, \pi_\alpha \rangle \colon (M^*_{\alpha+1}, E) \overset{*}{\to} (N^*_{\alpha+1}, F),$$

from which it follows that $\pi_{\alpha+1}$ is elementary. Let $c \subseteq \varepsilon(E)$ be finite. Let us take the case E does not have plus type; then $E_c = U_{I_c}$, and this gives us a Σ_1 definition of E_c over $M^*_{\alpha+1}$ from the parameter I_c, namely $X \in E_c$ iff

$$M^*_{\alpha+1} \models \exists \xi \exists Z (Z = \dot{F} \cap M||\xi \wedge \exists Y \in I_c \forall u \in Y (u, X) \in Z).$$

Similarly, $X \in F_{\pi_\alpha(c)}$ iff

$$N^*_{\alpha+1} \models \exists \xi \exists Z (Z = \dot{F} \cap M||\xi \wedge \exists Y \in J_{\pi_\alpha(c)} \forall u \in Y (u, X) \in Z).$$

But $\pi_\beta(I_c) = J_{\pi_\alpha(c)}$, so the Σ_1 definition of E_c is moved to a Σ_1 definition of $F_{\pi_\alpha(c)}$, as required. $\dashv$

Remark 4.5.18. The proof showed that if π is elementary, then whenever $\beta = T\text{-pred}(\alpha+1)$, then $\langle \pi_\beta, \pi_\alpha \rangle \colon (M^*_{\alpha+1}, E) \overset{*}{\to} (N^*_{\alpha+1}, F)$.

Unfortunately, we do need to copy plus trees under maps that are not elementary. One way to deal with this is to extend the definition of *plus tree* so as to allow gratuitous drops, and prove everything in that more general context. Another way is to eliminate gratuitous drops in $\pi\mathcal{T}$ by lifting it to a maximal tree $(\pi\mathcal{T})^+$ as we construct it. In this book we shall use the second method. The lifting here is a special case of a more general lifting procedure that we describe now.

Let $\mathcal{T}$ be a plus tree on the premouse M, and let $k = k(M)$. Let

$$\pi \colon M \to Q \trianglelefteq N$$

be nearly elementary; then we can lift $\mathcal{T}$ to a plus tree $\mathcal{U}$ on N as follows. $\mathcal{U}$ will have the same tree order as $\mathcal{T}$, so long as it is defined. Let M_α and N_α be the α-th models of $\mathcal{T}$ and $\mathcal{U}$, and E_α and F_α the α-th extenders. We shall have a nearly elementary

$$\pi_\alpha \colon M_\alpha \to Q_\alpha \trianglelefteq N_\alpha.$$

Here $\pi_0 = \pi$ and $Q_0 = Q$. We have the usual agreement and commutativity conditions:

(1) if $\beta \leq \alpha$, then $\pi_\alpha \restriction \varepsilon(E_\beta) = \pi_\beta \restriction \varepsilon(E_\beta)$ and $N_\alpha | \varepsilon(F_\beta) = N_\beta | \varepsilon(F_\beta)$, and
(2) if $\beta \leq_T \alpha$, then $\pi_\alpha \circ \hat{\imath}^{\mathcal{T}}_{\beta,\alpha} = \hat{\imath}^{\mathcal{U}}_{\beta,\alpha} \circ \pi_\beta$.

Drops in $\mathcal{T}$ of more than one degree will cause corresponding drops in $\mathcal{U}$. Drops of one degree may not. $\mathcal{U}$ may drop where $\mathcal{T}$ does not.

The successor step is the following. We are given E_α on M_α; set

$$F_\alpha = \pi_\alpha(E_\alpha),$$

or $F_\alpha = \dot{F}^{Q_\alpha}$ if $E_\alpha = \dot{F}^{M_\alpha}$. As above, our convention is that if $E_\alpha = E^+$ where E is on the M_α sequence, then $\pi_\alpha(E_\alpha) = \pi_\alpha(E^+) = \pi_\alpha(E)^+$. Let $\beta = T\text{-pred}(\alpha+1) =$ least ξ such that $\kappa < \hat{\lambda}(E_\xi)$, where $\kappa = \text{crit}(E_\alpha)$. By (1) above, $\beta = U\text{-pred}(\alpha+1)$ according to the rules of plus trees for $\mathcal{U}$. Let

$$M_{\alpha+1} = \text{Ult}(M^*_{\alpha+1}, E_\alpha),$$

and

$$N_{\alpha+1} = \text{Ult}(N^*_{\alpha+1}, F_\alpha),$$

where $M^*_{\alpha+1}$ and $N^*_{\alpha+1}$ are determined by the maximality of $\mathcal{T}$ and $\mathcal{U}$. Let

$$S = \pi_\beta(M^*_{\alpha+1}),$$

where as usual, if $M^*_{\alpha+1} = M_\beta \downarrow n$ then $S = Q_\beta \downarrow n$. Clearly $\pi_\beta \restriction M^*_{\alpha+1}$ is nearly elementary as a map into S, so $\text{crit}(F_\alpha)$ is a cardinal of S and $\text{crit}(F_\alpha) < \rho^-(S)$. It follows that

$$S \trianglelefteq N^*_{\alpha+1}.$$

Let $i^* \colon N^*_{\alpha+1} \to N_{\alpha+1}$ be the canonical embedding, and

$$Q_{\alpha+1} = i^*(S),$$

with the usual convention if $S = N^*_{\alpha+1} \downarrow n$ for some n. We obtain $\pi_{\alpha+1}$ by a variant of the Shift Lemma: let $R = M^*_{\alpha+1}$ and $k = k(R) = k(S)$. We obtain $\sigma \colon \text{Ult}_0(R^k, E_\alpha) \to Q^k_{\alpha+1}$ by setting

$$\sigma([a,f]^{R^k}_{E_\alpha}) = [\pi_\alpha(a), \pi_\beta(f)]^{N^*_{\alpha+1}}_{F_\alpha} = i^*(\pi_\beta(f))(\pi_\alpha(a)),$$

where the equivalence class on the right in line 1 is formed using functions appropriate to $\text{Ult}(N^*_{\alpha+1}, F_\alpha)$. By the proof of Lemma 2.5.19, σ is Σ_0 elementary and cardinal preserving map from $M^k_{\alpha+1}$ to $Q^k_{\alpha+1}$. We let $\pi_{\alpha+1}$ be its completion.

One can easily check that the inductive hypotheses are maintained. At limit λ we let Q_λ be the common value of $i^{\mathcal{U}}_{\alpha,\lambda}(Q_\alpha)$ for all $\alpha <_T \lambda$ sufficiently large. Note that $Q_{\alpha+1} \trianglelefteq i^{\mathcal{U}}_{\beta,\alpha+1}(Q_\beta)$, and we are assuming that N_λ is wellfounded, so there is such a common value. π_λ is defined using commutativity.

DEFINITION 4.5.19. Suppose that $\pi\colon M \to Q \trianglelefteq N$ is nearly elementary, and let $\mathcal{T}$ be a plus tree on M; then

(a) $(\pi\mathcal{T})^+$ is the plus tree $\mathcal{U}$ on N defined above. We call $(\pi\mathcal{T})^+$ the (π, Q) *lift of* $\mathcal{T}$ *to* N, or if $Q = N|\langle \nu, k\rangle$, the (π, ν, k)*-lift of* $\mathcal{T}$ *to* N. We call the map $\pi_\alpha\colon M_\alpha \to Q_\alpha$ defined above the α*-th lift map* associated to $(\pi\mathcal{T})^+$.

(b) When $M = Q$ and π is the identity, we let $\mathcal{T}^+ = (\pi\mathcal{T})^+$, and call $\mathcal{T}^+$ the *lift of* $\mathcal{T}$ *to* N.

Definition 4.5.19 extends to lifts of stacks of plus trees in the obvious way. One can extend the definition so as to allow non-maximal trees $\mathcal{T}$, and thereby obtain a natural reduction of arbitrary semi-normal trees to maximal ones.

Remark 4.5.20. The construction of $(\pi\mathcal{T})^+$ is a bit like the construction of a conversion system. In the conversion case we begin with $\pi\colon M \to Q \in N$, where $Q \in \text{lev}(\mathbb{C})$ for some construction $\mathbb{C}$ of N, instead of $\pi\colon M \to Q \trianglelefteq N$. The structure of $\mathbb{C}$ mediates the step from $\pi_\alpha(E_\alpha)$ to F_α. In both cases the lifting maps π_α are in general only nearly elementary, no matter how elementary the original π is. The reason is that the downstairs ultrapower is not just copied, its copy is embedded into an ultrapower formed by using more functions.

One can think of $(\pi\mathcal{T})^+$ as having been produced by the ordinary copying construction, which yields $\pi\mathcal{T}$ on Q, followed by applying the (id, ν, k)-lift to $\pi\mathcal{T}$, and obtaining $(\pi\mathcal{T})^+$. So our notation in (a) and (b) of Definition 4.5.19 is consistent with the earlier copying construction notation. We shall often write $\pi\mathcal{T}^+$ instead of $(\pi\mathcal{T})^+$. There is a possible confusion between $\pi\mathcal{U}$ for $\mathcal{U} = \mathcal{T}^+$ and $(\pi\mathcal{T})^+$ here, but context will resolve it.

Let us return now to copying plus trees, that is, the case that we have $\pi\colon M \to Q \trianglelefteq N$ where in fact $Q = N$.

LEMMA 4.5.21. *Let* $\pi\colon M \to N$ *be elementary and* $\mathcal{T}$ *be a plus tree on* M*; then* $\pi\mathcal{T} = \pi\mathcal{T}^+$.

PROOF. This is implicit in the “moreover” part of Proposition 4.5.17. ⊣

If π is only nearly elementary, then $\pi\mathcal{T} \neq \pi\mathcal{T}^+$ is possible, but the two are never out of step by more than one degree.

LEMMA 4.5.22. *Let* $\pi\colon M \to N$ *be nearly elementary, and* $\mathcal{T}$ *be a plus tree on* M*. Suppose that all models of* $\pi\mathcal{T}^+$ *are wellfounded, and let* $\pi_\alpha\colon M_\alpha \to Q_\alpha \trianglelefteq N_\alpha$ *be the associated* α*-th lift map. Then for any* α*, either*

(a) $Q_\alpha = N_\alpha$*, or*

(b) $Q_\alpha = N_\alpha^-$*,* π_α *is nearly elementary and exact, and* M_α *is stable.*

The proof is a routine induction.

Elementarity in various contexts. Here is a summary of elementarity in various situations we shall encounter. Let M be a pfs premouse.

(i) If M is solid and of type 1, then anticore map from $\mathfrak{C}(M)^-$ to M is elementary, cofinal, and exact.
(ii) The resurrection maps associated to a PFS construction are elementary, cofinal, and exact. (See Section 4.7.)
(iii) Fine ultrapower maps, and more generally, the maps $\hat{i}^{\mathcal{T}}_{\alpha,\beta}$ along branches of a quasi-normal iteration tree on M, are elementary. If M has type 1 and is strongly stable, then these maps are exact, and all the $M^{\mathcal{T}}_\alpha$ have type 1.
(iv) If $\pi\colon M \to N$ is nearly elementary, and $\mathcal{T}$ is a quasi-normal tree on M, then $\pi\mathcal{T}$ is semi-normal, and the copy maps $\pi_\alpha\colon \mathcal{M}^{\mathcal{T}}_\alpha \to \mathcal{M}^{\pi\mathcal{T}}_\alpha$ are nearly elementary. If π is elementary, then $\pi\mathcal{T}$ is quasi-normal, and all the π_α are elementary.
(v) The Dodd-Jensen and Weak Dodd-Jensen lemmas hold in the category of nearly elementary maps.

As with maps on ordinary premice, factor embeddings from one ultrapower to another that is formed using a larger class of functions can lead to maps that are nearly elementary but not elementary. See Examples 2.4.9 and 2.4.10. These include the embedding normalization maps σ_γ[41], the lifting maps associated to $\pi\mathcal{T}^+$ we defined above, and the lifting maps of a conversion system.

4.6. Iteration strategies and comparison

Iteration strategies acting on plus trees, or stacks of them, are what one would expect. If M is a pfs premouse, then $G^+(M,\theta)$ is the variant of $G^{\mathrm{qn}}(M,\theta)$ in which player II must pick cofinal wellfounded branches at limit steps as before, and given that $\mathcal{T}$ with $\mathrm{lh}(\mathcal{T}) = \alpha+1$ is the play so far, I must pick E_α such that E_α^- (which may or may not be equal to E_α) is on the $\mathcal{M}^{\mathcal{T}}_\alpha$ sequence, and such that $\hat{\lambda}(E_\beta) \leq \hat{\lambda}(E_\alpha)$ for all $\beta < \alpha$. Since $\mathcal{T}$ is to be maximal, this determines

$$T\text{-pred}(\alpha+1) = \text{least } \beta \text{ s.t. } \mathrm{crit}(E_\alpha) < \hat{\lambda}(E_\beta),$$

and for $\xi = T\text{-pred}(\alpha+1)$, the initial segment $\mathcal{M}^*_{\alpha+1}$ of $\mathcal{M}^{\mathcal{T}}_\xi$ such that

$$\mathcal{M}^{\mathcal{T}}_{\alpha+1} = \mathrm{Ult}(\mathcal{M}^*_{\alpha+1}, E_\alpha).$$

A plus tree on M is just a position in some $G^+(M,\theta)$ in which II has not yet lost.

The example from Remark 4.4.9 shows that there can be distinct normal plus trees by the same iteration strategy that have the same last model. The reason is that at some step, one tree might use E while the other uses E^+. What does hold is

[41] See Chapter 6.

LEMMA 4.6.1. *Let $\mathcal{T}$ and $\mathcal{U}$ be normal plus trees by the same iteration strategy, and having the same last model. Suppose that whenever $\alpha + 1 < \inf(\mathrm{lh}(\mathcal{T}), \mathrm{lh}(\mathcal{U}))$, then the plus case occurs at α in $\mathcal{T}$ iff the plus case occurs at α in $\mathcal{U}$; then $\mathcal{T} = \mathcal{U}$.*

We omit the simple proof.

For λ a limit ordinal or $\lambda = 1$, $G^+(M, \lambda, \theta)$ is the variant of $G^{\mathrm{qn}}(M, \lambda, \theta)$ whose output is a stack of plus trees on M of length λ. (So $G^+(M, 1, \theta) = G^+(M, \theta)$.) We allow a gratuitous drop at the beginning of each round. II wins iff all models reached are wellfounded, and if $\lambda > 1$, there are finitely many drops along the sequence of base models, and their direct limit is wellfounded.

An M-stack is a position in some $G^+(M, \lambda, \theta)$ in which II has not yet lost. Precisely,

DEFINITION 4.6.2. Let M be a premouse; then s is an *M-stack* iff

$$s = \langle (\nu_\alpha, k_\alpha, \mathcal{T}_\alpha) \mid \alpha < \beta \rangle,$$

and there are premice M_α for $\alpha < \beta$ such that

(1) $\mathcal{T}_\alpha$ is a plus tree on $M_\alpha | \langle \nu_\alpha, k_\alpha \rangle$,
(2) $M_0 = M$,
(3) if $\alpha < \beta$ and α is a limit ordinal, then M_α is the direct limit of the M_β for $\beta < \alpha$, and
(4) if $\gamma + 1 = \alpha < \beta$, then M_α is the last model of $\mathcal{T}_\gamma$

If each $\mathcal{T}_\alpha$ is normal, then we call s a *normal M-stack*. If $\langle \nu_\alpha, k_\alpha \rangle = l(M_\alpha)$ for all α, we say s is *maximal*.[42]

So a maximal stack is one with no gratuitous drops anywhere. It is normal iff its component trees are all normal.

DEFINITION 4.6.3. Let M be a pfs premouse; then a *complete θ-iteration strategy* for M is a winning strategy for player II in $G^+(M, \theta)$. A *complete (λ, θ)-iteration strategy* for M is a winning strategy for II in $G^+(M, \lambda, \theta)$.

DEFINITION 4.6.4. Let M be a pfs premouse; then M is *countably iterable* iff whenever N is countable and there is an elementary $\pi \colon N \to M$, then there is a complete $(\omega_1, \omega_1 + 1)$-iteration strategy for N.

The terminology and notation of Section 2.7 regarding tail strategies, pullback strategies, and positionality extends to strategies acting on stacks of plus trees in an obvious way.

DEFINITION 4.6.5. Let Ω be a winning strategy for II in $G^+(M, \lambda, \theta)$ and s be an M-stack according to Ω with $\mathrm{lh}(s) < \lambda$.

[42]We allow $k_\alpha = -1$, with the convention that $P | \langle \nu, -1 \rangle = P || \nu$.

(a) If $N = M_\infty(s)|\langle \nu, k\rangle$ for some ν, k, then $\Omega_{s,N}$ is the *tail strategy*

$$\Omega_{s,N}(t) = \Omega(s^\frown\langle N\rangle^\frown t).$$

We set $\Omega_s = \Omega_{s,M_\infty(s)}$.

(b) If $\pi\colon Q \to M$ is nearly elementary, then Ω^π is the *pullback strategy* for $G^+(Q,\lambda,\theta)$ given by

$$\Omega^\pi(s) = \Omega((\pi s)^+).$$

In part (b), $(\pi s)^+$ is the stack of plus trees that we get by copying and lifting so that each component tree in $(\pi s)^+$ is maximal. Thus if $\mathcal{T}$ is the first tree in s, then $(\pi\mathcal{T})^+$ is the first tree in $(\pi s)^+$. $(\pi s)^+$ does copy the gratuitous drops at the beginning of rounds in s. If π is elementary, then $(\pi s)^+ = \pi s$.[43]

Iterable pfs premice can be compared. In the most important case, the premice are strongly stable and of type 1.

THEOREM 4.6.6. *Let P and Q be strongly stable pfs premice of type 1 and of size $\leq \theta$, and suppose Σ and Ψ are complete $\theta^+ + 1$-iteration strategies for P and Q respectively; then there are normal, λ-tight plus trees $\mathcal{T}$ by Σ and $\mathcal{U}$ by Ψ of size θ, with last models R and S, such that either*

(a) $R \trianglelefteq S$, and P-to-R does not drop, or

(b) $S \trianglelefteq R$, and Q-to-S does not drop.

PROOF. The proof for ordinary premice works. (See Theorem 2.8.1.) We compare by iterating away least disagreements, so the comparison trees use only extenders on the sequence, with strictly increasing lengths. That is, they are λ-tight and normal. The standard reflection argument gives trees $\mathcal{T} = \mathcal{T}_\alpha$ by Σ and $\mathcal{U} = \mathcal{U}_\alpha$ by Ψ with last models R and S such that $R \trianglelefteq S$ or $S \trianglelefteq R$.

If $R \triangleleft S$, then R is sound, so by Lemma 4.3.11, the branch P-to-R did not drop, and we have conclusion (a). Similarly, if $S \triangleleft R$ we get conclusion (b). Thus we may assume $R = S$. It is now enough to show that one of the two branches P-to-R and Q-to-S did not drop. Assume otherwise, and let $X = M^{*,\mathcal{T}}_{\alpha+1}$ and $Y = M^{*,\mathcal{U}}_{\beta+1}$ be the last drops on the two branches. Since P and Q are strongly stable and we have dropped, both X and Y are sound type 1 pfs premice. (Cf. Lemma 4.4.6.) Let $i\colon X \to R$ and $j\colon Y \to S$ be the branch embeddings. By Lemma 4.3.11,

$$\rho(X) = \rho(R) = \rho(S) = \rho(Y),$$

and

$$i(p(X)) = p(R) = p(S) = j(p(Y)).$$

[43] If Ω were defined on stacks of non-maximal trees, we could have defined $\Omega^\pi(s) = \Omega(\pi s)$. We could also have set $\Omega^\pi(s) = \Omega((\pi s)^{++})$, where $(\pi s)^{++}$ is the maximal stack of plus trees that comes from eliminating gratuitous drops at the beginnings of rounds. In the abstract, these are different pullback strategies, but for the strategies Ω that we eventually construct, they are the same, because $\Omega^\pi = \Omega$ for each version of Ω^π.

Also, $\rho(X) < \text{crit}(i)$ and $\rho(Y) < \text{crit}(j)$ because X and Y are projectum solid, and

$$\mathfrak{C}(X) = \mathfrak{C}(R) = \mathfrak{C}(Y).$$

It follows that $X = \mathfrak{C}(R)^- = Y$, and i and j are the anticore map from $\mathfrak{C}(R)^-$ to R. So $i = j$, so the first extenders used i and j are compatible, and hence the same by the Jensen initial segment condition. This is a contradiction. ⊣

Phalanx comparisons work too, as we shall see later Sections 4.9 and 4.10. Those proofs require iteration strategies with the *Weak Dodd-Jensen property*. We need Weak Dodd-Jensen in the category of nearly elementary maps for some of them. In fact, we must go slightly beyond that in one case.

DEFINITION 4.6.7. Let M be a pfs premouse and $k = k(M) > 0$; then

$$B^k = \{\langle \varphi, b\rangle \mid \varphi \text{ is } \Sigma_1 \wedge b \in M||\rho_k \wedge M^{k-1} \models \varphi[b, p_k]\},$$

and

$$M_0^k = (M||\rho_k, B^k).$$

We call M_0^k the *reduct of* $\bar{\mathfrak{C}}_k(M)$.

M_0^k codes the strong core $\bar{\mathfrak{C}}_k(M)$. Any Σ_0 elementary map from $\pi\colon M_0^k \to N_0^k$ has a unique completion $\pi^*\colon \bar{\mathfrak{C}}_k(M) \to \bar{\mathfrak{C}}_k(N)$.

DEFINITION 4.6.8. Let M be a countable pfs premouse, and $\langle e_i \mid i < \omega\rangle$ enumerate the universe of M. A map $\pi\colon M \to N$ is *$\vec{e}$-minimal* just in case π is nearly elementary, and whenever $\sigma\colon M \to N|\langle \eta, k\rangle$ is nearly elementary, then $\langle \eta, k\rangle = l(N)$, and if $\sigma \neq \pi$, then for i least such that $\sigma(e_i) \neq \pi(e_i)$, we have $\pi(e_i) < \sigma(e_i)$ (in the order of construction).

DEFINITION 4.6.9. An iteration strategy Ω for M has the *Weak Dodd-Jensen property* relative to an enumeration $\vec{e}$ of its universe in order type ω iff whenever $N = M_\infty(s)$ for some stack s by Ω, then

(1) if there is a nearly elementary embedding from M to an initial segment of N, then the branch M-to-N of s does not drop, and the iteration map i^s is $\vec{e}$-minimal, and
(2) if M has type 1A, $k = k(M)$, and there is a Σ_0 elementary map from M_0^k to N_0^k, then the branch M-to-N of s does not drop in model.

LEMMA 4.6.10 (Weak Dodd-Jensen). *Let M be a pfs premouse, $\vec{e}$ be an enumeration of the universe of M in order type ω, and Ω a complete (ω_1, θ) iteration strategy for M; then there is a countable M-stack s by Ω having last model $N = M_\infty(s)$, and a nearly elementary $\pi\colon M \to N$, such that $(\Omega_{s,N})^\pi$ has the Weak Dodd-Jensen property relative to $\vec{e}$.*

The proof of Lemma 4.6.10 is essentially that of [39]. The main facts it uses are

(i) If $\pi\colon M \to N$ is nearly elementary, and $\mathcal{T}$ is a plus tree on M, then $(\pi\mathcal{T})^+$ is a plus tree on N.

(ii) The collection of nearly elementary maps from M to N is closed in the product topology on ${}^M N$.

There are some small additional elements needed to insure (2) in Definition 4.6.9.[44]

Comparing mice that are not strongly stable. Comparison by least disagreement works for iterable premice that are are not strongly stable. We shall not need such comparisons, but for the sake of completeness, we record the basic facts. The possible termination patterns are more complicated, because the side that comes out weaker may have a drop of one degree.

DEFINITION 4.6.11. Suppose that $\mathcal{T}$ is a plus tree on P. We say that $\mathcal{T}$ has a *small drop at* $\xi + 1$ iff $[0,\xi]_T \cap D^{\mathcal{T}} = \emptyset$, and for $k = k(P)$,

$$\mathcal{M}^{\mathcal{T}}_{\xi+1} = \mathrm{Ult}_{k-1}(\mathcal{M}^{\mathcal{T}}_{\xi}, D)$$

for some order zero D on the X sequence such that $\rho_k(X) \leq \mathrm{crit}(D) < \rho_{k-1}(X)$. In this case we say that $\hat{\imath}^{\mathcal{T}}_{0,\xi+1}$ is an *essentially* $r\Sigma_{k+1}$ *iteration map*.

THEOREM 4.6.12. *Let P and Q be stable pfs premice of type 1 and of size $\leq \theta$, and suppose Σ and Ψ are complete $\theta^+ + 1$-iteration strategies for P and Q respectively; then there are normal, λ-tight plus trees $\mathcal{T}$ by Σ and $\mathcal{U}$ by Ψ of size θ, with last models $R = \mathcal{M}^{\mathcal{T}}_{\xi}$ and $S = \mathcal{M}^{\mathcal{U}}_{\eta}$, such that either*

(a) *$R \trianglelefteq S$, and $[0,\xi]_T \cap D^{\mathcal{T}} = \emptyset$, or*
(b) *$R^- = S$, R has type 2, $[0,\xi]_T \cap D^{\mathcal{T}} = \emptyset$, and either $\mathcal{U}$ has a small drop at η, or $[0,\eta]_U$ drops in model, or*
(c) *$R = S$, $\mathcal{T}$ and $\mathcal{U}$ have small drops at ξ and η respectively, and $\mathcal{M}^{\mathcal{T}}_{\xi}$ and $\mathcal{M}^{\mathcal{U}}_{\eta}$ have type 2, or*
(d) *$S^- = R$, S has type 2, $[0,\eta]_U \cap D^{\mathcal{U}} = \emptyset$, and either $\mathcal{T}$ has a small drop at ξ, or $[0,\xi]_T$ drops in model, or*
(e) *$S \trianglelefteq R$, and $[0,\eta]_U \cap D^{\mathcal{U}} = \emptyset$.*

In (a)-(c), $\hat{\imath}^{\mathcal{T}}_{0,\xi}$ is an essentially $r\Sigma_{k+1}$ embedding of P into an initial segment of an iterate of Q, showing that P is no stronger than Q. In (c)-(e), $\hat{\imath}^{\mathcal{U}}_{0,\eta}$ is an essentially $r\Sigma_{k+1}$ embedding of Q into an initial segment of an iterate of P, so Q is no stronger than P.

In the case that they involve a small drop, $\hat{\imath}^{\mathcal{T}}_{0,\xi}$ and $\hat{\imath}^{\mathcal{U}}_{0,\eta}$ are very simple instances of Σ^* *elementary* maps that are not nearly elementary. In the wider context of

[44]Clause (2) of Definition 4.6.9 is a small step in the direction of minimality with respect to Σ^*-*preserving* maps. This is proved in [86, Lemma 9.2.5, Theorem 9.2.11] for the iteration strategies for ordinary Jensen premice of Chapter 3, and the proof adapts to pfs premice. To see how Σ^*-elementarity is a more refined notion than elementarity or near elementarity, suppose $k(M) = 1$, $N = \mathrm{Ult}_1(M,E)$, and $P = \mathrm{Ult}_0(N,F)$, where $\rho_1(N) \leq \mathrm{crit}(F)$. Let $i = i_F \circ i_E$. i is elementary as a map from M^- to P. According to our definitions, as a map from M to P it is not nearly elementary. On the other hand, i can be used to copy 1-ultrapowers of M by extenders with critical point $< \rho_1(M)$. In the terminology of [86], i is a Σ^* embedding from M to N. This makes copying possible, and leads to Dodd-Jensen in the broader category.

Σ^* elementary maps, copying does work and the Dodd-Jensen Lemmas hold, as shown in Zeman's book.[45] But we shall not need this generality. We do not need to compare premice that are not strongly stable, or to consider more than elementary and nearly elementary maps. The Weak Dodd-Jensen Lemma stated in Lemma 4.6.10 is enough for our purposes.

4.7. PFS constructions and their resurrection maps

We produce pfs premice in a background construction just as we did in Chapter 3, except that we take the cores that are appropriate to the pfs hierarchy. It is convenient to require that the set of eligible background extenders be part of a coherent pair.

DEFINITION 4.7.1. A PFS-construction above κ is a tuple

$$\mathbb{C} = \langle w, \mathcal{F}, \langle (M_{\nu,k}, F_\nu) | \langle \nu, k \rangle <_{\text{lex}} \text{lh}(\mathbb{C}) \rangle \rangle$$

such that

(a) $(w, \mathcal{F})$ is a coherent pair, and
(b) $\langle (M_{\nu,k}, F_\nu) | \langle \nu, k \rangle <_{\text{lex}} \text{lh}(\mathbb{C}) \rangle$ satisfies the properties in Definition 3.1.3, relative to $(w, \mathcal{F})$, except that for $\langle \nu, k \rangle <_{\text{lex}} \text{lh}(\mathbb{C}) \rangle$,
  (i) $M_{\nu,k}$ is a pfs premouse of type 1, and
  (ii) if $\langle \nu, k+1 \rangle <_{\text{lex}} \text{lh}(\mathbb{C})$, then $M_{\nu,k+1} = \mathfrak{C}(M_{\nu,k})$.

The background certificate requirements on F_ν in Definition 4.7.1 are the same as those in Definitions 3.1.2 and 3.1.3. Roughly, the background extenders are taken from $\mathcal{F}$, have strictly increasing strengths and critical point $> \kappa$, and are Mitchell minimal, then w-minimal, among such certificates.

We write $w^{\mathbb{C}}, \mathcal{F}^{\mathbb{C}}, M^{\mathbb{C}}_{\nu,k}$, and $F^{\mathbb{C}}_\nu$ for the objects associated to the construction $\mathbb{C}$. If $(w, \mathcal{F})$ can be understood from context, we may identify $\mathbb{C}$ with the sequence $\langle (M_{\nu,k}, F_\nu) | \langle \nu, k \rangle <_{\text{lex}} \text{lh}(\mathbb{C}) \rangle$ of premice and background extenders.

Since $(w, \mathcal{F})$ is coherent, we get

LEMMA 4.7.2. *Let $\mathbb{C}$ be a PFS construction above κ, $M^{\mathbb{C}}_{\nu,0} = (M^{<\nu}, F)$, and $F^* = F^{\mathbb{C}}_\nu$; then*

(a) $\text{lh}(F^*)$ is the least strongly inaccessible η such that $\lambda_F < \eta$ and $\forall \tau < \nu (\text{lh}(F^{\mathbb{C}}_\tau) < \eta)$,
(b) $\{\text{lh}(F^{\mathbb{C}}_\tau) \mid \tau < \nu\}$ is bounded in $\text{lh}(F^*)$, and
(c) $i^V_{F^*}(M^{<\nu}) \models \lambda_F$ is not measurable.

The proof is the same as that of Lemma 3.1.9.

We say $\mathbb{C}$ is *maximal* iff whenever $\langle \nu, 0 \rangle < \text{lh}(\mathbb{C})$ and there is an F such that $(M^{<\nu}, F)$ is a pfs premouse and F has a certificate $F^* \in \mathcal{F}^{\mathbb{C}}$, in the sense of

[45] See the last footnote.

Definition 3.1.2, then $\dot{F}^{M_{\nu,0}} \neq \emptyset$. We shall deal pretty much exclusively with maximal constructions.

It is convenient to give $(M^{<\nu}, \emptyset)^{\mathbb{C}}$ an index in $\mathbb{C}$.

DEFINITION 4.7.3. Let $\mathbb{C}$ be a PFS construction and $\langle \nu, 0\rangle \leq \text{lh}(\mathbb{C})$; then $M^{\mathbb{C}}_{\nu,-1} = (M^{<\nu}, \emptyset)$.

So for $\langle \nu, 0\rangle <_{\text{lex}} \text{lh}(\mathbb{C})$, $M^{\mathbb{C}}_{\nu,-1} = M^{\mathbb{C}}_{\nu,0}||\nu$.

A PFS construction can break down in various ways, all of which are ruled out by the countable iterability of its levels and associated bicephali and pseudo-premice.

DEFINITION 4.7.4. Let $\mathbb{C}$ be a PFS construction, and $\langle \nu, k\rangle < \text{lh}(\mathbb{C})$. We say that $\mathbb{C}$ is *good at* $\langle \nu, k\rangle$ iff

(a) if $k = -1$, and $\mathbb{C} \restriction \nu^\frown \langle (M^{<\nu}, F), F^*\rangle$ is a PFS construction, then F^* *certifies* F^+, in that
  (i) $F^+ \cap ([\lambda_F + 1]^{<\omega} \times M) = F^* \cap ([\lambda_F + 1]^{<\omega} \times M)$, and
  (ii) $\text{lh}(F)$ is a cardinal of $i_{F^*}(M)$,

(b) if $k = -1$, and $\mathbb{C} \restriction \nu^\frown \langle (M^{<\nu}, F), F^*\rangle$ and $\mathbb{C} \restriction \nu^\frown \langle (M^{<\nu}, G), G^*\rangle$ are PFS constructions, then $F = G$, and

(c) if $k \geq 0$, then $M_{\nu,k}$ is solid.

We say that $\mathbb{C}$ is *plus consistent at* ν iff (a) holds, and that $\mathbb{C}$ is *extender unique at* ν iff (b) holds. We say that $\mathbb{C}$ is *good* iff it is good at all $\langle \nu, k\rangle$ such that $\langle \nu, k\rangle < \text{lh}(\mathbb{C})$. We say that $\mathbb{C}$ *breaks down at* $\langle \nu, k\rangle$ iff $\mathbb{C}$ is good at all $\langle \eta, j\rangle <_{\text{lex}} \langle \nu, k\rangle$, but is not good at $\langle \nu, k\rangle$.

Plus consistency is important when we use the background extenders in $\vec{F}^{\mathbb{C}}$ to lift plus trees on some level of $\mathbb{C}$. The proof of (a) modulo iterability belongs to the same family of phalanx comparison arguments that yield solidity. More specifically, it resembles the proof of closure under initial segment in [35, §10], and it uses the λ-minimality property of certificates recorded in Lemma 4.7.2(c).

Extender uniqueness is needed in order to show that maximal constructions reach mice satisfying various large cardinal hypotheses. That (b) holds, granted iterability, is known as the *Bicephalus Lemma*. Item (c) says that cores behave well, so we can continue the construction, producing a next level $\mathfrak{C}(M_{\nu,k+1})$ that is a pfs premouse of type 1. It holds by definition unless $\langle \nu, k+1\rangle = \text{lh}(\mathbb{C})$.

The main theorem about PFS constructions is that, granted iterability, they are good at all $\langle \nu, k\rangle$. We shall prove this later in this chapter.[46]

The set of levels of a PFS construction $\mathbb{C}$ is $\text{lev}(\mathbb{C}) = \{M_{\nu,k} \mid \langle \nu, k\rangle <_{\text{lex}} \text{lh}(\mathbb{C})\}$, and $<_{\mathbb{C}}$ is the order on $\text{lev}(\mathbb{C})$ induced by $\mathbb{C}$. The order is well defined for the same reasons as before:

LEMMA 4.7.5. *Let $\mathbb{C}$ be a PFS construction, with levels $M_{\nu,k} = M^{\mathbb{C}}_{\nu,k}$.*

[46]See Theorem 4.11.4.

(a) Let $\langle\mu,l\rangle <_{\text{lex}} \langle\nu,k\rangle < \text{lh}(\mathbb{C})$, and suppose that whenever

$$\langle\mu,l\rangle \leq_{\text{lex}} \langle\eta,j\rangle \leq_{\text{lex}} \langle\nu,k\rangle,$$

then $\rho^-(M_{\mu,l}) \leq \rho^-(M_{\eta,j})$; then $M_{\mu,l} \lhd M_{\nu,k}$.

(b) Let $\gamma < o(M_{\nu,k})$ be a cardinal of $M_{\nu,k}$ such that $\gamma \leq \rho^-(M_{\nu,k})$, and suppose $P \trianglelefteq M_{\nu,k}$ is such that $\rho^-(P) = \gamma$; then

(i) there is a unique $\langle\mu,l\rangle \leq_{\text{lex}} \langle\nu,k\rangle$ such that $P = M_{\mu,l}$, moreover

(ii) if $P = M_{\mu,l}$, then $\gamma \leq \rho^-(M_{\eta,j})$ whenever $\langle\mu,l\rangle \leq_{\text{lex}} \langle\eta,j\rangle \leq_{\text{lex}} \langle\nu,k\rangle$.

COROLLARY 4.7.6. *Let $\mathbb{C}$ be a background construction; then for any premouse N, there is at most one $\langle\nu,k\rangle$ such that $N = M^{\mathbb{C}}_{\nu,k}$.*

We have also a parallel of Lemma 3.1.11 on the coherence of constructions. Recall that $\mathbb{C}\restriction\gamma = \langle (M_{\tau,k}, F_\tau) \mid \tau < \gamma \wedge k \leq \omega\rangle$, and $(M^{<\gamma}, \emptyset)$ is the last model of $\mathbb{C}^\frown\gamma$.

LEMMA 4.7.7. *Let $(w, \mathcal{F}, \mathbb{C})$ be a maximal PFS construction above κ, and suppose that $M^{\mathbb{C}}_{\nu,0} = (M^{<\nu}, F)$ where $F \neq \emptyset$, and let $F^* = F^{\mathbb{C}}_\nu$ and $\mathbb{D} = i_{F^*}(\mathbb{C})$; then*

(1) $\mathbb{D}\restriction\nu = \mathbb{C}\restriction\nu$,

(2) $M^{\mathbb{D}}_{\nu,0} \neq M^{\mathbb{C}}_{\nu,0}$; moreover if $\mathbb{C}$ is extender unique at ν, then $M^{\mathbb{D}}_{\nu,0} = (M^{<\nu}, \emptyset)$,

(3) $(M^{<\nu}, \emptyset) \lhd_0 i_{F^*}(M^{<\nu})$, and,

(4) if $\xi < \nu$, and $\mathbb{C}\restriction\xi$ has last model N such that $o(N) < \text{crit}(F^*)$, then $\mathbb{C}\restriction\xi \in V_{\text{crit}(F^*)}$.

PROOF. The proof of Lemma 3.1.11 goes through verbatim. ⊣

The point of the new premice and their constructions is that there is no case split in the definition of the resurrection maps. We resurrect $N \lhd Q$ from a successor level Q of $\mathbb{C}$ by resurrecting $\pi(N)$ from X, where $\pi\colon Q^- \to X$ is the anticore map. We resurrect from limit levels as before. As before, the resurrection maps satisfy

(∗) If $R <_{\mathbb{C}} Q$ and $\rho^-(R) \leq \rho(S)$ for all S such that $R \leq_{\mathbb{C}} S <_{\mathbb{C}} Q$, then

(i) $R \trianglelefteq Q$,

and for all $N \trianglelefteq R$ and Y such that $\text{Res}_R[N] \leq_{\mathbb{C}} Y \leq_{\mathbb{C}} R$,

(ii) $\text{Res}_{Q,Y}[N] = \text{Res}_{R,Y}[N]$, and

(iii) $\sigma_{Q,Y}[N] = \sigma_{R,Y}[N]$.

This enables us to resurrect from limit levels in an unambiguous way.

The formal definition goes by induction on Q, maintaining (∗) as we go.

(1) $\text{Res}_Q[Q] = Q$ and $\sigma_Q[Q] = \text{id}$.

(2) If $Q = M_{\nu,k+1}$, $X = M_{\nu,k}$, and $\pi\colon Q^- \to X$ is the anticore map, then

(a) $\text{Res}_{Q,X}[N] = \pi(N)$, and $\sigma_{Q,X}[N] = \pi$,

(b) $\text{Res}_Q[N] = \text{Res}_X[\text{Res}_{Q,X}[N]]$ and $\sigma_Q[N] = \sigma_X[\text{Res}_{Q,X}[N]] \circ \sigma_{Q,X}[N]$, and

(c) for Y such that $\text{Res}_Q[N] \leq_{\mathbb{C}} Y \leq_{\mathbb{C}} X$, $\text{Res}_{Q,Y}[N] = \text{Res}_{X,Y}[\text{Res}_{Q,X}[N]]$, and $\sigma_{Q,Y}[N] = \sigma_{X,Y}[\text{Res}_{Q,X}[N]] \circ \sigma_{Q,X}[N]$.

Since π is the identity on $\rho^-(Q)+1$, (∗) remains true.

(3) Suppose Q is a limit point in $\mathrm{lev}(\mathbb{C})$, that is, $k(Q) = 0$, and $N \lhd Q$. Let ρ be the minimum value of $\rho^-(R)$ for $N \unlhd R \lhd Q$, and let R be such that $N \unlhd R \lhd Q$ and $\rho^-(R) = \rho$. By $(*)$, we can set $\mathrm{Res}_{\mathrm{Q}}[N] = \mathrm{Res}_{\mathrm{R}}[N]$ and $\sigma_{\mathrm{Q}}[N] = \sigma_{\mathrm{R}}[N]$, and the results will be independent of our choice of R. Similarly, for Y such that $\mathrm{Res}_{\mathrm{Q}}[N] \leq_{\mathbb{C}} Y \leq_{\mathbb{C}} Q$, we let $\mathrm{Res}_{\mathrm{Q,Y}}[N]$ and $\sigma_{\mathrm{Q,Y}}[N]$ be the common values of $\mathrm{Res}_{\mathrm{R,Y}}[N]$ and $\sigma_{\mathrm{R,Y}}[N]$ for all such R.

In addition to $(*)$, we have the elementary properties of Proposition 3.2.2.

PROPOSITION 4.7.8. *Let* $\mathbb{C}$ *be a PFS construction, and* $N \unlhd Q \in \mathrm{lev}(\mathbb{C})$.

(i) $\mathrm{Res}_{\mathrm{Q}}[N]$ *is the* $<_{\mathbb{C}}$*-least* X *such that* $\mathrm{Res}_{\mathrm{Q,X}}[N]$ *is defined.*

(ii) $k(N) = k(\mathrm{Res}_{\mathrm{Q,X}}[N])$*, and* $\sigma_{\mathrm{Q,X}}[N]$ *is elementary and exact.*

(iii) *If* $P \lhd N$*, then* $\mathrm{Res}_{\mathrm{Q}}[P] <_{\mathbb{C}} \mathrm{Res}_{\mathrm{Q}}[N]$.

(iv) *If* $P \lhd N$ *and* $\mathrm{Res}_{\mathrm{Q,X}}[N]$ *is defined, then* $\mathrm{Res}_{\mathrm{Q,X}}[P] \lhd \mathrm{Res}_{\mathrm{Q,X}}[N]$.

(v) *Suppose that* $\mathrm{Res}_{\mathrm{Q}}[N] \leq_{\mathbb{C}} X \leq_{\mathbb{C}} Y \leq_{\mathbb{C}} Q$*; then*

(a) $\mathrm{Res}_{\mathrm{Q,X}}[N] = \mathrm{Res}_{\mathrm{Y,X}}[\mathrm{Res}_{\mathrm{Q,Y}}[N]]$*, and*

(b) $\sigma_{\mathrm{Q,X}}[N] = \sigma_{\mathrm{Y,X}}[\mathrm{Res}_{\mathrm{Q,Y}}[N]] \circ \sigma_{\mathrm{Q,Y}}[N]$.

(vi) *Suppose* $k(N) > 0$ *and* $\mathrm{Res}_{\mathrm{Q,X}}[N]$ *is defined; then* $\mathrm{Res}_{\mathrm{Q,X}}[N^-] = (\mathrm{Res}_{\mathrm{Q,X}}[N])^-$.

(vii) *If* $\mathrm{Res}_{\mathrm{Q}}[N] = M_{\nu,k+1}$*, then* $\mathrm{Res}_{\mathrm{Q}}[N^-] = M_{\nu,k}$*. Moreover, if* $\pi\colon (M_{\nu,k+1})^- \to M_{\nu,k}$ *is the anticore map, then* $\pi \circ \sigma_{\mathrm{Q}}[N] = \sigma_{\mathrm{Q}}[N^-]$.

Resurrection maps are exact because they are compositions of anticore maps, and anticore maps are elementary and exact by Lemma 4.3.6.

Our resurrection maps are now consistent with one another.

LEMMA 4.7.9. *Let* $\mathbb{C}$ *be a PFS construction,* $X <_{\mathbb{C}} Q$ *, and* $P \lhd N \lhd Q$*; then the* (Q,X) *resurrections of* P *and* N *are consistent, in that*

(a) $\mathrm{Res}_{\mathrm{Q,X}}[P] = \sigma_{\mathrm{Q,X}}[N](P)$*, and*

(b) $\sigma_{\mathrm{Q,X}}[P] = \sigma_{\mathrm{Q,X}}[N] \restriction P$.

The proposition and lemma are easy to prove by induction on $\mathrm{lev}(\mathbb{C})$. We also get a useful lemma on the agreement of resurrection maps.

LEMMA 4.7.10. *Let* $\mathbb{C}$ *be a PFS construction,* $X \unlhd K \unlhd Q$*, and* $\kappa < o(X)$*. Suppose that whenever* $X \unlhd S \lhd K$*, then* $\kappa \leq \rho(S)$*; then* $\sigma_{\mathrm{Q}}[X] \restriction \kappa^{+,X} = \sigma_{\mathrm{Q}}[K] \restriction \kappa^{+,X}$.

PROOF. Here we take $\kappa^{+,X} = o(X)$ if there are no cardinals of X strictly above κ. Letting $N = \mathrm{Res}_{\mathrm{Q}}[K]$ and $Y = \sigma_{\mathrm{Q}}[K](X)$, we have

$$\sigma_{\mathrm{Q}}[X] = \sigma_{\mathrm{N}}[Y] \circ \sigma_{\mathrm{Q}}[K] \restriction X$$

by the consistency of resurrections. Let $\mu = \sigma_{\mathrm{Q}}[K](\kappa)$. Resurrection maps are elementary, so whenever $Y \unlhd R \unlhd N$, $\rho^-(R) \geq \mu$. But this means that all the anticore maps at levels of $\mathbb{C}$ between $\mathrm{Res}_{\mathrm{N}}[Y]$ and N have critical point $> \mu^{+,Y}$. Thus $\sigma_{\mathrm{N}}[Y]$ is the identity on $\mu^{+,Y}$, so $\sigma_{\mathrm{Q}}[X] \restriction \kappa^{+,X} = \sigma_{\mathrm{Q}}[K] \restriction \kappa^{+,X}$. ⊣

The factoring of resurrection maps induced by dropdown sequences is simpler now.

DEFINITION 4.7.11. Let Q be a pfs premouse and $N \lhd Q$. The *N-dropdown sequence of Q* is given by

(1) $A_0 = N$,
(2) A_{i+1} is the least $B \unlhd Q$ such that $A_i \lhd B$ and $\rho^-(B) < \rho^-(A_i)$.

We write $A_i = A_i(Q,N)$, and let $n(Q,N)$ be the largest i such that A_i is defined. Let also $\kappa_i(Q,N) = \rho^-(A_i(Q,N))$.

Exact maps preserve dropdown sequences.

LEMMA 4.7.12. *Let Q and X be pfs premice, and $\pi\colon Q \to X$ be nearly elementary and exact, and $N \lhd Q$; then*

(1) $n(Q,N) = n(X,\pi(N))$, and
(2) for all $i \le n(Q,N)$,
   (a) $\pi(A_i(Q,N)) = A_i(X,\pi(N))$, and
   (b) $\pi(\kappa_i(Q,N)) = \kappa_i(X,\pi(N))$.

PROOF. Since π is exact,

$$\pi(\rho^-(A_i(Q,N))) = \rho^-(\pi(A_i(Q,N))$$

for all $i \le n(Q,N)$. One can then prove (2) by induction, starting with $i = n(Q,N)$ and working down to $i = 0$. ⊣

Lemma 4.7.12 is simpler than the version we get for inexact π. This makes the factoring of $\sigma_Q[N]$ induced by the (Q,N) dropdown sequence easier to describe.

LEMMA 4.7.13. *Let $\mathbb{C}$ be a PFS construction, $Q \in \mathrm{lev}(\mathbb{C})$, and $N \lhd Q$. Let $n = n(Q,N)$, and $A_i = A_i(Q,N)$; then*

(a) $A_n \le_{\mathbb{C}} Q$, and for all $P \unlhd A_n$, $\mathrm{Res}_{Q,A_n}[P] = P$ and $\sigma_{Q,A_n}[P] = id$. Thus $\mathrm{Res}_Q[A_n] = A_n$.

Moreover, if $n > 0$, then letting $X = \mathrm{Res}_Q[A_n^-]$ and $\pi = \sigma_{Q,X}[A_n^-]$,

(b) X is the immediate $<_{\mathbb{C}}$- predecessor of A_n, and $\pi\colon A_n^- \to X$ is the anticore map. Moreover, $n(X,\pi(N)) = n-1$, and for all $i \le n-1$,
   (i) $\mathrm{Res}_Q[A_i] = \mathrm{Res}_X[A_i(X,\pi(N))]$, and
   (ii) $\sigma_Q[A_i] = \sigma_X[A_i(X,\pi(N)] \circ \pi$.
(c) $\sigma_Q[N] = \pi_1 \circ \cdots \circ \pi_n \restriction N$, where π_i is the anticore map from $\mathrm{Res}_Q[A_i]$ to $\mathrm{Res}_Q[A_i^-]$.
(d) Let $\gamma = \kappa_i(Q,N)^{+,Q}$; then $\sigma_Q[A_i] \restriction \gamma = \sigma_Q[A_i^-] \restriction \gamma = \sigma_Q[N] \restriction \gamma$. In particular, $\sigma_Q[N]$ is the identity on $\kappa_n(Q,N)^{+,Q}$.

PROOF. (a) follows easily from property $(*)$. (b) follows easily from Lemma 4.7 12. Part (c) comes from applying (b) repeatedly until we reach $\mathrm{Res}_Q[A_1(Q,N)]^-$, and then applying (a).

For (d), note that $\sigma_Q[A_k] \restriction \gamma = \sigma_Q[A_k^-] \restriction \gamma$. This is because $\sigma_Q[A_k](\rho^-(A_k)) = \rho^-(\mathrm{Res}_Q[A_k])$, and the anticore map π_k is therefore the identity on $\sigma_Q[A_k](\gamma)$. But

$\sigma_Q[A_k^-] = \pi_k \circ \cdots \circ \pi_n$ and $\mathrm{crit}(\pi_1 \circ \cdots \circ \pi_{k+1}) > \sigma_Q[A_k^-](\gamma)$, so we get the second equality. ⊣

Lemma 4.7.13 is clearly simpler than its counterpart in [35, §11] for constructions of ordinary premice.

4.8. Conversion systems and induced strategies

We begin with a shift lemma for conversions.

DEFINITION 4.8.1. $\langle M, \psi, Q, \mathbb{C}, R\rangle$ is a *PFS conversion stage* iff

(1) R is a transitive model of ZFC, and $(R, \mathbb{C})$ is amenable,
(2) $(R, \mathbb{C}) \models$ "$\mathbb{C}$ is a maximal PFS construction", and
(3) M is a pfs premouse, $Q \in \mathrm{lev}(\mathbb{C})$, and $\psi\colon M \to Q$ is nearly elementary.

If $\langle M, \psi, Q, \mathbb{C}, R\rangle$ is a PFS conversion stage, then $(R, \in, w^{\mathbb{C}}, \mathcal{F}^{\mathbb{C}})$ is a coarse premouse, and $\mathbb{C}$ is the unique maximal PFS construction this coarse premouse determines. M may be of type 2, but Q has type 1 since $Q \in \mathrm{lev}(\mathbb{C})$. We have included the requirement that $\mathbb{C}$ be maximal because it is needed in Lemma 4.7.7 on the coherence properties of PFS constructions, and that lemma is useful in what we shall do later. Many of the basic lemmas about conversion stages and systems do not require it.

The constructions, conversion stages, and conversion systems that we use in the remainder of this book will be of the PFS variety, so we shall drop the qualifier "PFS" most of the time.

LEMMA 4.8.2 (Shift Lemma for Conversion Systems). *Let $\langle M, \psi, Q, \mathbb{C}, R\rangle$ be a conversion stage. Let E be an extender over M such that* $\mathrm{crit}(E) < \rho^-(M)$, *and let*

$$\nu = \begin{cases} \lambda(E) & \text{if } E \text{ is not of plus type,} \\ \lambda(E^-)+1 & \text{if } E \text{ is of plus type.} \end{cases}$$

Let E^* be an extender over R, and $\varphi\colon \mathrm{dom}(E) \cup \nu \to \mathrm{dom}(E^*) \cup \mathrm{lh}(E^*)$ be such that

(i) $(a, X) \in E$ iff $(\varphi``(a), \varphi(X)) \in E^*$, and
(ii) $\varphi \restriction \mathrm{dom}(E) = \psi \restriction \mathrm{dom}(E)$.

Let $i = i_E^M$ and $i^* = i_{E^*}^R$ be the ultrapower embeddings, and assume that $\mathrm{Ult}(R, E^*)$ *is wellfounded. There is then a nearly elementary map $\sigma\colon$* $\mathrm{Ult}(M, E) \to i^*(Q)$ *given by*[47]

$$\sigma([a, f]_E^M) = [\varphi(a), \psi(f)]_{E^*}^R,$$

[47] The definition of σ can be understood either by means of reducts, or by letting $\psi(f_{\tau,q}^M) = f_{\tau,\psi(q)}^Q$ for $\tau \in \mathrm{sk}_k$ and $q \in M$.

for all functions f used in $\mathrm{Ult}(M,E)$.[48] *Moreover*

(a) $\sigma \restriction \nu = \varphi \restriction \nu$,

(b) $\sigma \circ i = i^* \circ \psi$, *and*

(c) $\langle \mathrm{Ult}(M,E), \sigma, i^*(Q), i^*(\mathbb{C}), \mathrm{Ult}(R,E^*)\rangle$ *is a conversion stage.*

(d) *For all* $x \in Q$, $x \in \mathrm{ran}(\psi)$ *iff* $i^*(x) \in \mathrm{ran}(\sigma)$).

PROOF. This is a routine adaptation of the proof of Lemma 3.3.2. Let $k = k(M)$, $R = \mathrm{Ult}(M,E)$, and $S = i^*(Q)$. The additional thing we must show is that $\sigma(\hat{w}_k(R)) = \hat{w}_k(S)$. But this follows from the preservation properties of i, i^*, and ψ:

$$\sigma(\hat{w}_k(R)) = \sigma \circ i(\hat{w}_k(M)) = i^* \circ \psi(w_k(R)) = i^*(\hat{w}_k(Q)) = \hat{w}_k(S). \qquad \dashv$$

Remark 4.8.3. The proof of Lemma 4.8.2 is simpler than the proof of Lemma 3.3.2 because the notion of near elementarity has been simplified. Lemma 3.3.2 required the hypothesis that E is close to M, whereas Lemma 4.8.2 does not. The hypothesis came up in the proof that the copy map σ satisfies part (b) of Definition 2.5.14, concerning the relationship between η_{k-1} and ρ_k. The new definition of near elementarity replaces η_k by $\hat{\eta}_k$ in this context. k-ultrapower maps preserve $\hat{\eta}_k$ but may not preserve η_k, so this replacement simplifies things.

Now suppose that $c = \langle M, \psi, Q, \mathbb{C}, R\rangle$ is a conversion stage,[49] and $\mathcal{T}$ is a plus tree on M. We define a conversion system $\mathrm{lift}(\mathcal{T}, c)$ that lifts $\mathcal{T}$ to a nice, quasi-normal tree $\mathcal{T}^*$ on R. The definition is quite close to that in Section 3.4, so we shall skip some of the more detailed calculations done there. The dropping case in the inductive construction simplifies a bit because our new resurrection maps are simpler. If $E_\alpha^{\mathcal{T}}$ is of plus type, then the agreement between the lifting map at stage α and later maps is better now. (This is the motivation for plus trees.) On the other hand, if $E_\alpha^{\mathcal{T}}$ is of plus type, then the lifting map at $\alpha+1$ will map $\lambda(E_\alpha^{\mathcal{T}})$ strictly below $\mathrm{lh}(E_\alpha^{\mathcal{T}^*})$, with the result that $\mathcal{T}^*$ may be only quasi-normal.[50] Beyond these features, there is nothing new.

We shall use the same notation that we used in Section 3.4. So $\mathrm{lift}(\mathcal{T}, c) = \mathrm{lift}(\mathcal{T}, M, \psi, Q, \mathbb{C}, R)$, and

$$\mathrm{lift}(\mathcal{T}, c) = \langle \mathcal{T}^*, \langle c_\alpha \mid \alpha < \mathrm{lh}(\mathcal{T})\rangle\rangle,$$

where $c = c_0$, and $c_\alpha = \langle M_\alpha, \psi_\alpha, Q_\alpha, \mathbb{C}_\alpha, R_\alpha\rangle$ is a conversion stage. As before, we also write

$$\mathrm{lift}(\mathcal{T}, M, \psi, Q, \mathbb{C}, R) = \langle \mathcal{T}^*, \langle Q_\alpha \mid \alpha < \mathrm{lh}(\mathcal{T})\rangle, \langle \psi_\alpha \mid \alpha < \mathrm{lh}(\mathcal{T})\rangle\rangle,$$

since c_α is determined by the data displayed. The recycled notation should cause no trouble, because from now on we shall not be dealing with premice, constructions, and conversion systems in the sense of Chapter 3.

---

[48]The relevant functions are the boldface $r\Sigma_k^M$ functions, where $k = k(M)$.

[49]That is, a PFS conversion stage.

[50]See Remark 4.8.4.

Our first two induction hypotheses are the same as before.

$(1)_\alpha$ (a) $\mathcal{T}^* \upharpoonright \alpha+1$ is a nice, quasi-normal iteration tree on R with the same tree order as $\mathcal{T}$,

(b) for all $\nu \leq \alpha$, c_ν is a conversion stage, moreover, $M_\nu = \mathcal{M}_\nu^{\mathcal{T}}$, $R_\nu = \mathcal{M}_\nu^{\mathcal{T}^*}$, and $\mathbb{C}_\nu = i_{0,\nu}^{\mathcal{T}^*}(\mathbb{C})$.

The lifting maps commute appropriately with the embeddings of $\mathcal{T}$ and $\mathcal{T}^*$. Drops in model in $\mathcal{T}$ are mirrored by drops in the construction at the background level. Letting $i_{\xi,\nu} = i_{\xi,\nu}^{\mathcal{T}}$ and $i_{\xi,\nu}^* = i_{\xi,\nu}^{\mathcal{T}^*}$,

$(2)_\alpha$ Let $\xi <_T \nu \leq \alpha$; then

(a) $Q_\nu \leq_{\mathbb{C}_\nu} i_{\xi,\nu}^*(Q_\xi)$,

(b) $(\xi,\nu]_T$ drops in model or degree iff $Q_\nu <_{\mathbb{C}_\nu} i_{\xi,\nu}^*(Q_\xi)$, and

(c) if $(\xi,\nu]_T$ does not drop in model or degree, then $Q_\nu = i_{\xi,\nu}^*(Q_\xi)$ and $\psi_\nu \circ i_{\xi,\nu} = i_{\xi,\nu}^* \circ \psi_\xi$.

Having defined lift$(\mathcal{T} \upharpoonright \nu+1, c)$, where $\nu+1 < \text{lh}(\mathcal{T})$, we are given E_ν by $\mathcal{T}$, and we set

$$\begin{aligned} H_\nu &= \psi_\nu(E_\nu^-), \\ X_\nu &= Q_\nu | \text{lh}(H_\nu), \\ G_\nu &= \sigma_{Q_\nu}[X_\nu](H_\nu), \\ Y_\nu &= \text{Res}_{Q_\nu}[X_\nu], \\ G_\nu^* &= B^{\mathbb{C}_\nu}(G_\nu). \end{aligned}$$

Here $\sigma_{Q_\nu}[X_\nu]$ is the resurrection map of $\mathbb{C}_\nu$. E_ν^- is on the sequence of M_ν, but is of course different from E_ν if the plus case occurs at ν. H_ν is on the sequence of Q_ν, and is the last extender of X_ν. Its complete resurrection G_ν is the last extender of Y_ν. We let

$$E_\nu^{\mathcal{T}^*} = G_\nu^*.$$

Notice that

$$\lambda(G_\nu) < \text{lh}(G_\nu) < \text{lh}(G_\nu^*) < \lambda(G_\nu^*).$$

The ψ_ν will agree with one another in a way that lets us keep the conversion going. The agreement involves generator maps $\text{res}_\nu \in R_\nu$, defined when $\nu+1 < \text{lh}(\mathcal{T})$, that connect the generators of H_ν to those of G_ν and G_ν^*. These are defined just as in Section 3.4:

$$\text{res}_\nu = \sigma_{Q_\nu}[X_\nu]^{\mathbb{C}_\nu},$$

so that

$$\text{res}_\nu \colon X_\nu \to Y_\nu.$$

Let $\varepsilon_\nu = \varepsilon(E_\nu)$, that is

$$\varepsilon_\nu = \begin{cases} \text{lh}(E_\nu) & \text{if } E_\nu \text{ is of plus type,} \\ \lambda(E_\nu) & \text{otherwise,} \end{cases}$$

and let

$$\xi_\nu = \text{unique } \xi \text{ such that } Y_\nu = M_{\xi,0}^{\mathbb{C}_\nu}.$$

Since $\mathcal{T}$ is a plus tree, if $\nu < \gamma$ then $\varepsilon_\nu \leq \varepsilon_\gamma$, and if the plus case occurs at some η such that $\nu \leq \eta \leq \gamma$, then $\varepsilon_\nu < \varepsilon_\gamma$.

$(3)_\alpha$ If $\nu < \gamma \leq \alpha$, then

(a) $\text{res}_\nu \circ \psi_\nu \restriction \varepsilon_\nu = \psi_\gamma \restriction \varepsilon_\nu$,

(b) $V^{R_\nu}_{\text{lh}(G^*_\nu)} = V^{R_\gamma}_{\text{lh}(G^*_\nu)}$,

(c) $\mathbb{C}_\nu \restriction \xi_\nu = \mathbb{C}_\gamma \restriction \xi_\nu$, and $M^{\mathbb{C}_\gamma}_{\xi_\nu,0}$ is passive,

(d) $Y_\nu || o(Y_\nu) = Q_\gamma | o(Y_\nu)$, and

(e) $o(Y_\nu)$ is a cardinal of Q_γ, and $o(Y_\nu) \leq \rho^-(Q_\gamma)$.

Notice that the agreement recorded in (a) is better when the plus case occurs at ν, so that $\varepsilon_\nu = \text{lh}(E_\nu)$. In this case the maps actually agree on $\text{lh}(E_\nu)+1$. If the plus case does not occur at ν, then $\text{res}_\nu \circ \psi_\nu$ and ψ_γ disagree at $\varepsilon_\nu = \lambda(E_\nu)$.

$(4)_\alpha$ If $\nu < \gamma \leq \alpha$ and the plus case occurs at ν, then $\text{res}_\nu \circ \psi_\nu(\varepsilon_\nu) = \psi_\gamma(\varepsilon_\nu)$.

$(5)_\alpha$ If $\nu < \gamma \leq \alpha$ and the plus case does not occur at ν, then

(a) $\lambda(G^*_\nu) \leq \psi_\gamma(\varepsilon_\nu)$, and

(b) $\lambda(G^*_\nu)$ is a cardinal of Q_γ, and $\lambda(G^*_\nu) \leq \rho^-(Q_\gamma)$.

Notation. $(\dagger)_\alpha$ is the conjunction of $(1)_\alpha$ through $(5)_\alpha$.

$(\dagger)_\alpha$ involves objects that are associated to $\text{lift}(\mathcal{T} \restriction \alpha+1, c)$. Objects that are associated to E_α (such as H_α and G_α) do not play a role in it.

The step from α to $\alpha+1$ in the conversion process goes as follows. E_α determines $H_\alpha = \psi_\alpha(E_\alpha)$, res_α, etc., as above. Let

$$(E, H, X, Y, G, G^*) = (E_\alpha, H_\alpha, X_\alpha, Y_\alpha, G_\alpha, G^*_\alpha),$$

and

$$\beta = T\text{-pred}(\alpha+1).$$

We shall apply the Shift Lemma for Conversions with $\varphi = \text{res}_\alpha \circ \psi_\alpha$ as the embedding of $E_\alpha \restriction \varepsilon_\alpha$ into G^*. Notice that if E is of plus type, then $\text{res}_\alpha \circ \psi_\alpha$ embeds E into G^+, and since $\mathbb{C}_\alpha$ is good at $\langle \text{lh}(G), 0\rangle$, G^* backgrounds G^+.

Here are some elementary consequences of $(\dagger)_\alpha$.

CLAIM 1. *Assume $(\dagger)_\alpha$, and let $\nu < \gamma \leq \alpha$; then $\text{lh}(G_\nu) < \lambda(H_\gamma)$, and*

(a) if E_ν is not of plus type, then $\lambda(G^*_\nu) \leq \lambda(H_\gamma)$, and $\text{res}_\gamma \restriction \lambda(G^*_\nu)+1 = id$,

(b) if E_ν is of plus type, then $\text{res}_\gamma \restriction \text{lh}(G_\nu)+1 = id$,

(c) $Y_\nu \| o(Y_\nu) = Y_\gamma | o(Y_\nu)$,
(d) $\xi_\nu < \xi_\gamma$.

PROOF. Suppose first that E_ν is not of plus type. Then

$$\lambda(G^*_\nu) \le \psi_\gamma(\lambda(E_\nu)) \le \psi_\gamma(\lambda(E_\gamma)) = \lambda(H_\gamma).$$

The first inequality comes from $(5)_\alpha$. But $\mathrm{lh}(G_\nu) < \lambda(G^*_\nu)$, so $\mathrm{lh}(G_\nu) < \lambda(H_\gamma)$. Also, $\lambda(G^*_\nu) \le \rho^-(Q_\gamma)$ by $(5)_\alpha$, so $\mathrm{res}_\gamma = \sigma_{Q_\gamma}[X_\gamma]$ is the identity on $\lambda(G^*_\nu)+1$ by Lemma 4.7.13(d). Thus we have (a) of the claim.

Suppose next that E_ν is of plus type. Then $\mathrm{lh}(E_\nu) < \lambda(E_\gamma)$ by the rules of plus trees. So

$$\mathrm{lh}(G_\nu) = \psi_\gamma(\mathrm{lh}(E_\nu)) < \psi_\gamma(\lambda(E_\gamma)) = \lambda(H_\gamma).$$

The first equality comes from $(4)_\alpha$. Also, $\mathrm{lh}(G_\nu) \le \rho^-(Q_\gamma)$ by $(3)_\alpha$, so by Lemma 4.7.13(d), $\mathrm{res}_\gamma = \sigma_{Q_\gamma}[X_\gamma]$ is the identity on $\mathrm{lh}(G_\nu)+1$. This proves (b) of the claim, and we have shown $\mathrm{lh}(G_\nu) < \lambda(H_\gamma)$ in both cases.

For (c): we have $Y_\nu \| o(Y_\nu) = Q_\gamma | o(Y_\nu)$ by $(3)_\alpha$. But $o(Y_\nu) = \mathrm{lh}(G_\nu)$, so we have just shown that $o(Y_\nu) < o(X_\gamma)$, and res_γ is the identity on $o(Y_\nu)$. Hence $Y_\nu \| o(Y_\nu) = Y_\gamma | o(Y_\nu)$.

For (d): $\mathbb{C}_\nu \restriction \xi_\nu = \mathbb{C}_\gamma \restriction \xi_\nu$ has last model $Y_\nu \| o(Y_\nu)$. Since $\mathrm{lh}(G_\nu) < \lambda(H_\gamma) < \mathrm{lh}(G_\gamma)$, $\xi_\nu \ne \xi_\gamma$. If $\xi_\gamma < \xi_\nu$, then since the last model $Y_\gamma \| o(Y_\gamma)$ of $\mathbb{C}_\nu \restriction \xi_\gamma$ is not an initial segment of $Y_\nu \| o(Y_\nu)$, there is $\kappa < o(Y_\nu)$ such that κ is a cardinal in Y_γ but not in Y_ν. (Take $\kappa = \rho^{+,Y_\gamma}$, where ρ is the smallest projectum associated to a stage of $\mathbb{C}_\nu$ between ξ_γ and ξ_ν.) This contradicts (c). ⊣

Let us show that we obtain a quasi-normal extension of $\mathcal{T}^* \restriction \alpha+1$ by setting $\beta = T^*\text{-pred}(\alpha+1)$. Let

$$\kappa = \mathrm{crit}(E),$$
$$\kappa^* = \mathrm{crit}(G) = \mathrm{res}_\alpha \circ \psi_\alpha(\kappa).$$

CLAIM 2. *(1) Suppose $\gamma < \alpha$; then*
(a) $\mathrm{lh}(G^*_\gamma) \le \mathrm{lh}(G^*)$, *and* $\mathrm{lh}(G^*_\gamma) < \mathrm{lh}(G^*)$ *if E_γ is not of plus type.*
(b) $\kappa < \lambda(E_\gamma)$ *iff* $\kappa^* < \mathrm{lh}(G^*_\gamma)$.
(2) $\mathcal{T}^* \restriction \alpha+2$ *is quasi-normal.*

PROOF. If E_γ is not of plus type, then

$$\mathrm{lh}(G^*_\gamma) < \lambda(G^*_\gamma) \le \lambda(H) \le \lambda(G) < \mathrm{lh}(G^*),$$

by Claim 1(a). If E_γ is of plus type, then $\mathrm{lh}(E_\gamma) < \lambda(E_\alpha)$, so

$$\mathrm{lh}(G_\gamma) = \psi_\alpha(\mathrm{lh}(E_\gamma)) < \mathrm{res}_\alpha \circ \psi_\alpha(\lambda(E_\alpha)) = \lambda(G_\alpha).$$

Let $\eta = \mathrm{lh}(G^*_\gamma)$. By Lemma 4.7.7, in R_γ, η is the least strongly inaccessible such that $\lambda(G_\gamma) < \eta$ and $\forall \tau < \xi_\gamma(\mathrm{lh}(F^{\mathbb{C}_\gamma}_\tau) < \eta)$. Our agreement hypotheses $(3)_\alpha$(b),(c) imply that in R_α, η is the least strongly inaccessible such that $\lambda(G_\gamma) < \eta$ and $\forall \tau < \xi_\gamma(\mathrm{lh}(F^{\mathbb{C}_\alpha}_\tau) < \eta)$. But $\lambda(G_\gamma) < \lambda(G_\alpha)$ and $\xi_\gamma < \xi_\alpha$, and in R_α, $\mathrm{lh}(G^*)$ is the least strongly inaccessible μ such that $\lambda_G < \mu$ and $\forall \tau < \xi_\alpha(\mathrm{lh}(F_\tau^{\mathbb{C}_\alpha}) < \mu)$. It follows that $\eta \leq \mathrm{lh}(G^*)$, so again we have (1)(a).

Remark 4.8.4. The argument of the last paragraph does not seem to give $\mathrm{lh}(G^*_\gamma) < \mathrm{lh}(G^*_\alpha)$ when E_γ is of plus type. This is why we must allow $\mathcal{T}^*$ to be merely quasi-normal.

For (1)(b), suppose first $\kappa < \lambda(E_\gamma)$; then $\mathrm{res}_\gamma \circ \psi_\gamma(\kappa) < \lambda(G_\gamma)$, so $\psi_\alpha(\kappa) < \lambda(G_\gamma)$ by $(3)_\alpha$. But $\mathrm{res}_\alpha \restriction \mathrm{lh}(G_\gamma) = \mathrm{id}$ by Claim 1. So $\kappa^* < \lambda(G_\gamma) < \mathrm{lh}(G^*_\gamma)$, as desired.

Suppose next $\lambda(E_\gamma) \leq \kappa$ and E_γ is not of plus type. Then $\psi_\alpha(\lambda(E_\gamma)) \leq \psi_\alpha(\kappa)$, so

$$\lambda(G^*_\gamma) \leq \psi_\alpha(\kappa) \leq \mathrm{res}_\alpha \circ \psi_\alpha(\kappa) = \kappa^*,$$

as desired.

Suppose finally $\lambda(E_\gamma) \leq \kappa$ and E_γ is of plus type. By the rules of plus trees, $\mathrm{lh}(E_\gamma) < \kappa$. So

$$\begin{aligned}\mathrm{lh}(G_\gamma) &= \mathrm{res}_\gamma \circ \psi_\gamma(\mathrm{lh}(E_\gamma)) = \psi_\alpha(\mathrm{lh}(E_\gamma))\\ &< \mathrm{res}_\alpha \circ \psi_\alpha(\kappa) = \kappa^*.\end{aligned}$$

By part (4) of Lemma 4.7.7, $\mathbb{C}_\gamma \restriction \xi_\gamma \in V^{R_\alpha}_{\kappa^*}$. But κ^* is measurable in R_α, and $\mathrm{lh}(G^*_\gamma)$ is the least inaccessible τ in R_γ, equivalently R_α, such that $\mathbb{C}_\gamma \restriction \xi_\gamma \in V_{\tau+1}$. It follows that $\mathrm{lh}(G^*_\gamma) < \kappa^*$, as desired. ⊣

CLAIM 3.

(a) $\mathrm{res}_\alpha \circ \psi_\alpha \restriction \varepsilon(E_\beta) = \mathrm{res}_\beta \circ \psi_\beta \restriction \varepsilon(E_\beta)$.

(b) If $\alpha + 1 \notin D^{\mathcal{T}}$, *then*

  (i) $\psi_\beta \restriction \mathrm{dom}(E) + 1 = \psi_\alpha \restriction \mathrm{dom}(E) + 1$, *and*

  (ii) res_β *and* res_α *are the identity on* $\psi_\beta(\mathrm{dom}(E) + 1)$.

PROOF. For (a): this is clear if $\beta = \alpha$, so assume $\beta < \alpha$. Then $(3)_\alpha$ implies that ψ_α agrees with $\mathrm{res}_\beta \circ \psi_\beta$ on $\varepsilon(E_\beta)$. By (e) of $(3)_\alpha$, $\mathrm{lh}(G_\beta) \leq \rho^-(Q_\alpha)$, so by Lemma 4.7.13(d) we get that res_α is the identity on $\mathrm{lh}(G_\beta)$, and hence on $\psi_\beta(\varepsilon(E_\beta))$. This yields (a).

For (b): Note that $\mathrm{dom}(E) < \hat{\lambda}(E_\beta)$, so $\psi_\beta(\mathrm{dom}(E)) < \hat{\lambda}(H_\beta)$. Since we are not dropping in $\mathcal{T}$, E is total on M_β and $\mathrm{dom}(E) \leq \rho^-(M_\beta)$, so since ψ_β is nearly elementary, $\psi_\beta(\mathrm{dom}(E)) \leq \rho_k(Q_\beta)$ and $\psi_\beta(\mathrm{dom}(E))$ is a cardinal initial segment of Q_β. Thus for $n = n(Q_\beta, X_\beta)$, $\psi_\beta(\mathrm{dom}(E)) \leq \kappa_n(Q_\beta, X_\beta)$. So by Lemma 4.7.13(d), $\sigma_{Q_\beta}[X_\beta] \restriction \psi_\beta(\mathrm{dom}(E)) = \mathrm{id}$. If $\beta = \alpha$, we have (b)(ii). If $\beta < \alpha$, then res_α is the identity on $\mathrm{lh}(G_\beta)$, as we saw in the last paragraph. But $\psi_\beta(\mathrm{dom}(E) + 1) < \hat{\lambda}(G_\beta)$. Thus we have (b)(ii) in either case. From this and (a) we get (b)(i). ⊣

We define $\psi_{\alpha+1}$ and $Q_{\alpha+1}$ by cases.

The non-dropping case: $\alpha+1 \notin D^{\mathcal{T}}$.

We are in case (b) of Claim 3. So ψ_α agrees with ψ_β on $\text{dom}(E)$, $\psi_\beta(\text{dom}(E)) = \text{dom}(H)$, and res_α is the identity on $\text{dom}(H)$, so that $\text{dom}(H) = \text{dom}(G)$. This means we can apply Lemma 4.8.2, the Shift Lemma for Conversions, with its inputs being $\langle M_\beta, \psi_\beta, Q_\beta, \mathbb{C}_\beta, R_\beta\rangle$ and $\varphi = \text{res}_\alpha \circ \psi_\alpha$. That is, we set

$$Q_{\alpha+1} = i^*_{\beta,\alpha+1}(Q_\beta)$$

and for $a \in [\varepsilon_\alpha]^{<\omega}$ and appropriate f,[51]

$$\psi_{\alpha+1}([a,f]^{M_\beta}_E)) = [\text{res}_\alpha \circ \psi_\alpha(a), \psi_\beta(f)]^{R_\beta}_{G^*}.$$

Since G^* backgrounds G^+, this makes sense even if E is of plus type. By Lemma 4.8.2, $\langle M_{\alpha+1}, \psi_{\alpha+1}, Q_{\alpha+1}, \mathbb{C}_{\alpha+1}, R_{\alpha+1}\rangle$ is a conversion stage.

We have the diagram

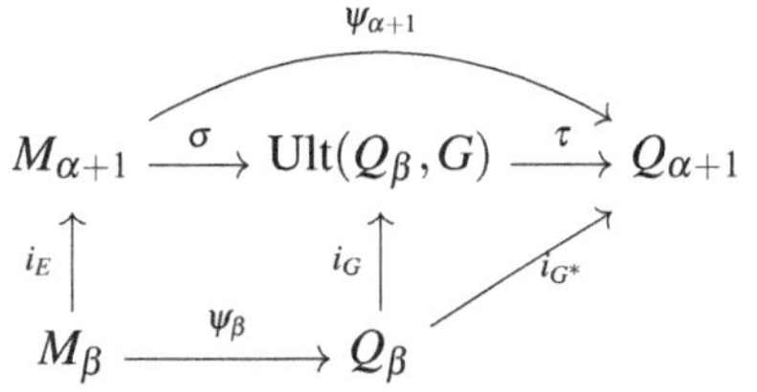

Here σ is the copy map, τ is the factor map into the larger ultrapower using all functions in R_β, and $\psi_{\alpha+1} = \tau \circ \sigma$.

Let us check that our induction hypotheses continue to hold.

CLAIM 4. *In the non-dropping case,* $(\dagger)_{\alpha+1}$ *holds.*

PROOF. We have already verified (1) of $(\dagger)_{\alpha+1}$. The commutativity condition (2) is easy based on the diagram above.

Let us now check the agreement hypotheses $(3)_{\alpha+1}$. $\psi_{\alpha+1}$ agrees with $\text{res}_\alpha \circ \psi_\alpha$ on ε_α by the Shift Lemma. If $\nu < \alpha$, then $(3)_\alpha$ implies that $\text{res}_\nu \circ \psi_\nu$ agrees with ψ_α on ε_ν, and hence with $\text{res}_\alpha \circ \psi_\alpha$ on ε_ν. Thus $\text{res}_\nu \circ \psi_\nu$ agrees with $\psi_{\alpha+1}$ on ε_α, as desired. So we have (a). Clause (b) is a simple consequence of the quasi-normality of $\mathcal{T}^* \restriction \alpha+2$.

For (c), it is enough to show $\mathbb{C}_\alpha \restriction \xi_\alpha = \mathbb{C}_{\alpha+1} \restriction \xi_\alpha$, and $M^{\mathbb{C}_{\alpha+1}}_{\xi_\alpha,0}$ is passive, since the rest of (c) then follows from $(3)_\alpha$(c). But letting $\mathbb{D} = i^{R_\alpha}_{G^*}(\mathbb{C}_\alpha)$, $\mathbb{C}_\alpha \restriction \xi_\alpha = \mathbb{D} \restriction \xi_\alpha$ and $M^{\mathbb{D}}_{\xi_\alpha,0}$ is passive, by Lemma 4.7.7. Thus we are done if $\beta = \alpha$, so assume $\beta < \alpha$. This implies $\kappa^* = \text{crit}(G^*) < \text{lh}(G^*_\beta)$, so $\kappa^* < \xi_\beta$, so $\mathbb{C}_\beta \restriction \kappa^* = \mathbb{C}_\alpha \restriction \kappa^*$, so $\mathbb{C}_{\alpha+1} \restriction i_{G^*}(\kappa^*) = \mathbb{D} \restriction i_{G^*}(\kappa^*)$. But $\xi_\alpha < i_{G^*}(\kappa^*)$, so we are done.

---

[51]Let $M_{\alpha+1} = \text{Ult}_k(M_\beta, E)$, where $k = k(M_\beta)$. As usual, the appropriate f are those that are $r\Sigma_k^{M_\beta}$ in some q. $\psi_\beta(f)$ is defined over Q_β from $\psi_\beta(q)$ via the same $r\Sigma_k$ formula.

For (d), it is enough again to consider the case $\nu = \alpha$, since the case $\nu < \alpha$ then follows from $(3)_\alpha$. So we must show that $Y_\alpha || o(Y_\alpha) = Q_{\alpha+1} | o(Y_\alpha)$. By Lemma 4.7.7, this is true if we replace $Q_{\alpha+1}$ with $i_{G^*}(Y_\alpha)$. But $Q_\beta | \kappa^* = Y_\beta \restriction \kappa^* = Q_\alpha \restriction \kappa^* = Y_\alpha \restriction \kappa^*$ by Claim 3. Since $o(Y_\alpha) < i_{G^*}(\kappa^*)$, we get (d). The same proof shows that $o(Y_\alpha)$ is a cardinal in $Q_{\alpha+1}$.

For (e), note that $\kappa^* < \rho^-(Q_\beta)$, so $\lambda(G^*) < \rho^-(Q_{\alpha+1})$. Hence $o(Y_\alpha) < \rho^-(Q_{\alpha+1})$, and we just observed that it is a cardinal in $Q_{\alpha+1}$. This gives us $(3)_{\alpha+1}$(e) when $\nu = \alpha$, and the case $\nu < \alpha$ then follows because $o(Y_\nu) < o(X_\alpha)$, $o(Y_\nu)$ is a cardinal of Q_α, and $\mathrm{res}_\alpha \restriction o(Y_\nu) + 1 = \mathrm{id}$.

$(4)_{\alpha+1}$ and $(5)_{\alpha+1}$(a) follow at once from the Shift Lemma for conversions. $(5)_{\alpha+1}$(b) holds for $\nu = \alpha$ by the argument of the last paragraph. The case $\nu < \alpha$ then follows because $\lambda(G^*_\nu) < o(X_\alpha)$, $\lambda(G^*_\nu)$ is a cardinal of Q_α, and res_α is the identity on $\lambda(G^*_\nu)$. ⊣

The dropping case: $\alpha + 1 \in D^{\mathcal{T}}$.

Let $J = M^{*,\mathcal{T}}_{\alpha+1}$, so that $J \lhd M_\beta$ and

$$M_{\alpha+1} = \mathrm{Ult}(J, E),$$

and let

$$K = \psi_\beta(J).$$

Here if $J = M_\beta \downarrow n$, then we understand K to be $Q_\beta \downarrow n$. The conversion stage that we shall move up to $c_{\alpha+1}$ via $i^{R_\beta}_{G^*}$ is

$$d = \langle J, \sigma_{Q_\beta}[K] \circ \psi_\beta, \mathrm{Res}_{Q_\beta}[K], \mathbb{C}_\beta, R_\beta \rangle.$$

In order to do that we must see that $\mathrm{res}_\alpha \circ \psi_\alpha$ agrees with $\sigma_{Q_\beta}[K] \circ \psi_\beta$ on $\mathrm{dom}(E)$. But $\mathrm{res}_\alpha \circ \psi_\alpha$ agrees with $\mathrm{res}_\beta \circ \psi_\beta$ on $\mathrm{dom}(E)$, so it is enough to show

CLAIM 5. res_β *agrees with* $\sigma_{Q_\beta}[K]$ *on* $\mathrm{dom}(H)$.

PROOF. Since $\mathcal{T}$ is maximal, J is the first initial segment of M_β past $\mathrm{lh}(E_\beta)$ with projectum $\rho(J) \le \mathrm{crit}(E)$. Since ψ_β is nearly elementary and $J \lhd M_\beta$,

$$\psi_\beta(\rho^-(J)) = \rho^-(K).$$

Moreover, $\forall R[M_\beta | \mathrm{lh}(E_\beta) \unlhd R \unlhd J \Longrightarrow \mathrm{dom}(E) \le \rho^-(R)]$, so Lemma 4.7.10 then implies that $\sigma_{Q_\beta}[K]$ agrees with $\sigma_{Q_\beta}[X_\beta]$ on $\mathrm{dom}(H)^{+,X_\beta}$. This is what we want. ⊣

By the Shift Lemma for conversion stages, letting

$$\psi_{\alpha+1}([a,f]^J_E) = [\mathrm{res}_\alpha \circ \psi_\alpha(a), \sigma_{Q_\beta}[K] \circ \psi_\beta(f)]^{R_\beta}_{G^*},$$

and

$$Q_{\alpha+1} = i_{G^*}(\mathrm{Res}_{Q_\beta}[K]),$$

we get the next conversion stage $c_{\alpha+1}$. The induction hypotheses $(\dagger)_{\alpha+1}$ are easy to verify.[52]

This completes the successor step in our inductive definition of $\text{lift}(\mathcal{T}, c)$. Now suppose γ is a limit ordinal $< \text{lh}(\mathcal{T})$. We define $\mathcal{T}^* \restriction \gamma+1$ by setting $[0,\gamma]_{T^*} = [0,\gamma]_T$. If this results in $\mathcal{M}_\gamma^{\mathcal{T}^*}$ being illfounded, then we stop the conversion. So suppose that $\mathcal{M}_\gamma^{\mathcal{T}^*}$ is wellfounded. Induction hypothesis (2) then tells us that $D^{\mathcal{T}} \cap [0,\gamma)_T$ is finite. Let $\alpha <_T \gamma$ be large enough that $D^{\mathcal{T}} \cap \gamma \subseteq \alpha$. By (2) we have $i^*_{\alpha,\xi}(Q_\alpha) = Q_\xi$ for all $\xi \in [\alpha,\gamma)_T$. We set

$$Q_\gamma = i^*_{\alpha,\gamma}(Q_\alpha),$$

and define $\psi_\gamma \colon M_\gamma \to Q_\gamma$ by letting

$$\psi_\gamma(i^{\mathcal{T}}_{\xi,\gamma}(x)) = i^*_{\xi,\gamma}(\psi_\xi(x))$$

for all $\xi \in [\alpha,\gamma)_T$. By (2), ψ_γ is well-defined. It is now easy to check that $(\dagger)_\gamma$ holds.

DEFINITION 4.8.5. Let $c = \langle M, \psi, Q, \mathbb{C}, R\rangle$ be a PFS conversion stage, and let $\mathcal{T}$ be a plus tree on M; then

(1) $\text{lift}(\mathcal{T}, c) = \langle \mathcal{T}^*, \langle c_\alpha \mid \alpha < \text{lh}(\mathcal{T})\rangle\rangle$. is the conversion system defined above. We write $\mathcal{T}^* = \text{lift}(\mathcal{T}, c)_0$ for its tree component, and $\mathbb{C}_\xi = i^{\mathcal{T}^*}_{0,\xi}(\mathbb{C})$.
(2) $\text{res}_\xi(\mathcal{T}, c) = \text{res}_\xi = \sigma_{Q_\xi}[Q_\xi | \text{lh}(\psi_\xi(E^{\mathcal{T}}_\xi))]^{\mathbb{C}_\xi}$. We call res_ξ the ξ-th *generator map* associated to $\text{lift}(\mathcal{T}, c)$.
(3) $\text{stg}(\mathcal{T}, c, \alpha) = \langle P_\alpha, \psi_\alpha, Q_\alpha, \mathbb{C}_\alpha, R_\alpha\rangle$ is the conversion stage c_α occurring at α in the construction of $\text{lift}(\mathcal{T}, c)$.

We may sometimes display the components of the conversion stages by writing $\text{lift}(\mathcal{T}, M, \psi, Q, \mathbb{C}, R) = \langle \mathcal{T}^*, \langle Q_\xi \mid \xi < \text{lh}(\mathcal{T})\rangle, \langle \psi_\xi \mid \xi < \text{lh}(\mathcal{T})\rangle\rangle$.

DEFINITION 4.8.6. In the special case of Definition 4.8.5 that $M = Q$ and $\psi = \text{id}$, we set

$$\text{lift}(\mathcal{T}, M, \mathbb{C}, R) = \text{lift}(\mathcal{T}, M, \text{id}, M, \mathbb{C}, R).$$

We also let

$$\text{lift}(\mathcal{T}, M, \mathbb{C}) = \text{lift}(\mathcal{T}, M, \mathbb{C}, V)$$

in the case that $R = V$ (the universe of all sets).

[52] If $J \lhd M^-_\beta$, then $\rho(K) \le \text{crit}(H)$. But it is possible that $J = M^-_\beta$ and $\text{dom}(H) \le \rho(K) = \rho^-(Q_\beta)$. In that case, setting $Q_{\alpha+1} = i^*(Q_\beta)$ instead of $Q_{\alpha+1} = i^*(K)$ would give a superficially different conversion system. We believe that because PFS resurrection maps are consistent with one another, this other system would be equivalent to the one we are defining.

Induced strategies. We define induced strategies just as in Chapter 3. Suppose that $c = \langle M, \psi, Q, \mathbb{C}, R\rangle$ is a conversion stage, and that Σ^* is a $(\theta, \vec{F}^{\mathbb{C}})$ iteration strategy for the background universe R; then Σ^* induces a complete θ-iteration strategy Σ for M as follows: for $\mathcal{T}$ a plus tree on M,

$$\mathcal{T} \text{ is by } \Sigma \iff \text{lift}(\mathcal{T}, c)_0 \text{ is by } \Sigma^*.$$

We write

$$\Sigma = \Omega(c, \Sigma^*)$$

for this induced strategy. When $M \in \text{lev}(\mathbb{C})$, we set

$$\Omega(\mathbb{C}, M, R, \Sigma^*) = \Omega(\langle M, \text{ id}, M, \mathbb{C}, R\rangle, \Sigma^*).$$

We write $\Omega(\mathbb{C}, M, \Sigma^*)$ when R can be understood from context. We may occasionally use the notation $\text{lift}(\mathcal{T}, c, \Sigma^*)$ for the largest initial segment of $\text{lift}(\mathcal{T}, c)$ that is by Σ^*. So $\mathcal{T}$ is by $\Omega(\mathbb{C}, M, \Sigma^*)$ iff $\text{lift}(\mathcal{T}, c) = \text{lift}(\mathcal{T}, c, \Sigma^*)$. We have shown above that the lifted tree $\mathcal{T}^*$ is quasi-normal, so Σ^* need only be defined on nice quasi-normal iteration trees.

If Σ^* is defined on stacks of quasi-normal trees, of any length, then we can extend the lifting process and the induced strategy Σ for M so that it is defined on stacks of plus trees of the same length. For example, let

$$c = \langle M, \psi, Q, \mathbb{C}, R\rangle$$

be a conversion stage, and Σ^* an $(\eta, \theta, \vec{F}^{\mathbb{C}})$ iteration strategy for R, where $\eta > 1$. Let $\Omega = \Omega(c_0, \Sigma^*)$, and $\mathcal{T}$ be a plus tree on M by Ω having last model $M_\alpha^{\mathcal{T}}$, and let $N \trianglelefteq M_\alpha^{\mathcal{T}}$. We get a tail strategy for plus trees on N as follows. Letting

$$\text{stg}(\mathcal{T}, c, \alpha) = \langle M_\alpha, \psi_\alpha, Q_\alpha, \mathbb{C}_\alpha, R_\alpha\rangle,$$

we set

$$d = \langle N, \sigma_{Q_\alpha}[\psi_\alpha(N)] \circ \psi_\alpha, \text{Res}_{Q_\alpha}[\psi_\alpha(N)], \mathbb{C}_\alpha, R_\alpha\rangle,$$

and define the tail strategy $\Omega_{\mathcal{T},N}$ on plus trees of length $< \theta$ by

$$\mathcal{U} \text{ is by } \Omega_{\mathcal{T},N} \iff \text{lift}(\mathcal{U}, d)_0 \text{ is by } \Sigma^*_{\mathcal{T}^*, R_\alpha},$$

where $\mathcal{T}^* = \text{lift}(\mathcal{T}, c)_0$. Clearly we can continue this process so as to define a tail strategy $\Omega_{\mathcal{T},N,\mathcal{U},P}$, for any P that is an initial segment of the last model of $\mathcal{U}$, and so on.

DEFINITION 4.8.7. Let $c = \langle M, \psi, Q, \mathbb{C}, R\rangle$ be a conversion stage, and let Σ^* be a $(\lambda, \theta, \mathcal{F}^{\mathbb{C}})$-iteration strategy for R; then $\Omega(c, \Sigma^*)$ is the complete (λ, θ)-iteration strategy induced by Σ^* as above.

Again, when $M \in \text{lev}(\mathbb{C})$ and R can be understood from context, we write

$$\Omega(\mathbb{C}, M, \Sigma^*) = \Omega(\langle M, \text{ id}, M, \mathbb{C}, R\rangle, \Sigma^*).$$

We shall show in Corollary 5.1.4 that all strategies induced by $\mathbb{C}$ and Σ^* are pullbacks of induced strategies for levels of $\mathbb{C}$. For now, notice

LEMMA 4.8.8. *Let $c = \langle M, id, M, \mathbb{C}, R\rangle$ be a conversion stage, let Σ^* be a $(\lambda, \theta, \mathcal{F}^{\mathbb{C}})$-iteration strategy for R, and let $N \trianglelefteq M$; then*

$$\Omega(\mathbb{C}, M, \Sigma^*)_N = \Omega(\mathbb{C}, \mathrm{Res_M}[N], \Sigma^*)^{\sigma_M[N]}$$

PROOF. This is immediate from the definitions. $\Omega(\mathbb{C}, M, \Sigma^*)_N$ is the tail of $\Omega(\mathbb{C}, M, \Sigma^*)$ after the empty tree followed by a drop to N. The only model in the empty tree is M, the lifting map is $\psi = \mathrm{id}$, and the new background universe and construction $\mathbb{D}$ are the same as the old ones. So by definition

$$\begin{aligned}\Omega(\mathbb{C}, M, \Sigma^*)_N &= \Omega(\mathbb{D}, \mathrm{Res_M}[N], \Sigma^*)^{\sigma_M[N]\circ\psi}\\ &= \Omega(\mathbb{C}, \mathrm{Res_M}[N], \Sigma^*)^{\sigma_M[N]}.\end{aligned}$$ ⊣

Mild positionality makes perfect sense in the context of plus trees on pfs premice:

DEFINITION 4.8.9. Let Ω be a (λ, θ)-iteration strategy for a pfs premouse M; then Ω is *mildly positional* iff

(a) $\Omega = \Omega_M$, and
(b) whenever s is a stack by Ω and $P \trianglelefteq N \trianglelefteq M_\infty(s)$, then $(\Omega_{s,N})_P = \Omega_{s,P}$.

Because our resurrection maps are consistent, we get

LEMMA 4.8.10. *Let $c = \langle M, \pi, Q, \mathbb{C}, R\rangle$ be a conversion stage and let Σ^* be a $(\lambda, \theta, \mathcal{F}^{\mathbb{C}})$-iteration strategy for R; then $\Omega(c, \Sigma^*)$ is mildly positional.*

PROOF. Let $\Lambda = \Omega(\langle Q, \mathrm{id}, Q, \mathbb{C}, R\rangle, \Sigma^*)$. By Corollary 5.1.4, $\Omega(c, \Sigma^*) = \Lambda^\pi$. Mild positionality is preserved by pullbacks, so it is enough to show that Λ is mildly positional. We prove (b) in the case $s = \emptyset$; the general case is the same.

Let $P \trianglelefteq N \trianglelefteq Q$, $N_1 = \mathrm{Res_Q}[N]$, $P_1 = \mathrm{Res_Q}[P]$, and let $\tau\colon N \to N_1$ and $\sigma\colon P \to P_1$ be the two resurrection maps. Let Ψ and Φ be the strategies for N_1 and P_1 induced by $\mathbb{C}$ and Σ^*. Let $\theta = \sigma_{\mathrm{N_1}}[\tau(P)]$ resurrect $\tau(P)$ from N_1. By resurrection consistency, θ maps into $\mathrm{Res_{N_1}}[\tau(P)] = P_1$, and

$$\sigma = \theta \circ \tau.$$

Thus

$$\begin{aligned}(\Lambda_N)_P &= (\Psi_{\tau(P)})^\tau\\ &= (\Phi^\theta)^\tau\\ &= \Phi^\sigma\\ &= \Lambda_P.\end{aligned}$$ ⊣

In our case of interest, the background universe strategy Σ^* chooses unique wellfounded branches. It follows from Lemma 4.4.12 that Σ^* does not distinguish between a nice tree $\mathcal{T}$ and its normal companion $\mathcal{T}^{\mathrm{nrm}}$.[53] If $\mathcal{F}^{\mathbb{C}}$ is coarsely coherent,

[53] We could have defined our conversion systems so that $\mathrm{lift}(\mathcal{T}, c)_0$ is always normal, but the price would be that $\mathcal{T}$ and $\mathrm{lift}(\mathcal{T}, c)_0$ might have different lengths, which would be a nuisance.

then by Lemma 2.9.12, $\mathcal{T}^{\text{nrm}}$ is then the unique normal $\mathcal{F}^{\mathbb{C}}$-tree with last model $\mathcal{M}_\infty^{\mathcal{T}^{\text{nrm}}}$.

In general, we wo not assume that $\mathcal{F}^{\mathbb{C}}$ is coarsely coherent. Also, $\text{lift}(\mathcal{T}, c)_0$ may fail to be normal, and in general, quasi-normal iterations are not determined by their last model, because one can insert delays. We do have a uniqueness lemma, however.

DEFINITION 4.8.11. We say that $\langle F_\alpha \mid \alpha < \theta \rangle$ is *mildly coherent* iff for all $\alpha < \theta$

(1) F_α is a nice extender,
(2) $\alpha < \beta < \theta \Longrightarrow \text{lh}(F_\alpha) < \text{lh}(F_\beta)$, and
(3) $i_{F_\alpha}(\vec{F}) \restriction \alpha = \vec{F} \restriction \alpha$.

It follows at once from Lemma 4.7.7(1) that if $\mathbb{C}$ is a maximal PFS construction, then the sequence $\langle F_\nu^{\mathbb{C}} \mid \langle \nu, 0 \rangle < \text{lh}(\mathbb{C}) \rangle$ of background extenders it actually uses is mildly coherent. We do have

LEMMA 4.8.12. *Let $\vec{F}$ be mildly coherent in M, and let Σ be an $\vec{F}$-iteration strategy for M; then for any N, there is at most one quasi-normal $\vec{F}$-iteration tree played according to Σ whose last model is N.*

PROOF. Let $\mathcal{T}$ and $\mathcal{U}$ be distinct such trees. Suppose that $\mathcal{T} \restriction \beta + 1 = \mathcal{U} \restriction \beta + 1$, but $G \neq H$, where $G = E_\beta^{\mathcal{T}}$ and $H = E_\beta^{\mathcal{U}}$. Both G and H are taken from $i(\vec{F})$, where $i = i_{0,\beta}^{\mathcal{T}} = i_{0,\beta}^{\mathcal{U}}$. Say G occurs before H in $i(\vec{F})$, or equivalently, $\text{lh}(G) < \text{lh}(H)$. Then $G \in \mathcal{M}_{\beta+1}^{\mathcal{U}}$, so $G \in N$ because $\mathcal{U}$ is length non-decreasing. But $G \notin N$ because $G \notin \mathcal{M}_{\beta+1}^{\mathcal{T}}$, and $\mathcal{T}$ is length non-decreasing. ⊣

It follows that if c is a conversion stage whose construction is $\mathbb{C}$, and $\mathcal{T}$ is by $\Omega(c, \Sigma^*)$, then $\text{lift}(\mathcal{T}, c)_0$ is the unique quasi-normal $\vec{F}^{\mathbb{C}}$-tree having the last model it has, and played by Σ^*.

4.9. Backgrounds for plus extenders

PFS constructions can break down by reaching some $M_{\nu,k}$ such that either $M_{\nu,k}$ is not solid, or its last extender is not unique, or its last extender F is not properly certified, in that F^* does not certify F^+. Granted iterability assumptions, we can prove none of that happens. In this section, we rule out the last possibility, and in the next two sections, we rule out the others.

It might seem that we could define away the last problem, by simply restricting our attention to constructions in which F^* always certifies F^+. The trouble with that approach is that such background constructions may not produce enough mice. Our existence proofs for pfs mice, and later for strategy mice, would all have a gap.[54] Proposition 3.1.9 implies that the requirements on certificates in Section 3.1

[54] A new gap, beyond the lack of a general iterability theorem.

do not restrict the certified extenders in any way that matters for the existence theorems in this book. Theorem 4.9.1 says that the requirement that E^* background E^+ is also not restrictive, because in fact it follows from the other requirements.

THEOREM 4.9.1. *Let $\mathbb{C}$ be a maximal PFS-construction, and assume that V is countably $\mathcal{F}^{\mathbb{C}}$-iterable. Let ν be an extender-active stage of $\mathbb{C}$, let $M^{\mathbb{C}}_{\nu,0} = (M^{<\nu}, F)$, and let $F^* = F^{\mathbb{C}}_\nu$ be the certificate for F; then $F^* \cap ([\lambda_F + 1]^{<\omega} \times M^{<\nu}) = F^+$, and $\mathrm{lh}(F)$ is a cardinal of $i^V_{F^*}(M)$.*

PROOF. The proof resembles the proof of closure under initial segment in Section 10 of [35]. Let $F^* = F^*_\nu$ be the certificate for F in the construction $\mathbb{C}$, where F is the last extender of $M = M^{\mathbb{C}}_{\nu,0}$. Let $\kappa = \mathrm{crit}(F)$ and

$$\phi = i^V_{F^*} \restriction M$$

and let

$$G = \begin{cases} E_\phi \restriction \mathrm{lh}(F) & \text{if } \mathrm{lh}(F) = \phi(\kappa^{+,M}) \\ E_\phi \restriction \mathrm{lh}(F)+1 & \text{if } \mathrm{lh}(F) < \phi(\kappa^{+,M}), \end{cases}$$

$$N = \mathrm{Ult}(M,G)||(\lambda_F^+)^{\mathrm{Ult}(M,G)}.$$

One of our goals is to show that $\mathrm{lh}(G) = \mathrm{lh}(F) = o(N)$. Since $\lambda_F < \lambda_{F^*}$, λ_F is a generator of E_ϕ, a limit cardinal of N, and the largest cardinal of N. $\mathrm{lh}(F)$ is the next potential generator of E_ϕ, and it is a generator iff $\mathrm{lh}(F)$ is not a cardinal of $i^V_{F^*}(M)$ iff $\mathrm{lh}(F) < o(N)$. The factor embedding from $\mathrm{Ult}(M,G)$ to $i_{F^*}(M)$ has critical point $\geq o(N)$, so $N \trianglelefteq_0 i_{F^*}(M)$. For all we know at the moment, $\mathrm{lh}(F)$ may be an active stage in N, but by Lemma 4.7.7(2), it cannot index F in N.

For $\eta < \kappa^{+,M}$, the fragment $G_\eta = G \cap ([\lambda_F \cup \{\lambda_F, \mathrm{lh}(F)\}]^{<\omega} \times M|\eta)$ belongs to N, by the usual Kunen argument.[55] The G_η are constructed cofinally in $o(N)$, so we can code G by a predicate $\hat{G}$ that is amenable to N. (See [35, p.13].)

For any extender H over some M and $\eta < \mathrm{lh}(H)$, let us write

$$\gamma(H,\eta) = (\eta^+)^{\mathrm{Ult}(M,H\restriction\eta)}.$$

We care about the case that $H \restriction \eta$ is whole, so that $\gamma(H,\eta) = i_{H\restriction\eta}(\mathrm{dom}(H))$.

The following facts are captured by first order sentences in the theory of $(N,\hat{G})$:

(∗) There is a largest cardinal ν, moreover
  (1) ν is a limit cardinal, $\nu = \lambda(G\restriction\nu)$, and ν is a generator of G,
  (2) ν is not measurable by the $\mathrm{Ult}(N,G)$-sequence,
  (3) letting $\gamma = \gamma(G,\nu)$, it is not the case that $\gamma < o(N)$ and E^N_γ is the trivial completion of $G\restriction\nu$.

Clause (2) holds because F^* is λ-minimal. Clause (3) holds by Lemma 4.7.7(2).

[55] Let $f\colon \kappa \to M|\eta$ and $f \in M$. Then $i_G(f) \in N$, and for $a \in [\lambda_F \cup \{\lambda_F, \mathrm{lh}(F)\}]^{<\omega}$ and $\alpha < \kappa$, $(a, f(\alpha)) \in G$ iff $a \in i_G(f)(\alpha)$.

Let us call a structure satisfying the sentences that capture $(*)$ a *pseudo-premouse*. If (P,B) is a pseudo-premouse, then

$$\nu(P,B) = \text{largest cardinal of } P,$$
$$G(P,B) = \text{extender over } P \text{ coded by } B,$$

and

$$\gamma(P,B) = \gamma(G(P,B), \nu(P,B)).$$

Let also

$$F(P,B) = \text{Jensen completion of } G(P,B) \restriction \nu(P,B).$$

Thus $\gamma(P,B) = \text{lh}(F(P,B))$.

If (M,A) is another pseudo-premouse, then we say that (P,B) is an *initial segment of* (M,A) iff $P = M||o(P)$ and $G(P,B) = G(M,A) \restriction o(P)$.

Our goal is to show that $G = F^+ \restriction \text{lh}(F)$. Let us say that a pseudo-premouse (P,B) is *bad* iff $G(P,B) \neq F(P,B)^+ \restriction \gamma(P,B)$. Thus our goal is to show that $(N,\hat{G})$ is not bad. If $\gamma(P,B) < o(P)$ then (P,B) is bad, and if $\gamma(P,B) = o(P)$ then (P,B) is bad iff $G(P,B) \neq F(P,B)^+$.

$(N,\hat{G})$ also has an iteration strategy Σ that we get from $\mathbb{C}$. Along non-dropping branches of an iteration tree $\mathcal{T}$ on $(N,\hat{G})$ the ultrapowers taken are all Σ_0 ultrapowers, so the canonical embeddings are cofinal and Σ_1 elementary.

CLAIM 0. *If (M,A) is a pseudo-premouse, E is an extender over M, and $(P,B) = \text{Ult}_0((M,A),E)$ is wellfounded, then*

(a) (P,B) is a pseudo-premouse,
(b) $\gamma(P,B) = \sup i_E``\gamma(M,A)$,
(c) $\gamma(M,A) = o(M)$ iff $\gamma(P,B) = o(P)$,
(d) if (M,A) is bad, then (P,B) is bad.

PROOF. Let $i = i_E^M$, $\nu = \nu(M,A)$, and $\gamma = \gamma(M,A)$. i is cofinal and Σ_1 elementary, so $i(\nu)$ is the largest cardinal of P. Let $H = G(M,A)$ and $K = G(P,B)$. i_E maps $\text{dom}(H \restriction \nu)$ cofinally into $\text{dom}(K \restriction i(\nu))$, so i_E maps $i_{H\restriction\nu}``\text{dom}(H)$ cofinally into $i_{K\restriction i(\nu)}``\text{dom}(K)$. This gives us (b). Since i_E maps $o(M)$ cofinally into $o(P)$, we also get (c).

$(*)(1)$ and $(*)(2)$ are $\Pi_1^{(M,A)}$ facts about ν, and hence they hold of $i(\nu) = \nu(P,B)$. If $\gamma(M,A) = o(M)$, then $\gamma(P,B) = o(P)$, so $(*)(3)$ for (P,B) is vacuous. Suppose $\gamma = \gamma(M,A) < o(M)$. Since $E_\gamma^M \restriction \nu \neq K \restriction \nu$, and this is a $\Sigma_0^{(M,A)}$ fact about E_γ^N and some $(a,X) \in E_\gamma^N$, we have that $E_{i(\gamma)}^P \restriction i(\nu) \neq H \restriction i(\nu)$, so $(*)(3)$ holds if $i(\gamma) = \gamma(P,B)$. But otherwise $\gamma(P,B) = \sup i``\gamma < i(\gamma)$, so $\text{cof}_0^P(\sup i``\gamma) = \text{crit}(E)$ is a limit cardinal in P, and hence not the index of an extender on the P-sequence. Thus $(*)(3)$ holds in any case, and (P,B) is a pseudo-premouse.

Finally, suppose (M,A) is bad. If it is bad because $\gamma < o(M)$, then (P,B) is also bad by (c), so suppose $\gamma = o(M)$ and $G(M,A) \neq F(M,A)^+$. Let

$$(a,X) \in G(M,A) \triangle F(M,A)^+,$$

where $a \subseteq \nu(M,A)+1$ is finite and $X \in \mathrm{dom}(G)$. This is a $\Sigma_1^{(M,A)}$ property of (a,X), so

$$(i(a),i(X)) \in G(P,B) \triangle F(P,B)^+.$$

Thus (P,B) is bad, as desired. ⊣

Let us now assume toward contradiction that $(N,\hat{G})$ is bad. By the claim, all non-dropping iterates of $(N,\hat{G})$ are also bad pseudo-premice. We may assume that $(N,\hat{G})$ is countable, as otherwise we can just replace $(N,\hat{G})$ with a countable elementary submodel of itself, and Σ by its pullback under the anticollapse map. Let $\vec{e}$ be an enumeration of N in order type ω. By the proof of Lemma 4.6.10, we may assume that Σ has the Weak Dodd-Jensen property relative to $\vec{e}$, in the following sense:

(†) If $(M,\hat{H})$ is a non-dropping Σ-iterate of $(N,\hat{G})$ with iteration map $i\colon (N,\hat{G}) \to (M,\hat{H})$, and (P,B) is an initial segment of $(M,\hat{H})$, and $\pi\colon (N,\hat{G}) \to (P,B)$ is cofinal and Σ_1 elementary, then
 (a) $(P,B) = (M,\hat{H})$, and
 (b) for any n, if $i(e_k) = \pi(e_k)$ for all $k < n$, then $i(e_n) \leq_P \pi(e_n)$.
 Here $\leq_P$ is the order of construction in P.

Now let

$$P_0 = Q_0 = (N,\hat{G}),$$

and

$$P_1 = \mathrm{Ult}(P_0, G \restriction \nu).$$

We are going to compare the phalanx (P_0,P_1,ν) with Q_0. The resulting tree on the phalanx we call $\mathcal{T}$, with models $P_\xi = \mathcal{M}_\xi^{\mathcal{T}}$, and the tree on Q_0 we call $\mathcal{U}$, with models $Q_\xi = \mathcal{M}_\xi^{\mathcal{U}}$. The trees $\mathcal{T}$ and $\mathcal{U}$ will be λ-tight. At the same time, we lift $\mathcal{T}$ to a λ-tight tree $\mathcal{T}^*$ with models P_ξ^*, and embeddings $\pi_\xi : P_\xi \to P_\xi^*$. Here $\pi_0 = \mathrm{id}$, and

$$P_1^* = \mathrm{Ult}(P_0, G),$$

with π_1 being the natural factor map. π_1 is cofinal and Σ_1 elementary, and $\pi_1 \restriction \nu$ is the identity, so we can indeed lift $\mathcal{T}$ by (π_0,π_1), the construction being the same as the one that produced $\pi\mathcal{T}$ in Section 4.5. Since $\nu = i_{G\restriction\nu}(\kappa)$, $\pi_1(\nu) = i_G(\kappa) > \nu$. The trees $\mathcal{T}^*$ and $\mathcal{U}$ are according to Σ.

$\mathcal{T}$ is not literally an iteration tree on P_0, since $G \restriction \nu$ is not on the P_0 sequence, but we shall use iteration tree notation for it. In particular, $0 <_T 1$, and $i_{0,1}^{\mathcal{T}} = i_{G\restriction\nu}$.[56]

[56] [35] calls $\mathcal{T}$ a *pseudo-iteration tree*.

Notice that ν is a limit cardinal in P_0, and $k(P_0) = 0$, so that if T-pred$(\xi + 1) = 0$, then $P_{\xi+1} = \text{Ult}_0(P_0, E_\xi^{\mathcal{T}})$. In other words, we never drop when an extender in $\mathcal{T}$ is applied to P_0. This means that certain anomalous cases that occur in more delicate phalanx comparisons do not occur here.[57]

The non-dropping iterates of P_0 in the trees $\mathcal{T}, \mathcal{T}^*$, and $\mathcal{U}$ are all pseudo-premice. If P_0-to-P_ξ does not drop, then π_ξ is cofinal and Σ_1 elementary. If P_0-to-P_ξ does drop, then P_ξ and P_ξ^* are type 1 pfs premice[58] and π_ξ is elementary.[59]

P_0 also satisfies the "weak initial segment condition", in that whenever H is a whole proper initial segment of $G \restriction \nu$, then the completion of H is indexed on the P_0 sequence. One of our problems is that the weak initial segment condition can fail in iterates of P_0 below the image of ν if the iteration map is discontinuous at ν.

CLAIM 1. *Suppose that $[0, \xi]_T \cap D^{\mathcal{T}} = \emptyset$, and let (Q, C) be a pseudo-premouse that is a proper initial segment of P_ξ; then there is a proper initial segment (R, D) of P_ξ^* such that $\pi_\xi \restriction Q$ is cofinal and Σ_1 elementary as a map from (Q, C) to (R, D).*

PROOF. Let $\eta = \nu(Q, C)$ and $\delta = \pi_\xi(\eta)$. Let $P_\xi = (P, A)$, $H = G(P, A)$ be its last extender, $P_\xi^* = (P^*, B)$, and $H^* = G(P^*, B)$. We are given that $Q \lhd P$ and $G(Q, C) = H \restriction o(Q)$. We set

$$R = P_\xi^* || \sup \pi_\xi``o(Q),$$
$$\sigma = \pi_\xi \restriction Q,$$

and

$$D = \bigcup_{\alpha < o(Q)} \sigma(C \cap \alpha).$$

Clearly σ is cofinal and Σ_1 elementary as a map from (Q, C) to (R, D), so we just need to see that $G(R, D) = H^* \restriction o(R)$.

$G(Q, C)$ and $G(R, D)$ are determined by looking at the extender fragments coded by C and D. Let γ be the largest generator of $H \restriction o(Q)$, that is, $\gamma = \eta$ if $\gamma(Q, C) = o(Q)$ and $\gamma = \gamma(Q, C)$ otherwise. For $\beta < \text{dom}(H)$, let

$$H_\beta = H \cap ([\nu(P, A)]^{<\omega} \times P|\beta),$$

and for $\beta < \text{dom}(H^*)$ let

$$H_\beta^* = H^* \cap ([\nu(P^*, B)]^{<\omega} \times P^*|\beta).$$

Then for all $\beta < \text{dom}(H)$

$$\pi_\xi(H_\beta) = H_{\pi_\xi(\beta)}^*.$$

---

[57] See Remark 4.10.4.

[58] Type 1 because this is preserved after a drop in plus trees.

[59] That the maps are elementary relies on arguments from [47] and [86]. See Lemma 4.5.21. One could just make do with Lemma 4.5.22 and near elementarity for the π_ξ at this point.

Moreover, $\operatorname{ran}(\pi_\xi)$ is cofinal in $\operatorname{dom}(H^*)$ because $[0,\xi]_T$ does not drop, and $i^{\mathcal{T}}_{0,\xi}$ and $i^{\mathcal{T}^*}_{0,\xi}$ are continuous at $\operatorname{dom}(G)$. It follows that

$$\bigcup_{\beta<\operatorname{dom}(H)} \sigma(H_\beta\restriction\gamma+1) = H^*_\beta\restriction\sigma(\gamma+1).$$

But this just means that $G(R,D) = H^*\restriction o(R)$, as desired. ⊣

The comparison of (P_0,P_1,ν) with Q_0 proceeds by iterating away least disagreements. Let

$$\varepsilon^{\mathcal{T}}_\beta = \begin{cases} i^{\mathcal{T}}_{0,\beta}(\nu) & \text{if } E^{\mathcal{T}}_\beta \text{ is coded by the image of } \hat{G} \text{ along } [0,\beta]_T,\\ \lambda(E^{\mathcal{T}}_\beta) & \text{otherwise.}\end{cases}$$

Similarly for $\varepsilon^{\mathcal{U}}_\beta$. The normality rules are that T-pred$(\xi+1)$ is the least β such that $\operatorname{crit}(E^{\mathcal{T}}_\xi) < \varepsilon^{\mathcal{T}}_\beta$, and similarly for $\mathcal{U}$. Notice that in the case that $E^{\mathcal{T}}_\beta$ is coded by the image of $\hat{G}$ along $[0,\beta]_T$, $\operatorname{crit}(E^{\mathcal{T}}_\xi) \neq i^{\mathcal{T}}_{0,\beta}(\nu)$ by property (∗)(2) of pseudo-premice. Thus our normality rules do prevent the generators of $E^{\mathcal{T}}_\beta$ from being moved along branches where it has been used.

CLAIM 2. *The comparison terminates.*

PROOF. This is not completely routine, because the weak initial segment condition may fail for iterates of $(N,\hat{G})$. Important generators are not moved along branches, so the usual proof gives us some countable α and $\eta+1,\xi+1$ such that

$$\alpha = T\text{-pred}(\eta+1) = U\text{-pred}(\xi+1),$$

and for $H = E^{\mathcal{T}}_\eta$ and $K = E^{\mathcal{U}}_\xi$, $\operatorname{dom}(H) = \operatorname{dom}(K)$ and H and K are compatible. This is impossible unless one of H and K is coded by the image of $\hat{G}$ along the branch to its model.

Case 1. $[0,\eta]_T$ does not drop, and H is coded by the top predicate of P_η.
Let $\mu = i^{\mathcal{T}}_{0,\eta}(\nu)$ be the largest cardinal of P_η. We have that μ is a cutpoint of H, a generator of H, and for $\gamma = \gamma(P_\eta,\hat{H})$ and $J = E^{\operatorname{Ult}(P_\eta,H\restriction\mu)}_\gamma$, $J\restriction\mu \neq H\restriction\mu$.

Subcase 1A. $[0,\xi]_U$ does not drop, and K is coded by the top predicate of Q_ξ.
$H \neq K$, because otherwise the comparison was finished before we used them. Suppose K is a proper initial segment of H, so that Q_ξ is a proper initial segment of P_η in the pseudo-premouse sense. $i^{\mathcal{U}}_{0,\xi}$ is cofinal and Σ_1 elementary, so by Claim 0, $\pi_\eta \circ i^{\mathcal{U}}_{0,\xi}$ is a cofinal, Σ_1 elementary embedding from P_0 to a proper initial segment of P^*_ξ, contrary to the Weak Dodd-Jensen property of Σ. If H is a proper initial segment of K, then $i^{\mathcal{T}}_{0,\eta}$ is a cofinal, Σ_1 elementary map from from $P_0 = Q_0$ into a proper initial segment of Q_ξ, which again contradicts the Weak Dodd-Jensen property of Σ.

Subcase 1B. Subcase 1A does not hold.

We then have that $Q_\xi | \mathrm{lh}(K)$ is a pfs premouse. If $\mu < \lambda_K$, then $H \restriction \mu$ is a whole proper initial segment of K, so by the initial segment condition its trivial completion I is indexed on the Q_ξ-sequence at $\gamma = \gamma(P_\eta, \hat{H})$. But then $I = E_\gamma^{Q_\xi} = E_\gamma^{Q_{\omega_1}} = E_\gamma^{P_{\omega_1}} = E_\gamma^{P_\eta} = J$, contrary to $J \restriction \mu \neq H \restriction \mu$.

Suppose next that $\lambda_K < \mu$. Let $i = i^{\mathcal{T}}_{0,\eta}$. For any $\tau < \nu$ such that $G \restriction \tau$ is whole, the Jensen completion of $G \restriction \tau$ is on the N sequence. It follows that for any $\tau < \sup i``\nu$ such that $H \restriction \tau$ is whole, the Jensen completion of $H \restriction \tau$ is on the P_η sequence.[60] Since K is not on the P_η sequence, we must have

$$\sup i``\nu \leq \lambda_K < \mu = i(\nu).$$

Thus ν is singular in N, and since ν is regular in $N||\mathrm{lh}(F) = \mathrm{Ult}(N, G \restriction \nu)||\mathrm{lh}(F)$,

$$\mathrm{lh}(F) = \gamma(N, \hat{G}) < o(N).$$

Let S be the first level of N above $\mathrm{lh}(F)$ that projects to ν. For any $X \subset \kappa$ such that $X \in N$, we have some $\beta < \nu$ such that

$$i_{G \restriction \nu}(X) = h_S(\beta, p(S)),$$

where h_S is the canonical Skolem function and $p(S)$ is the standard parameter. This fact is preserved by i, so $i_{H \restriction \mu}(i(X)) = h_{i(S)}(i(\beta), p(i(S)))$. But this means

$$i_K(i(X)) = h_{i(S)}(i(\beta), p(i(S))) \cap \lambda_K.$$

Noting that $i(\beta) < \lambda_K$ and $\mathrm{ran}(i)$ is cofinal in $\mathrm{dom}(H) = \mathrm{dom}(K)$, we see that

$$\mathrm{lh}(K) \subseteq \mathrm{Hull}^{i(S)}_{k(S)+1}(\sup i``\nu \cup p(i(S))),$$

so $\mathrm{lh}(K)$ has cardinality λ_K in P_η. But K was used in $\mathcal{U}$ before we reached P_η, so $\mathrm{lh}(K)$ is a cardinal in the lined up part of P_η, and hence in P_η. This is a contradiction.

Thus we must have $\lambda_K = \mu$. Also

$$\mathrm{lh}(K) = \mu^{+,Q_{\omega_1}} = \mu^{+,P_{\omega_1}} = \mu^{+,P_{\eta+1}} = o(P_\eta),$$

so $\gamma(H, \mu) = o(P_\eta)$. Thus P_η is bad because

$$H \neq F(H, \mu)^+.$$

But let $l = i^{\mathcal{T}}_{\eta+1,\omega_1}$ and $j = i^{\mathcal{U}}_{\xi+1,\omega_1}$ be the branch tails. μ is not measurable by the $\mathrm{Ult}(P_\eta, H)$-sequence by $(*)$, so μ is not measurable by the sequence in $\mathrm{Ult}(P_\alpha, H) = P_{\eta+1}$, since the two ultrapowers agree to $i_H(\mathrm{crit}(H))$. Thus $\mu < \mathrm{crit}(l)$ and μ is not measurable in P_{ω_1}. But $P_{\omega_1}|\omega_1 = Q_{\omega_1}|\omega_1$, so μ is not measurable by the sequence in Q_{ω_1}.

On the other hand, $\mu = \lambda_K$ is measurable by the $Q_{\xi+1}$-sequence. It follows that $E^{\mathcal{U}}_{\xi+1}$ is the order zero measure on λ_K, and $\xi + 1 <_U \xi + 2 <_U \omega_1$, so that

[60] This is very easy if ν is a limit of τ such that $G \restriction \tau$ is whole. If $\delta < \nu$ is largest such that $G \restriction \delta$ is whole, then one can show that $\nu = \sup\{i_G(f)(a) \mid f \colon \kappa_G \to \kappa_G \wedge a \in [\delta+1]^{<\omega}\}$. This then implies that there are no τ such that $i(\delta) < \tau < \sup i``\nu$ and $H \restriction \tau$ is whole.

K-then-$E^{\mathcal{U}}_{\xi+1}$ is the initial segment of the extender of $i^{\mathcal{T}}_{\alpha,\omega_1}$ with generators $\mu+1$. This implies that

$$H = K\text{-then-}E^{\mathcal{U}}_{\xi+1},$$

so $H = F(H,\mu)^+$, a contradiction.

This finishes our termination proof in case 1.

Case 2. $[0,\xi]_U$ does not drop, and K is coded by the top predicate of $\mathcal{M}^{\mathcal{U}}_{\xi}$.

This case is completely parallel to Case 1.

This proves Claim 2. ⊣

Now let $\theta+1 = \mathrm{lh}(\mathcal{T})$ and $\tau+1 = \mathrm{lh}(\mathcal{U})$.

CLAIM 3. *$P_\theta = Q_\tau$, neither $[0,\theta]_T$ nor $[0,\tau]_U$ drops, and $i^{\mathcal{T}}_{0,\theta} = i^{\mathcal{U}}_{0,\tau}$.*

PROOF. By standard Weak Dodd-Jensen arguments, using of course $\pi_\theta \colon P_\theta \to P^*_\theta$ at various points. ⊣

CLAIM 4. $1 \leq_T \theta$.

PROOF. Suppose not. Let $\eta+1 \leq_T \theta$ with T-pred$(\eta+1) = 0$, and $\xi+1 \leq_U \tau$ with U-pred$(\xi+1) = 0$. Let $H = E^{\mathcal{T}}_{\eta}$ and $K = E^{\mathcal{U}}_{\xi}$. We reach the same contradictions we reached in the proof that the comparison process terminates. ⊣

By Claims 3 and 4, $i^{\mathcal{U}}_{0,\tau}$ is not the identity, so $\tau > 0$. Let $\xi+1 \leq_U \tau$ and U-pred$(\xi+1) = 0$, and let $K = E^{\mathcal{U}}_{\xi}$; then

$$K \restriction \nu = G \restriction \nu = F \restriction \lambda_F.$$

(It is easy to see $\nu \leq \lambda_K$.) Now $P_1|\mathrm{lh}(F) = \mathrm{Ult}(M,F)|\mathrm{lh}(F) = M||\mathrm{lh}(F)$, and $Q_0||\mathrm{lh}(F) = M||\mathrm{lh}(F)$ by the properties of F^* recorded in Lemma 4.7.7. We were iterating away disagreements, so $\mathrm{lh}(K) \geq \mathrm{lh}(F)$. But $K \neq F$, since otherwise F is on the sequence of $Q_0 = (N,\hat{G})$, contrary to $(*)(3)$. Thus $\mathrm{lh}(K) > \mathrm{lh}(F)$, and K is not an ordinary extender on the Q_ξ sequence, as otherwise the Jensen completion of $K \restriction \nu$, which is F, would be on the Q_ξ sequence, and hence on the Q_0 sequence.

It follows that K is coded by $i^{\mathcal{U}}_{0,\xi}(\hat{G})$. We claim that $\xi = 0$. Suppose not. Since $\mathrm{crit}(K) = \kappa = \mathrm{crit}(G)$, $\mathrm{crit}(i^{\mathcal{U}}_{0,\xi}) > \kappa$. Suppose first that $\mathrm{crit}(i^{\mathcal{U}}_{0,\xi}) < \nu$. If there is a β such that $\mathrm{crit}(i^{\mathcal{U}}_{0,\xi}) \leq \beta < \nu$ and $G \restriction \beta$ is whole, then since G had the weak initial segment condition below ν, $G \restriction \beta \in Q_0$, so $K \restriction i^{\mathcal{U}}_{0,\xi}(\beta) \in Q_\xi$, so $K \restriction \nu \in Q_\xi$, contradiction. Thus $\beta < \mathrm{crit}(i^{\mathcal{U}}_{0,\xi})$, where β is largest such that $G \restriction \beta$ is whole. As above, this implies that $K \restriction \gamma$ is not whole for all $\gamma \in (\beta, \sup i^{\mathcal{U}}_{0,\xi}``\nu)$, so $K \restriction \nu$ is not whole, contradiction. Thus $\nu \leq \mathrm{crit}(i^{\mathcal{U}}_{0,\xi})$. But ν is the largest cardinal of Q_0 and $[0,\xi]_U \cap D^{\mathcal{U}} = \emptyset$, so $\mathrm{crit}(i^{\mathcal{U}}_{0,\xi}) \geq \nu$ is impossible.

It follows that $\xi = 0$, and $K = G$. Note that $\mathrm{crit}(i^{\mathcal{U}}_{1,\tau}) > \nu$ and $\mathrm{crit}(i^{\mathcal{T}}_{1,\theta}) \geq \nu$, so

$$P_1|\nu^{+,P_1} = P_\theta|\nu^{+,P_\theta} = Q_\tau|\nu^{+,Q_\tau} = Q_1|\nu^{+,Q_1} = N.$$

Thus $\mathrm{lh}(F) = \gamma(G,\nu) = o(N)$, and the badness of $(N,\hat{G})$ consists in the fact that $G \neq F^+$.

ν is not measurable in $\mathrm{Ult}(Q_0,G)$, so ν is not measurable in P_θ. But ν is measurable in P_1, so $E_1^{\mathcal{T}}$ must be the order zero measure on ν. By the rules of $\mathcal{T}$, it is applied to P_1, and since it has order zero, the fact that $1 \leq_T \theta$ implies $2 \leq_T \theta$. That is, $G \upharpoonright \nu$-then-$E_1^{\mathcal{T}}$ is an initial segment of the extender of $i_{0,\theta}^{\mathcal{T}}$. But G is an initial segment of the extender of $i_{0,\tau}^{\mathcal{U}} = i_{0,\theta}^{\mathcal{T}}$, so

$$G = G \upharpoonright \nu\text{-then-}E_1^{\mathcal{T}},$$

so G is of plus type, contradiction. ⊣

4.10. Solidity in PFS constructions

We begin with some consequences of amenable closure for stability and projectum solidity in PFS constructions. The proofs here are identical to the proofs of the corresponding facts in Theorems 3.7.1 and 3.8.2.

LEMMA 4.10.1. *Let $\mathbb{C}$ be a maximal PFS-construction and $M = M_{\nu,k}^{\mathbb{C}}$, where $0 \leq k < \omega$. Suppose that V is countably $\vec{F}^{\mathbb{C}}$-iterable; then*

(1) $\rho_{k+1}(M)$ is not the critical point of an M-total extender on the M sequence, and

(2) if $\rho_{k+1}(M) \leq \eta_k^M$, then η_k^M is not the critical point of an M-total extender on the M sequence.

PROOF. The amenable closure argument for part (1) of Theorem 3.7.1 goes over verbatim, and yields (1) above. The proof of Theorem 3.8.2 yields (2), but let us go through it again.

Let $\rho = \rho_{k+1}^M$, $\eta = \eta_k^M$, and assume toward contradiction that $\rho \leq \eta$ and η is measurable in M. By part (1), $\rho < \eta$, and since M is a pfs premouse, $\eta < \rho_k(M)$.

We claim that η is measurable in V. For let E be a total-on-M extender from the M sequence; then

$$\sigma_{\mathrm{M}}[M|\mathrm{lh}(E)] \upharpoonright \mathrm{dom}(E) = \mathrm{id},$$

because $\mathrm{dom}(E)$ is a cardinal of M and $\rho^-(M) \geq \mathrm{dom}(E)$. (See Lemma 4.7.5.) Thus the background extender $B^{\mathbb{C}}(E)$ has critical point η, and η is measurable in V.

But then $\mathrm{cof}^V(\rho_k(M)) = \eta$. On the other hand, $\mathrm{cof}^V(\rho_k(M)) \leq \rho$, because the new $\Sigma_1^{M^k}$ subset of ρ generates a partial $\Sigma_1^{M^k}$ map from ρ cofinally into $o(M^k)$. This is a contradiction. ⊣

As we saw in Section 3.7, the remaining clause in projectum solidity, concerning the relationship between $\bar{\mathfrak{C}}(M)$ and $\mathfrak{C}(M)$, is a corollary to the proof of parameter solidity. The proof of parameter solidity is essentially the same as that in [35], but there are new problems that arise from the fact that ultrapowers of type 1 premice can have type 2. Our solution to these problems is constrained by the need to generalize it to a proof of parameter solidity for strategy mice.

If M is strongly stable, then the issue of type 2 ultrapowers does not arise, and our proof is essentially the same as that in [35, §8]. We begin with this case.

LEMMA 4.10.2. *Let M be a strongly stable, countably iterable pfs premouse of type 1, and $k = k(M)$; then*

(a) M is parameter solid, and

(b) if $\rho_{k+1}(M)$ is not measurable by the M-sequence, then M is projectum solid.

PROOF. The proof is based on comparing phalanxes of the form (M, H, α) with M. M is strongly stable, so soundness in plus trees on M behaves according to the familiar pattern of Lemma 4.4.6 and comparison works as in Theorem 4.6.6. All models are type 1, and all branch embeddings are exact.

We wish to prove that M satisfies certain sentences, so we may assume that M is countable. By Lemma 4.6.10 we can fix an enumeration $\vec{e}$ of M and an $(\omega_1, \omega_1 + 1)$ iteration strategy Σ for M with the Weak Dodd-Jensen property relative to $\vec{e}$. Let $k = k(M)$, and

$$\rho = \rho_1(M^k) = \rho_{k+1}(M),$$
$$r = p_1(M^k) = p_{k+1}(M).$$

We choose $\vec{e}$ so that $r = \{e_0, \ldots, e_l\}$, where $e_0 > e_1 > \cdots > e_l$. Let q be the longest solid initial segment of r in this decreasing enumeration, and let

$$r = s \cup q,$$

where either $s = \emptyset$ or $\max(s) < \min(q)$. Let

$$\begin{aligned} \alpha_0 &= \text{least } \beta \text{ such that } \mathrm{Th}_1^{M^k}(\beta \cup q) \notin M. \\ &= \text{least } \beta \text{ such that } \mathrm{Th}_{k+1}^{M}(\beta \cup q) \notin M. \end{aligned}$$

We may assume that $\alpha_0 \in M^k$, as otherwise $r = \emptyset$ and $\alpha_0 = \rho_{k+1}(M) = \rho_k(M)$, in which case the theorem is trivially true. If r is solid, then $\alpha_0 = \rho_{k+1}(M)$. Let

$$\begin{aligned} H &= \mathrm{cHull}_{k+1}^{M}(\alpha_0 \cup q) \\ &= \mathrm{Dec}(\mathrm{cHull}_1^{M^k}(\alpha_0 \cup q)), \end{aligned}$$

and let

$$\pi \colon H \to M$$

be the anticollapse map. Note that $k(H) = k(M) = k$, and π is elementary by the Downward Extension Lemma.[61] Part of elementarity is that $\pi(w_k(H)) = w_k(M)$,

[61] Let $\sigma\colon (P, B) \to M^k$ be the anticollapse map. σ is Σ_1 elementary, and it is cofinal because $\mathrm{Th}_1^{M^k}(\alpha_0 \cup q) \notin M^k$.

which is true because M^k has a name for $w_k(M)$. Since $\pi(\eta_k^H) = \eta_k^M$ and M is strongly stable, H is strongly stable.

CLAIM 0. *(a) If $q = r$, then $\rho = \alpha_0$.*
(b) If $q \neq r$, then $\pi \neq$ id, and $\rho < \alpha_0 \leq \mathrm{crit}(\pi) \leq \max(s)$.
(c) $H \models \alpha_0$ is a cardinal.

PROOF. (a) is clear. For (b), let

$$W = \mathrm{cHull}_1^{M^k}(\max(s) \cup q)$$

be the solidity witness for $q \cup \{\max(s)\}$. We are assuming $W \notin M$. This implies that $\mathrm{Th}_1^{M^k}(\max(s) \cup q) \notin M$. [Proof: Suppose $T = \mathrm{Th}_1^{M^k}(\max(s) \cup q)$ is in M. Note $\max(s)$ is a cardinal of W, and $\max(s) = \mathrm{crit}(\psi)$, where $\psi \colon W \to M^k$ is the anti-collapse. So $T \in M|\psi(\max(s))$, and $M|\pi(\max(s)) \models \mathsf{KP}$. So $W \in M|\psi(\max(s))$.] Thus $\alpha_0 \leq \max(s)$. But then if $\max(s) < \mathrm{crit}(\pi)$, then $\max(s) = h^1_{M^k}(\beta, q)$ for some $\beta < \alpha_0$, which easily implies that r is not minimal in the parameter order among parameters defining a new $\Sigma_1^{M^k}$ subset of ρ. So $\alpha_0 \leq \mathrm{crit}(\pi) \leq \max(s)$.

We have $\rho < \alpha_0$ because otherwise $p(M) = q$. So we have (b).

(c) is clear if $\alpha_0 = \rho$. So we may assume $q \neq r$, hence $\pi \neq$ id. (c) is clear if $\alpha_0 = \mathrm{crit}(\pi)$, so we may assume $\alpha_0 < \mathrm{crit}(\pi)$. Suppose $f \colon \beta \to \alpha_0$ is a surjection, with $\beta < \alpha_0$ and $f \in H$. Let $\beta < \gamma < \alpha_0$ be such that $\pi(f)$ is $\Sigma_1^{M^k}$ definable from parameters in $\gamma \cup q$. Then from $\mathrm{Th}_1^{M^k}(\gamma \cup q)$ one can easily compute $\mathrm{Th}_1^{M^k}(\alpha_0 \cup q)$, so $\mathrm{Th}_1^{M^k}(\gamma \cup q) \notin M$, contrary to the minimality of α_0. ⊣

In view of Claim 0, we may assume that $\pi \neq$ id, and

$$\mathrm{crit}(\pi) < \rho_k(H).$$

For if $\pi \restriction H^k = $ id, then H^k is an initial segment of M^k. It cannot be a proper initial segment because $\mathrm{Th}_1^{M^k}(\alpha_0 \cup q) \notin M$. But if $H^k = M^k$ and $\pi = $ id, then r is solid and universal over M^k. Moreover $\alpha_0 = \rho$, so M is its own strong core, so the collapse of r is solid and universal over $\bar{\mathfrak{C}}(M)^-$.

We show now that if $q \neq r$, then $\mathrm{Th}_{k+1}^M(\alpha_0 \cup q) \in M$. This implies $q = r$, so r is solid over M^k and $H = \bar{\mathfrak{C}}(M)^-$. We then show that H agrees with M up to $\rho^{+,M}$. The argument is based on comparing the phalanx (M, H, α_0) with M.

The comparison proceeds by iterating away least disagreements. On the M side it produces a normal, λ-tight iteration tree $\mathcal{U}$ that is according to Σ. On the phalanx side we get a normal, λ-tight "pseudo-iteration tree" $\mathcal{T}$. The first two models of $\mathcal{T}$ are $M_0^{\mathcal{T}} = M$ and $M_1^{\mathcal{T}} = H$, and from there we proceed as if these were the first two models in an ordinary tree in which $\lambda(E_0^{\mathcal{T}}) = \alpha_0$. Since $H||\alpha_0 = M||\alpha_0$, all extenders used in $\mathcal{T}$ or $\mathcal{U}$ have length greater than or equal to α_0.

Let $\pi_0 = $ id and $\pi_1 = \pi$. We can copy the pseudo-tree $\mathcal{T}$ on (M, H, α_0) to a normal, λ-tight tree

$$\mathcal{T}^* = (\pi_0, \pi_1)\mathcal{T}$$

on M. The construction is similar to the lifting of an iteration tree on a phalanx in Section 4.9, and to the construction of $\pi\mathcal{T}^+$ in Section 4.5. We are given the first two copy maps at the outset, they are nearly elementary (in fact, elementary), and they agree up to the relevant exchange ordinal α_0, so the copying can continue from there. Σ induces a pullback strategy $\Sigma^{(\pi_0,\pi_1)}$ for (M,H,α_0), and we use this strategy to choose branches of our comparison tree $\mathcal{T}$ at limit steps. The construction thereby produces

$$\begin{aligned} P_\xi &= \mathcal{M}_\xi^{\mathcal{T}} &&\text{and} && i_{\xi,\gamma} = i_{\xi,\gamma}^{\mathcal{T}}, \\ P_\xi^* &= \mathcal{M}_\xi^{\mathcal{T}^*} &&\text{and} && i_{\xi,\gamma}^* = i_{\xi,\gamma}^{\mathcal{T}^*}, \\ Q_\xi &= \mathcal{M}_\xi^{\mathcal{U}} &&\text{and} && j_{\xi,\gamma} = i_{\xi,\gamma}^{\mathcal{U}}, \end{aligned}$$

and copy/lifting maps

$$\pi_\xi : P_\xi \to N_\xi \trianglelefteq P_\xi^*.$$

Except in some anomalous cases discussed below, P_ξ, P_ξ^*, and Q_ξ are pfs premice, the branch embeddings of $\mathcal{T}$, $\mathcal{T}^*$, and $\mathcal{U}$ are elementary, $N_\xi = P_\xi^*$, and π_ξ is elementary. $P_0 = M$, $P_1 = H$, and $P_0^* = P_1^* = M$. $\mathcal{T}^*$ is a "padded" iteration tree, in that the first node is indexed twice, for bookkeeping purposes.

The fact that the initial models in our phalanx (M,H,α_0) do not come from a single iteration tree can lead to fine structural anomalies when an extender in $\mathcal{T}$ is applied to a proper initial segment of M. The next two remarks deal with these anomalous cases. The reader who is only looking for the main idea of the proof should probably skip them, and just assume the cases they deal with do not arise.[62]

Remark 4.10.3. Suppose that $\operatorname{crit}(E_\gamma^{\mathcal{T}}) < \alpha_0$ and $E_\gamma^{\mathcal{T}}$ is not total on M. In this case, letting $E = E_\gamma^{\mathcal{T}}$ and $\mu = \operatorname{crit}(E)$, we must have $\alpha_0 = \mu^{+,H} < \mu^{+,M}$ and $\operatorname{dom}(E) = M||\alpha_0$. Let N be the first $S \trianglelefteq M$ such that $\rho(S) \leq \mu$. μ is a cardinal of $M||\alpha_0$, and hence a cardinal of M because $\pi_0(\mu) = \mu$. Thus $\rho(N) = \mu$. $\mathcal{T}$ is following the rules of normal plus trees, so we shall set $0 = \mathcal{T}\text{-pred}(\gamma+1)$. We would like to set $P_{\gamma+1} = \operatorname{Ult}(N,E)$, but $\operatorname{Ult}(N,E)$ might have type 2, and we want to avoid type 2 premice in $\mathcal{T}$ because they complicate the way comparisons terminate. So letting $k = k(N)$, we set

$$P_{\gamma+1} = \begin{cases} \operatorname{Ult}_k(N,E) & \text{if } \operatorname{Ult}_k(N,E) \text{ has type 1}, \\ \operatorname{Ult}_k(\bar{\mathfrak{C}}_k(N),E) & \text{if } \operatorname{Ult}(N,E) \text{ has type 2}. \end{cases}$$

The second case occurs iff N has type 1B and $\operatorname{crit}(E) = \eta_k^N$.

In both cases, letting $E^* = \pi_\gamma(E)$, $\operatorname{crit}(E^*) = \mu$ and $\operatorname{dom}(E^*) = M||\mu^{+,M}$ by the agreement between π_γ and π_0. Thus E^* is total on M and we shall apply E^* to

[62] These cases did not arise in Section 4.9 because the exchange ordinal α_0 was a cardinal of M in that situation.

M in $\mathcal{T}^*$. As in the definition of $(\pi\mathcal{T})^+$, we then have a natural map

$$\pi_{\gamma+1}\colon P_{\gamma+1} \to i^M_{E^*}(N) = N_{\gamma+1}$$

or

$$\pi_{\gamma+1}\colon P_{\gamma+1} \to i^M_{E^*}(\bar{\mathfrak{C}}_k(N)) = N_{\gamma+1},$$

depending on which of the two cases define $P_{\gamma+1}$. In the first case, $\pi_{\gamma+1}$ is nearly elementary, not necessarily elementary, and it maps $\mathcal{M}^{\mathcal{T}}_{\gamma+1}$ to the proper initial segment $N_{\gamma+1}$ of $P^*_{\gamma+1}$. In the second case $N_{\gamma+1}$ is the strong core of a proper initial segment of $P^*_{\gamma+1}$, and $\pi_{\gamma+1}$ is nearly elementary in that it completes a Σ_0 elementary map on the reducts $(P_{\gamma+1})^k_0$ and $(N_{\gamma+1})^k_0$ coding $P_{\gamma+1}$ and $N_{\gamma+1}$.[63] This is good enough to continue lifting $\mathcal{T}$ to $\mathcal{T}^*$. So in this respect, the construction of $\mathcal{T}^*$ is like that of $\pi\mathcal{T}^+$, rather than that of $\pi\mathcal{T}$.

Remark 4.10.4. Continuing with the last remark, there is a case when $\mathrm{Ult}(N,E)$ is not a premouse of any sort. The method for dealing with it is due to Schindler and Zeman. (See [53].) This case occurs when $\alpha_0 = \mathrm{lh}(F)$, for some extender F from the M-sequence, and $\mathrm{crit}(E) = \lambda_F$. The collapsing structure in M for α_0 is then just $N = M|\alpha_0$, and $\mathcal{M}^{\mathcal{T}}_{\gamma+1} = \mathrm{Ult}_0(M|\mathrm{lh}(F),E)$. The trouble is that $\mathcal{M}^{\mathcal{T}}_{\gamma+1}$ is not a premouse at all, because F is a missing whole initial segment of $i^{\mathcal{T}}_{0,\gamma+1}(F)$.[64] But this is ok. The next disagreement will force us to apply $i^{\mathcal{T}}_{0,\gamma+1}(F)$ to M, and that will produce a pfs premouse; moreover, $\lambda(E^{\mathcal{T}}_\gamma) = \lambda(i^{\mathcal{T}}_{0,\gamma+1}(F))$, so there will be no ξ such that $T\text{-pred}(\xi) = \gamma+1$. ($\gamma+1$ is a "dead node" in $\mathcal{T}$.) One can cope with the fact that $i^{\mathcal{T}}_{\xi,\gamma+1}(F)$ has a missing whole initial segment in the termination arguments; the argument is the same as that of Schindler-Zeman.

We shall call $\gamma+1$ a $\mathcal{T}$-*anomaly* in the case that $P_{\gamma+1}$ is not a pfs premouse for the reason just described.[65] In order to simplify the exposition a bit, let us assume in the proof to follow that there are no $\mathcal{T}$-anomalies.

In sum, $N_\xi = P^*_\xi$ and π_ξ is elementary unless we are in the situation covered by Remarks 4.10.3 and 4.10.4. We are assuming the situation of Remark 4.10.4 does not arise. In the situation of Remark 4.10.3, $0 <_T \xi$, $N_\xi \in P^*_\xi$, and π_ξ is nearly elementary, perhaps in the sense appropriate to maps on strong cores.[66]

The proof of the Comparison Theorem for pfs premice works here, so we have last models P_θ for $\mathcal{T}$ and Q_δ for $\mathcal{U}$ such that $P_\theta \trianglelefteq Q_\delta$ or $Q_\delta \trianglelefteq P_\theta$.

---

[63] See Definition 4.6.7. The names for η_k and ρ_k are removed from the language of the reducts.

[64] As we have seen, plus trees do not involve taking such ultrapowers. However, phalanx iterations might.

[65] This is called a *strong anomaly* in [86]. Lemma 9.2.8 of [86] shows that π_ξ is elementary from $\mathcal{M}^{\mathcal{T}}_\xi$ to $\mathcal{M}^{\mathcal{T}^*}_\xi$ as long as there is no anomaly of either sort along the branch of $\mathcal{T}$ to ξ.

[66] For this last part one must show that $\langle \pi_\beta, \pi_\alpha \rangle$ is a Σ_1 embedding of $(Q, E^{\mathcal{T}}_\alpha)$ into $(R, E^{\mathcal{T}^*}_\alpha)$, where Q and R are the initial segments of P_β and P^*_β to which $E^{\mathcal{T}}_\alpha$ and $E^{\mathcal{T}^*}_\alpha$ are applied. This is done as in [86, Lemmas 9.2.7].

We show first that P_θ is above H in $\mathcal{T}$.

CLAIM 1. $1 \le_T \theta$.

PROOF. Suppose that P_θ is above M, i.e., $0 \le_T \theta$. We shall derive a contradiction using the Weak Dodd-Jensen property of Σ.

Case (a). $P_\theta \trianglelefteq Q_\delta$ and $[0,\theta]_T \cap D^{\mathcal{T}} = \emptyset$.

PROOF. If $P_\theta \triangleleft Q_\delta$ or $[0,\delta]_U \cap D^{\mathcal{U}} \neq \emptyset$, then $i_{0,\theta}$ is an elementary map from M to an initial segment of a Σ-iterate of M of the sort that is forbidden by the Weak Dodd-Jensen property (Definition 4.6.9) of Σ. Suppose then that $Q_\delta = P_\theta$ and $[0,\delta]_U \cap D^{\mathcal{U}} = \emptyset$. Thus we have elementary iteration maps

$$i = i_{0,\theta} \colon M \to P_\theta,$$
$$j = j_{0,\delta} \colon M \to Q_\delta.$$

We claim that $i = j$. Otherwise, let n be least such that $i(e_n) \neq j(e_n)$. If $i(e_n) < j(e_n)$ (in the order of construction), then i is an elementary map that is $<_{\text{lex}}$ the iteration map j, contrary to the Weak Dodd-Jensen property of Σ. If $j(e_n) < i(e_n)$, then

$$\pi_\theta \circ i(e_n) < \pi_\theta \circ j(e_n).$$

But $\pi_\theta \circ i = i^*_{0,\theta}$, so $\pi_\theta \circ j$ is $<_{\text{lex}}$ the iteration map $i^*_{0,\theta}$, contrary to the Weak Dodd-Jensen property of Σ. This is a contradiction. Now let E and F be the first extenders used along the branches $[0,\theta]_T$ and $[0,\delta]_U$. Since $i = j$ and generators are not moved, E is an initial segment of F or vice-versa. This leads to the same contradiction we got in the proof that the comparison process terminates.[67] ⊣

Case (b). $Q_\delta \trianglelefteq P_\theta$ and $[0,\delta]_U \cap D^{\mathcal{U}} = \emptyset$.

PROOF. Suppose first that $\pi_\theta \circ j_{0,\delta}$ is nearly elementary. If $Q_\delta \triangleleft P_\theta$ or $[0,\theta]_T \cap D^{\mathcal{T}} \neq \emptyset$, then $\pi_\theta \circ j_{0,\delta}$ is a nearly elementary map from M to an initial segment of P^*_θ of the sort that is forbidden by clause (1) of the Weak Dodd-Jensen property. This leads to $i_{0,\theta} = j_{0,\delta}$ and the same contradiction as in case (a).

If $\pi_\theta \circ j_{0,\delta}$ is not nearly elementary, then $Q_\delta = P_\theta$ and π_θ is not nearly elementary. Thus we must be in the anomalous case described in Remark 4.10.3. In this case, the branch M-to-P_θ of $\mathcal{T}$ has dropped in model, but only at the first step along it, P_θ has type 1A, and π_θ completes a Σ_0 elementary map from $(P_\theta)^k_0$ to $(N_\theta)^k_0$. Since $Q_\delta = P_\theta$, Q_δ has type 1A, so M has type 1A. But then $\pi_\theta \circ j_{0,\delta}$ completes a Σ_0 elementary map of M^k_0 to $(N_\theta)^k_0$ and $[0,\theta]_{T^*}$ has dropped in model, contrary to Clause (2) of the Weak Dodd-Jensen property. ⊣

This proves Claim 1. ⊣

Now that we know $1 \le_T \theta$, the anomalous cases in Remarks 4.10.3 and 4.10.4 are no longer relevant. $N_\theta = P^*_\theta$ and π_θ is elementary.

[67]This case used only Weak Dodd-Jensen in the category of nearly elementary maps.

CLAIM 2. *$D^{\mathcal{T}} \cap [1,\theta]_T = \emptyset$, and $P_\theta \trianglelefteq Q_\delta$. moreover, $i_{1,\theta}(\rho_k(H)) = \rho_k(P_\theta)$.*

PROOF. If $[1,\theta]_T \cap D^{\mathcal{T}} \neq \emptyset$, then $[0,\delta]_U \cap D^{\mathcal{U}} = \emptyset$, so $\pi_\theta \circ j_{0,\delta}$ is a nearly elementary map of M into P^*_θ and $[1,\theta]_{T^*}$ has dropped, contrary to Weak Dodd-Jensen. So $[1,\theta]_T$ does not drop. If $Q_\delta \triangleleft P_\theta$, then $[0,\delta]_U \cap D^{\mathcal{U}} \neq \emptyset$, and $\pi_\theta \circ j_{0,\delta}$ is a nearly elementary map of M into a proper initial segment of P^*_θ, contrary to Weak Dodd-Jensen. ⊣

We want to show $P_\theta = Q_\delta$ and $[0,\delta]_U$ does not drop. For that we need some simple facts about definability over the models in $\mathcal{T}$ and $\mathcal{U}$.

CLAIM 3. *Suppose $D^{\mathcal{U}} \cap [0,\eta]_U = \emptyset$; then*

(a) for any $\beta < \alpha_0$, $Th^{Q_\eta}_{k+1}(j_{0,\eta}(\beta) \cup j_{0,\eta}(q)) \in Q_\eta$,

(b) $\sup j_{0,\eta}$"$\rho = \rho_1(Q^k_\eta)$, and

(c) if $q \neq r$, then $Th^{Q_\eta}_{k+1}(\rho(Q_\eta) \cup j_{0,\eta}(q)) \in Q_\eta$.

PROOF. Part (a) holds because $j_{0,\eta}(\mathrm{Th}^{M^k}_1(\beta \cup q))$ can be used to compute $\mathrm{Th}^{(Q_\eta)^k}_1(j_{0,\eta}(\beta) \cup j_{0,\eta}(q))$ by the usual proof for solidity witnesses. Part (b) can be proved by induction along the branch from 0 to η, using the fact that if E is applied to Q^k_ξ along this branch, then $\mathrm{crit}(E) < \rho_1(Q^k_\xi)$, so E is very close to Q_ξ.[68] If $q \neq r$, then $\rho < \alpha_0$. But $\rho(Q_\eta) \leq j_{0,\eta}(\rho)$, so we get (c) by using (a) with $\beta = \rho$. ⊣

Set $\mu = \rho(H)$, and $t = \pi^{-1}(q)$.

CLAIM 4. *Either*

(i) $\mu = \alpha_0$, or

(ii) $\mu < \alpha_0 = \mathrm{crit}(\pi) = (\mu^+)^H$.

PROOF. $\mathrm{Th}^H_{k+1}(\alpha_0 \cup t) \notin H$, and therefore $\mu \leq \alpha_0$.

Suppose $\mu < \alpha_0$. We can then find some finite $p \subset \alpha_0$ such that $R = \mathrm{Th}^H_{k+1}(\mu \cup p \cup t) \notin H$. Since $\max(p) < \alpha_0$, our minimality hypothesis on α_0 implies that $R \in M$. Thus $P(\mu)^H \neq P(\mu)^M$, and since $\mathrm{crit}(\pi) > \mu$, we get (ii). ⊣

CLAIM 5. *$\mu = \rho(P_\theta)$.*

PROOF. This follows easily from the fact that all extenders used in $[1,\theta]_T$ are close to the model to which they are applied, and $\mathrm{crit}(i_{1,\theta}) \geq \alpha_0$. ⊣

CLAIM 6. *(i) For all $\eta \leq \delta$, $P(\alpha_0) \cap Q_\eta \subseteq M$.*

(ii) $P_\theta = Q_\delta$, and $D^{\mathcal{U}} \cap [0,\delta]_U = \emptyset$.

(iii) $i_{0,\delta}(w_k(M)) = w_k(Q_\delta) = i_{1,\theta}(w_k(H))$.

PROOF. For (i), clearly we may assume $\eta > 0$. Let $E = E^{\mathcal{U}}_0$; then $\mathrm{lh}(E)$ is a cardinal in Q_η, and $Q_\eta || \mathrm{lh}(E) = M || \mathrm{lh}(E)$, so if $\alpha_0 < \mathrm{lh}(E)$ we are done. The

[68] See Lemma 9.6.1 for a stronger result of this form.

alternative is that $\alpha_0 = \text{lh}(E)$. In that case, $P(\alpha_0) \cap Q_\eta \subseteq Q_1$ by the argument just given. But $Q_1 = \text{Ult}(M,E)$, moreover $\pi(\text{crit}(\pi))$ is a cardinal of M above α_0, so $P(\alpha_0) \cap \text{Ult}(M,E) \subseteq M$.

Let us prove (ii). Let

$$A = \text{Th}^H_{k+1}(\alpha_0 \cup t) = \text{Th}^M_{k+1}(\alpha_0 \cup q),$$

coded as a subset of α_0. Since $[1,\theta]_T$ does not drop and $\text{crit}(i_{1,\theta}) \geq \alpha_0$, A is $\Sigma_1^{P^k_\theta}$ in the parameter $i_{1,\theta}(t)$. If $A \in Q_\delta$, then by (i), $A \in M$, contradiction. It follows that $\hat{P}_\theta = \hat{Q}_\delta$.

Suppose toward contradiction that $[0,\delta]_U \cap D^{\mathcal{U}} \neq \emptyset$, and let $\xi + 1$ be largest in $[0,\theta]_U \cap D^{\mathcal{U}}$. Let $\beta = U\text{-pred}(\xi + 1)$ and

$$J = \mathcal{M}^{*,\mathcal{U}}_{\xi+1},$$

and let $n = k(J) = k(Q_\delta)$. We have that Q_δ is not $n+1$-sound, by Lemma 4.4.6. Since $\hat{Q}_\delta = \hat{P}_\theta$ and P_θ is k-sound, $k \leq n$. But then

$$\rho_{n+1}(J) = \rho_{n+1}(Q_\delta) \leq \rho_{k+1}(P_\theta) \leq \alpha_0.$$

We claim that $\beta = 0$. For otherwise, let G be the first extender used on the branch from M to Q_β; then by Corollary 4.4.14, $\text{lh}(G)$ is a cardinal of Q_β and $\text{lh}(G) \leq \rho^-(Q_\beta)$. Since $J \lhd Q_\beta$, $\text{lh}(G) \leq \rho(J)$, so

$$\alpha_0 \leq \text{lh}(G) \leq \rho_{n+1}(J) \leq \alpha_0.$$

Thus $\text{lh}(G) = \alpha_0$ and $G = E^{\mathcal{U}}_0 \in M$. But α_0 is a cardinal of H, so $\text{crit}(\pi) = \alpha_0$ and $\alpha_0 = \nu^{+,H}$ for some cardinal ν of M. It follows that G is total on M, and since $\alpha_0 < \rho_k(M)$,

$$\rho^-(Q_\beta) \geq \rho^-(\text{Ult}(M,G)) = i_G(\rho^-(M)) > \alpha_0.$$

This is a contradiction.

So $\beta = 0$ and $J \lhd M$. Let $G = E^{\mathcal{U}}_\xi$, and suppose first that $\alpha_0 \leq \text{crit}(G)$. If $J = M \downarrow i$ for some $i < k$ then Q_δ is not k-sound, contradiction. Thus $J \in M$. But A is boldface $r\Sigma_{n+1}$ over Q_δ, so by Lemma 4.3.11 A is boldface $r\Sigma_{n+1}$ over J. Hence $A \in M$, contradiction.

Thus $\text{crit}(G) < \alpha_0$. Let $\kappa = \text{crit}(G)$. Since $\kappa < \rho_k(M)$ and $J \lhd M$, we have $J \in M$. Since $\text{lh}(E^{\mathcal{U}}_0) \geq \alpha_0$, $o(J) \geq \alpha_0$, and thus J collapses α_0 to κ in M. Since α_0 is a cardinal of H and not of M, $\text{crit}(\pi) = \alpha_0$ and there is a largest cardinal of M strictly less than α_0, which must then be κ; moreover $\alpha_0 = \kappa^{+,H}$. (We are not claiming that $\kappa = \mu$.)

We now show again that A is boldface $r\Sigma_{n+1}$ over J. Since $\text{crit}(j_{\xi+1,\delta}) \geq \alpha_0$, Lemma 4.3.11 implies that A is boldface $r\Sigma_{n+1}$ over $\text{Ult}_n(J,G)$, that is, boldface Σ_1 over $\text{Ult}_0(J^n,G)$. Let φ be a Σ_0 formula and $[a,f] \in \text{Ult}_0(J^n,G)$ be such that for all $\beta < \alpha_0$

$$\beta \in A \text{ iff } \text{Ult}_0(J^n,G) \models \exists v \varphi[v,\beta,[a,f]].$$

We may assume that κ is the least element of a. Then if X is a wellorder of κ and $X \in J$, and $|X|$ is its order type,

$$|X| \in A \text{ iff } \exists Z \in G_a(J^n \models \exists g \forall u \in Z \varphi[g(u), |X \cap (u_0 \times u_0)|, f(u)]).$$

But G is close to J, so G_a is definable over J from parameters. Thus A is definable over J, so $A \in M$, contradiction.

This proves (ii) of Claim 6. Part (iii) follows from the fact that both H and M are strongly stable. ⊣

Remark 4.10.5. The fact that M is strongly stable is used at precisely this point, in proving that $i_{1,\theta}(\rho_k(H)) = j_{0,\delta}(\rho_k(M))$.

CLAIM 7. $i_{1,\theta}(t) = j_{0,\delta}(q)$.

PROOF. Let β be the first (i.e., largest) element of q such that $j_{0,\delta}(\beta) \neq i_{1,\theta} \circ \pi^{-1}(\beta)$. If

$$j_{0,\delta}(\beta) < i_{1,\theta} \circ \pi^{-1}(\beta),$$

then

$$\pi_\theta \circ j_{0,\delta}(\beta) < \pi_\theta \circ i_{1,\theta} \circ \pi^{-1}(\beta) = i^*_{1,\theta}(\beta).$$

(Recall here that $\pi = \pi_1$, and $i^*_{1,\theta} \circ \pi_1 = \pi_\theta \circ i_{1,\theta}$.) The maps on the two sides above agree at all earlier elements of q, and $\vec{e}$ started out with r, so this contradicts the Weak Dodd-Jensen property of Σ relative to $\vec{e}$. On the other hand, suppose

$$j_{0,\delta}(\beta) > i_{1,\theta} \circ \pi^{-1}(\beta).$$

Let $\bar{\beta} = \pi^{-1}(\beta)$, and $u = t - (\bar{\beta} + 1)$. Since q is solid at β and $j_{0,\delta}(w_k(M)) = w_k(Q_\delta)$,

$$\mathrm{Th}^{Q_\delta}_{k+1}(j_{0,\delta}(\beta) \cup j_{0,\delta}(q - (\beta + 1))) \in Q_\delta.$$

But $i_{1,\theta}(u) = j_{0,\delta}(q - (\beta + 1))$ and $i_{1,\theta}(\bar{\beta}) < j_{0,\delta}(\beta)$, so

$$\mathrm{Th}^{P_\theta}_{k+1}(i_{1,\theta}(\bar{\beta} + 1 \cup u)) \in P_\theta.$$

It follows that $\mathrm{Th}^{P_\theta}_{k+1}(\alpha_0 \cup i_{1,\theta}(t)) \in P_\theta$. The theory is essentially a subset of α_0, and it is equal to $\mathrm{Th}^{H}_{k+1}(\alpha_0 \cup t)$ because $i_{1,\theta}(w_k(H)) = w_k(P_\theta)$. So $\mathrm{Th}^{H}_{k+1}(\alpha_0 \cup t) \in M$ contradiction. ⊣

CLAIM 8. *If* $\delta > 0$, *then* $\alpha_0 \leq \mathrm{crit}(j_{0,\delta})$.

PROOF. Suppose U-pred$(\eta + 1) = 0$ and $\eta + 1 \leq_U \delta$, and let $E = E^{\mathcal{U}}_\eta$. Let $\kappa = \mathrm{crit}(E)$, and suppose $\kappa < \alpha_0$.

If $\rho \leq \kappa$, then $\rho = \rho(Q_\delta)$, so $\rho = \mu$, and so we have $\mu < \alpha_0$, and thus (ii) of Claim 4 holds, and $\mu^{+,M} > \alpha_0$. But then

$$\alpha_0 = \mu^{+,H} = \mu^{+,P_\theta} = \mu^{+,Q_\delta} < \mu^{+,M},$$

and $\mu \leq \mathrm{crit}(E)$, so $\eta + 1 \in D^{\mathcal{U}}$, contrary to Claim 6.

Thus $\kappa < \rho$. But then

$$\alpha_0 \leq \sup i_E``\kappa^{+,M} \leq \rho(Q_\delta) = \mu \leq \alpha_0,$$

so $\alpha_0 = \mu = \text{lh}(E)$. If $q \neq r$, then (c) of Claim 3, applied with $\eta = \delta$, implies that $\text{Th}_{k+1}^{Q_\delta}(\alpha_0 \cup j_{0,\delta}(q)) \in Q_\delta$. Hence $\text{Th}_{k+1}^{H}(\alpha_0 \cup t) \in H$, a contradiction. On the other hand, if $q = r$, then $\alpha_0 = \rho$ is a cardinal of M, contrary to $\alpha_0 = \text{lh}(E)$.

Thus $\alpha_0 \leq \kappa$, as desired. ⊣

CLAIM 9. *r is solid; that is, $q = r$.*

PROOF. Suppose $q \neq r$, so that $\rho(M) < \rho(H)$. Since $\text{crit}(j_{0,\delta}) \geq \alpha_0 \geq \rho(H)$, we then have

$$\rho(M) = \rho(Q_\delta) = \rho(P_\theta) = \rho(H) > \rho(M),$$

a contradiction. ⊣

By Claim 9, $\rho = \alpha_0$, so $H = \bar{\mathfrak{C}}(M)^-$. Let us prove that r is universal.

CLAIM 10. *(i) $H|\rho^{+,M} = M|\rho^{+,M}$,*

(ii) If $A \subseteq \rho$ and A is $\Sigma_1^{M^k}$ in parameters, then A is $\Sigma_1^{H^k}$ in parameters.

PROOF. Since $\text{crit}(j_{0,\delta}) \geq \alpha_0$ and $\text{crit}(i_{1,\theta}) \geq \alpha_0$,

$$M|\rho^{+,M} = Q_\delta|\rho^{+,Q_\delta} = P_\theta|\rho^{+,P_\theta} = H|\rho^{+,H}.$$

Moreover, if $A \subseteq \rho$ and A is $\Sigma_1^{M^k}$, then A is $\Sigma_1^{Q_\delta^k}$, so A is $\Sigma_1^{P_\theta^k}$, so A is $\Sigma_1^{H^k}$.[69] ⊣

Claims 1 through 10 show that r is solid and universal over M^k. To finish the proof of Lemma 4.10.2, we must also show that t is solid and universal over H^k. But H is itself a strongly stable, countably iterable pfs premouse, so the argument we just gave applies to it, and shows that t is solid over H^k. (Universality is then trivial.) Thus M is parameter solid.

Finally,

CLAIM 11. *If ρ is not measurable by the M-sequence, then M is projectum solid.*

PROOF. The proof is the same as that of Theorem 3.7.1. If $\text{crit}(\pi) > \rho$ then $\bar{\mathfrak{C}}(M) = \mathfrak{C}(M)$, as required. Suppose then that $\text{crit}(\pi) = \rho$. We have shown that $i_{1,\theta}(t) = j_{0,\delta}(q)$ and $j_{0,\delta} \restriction \rho = \text{id}$. It follows that

$$\pi = j_{0,\delta}^{-1} \circ i_{1,\theta}.$$

Since ρ is not measurable by M, $\text{crit}(j_{0,\delta}) > \rho$, and thus $\rho = \text{crit}(\pi) = \text{crit}(i_{1,\theta})$. Let D be the first extender used in $i_{1,\theta}$. Since ρ is not measurable by M and $\text{crit}(j_{0,\delta}) > \rho$, ρ is not measurable by $Q_\delta = P_\theta$, and hence ρ is not measurable by $\text{Ult}(H,D)$. Thus D is the order zero measure of H on ρ and $\text{Ult}(H,D) = P_2$.

[69] These last statements follow by induction from the fact that along both branches, the extenders used are close to the models to which they are applied.

Finally, letting $\tau\colon \mathfrak{C}(M) \to M$ be the anticore map, $\mathrm{Ult}(H,D)$ is isomorphic to $\mathfrak{C}(M)$ via ψ, where

$$\psi = \tau^{-1} \circ j_{0,\delta}^{-1} \circ i_{2,\theta}.$$

To see that ψ is an isomorphism, note that $\psi \restriction \rho + 1$ is the identity, and

$$\psi(p(\mathrm{Ult}(H,D))) = \psi(i_{1,2}(t)) = \tau^{-1} \circ j_{0,\delta}^{-1}(i_{1,\theta}(t)) = \tau^{-1}(q). \qquad \dashv$$

This finishes the proof of Lemma 4.10.2. ⊣

Suppose that M is stable, but not strongly stable. This implies that $k = k(M) > 0$. The problem with the argument we just gave is that p_{k+1} has been defined in such a way that it depends on ρ_k and η_k. For this reason, our proof of Claim 7 needed part (iii) of Claim 6, that $j_{0,\delta}(w_k(M)) = i_{1,\theta}(w_k(H))$.

What the proof in Case 1 does give is solidity and universality for the variant of p_{k+1} defined without reference to w_k. That variant is essentially the usual standard parameter. Namely, recall that if N is a pfs premouse of type 1 such that $k = k(N) > 0$, then

$$B^k = \{\langle \varphi, b \rangle \mid \varphi \text{ is } \Sigma_1 \wedge b \in N||\rho_k(N) \wedge N^{k-1} \models \varphi[b, p_k]\}$$

and

$$N_0^k = (N||\rho_k(N), B^k).$$

N_0^k codes the strong core $\bar{\mathfrak{C}}_k(N)$. It is easy to see that $\rho_{k+1}(N) = \rho_1(N_0^k)$.[70] Let

$$r_{k+1}(N) = p_1(N_0^k).$$

$r_{k+1}(N)$ is essentially the usual standard parameter of N.

DEFINITION 4.10.6. For $x \in M||\rho_k(M)$ and φ a Σ_1 formula, let

$$d_0^k(\langle x, \varphi \rangle) = d^{k-1} \circ h^1_{M_0^{k-1}}(x, p_k(M)),$$

and

$$\mathfrak{D}_{k+1}(M) = \text{transitive collapse of } d_0^k\text{“}\mathrm{Hull}_1^{M_0^k}(\rho_{k+1}(M) \cup r_{k+1}(M)).$$

$\bar{\mathfrak{C}}_k(M)$ is the transitive collapse of $\mathrm{ran}(d_0^k)$. $\mathfrak{D}_{k+1}$ is essentially the usual $k+1$-st core of M. Letting $\pi\colon \mathfrak{D}_{k+1}(M) \to M$ be the natural map, it is possible that η_k^M and $\rho_k(M)$ are not in $\mathrm{ran}(\pi)$.

Our plan is the following. Given M that is stable but not strongly stable, we shall replace M with $N = \mathrm{Ult}_k(\bar{\mathfrak{C}}_k(M), D)$, where $k = k(M)$ and D is the order zero measure of M on η_k^M. N has type 1A, $\eta_k^N < \rho_{k+1}(N)$, and N is strongly stable. We shall use these facts and the proof of Lemma 4.10.2 to show that $r_{k+1}(N)$ behaves well, and generates $\rho_k(N)$ as a point. We then pull this back to M and $r_{k+1}(M)$ using i_D. Finally, we shall use what we have proved about $r_{k+1}(M)$ to show that

[70] In one direction, B^k is clearly Σ_0 over N^k. In the other, A^k is Σ_0 over N_0^k in any $\gamma < \rho_k(N)$ such that $h^k_{\bar{\mathfrak{C}}_k(N)}(\gamma, p_k(\bar{\mathfrak{C}}_k(N))) = \rho_k(N)$.

$p_{k+1}(M)$ behaves well. The fact that $r_{k+1}(M)$ generates $\rho_k(M)$ as a point comes in at this point.

The proof of Lemma 4.10.2 yields the following.

LEMMA 4.10.7. *Let M be a strongly stable, countably iterable pfs premouse of type 1A. Let $k = k(M)$, and suppose that $\eta_k^M < \rho_{k+1}(M)$ and $\rho_{k+1}(M)$ is not measurable by the M-sequence. Let $\pi\colon \mathfrak{D}_{k+1}(M) \to M$ be the anticore map; then*

(a) $p_1(M_0^k)$ is solid and universal over M_0^k,
(b) if $\operatorname{crit}(\pi) = \rho_{k+1}(M) = \rho$, then letting $D = (E_\pi)_\rho$, D is the order zero measure of $\mathfrak{D}_{k+1}(M)$ on ρ,
(c) $\rho_k(M) = \pi(\rho_k(\mathfrak{D}))$, and
(d) $\eta_k^M = \eta_k^{\mathfrak{D}}$, where $\mathfrak{D} = \mathfrak{D}_{k+1}(M)$.

PROOF. The proof of parameter solidity in Lemma 4.10.2 goes over to r_{k+1} nearly verbatim. Let

$$\rho = \rho_1(M_0^k) = \rho_1(M^k),$$
$$r = p_1(M_0^k).$$

We choose $\vec{e}$ so that $r = \{e_0, \ldots, e_l\}$, where $e_0 > e_1 > \cdots > e_l$. Let q be the longest solid initial segment of r in this decreasing enumeration, where solidity is interpreted relative to the Σ_1 theory of M_0^k, and let

$$r = s \cup q,$$

where either $s = \emptyset$ or $\max(s) < \min(q)$. Let

$$\alpha_0 = \text{least } \beta \text{ such that } \mathrm{Th}_1^{M_0^k}(\beta \cup q) \notin M.$$

α_0 may not be the least β such that $\mathrm{Th}_{k+1}^M(\beta \cup q) \notin M$, since $k+1$ theories have access to w_k. We may assume that $\alpha_0 < \rho_k(M)$, as otherwise the theorem is trivially true. Let

$$\begin{aligned} H &= \mathrm{Dec}(\mathrm{cHull}_1^{M_0^k}(\alpha_0 \cup q)) \\ &= \text{transitive collapse of } d_0^k\text{``}\mathrm{Hull}_1^{M_0^k}(\alpha_0 \cup q), \end{aligned}$$

and let

$$\pi\colon H \to M$$

be the anticollapse map. Note here that $\bar{\mathfrak{C}}_k(M) = M$. $\pi \restriction H_0^k\colon H_0^k \to M_0^k$ is cofinal and Σ_1 elementary. We shall eventually show that $\pi(\eta_k^H) = \eta_k^H = \eta_k^M$ and $\pi(\rho_k(H)) = \rho_k(M)$, but at the moment we do not know either. (Both statements require that an appropriate parameter be in $\operatorname{ran}(\pi)$.) For all we know now, $\alpha_0 \le \eta_k^H < \rho_k(H)$ and η_k^H is measurable by H. For all we know now, $\rho_k(H)$ is measurable by H, so H is not a pfs premouse of degree k.

We now compare the phalanx (M, H, α_0) with M, just as in the proof of Lemma 4.10.2. In this process, we take k-ultrapowers of H by extenders E such that

$\text{crit}(E) < \rho_k(H)$ in the usual way, by decoding $\text{Ult}_0(H_0^k, E)$, or equivalently, decoding $\text{Ult}_1(H^{k-1}, E)$. (H^- is a pfs premouse of degree $k-1$.) Let us adopt all the notation of Lemma 4.10.2: $\mathcal{T}$, $\mathcal{T}^*$, and $\mathcal{U}$ are the trees that arise, P_ξ, P_ξ^*, and Q_ξ are their models, and so on. $\mu = \rho_{k+1}(H)$ and $t = \pi^{-1}(q)$.

Claims 0 and 1 go through with no change. (Claim 1 concerns the possibility that $\mathcal{T}$ terminates above M, so the fact that we have a different sort of H now is irrelevant.) So do Claims 3-5.

The counterpart of Claims 2 and 6 is

CLAIM 12. *(i) For all $\eta \le \delta$, $P(\alpha_0) \cap Q_\eta \subseteq M$.*
(ii) $P_\theta = Q_\delta$,
(iii) $D^{\mathcal{T}} \cap [1,\theta]_T = \emptyset$ and $D^{\mathcal{U}} \cap [0,\delta]_U = \emptyset$.

PROOF. Part (i) is proved just as in Claim 6 above. If $[1,\theta]_T$ drops in model or $Q_\xi \lhd P_\theta \downarrow 0$, then $\pi_\theta \circ j_{0,\delta} \colon M \to P_\theta^*$ is a map of the sort that is ruled out by the Weak Dodd-Jensen property of Σ. If $P_\theta \lhd Q_\delta \downarrow 0$, then $\text{Th}_1^{H_0^k}(\alpha_0 \cup t) \in Q_\xi$ for some ξ, so $\text{Th}_1^{H_0^k}(\alpha_0 \cup t) \in M$, contradiction. Thus $\hat{P}_\theta = \hat{Q}_\delta$, and neither branch drops in model. Using the fact that M^- and H^- are strongly sound (that is, $M = \bar{\mathfrak{C}}_k(M)$ and $K = \bar{\mathfrak{C}}_k(H)$), one can show that neither side drops in degree. Thus we have (i) and (iii). ⊣

We are now where we were after Claim 6, except that we are missing the information in Claim 6 that $i_{1,\theta}(\rho_k(H)) = \rho_k(P_\theta) = j_{0,\delta}(\rho_k(M))$. However, this is no longer relevant to the proof of Claim 7, because the solidity witnesses for q are theories in a language without names for ρ_k or η_k. So the proof of Claim 7 goes through.

The proofs of Claims 8-11 now go through without change. We have $\text{crit}(j_{0,\delta}) \ge \alpha_0$ or $j_{0,\delta} = \text{id}$, r is solid and universal, and if $\rho_{k+1}(M)$ is not measurable in M, then M is projectum solid in the sense of (b). $\mathfrak{D} = H$. Moreover, $\pi = j_{0,\delta}^{-1} \circ i_{1,\theta}$. Thus we have proved (a) and (b) of Lemma 4.10.7.

CLAIM 13. *$\pi(\rho_k(H)) = \rho_k(M)$ and $\eta_k^H = \eta_k^M$.*

PROOF. We have that $\eta_k^M < \rho_{k+1}(M) \le \text{crit}(j_{0,\delta})$, so

$$\eta_k^{Q_\delta} = \eta_k^M$$

and

$$\rho_k(Q_\delta) = j_{0,\delta}(\rho_k(M)) = \sup j_{0,\delta}``\rho_k(M).$$

But then $\eta_k^H < \alpha_0$, for otherwise $\alpha_0 \le \eta_k^H < \rho_k(H)$ and $\alpha_0 \le \sup i_{1,\theta}``\eta_k^H = \eta_k^{P_\theta} = \eta_k^{Q_\delta}$. Since $\eta_k^H < \alpha_0$, $i_{1,\theta}$ must be continuous at $\rho_k(H)$, and

$$i_{1,\theta}(\rho_k(H)) = \rho_k(P_\theta) = j_{0,\delta}(\rho_k(M))$$

and

$$\eta_k^H = i_{1,\theta}(\eta_k^H) = \eta_k^{P_\theta} = \eta_k^M.$$

The fact that $\pi = j_{0,\delta}^{-1} \circ i_{1,\theta}$ now yields our claim. ⊣

Claim 13 yields (c) and (d) of Lemma 4.10.7. ⊣

Now we use Lemma 4.10.7 by pulling back its conclusions under an ultrapower map.

LEMMA 4.10.8. *Let M be a stable, countably iterable pfs premouse of type 1, $k = k(M)$, and suppose that $\rho_{k+1}(M)$ is not measurable by the M-sequence. Suppose that M is not strongly stable. Let $\pi\colon \mathfrak{D} \to \bar{\mathfrak{C}}_k(M)$ be the anticore map, where $\mathfrak{D} = \mathfrak{D}_{k+1}(M)$; then*

(a) $p_1(M_0^k)$ is solid and universal over M_0^k,
(b) if $\operatorname{crit}(\pi) = \rho_{k+1}(M) = \rho$, then $(E_\pi)_\rho$ is the order zero measure of $\mathfrak{D}$ on ρ,
(c) $\pi(\rho_k(\mathfrak{D})) = \rho_k(M)$, and
(d) $\eta_k^{\mathfrak{D}} = \eta_k^M$.

PROOF. Let $\eta = \eta_k^M$, and let D be the order zero measure of M on η. Since M is stable, $\eta < \rho_{k+1}(M)$. We may assume that $\rho_{k+1}(M) < \rho_k(M)$, and hence $D \in M^k$. Let

$$N = \mathrm{Ult}_k(\bar{\mathfrak{C}}_k(M), D),$$

$$i = i_D^{\bar{\mathfrak{C}}_k(M)},$$

$$\rho = \rho_{k+1}(M) = \rho_1(M_0^k),$$

and

$$r = p_1(M_0^k).$$

In terms of reducts, $N_0^k = \mathrm{Ult}_0(M_0^k, D)$ and $i \restriction M_0^k$ is the canonical embedding. $\rho_k(N) = \sup i``\rho_k(M) < i(\rho_k(M))$, and $\eta_k^N = \eta$ is not measurable in N. Thus N is a strongly stable pfs premouse of type 1A, and N is countably iterable because M is, so N satisfies the hypotheses of Lemma 4.10.7.

CLAIM 1. *Let $\alpha < \rho_k(M)$ and $q \in M_0^k$; then*

$$Th_1^{M_0^k}(\alpha \cup \{q\}) \in M \text{ iff } Th_1^{N_0^k}(\sup i``\alpha \cup \{i(q)\}) \in N.$$

PROOF. The usual proof for solidity witnesses shows that if $\mathrm{Th}_1^{M_0^k}(\alpha \cup \{q\}) \in M$ then $\mathrm{Th}_1^{N_0^k}(\sup i``\alpha \cup \{i(q)\}) \in N$. Conversely, suppose $\mathrm{Th}_1^{N_0^k}(\sup i``\alpha \cup \{i(q)\}) \in N$, and let $\mathrm{Th}_1^{N_0^k}(\sup i``\alpha \cup \{i(q)\}) = [a, f]_D^{M_0^k}$. For φ a Σ_1 formula and $\beta < \alpha$,

$$\begin{aligned} M_0^k \models \varphi[\beta, q] &\text{ iff } N_0^k \models \varphi[i(\beta), i(q)] \\ &\text{ iff } \langle \varphi, i(\beta), i(q) \rangle \in [a, f]_D^{M_0^k} \\ &\text{ iff for } D_a \text{ a.e. } u,\ \langle \varphi, \beta, q \rangle \in f(u). \end{aligned}$$

Since $D_a \in M$, $\mathrm{Th}_1^{M_0^k}(\alpha \cup \{q\}) \in M$. ⊣

CLAIM 2. $\rho_{k+1}(N) = \sup i``\rho$.

PROOF. $\mathrm{Th}_1^{M_0^k}(\rho \cup \{r\}) \notin M$, so $\mathrm{Th}_1^{N_0^k}(\sup i``\rho \cup \{i(r)\}) \notin N$, so $\rho(N) \leq \sup i``\rho$. For the other inequality, let $\beta < \rho$ and $q = i(f)(a) \in N_0^k$, where $a \in [\varepsilon(D)]^{<\omega}$. Since $\mathrm{Th}_1^{M_0^k}(\beta \cup \{f\}) \in M^k$, we have $\mathrm{Th}_1^{N_0^k}(i(\beta) \cup \{i(f)\}) \in N^k$. Choosing $\beta \geq \mathrm{crit}(D)$, we get that $a \subset i(\beta)$ so using $\mathrm{Th}_1^{N_0^k}(i(\beta) \cup \{i(f)\})$ we can compute $\mathrm{Th}_1^{N_0^k}(i(\beta) \cup \{q\})$ in N, as desired. ⊣

CLAIM 3. *r is solid.*

PROOF. Let q be the longest solid initial segment of r, and suppose that $q \neq r$. Let $\gamma = \max(q - r)$, and let

$$\alpha = \text{least } \xi \text{ such that } \mathrm{Th}_1^{M_0^k}(\xi \cup q) \notin M.$$

Since $q \neq r$, $\alpha \leq \gamma$.

By Claims 1 and 2, $\mathrm{Th}_1^{N_0^k}(\rho(N) \cup i(r)) \notin N$, so $p_1(N_0^k) <^* i(r)$, where $<^*$ is the parameter order. But $i(q)$ is solid over N_0^k, so $i(q)$ is a proper initial segment of $p_1(N_0^k)$. Let

$$\beta = \text{least } \xi \text{ in } p_1(N_0^k) - i(q).$$

If $\beta < \sup i``\alpha$, then $\mathrm{Th}_1^{N_0^k}(\beta + 1 \cup i(q)) \in N$ by Claim 1 and our choice of $\alpha$, contradiction. Thus $\sup i``\alpha \leq \beta$. But then $\mathrm{Th}_1^{N_0^k}(\beta \cup i(q)) \notin N$ by Claim 1, so $p_1(N_0^k)$ is not solid, contradiction. ⊣

It follows from Claims 1-3 that $p_1(N_0^k) = i(r)$.

CLAIM 4. *r is universal.*

PROOF. Let $X \subseteq \rho$ and $X \in M$. By the universality of $i(r)$, we have $\gamma < \sup i``\rho$ such that

$$i(X) \cap \sup i``\rho = h^1_{N_0^k}(\gamma, i(r)) \cap \sup i``\rho.$$

Let $\gamma = [\{\kappa\}, g]_D^{M_0^k}$. Since $\gamma < \sup i``\rho$, we may assume that $g \in M||\rho$. But then for $\xi < \rho$, $\xi \in X$ iff there is a $Z \in D$ and a $\theta < \rho_k(M)$ such that for all $u \in Z$, $h^1_{M_0^k||\theta}(g(u), r)$ is defined and $\xi \in h^1_{M_0^k||\theta}(g(u), r)$. This shows that $X \in \mathrm{Hull}_1^{M_0^k}(\rho \cup r)$, as desired. Part (b) of universality (Definition 4.1.7) can be proved similarly. ⊣

CLAIM 5. *If $\mathrm{crit}(\pi) = \rho$, then $(E_\pi)_\rho$ is the order zero measure of $\mathfrak{D}_{k+1}(M)$ on ρ.*

PROOF. Suppose that $\mathrm{crit}(\pi) = \rho$. Since r is universal, ρ is regular in M. But then since $\rho < \rho_k(M)$, $\mathrm{cof}_k^M(\rho) = \rho$, so that i is continuous at ρ, and $i(\rho) =$

$\rho_{k+1}(N)$. Thus $\rho_{k+1}(N)$ is not measurable by the N-sequence. Also, the fact that $\rho \notin \mathrm{Hull}_1^{M_0^k}(\rho \cup \{r\})$ is Π_1 over M_0^k, so

$$i(\rho) \notin \mathrm{Hull}_1^{N_0^k}(\rho \cup \{i(r)\}).$$

Letting $Q = \mathfrak{D}_{k+1}(N)$ and $\tau\colon Q \to N$ be the anticore map, and $F = (E_\tau)_{i(\rho)}$, we have by Lemma 4.10.7(b) that F is the order zero measure of Q on $i(\rho)$. Letting $P = \mathfrak{D}_{k+1}(M)$ and $j = \tau^{-1} \circ i \circ \pi$. we have the diagram

$$\begin{array}{ccc} Q = \mathfrak{D}_{k+1}(N) & \xrightarrow{\ \tau\ } & N \\ \uparrow{\scriptstyle j} & & \uparrow{\scriptstyle i} \\ P = \mathfrak{D}_{k+1}(M) & \xrightarrow{\ \pi\ } & \bar{\mathfrak{C}}_k(M) \end{array}$$

It is easy to see that j is well defined, and letting $G = j^{-1}(F)$, i_G^P factors into π the way that i_F^Q factors into τ. This completes the proof of Claim 5. ⊣

Finally, we prove parts (c) and (d).

CLAIM 6. *$\pi(\rho_k(\mathfrak{D})) = \rho_k(M)$ and $\eta_k^{\mathfrak{D}} = \eta$.*

PROOF. By Lemma 4.10.7(d) there is some $z < i(\rho)$ such that $\mathrm{Hull}_1^{N^{k-1}}(\eta \cup \{i(r), p_k(N), z\})$ is cofinal in $\rho_k(N)$. Let $z = [\{\eta\}, g]_D^{M||\rho_k(M)}$; then since $\mathrm{ran}(i)$ is cofinal in $\rho_k(N)$, we get that $\mathrm{Hull}_1^{M^{k-1}}(\eta \cup \{r, p_k(M), g\})$ is cofinal in $\rho_k(M)$. But $\mathrm{ran}(\pi)$ is cofinal in $\rho_k(M)$, so $\mathrm{Hull}_1^{\mathfrak{D}^{k-1}}(\eta \cup \{\pi^{-1}(r), p_k(\mathfrak{D}), g\})$ is cofinal in $\rho_k(\mathfrak{D})$. Thus $\eta_k^{\mathfrak{D}} \leq \eta$. It is easy to see that $\eta \leq \eta_k^{\mathfrak{D}}$.

Thus $\eta_k^{\mathfrak{D}} = \eta = \eta_k^M$. If f is a nice witness that $\mathrm{cof}_k^{\mathfrak{D}}(\rho_k(\mathfrak{D})) = \eta$, then $\pi(f)$ is a nice witness that $\mathrm{cof}_k^M(\pi(\rho_k(\mathfrak{D}))) = \eta$, because $\eta < \mathrm{crit}(\pi)$ and π is Π_2-elementary as a map from $\mathfrak{D}^{k-1}$ to M^{k-1}. It follows that π is continuous at $\rho_k(\mathfrak{D})$, and thus $\pi(\rho_k(\mathfrak{D})) = \rho_k(M)$. ⊣

This completes the proof of Lemma 4.10.8. ⊣

Let us put the pieces together.

THEOREM 4.10.9. *Let M be a countably iterable pfs premouse of type 1, and $k = k(M)$. Suppose that M is stable and $\rho_{k+1}(M)$ is not measurable M by the M-sequence; then M is parameter solid.*

PROOF. By Lemma 4.10.2 we may assume that M is not strongly stable. Thus M has the properties enumerated in Lemma 4.10.8. Let

$$\begin{aligned} r &= r_{k+1}(M) \\ &= p_1(M_0^k) \\ &= \langle e_0, \ldots, e_l \rangle, \end{aligned}$$

and

$$\begin{aligned} p &= p_{k+1}(M) \\ &= p_1(M^k) \\ &= \langle d_0, \ldots, d_m \rangle, \end{aligned}$$

where the enumerations are in decreasing order. r is solid over M_0^k. We shall use this to show that p is solid over M^k.

We may assume that $\rho_k(M) < \rho_{k-1}(M)$. For letting $\pi\colon \bar{\mathfrak{C}}_k(M) \to M$ be the anticore map, we have $\rho_{k-1}(M) \in \operatorname{ran}(\pi)$, so if $\rho_k(M) = \rho_{k-1}(M)$, then π is the identity. This means that M_0^k and M^k are essentially equivalent, so $p = r$. Solidity for r over M_0^k then implies solidity for p over M^k.

So we assume $\rho_k(M) < \rho_{k-1}(M)$. Let

$$\begin{aligned} \rho &= \rho_{k+1}(M), \\ \rho_k &= \rho_k(M), \\ N &= \bar{\mathfrak{C}}_k(M), \end{aligned}$$

and if M has type 1B,

$$D = \text{order zero measure of } N \text{ on } \rho_k.$$

If M has type 1B, let $i = i_D^N$. If M has type 1A, let $N = M$ and i be the identity. Our plan is to translate between r and p, and the key here is that by Lemma 4.10.8(c), there is an M_0^k-name for D in $\operatorname{Hull}_1^{M_0^k}(\rho \cup r)$. The name is ε, where

$$\varepsilon = \begin{cases} \text{least } \gamma \text{ s.t. } \rho_k = h^1_{M^{k-1}}(\gamma, p_k(M)) & \text{if } M \text{ has type 1A,} \\ \text{least } \gamma \text{ s.t. } i(D) = h^1_{M^{k-1}}(\gamma, p_k(M)) & \text{if } M \text{ has type 1B.} \end{cases}$$

CLAIM 1. *$\varepsilon \in \operatorname{Hull}_1^{M^k}(\emptyset)$, and $\varepsilon \in \operatorname{Hull}_1^{M_0^k}(\rho \cup r)$.*

PROOF. Let

$$z = \begin{cases} \rho_k & \text{if } M \text{ has type 1A,} \\ D & \text{if } M \text{ has type 1B.} \end{cases}$$

Let $\tau\colon \mathfrak{D}_{k+1}(M) \to \bar{\mathfrak{C}}_k(M)$ be the anticore map. By Lemma 4.10.8(c), $z \in \operatorname{ran}(\tau)$, so there is a $\gamma < \rho_k$ in $\operatorname{ran}(\tau)$ such that $h^1_{N^{k-1}}(\gamma, p_k(N)) = z$, or equivalently, $h^1_{M^{k-1}}(\gamma, p_k(M)) = i(z)$. But then the least such γ is in $\operatorname{ran}(\tau)$.[71] So $\varepsilon \in \operatorname{ran}(\tau) \cap M||\rho_k = \operatorname{Hull}_1^{M_0^k}(\rho \cup r)$.

We claim that $i(\rho_k) \in \operatorname{Hull}_1^{M^{k-1}}(\{\rho_k, p_k(M)\})$. For by elementarity, there is an $\xi \geq \rho_k$ in $\operatorname{Hull}_1^{M^{k-1}}(\{\rho_k, p_k(M)\})$ such that $\xi = h^1_{M^{k-1}}(\gamma, p_k(M))$ for some $\gamma < \rho_k$. By elementarity again, the least such ξ is in $\operatorname{Hull}_1^{M^{k-1}}(\{\rho_k, p_k(M)\})$. But the least such ξ is $i(\rho_k)$.

---

[71]Let γ be least in $\operatorname{ran}(\tau)$ such that $h^1_{N^{k-1}}(\gamma, p_k(N)) = z$. If $\exists \delta < \gamma (h^1_{N^{k-1}}(\delta, p_k(N)) = z)$, then by elementarity $\exists \delta < \gamma (\delta \in \operatorname{ran}(\tau) \wedge h^1_{N^{k-1}}(\delta, p_k(N)) = z)$, contradiction.

It follows that $i(z) \in \mathrm{Hull}_1^{M^{k-1}}(\{\rho_k, p_k(M)\})$. But

$$\varepsilon = \text{least } \gamma \text{ such that } i(z) = h^1_{M^{k-1}}(\gamma, p_k(M)).$$

Since $i(z) \in \mathrm{Hull}_1^{M^{k-1}}(\{\rho_k, p_k(M)\})$, we get that $\varepsilon \in \mathrm{Hull}_1^{M^{k-1}}(\{\rho_k, p_k(M)\})$.[72] But $\varepsilon < \rho_k$, so $\varepsilon \in \mathrm{Hull}_1^{M^k}(\emptyset)$. ⊣

We shall show that $p \subseteq r$, and that the solidity witnesses for r yield solidity witnesses for p. This involves showing by induction that their initial segments are Σ_1-interdefinable over M_0^k, modulo parameters in $\rho_{k+1}(M) \cup \{\varepsilon\}$. The proof is essentially the same as Zeman's proof in [87] that in Jensen premice, the standard parameter is intertranslatable with the Dodd parameter.[73]

Let $h_B = h^1_{M_0^k}$. For the remainder of this proof, let us say that *x is generated by y* iff $x \in \mathrm{Hull}_1^{M_0^k}(\rho \cup \{y\})$, or equivalently, $x = h_B(y, a)$ for some finite $a \subseteq \rho$. By Claim 1, ε is generated by r.

CLAIM 2. *If ε is generated by x, then for any $\gamma < \rho_k(M)$, $Th_1^{M^k}(\gamma \cup \{x\})$ is rudimentary in $Th_1^{M_0^k}(\gamma \cup \{x\})$. In particular, if the latter belongs to M, then so does the former.*

PROOF. We claim first that there is a recursive function $\varphi \mapsto \varphi^*$ defined on Σ_1 formulae φ such that whenever $\beta < \rho_k$,

$$M^{k-1} \models \varphi[\beta, \rho_k, p_k(M)] \text{ iff } M^{k-1} \models \varphi^*[\beta, \varepsilon, p_k(M)].$$

If M has type 1A this is obvious, so suppose M has type 1B. We let

$$\varphi^*(u, v, w) = \exists E (E = h^1(v, w) \wedge \exists \gamma \exists X \in E \, \forall \alpha \in X \, (M^{k-1} || \gamma \models \varphi[u, \alpha, w])).$$

This works because by Łoś's Theorem, for $\beta < \rho_k$,

$$\begin{aligned} M^{k-1} \models \varphi[\beta, \rho_k, p_k(M)] &\text{ iff } \exists \gamma \exists X \in D \forall \alpha \in X (N^{k-1} || \gamma \models \varphi[\beta, \alpha, p_k(N)]) \\ &\text{ iff } \exists \gamma \exists X \in i(D) \forall \alpha \in X (M^{k-1} || \gamma \models \varphi[\beta, \alpha, p_k(M)]). \end{aligned}$$

For $\gamma \leq \rho_k$, let

$$B_\gamma = B^k \cap M || \gamma$$

and

$$A_\gamma = A^k \cap M || \gamma.$$

Using the map $\varphi \mapsto \varphi^*$, it is easy to construct a Σ_0 formula θ such that whenever $\varepsilon < \gamma$

$$\theta(B_\gamma, \varepsilon, Z) \text{ iff } Z = A_\gamma.$$

---

[72] See the last footnote.

[73] Our r is analogous the Dodd parameter in Zeman's proof, p is analogous to the standard parameter, and ε is analogous to the index of the longest proper initial segment of the top extender.

Now suppose x generates ε, and let $h_B(a,x) = \varepsilon$ where $a \subseteq \rho$ is finite. Let $\rho \leq \gamma < \rho_k$, and suppose that $\mathrm{Th}_1^{M_0^k}(\gamma \cup \{x\}) \in M$. Then for φ a Σ_1 formula and $\delta < \gamma$

$$M^k \models \varphi[\delta,x] \text{ iff } M_0^k \models \exists T \exists \xi \exists A \exists \varepsilon (h^1(a,x) = \varepsilon \wedge T = B_\xi \wedge \theta(T,\varepsilon,A) \wedge (M||\xi,A) \models \varphi[\delta,x]).$$

The right hand side is of the form $\langle \psi, \delta, x \rangle \in \mathrm{Th}_1^{M_0^k}(\gamma \cup \{x\})$, where ψ is obtained recursively from φ. This yields the claim. ⊣

Let $<^*$ be the parameter order, that is, the lexicographic order of descending sequences of ordinals. Since the predicate B^k of M_0^k is lightface Σ_0 over M^k, $p \leq^* r$.

CLAIM 3. *For all $i \leq m$,*

(a) $d_i \in r$,

(b) $\langle p-(d_i+1), \varepsilon \rangle$ generates $r-(d_i+1)$, and

(c) $Th_1^{M^k}(d_i \cup p-(d_i+1)) \in M$.

PROOF. If ε is generated by $\emptyset$, then by Claim 2, $p = r$, and the solidity witnesses for p are rudimentary in the solidity witnesses for r, so we have Claim 3. So let us fix $n \geq 0$ least such that $\langle e_0, \ldots, e_n \rangle$ generates ε.

We prove the claim by induction on i. Let $i = 0$. Since $p \leq^* r$, $d_0 \leq e_0$. We show first by induction on j that if $j \leq n$ and $e_j > d_0$, then

$$e_j = \text{least } \xi \text{ s.t. for some finite } s \subseteq \xi,\ \{r \restriction j, \xi, s\} \text{ generates } \varepsilon.$$

For this, since r generates ε, it is enough to see that there is no $\xi < e_j$ as on the right. Suppose there were, and let $\alpha = \max(\xi, d_0) + 1$. Since $\alpha < e_j$ and r is solid, $\mathrm{Th}_1^{M_0^k}(\alpha \cup r \restriction j) \in M$, so by Claim 2, $\mathrm{Th}_1^{M^k}(\alpha \cup r \restriction j) \in M$. But $p \subseteq \alpha$, contradiction.

The formula above gives a $\Sigma_1^{M_0^k}$ definition of e_j from $\langle e_0, \ldots, e_{j-1} \rangle$ and ε.[74] Thus if $j \leq n$ and $e_j > d_0$, then e_j is generated by ε. This proves (b) of the claim in the case $i = 0$.

Suppose now that $e_n > d_0$. Then just as in the case that $\varepsilon = \emptyset$, $p = r \cap e_n$ and the solidity witness for $p \restriction l$ can be computed from the solidity witness for $r \restriction n+1+l$. Thus we may assume $e_n \leq d_0$.

We claim that $d_0 \in r$. If not, then $r - d_0$ is generated by ε, so $\mathrm{Th}_1^{M^k}(\rho \cup r \cap d_0)$ can be used to compute $\mathrm{Th}_1^{M^k}(\rho \cup r)$, so $\mathrm{Th}_1^{M^k}(\rho \cup r \cap d_0) \notin M$. But $r \cap d_0 <^* p$, contradiction. Thus $d_0 \in r$.

For solidity, note first that there is a finite $s \subseteq d_0$ such that $\{r-(d_0+1), s\}$ generates ε, since otherwise d_0 is the least ξ such that for some finite $t \subseteq \xi$, $\{r-(d_0-1), \xi, t\}$ generates ε. This implies that ε generates d_0, which is impossible since $d_0 \in p$. So ε is generated by $d_0 \cup r-(d_0+1)$, and by Claim 2, the r-solidity

[74] "Least" may seem to introduce a Π_1 element, but see the proof of Claim 1.

witness $\mathrm{Th}_1^{M_0^k}(d_0 \cup r - (d_0+1))$ can be used to compute the p-solidity witness $\mathrm{Th}_1^{M^k}(d_0)$. This completes the base case $i = 0$.

If $e_n = d_0$, then $r \cap e_n = p \cap d_0$, and the solidity witnesses for r can be used to compute solidity witnesses for p, so we are done. Thus we may assume $e_n < d_0$, and go on to $i = 1$.

The induction step is very similar. Suppose we have (a)-(c) at i, and that d_{i+1} exists. Suppose also that $e_n < d_i$, since otherwise we have $r \cap e_n = p \cap e_n$ and the solidity witnesses can be translated, as above. By the argument above, we get that whenever $j \le n$ and $d_i > e_j > d_{i+1}$, then

$$e_j = \text{least } \xi \text{ s.t. for some finite } s \subseteq \xi,\ \{r \restriction j, \xi, s\} \text{ generates } \varepsilon.$$

For otherwise, there is a $\xi < e_j$ as on the right, and setting $\alpha = \max(\xi, d_{i+1}) + 1$, $\mathrm{Th}_1^{M_0^k}(\alpha \cup r \restriction j) \in M$ by solidity for r, so $\mathrm{Th}_1^{M^k}(\alpha \cup r \restriction j) \in M$ by Claim 2. Since $p \subseteq \{e_0, \ldots, e_{j-1}\} \cup \alpha$, this is a contradiction. The displayed formula implies that e_j is generated by $\{d_0, \ldots, d_i, \varepsilon\}$, so we have (b).

If $e_n > d_{i+1}$, then $p \cap e_n = r \cap e_n$ and the solidity witnesses for r yield witnesses for p. Here we use that $\{d_0, \ldots, d_i\} \subseteq r$, so $\{e_0, \ldots, e_n\}$ generates $\{d_0, \ldots, d_i, \varepsilon\}$. So we may assume $e_n \le d_{i+1}$.

Now we get that $d_{i+1} \in r$ and $\mathrm{Th}_1^{M^k}(d_{i+1} \cup \{d_0, \ldots, d_i\}) \in M$ just as in the case that $i = 0$. If $e_n = d_{i+1}$ then $r \cap e_n = p \cap d_{i+1}$ and solidity witnesses translate, so we are done. If $e_n < d_{i+1}$ we go on to $i+2$. ⊣

By Claim 3, $p_{k+1}(M)$ is solid. Moreover, $\mathrm{Hull}_1^{M^k}(\rho_{k+1}(M) \cup p_{k+1}(M)) = \mathrm{Hull}_1^{M_0^k}(\rho_{k+1}(M) \cup r)$, so $\bar{\mathfrak{C}}_{k+1}(M) = \mathfrak{D}_{k+1}(M)$. By Lemma 4.10.8, M is solid.

This finishes the proof of Theorem 4.10.9. ⊣

Our proof of solidity also yields a useful condensation theorem. The following is a simplified version of Theorem 9.3.2 of [86].

THEOREM 4.10.10 (Condensation). *Let M be a strongly stable, sound, countably iterable pfs premouse, and let H be a sound pfs premouse, and $\pi\colon H \to M$ be nearly elementary, with $\rho(H) \le \mathrm{crit}(\pi)$. Suppose that $H \in M$; then either*

(a) $H \lhd M$, or

(b) $H \unlhd \mathrm{Ult}(M, E)$, where E is on the M sequence and $\mathrm{lh}(E) = \mathrm{crit}(\pi) = \rho(H)$.

PROOF SKETCH. We may assume M is countable and enumerated by $\vec{e}$, and that Σ is an iteration strategy for M with the Weak Dodd-Jensen property relative to $\vec{e}$. Let $\alpha = \mathrm{crit}(\pi)$. We compare the phalanx (M, H, α) with M as in Lemma 4.10.2, using (id, π) to lift trees on (M, H, α) to trees on M, and using Σ to iterate M.[75] This yields $\mathcal{T}$ on (M, H, α), $\mathcal{T}^* = (\mathrm{id}, \pi)\mathcal{T}$ on M, and $\mathcal{U}$ on M.

CLAIM 1. *The last model P of $\mathcal{T}$ must be above H in $\mathcal{T}$.*

[75] It would also work to compare $(M, H, \rho(H))$ with M.

PROOF. Exactly as in the proof of Lemma 4.10.2. ⊣

Let Q be the last model of $\mathcal{U}$.

CLAIM 2. *$P \trianglelefteq Q$ and H-to-P does not drop.*

PROOF. Let P^* be the last model of $\mathcal{T}^*$, and $\pi^*: P \to P^*$ be the copy map. If $Q \triangleleft P$, then M-to-Q does not drop in $\mathcal{U}$, and letting j be the branch embedding, $\pi^* \circ j$ maps M to a proper initial segment of P^*, contrary to Weak Dodd-Jensen. So $P \trianglelefteq Q$. Similarly, H-to-P does not drop, as otherwise $\pi^* \circ j$ maps M to an initial segment of a dropping iterate of M. ⊣

CLAIM 3. $H = P$.

PROOF. Suppose not, and let $i: H \to P$ be the branch embedding. We have assumed that H is sound, so $\text{crit}(i) > \rho(H)$,

$$H = \mathfrak{C}(P)^-,$$

and

$$i = \text{anticore map}.$$

Thus P is not sound, so $P = Q$. Since M is sound and Q is not, the branch M-to-Q in $\mathcal{U}$ dropped. Let $Q = Q_\delta$, $\eta <_U \delta$ be largest in $D^{\mathcal{U}}$, and $K = \mathcal{M}^{*,\mathcal{U}}_{\xi+1}$ where $\xi + 1 \leq_U \delta$ and U-pred$(\xi+1) = \eta$. Let $j = i^{\mathcal{U}}_{\xi+1,\delta} \circ i^{*,\mathcal{U}}_{\xi+1}$ be the branch embedding. Then

$$K = \mathfrak{C}(Q)^-,$$

and

$$j = \text{anticore map}.$$

Thus $H = K$ and $i = j$. This implies that the first extenders used in i and j are the same, contrary to the fact that we were iterating away disagreements. ⊣

If $Q = M$, then $H \triangleleft M$ and we are done. Otherwise, let G be the first extender used in M-to-Q. We have that $\alpha \leq \text{lh}(G) < o(H)$, and $\text{lh}(G)$ is a cardinal of Q. Suppose toward contradiction that $\rho(H) < \text{lh}(G)$. Then H collapses $\text{lh}(G)$ via a $r\Sigma^H_{k(H)+1}$ function, and hence $H \notin Q$, so $Q = H$. But G is used in $\mathcal{U}$, and Q is the last model of $\mathcal{U}$, so $\rho(Q)$ is not in the interval $(\text{crit}(G), \text{lh}(G))$. Thus $\rho(H) < \text{crit}(G)$. But this is impossible because H is sound.

Thus

$$\alpha \leq \text{lh}(G) \leq \rho(H) \leq \alpha.$$

We just need to see that $H \triangleleft \text{Ult}(M, G)$. But if not, there is a second extender K that is used in $\mathcal{U}$, and $\text{lh}(G) < \text{lh}(K) < o(H)$. Since $\text{lh}(K)$ is a cardinal of Q and H collapses $\text{lh}(K)$, we get $H = Q$. As in the last paragraph, this leads to a contradiction. ⊣

For a simple application of Condensation, suppose that N is a countable iterable pfs premouse, $N \models \mathsf{ZFC}$, and κ is a regular cardinal of N. Let $M \trianglelefteq N$ and $\rho(M) = \kappa$. Working inside N we can find club many $\alpha < \kappa$ such that $\mathrm{Hull}_\omega^M(\alpha) \cap \kappa = \alpha$. Theorem 4.10.10 implies that for such α, letting $H = \mathrm{cHull}_\omega^M(\alpha)$, either $H \triangleleft M$ or $H \triangleleft \mathrm{Ult}(M, \dot{E}_\alpha^M)$.

The two possibilities (a) and (b) in the conclusion of Theorem 4.10.10 are mutually exclusive. Alternative (b) is sometimes realized. For example, in the last paragraph, if $\kappa = \mu^+$ and μ is subcompact in N, then there are stationarily many $\alpha < \kappa$ such that $\alpha = \mathrm{lh}(E)$ for some $E \in \dot{E}^M$.[76] For such α, alternative (b) of Theorem 4.10.10 must apply.

A variant of the condensation argument yields weak ms-solidity.

LEMMA 4.10.11. *Let N be a countably iterable pfs premouse such that $k(N) = 0$; then N is weakly ms-solid.*

PROOF. We may assume that N is active. Let $M = \mathfrak{C}_1(N)^-$ be the first core of N, but with degree zero, and let $E = \dot{F}^M$; we must show that E has the weak ms-ISC. Let $\kappa = \mathrm{crit}(E)$ and let F be the Jensen completion of $E_{\{\kappa\}}$, and let

$$H_0 = \mathrm{Ult}_0(M|\kappa^{+,M}, F).$$

We have the diagram

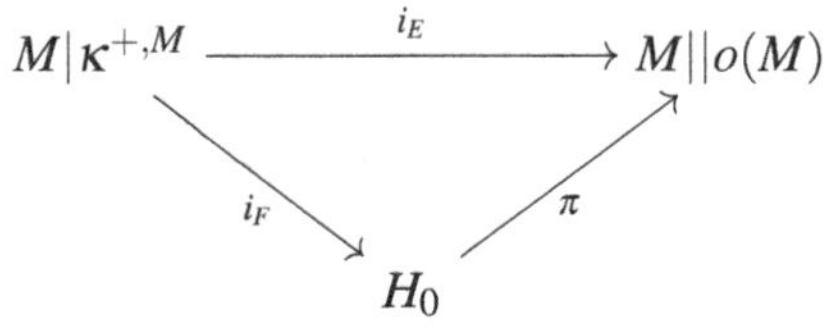

Here $\pi(i_F(g)(\kappa)) = i_E(g)(\kappa)$. It is not hard to see that for all $\alpha < \kappa^{+,M}$, $\pi(i_F \restriction \alpha) = i_E \restriction \alpha$, that is, π maps the fragments of F to the corresponding fragments of E. π is cofinal, so letting

$$H = (H_0, F),$$

we have that

$$\pi : H \to M$$

is a cofinal and Σ_1 elementary. Clearly $\kappa^{+,H} = \kappa^{+,M}$; let us write κ^+ for the common value.

CLAIM 1. *If $H \notin M$, then $H = M$.*

PROOF. Suppose $H \notin M$. We must have $\rho_1(M) \leq \kappa^+$, as otherwise $E_{\{\kappa\}} \in M$, so $E_{\{\kappa\}} \in M|\lambda_E$, so $H \in M$. Similarly $p_1(M) \subseteq \kappa^+$, for if $\gamma \in p_1(M) - \kappa^+$, then the solidity witness for γ can be used to compute $E_{\{\kappa\}}$ inside M, so again $H \in M$.

[76] This is due to Jensen, who showed that in extender models, it is equivalent to subcompactness. See [49].

It follows that $\rho_1(M) \cup p_1(M) \subseteq \text{ran}(\pi)$, so since π is elementary, $M = \text{ran}(\pi)$, as desired. ⊣

So if $H \notin M$, then $E = F$, and we are done. Thus we may assume $H \in M$ and $\pi \neq \text{id}$. Clearly $o(H)$ has cardinality κ^+ in M, so

$$\alpha = \text{crit}(\pi) = \kappa^{++,H}.$$

Note also that $F_{\{\kappa\}} \notin H$, so that $\rho(H) \leq \kappa^+$. There is a lightface Σ_1^H map from κ^+ onto $o(H)$, so $p(H) \subseteq \kappa^+$.

We now compare (M, H, α) with M just as in the condensation proof. Here $k(M) = k(H) = 0$, so M and H are strongly stable. Since $\alpha = \kappa^{++,M|\alpha}$, the anomalies described in Remarks 4.10.3 and 4.10.4 cannot arise. The proof of condensation did use that H was sound at a couple points, and we do not have that to work with now.

Let us adopt the notation from the proof of Theorem 4.10.10. We assume M is countable and enumerated by $\vec{e}$, and that Σ is an iteration strategy for M with the Weak Dodd-Jensen property relative to $\vec{e}$. We compare the phalanx (M, H, α) with M, obtaining trees $\mathcal{T}$ on (M, H, α), $\mathcal{T}^* = (\text{id}, \pi)\mathcal{T}$ on M, and $\mathcal{U}$ on M.

CLAIM 2. *The last model P of $\mathcal{T}$ is above H in $\mathcal{T}$.*

PROOF. As before. ⊣

Let Q be the last model of $\mathcal{U}$.

CLAIM 3. *$P \trianglelefteq Q$ and H-to-P does not drop.*

PROOF. By the Weak Dodd Jensen property, as in the proof of Theorem 4.10.10, Claim 2. ⊣

CLAIM 4. $H = P$.

PROOF. Suppose not, and let $i\colon H \to P$ be the branch embedding. $\rho(H) \leq \kappa^+ < \text{crit}(i)$, so P is not 1-sound, so $P = Q$. Since $F_{\{\kappa\}} \notin H$, $F_{\{\kappa\}} \notin P$. But $F_{\{\kappa\}} \in M$ because $H \in M$, so the branch M-to-Q in $\mathcal{U}$ dropped. Let $Q = Q_\delta$, $\eta <_U \delta$ be largest in $D^{\mathcal{U}}$, and $K = \mathcal{M}^{*,\mathcal{U}}_{\xi+1}$ where $\xi + 1 \leq_U \delta$ and $U\text{-pred}(\xi+1) = \eta$. We have

$$\dot{F}^H_{\{\kappa\}} = \dot{F}^P_{\{\kappa\}} = \dot{F}^Q_{\{\kappa\}} = \dot{F}^K_{\{\kappa\}}.$$

Let $j = i^{\mathcal{U}}_{\xi+1,\delta} \circ i^{*,\mathcal{U}}_{\xi+1}$ be the branch embedding. Since $\kappa \in \text{ran}(j)$, $\text{crit}(j) > \kappa^+$. Moreover, $p_1(H) \subset \kappa^+$, so $p_1(H) = p_1(Q) = p_1(K)$. It follows that

$$H = \text{cHull}^P(\kappa^+),$$
$$i = \text{anticollapse map},$$

and

$$K = \text{cHull}^Q(\kappa^+),$$
$$j = \text{anticollapse map}.$$

Thus $H = K$ and $i = j$. This implies that the first extenders used in i and j are the same, contrary to the fact that we were iterating away disagreements. $\dashv$

CLAIM 5. $M = Q$.

PROOF. Otherwise, let G be the first extender used in M-to-Q. We have that $\alpha \leq \text{lh}(G) < o(H)$, and $\text{lh}(G)$ is a cardinal of Q. Since H collapses $\text{lh}(G)$, $H \notin Q$, so $H = Q$. If $\text{crit}(G) < \kappa^+$, then G is total on M, and letting $N = \text{Ult}_0(M, G)$, $\text{crit}(\dot{F}^N) > \kappa$. Since $\text{crit}(\dot{F}^Q) = \kappa$, there must have been a drop on the branch N-to-Q, and this implies $\text{lh}(G) \leq \rho(Q)$. But $\rho(Q) \leq \kappa^+$, contradiction. Thus $\kappa^+ < \text{crit}(G)$, so $\alpha < \text{crit}(G)$, and M-to-Q dropped.

But again, let $\eta <_U \delta$ be largest in $D^{\mathcal{U}}$, and $K = \mathcal{M}^{*,\mathcal{U}}_{\xi+1}$ where $\xi + 1 \leq_U \delta$ and $U\text{-pred}(\xi + 1) = \eta$. Let $j \colon K \to Q$ be the iteration map. We have $\alpha < \text{crit}(j)$, so $\text{crit}(j) \notin \text{Hull}^Q_1(\kappa^+)$, contrary to $Q = H$. This is a contradiction. $\dashv$

Claims (2)-(5) imply that if $H \in M$, then $H \lhd M$. This completes the proof of Lemma 4.10.11. $\dashv$

Putting Lemma 4.10.1, Lemma 4.10.11, and Theorem 4.10.9 together, we get

THEOREM 4.10.12. *Let $\mathbb{C}$ be a maximal PFS-construction, and assume that V is countably $\mathcal{F}^{\mathbb{C}}$-iterable; then for all $\langle \nu, k \rangle < \text{lh}(\mathbb{C})$ such that $k \geq 0$, $M^{\mathbb{C}}_{\nu,k}$ is solid.*

4.11. The Bicephalus Lemma

The final thing we want of our constructions is that at any given stage, there is at most one extender that can be added. This follows, modulo iterability, from the Bicephalus Lemma.

DEFINITION 4.11.1. An *bicephalus* is a structure $\mathcal{B} = (B, F, G)$ such that both (B, F) and (B, G) are extender-active pfs premice of degree 0. We say that $\mathcal{B}$ is *nontrivial* iff $F \neq G$.

We think of $\mathcal{B}$ as a structure in the language with predicate symbols $\dot{E}, \dot{F}$, and $\dot{G}$ for the extender sequence of B, and the two last extenders F and G. The degree of $\mathcal{B}$ is zero, i.e., $k(\mathcal{B}) = 0$. For $\nu < o(\mathcal{B}) = \hat{o}(\mathcal{B})$, we set $\mathcal{B}|\langle \nu, l \rangle = B|\langle \nu, l \rangle$. The extender sequence of $\mathcal{B}$ is $\dot{E}^{\mathcal{B}}$ together with $\dot{F}^{\mathcal{B}}$ and $\dot{G}^{\mathcal{B}}$; it is not actually a sequence.

We need only consider normal, λ-tight iteration trees on $\mathcal{B}$. These are iteration trees $\mathcal{T}$ such that $\mathcal{M}^{\mathcal{T}}_0 = \mathcal{B}$, the extenders used in $\mathcal{T}$ are length-increasing and nonoverlapping along branches, and $E^{\mathcal{T}}_\alpha$ comes from the sequence of $\mathcal{M}^{\mathcal{T}}_\alpha$. If $\mathcal{M}^{\mathcal{T}}_\alpha$ is a bicephalus, this means that the extenders from $\dot{E}^{\mathcal{M}_\alpha}$ together with $\dot{F}^{\mathcal{M}_\alpha}$ and $\dot{G}^{\mathcal{M}_\alpha}$ are eligible. A θ-iteration strategy is an iteration strategy defined on all normal trees of length $< \theta$. $\mathcal{B}$ is *countably iterable* iff every countable elementary submodel of $\mathcal{B}$ has an $\omega_1 + 1$-iteration strategy.

The main theorem about bicephali is that the iterable ones are trivial. As befits such a basic result, the proof is simple and natural.[77]

THEOREM 4.11.2. *Let (B, F, G) be a countably iterable bicephalus; then $F = G$.*

PROOF SKETCH. Suppose $F \neq G$. This is a first order fact, so it passes to Skolem hulls of (B, F, G). Thus we may assume B is countable. Let Σ be an $\omega_1 + 1$-iteration strategy for (B, F, G). We now compare (B, F, G) with itself, by iterating least disagreements, producing normal trees $\mathcal{T}$ and $\mathcal{U}$ on (B, F, G) that are by Σ.

There will always be a disagreement, because $F \neq G$. For example, $E_0^{\mathcal{T}} = F$ and $E_0^{\mathcal{U}} = G$ is a legitimate first step. In general, if $\mathcal{M}_\alpha^{\mathcal{T}} = (B_\alpha, F_\alpha, G_\alpha)$ is a non-dropping iterate of (B, F, G), then either some extender on $\dot{E}^{B_\alpha}$ disagrees with its counterpart on the sequence of the current model of $\mathcal{U}$, or B_α is an initial segment of that model, and one of F_α and G_α disagrees with its counterpart (because $F_\alpha \neq G_\alpha$).

Thus this comparison yields $\mathcal{T}$ and $\mathcal{U}$ of length $\omega_1 + 1$. The usual termination argument now leads to a contradiction. ⊣

COROLLARY 4.11.3. *Suppose that*

$$\mathbb{C} \restriction \nu^\frown \langle (M^{<\nu}, F), F^* \rangle \quad \textit{and} \quad \mathbb{C} \restriction \nu^\frown \langle (M^{<\nu}, G), G^* \rangle$$

are maximal PFS constructions, and that V is countably $\vec{F} \cup \{F^*, G^*\}$ iterable; then $F = G$.

Putting things together, we have

THEOREM 4.11.4. *Suppose that V is countably iterable, and let $\mathbb{C}$ be a maximal PFS construction; then $\mathbb{C}$ is good at all $\langle \nu, k \rangle < \mathrm{lh}(\mathbb{C})$.*

PROOF. This follows from Theorems 4.9.1, 4.10.12, and Corollary 4.11.3. ⊣

We have shown that maximal PFS constructions do not break down, granted iterabilty for V. But do they reach anything interesting? We shall show in Section 10.4 that under certain hypotheses they do, but the following simple question is open. Suppose V is strongly uniquely (θ, θ) iterable for all θ. Let δ be a Woodin cardinal. Must there be a PFS construction $\mathbb{C}$ such that $\mathcal{F}^{\mathbb{C}}$ consists of nice extenders over V and $L[M_{\delta,0}^{\mathbb{C}}] \models$ "δ is Woodin"? If we had adopted ms-indexing and its corresponding background certificate requirement, the answer would be yes, essentially by [35, §11]. But we have adopted Jensen indexing. Our background certificate requirement is sufficiently liberal that we can prove the results of Section 10.4, but we do not see that it yields a positive answer to this question.

[77] It is due to Jensen and Mitchell. See [86, §4.6] and [33, Theorem 5].

Chapter 5

SOME PROPERTIES OF INDUCED STRATEGIES

In this chapter we show that certain internal consistency properties pass from an iteration strategy Σ^* for a coarse premouse to the iteration strategies that Σ^* induces via PFS constructions. These results are preliminary. We shall return to the topic in Chapter 7, where we shall prove much stronger results along the same lines.

Our results in the rest of this book have to do with pfs premice and constructions. The strategy mice that we study later are built upon the projectum-free-spaces fine structure. So from here on, we shall often drop the qualifier "pfs". Pure extender premice are pfs premice, not Jensen premice, unless otherwise specified.

5.1. Copying commutes with conversion

Let us show that copying commutes with conversion. The proof is completely routine, but it has the structure of less routine inductions we shall do later, so we give it here. We shall use the result later.

THEOREM 5.1.1. *Let R and S be transitive models of* ZFC, $R \models$ "$\mathbb{C}$ *is a PFS construction", and let* $\sigma : R \to S$ *be elementary with* $\sigma(\mathbb{C}) = \mathbb{D}$. *Let* $c = \langle M, \varphi, P, \mathbb{C}, R\rangle$ *and* $d = \langle N, \psi, Q, \mathbb{D}, S\rangle$ *be conversion stages, and suppose that* $\pi : M \to N$ *is nearly elementary, and* $\psi \circ \pi = \sigma \circ \varphi$; *then for any plus tree* $\mathcal{T}$ *on* M, *if all the models in* $\mathrm{lift}(\pi\mathcal{T}, d)_0$ *are wellfounded, then so are those in* $\mathrm{lift}(\mathcal{T}, c)_0$, *and*

$$\sigma\,\mathrm{lift}(\mathcal{T}, c)_0 = \mathrm{lift}(\pi\mathcal{T}^+, d)_0.$$

PROOF. We assume first that π is elementary, so that by Lemma 4.5.21, $\pi\mathcal{T}^+ = \pi\mathcal{T}$, and all the copy maps associated to $\pi\mathcal{T}$ are elementary. The general case is almost the same, but the notation and diagrams are less tidy. We discuss it at the end of the proof. Here is a diagram of our starting position:

$$\begin{array}{ccccc} N & \xrightarrow{\ \psi\ } & Q & \qquad & S \\ {\scriptstyle\pi}\uparrow & & \uparrow{\scriptstyle\sigma} & & \uparrow{\scriptstyle\sigma} \\ M & \xrightarrow[\ \varphi\ ]{} & P & & R \end{array}$$

Let $\mathcal{U} = \pi\mathcal{T}$, $\mathcal{T}^* = \text{lift}(\mathcal{T}, c)_0$ and $\mathcal{U}^* = \text{lift}(\mathcal{U}, d)_0$. We must see that $\mathcal{U}^* = \sigma\mathcal{T}^*$. Let $M_\xi = \mathcal{M}_\xi^{\mathcal{T}}$, $N_\xi = \mathcal{M}_\xi^{\mathcal{U}}$, and

$$\pi_\xi : M_\xi \to N_\xi$$

be the elementary copy map.

Let

$$\text{stg}(\mathcal{T}, c, \xi) = \langle M_\xi, \varphi_\xi, P_\xi, \mathbb{C}_\xi, R_\xi \rangle$$

and

$$\text{stg}(\mathcal{U}, d, \xi) = \langle N_\xi, \psi_\xi, Q_\xi, \mathbb{D}_\xi, S_\xi \rangle$$

be the conversion stages associated to the two conversion systems. We shall define $\sigma_\xi : R_\xi \to S_\xi$ by induction on ξ, maintaining by induction on ξ

(a) $\mathcal{U}^* \upharpoonright \xi + 1 = \sigma\mathcal{T}^* \upharpoonright \xi + 1$, and for all $\alpha \leq \xi$, σ_α is the associated copy map,
(b) $\sigma_\xi(P_\xi) = Q_\xi$, and
(c) $\sigma_\xi \circ \varphi_\xi = \psi_\xi \circ \pi_\xi$.

Let $(\dagger)_\xi$ be the conjunction of (a)-(c). Setting $\sigma_0 = \sigma$, $(\dagger)_0$ is just the hypothesis of the theorem.

Now suppose that $(\dagger)_\xi$ holds. Let $E = E_\xi^{\mathcal{T}}$ and $F = E_\xi^{\mathcal{U}}$. For simplicity, let us assume that E is not of plus type, that is, E is on the M_ξ sequence. (The other case is almost the same.) The map that resurrects $\varphi_\xi(E)$ inside $\mathbb{C}_\xi$ is

$$\rho_\xi = \sigma_{P_\xi}[P_\xi \,|\, \text{lh}(\varphi_\xi(E))]^{\mathbb{C}_\xi}.$$

Similarly, the resurrection map for $\psi_\xi(F)$ is

$$\tau_\xi = \sigma_{Q_\xi}[Q_\xi \,|\, \text{lh}(\psi_\xi(F))]^{\mathbb{D}_\xi}.$$

By $(\dagger)_\xi$(a), $\sigma_\xi(\mathbb{C}_\xi) = \mathbb{D}_\xi$. By (b), $\sigma_\xi(P_\xi) = Q_\xi$, and by (c), $\sigma_\xi(\varphi_\xi(E)) = \psi_\xi(F)$. It follows that

$$\sigma_\xi(\rho_\xi) = \tau_\xi.$$

Recall that $B^{\mathbb{C}}(G)$ is the background extender assigned by $\mathbb{C}$ to G. We have

$$E_\xi^{\mathcal{T}^*} = B^{\mathbb{C}_\xi} \circ \rho_\xi \circ \varphi_\xi(E_\xi^{\mathcal{T}}),$$

and

$$E_\xi^{\mathcal{U}^*} = B^{\mathbb{D}_\xi} \circ \tau_\xi \circ \psi_\xi(E_\xi^{\mathcal{U}}).$$

Let $E^* = E_\xi^{\mathcal{T}^*}$ and $F^* = E_\xi^{\mathcal{U}^*}$; then

$$\begin{aligned}
\sigma_\xi(E^*) &= \sigma_\xi(B^{\mathbb{C}_\xi} \circ \rho_\xi \circ \varphi_\xi(E)) \\
&= B^{\mathbb{D}_\xi} \circ \tau_\xi(\sigma_\xi(\varphi_\xi(E))) \\
&= B^{\mathbb{D}_\xi} \circ \tau_\xi \circ \psi_\xi \circ \pi_\xi(E) \\
&= F^*.
\end{aligned}$$

Line 2 comes from the fact that $\sigma_\xi(\rho_\xi) = \tau_\xi$, and line 3 comes from (c). Since $\sigma_\xi(E^*) = F^*$, we get that F^* is the next extender used in $\pi\mathcal{T}^*$, and thus $\pi\mathcal{T}^* \restriction \xi + 2 = \mathcal{U}^* \restriction \xi + 2$. We let $\sigma_{\xi+1}$ be the copy map,

$$\sigma_{\xi+1}([a,f]_{E^*}^{\mathcal{M}_\beta^{\mathcal{T}^*}}) = [\sigma_\xi(a), \sigma_\beta(f)]_{F^*}^{\mathcal{M}_\beta^{\mathcal{U}^*}},$$

where $\beta = T\text{-pred}(\xi+1)$ is the predecessor of $\xi+1$ in all our trees.

We must verify (b) and (c) of $(\dagger)_{\xi+1}$. Suppose first that $\xi+1$ is not a drop in $\mathcal{T}$. It is then not a drop in $\mathcal{U}$ either, so $P_{\xi+1} = i^{\mathcal{T}^*}_{\beta,\xi+1}(P_\beta)$ and $Q_{\xi+1} = i^{\mathcal{U}^*}_{\beta,\xi+1}(Q_\beta)$. But then

$$\begin{aligned}\sigma_{\xi+1}(P_{\xi+1}) &= \sigma_{\xi+1} \circ i^{\mathcal{T}^*}_{\beta,\xi+1}(P_\beta)\\ &= i^{\mathcal{U}^*}_{\beta,\xi+1} \circ \sigma_\beta(P_\beta)\\ &= i^{\mathcal{U}^*}_{\beta,\xi+1}(Q_\beta)\\ &= Q_{\xi+1},\end{aligned}$$

so we have (b). For (c), let us consider the diagram

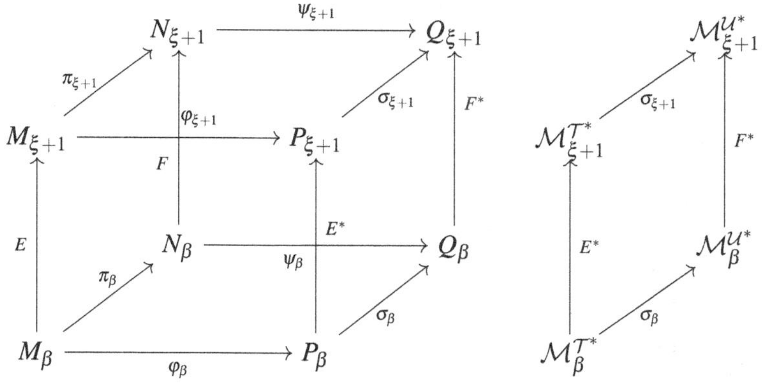

We are asked to show that $\sigma_{\xi+1} \circ \varphi_{\xi+1} = \psi_{\xi+1} \circ \pi_{\xi+1}$, that is, that the rectangle on the top face of the cube commutes. We are given that all other faces of the cube commute, so we have that $\sigma_{\xi+1} \circ \varphi_{\xi+1}$ agrees with $\psi_{\xi+1} \circ \pi_{\xi+1}$ on $\text{ran}(i^{\mathcal{T}}_{\beta,\xi+1})$. Since $M_{\xi+1}$ is generated by $\text{ran}(i^{\mathcal{T}}_{\beta,\xi+1}) \cup \varphi(E)$, it is enough to show that $\sigma_{\xi+1} \circ \varphi_{\xi+1}$ agrees with $\psi_{\xi+1} \circ \pi_{\xi+1}$ on $\varphi(E)$. But on $\varphi(E)$, $\sigma_{\xi+1} \circ \varphi_{\xi+1}$ agrees with $\sigma_\xi \circ \varphi_\xi$ and $\psi_{\xi+1} \circ \pi_{\xi+1}$ agrees with $\psi_\xi \circ \pi_\xi$, by the Shift Lemma. Hence our induction hypothesis $(\dagger)_\xi(\text{c})$ gives us what we want.

The case that $\mathcal{T}$ drops at $\xi+1$ is similar. Suppose the drop is to $J \lhd \mathcal{M}_\beta$. Then since π_β is elementary, $K = \pi_\beta(J)$ is what $\mathcal{U}$ drops to at $\xi+1$. Let $L = \varphi_\beta(J)$ and $N = \psi_\beta(K)$. Thus $\sigma_\beta(L) = N$ by $(\dagger)_\beta(\text{c})$. To get to $P_{\xi+1}$ and $Q_{\xi+1}$ we must resurrect our drop. Let $Y = \text{Res}^{\mathbb{C}_\beta}_{P_\beta}[L])$ and $t = \sigma_{P_\beta}[L]^{\mathbb{C}_\beta}$. Similarly, let

$Z = \mathrm{Res}^{\mathbb{D}_\beta}_{Q_\beta}[N])$ and $u = \sigma_{Q_\beta}[N]^{\mathbb{D}_\beta}$. From the definition of a conversion system, we see that

$$P_{\xi+1} = i^{\mathcal{T}^*}_{\beta,\xi+1}(Y)$$

and

$$Q_{\xi+1} = i^{\mathcal{U}^*}_{\beta,\xi+1}(Z).$$

But $\sigma_\beta(L) = N$, so $\sigma_\beta(Y) = Z$ by elementarity, so $\sigma_{\xi+1}(P_{\xi+1}) = Q_{\xi+1}$. This gives us (b) of $(\dagger)_{\xi+1}$. The reader can easily check (c) using a diagram like the one above. Note here that $\sigma_\beta(t) = u$.

Now let us consider the general case, when π is only nearly elementary. Let $\mathcal{U} = \pi\mathcal{T}^+$, with models $N_\xi = \mathcal{M}^{\mathcal{U}}_\xi$. Now our copy maps have the form

$$\pi_\xi : M_\xi \to J_\xi \trianglelefteq N_\xi.$$

By Lemma 4.5.22, π_ξ is nearly elementary, and either $J_\xi = N_\xi$, or $J_\xi = N^-_\xi$ and π_ξ is exact. Let us keep the notation for lift$(\mathcal{T}, c)$ and lift$(\mathcal{U}, d)$ that we had. The general version of our induction hypothesis is captured by the diagram

$$\begin{array}{ccccccc} N_\xi & \xrightarrow{\psi_\xi} & Q_\xi & \overset{\tau_\xi}{\dashrightarrow} & L_\xi & \qquad & S_\xi \\ {\scriptstyle \pi_\xi}\uparrow & & {\scriptstyle \sigma_\xi}\uparrow & & {\scriptstyle \sigma_\xi}\uparrow & & \uparrow{\scriptstyle \sigma_\xi} \\ M_\xi & \longrightarrow & K_\xi & \longrightarrow & P_\xi & & R_\xi \end{array}$$

(with a curved arrow φ_ξ from M_ξ to P_ξ)

Here we have suppressed J_ξ; it can be that π_ξ is only nearly elementary as a map from M_ξ to N^-_ξ. This happens iff $k(N_\xi) = k(M_\xi) + 1$. In that case, letting $k = k(M_\xi)$, we set $K_\xi = \mathfrak{C}_{k+1}(P_\xi)$ and we have $\sigma_\xi(K_\xi) = Q_\xi$ and the diagram above, with $L_\xi = \sigma_\xi(P_\xi)$. If $k(M_\xi) = k(N_\xi)$, then $K_\xi = P_\xi$ and $Q_\xi = L_\xi$, and we have the diagram from before.

So the general version of $(\dagger)_\xi$ is

(a) $\mathcal{U}^* \restriction \xi + 1 = \sigma\mathcal{T}^* \restriction \xi + 1$, and for all $\alpha \leq \xi$, σ_α is the associated copy map.
(b) If $k(N_\xi) = k(M_\xi)$, then
  (i) $\sigma_\xi(P_\xi) = Q_\xi$, and
  (ii) $\sigma_\xi \circ \varphi_\xi = \psi_\xi \circ \pi_\xi$.
(c) If $k(N_\xi) = k(M_\xi) + 1$, then
  (i) $\sigma_\xi(\mathfrak{C}(P_\xi)) = Q_\xi$, and
  (ii) $\sigma_\xi \circ \varphi_\xi = \tau \circ \psi_\xi \circ \pi_\xi$, where $\tau\colon Q^-_\xi \to \sigma_\xi(P_\xi)$ is the anticore map.

The need for clause (c) in $(\dagger_\xi)$ arises as follows. Suppose we are at a successor step $\xi + 1$, where T-pred$(\xi + 1) = \beta$. Let $E = E^{\mathcal{T}}_\xi$ and $F = E^{\mathcal{U}}_\xi$, and suppose we are dropping in $\mathcal{T}$ at $\xi + 1$, so that

$$M_{\xi+1} = \mathrm{Ult}(J, E),$$

where $J \lhd M_\beta$. If $J \lhd M_\beta^-$ then F is applied to $\pi_\beta(J)$ in $\mathcal{U}$ because $\pi_\beta(\rho(J)) = \rho(\pi_\beta(J))$. In this case we can proceed as before. If $J = M_\beta^-$ and $k(N_\beta) = k(M_\beta) + 1$, then $\pi_\beta(\rho(J)) = \rho(\pi_\beta(J))$ because π_β is exact, so once again F is applied to $\pi_\beta(J)$ in $\mathcal{U}$, and we can proceed as before. We are left with the case

$$J = M_\beta^-$$

and $k(M_\beta) = k(N_\beta)$. In this case $\pi_\beta(\rho(J)) < \text{crit}(F) < \rho(N_\beta^-)$ is possible, so that

$$N_{\xi+1} = \text{Ult}(N_\beta, F).$$

If we now trace through the relevant diagram, we see that it leads to $(\dagger)_{\xi+1}(\text{c})$. Letting $P_\beta = \mathfrak{C}(X)$, we shall have $P_{\xi+1} = i_{E^*}(X)$ and $Q_{\xi+1} = i_{F^*}(Q_\beta)$, so $Q_{\xi+1}$ will be the core of $\sigma_{\xi+1}(P_{\xi+1})$. Here is the diagram.

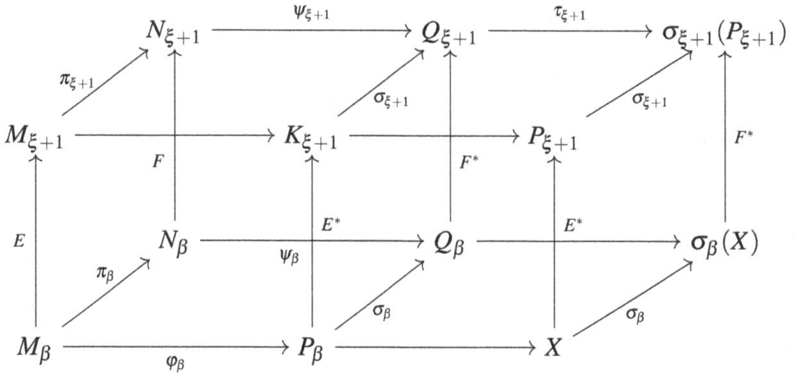

⊣

Remark 5.1.2. Section 5.4 extends the argument at the end of the proof.

We get at once a coarse condensation theorem for induced strategies.

COROLLARY 5.1.3. *Let R and S be transitive models of* ZFC, $R \models$ "$\mathbb{C}$ *is a PFS construction", and let* $\sigma : R \to S$ *be elementary with* $\sigma(\mathbb{C}) = \mathbb{D}$. *Let* $P \unlhd M \in \text{lev}(\mathbb{C})$ *and* $\langle Q, N\rangle = \sigma(\langle P, M\rangle)$. *Let* Σ *be a* (λ, θ)*-iteration strategy for* $(S, w^{\mathbb{D}}, \mathcal{F}^{\mathbb{D}})$*; then*

$$\Omega(\mathbb{C}, M, \Sigma^\sigma)_P = (\Omega(\mathbb{D}, N, \Sigma)_Q)^\sigma.$$

PROOF. Let $\varphi = \sigma_M[P]^{\mathbb{C}}$ and $\psi = \sigma_N[Q]^{\mathbb{D}}$, so that $\sigma(\varphi) = \psi$. The initial conversion stage in the definition of $\Omega(\mathbb{C}, M, \Sigma^\sigma)_P$ is

$$c = \langle P, \varphi, \text{Res}_M[P]^{\mathbb{C}}, \mathbb{C}, R\rangle,$$

and the initial stage in the definition of $\Omega(\mathbb{D}, N, \Sigma)_Q$ is

$$d = \langle Q, \psi, \text{Res}_N[Q]^{\mathbb{D}}, \mathbb{D}, S\rangle.$$

We apply Theorem 5.1.1 with $\pi = \sigma \restriction P$. We get that for any plus tree $\mathcal{T}$ on P,

$$\sigma\,\text{lift}(c, \mathcal{T})_0 = \text{lift}(d, \sigma\mathcal{T})_0,$$

so $\text{lift}(c,\mathcal{T})_0$ is by Σ^σ iff $\text{lift}(d,\sigma\mathcal{T})_0$ is by Σ, that is, $\mathcal{T}$ is by $\Omega(c,\Sigma^\sigma)$ iff $\sigma\mathcal{T}$ is by $\Omega(d,\Sigma)$. The argument easily extends to stacks of plus trees. ⊣

Another elementary consequence is

COROLLARY 5.1.4. *Let $\langle M,\pi,N,\mathbb{C},R\rangle$ be a conversion stage, and Σ be a (λ,θ)-iteration strategy for $(R,w^{\mathbb{C}},\mathcal{F}^{\mathbb{C}})$; then*

$$\Omega(\langle M,\pi,N,\mathbb{C},R\rangle,\Sigma) = \Omega(\mathbb{C},N,\Sigma)^\pi.$$

PROOF. Let $c = \langle M,\pi,N,\mathbb{C},R\rangle$ and $d = \langle N,\text{id},N,\mathbb{C},R\rangle$. We apply Theorem 5.1.1 with $R = S$ and $\sigma = \text{id}$. We get that for any plus tree $\mathcal{T}$ on M, $\text{lift}(c,\mathcal{T})_0 = \text{lift}(d,\pi\mathcal{T}^+)_0$, so $\mathcal{T}$ is by $\Omega(\langle M,\pi,N,\mathbb{C},R\rangle,\Sigma)$ iff $\pi\mathcal{T}^+$ is by $\Omega(\mathbb{C},N,\Sigma)$. The argument easily extends to stacks of plus trees. ⊣

Thus if Σ is an iteration strategy for some model R of ZFC, and $\mathbb{C}$ is a maximal PFS construction in the sense of R, then the strategies Σ induces via $\mathbb{C}$ are the strategies it induces for the levels of $\mathbb{C}$, together with their pullbacks. Pretty much all regularity properties of iteration strategies are preserved under pullbacks, so if they hold for all strategies of the form $\Omega(\mathbb{C},M,\Sigma)$, where $M \in \text{lev}(\mathbb{C})$, then they hold for all strategies induced by Σ and $\mathbb{C}$.

5.2. Positionality and strategy coherence

Let us define positionality in our new context.

DEFINITION 5.2.1. Let M be a pfs premouse, and Ω be a complete (λ,θ) iteration strategy, then Ω is *positional* iff whenever s and t are M-stacks by Ω of length $< \lambda$, and $N \trianglelefteq M_\infty(s)$ and $N \trianglelefteq M_\infty(t)$, then $\Omega_{s,N} = \Omega_{t,N}$.

We are equipped now to prove some instances of positionality. For example, there is the trivial instance we discussed in Section 3.6.

PROPOSITION 5.2.2. *Let $c = \langle M,\psi,K,\mathbb{C},S\rangle$ be a conversion stage, and let Σ^* be a (λ,θ) iteration strategy for $(S,w^{\mathbb{C}},\mathcal{F}^{\mathbb{C}})$. Let $P \trianglelefteq N \trianglelefteq M$; then $(\Omega(c,\Sigma^*)_N)_P = \Omega(c,\Sigma^*)_P$.*

PROOF. Suppose first that $M = K$ and $\psi = \text{id}$. By Lemma 4.7.9, the resurrections of N and P from M in $\mathbb{C}$ are consistent. So letting $Q = \text{Res}_M[N]$ and $\sigma = \sigma_M[N]$, and $R = \text{Res}_Q[\sigma(P)]$ and $\tau = \sigma_Q[\sigma(P)]$, we have that $R = \text{Res}_M[P]$ and $\tau \circ \sigma = \sigma_M[P]$. Setting $\Omega = \Omega(\mathbb{C},M,\Sigma^*)$, we get

$$\Omega_P = (\Omega_R)^{\tau\circ\sigma} = ((\Omega_R)^\tau)^\sigma = (\Omega_Q)^\sigma = \Omega_N.$$

Here $\Omega_N = \Omega_{\emptyset,N} = (\Omega_Q)^\sigma$ by our definition of induced tail strategies, and similarly for the others.

In the general case, $\Omega(c,\Sigma^*) = \Lambda^\psi$, where $\Lambda = \Omega(\mathbb{C},K,\Sigma^*)$. Since $(\Lambda_{\psi(N)})_{\psi(P)} = \Lambda_{\psi(P)}$, we get by copying empty trees that $(\Omega(c,\Sigma^*)_N)_P = \Omega(c,\Sigma^*)_P$. ⊣

Again, we know of no direct proof of this proposition for the premice and constructions of Chapter 3. It seems like the sort of simple fact whose proof ought to be routine. In the context of pfs premice and constructions, that is true.

Strategy coherence is a more useful consequence of positionality. For coarse strategies, the definition is

DEFINITION 5.2.3. Let $(R, w, \mathcal{F})$ be a coarse extender premouse, and Σ be a (λ, θ) iteration strategy for $(R, w, \mathcal{F})$; then $((R, w, \mathcal{F}), \Sigma)$ is *strategy coherent* iff whenever $s^\frown\langle\mathcal{T}\rangle$ and $s^\frown\langle\mathcal{U}\rangle$ are stacks by Σ, and N is an initial segment of both last models, then $\Sigma_{s^\frown\langle\mathcal{T}\rangle,N} = \Sigma_{s^\frown\langle\mathcal{U}\rangle,N}$.

It is clear that if Σ witnesses the strong unique iterability of $(R, w, \mathcal{F})$, then $((R, w, \mathcal{F}), \Sigma)$ is strategy coherent.

If $((R, w^{\mathbb{C}}, \mathcal{F}^{\mathbb{C}}), \Sigma)$ is strategy coherent, where $\mathbb{C}$ is a maximal PFS construction of R, then the strategies for levels of $\mathbb{C}$ induced by Σ are induced locally, in the following sense. Let M be the last model of $\mathbb{C}\restriction\nu$, and let α be an inaccessible cardinal of R such that $\mathbb{C}\restriction\nu \subseteq V_\alpha^R$. Let $S = V_\alpha^R$; then

$$\Omega(\mathbb{C}, M, \Sigma) = \Omega(\mathbb{C}\restriction\xi, M, \Sigma_{(S, w^{\mathbb{C}}\cap S, \mathcal{F}^{\mathbb{C}}\cap S)}).$$

This simple fact will play a role in various arguments to come.

The natural fine structural form of strategy coherence is

DEFINITION 5.2.4. Let M be a pfs premouse and Σ be a complete iteration strategy for M; then (M, Σ) is *strategy coherent* iff whenever $s^\frown\langle\mathcal{T}\rangle$ and $s^\frown\langle\mathcal{U}\rangle$ are stacks by Σ, and N is an initial segment of both last models, then $\Sigma_{s^\frown\langle\mathcal{T}\rangle,N} = \Sigma_{s^\frown\langle\mathcal{U}\rangle,N}$.

Notice that in both the fine and coarse cases, $\mathcal{T}$ and $\mathcal{U}$ are quasi-normal, so they have a common initial segment $\mathcal{W}$ such that $N \trianglelefteq \mathcal{M}_\infty^{\mathcal{W}}$. So strategy coherence is equivalent to the assertion that if $s^\frown\langle\mathcal{T}\rangle$ is by Σ, and $N \trianglelefteq \mathcal{M}_\infty^{\mathcal{T}}$, and $o(N) \leq \mathrm{lh}(E_\alpha^{\mathcal{T}})$, then $\Sigma_{s^\frown\langle\mathcal{T}\restriction\alpha+1\rangle,N} = \Sigma_{s^\frown\langle\mathcal{T}\rangle,N}$.

Ultimately we shall show that if Σ^* is a strongly unique iteration strategy for V, and $\Sigma = \Omega(\mathbb{C}, M, \Sigma^*)$, then (M, Σ) is strategy coherent. The background coherence problem prevents us from showing this now, but we can prove the following approximation.

THEOREM 5.2.5. *Let $c = \langle M, \pi, Q, \mathbb{C}, R\rangle$ be a conversion stage, let Σ^* be a strongly unique (λ, θ) iteration strategy for $(R, w^{\mathbb{C}}, \mathcal{F}^{\mathbb{C}})$, and let $\Sigma = \Omega(c, \Sigma^*)$. Let $\mathcal{T}$ be a plus tree on M by Σ and let N be an initial segment of its last model. Let $\nu + 1 < \mathrm{lh}(\mathcal{T})$, and suppose that either*

(a) $o(N) < \hat{\lambda}(E_\nu^{\mathcal{T}})$, or

(b) $E_\nu^{\mathcal{T}}$ is of plus type, and $o(N) \leq \mathrm{lh}(E_\nu^{\mathcal{T}})$;

then $\Sigma_{\mathcal{T}\restriction\nu+1,N} = \Sigma_{\mathcal{T},N}$.

PROOF. For $\eta < \mathrm{lh}(\mathcal{T})$, let $c_\eta = \mathrm{stg}(\mathcal{T}, c, \eta) = \langle M_\eta, \psi_\eta, Q_\eta, \mathbb{C}_\eta, R_\eta\rangle$. Let $\gamma + 1 = \mathrm{lh}(\mathcal{T})$. We need to see that c_ν and c_γ induce the same iteration strategy for

N. For this we use the agreement properties (3),(4), and (5) in the definition of conversion systems. Our hypothesis on $o(N)$ guarantees that they suffice.

Adopting the notation of Section 4.8, let

$$\begin{aligned} E &= (E_\nu^{\mathcal{T}})^-, & H &= \psi_\nu(E), \\ X &= Q_\nu | \mathrm{lh}(H), & G &= \sigma_{Q_\nu}[X](H), \\ Y &= \mathrm{Res}_{Q_\nu}[X], & G^* &= B^{\mathbb{C}_\nu}(G). \end{aligned}$$

(The resurrections are in $\mathbb{C}_\nu$, of course.) Let ξ be the unique η such that $Y = M_{\eta,0}^{\mathbb{C}_\nu}$. We have that $V_{\mathrm{lh}(G^*)}^{R_\nu} = V_{\mathrm{lh}(G^*)}^{R_\gamma}$ and $\mathbb{C}_\nu \restriction \xi = \mathbb{C}_\gamma \restriction \xi$, moreover $Y || o(Y)$ is the last model of $\mathbb{C}_\nu \restriction \xi$. Let

$$(S, \mathcal{G}) = (V_{\mathrm{lh}(G^*)}^{R_\nu}, \vec{F}^{\mathbb{C}_\nu \restriction \xi}) = (V_{\mathrm{lh}(G^*)}^{R_\gamma}, \vec{F}^{\mathbb{C}_\gamma \restriction \xi});$$

then $\Sigma^*_{\mathcal{T}^* \restriction \nu+1,(S,\mathcal{G})} = \Sigma^*_{\mathcal{T}^* \restriction \gamma+1,(S,\mathcal{G})}$ because (R, Σ^*) is strategy coherent.

This is the agreement of background strategies we need; let us calculate the agreement of lifting maps. Set $N_1 = \psi_\nu(N)$, $N_2 = \mathrm{Res}_\nu(N_1) = \mathrm{Res}_{Q_\nu, Y}[N_1]$, and $N_3 = \mathrm{Res}_Y[N_2]$. The resurrections are all in $\mathbb{C}_\nu \restriction \xi$ because $o(N) < \mathrm{lh}(E)$. Let $\sigma_1 = \sigma_{Q_\nu, Y}[N_1]$ and $\sigma_2 = \sigma_Y[N_2]$ be the associated resurrection maps. We have the diagram

$$\begin{array}{ccccc} N & \xrightarrow{\psi_\nu} & N_1 & \xrightarrow{\qquad\sigma_3\qquad} & N_3 \\ & & & \searrow^{\sigma_1} \quad {}^{\sigma_2}\nearrow & \\ & & & N_2 & \end{array}$$

Here $\sigma_3 = \sigma_{Q_\nu}[N_1] = \sigma_2 \circ \sigma_1$. All resurrections here are in the sense of $\mathbb{C}_\nu$. The resurrection map σ_2 is determined by $\mathbb{C}_\nu \restriction \xi$, so $\sigma_2 = \sigma_Y[N_2]^{\mathbb{C}_\gamma}$. But $o(Y)$ is a cardinal of Q_γ and $o(Y) \leq \rho^-(Q_\gamma)$, so by Lemma 4.7.10, $\sigma_2 = \sigma_{Q_\gamma}[N_2]^{\mathbb{C}_\gamma}$. Our hypothesis (2) on $o(N)$ guarantees that

$$\mathrm{res}_\nu \circ \psi_\nu \restriction o(N) + 1 = \psi_\gamma \restriction o(N) + 1,$$

so $N_2 = \psi_\gamma(N)$. Thus we can calculate

$$\begin{aligned} \Sigma_{\mathcal{T} \restriction \nu+1,N} &= \Omega(N, \sigma_3 \circ \psi_\nu, N_3, \mathbb{C}_\nu \restriction \xi, S, \Sigma^*_{\mathcal{T}^* \restriction \nu+1,(S,\mathcal{G})}) \\ &= \Omega(N, \sigma_2 \circ \mathrm{res}_\nu \circ \psi_\nu, N_3, \mathbb{C}_\nu \restriction \xi, S, \Sigma^*_{\mathcal{T}^* \restriction \nu+1,(S,\mathcal{G})}) \\ &= \Omega(N, \sigma_2 \circ \psi_\gamma, N_3, \mathbb{C}_\gamma \restriction \xi, S, \Sigma^*_{\mathcal{T}^* \restriction \gamma+1,(S,\mathcal{G})}) \\ &= \Sigma_{\mathcal{T} \restriction \gamma+1,N}, \end{aligned}$$

as desired. $\dashv$

From this we get strategy coherence within plus trees that are λ-separated.

COROLLARY 5.2.6. *Let* $\Sigma = \Omega(M, \pi, Q, \mathbb{C}, R, \Sigma^*)$, *where* Σ^* *is a strongly unique* (λ, θ) *iteration strategy for* $(R, w^{\mathbb{C}}, \mathcal{F}^{\mathbb{C}})$. *Let* $s^\frown\langle\mathcal{T}\rangle$ *and* $s^\frown\langle\mathcal{U}\rangle$ *are stacks by* Σ, *and* N *is an initial segment of both last models. Suppose that* $\mathcal{T}$ *and* $\mathcal{U}$ *are* λ*-separated; then* $\Sigma_{s^\frown\langle\mathcal{T}\rangle,N} = \Sigma_{s^\frown\langle\mathcal{U}\rangle,N}$.

PROOF. We may assume $s = \emptyset$ without loss of generality. Let ν be least such that $N \trianglelefteq \mathcal{M}^{\mathcal{T}}_{\nu}$; then $\mathcal{T} \restriction \nu+1 = \mathcal{U} \restriction \nu+1$ by the uniqueness of normal iterations by a fixed strategy. (Recall that λ-separated trees are length-increasing.) By the symmetry of the situation, it will suffice to show that $\Sigma_{\mathcal{T}\restriction\nu+1,N} = \Sigma_{\mathcal{T},N}$. If $\nu+1 = \mathrm{lh}(\mathcal{T})$ this is true. If $\nu+1 < \mathrm{lh}(\mathcal{T})$, then $o(N) \le \mathrm{lh}(E^{\mathcal{T}}_{\nu})$, since otherwise $E^{\mathcal{T}}_{\nu}$ is on the N sequence, so $N \ntrianglelefteq \mathcal{M}^{\mathcal{T}}_{\infty}$. Since $E^{\mathcal{T}}_{\nu}$ has plus type, Theorem 5.2.5(b) then tells us that $\Sigma_{\mathcal{T}\restriction\nu+1,N} = \Sigma_{\mathcal{T},N}$. ⊣

This corollary and stronger results along its lines are the reason we are giving special attention to λ-separated trees.

5.3. Pullback consistency

Roughly speaking, an iteration strategy is *pullback consistent* if it pulls back to itself under its own iteration maps. We shall show that any background-induced strategy is pullback consistent, provided that the strategy inducing it is pullback consistent. This is a warm-up for Chapter 7, where we shall prove stronger results along the same lines.

DEFINITION 5.3.1. Let Ω be a complete (λ, θ) iteration strategy for a premouse M. We say that Ω is *pullback consistent* iff whenever $s^\frown\langle P, \mathcal{T}\rangle$ is an M-stack by Ω, $\alpha <_T \beta$, $K \trianglelefteq \mathcal{M}^{\mathcal{T}}_{\alpha}$, and $L = \hat{\imath}^{\mathcal{T}}_{\alpha,\beta}(K)$, then

$$\Omega_{s^\frown\langle P,\mathcal{T}\restriction\alpha+1\rangle,K} = (\Omega_{s^\frown\langle P,\mathcal{T}\restriction\beta+1\rangle,L})^{\hat{\imath}^{\mathcal{T}}_{\alpha,\beta}}.$$

The definition applies even if there are drops along the branch of $\mathcal{T}$ from α to β, so long as K is in the domain of the partial iteration map $\hat{\imath} = \hat{\imath}^{\mathcal{T}}_{\alpha,\beta}$. Indeed $K = \mathrm{dom}(\hat{\imath})$ is possible, in which case $L = \mathcal{M}^{\mathcal{T}}_{\beta}$.[1]

Pullback consistency for the iteration strategies described in Chapter 3 cannot be proved directly, so far as we can see. For example, let $\Omega = \Omega(\mathbb{C}, M, \Sigma^*)$, where $\mathbb{C}$ is a construction in the sense of Chapter 3, and suppose $P \triangleleft N \triangleleft M$. We can think Ω_P as the pullback of $(\Omega_N)_P$ under the identity map, which is indeed the iteration map associated to the empty tree on N. So $\Omega_P = (\Omega_N)_P$ is an instance of pullback

[1]The branch embeddings in an M-stack are elementary, so pullback consistency only concerns pullbacks under elementary maps. The possible difference between $\pi\mathcal{T}$ and $\pi\mathcal{T}^+$ when π is only nearly elementary is not relevant.

consistency. But as we saw in Section 3.6, the attempt to prove $\Omega_P = (\Omega_N)_P$ directly is blocked by the possibility of resurrection inconsistencies.[2]

We have stated pullback consistency for pullbacks within a single normal tree $\mathcal{T}$, but this implies we can pull back consistently from one normal tree in a stack into any previous one, step by step. This is simply because $\Omega^{i \circ j} = (\Omega^i)^j$.

As one might guess, pullback consistency passes from Ω to its pullbacks.

LEMMA 5.3.2. *Let $\pi \colon M \to N$ be nearly elementary, and let Ω be a pullback consistent iteration strategy for N; then Ω^π is pullback consistent.*

PROOF. Let $s^\frown \langle P, \mathcal{T} \rangle$ be an M-stack by Ω^π, and $\alpha <_T \beta$. To simplify the notation, let us assume $s = \emptyset$ and $P = M$; the general case is no different. Let

$$i = \hat{\imath}^{\mathcal{T}}_{\alpha,\beta},$$

and let $K \trianglelefteq \mathcal{M}^{\mathcal{T}}_\alpha$, and

$$i(K) = L.$$

We must see that $\Omega^\pi_{\mathcal{T} \restriction \alpha+1, K} = (\Omega^\pi_{\mathcal{T} \restriction \beta+1, L})^i$.

Let $\mathcal{U} = \pi \mathcal{T}$, let $N_\xi = \mathcal{M}^{\mathcal{U}}_\xi$, and let $\pi_\xi : M_\xi \to N_\xi$ be the copy map. π_ξ is nearly elementary. Let $R = \pi_\alpha(K)$, $j = \hat{\imath}^{\mathcal{U}}_{\alpha,\beta}$, and $S = j(R)$. Here if $K = \mathcal{M}^{\mathcal{T}}_\alpha \downarrow n$, then $R = \mathcal{M}^{\mathcal{U}}_\alpha \downarrow n$, as usual. Similarly, $\hat{R} = \operatorname{dom}(j)$ is possible.

Tracing through the appropriate diagram, we get

$$\begin{aligned} \Omega^\pi_{\mathcal{T} \restriction \alpha+1, K} &= (\Omega_{\mathcal{U} \restriction \alpha+1, R})^{\pi_\alpha} \\ &= (\Omega_{\mathcal{U} \restriction \beta+1, S})^{j \circ \pi_\alpha} \\ &= (\Omega_{\mathcal{U} \restriction \beta+1, S})^{\pi_\beta \circ i} \\ &= (\Omega^\pi_{\mathcal{T} \restriction \beta+1, L})^i, \end{aligned}$$

as desired. Lines 1 and 4 come from the definition of Ω^π. Line 2 comes from the pullback consistency of Ω, and line 3 from the fact that $j \circ \pi_\alpha = \pi_\beta \circ i$. ⊣

Pullback consistency also makes sense for coarse iteration strategies. Like the other regularity properties for coarse strategies we shall consider, it holds for strategies that witness strong unique iterability.

THEOREM 5.3.3. *Let $(N, \in, w, \mathcal{F})$ be a coarse extender premouse, and suppose that Σ witnesses that N is strongly uniquely $(\lambda, \theta, \mathcal{F})$-iterable; then Σ is pullback consistent.*

PROOF. Let $s^\frown \langle P, \mathcal{T} \rangle$ be by Σ. Let $\alpha <_T \beta$, $Q = \mathcal{M}^{\mathcal{T}}_\alpha$, $R = \mathcal{M}^{\mathcal{T}}_\beta$, and $\pi = i^{\mathcal{T}}_{\alpha,\beta}$; we must show that $\Sigma_{s^\frown \langle \mathcal{T} \restriction \alpha+1 \rangle, Q} = (\Sigma_{s^\frown \langle \mathcal{T} \restriction \beta+1 \rangle, R})^\pi$.

---

[2]Pullback consistency in the non-dropping case can be proved directly for the induced strategies for Jensen premice described in Chapter 3. In the example just given, the empty tree from N to P has a drop along its (trivial) main branch.

But if $\mathcal{U}$ is by $(\Sigma_{s^\frown\langle \mathcal{T}\restriction\beta+1\rangle,R})^\pi$, then its models are wellfounded, because they embed by copy maps into models of $\pi\mathcal{U}$, and these are wellfounded. Since $\Sigma_{s^\frown\langle \mathcal{T}\restriction\alpha+1\rangle,Q}$ chooses unique wellfounded branches, $\mathcal{U}$ must be by $\Sigma_{s^\frown\langle \mathcal{T}\restriction\alpha+1\rangle,Q}$, as desired. ⊣

The proof that pullback consistency passes to pullback strategies shows that it passes to background-induced strategies.

THEOREM 5.3.4. *Let $c = \langle M,\psi,Q,\mathbb{C},R\rangle$ be a conversion stage, and suppose that Σ^* is a pullback consistent (λ,θ) iteration strategy for $(R,\in w,\mathcal{F}^{\mathbb{C}})$; then $\Omega(c,\Sigma^*)$ is pullback consistent.*

PROOF. Let $\Sigma = \Omega(c,\Sigma^*)$. For notational simplicity, we consider pullbacks within a plus tree $\mathcal{T}$ by Σ. The argument we give applies equally well to $\mathcal{T}$ that are by a tail strategy $\Sigma_{s,P}$. Suppose that $\mathcal{T}$ is a plus tree on M and $\alpha <_T \beta$. Let $i = \hat{\imath}^{\mathcal{T}}_{\alpha,\beta}$, and suppose that $i(K) = L$ (with the usual understandings if $\hat{K} = \mathrm{dom}(i)$.) We must see that $\Sigma_{\mathcal{T}\restriction\alpha+1,K} = (\Sigma_{\mathcal{T}\restriction\beta+1,L})^i$.

Let $\mathrm{lift}(\mathcal{T},c)_0 = \mathcal{T}^*$ and $\mathrm{stg}(\mathcal{T},c,\eta) = \langle M_\eta,\psi_\eta,Q_\eta,\mathbb{C}_\eta,R_\eta\rangle$. Letting $\gamma+1 \leq_T \beta$ be such that $T\text{-pred}(\gamma+1) = \alpha$, we may assume without loss of generality that $D^{\mathcal{T}} \cap (\gamma+1,\beta]_T = \emptyset$. (If there is more than one drop between α and β, we then just pull back successively across one drop at a time.) Let $J = \mathcal{M}^{*,\mathcal{T}}_{\gamma+1}$; then $K \trianglelefteq J = \mathrm{dom}(i) \trianglelefteq M_\alpha$.

Let

- $J_1 = \psi_\alpha(J)$ and $K_1 = \psi_\alpha(K)$,
- $J_2 = \mathrm{Res}_{Q_\alpha}[J_1]^{\mathbb{C}_\alpha}$,
- $\sigma = \sigma_{Q_\alpha}[J_1]^{\mathbb{C}_\alpha}$ and $K_2 = \sigma(K_1)$, and
- $L_1 = \psi_\beta(L)$.

Here is a diagram of the situation.

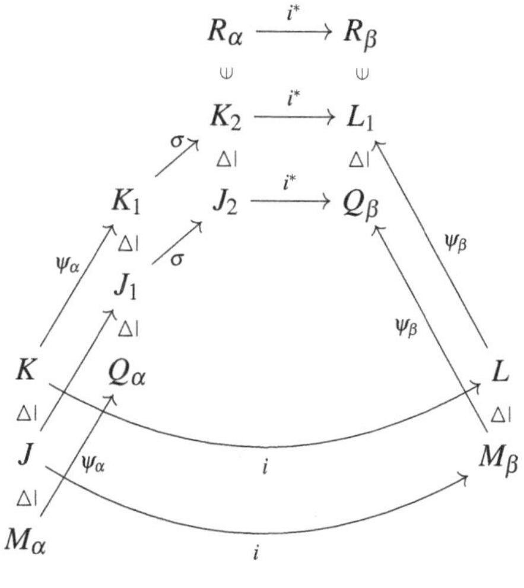

Here $i^* = i^{\mathcal{T}^*}_{\alpha,\beta}$. We get the desired conclusion by pulling back the induced strategy for L_1 to K, along $\psi_\beta \circ i$ and $i^* \circ \sigma \circ \psi_\alpha$. Let

$$\Lambda = \Omega(\mathbb{C}_\beta, Q_\beta, R_\beta, \Sigma^*_{\mathcal{T}^* \restriction \beta+1})_{L_1},$$
$$\Gamma = \Omega(\mathbb{C}_\alpha, J_2, R_\alpha, \Sigma^*_{\mathcal{T}^* \restriction \alpha+1})_{K_2}.$$

Since Σ^* is pullback consistent, $\Sigma^*_{\mathcal{T}^* \restriction \alpha+1}$ is the pullback of $\Sigma^*_{\mathcal{T}^* \restriction \beta+1}$ under i^*, so by Corollary 5.1.3, $\Gamma = \Lambda^{i^*}$. We can therefore calculate

$$\Sigma_{\mathcal{T} \restriction \alpha+1,K} = \Gamma^{\sigma \circ \psi_\alpha} = (\Lambda^{i^*})^{\sigma \circ \psi_\beta} = \Lambda^{i^* \circ \sigma \circ \psi_\alpha} = \Lambda^{\psi_\beta \circ i} = (\Lambda^{\psi_\beta})^i = (\Sigma_{\mathcal{T} \restriction \beta+1,L})^i,$$

as desired. ⊣

5.4. Internal lift consistency

Given $\pi\colon P \to R$ nearly elementary, we can copy a P-stack s to an R-stack πs, until we reach an illfounded model on the πs side. In Section 4.5 we extended the copying construction slightly, so as to allow stronger ultrapowers on the R side than the copied ones. This leads to a more general way to pull back iteration strategies.

DEFINITION 5.4.1. Suppose that $\pi\colon P \to Q|\langle \nu, k\rangle$ is nearly elementary and Ω is a strategy for Q defined on plus trees; then $\Omega^{(\pi,\nu,k)}$ is the strategy on plus trees given by pulling back:

$$\Omega^{(\pi,\nu,k)}(\mathcal{T}) = \Omega((\pi\mathcal{T})^+).$$

When $P = Q|\langle \nu, k\rangle$ and $\pi = \mathrm{id}$, we write Ω^+_P for $\Omega^{(\pi,\nu,k)}$.

We show now that background-induced strategies pull back to themselves when lifted under the identity map. Again, resurrection consistency plays a role, and we know of no direct proof of the corresponding fact for the induced strategies of Chapter 3. The main lemma we need is that lifting a stack of plus trees from some initial segment of Q to Q itself under the identity map commutes with conversion. In order to prove this for stacks, we need to consider (π, ν, k) lifts where π is not the identity.

LEMMA 5.4.2. *Let $d = \langle Q, \psi, N, \mathbb{C}, R\rangle$ be a conversion stage, $\pi\colon P \to J \lhd Q$ be nearly elementary, and $c = \langle P, \sigma_{\mathrm{N}}[\psi(J)] \circ \psi \circ \pi, \mathrm{Res}_{\mathrm{N}}[\psi(J)], \mathbb{C}, R\rangle$. Suppose that $\mathcal{U}$ is a plus tree on P, and let $\pi\mathcal{U}^+$ be the (π, ν, k)-lift of $\mathcal{U}$ to Q, where $J = Q|\langle \nu, k\rangle$; then*

$$\mathrm{lift}(\mathcal{U}, c)_0 = \mathrm{lift}(\pi\mathcal{U}^+, d)_0.$$

PROOF. Let $\mathcal{V}^* = \mathrm{lift}(\mathcal{U}, c)_0$ and $\mathcal{W}^* = \mathrm{lift}(\pi\mathcal{U}^+, d)_0$. We mean of course that $\mathcal{V}^*$ and $\mathcal{W}^*$ are the same up to and at the first point, if one exists, that they reach an illfounded model. Let also

$$\operatorname{stg}(\mathcal{U},c,\alpha) = \langle P_\alpha, \theta_\alpha, M_\alpha, \mathbb{C}_\alpha, R_\alpha \rangle,$$
$$\operatorname{stg}(\pi\mathcal{U}^+,d,\alpha) = \langle Q_\alpha, \psi_\alpha, N_\alpha, \mathbb{D}_\alpha, S_\alpha \rangle.$$

We shall show by induction that $\mathcal{V}^* \restriction \alpha+1 = \mathcal{W}^* \restriction \alpha+1$, and hence $R_\alpha = S_\alpha$ and $\mathbb{C}_\alpha = \mathbb{D}_\alpha$.

We have $P_\alpha = \mathcal{M}^{\mathcal{U}}_\alpha$ and $Q_\alpha = \mathcal{M}^{\pi\mathcal{U}^+}_\alpha$. Let

$$\pi_\alpha \colon P_\alpha \to J_\alpha \trianglelefteq Q_\alpha$$

be the lifting map from the definition of $\pi\mathcal{U}^+$. Thus $J_0 = J$ and $\pi_0 = \pi$. Let $\psi_\alpha(J_\alpha) = K_\alpha \trianglelefteq N_\alpha$.

We shall show that the following diagram commutes:

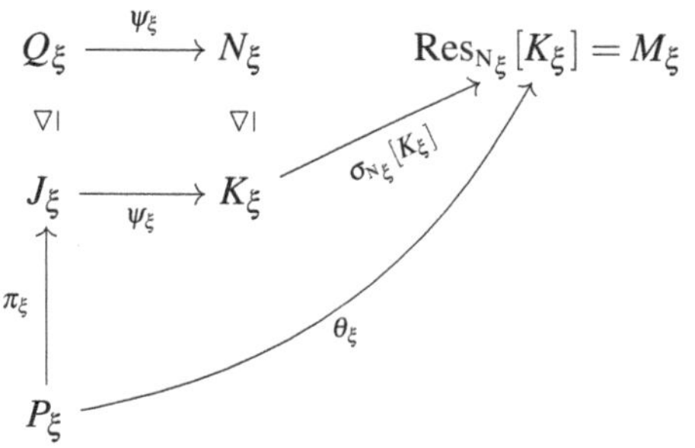

Here the resurrection in the upper right corner is taking place in the construction $\mathbb{C}_\xi = \mathbb{D}_\xi$ of $\mathcal{M}^{\mathcal{W}^*}_\xi = \mathcal{M}^{\mathcal{V}^*}_\xi$. Our induction hypotheses are

(1) $\mathcal{V}^* \restriction \xi+1 = \mathcal{W}^* \restriction \xi+1$,
(2) $M_\xi = \operatorname{Res}_{N_\xi}[K_\xi]$, and
(3) $\theta_\xi = \sigma_{N_\xi}[K_\xi] \circ \psi_\xi \circ \pi_\xi$.

They hold at $\xi = 0$ because we defined M_0 as $\operatorname{Res}_{N_0}[K_0]$ and θ_0 as $\sigma_{N_0}[K_0] \circ \psi_0 \circ \pi$.

Now suppose (1)-(3) hold at all $\eta \le \xi$. Let

$$E = E^{\mathcal{U}}_\xi, \qquad F = \pi_\xi(E) = E^{\pi\mathcal{U}^+}_\xi,$$
$$G = \psi_\xi(F), \qquad H = \theta_\xi(E).$$

Letting $\mathbb{D} = \mathbb{C}_\xi = \mathbb{D}_\xi$ and interpreting the resurrection as being in $\mathbb{D}$, (3) implies that $\sigma_{N_\xi}[K_\xi](G) = H$. Thus

$$\begin{aligned} E^{\mathcal{V}^*}_\xi &= B^{\mathbb{D}} \circ \sigma_{M_\xi}[M_\xi | \operatorname{lh}(H)] \circ \theta_\xi(E) \\ &= B^{\mathbb{D}} \circ \sigma_{M_\xi}[M_\xi | \operatorname{lh}(H)](H) \\ &= B^{\mathbb{D}} \circ \sigma_{M_\xi}[M_\xi | \operatorname{lh}(H)] \circ \sigma_{N_\xi, M_\xi}[K_\xi](G) \\ &= B^{\mathbb{D}} \circ \sigma_{M_\xi}[M_\xi | \operatorname{lh}(H)] \circ \sigma_{N_\xi, M_\xi}[K_\xi | \operatorname{lh}(G)](G) \\ &= E^{\mathcal{W}^*}_\xi, \end{aligned}$$

so $\mathcal{V}^* \restriction \xi+2 = \mathcal{W}^* \restriction \xi+2$. The step from line 3 to line 4 uses resurrection consistency.

Let us check that (2) and (3) continue to hold at $\xi+1$. Let $\beta = U\text{-pred}(\xi+1)$ and $i^* = i^{\mathcal{V}^*}_{\beta,\xi+1} = i^{\mathcal{W}^*}_{\beta,\xi+1}$, so that $N_{\xi+1} = i^*(N_\beta)$ and $M_{\xi+1} = i^*(M_\beta)$. Let

$$i = \hat{i}^{\mathcal{U}}_{\beta,\xi+1} : P^*_{\xi+1} \to P_{\xi+1}$$

and

$$j = \hat{i}^{\pi\mathcal{U}^+}_{\beta,\xi+1} : Q^*_{\xi+1} \to Q_{\xi+1}$$

be the canonical embeddings. Suppose for simplicity that $\xi+1 \notin D^{\mathcal{U}}$; we leave the dropping case to the reader. Here is the relevant diagram:

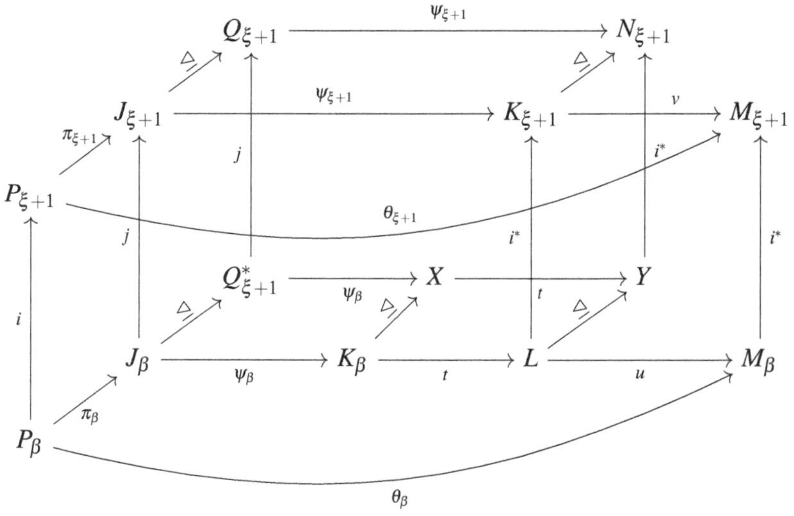

Here $X = \psi_\beta(Q^*_{\xi+1})$, $Y = \mathrm{Res}_{N_\beta}[X]$, $t = \sigma_{N_\beta}[X]$, $L = t(K_\beta)$, and $u = \sigma_Y[L]$, the resurrections being in $\mathbb{C}_\beta$.

We have $K_{\xi+1} = i^*(L)$, and $N_{\xi+1} = i^*(Y)$ by the definition of $\mathrm{lift}(\mathcal{U}^+, d)$. Setting $v = i^*(u)$, this implies that $v = \sigma_{N_{\xi+1}}[K_{\xi+1}]$, the resurrection being in $\mathbb{C}_{\xi+1}$. Thus $M_{\xi+1} = \mathrm{Res}_{N_{\xi+1}}[K_{\xi+1}]$, as desired for (2).

Let us verify that $\theta_{\xi+1} = v \circ \psi_{\xi+1} \circ \pi_{\xi+1}$. We have that $\pi_{\xi+1} \circ i = j \circ \pi_\beta$ by the way lifting from $\mathcal{U}$ to $\pi\mathcal{U}^+$ works, and $i^* \circ t \circ \psi_\beta = \psi_{\xi+1} \circ j$ by the way the conversion of $\pi\mathcal{U}^+$ works. The bottom face of the slab commutes by our induction hypothesis. Thus all parts of the diagram that do not involve $\theta_{\xi+1}$ commute. We have also $\theta_{\xi+1} \circ i = i^* \circ \theta_\beta$ by the way $\mathcal{U}$ converts to $\mathcal{W}^*$. So the front face commutes.

It follows that $\theta_{\xi+1}$ agrees with $v \circ \psi_{\xi+1} \circ \pi_{\xi+1}$ on $\mathrm{ran}(i)$. But $P_{\xi+1}$ is generated from $\varepsilon(E) \cup \mathrm{ran}(i)$, where $\varepsilon(E) = \mathrm{lh}(E)$ if E has plus type, and $\varepsilon(E) = \lambda(E)$ otherwise. So it is enough to show that $\theta_{\xi+1}$ agrees with $v \circ \psi_{\xi+1} \circ \pi_{\xi+1}$ on $\varepsilon(E)$.

To see this, let

$$Y_\xi = \mathrm{Res}_{M_\xi}[M_\xi \,|\, \mathrm{lh}(H)],$$
$$s = \sigma_{M_\xi}[M_\xi \,|\, \mathrm{lh}(H)].$$

s is res_ξ for the lift$(\mathcal{U}, c)$ conversion, and $s \circ \sigma_{N_\xi}[K_\xi]$ is res_ξ for the lift$(\pi\mathcal{U}^+, d)$ conversion. We have

$$\begin{aligned}
\theta_{\xi+1} \restriction \varepsilon(E) &= s \circ \theta_\xi \restriction \varepsilon(E) \\
&= s \circ \sigma_{N_\xi}[K_\xi] \circ \psi_\xi \circ \pi_\xi \restriction \varepsilon(E) \\
&= \mathrm{res}_\xi^{\mathrm{lift}(\pi\mathcal{U}^+,d)} \circ \psi_\xi \circ \pi_\xi \restriction \varepsilon(E) \\
&= \psi_{\xi+1} \circ \pi_{\xi+1} \restriction \varepsilon(E) \\
&= v \circ \psi_{\xi+1} \circ \pi_{\xi+1} \restriction \varepsilon(E).
\end{aligned}$$

Line 1 comes from the way lift$(\mathcal{U}, c)$ works. Line 2 comes from our induction hypothesis. Line 4 comes from the fact that $\pi_{\xi+1}$ agrees with π_ξ on $\mathrm{lh}(E)$[3] and the fact that $\psi_{\xi+1}$ agrees with $\mathrm{res}_\xi \circ \psi_\xi$ on $\varepsilon(F)$ by the agreement of maps in lift$(\pi\mathcal{U}^+, d)$. Finally, line 5 holds because $o(Y_\xi)$ is a cardinal of $N_{\xi+1}$, and $o(Y_\xi) \leq \rho^-(N_{\xi+1})$[4], so $v \restriction o(Y_\xi)$ is the identity by the properties of resurrection maps.

This completes the step from ξ to $\xi + 1$ in our induction. The limit step is easy. ⊣

COROLLARY 5.4.3. *Under the hypotheses of Lemma 5.4.2, if $\vec{\mathcal{U}}$ is a stack of plus trees on P, and $\pi\vec{\mathcal{U}}^+$ is its (π, v, k)-lift to a stack on Q, then* lift$(\vec{\mathcal{U}}, c)_0 =$ lift$(\pi\vec{\mathcal{U}}^+, d)_0$.

PROOF. This is really a corollary to the proof. Let $\mathcal{U}$ be the first plus tree in $\vec{\mathcal{U}}$, and $\xi + 1 = \mathrm{lh}(\mathcal{U})$. Our induction hypothesis tells us that $c_\xi = \mathrm{stg}(\mathcal{U}, c, \xi)$ is related to $d_\xi = \mathrm{stg}(\pi\mathcal{U}^+, d, \xi)$ by π_ξ in the same way that c was related to d by π, that is

$$d_\xi = \langle Q_\xi, \psi_\xi, N_\xi, \mathbb{C}_\xi, R_\xi \rangle,$$

and

$$c_\xi = \langle P_\xi, \sigma_{\mathrm{N}_\xi}[\psi_\xi(J_\xi)] \circ \psi_\xi \circ \pi_\xi, \mathrm{Res}_{\mathrm{N}_\xi}[\psi_\xi(J_\xi)], \mathbb{C}_\xi, R_\xi \rangle,$$

where R_ξ is the common last model $\mathcal{M}_\xi^{\mathcal{U}^*}$ of the two lifts, and $\mathbb{C}_\xi$ is its construction. So if $\mathcal{U}_1$ is the next plus tree in $\vec{\mathcal{U}}$, then it lifts under $\sigma_{\mathrm{N}_\xi}[\psi_\xi(J_\xi)] \circ \psi_\xi \circ \pi_\xi, \mathrm{Res}_{\mathrm{N}_\xi}[\psi_\xi(J_\xi)]$ the same way that $\pi_\xi\mathcal{U}_1^+$ lifts under ψ_ξ. And so on. ⊣

---

[3]It may agree less with π_α for $\alpha > \xi + 1$.

[4]See $(3)_\xi$(d)(e) in the definition of conversion.

DEFINITION 5.4.4. Let Ω be a complete (λ,θ) iteration strategy for a premouse M. We say that Ω is *internally lift consistent* iff whenever s is a stack of plus trees by Ω and $P \trianglelefteq Q \trianglelefteq \mathcal{M}_\infty(s)$, then $\Omega_{s,P} = (\Omega_{s,Q})_P^+$.

We get at once

THEOREM 5.4.5. *Let $d = \langle M, \psi, N, \mathbb{C}, R\rangle$ be a conversion stage, Σ^* be a (λ,θ) iteration strategy for $(R, w^{\mathbb{C}}, \mathcal{F}^{\mathbb{C}})$, and $\Omega = \Omega(d,\Sigma^*)$; then Ω is internally lift consistent.*

PROOF. Let us take the case that $s = \emptyset$. Let $P \trianglelefteq Q = M$. In that case, we must show that $\Omega_P = \Omega_P^+$. Let $c = \langle P, \tau\circ\psi, K, \mathbb{C}, R\rangle$, where $K = \mathrm{Res}_N[\psi(P)]$ and $\tau = \sigma_N[\psi(P)]$. Then $\vec{\mathcal{U}}$ is by Ω_P iff $\mathrm{lift}(\vec{\mathcal{U}},c)_0$ is by Σ^* iff $\mathrm{lift}(\vec{\mathcal{U}}^+,d)_0$ is by Σ^* iff $\vec{\mathcal{U}}^+$ is by Ω iff $\vec{\mathcal{U}}$ is by Ω_P^+.

The general case, when s is arbitrary and $P \trianglelefteq Q \trianglelefteq \mathcal{M}_\infty(s)$, is similar. ⊣

It is not hard to see that internal lift consistency passes to pullback strategies. Another simple diagram shows that the action of an internally lift consistent strategy on stacks of plus trees is determined by its action on maximal stacks of plus trees.

5.5. A reduction to λ-separated trees

It is easy to show that background induced iteration strategies are determined by their action on λ-separated trees. First, there is a natural minimal λ-separation of a given plus tree:

DEFINITION 5.5.1. Let $\mathcal{T}$ be a plus tree on M. We define a λ-separated tree $\mathcal{U} = \mathcal{T}^{\mathrm{sep}}$, along with elementary maps

$$\pi_\alpha : \mathcal{M}_\alpha^{\mathcal{T}} \to \mathcal{M}_\alpha^{\mathcal{U}}$$

by: $\pi_0 = \mathrm{id}$, and

$$E_\alpha^{\mathcal{U}} = \begin{cases} \pi_\alpha(E_\alpha^{\mathcal{T}}) & \text{if } E_\alpha^{\mathcal{T}} \text{ is of plus type,} \\ \pi_\alpha(E_\alpha^{\mathcal{T}})^+ & \text{otherwise,} \end{cases}$$

and $\pi_{\alpha+1}$ is the natural copy map from $\mathrm{Ult}(P, E_\alpha^{\mathcal{T}})$ to $\mathrm{Ult}(\pi_\beta(P), E_\alpha^{\mathcal{U}})$, where $\beta = T\text{-pred}(\alpha+1) = U\text{-pred}(\alpha+1)$, and $P \trianglelefteq \mathcal{M}_\beta^{\mathcal{T}}$ is what $E_\alpha^{\mathcal{T}}$ is applied to.

We then have

THEOREM 5.5.2. *Let $c = \langle M, \psi, Q, \mathbb{C}, R\rangle$ be a conversion stage, and $\mathcal{T}$ be a plus tree on M; then*

(a) $\mathrm{lift}(\mathcal{T},c)_0 = \mathrm{lift}(\mathcal{T}^{\mathrm{sep}},c)_0$, *and*
(b) if Σ^* is a complete (λ,θ) iteration strategy for $(R, w^{\mathbb{C}}, \mathcal{F}^{\mathbb{C}})$, then $\mathcal{T}$ is by $\Omega(c,\Sigma^*)$ iff $\mathcal{T}^{\mathrm{sep}}$ is by $\Omega(c,\Sigma^*)$.

PROOF. Let

$$\text{stg}(\mathcal{T},c,\alpha) = \langle M_\alpha, \psi_\alpha, P_\alpha, \mathbb{C}_\alpha, R_\alpha \rangle$$

and

$$\text{stg}(\mathcal{T}^{\text{sep}},c,\alpha) = \langle N_\alpha, \varphi_\alpha, Q_\alpha, \mathbb{D}_\alpha, S_\alpha \rangle$$

be the conversion stages in the two liftings. Let $\pi_\alpha : M_\alpha \to N_\alpha$ be the separation map described in Definition 5.5.1. A completely routine induction shows that for all α, $\langle P_\alpha, \mathbb{C}_\alpha, R_\alpha \rangle = \langle Q_\alpha, \mathbb{D}_\alpha, S_\alpha \rangle$, and $\psi_\alpha = \varphi_\alpha \circ \pi_\alpha$.

This proves (a). Part (b) follows at once. $\dashv$

Chapter 6

NORMALIZING STACKS OF ITERATION TREES

In this chapter, we shall show how one can re-order the use of extenders in a finite stack s of normal plus trees, so as to produce a single normal plus tree $W(s)$ such that the last model of s embeds into the last model of $W(s)$. We call this process *embedding normalization*. Our goal here is to give some basic definitions and prove some elementary theorems that help one deal with the complexities of normalizing and quasi-normalizing. In Chapter 8 we shall apply the resulting theory to the comparison of iteration strategies.[1]

We shall assume for most of the chapter that the stack s to be normalized is finite and maximal, and consists of normal trees. In Section 6.7 we consider arbitrary finite M-stacks, but that more general case is not needed for strategy comparison.

The results of this chapter have the pleasant feature that one need only understand the basic facts about iteration trees and premice in order to follow their proofs. Indeed, it seems to us that this is a place where someone with minimal background knowledge could get a feel for iteration tree combinatorics. With that in mind, we have gone more slowly, including more examples and variant proofs than a more advanced reader would require.

In that spirit, we begin in Section 6.1 by considering the simplest possible case, normalizing a stack of length two in which each component tree uses only one extender. The results of this section are not used later, but they do help give a feel for what's going on. We also show in Section 6.1 that these simple stacks can be fully normalized, in that, granted an iterability assumption, one can find a normal tree $X(s)$ whose last model is equal to the last model of s.

In Section 6.2 we consider the special case of stacks $\langle \mathcal{T}, \mathcal{U} \rangle$ in which $\mathcal{U}$ uses only one extender, and in Section 6.5 we define $W(\langle \mathcal{T}, \mathcal{U} \rangle) = W(\mathcal{T}, \mathcal{U})$ for the general maximal stack of length two. We do use some of the definitions of Section 6.2 in Section 6.5.

In Section 6.3 we introduce *extender trees*, which are simple re-packagings of iteration trees that are sometimes helpful. In Section 6.4 we introduce something

[1] Some of our work on normalization was done earlier (but never written up) with Itay Neeman, and then later with Grigor Sargsyan. Fuchs, Neeman and Schindler ([13]) and Mitchell (unpublished), and probably others, have considered the question. Much of what seems to be new in this chapter was done independently, and at roughly the same time, by Farmer Schlutzenberg. (See [57].)

much more important, the notion of a *tree embedding*.[2] This notion is absolutely central to our work here. A key part of what makes an iteration strategy Σ comparable with other strategies is that if $\mathcal{U}$ is by Σ, and $\mathcal{T}$ is tree-embeddable into $\mathcal{U}$, then $\mathcal{T}$ is by Σ. We call this property of Σ *strong hull condensation*. Tree embeddings play an important role in the definition of $W(\mathcal{T},\mathcal{U})$, as we shall see.

Sections 6.6 and 6.8 are devoted to elementary facts about $W(\mathcal{T},\mathcal{U})$. The most substantial result here concerns the way branches of $W(\mathcal{T},\mathcal{U})$ correspond in one-one fashion with pairs consisting of a branch of $\mathcal{T}$ and a branch of $\mathcal{U}$. In Section 6.9 we describe the normalization of stacks of arbitrary finite length, and we say a few words about normalizing stacks of infinite length.

Finally, in Section 6.7 we describe the quasi-normalization $V(s)$ of a stack s. If the components of s are λ-separated, then $V(s) = W(s)$, so if one were willing to restrict all iteration strategies to stacks of λ-separated trees, then one could ignore quasi-normalization. There seems to be no great loss in doing that. In general, if s is maximal and normal, $W(s)$ is the normal companion of $V(s)$. Quasi-normalization is needed in showing that background induced strategies embedding normalize well[3] on stacks whose components may not be λ-separated.

In general, there are two sorts of base models M for the iteration trees we deal with in this book: coarse premice and fine-structural premice. Both sorts divide further into pure extender and strategy premice. The definition of $W(\mathcal{T},\mathcal{U})$ will make sense in both cases. In this chapter we shall focus on the case that M is a pfs premouse. Until we get to Chapter 9, this is what we mean by the unqualified *premouse*. In the most important case, M has type 1 and is strongly stable. We do also need to define $W(\mathcal{T},\mathcal{U})$ in the coarse structural case as well, and we shall indicate how to do so as we proceed. But then we are just talking about ultrapowers of models of ZFC by nice extenders, so various things simplify.

The construction of $W(\mathcal{T},\mathcal{U})$ does not require that any iteration strategy for M be fixed; however, it may break down by reaching illfounded models, even if the models of $\mathcal{T}^\frown\mathcal{U}$ are wellfounded. In the case we care about, M has an iteration strategy Σ , $\langle\mathcal{T},\mathcal{U}\rangle$ is played according to Σ, and the initial segment of $W(\mathcal{T},\mathcal{U})$ up to our point of interest is also played by Σ. We shall eventually show that if Σ has been properly induced, then $W(\mathcal{T},\mathcal{U})$ is also by Σ, and hence the construction of $W(\mathcal{T},\mathcal{U})$ does not break down.

6.1. Normalizing trees of length 2

We begin by looking closely at stacks of the form $\langle\langle E\rangle,\langle F\rangle\rangle$.

Let M be a pfs premouse, E on the extended sequence[4] of M, $\mathrm{crit}(E) < \rho_{k(M)}(M)$, and $N = \mathrm{Ult}(M,E)$. Let F be on the extended sequence of N, and

[2]Tree embeddings were isolated independently by Schlutzenberg and the author. See [57].

[3]Meaning s is by Σ iff $W(s)$ is by Σ.

[4]That is, E^- is on the sequence; see Definition 4.4.2.

$\operatorname{crit}(F) < \hat{\lambda}(E)$. It follows that $k(M) = k(N)$ and $\operatorname{crit}(F) < \rho_{k(N)}(N)$, so that $\operatorname{Ult}(N,F)$ makes sense, and both ultrapowers are n-ultrapowers, where $n = k(M)$.

Let

$$\kappa = \operatorname{crit}(E), \quad \mu = \operatorname{crit}(F), \quad \text{and } Q = \operatorname{Ult}(N,F).$$

Let $\mathcal{T}$ be the iteration tree such that $E_0^{\mathcal{T}} = E$, $E_1^{\mathcal{T}} = F$, $\mathcal{M}_0^{\mathcal{T}} = M$, $\mathcal{M}_1^{\mathcal{T}} = N$, and $\mathcal{M}_2^{\mathcal{T}} = Q$. Since $\mu < \hat{\lambda}(E)$, $\mathcal{T}$ is not normal. We show how to normalize it. There are two cases.

Case 1. $\operatorname{crit}(F) \le \operatorname{crit}(E)$.

Since $\mu \le \kappa$ and E is an extender over M (that is, over the reduct M^n, for $n = k(M)$), F is also an extender over M. Let $P = \operatorname{Ult}(M,F)$, and $i_F^M : M \to P$ be the canonical embedding. We have the diagram

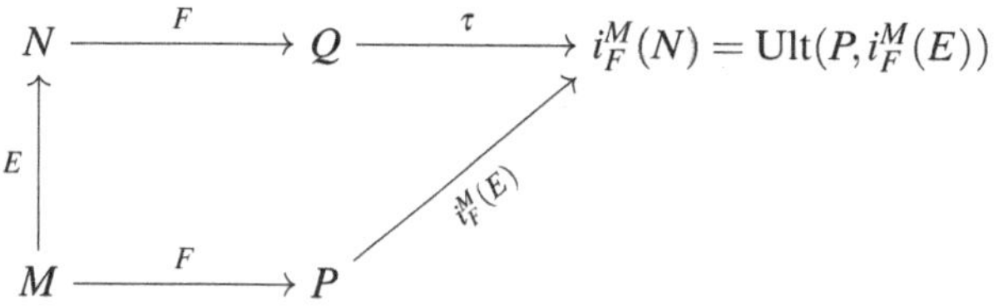

Suppose first that $M \models \mathsf{ZFC}$; then N is definable over M from E, and i_F^M moves the fact that $N = \operatorname{Ult}(M,E)$ over to the fact that $i_F^M(N) = \operatorname{Ult}(P, i_F^M(E))$. τ is the natural embedding from $i_F^N(N)$ to $i_F^M(N)$. That is,

$$\tau([a,g]_F^N) = [a,g]_F^M$$

for $g : [\mu]^{|a|} \to N$, with $g \in N$. The tree $\mathcal{U}$ with models

$$\mathcal{M}_0^{\mathcal{U}} = M,\ \mathcal{M}_1^{\mathcal{U}} = N,\ \mathcal{M}_2^{\mathcal{U}} = P,\ \mathcal{M}_3^{\mathcal{U}} = \operatorname{Ult}_0(\mathcal{P}, i_F^M(E))$$

and extenders

$$E_0^{\mathcal{U}} = E,\ E_1^{\mathcal{U}} = F,\ E_2^{\mathcal{U}} = i_F^M(E),$$

is normal. We call $\mathcal{U}$ the *embedding normalization* of $\mathcal{T}$.

Remark 6.1.1. This implicitly assumes $\operatorname{lh}(E) < \operatorname{lh}(F)$. If $\operatorname{lh}(F) < \operatorname{lh}(E)$, then F is already on the M-sequence, and the extenders of $\mathcal{U}$ would be $E_0^{\mathcal{U}} = F$, $E_1^{\mathcal{U}} = i_F^M(E)$. The diagrams and calculations above do not change, however.

The proof just given was based on N being definable over M as its E-ultrapower and i_F^M acting elementarily on this definition. But of course, $\mathrm{OR}^N > \mathrm{OR}^M$ is possible, and anyway, we need to know i_F^M has enough elementarity. If $M \models \mathsf{ZFC}$, all is fine. We now give a more careful proof that works in general.

We assume $k(M) = k(N) = 0$ so that we can avoid the details of ultrapowers of reducts and their decodings. The general case is similar. So every $x \in Q$ has the form $i_F^N(g)(b)$ for $g \in N$ and $b \in [\mathrm{lh}(F)]^{<\omega}$. We can write $g = i_E^M(h)(a)$, where $h \in M$ and $a \in [\mathrm{lh}(E)]^{<\omega}$. So

$$\begin{aligned} x &= i_F^N(i_E^M(h)(a))(b) \\ &= i_F^N \circ i_E^M(h)(i_F^N(a))(b), \end{aligned}$$

with b, $i_F^N(a) \in [\sup i_F^N\text{“}\,\mathrm{lh}(E)]^{<\omega}$. Let

$$G = (\text{extender of } i_F^N \circ i_E^M) \restriction \sup i_F^N\text{“}(\mathrm{lh}(E)),$$

so that

$$Q = \mathrm{Ult}(M, G).$$

The space of G is κ, and its critical point is μ. Notice that $\mathrm{lh}(E)$ is regular in N, and $\rho^-(N) > \mathrm{lh}(E)$, so i_F^N is continuous at $\mathrm{lh}(E)$. Let us write

$$\begin{aligned} R &= \mathrm{Ult}_0(P, i_F^M(E)) \\ H &= (\text{extender of } i^P_{i_F^M(E)} \circ i_F^M) \restriction \sup i_F^M(\mathrm{lh}(E)). \end{aligned}$$

It is easy to see that

$$R = \mathrm{Ult}(M, H).$$

$P = \mathrm{Ult}_1(M, F)$ iff $R = \mathrm{Ult}_1(M, H)$. But now we can calculate that G is a subextender of H. For let $b \in [\mathrm{lh}(F)]^{<\omega}$ and $g : [\mu]^{|b|} \to [\mathrm{lh}(E)]^l$ with $g \in N$. Let $A \subseteq [\mathrm{crit}(E)]^l$ with $A \in N$. (Equivalently, $A \in M$.) We have

$$\begin{aligned} ([b,g]_F^N, A) \in G &\text{ iff } [b,g]_F^N \in i_F^N \circ i_E^M(A) \\ &\text{ iff for } F_b \text{ a.e. } \bar{\mu},\ g(\bar{\mu}) \in i_E^M(A) \\ &\text{ iff for } F_b \text{ a.e. } \bar{\mu},\ (g(\bar{\mu}), A) \in E \\ &\text{ iff } ([b,g]_F^M, i_F^M(A)) \in i_F^M(E) \\ &\text{ iff } [b,g]_F^M \in i^P_{i_F^M(E)} \circ i_F^M(A) \\ &\text{ iff } ([b,g]_F^M, A) \in H. \end{aligned}$$

Let $S = N||\mathrm{lh}(E) = M||\mathrm{lh}(E)$, and let $\sigma : i_F^N(S) \to i_F^M(S)$ be given by

$$\sigma([b,g]_F^N) = [b,g]_F^M.$$

σ is nearly elementary, and maps $\mathrm{lh}(G)$ into $\mathrm{lh}(H)$. We have just shown that

$$(a, A) \in G \text{ iff } (\sigma(a), A) \in H,$$

so G is a subextender of H under σ. (P may have been constructed using functions that are Σ_1^M, but that certainly includes g.) We can therefore define τ from Q into R by

$$\tau([a,f]_G^M) = [\sigma(a), f]_H^M.$$

Notice that τ agrees with σ on $i_F^N(S)$, and $\tau \restriction \mathrm{lh}(F) = \sigma \restriction \mathrm{lh}(F) =$ identity. One can easily show that in the case $M \models \mathsf{ZFC}$, our current definition of τ coincides with the earlier one.

Here is another way to obtain τ, one that is closer to the way we shall handle the general case below. Let $\psi\colon \mathrm{Ult}(M,E) \to \mathrm{Ult}(P,E^*)$ be the Shift Lemma map, where $E^* = i_F^M(E)$. That is,

$$\psi([a,f]_E^M) = [i_F^M(a), i_F^M(f)]_{E^*}^P.$$

By the Shift Lemma, ψ agrees with i_F^M on $\mathrm{lh}(E)$. It follows that F is an initial segment of E_ψ, the extender of ψ. Let θ be the factor embedding from $\mathrm{Ult}(N,F)$ to $\mathrm{Ult}(N,E_\psi)$, given by

$$\theta([a,g]_F^N) = [a,g]_{E_\psi}^N = \psi(g)(a),$$

for all $a \in [\mathrm{lh}(F)]^{<\omega}$. We claim that $\theta = \tau$.

To see this, note that θ is the unique map π from Q to $\mathrm{Ult}(P,E^*)$ such that $\psi = \pi \circ i_F^N$ and $\pi \restriction \mathrm{lh}(F)$ is the identity. Clearly $\tau \restriction \mathrm{lh}(F) = \sigma \restriction \mathrm{lh}(F) =$ identity, so we must see that $\psi = \tau \circ i_F^N$. Now both θ and τ make the diagram

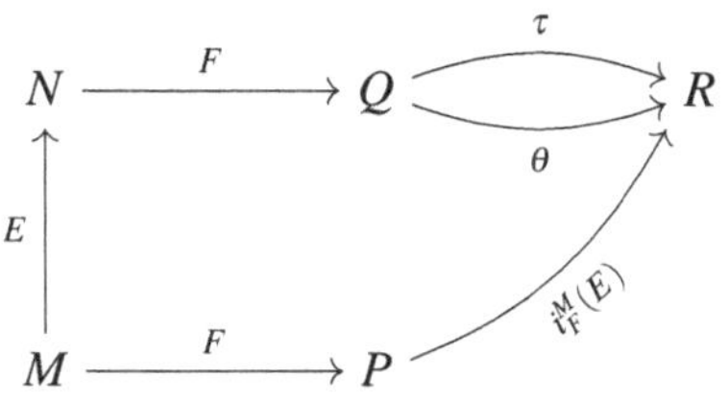

commute, where $R = \mathrm{Ult}(P, i_F^M(E))$, so ψ agrees with $\tau \circ i_F^N$ on $\mathrm{ran}(i_E^M)$. Thus it is enough to see that ψ agrees with $\tau \circ i_F^N$ on the generators of E, that is, on $\mathrm{lh}(E)$. But for $a \in [\mathrm{lh}(E)]^{<\omega}$,

$$\begin{aligned} \tau \circ i_F^N(a) &= \sigma \circ i_F^N(a) \\ &= i_F^M(a) \\ &= \psi(a), \end{aligned}$$

by the definitions of ψ and τ. This completes our proof that $\tau = \theta$.

Another diagram of the situation is shown in Figure 6.1.1, where F is an initial segment of the extender of ψ, and τ is the factor map. N is generated by $M || \mathrm{lh}(E) \cup \mathrm{ran}(i_E)$, and Q is generated by $Q || \mathrm{lh}(G) \cup \mathrm{ran}(i_G)$. τ is the unique map that agrees with σ on $\mathrm{lh}(G)$ and makes the diagram commute.

Remark 6.1.2. τ is Σ_0 as a map from Q^n to $\mathrm{Ult}(P, i_F^M(E))^n$, so using the fact that the diagram commutes, we see that τ is nearly elementary. If all the ultrapowers in the diagram are n-ultrapowers, for $n = k(M)$, then all their maps are cofinal, so by commutativity τ is cofinal, and hence also elementary. But in general, τ behaves

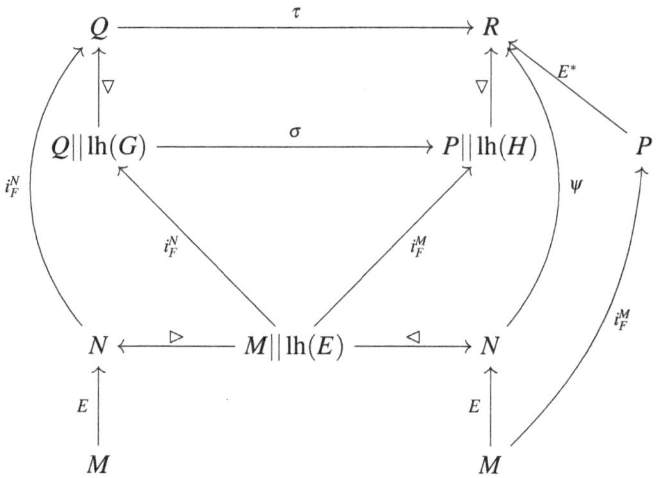

FIGURE 6.1.1

like any factor map from one ultrapower to a larger one: it is nearly elementary, but may not be elementary.

Case 2. $\text{crit}(E) < \text{crit}(F)$.

Let $\mu = \text{crit}(F)$ and $\kappa = \text{crit}(E)$. We have assumed $\mu < \hat{\lambda}(E)$, as otherwise $\mathcal{T}$ is already normal. Let

$$S = M||\text{lh}(E) = N||\text{lh}(E),$$
$$J = M|\langle \xi, k\rangle, \text{where } \langle \xi, k\rangle \text{ is lex least such that } \rho(M|\langle \xi, k\rangle) \leq \mu,$$
$$P = \text{Ult}(J, F).$$

Let $N = \text{Ult}(M, E)$ and $Q = \text{Ult}(N, F)$.

The embedding normalization of $\mathcal{T}$ continues from $M_0 = M$, $M_1 = N$ (assuming $\text{lh}(E) < \text{lh}(F)$), $M_2^* = J$, and $M_2 = P$ by using $i_F^J(E)$ now. Note $i_F^J(E)$ should be applied to M, not P, in a normal tree. So let

$$R = \text{Ult}(M, i_F^J(E)).$$

Since $\text{crit}(i_F^J(E)) = \kappa$, the ultrapowers producing N and R have the same degree. We assume again for simplicity that it is zero. Let G be the extender of $i_F^N \circ i_E^M$, and notice that G is short, with $\lambda(G) = i_F^N(\lambda(E)) = \sup i_F^N``\lambda(E)$. Let

$$\sigma : \text{Ult}(S, F) \to i_F^J(S)$$

be given by

$$\sigma([b, g]_F^N) = [b, g]_F^J$$

for $g : [\mu]^{|b|} \to \lambda(E)$ with $g \in N$. (Note that for $n = k(M) = k(N)$, we have $\kappa < \rho_n(M)$, so $\mathrm{lh}(E) < \rho_n(N)$, so every $r\Sigma_n^N$ such function g belongs to S. That is, $\mathrm{Ult}(S,F) = i_F^N(S)$.) We claim that

CLAIM 6.1.3. *G is a subextender of $i_F^J(E)$ under σ.*

Remark 6.1.4. In this case, G and $i_F^J(E)$ are short, and σ is the identity on their common domain.

PROOF. Let $a \subseteq i_F^N(\mathrm{lh}(E))$ be finite, and let $A \subseteq [\kappa]^{|a|}$ be in M. Let $a = [b,g]_F^N$, where $g \in N$ and $g : [\mu]^{|b|} \to [\nu(E)]^{|a|}$. Then

$$\begin{aligned}(a,A) \in G \text{ iff } & ([b,g]_F^N, A) \in G \\ \text{iff } & [b,g]_F^N \in i_F^N \circ i_E^M(A) \\ \text{iff } & \text{for } F_b \text{ a.e. } \bar{\mu},\ g(\bar{\mu}) \in i_E^M(A) \\ \text{iff } & \text{for } F_b \text{ a.e. } \bar{\mu},\ (g(\bar{\mu}),A) \in E \\ \text{iff } & ([b,g]_F^J, A) \in i_F^J(E) \\ \text{iff } & (\sigma(a),A) \in i_F^J(E).\end{aligned}$$

⊣

Thus we have a factor map $\tau : Q \to R$ from $Q = \mathrm{Ult}(M,G)$ to $\mathrm{Ult}(M, i_F^J(E))$ given by

$$\tau([a,f]_G^M) = [\sigma(a),f]_{i_F^J(E)}^M.$$

Assuming $\mathrm{lh}(E) < \mathrm{lh}(F)$, the embedding normalization of $\mathcal{T}$ is then $\mathcal{U}$, where

$$E_0^{\mathcal{U}} = E,\ E_1^{\mathcal{U}} = F,\ E_2^{\mathcal{U}} = i_F^M(E).$$

If $\mathrm{lh}(F) < \mathrm{lh}(E)$, it is $E_0^{\mathcal{U}} = F$, $E_1^{\mathcal{U}} = i_F^M(E)$.

Notice that E is an amenable class of $N||\mathrm{lh}(E)$, so we can make sense of $i_F^N(E)$ as the union of all $i_F^N(E \cap x)$ for $x \in N||\mathrm{lh}(E)$. The proof of the claim showed

$$G = i_F^N(E).$$

So the situation in Case 2 is summarized by the diagram

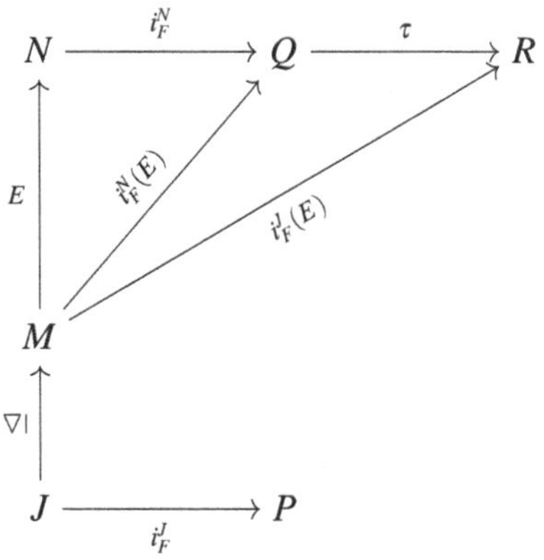

We have assumed here $k = 0$ to remove some clutter. As in Case 1, τ is nearly elementary.

Remark 6.1.5. If $J = M|\mathrm{lh}(E)$, then $i_F^J = i_F^N \upharpoonright N|\mathrm{lh}(E)$, so $i_F^N(E) = i_F^J(E)$, and $Q = R$. This is what happens if $\nu(E) \le \mathrm{crit}(F) < \lambda(E)$. The original $\mathcal{T}$ is ms-normal but not Jensen normal. Its embedding normalization is Jensen normal, and has the same last model as $\mathcal{T}$.

If $J = M$, then the diagram simplifies to

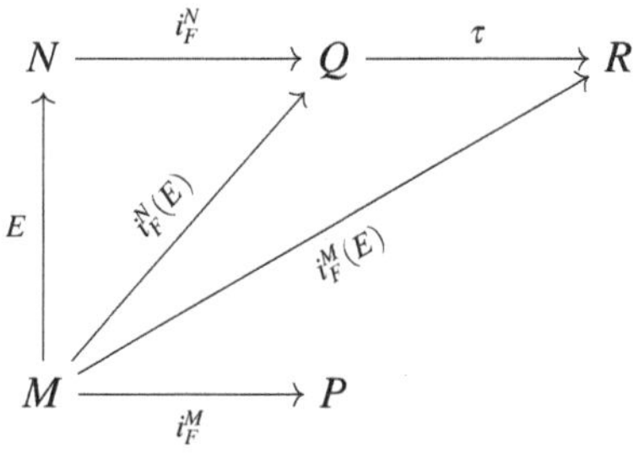

If $\mu < \nu(E)$ and $\nu(E)$ is a cardinal of M and $J = M$, then $i_F^N(E)$ is the trivial completion of $i_F^M(E) \upharpoonright \sup i_F^N``\nu(E)$. In this case, $Q = R$ iff $\mathrm{cof}^M(\nu(E)) \neq \mu$, and if $Q \neq R$, then $\mathrm{crit}(\tau) = \sup i_F^N``\nu(E)$.

Full normalization. Suppose we are in the situation above: E is on the extended M-sequence, $N = \mathrm{Ult}(M,E)$, F is on the extended N-sequence, and $Q = \mathrm{Ult}(N,F)$. Let us consider the problem of fully normalizing E-then-F. That is, we seek a normal tree on M whose last model is literally equal to Q.

We saw in Case 2 that $Q = \mathrm{Ult}(M, i_F^N(E))$, and it is not hard to check that $Q = \mathrm{Ult}(P, i_F^N(E))$ in Case 1, where $P = \mathrm{Ult}(M,F)$. So it would be enough to show that $i_F^N(E)$ is on the extended P-sequence. $i_F^N(E)$ is a subextender of $i_F^M(E)$ under the map σ that we identified in the proof of embedding normalization. $i_F^M(E)$ is on the extended P-sequence, so perhaps we can apply condensation to σ and conclude that $i_F^N(E)$ is on the extended P-sequence. We shall sketch here a proof that this can be done.

Full normalization is not important in this book, but it is very useful in its sequels, for example [73] and [80]. The paper [64] proves a general theorem on the existence of full normalizations for stacks of normal trees on premice. The argument we are about to give contains one of the main ideas in that proof.

Let us assume that E and F are not of plus type, so that we do not have to bother with extended sequences. Let us assume $\mathrm{crit}(F) \le \mathrm{crit}(E)$ as well. The proof we give easily generalizes to the other cases. In the course of the proof we shall assume that certain ultrapowers produce type 1 pfs premice. That assumption, which is less easily removed, has the effect of making the Condensation Theorem 4.10.10 adequate to our task. One can avoid it by going further into the fine structure of pfs premice and proving a stronger condensation theorem. We shall not do that here.

Remark 6.1.6. Let us consider the case that $\nu(E)$ is a cardinal in M. Then $({}^{\mu}\alpha)^M = ({}^{\mu}\alpha)^N$ for all $\alpha < \nu(E)$, so for σ as above, $\sigma \restriction \sup i_F^N \text{``} \nu(E) = \text{identity}$. Thus $i_F^N(E)$ is the trivial completion of $i_F^M(E) \restriction \sup i_F^M \text{``} \nu(E)$. If i_F^M is continuous at $\nu(E)$ (i.e., $\text{cof}^M(\nu(E)) \neq \mu$), then $i_F^N(E) = i_F^M(E)$ and $Q = R$. If i_F^M is discontinuous at $\nu(E)$ (i.e., $\text{cof}^M(\nu(E)) = \mu$), then $Q \neq R$, and in fact $\text{crit}(\tau) = \sup i_F^M \text{``} \nu(E)$.

So in this case, the embedding normalization of $\mathcal{T}$ uses $i_F^M(E)$ to continue from P, while the full normalization may use a proper initial segment of $i_F^M(E)$ to continue from P.

Clearly, a full normalization of $\mathcal{T}$ must start with E and then F. We are now at the model P, and to get to Q, we must replace $i_F^M(E)$ by $i_F^N(E)$. In order to do that, we look at $i_F^J(E)$ for all those J in the $(M, M|\text{lh}(E))$ dropdown sequence. Assuming for simplicity that the $\text{Ult}(J,F)$ have type 1, we show inductively that each such $i_F^J(E)$ is on the P-sequence, starting with $J = M$, where this is clearly true, and working down to $J = M|\text{lh}(E)$, where $i_F^J(E) = i_F^N(E)$. At each step of the induction we apply Theorem 4.10.10.

Let us check that $i_F^N(E)$ is indeed the extender we want.

CLAIM 6.1.7. *Let $J = M||\text{lh}(E)$; then $Q = \text{Ult}(P, i_F^J(E))$.*

PROOF. $\text{lh}(E)$ is a regular cardinal in N, and $\text{lh}(E) \leq \rho^-(N)$, so

$$i_F^J = i_F^N \restriction J,$$

and i_F^N is continuous at $\text{lh}(E)$. Let

$$L = (\text{extender of } i^P_{i_F^J(E)} \circ i_F^M) \restriction i_F^N(\text{lh}(E)),$$

then it is easy to see that

$$\text{Ult}(P, i_F^J(E)) = \text{Ult}(M, L).$$

Recall that G was the extender of length $i_F^N(\nu(E))$ given by $i_F^N \circ i_E^M$. As before, we get $\bar{\sigma} : \text{lh}(G) \to \text{lh}(L)$ by

$$\bar{\sigma}([b,g]_F^N) = [b,g]_F^{M||\text{lh}(E)},$$

defined for $b \in [\nu(E)]^{<\omega}$ and $g : [\mu]^{|b|} \to \nu(E)$ with $g \in N$. (We assume here $k(M) = k(N) = 0$; otherwise replace M and N by their $k(M)$-reducts.) But all such g are in $M||\text{lh}(E)$, so

$$\bar{\sigma} = \text{identity}.$$

As before, we get that G is a subextender of L under $\bar{\sigma}$, but this just means that $G = L$, proving Claim 6.1.7. ⊣

Now let J and K be successive elements of the $(M, M|\mathrm{lh}(E))$ dropdown sequence, that is, $J = A_i(M, M|\mathrm{lh}(E))$ and $K = A_{i+1}(M, M|\mathrm{lh}(E))$ for some i. Let $m = k(J)$ and $n+1 = k(K)$. Let

$$X = \mathrm{Ult}_m(J,F),$$
$$Y = \mathrm{Ult}_n(K,F),$$
$$Z = \mathrm{Ult}_{n+1}(K,F).$$

Notice here that we are not guaranteed that X, Y, or Z are type 1 pfs premice, or equivalently, that X^-, Y^-, or Z^- are sound. For example, if $\eta^J_m = \mu$, where $\mu = \mathrm{crit}(F)$, and J has type 1B, then X^- is not sound. This is a complication because our condensation theorem applies to sound mice. Rather than go further into pfs fine structure, we shall now simply assume that X, Y, and Z are type 1 pfs premice. The full proof is given in [64].[5]

CLAIM 6.1.8. $X \trianglelefteq Y \trianglelefteq Z$.

PROOF. Let $i\colon J \to X$, $k\colon K^- \to Y$, and $l\colon K \to Z$ be the canonical embeddings. We have a factor map

$$\psi\colon X \to k(J)$$

given by

$$\psi([a,f]^J_F) = [a,f]^{K^-}_F.$$

Let $\gamma = \rho_m(J)$. γ is a cardinal of K, and $\gamma \le \rho_n(K)$, so every $r\Sigma^K_n$ function f with domain μ and range bounded in γ belongs to J. Thus

$$\rho_m(X) = \sup i``\gamma \le \mathrm{crit}(\psi).$$

We may assume that ψ is not the identity, as otherwise $X = k(J)$, so $X \trianglelefteq Y$ as desired. But then ψ witnesses that the reduct X^m is a proper initial segment of $k(J)^m$, so $X^m \in k(J)^m$, so $X \in k(J)$. This means that the Condensation Theorem 4.10.10 applies[6], and we have that either $X \trianglelefteq k(J)$ or $X \trianglelefteq \mathrm{Ult}_0(k(J),D)$, where $\mathrm{lh}(D) = \mathrm{crit}(\psi) = \rho_m(X)$, D is on the $k(J)$ sequence. By the display above, $\mathrm{crit}(\psi) = \sup i``\gamma$ in this latter case.

Suppose toward contradiction that D witnesses the latter "one ultrapower away" possibility. Since i is discontinuous at γ, then $\mathrm{cof}^J_m(\gamma) = \mu$, so $\mathrm{cof}^{k(J)}_m(\sup i``\gamma) = \mu$, so $\mathrm{cof}^{k(J)}_m(\mathrm{dom}(D)) = \mu$. But $\mathrm{dom}(D)$ is a successor cardinal of $k(J)$, and $\mathrm{dom}(D) < \rho_m(k(J))$, so this is impossible.

[5]If X has type 2, then $\mathfrak{C}_m(X) \trianglelefteq Y$, and $X^- = \mathrm{Ult}_{m-1}(\mathfrak{C}_m(X)^-, D)$ for D the order zero measure of $\mathfrak{C}_m(X)$ on $\hat{\rho}_m(X)$. The end result is that $i^N_F(E)$ may not be on the sequence of $\mathrm{Ult}(M,F)$, but instead be an ultrapower away from it by an order zero measure.

[6]In the notation there, $H = X^-$ and $M = k(J)^-$. We are assuming they are sound, not just almost sound, so the Condensation Theorem 4.10.10 applies.

So we have shown that $X \trianglelefteq Y$. The proof that $Y \trianglelefteq Z$ is similar. Let $\delta = \rho_{n+1}(K)$, and let $\theta : Y \to Z^-$ be given by

$$\theta([a,f]_F^{K^-}) = [a,f]_F^K.$$

The ultrapower on the left uses $r\Sigma_n^K$ functions and that on the right uses $r\Sigma_{n+1}^K$ functions. These are the same functions if the range is bounded in δ, so

$$\rho_{n+1}(Z) = \sup l``\delta = \sup k``\delta \leq \mathrm{crit}(\psi).$$

Let $\rho = \rho_{n+1}(Z)$. For simplicity, let us assume that $K = \bar{\mathfrak{C}}_{n+1}(K)$ is its own strong $n+1$-core. Letting $r = p_{n+1}(K)$, we get that $K^n = h^1_{K^n}``(\delta \cup r)$, so that $h^1_{Y^n}``(\rho \cup k(r)) = Y^n$. This and the existence of solidity witnesses in Y^n implies that $\rho = \rho_{n+1}(Y)$ and $k(r) = p_{n+1}(Y)$.

Let us take the case $n = 0$ now, just to be more concrete. $\theta : Y \to Z$ is thus Σ_0 elementary, and the identity on $\rho_1(Y) = \rho_1(Z)$. If θ is cofinal, then $Y = Z$. If θ is not cofinal, then $Y \in Z$, so the Condensation Theorem 4.10.10 shows that $Y \lhd Z$, or $Y \lhd \mathrm{Ult}_0(Z,D)$ where $\mathrm{lh}(D) = \rho_1(Z)$. The latter can be ruled out in the same way we did above. Thus $Y \lhd Z$.

If $K \neq \bar{\mathfrak{C}}_{n+1}(K)$, then it is possible that Y is not $n+1$-sound, because $k(\delta)$ is not in the appropriate hull. In this case we would need to replace Y by $\mathrm{Ult}_n(\bar{\mathfrak{C}}_{n+1}(K))$ in the argument above. This leads further into the condensation properties of pfs premice. See [64] for a full account. ⊣

Now let $\langle A_0, \ldots, A_n\rangle$ be the $(M, \mathrm{lh}(E))$ dropdown sequence, let $X_i = \mathrm{Ult}_m(A_i, F)$, where $m = k(A_i)$. The two claims clearly imply that, under the simplifying assumption that each X_i is a sound pfs premouse, $X_0 \trianglelefteq X_n$. But $X_n = \mathrm{Ult}(M,F)$, and $i_F^N(E)$ is the top extender of X_0. So $i_F^N(E)$ is on the sequence of $\mathrm{Ult}(M,F)$, as desired.

Remark 6.1.9. If $M \in \mathrm{lev}(\mathbb{C})$, then $\mathbb{C}$ will associate background extenders G and H to $i_F^N(E)$ and $i_F^M(E)$ via the conversion process. H is just the image at the background level of the background originally assigned to E. On the other hand, there is no useful connection between G and the original background for E. This is the main reason that embedding normalization is more useful than full normalization in this book. Embedding normalization commutes with conversion, but full normalization does not.

6.2. Normalizing $\mathcal{T}^\frown\langle F\rangle$

Let M be a premouse, let $\mathcal{T}$ a normal plus tree on M having last model $\mathcal{M}_\theta^\mathcal{T}$, and let F be on the extended sequence of $M_\theta^\mathcal{T}$. Let Q be the longest initial segment of $\mathcal{M}_\theta^\mathcal{T}$ such that $\mathrm{Ult}(Q,F)$ makes sense, that is, such that F is total on Q and $\mathrm{crit}(F) < \rho_{k(Q)}(Q)$. We construct a normal plus tree $\mathcal{W}$ on M such that $\mathrm{Ult}(Q,F)$ embeds into the last model of $\mathcal{W}$ via a nearly elementary map. We call $\mathcal{W}$ the *embedding normalization of* $\mathcal{T}^\frown\langle F\rangle$, and write

$$\mathcal{W} = W(\mathcal{T}, F).$$

The reader can find some diagrams which may help visualize the construction of $\mathcal{W}$ at the end of this section.

Let $M_\nu = \mathcal{M}_\nu^{\mathcal{T}}$, $M_\nu^* = M_\nu^{*,\mathcal{T}}$, and $E_\nu = E_\nu^{\mathcal{T}}$ be the models and extenders of the given $\mathcal{T}$, and let $N_\nu = \mathcal{M}_\nu^{\mathcal{W}}, N_\nu^* = N_\nu^{*,\mathcal{W}}$, and $F_\nu = E_\nu^{\mathcal{W}}$ be the models and extenders of the desired $\mathcal{W}$. Let α be least such that F is on the extended M_α-sequence. Then M_α agrees with Q up to $\text{lh}(F)+1$, and Q agrees with $\text{Ult}(Q,F)$ up to $\text{lh}(F)$, but not $\text{lh}(F)+1$. (Recall our convention that $\text{lh}(F) = \text{lh}(F^-)$.) We set

$$\mathcal{W}\restriction(\alpha+1) = \mathcal{T}\restriction(\alpha+1).$$

This does *not* imply $F_\alpha = E_\alpha$, just $N_\alpha = M_\alpha$. We set

$$F_\alpha = F,$$

and the rest of $\mathcal{W}\restriction\alpha+2$ is dictated by normality. Let $\mu = \text{crit}(F)$, and let $\beta \leq \alpha$ be least such that either $\mu < \hat{\lambda}(E_\beta)$, or $\beta = \alpha$. F must be applied to an initial segment of $N_\beta = M_\beta$ in $\mathcal{W}$. That is

$$W\text{-pred}(\alpha+1) = \beta,$$

and

$$N_{\alpha+1}^* = M_\beta|\langle\xi_0, k_0\rangle$$

where $\langle\xi_0,k_0\rangle$ is least such that $\rho(M_\beta|\langle\xi_0,k_0\rangle) \leq \mu$ or $\langle\xi_0,k_0\rangle = l(M_\beta)$, and

$$N_{\alpha+1} = \text{Ult}(N_{\alpha+1}^*, F).$$

This defines $\mathcal{W}\restriction(\alpha+2)$.

Case 1. $Q \neq M_\theta$.
If $\beta+1 < \text{lh}(\mathcal{T})$, then Q is a proper initial segment of $M_\beta|\text{lh}(E_\beta)$, by the following claim.

CLAIM 6.2.1. *Let $\mathcal{T}$ be a normal plus tree, $\beta+1 < \text{lh}(\mathcal{T})$, and $\mathcal{M}_\beta^{\mathcal{T}}|\text{lh}(E_\beta^{\mathcal{T}}) \trianglelefteq R \trianglelefteq \mathcal{M}_\nu^{\mathcal{T}}$ for some $\nu \geq \beta+1$; then $\text{lh}(E_\beta^{\mathcal{T}}) \leq \rho^-(R)$.*

PROOF. Let $S = \mathcal{M}_\theta^{\mathcal{T}}$. It is easy to see that $\rho^-(S) \geq \text{lh}(G)$ for all extenders G used in the branch $[0,\theta)_T$. Since some G with $\text{lh}(G) \geq \text{lh}(E_\beta^{\mathcal{T}})$ was used in $[0,\nu)_T$, we are done if $R = S$. If $\hat{o}(R) = \hat{o}(S)$ but $k(R) < k(S)$, then $\rho^-(S) \leq \rho^-(R)$, so again we are done. Finally, if $\hat{o}(R) < \hat{o}(S)$, then $R \in S$, so $\rho^-(R) < \text{lh}(E_\beta^{\mathcal{T}}) \leq o(R)$ implies that $\text{lh}(E_\beta^{\mathcal{T}})$ is not a cardinal in S. This is a contradiction. ⊣

If $\beta < \theta$, then we apply the claim to $R = Q^+$. We have $Q \lhd M_\theta$, so $R \unlhd M_\theta$. We have $\rho(Q) = \rho^-(R) \leq \mu < \mathrm{lh}(E_\beta^{\mathcal{T}})$, so it follows from the claim that $R \lhd M_\beta | \mathrm{lh}(E_\beta)$. Thus Q is a proper initial segment of $M_\beta | \mathrm{lh}(E_\beta)$.

So if $\beta + 1 < \mathrm{lh}(\mathcal{T})$, then $\alpha = \beta$, $N^*_{\alpha+1} = Q$, and and $N_{\alpha+1} = \mathrm{Ult}(Q, F)$. These conclusions hold trivially if $\beta + 1 = \mathrm{lh}(\mathcal{T})$, so in either case we set

$$\begin{aligned} W(\mathcal{T}, F) &= \mathcal{W} \restriction (\alpha+2) \\ &= \mathcal{T} \restriction (\beta+1)^\frown\langle F\rangle. \end{aligned}$$

We call this the *dropping case* in the definition of $W(\mathcal{T}, F)$. In this case, $\mathrm{Ult}(Q, F)$ is actually equal to the last model of $W(\mathcal{T}, F)$.

Case 2. $Q = M_\theta$, and $\theta = \beta$.

In this case $\alpha = \beta$, and again

$$\begin{aligned} W(\mathcal{T}, F) &= \mathcal{W} \restriction (\alpha+2) \\ &= \mathcal{T} \restriction (\beta+1)^\frown\langle F\rangle. \end{aligned}$$

Again, $\mathrm{Ult}(Q, F)$ is actually equal to the last model of $W(\mathcal{T}, F)$. The difference between this and the previous case is just that we did not drop when we applied F to $\mathcal{T}$.

Case 3. $Q = M_\theta$, and $\theta > \beta$.

In this case, $\mathrm{Ult}(M_\theta, F)$ makes sense, so $M_\beta | \mathrm{lh}(E_\beta) \unlhd N^*_{\alpha+1}$. In fact, if $\beta < \eta \leq \theta$, then $\mathrm{Ult}(M_\eta, F)$ makes sense, because $\mathrm{lh}(E_\beta)$ is a cardinal of M_η and $\mathrm{lh}(E_\beta) \leq \rho^-(M_\eta)$.

For $\eta \leq \theta$, set

$$u(\eta) = \begin{cases} \eta & \text{if } \eta < \beta; \\ (\alpha+1) + (\eta - \beta) & \text{if } \eta \geq \beta. \end{cases}$$

So $u : [0, \mathrm{lh}(\mathcal{T})) \cong [0, \beta) \cup [\alpha+1, (\alpha+1) + (\theta + 1 - \beta))$ order-preservingly. We shall define $N_{u(\eta)}$, and an elementary map

$$t_\eta : M_\eta \to N_{u(\eta)}.$$

For $\eta < \beta$, $u(\eta) = \eta$ and $M_\eta = N_\eta$ and $t_\eta =$ identity. We let

$$t_\beta = \text{canonical embedding of } N^*_{\alpha+1} \text{ into } N_{\alpha+1}.$$

(So the display above is a bit off; for $\eta = \beta$, t_η may not act on all of M_η. For $\eta \neq \beta$, t_η will act on all of M_η.) Note that F is close to $N^*_{\alpha+1}$ because it arose in a later model of $\mathcal{T}$, so t_β is cofinal and elementary.

We define t_η and $N_{u(\eta)}$ for $\eta \geq \beta + 1$ by induction.

For $\eta = \beta + 1$, we let

$$F_{u(\beta)} = t_\beta(E_\beta),$$

and let $\tau \leq \beta$ be least such that $\text{crit}(F_{u(\beta)}) < \hat{\lambda}(F_\tau)$, and $\langle \gamma, k\rangle$ be least such that $\text{crit}(F_{u(\beta)}) \geq \rho_{k+1}(N_\tau|\gamma)$, and set

$$N_{u(\beta+1)} = \text{Ult}(N_\tau|\langle \gamma, k\rangle, E^{\mathcal{W}}_{u(\beta)}),$$

as required by normality. We get $t_{\beta+1}$ from the Shift Lemma. There are two cases, based on the location of $\mu = \text{crit}(F)$.

Case A. $\mu \leq \text{crit}(E_\beta)$.

Since $t_\beta = i_F^{M_\beta|\langle \xi_0, k_0\rangle}$, $\text{crit}(t_\beta(E_\beta)) \geq \hat{\lambda}(F)$. But $F = F_\alpha$. Thus $F_{u(\beta)}$ is applied to $N_{\alpha+1}$, or an initial segment of it. That is

$$\tau = u(\beta) = \alpha + 1$$

in this case. In $\mathcal{T}$, we must have

$$T\text{-pred}(\beta+1) = \beta,$$

because β was the least ξ such that $\mu < \hat{\lambda}(E_\xi)$. Similarily, the case hypothesis implies that

$$M_{\beta+1} = \text{Ult}(M_\beta|\langle \xi_1, k_1\rangle, E_\beta)$$

where $\langle \xi_1, k_1\rangle \leq_{\text{lex}} \langle \xi_0, k_0\rangle$. We have that $t_\beta : M_\beta|\langle \xi_1, k_1\rangle \to t_\beta(M_\beta|\langle \xi_1, k_1\rangle)$ is elementary, so the Shift Lemma applies, and we can set

$$t_{\beta+1} = \text{copy map associated to } (t_\beta, t_\beta, E_\beta).$$

We are copying an internal ultrapower under an elementary map, so $t_{\beta+1}$ is elementary. (See Remark 2.5.21.)

Case B. $\text{crit}(E_\beta) < \mu$.

Then $\text{crit}(t_\beta(E_\beta)) = \text{crit}(E_\beta)$, so $\tau = T\text{-pred}(\beta+1) = W\text{-pred}(u(\beta+1))$. It is clear that E_β and $t_\beta(E_\beta)$ are applied to the same initial segment S of $M_\tau = N_\tau$. The Shift Lemma applies to $t_\beta : M_\beta \to N_{u(\beta)}$ and $\text{id} : S \to S$, and we let

$$t_{\beta+1} = \text{copy map associated to } (\text{id}, t_\beta, E_\beta).$$

Again, $t_{\beta+1}$ is elementary[7], and $t_{\beta+1}$ agrees with t_β on $\text{lh}(E_\beta)+1$.

Remark 6.2.2. In Case A, $u(T\text{-pred}(\beta+1)) = W\text{-pred}(u(\beta+1))$, while in Case B, this fails, and in fact $T\text{-pred}(\beta+1) = W\text{-pred}(\beta+1)$. It is because u may not preserve point-of-application for extenders that $\mathcal{T}$ may not be a hull of $\mathcal{W}$, under u and the t_η's, in the sense of Sargsyan's thesis [42]. In fact, $\mathcal{T}$ will be such a hull iff $\text{crit}(E_\eta) \geq \mu$ for all $\eta \geq_T \beta$. For example, this happens when $\mathcal{T}$ factors as $\mathcal{T}\restriction(\beta+1)^\frown\mathcal{S}$, where $\mathcal{S}$ is a tree on M_β with all critical points $\geq \mu$.

The successor case when $\eta > \beta$ is similar. Suppose by induction that whenever $\xi, \delta \leq \eta$:

---

[7]It is easy to see that $\langle \text{id}, t_\beta\rangle : (S, E_\beta) \overset{*}{\to} (S, \pi_\beta(E_\beta))$, so Corollary 2.5.20 apllies here.

(1) $F_{u(\delta)} = t_\delta(E_\delta)$.

(2) If $\delta \neq \beta$, then t_δ is an elementary embedding from M_δ to $N_{u(\delta)}$. (t_β is elementary from $M_\beta|\langle\xi_0,k_0\rangle$ to $N_{u(\beta)}$.)

(3) If $\xi < \delta$, then t_δ agrees with t_ξ on $\mathrm{lh}(E_\xi)+1$.

(4) (a) If T-pred$(\delta) \neq \beta$, then $u(T\text{-pred}(\delta)) = W\text{-pred}(u(\delta))$.
 (b) If T-pred$(\delta) = \beta$, then
  (i) $\mathrm{crit}(E_{\delta-1}) \geq \mu \implies u(T\text{-pred}(\delta)) = W\text{-pred}(u(\delta))$.
  (ii) $\mathrm{crit}(E_{\delta-1}) < \mu \implies W\text{-pred}(u(\delta)) = \beta$.
 (c) (i) If $\delta \neq \beta$, then ($\delta\, T\, \xi$ iff $u(\delta)\, W\, u(\xi)$).
  (ii) $\beta\, T\, \xi \implies u(\beta)\, W\, u(\xi)$ iff the first extender used in $(\beta,\xi]_T$ has critical point $\geq \mu$.

(5) (a) if $\delta \neq \beta$, then $\delta \in D^{\mathcal{T}}$ iff $u(\delta) \in D^{\mathcal{W}}$, and
 (b) if $\delta \neq \beta$, $\delta\, T\, \xi$, and $D^{\mathcal{T}} \cap (\xi,\delta]_T = \varnothing$, then $t_\xi \circ i^{\mathcal{T}}_{\delta,\xi} = i^{\mathcal{W}}_{u(\delta),u(\xi)} \circ t_\delta$.

We then define $t_{\eta+1}: M_{\eta+1} \to N_{u(\eta+1)}$ so as to maintain those conditions. Namely,

$$F_{u(\eta)} = t_\eta(E_\eta),$$

and letting τ be least such that $\mathrm{crit}(F_{u(\eta)}) < \hat{\lambda}(E_\tau)$, and $\langle\gamma,k\rangle$ be appropriate for normal trees,

$$N_{u(\eta+1)} = \mathrm{Ult}(N_\tau|\langle\gamma,k\rangle, F_{u(\eta)}).$$

We get $t_{\eta+1}$ from the Shift Lemma, with two cases, as before.

Case A. $\mu \leq \mathrm{crit}(E_\eta)$.

Let $\sigma = T$-pred$(\eta+1)$, i.e., σ is least such that $\mathrm{crit}(E_\eta) < \hat{\lambda}(E_\sigma)$. Clauses (1) and (3) above tell us that $u(\sigma)$ is the least θ in $\mathrm{ran}(u)$ such that $\mathrm{crit}(F_{u(\eta)}) < \hat{\lambda}(F_\theta)$. But $\tau \geq u(\beta)$ by our case hypotheses, so $\tau \in \mathrm{ran}(u)$, so $\tau = u(\sigma)$. We leave it to the reader to show that if

$$M_{\eta+1} = \mathrm{Ult}(M_\sigma|\langle\lambda,i\rangle, E_\eta),$$

then in fact $i = k$, and $t_\sigma(\lambda) = \gamma$.

We claim that the Shift Lemma applies, in that

$$\langle t_\sigma, t_\eta\rangle\colon (M_\sigma|\langle\lambda,i\rangle, E_\eta) \xrightarrow{*} (N_{u(\sigma)}|\langle t(\lambda),i\rangle, t_\eta(E_\eta)).$$

The proof is the same as that in the successor step of the Copy Lemma 4.5.17. Let $P = M_\sigma|\langle\lambda,i\rangle$, $Q = N_{u(\sigma)}|\langle t(\lambda),i\rangle$, $E = E_\eta$, and $F = t_\eta(E_\eta)$. Our inductive agreement hypotheses imply that $\langle t_\sigma,t_\eta\rangle\colon (P,E) \to (Q,F)$, so we just need to see that this is a Σ_1 embedding of the extenders. If $\sigma = \eta$, that follows from Remark 2.5.21. If E is very close to both $M_\eta|\mathrm{lh}(E)$ and P, then Lemma 4.5.15 shows that in fact $\langle t_\sigma,t_\eta\rangle\colon (P,E) \xrightarrow{**} (Q,F)$. Finally, we have the case that η is special in $\mathcal{T}$. This implies that $u(\eta)$ is special in $\mathcal{U}$. Let $I = I^{\mathcal{T}}_{\sigma,\eta}$ and $J = I^{\mathcal{W}}_{u(\sigma),u(\eta)}$ be the two well supported branch extenders. By Lemma 4.5.8 they are very close to P and Q respectively, and by Lemma 4.5.16,

$$\langle t_\sigma, t_\eta \rangle \colon (P,I) \stackrel{**}{\to} (Q,J).$$

But then for any finite $c \subseteq \varepsilon(E)$, I_c is a good code of E_c over P, and $J_{t_\eta(c)}$ is a good code of $F_{t_\eta(c)}$ over Q, moreover, $t_\sigma(I_c) = J_{t_\eta(c)}$. As in the proof of the Copy Lemma 4.5.17, this implies that $\langle t_\sigma, t_\eta \rangle \colon (P,E) \stackrel{*}{\to} (Q,F)$, which is our claim.

Since the Shift Lemma applies, we may set

$$t_{\eta+1} = \text{copy map associated to } (t_\sigma, t_\eta, E_\eta),$$

and everything works out so that (1)-(5) still hold.

Case B. $\text{crit}(E_\eta) < \mu$.

Again, let $\sigma = T\text{-pred}(\eta+1)$. So $\sigma \leq \beta$. Since $t_\eta \restriction \text{lh}(E_\beta) = t_\beta \restriction \text{lh}(E_\beta)$, $t_\eta \restriction \mu =$ identity, so $\text{crit}(E_\eta) = \text{crit}(F_{u(\eta)})$. Thus $\sigma = \tau$. One can show that E_η and $F_{u(\eta)}$ are applied to the same initial segment S of $M_\tau = N_\tau$, via ultrapowers of the same degree. The Shift Lemma applies to $(\text{id}_S, t_\eta, E_\eta)$, and in fact

$$\langle \text{id}, t_\eta \rangle \colon (S, E_\eta) \stackrel{*}{\to} (S, t_\eta(E_\eta)),$$

by the proof given in Case A. We let

$$t_{\eta+1} = \text{copy map associated to } (t_\eta, \text{id}_S,, E_\eta),$$

and $t_{\eta+1}$ is elementary, and (1)-(5) still hold.

This finishes the definition of $t_{\eta+1}$. For λ a limit, $N_{u(\lambda)}$ and $t_\lambda : M_\lambda \to N_{u(\lambda)}$ are defined by

$$N_{u(\lambda)} = \text{dirlim of } N_{u(\alpha)} \text{ for } \alpha\, T\, \lambda \text{ sufficiently large},$$
$$t_\lambda(i^{\mathcal{T}}_{\alpha\lambda}(x)) = i^{\mathcal{W}}_{u(\alpha),u(\lambda)}(t_\alpha(x)), \text{ for } \alpha\, T\, \lambda \text{ sufficiently large}.$$

(1)-(5) imply this makes sense, and that (1)-(5) continue to hold. This completes our description of the embedding normalization of $\mathcal{T}^\frown\langle F\rangle$.

We must see that $\text{Ult}(Q,F)$ embeds into the last model of $\mathcal{W}$. If $Q \lhd M_\theta$, then we are in Case 1, and $\text{Ult}(Q,F)$ *is* the last model of $\mathcal{W}$, so let us assume that $Q = M_\theta$.

LEMMA 6.2.3. *For any $\gamma \geq \beta$, F is an initial segment of the extender of t_γ.*

PROOF. F is the extender of t_β. Since $t_\beta \restriction (\mu^+)^{M_\beta|\xi} = t_\gamma \restriction (\mu^+)^{M_\beta|\xi}$ (because $(\mu^+)^{M_\beta|\xi} < \text{lh}(E_\beta)$), we are done. ⊣

Thus there is a natural factor embedding τ from $\text{Ult}(Q,F)$ into R, where $R = N_{u(\theta)}$. Letting $n = k(Q)$, we have that $t_\theta \colon Q^n \to R^n$ is Σ_1 elementary. τ completes τ_0, where

$$\tau_0([a,f]^{Q^n}_F) = t_\theta(f)(a).$$

Here f ranges over functions belonging to Q^n.[8]

LEMMA 6.2.4. *τ is nearly elementary.*

PROOF. Let G be the extender of t_θ, so that F is an initial segment of G, and τ is the natural map from $\mathrm{Ult}(Q,F)$ to $\mathrm{Ult}(Q,G)$. As in the proof of the Shift Lemma for Conversion Systems, τ is nearly elementary. ⊣

Remark 6.2.5. There is an analogous construction that starts with an ms-normal tree $\mathcal{T}$ on M, and an extender F on the sequence of its last model N, and produces an ms-normal tree $\mathcal{W}^{\mathrm{ms}}(\mathcal{T},F)$ such that $\mathrm{Ult}(N,F)$ embeds into its last model.

DEFINITION 6.2.6. For $\mathcal{U}$ a normal iteration tree on M, let

$$\mathcal{U}^{<\gamma} = \mathcal{U}\restriction(\alpha+1), \text{ where } \alpha \text{ is least such that } \mathrm{lh}(E^{\mathcal{U}}_\alpha) \geq \gamma,$$

and $\mathcal{U}^{<\gamma} = \mathcal{U}$ if there is no such α. Let

$$\mathcal{U}^{>\gamma} = \langle \mathcal{M}^{\mathcal{U}}_\eta \mid E^{\mathcal{U}}_\eta \text{ exists} \wedge \gamma < \lambda(E^{\mathcal{U}}_\eta)\rangle.$$

DEFINITION 6.2.7. Let M, $\mathcal{T}$, F and $\mathcal{W}$ be as above, then we write

$$W(\mathcal{T},F) = \mathcal{T}^{<\mathrm{lh}(F)}{}^\frown\langle F\rangle^\frown i_F\text{``}\mathcal{T}^{>\mathrm{crit}(F)}$$

for the embedding normalization of $\mathcal{T}^\frown\langle F\rangle$ just defined. We write $\alpha^{\mathcal{T},F}$, $\beta^{\mathcal{T},F}$, $u^{\mathcal{T},F}$, and $t_\xi^{\mathcal{T},F}$ for the auxiliary objects α,β,u,t_ξ that we defined above.

Thus $\alpha(\mathcal{T},F)$ is the least γ such that F^- is on the $\mathcal{M}^{\mathcal{T}}_\gamma$-sequence, and $\beta(\mathcal{T},F)$ is the least γ such that $\mathrm{crit}(F) < \hat{\lambda}(E^{\mathcal{T}}_\gamma)$ or $\gamma = \xi$.

Remark 6.2.8. There is nothing guaranteeing that the models of $W(\mathcal{T},F)$ are wellfounded. In our context of interest, $\mathcal{T}$ is played according to an iteration strategy Σ. Part of "normalizing well" for Σ will then be that $W(\mathcal{T},F)$ is according to Σ.

Here are some illustrations related to $W(\mathcal{T},F)$ that the reader may or may not find helpful. Let $\mathcal{T}$ be normal on M of length $\theta+1$, F on the sequence of $\mathcal{M}^{\mathcal{T}}_\theta$, $\mu = \mathrm{crit}(F)$, β least such that $\mu < \hat{\lambda}(E^{\mathcal{T}}_\beta)$, and α least such that F is on the sequence of $\mathcal{M}^{\mathcal{T}}_\alpha$, as above. We assume in the diagram that $\beta < \theta$, and that $\mathrm{Ult}(\mathcal{M}^{\mathcal{T}}_\theta,F)$ makes sense. Let $u : \theta \cong [0,\beta)\cup[\alpha+1,(\alpha+1)+(\theta-\beta)]$ be the order-isomorphism as above.

In Figure 6.2.1, we illustrate first the embedding of $\mathcal{T}$ into $\mathcal{W}(\mathcal{T},F)$, as it appears in the agreement diagrams. We draw them as if $\beta < \alpha$, although $\beta = \alpha$ is possible.

[8] Alternatively, $\tau([a,f]^Q_F) = t_\theta(f)(a)$, where f is $r\Sigma^Q_n$.

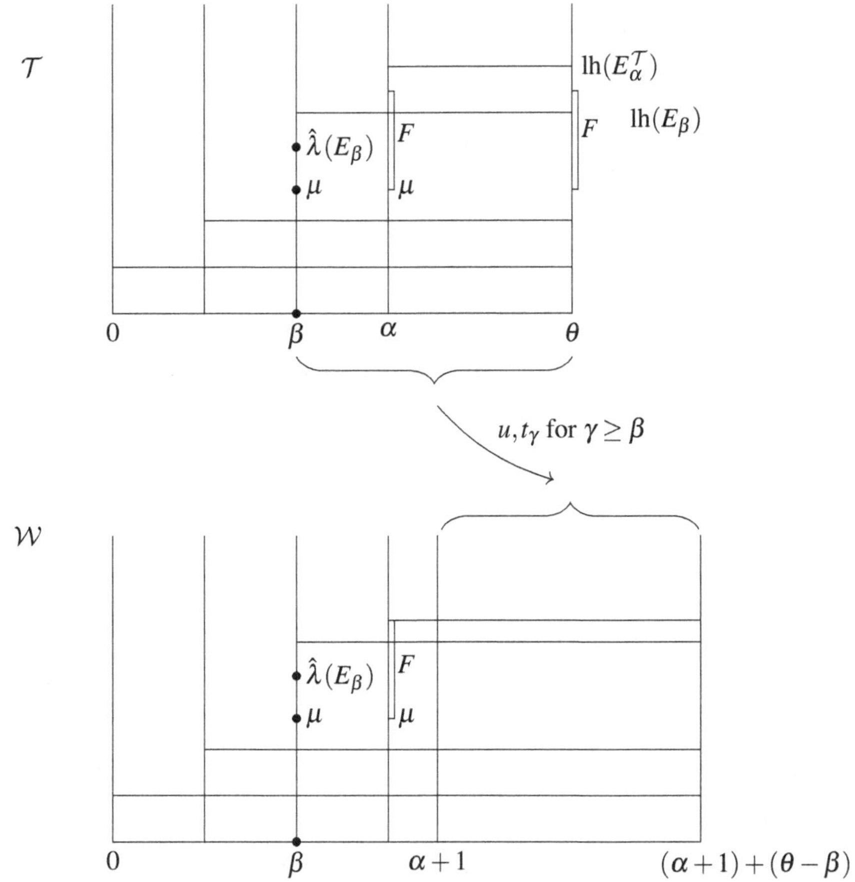

FIGURE 6.2.1

We have

$$\mathcal{T}\restriction(\alpha+1) = \mathcal{W}\restriction(\alpha+1),$$
$$F = E_\alpha^{\mathcal{W}},$$

and

$$i_F\text{“}\mathcal{T}^{>\mu} = \text{remainder of } \mathcal{W}.$$

The next diagram shows how u may fail to preserve tree order. By (4)(c) above, we can have $\delta \leq_T \xi$ but $u(\delta) \not\leq_W u(\xi)$ iff $\delta = \beta$, and the first extender G used in $(0, \xi)_T$ such that G is applied to an initial segment of $\mathcal{M}_\beta^{\mathcal{T}}$ satisfies $\text{crit}(G) < \mu$. Let $S^{<\mu}$ be the set of such $\xi >_{\mathcal{T}} \beta$, and $S^{\geq\mu}$ the remaining $\xi >_{\mathcal{T}} \beta$. The picture is

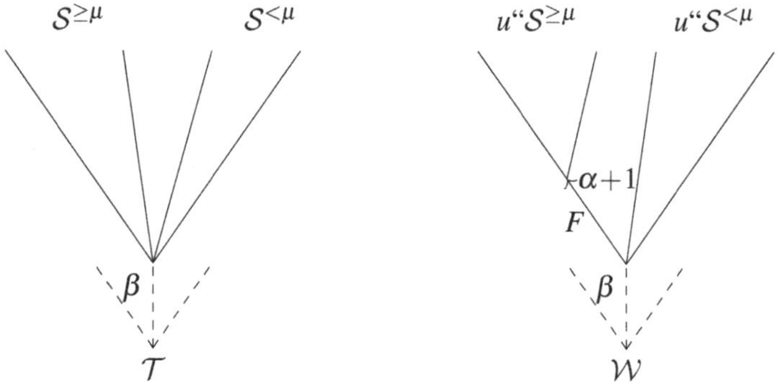

Finally, we illustrate the relationship between the branch extenders of $[0,\xi)_T$ and $[0,u(\xi))_W$. If $\xi < \beta$, they are equal. For $\xi = \beta$, the picture is

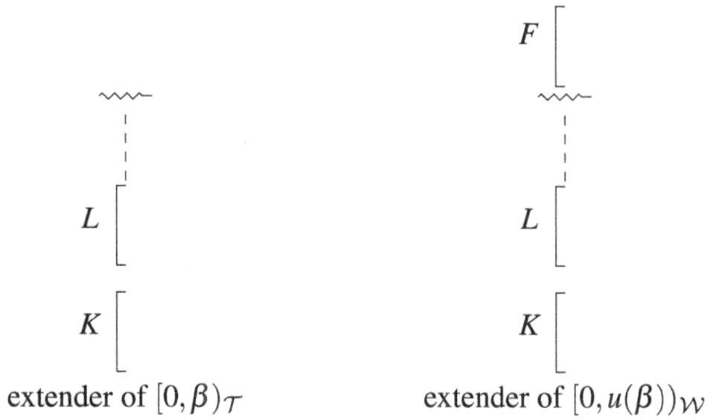

because $[0,\beta)_T \subseteq [0,u(\beta))_W$, and just the one additional extender F is used.

For $\xi > \beta$, let G be the first extender used in $[0,\xi)_T$ such that $\hat{\lambda}(G) \geq \hat{\lambda}(E_\beta^\mathcal{T})$. The picture depends on whether $\mu \leq \text{crit}(G)$. If $\mu \leq \text{crit}(G)$, it is

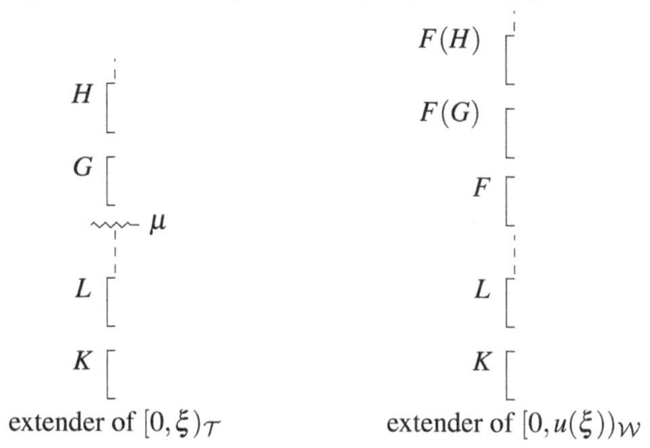

In this case, F is used on $[0,u(\xi))_W$, and the remaining extender used are the images of old ones under copy maps.

If $\mathrm{crit}(G) < \mu < \lambda(G)$, the picture is

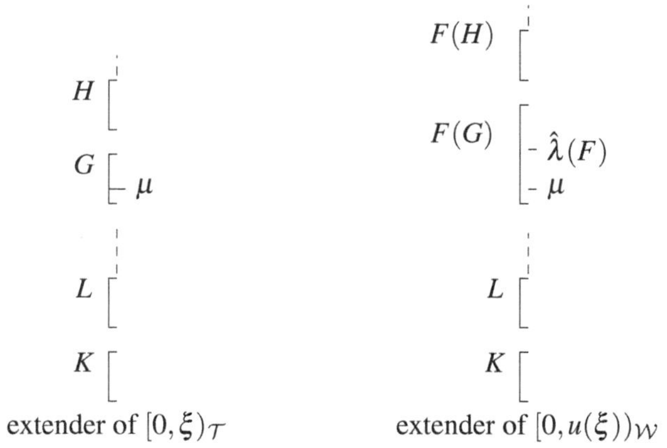

In this case, the two branches use the same extenders until G is used on $[0,\xi)_T$. At that point and after, $[0,u(\xi))_W$ uses the images of extenders under the copy maps.

Notice that in either case, there is an L used in $[0,u(\xi))_W$ such that $\mathrm{crit}(L) \leq \mathrm{crit}(F) < \hat{\lambda}(F) \leq \hat{\lambda}(L)$. This will be important later.

Full normalization. The definition of $W(\mathcal{T},F)$ makes perfect sense in the coarse case, in which $\mathcal{T}$ is a nice, quasi-normal tree on some M satisfying ZFC, and F is a nice extender in the last model Q of $\mathcal{T}$. In this case, we set

$$\alpha(\mathcal{T},F) = \text{least } \eta \text{ such that } F \in \mathcal{M}_\eta^\mathcal{T},$$

and

$$\beta(\mathcal{T},F) = \text{ least } \eta \text{ such that } \mathrm{crit}(F) < \mathrm{lh}(E_\eta^\mathcal{T}) \text{ or } \eta+1 = \mathrm{lh}(\mathcal{T}).$$

As in the fine case, we set $u(\xi) = \xi$ for $\xi < \beta$, and $u(\xi) = \alpha+1+(\xi-\beta)$ if $\beta \leq \xi \leq \theta$. The construction gives us fully elementary maps $t_\xi : \mathcal{M}_\xi^\mathcal{T} \to \mathcal{M}_{u(\xi)}^\mathcal{W}$.

In the coarse case, $\mathrm{Ult}(Q,F)$ is equal to the last model of $W(\mathcal{T},F)$, and the factor map τ that we defined in the fine case is the identity.

PROPOSITION 6.2.9. *Let $M \models$ ZFC, and let $\mathcal{T}$ be a nice, quasi-normal tree on M of length $\theta+1$. Let F be a nice extender in $\mathcal{M}_\theta^\mathcal{T}$, and let $\mathcal{W} = W(\mathcal{T},F)$; then for all γ such that $\beta(\mathcal{T},F) \leq \gamma \leq \theta$,*

$$\mathcal{M}_{u(\gamma)}^\mathcal{W} = \mathrm{Ult}(\mathcal{M}_\gamma^\mathcal{T},F),$$

and the embedding normalization map t_γ is the same as the F-ultrapower map.

PROOF. We show this by induction on γ. For $\gamma = \beta$, this is the definition of $\mathcal{M}_{u(\beta)}^\mathcal{W}$ and t_β. Suppose it holds for all $\gamma \leq \eta$, we must show it holds at $\eta+1$. Let $E = E_\eta^\mathcal{T}$ and $E^* = t_\eta(E) = E_{u(\eta)}^\mathcal{W}$. Let $\sigma = T\text{-pred}(\eta+1)$.

Case 1. $\mu \leq \operatorname{crit}(E)$.

Then $\sigma \geq \beta$, and $u(\sigma) = W\text{-pred}(u(\eta+1))$. Let $S = \operatorname{Ult}(\mathcal{M}^{\mathcal{T}}_{\eta+1}, F)$, and let $i_F^{\mathcal{M}^{\mathcal{T}}_{\eta+1}}$ be the canonical embedding. We have the diagram

$$\begin{array}{ccccc}
\mathcal{M}^{\mathcal{T}}_{\eta+1} & \xrightarrow{\;i_F^{\mathcal{M}^{\mathcal{T}}_{\eta+1}}\;} & S & \xrightarrow{\;\tau\;} & \mathcal{M}^{\mathcal{W}}_{u(\eta+1)} \\
\uparrow E & & & \nearrow E^* & \\
\mathcal{M}^{\mathcal{T}}_{\sigma} & \xrightarrow[\;i_F^{\mathcal{M}^{\mathcal{T}}_{\sigma}} = t_\sigma\;]{} & \mathcal{M}^{\mathcal{W}}_{u(\sigma)} & &
\end{array}$$

Here τ comes from the argument in Case 1 of two-step normalization. Namely, let G be the extender of $i_F^{\mathcal{M}^{\mathcal{T}}_{\eta+1}} \circ i_E^{\mathcal{M}^{\mathcal{T}}_{\sigma}}$, and H be the extender of $i_{E^*}^{\mathcal{M}^{\mathcal{W}}_{u(\sigma)}} \circ i_F^{\mathcal{M}^{\mathcal{T}}_{\sigma}}$. Note $\nu(G) = \sup i_F^{\mathcal{M}^{\mathcal{T}}_{\eta+1}}``\operatorname{lh}(E)$ and $\nu(H) = \sup i_F^{\mathcal{M}^{\mathcal{T}}_{\eta}}``\operatorname{lh}(E)$. Here we use that $\mathcal{T}$ is nice, so $\operatorname{lh}(E)$ is inaccessible in $\mathcal{M}^{\mathcal{T}}_{\eta}$, so $\operatorname{lh}(E^*) = \sup i_F^{\mathcal{M}^{\mathcal{T}}_{\eta}}``\operatorname{lh}(E) = \sup t_\eta``\operatorname{lh}(E)$.[9]

CLAIM 6.2.10. *G is a subextender of H under the map ψ, where*

$$\psi([b,g]_F^{\mathcal{M}^{\mathcal{T}}_{\eta+1}}) = [b,g]_F^{\mathcal{M}^{\mathcal{T}}_{\eta}},$$

for $b \in [\operatorname{lh}(F)]^{<\omega}$ and $g : [\mu]^{|b|} \to \operatorname{lh}(E)$, $g \in \mathcal{M}^{\mathcal{T}}_{\eta+1}$.

PROOF. We calculate as before: for b, g as above and $A \subseteq [\operatorname{crit}(E)]^{<\omega}$ with $A \in \mathcal{M}^{\mathcal{T}}_{\sigma}$,

$$\begin{aligned}
([b,g]_F^{\mathcal{M}^{\mathcal{T}}_{\eta+1}}, A) \in G \text{ iff } & [b,g]_F^{\mathcal{M}^{\mathcal{T}}_{\eta+1}} \in i_F^{\mathcal{M}^{\mathcal{T}}_{\eta+1}} \circ i_E^{\mathcal{M}^{\mathcal{T}}_{\sigma}}(A) \\
\text{iff } & \text{for } F_b \text{ a.e. } u,\ g(u) \in i_E^{\mathcal{M}^{\mathcal{T}}_{\sigma}}(A)
\end{aligned}$$

(by Łoś's Theorem for $\operatorname{Ult}(\mathcal{M}^{\mathcal{T}}_{\eta+1}, F)$)

$$\begin{aligned}
\text{iff } & \text{for } F_b \text{ a.e. } u,\ (g(u), A) \in E \\
\text{iff } & ([b,g]_F^{\mathcal{M}^{\mathcal{T}}_{\eta}}, i_F^{\mathcal{M}^{\mathcal{T}}_{\eta}}(A)) \in E^*
\end{aligned}$$

(by Łoś's Theorem for $\operatorname{Ult}(\mathcal{M}^{\mathcal{T}}_{\eta}, F)$)

$$\text{iff } [b,g]_F^{\mathcal{M}^{\mathcal{T}}_{\eta}} \in i_{E^*}^{\mathcal{M}^{\mathcal{T}}_{u(\sigma)}}(i_F^{\mathcal{M}^{\mathcal{T}}_{\eta}}(A))$$

(since $i_{E^*}^{\mathcal{M}^{\mathcal{T}}_{u(\sigma)}}$ and $i_{E^*}^{\mathcal{M}^{\mathcal{T}}_{\eta}}$ agree on subsets of $\operatorname{crit}(E^*)$)

$$\text{iff } [b,g]_F^{\mathcal{M}^{\mathcal{T}}_{\eta}} \in i_{E^*}^{\mathcal{M}^{\mathcal{T}}_{u(\sigma)}}(i_F^{\mathcal{M}^{\mathcal{T}}_{\sigma}}(A))$$

---

[9] We do not really need that $\mathcal{T}$ is nice; the proposition is true under the weaker assumption that $\mathcal{M}^{\mathcal{T}}_{\eta} \models \operatorname{cof}(\operatorname{lh}(E_\eta)^{\mathcal{T}}) \neq \mu$ for all η, with $\operatorname{lh}(E^{\mathcal{T}}_{\eta})$ being both the strength and the sup of generators of $E^{\mathcal{T}}_{\eta}$ in $\mathcal{M}^{\mathcal{T}}_{\eta}$.

(since $i_F^{\mathcal{M}_\eta^{\mathcal{T}}}$ agrees with t_η, hence t_γ, hence $i_F^{\mathcal{M}_\sigma^{\mathcal{T}}}$ on subsets of $\mathrm{crit}(E)$)

$$\text{iff } ([b,g]_F^{\mathcal{M}_\eta^{\mathcal{T}}}, A) \in H. \qquad \dashv$$

But now $\mathcal{M}_\eta^{\mathcal{T}}$ and $\mathcal{M}_{\eta+1}^{\mathcal{T}}$ have the same functions $g : [\mu]^{<\omega} \to \mathrm{lh}(E)$, by our "coarseness" assumptions. So $\psi =$ identity, and $G = H$, and $S = \mathcal{M}_{u(\eta+1)}^{\mathcal{W}}$. So our diagram is

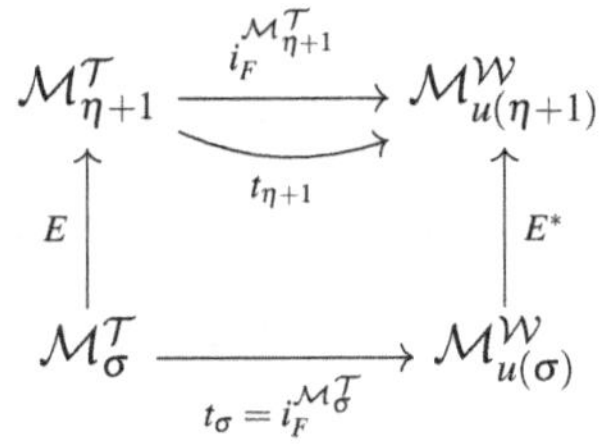

It remains to show $i_F^{\mathcal{M}_{\eta+1}^{\mathcal{T}}} = t_{\eta+1}$. Since both maps make the diagram commute, it is enough to show $i_F^{\mathcal{M}_{\eta+1}^{\mathcal{T}}} \restriction \mathrm{lh}(E) = t_{\eta+1} \restriction \mathrm{lh}(E)$. But $t_{\eta+1} \restriction \mathrm{lh}(E) = t_\eta \restriction \mathrm{lh}(E)$ by the Shift Lemma, and $t_\eta \restriction \mathrm{lh}(E) = i_F^{\mathcal{M}_\eta^{\mathcal{T}}} \restriction \mathrm{lh}(E)$ by induction, and $i_F^{\mathcal{M}_\eta^{\mathcal{T}}} \restriction \mathrm{lh}(E) = i_F^{\mathcal{M}_{\eta+1}^{\mathcal{T}}} \restriction \mathrm{lh}(E)$ because $\mathcal{M}_\eta^{\mathcal{T}}$ and $\mathcal{M}_{\eta+1}^{\mathcal{T}}$ have the same functions $g : [\mu]^{<\omega} \to \mathrm{lh}(E)$.

Case 2. $\mathrm{crit}(E) < \mu$.
Let $\sigma = T\text{-pred}(\eta+1)$. Then in this case, $\sigma = W\text{-pred}(\eta+1)$. Let

$$S = \mathrm{Ult}(\mathcal{M}_{\eta+1}^{\mathcal{T}}, F).$$

We have the diagram

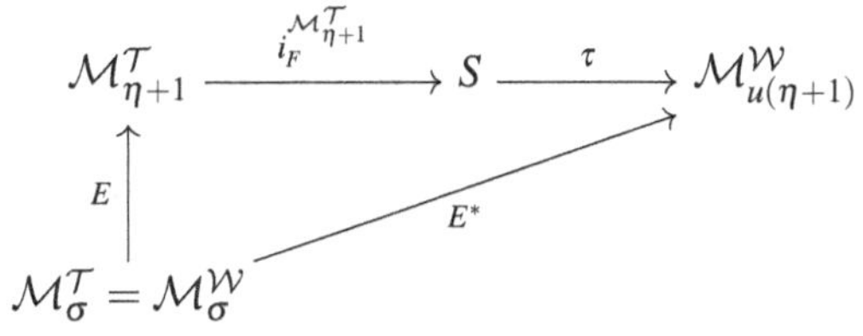

We show that $S = \mathcal{M}_{\phi(\eta+1)}^{\mathcal{W}}$ and $i_F^{\mathcal{M}_{\eta+1}^{\mathcal{T}}} = t_{\eta+1}$ by the calculations in Case 2 of two-step normalization. $\dashv$

Let us return briefly to the fine case. The full normalization $X(\mathcal{T},F)$ of $\mathcal{T}^\frown\langle F\rangle$ can be obtained as follows. We assume that $\mathcal{T}$ is normal on M, N is the last model of $\mathcal{T}$, F is on the N sequence, and $\mathrm{crit}(F) < \rho_n(N)$, for $n = k(N)$. Let

$$\mathcal{W} = \mathcal{T}^{<\mathrm{lh}F\frown}\langle F\rangle^\frown i_F``\mathcal{T}^{>\mathrm{crit}(F)}$$

be the embedding normalization. Let $\mathcal{T}^{<\text{lh}F} = \mathcal{T}\upharpoonright(\alpha+1)$, $\beta = W\text{-pred}(\alpha+1)$, and $u : \text{lh}\,\mathcal{T} \to \text{lh}\,W$ be as above. The full normalization is $\mathcal{X}$, where

$$\mathcal{X}\upharpoonright(\alpha+2) = \mathcal{W}\upharpoonright(\alpha+2)$$

and

$$\mathcal{M}^{\mathcal{X}}_{u(\eta)} = \text{Ult}(\mathcal{M}^{\mathcal{T}}_{\eta}, F) \text{ for } \eta > \beta.$$

(Note that if $\eta > \beta$, then some G such that $\text{crit}(F) = \mu < \hat{\lambda}(G)$ was used on the branch to $\mathcal{M}^{\mathcal{T}}_{\eta}$, so for $k = k(\mathcal{M}^{\mathcal{T}}_{\eta})$, $\mu < \rho_k(\mathcal{M}^{\mathcal{T}}_{\eta})$.) The tree order of $\mathcal{X}$ is the same as that of $\mathcal{W}$. We have

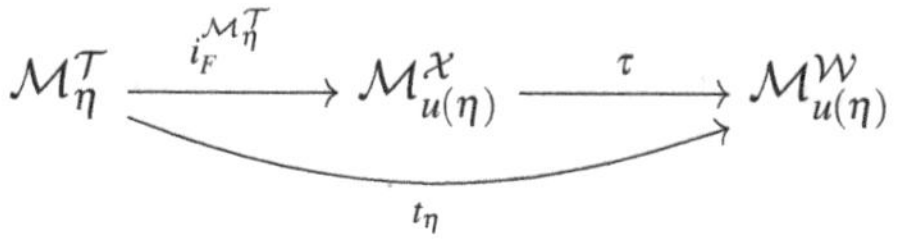

where τ is the natural factor map. What remains is to find the extenders $E^{\mathcal{X}}_{u(\eta)}$ that make $\mathcal{X}$ into a normal iteration tree. For this, let $E = E^{\mathcal{T}}_{\eta}$, and

$$t : \mathcal{M}^{\mathcal{T}}_{\eta}|\langle \text{lh}(E), 0\rangle \to \text{Ult}(\mathcal{M}^{\mathcal{T}}_{\eta}|\langle \text{lh}(E), 0\rangle, F)$$

be the canonical embedding. One can show using condensation that $t(E)$ is on the sequence of $\mathcal{M}^{\mathcal{X}}_{u(\eta)}$. Moreover, for $\sigma = W\text{-pred}(\eta+1)$,

$$\mathcal{M}^{\mathcal{X}}_{u(\eta+1)} = \text{Ult}(\mathcal{M}^{\mathcal{W}}_{\sigma}|\langle \xi, n\rangle, \pi(E)),$$

where $n = k(\mathcal{M}^{\mathcal{W}}_{\eta+1}) = k(\mathcal{M}^{\mathcal{T}}_{\eta+1})$ and ξ is appropriate. The details here are like those in the two-step case. Since we do not actually need full normalization in comparing iteration strategies, we give no further detail here. There is a much more careful discussion in [64]. Here is a diagram of the situation.

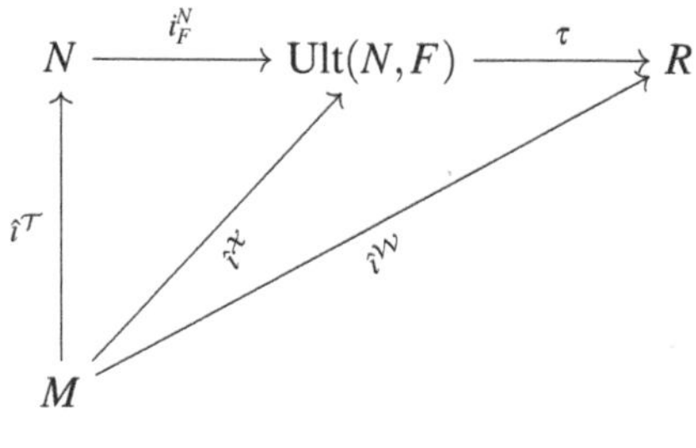

6.3. The extender tree $\mathcal{V}^{\text{ext}}$

The fact that $u^{\mathcal{T},F}$ does not fully preserve tree order or tree predecessor is awkward. Here is another way to visualize our embedding of $\mathcal{T}$ into $W(\mathcal{T}, F)$ given by u and the t_ξ's.

For $\mathcal{V}$ a quasi-normal plus tree, let

$$\text{Ext}(\mathcal{V}) = \{E_\alpha^\mathcal{V} \mid \alpha + 1 < \text{lh}(\mathcal{V})\}$$

be the set of extenders used. Note that $\text{Ext}(\mathcal{V})$ determines $\mathcal{V}$ modulo a strategy Σ for the base model of $\mathcal{V}$, by normality. For $\gamma < \text{lh}(\mathcal{V})$,

$$e_\gamma^\mathcal{V} = \text{increasing enumeration of } \{E_\alpha^\mathcal{V} \mid \alpha + 1 \leq_V \gamma\},$$

increasing in order of use (index, length). Set

$$\mathcal{V}^{\text{ext}} = \{e_\gamma^\mathcal{V} \mid \gamma < \text{lh}\,\mathcal{V}\}.$$

$\mathcal{V}^{\text{ext}}$ determines $\mathcal{V}$. The structure $(\mathcal{V}^{\text{ext}}, \subseteq)$ is the *extender-tree* of $\mathcal{V}$.

If F and G are extenders, then F and G *overlap* iff

$$[\text{crit}(F), \hat{\lambda}(F)) \cap [\text{crit}(G), \hat{\lambda}(G)) \neq \varnothing.$$

We say F and G are *compatible* iff $\exists\alpha(F = G \restriction \alpha \text{ or } G = F \restriction \alpha)$. Here are two elementary facts:

PROPOSITION 6.3.1. *Let $\mathcal{V}$ be a quasi-normal plus tree; then*

(1) if $s^\frown\langle F\rangle \in \mathcal{V}^{\text{ext}}$ and $s^\frown\langle G\rangle \in \mathcal{V}^{\text{ext}}$, then F and G overlap, and
(2) if $s, t \in \mathcal{V}^{\text{ext}}$ and $s(i)$ is compatible with $t(k)$, then $i = k$ and $s\restriction(i+1) = t\restriction(i+1)$.

Now let $\mathcal{T}$ be normal on M, and $\mathcal{W} = W(\mathcal{T}, F)$. Let $u = u^{\mathcal{T},F}$, $t_\xi = t_\xi^{\mathcal{T},F}$, etc. We define a partial map

$$p_{\mathcal{T},F} : \text{Ext}(\mathcal{T}) \to \text{Ext}(\mathcal{W})$$

by

$$p_{\mathcal{T},F}(E_\xi^\mathcal{T}) = t_\xi(E_\xi^\mathcal{T}) = E_{u(\xi)}^\mathcal{W}.$$

So $p_{\mathcal{T},F}(E_\xi^\mathcal{T})\downarrow$ iff $\xi \in \text{dom}\, u$, and either $\xi \neq \beta$, or $\xi = \beta$ and $\mathcal{M}_\beta^\mathcal{T}|\text{lh}(E_\beta^\mathcal{T}) \trianglelefteq \mathcal{M}_{\alpha+1}^{*,\mathcal{W}}$.

We can view p as acting on branch extenders. For $s \in \mathcal{T}^{\text{ext}}$, let

$$i_s^F = i_s = \begin{cases} \text{least } i \text{ such that } \text{crit}(F) < \hat{\lambda}(s(i)) & \text{if this exists,} \\ \text{undefined} & \text{otherwise.} \end{cases}$$

Let $\xi \in \text{dom}\, u$ and $s = e_\xi^\mathcal{T}$; then if $\text{dom}(u) = \beta + 1$, we have

$$e_{u(\xi)}^\mathcal{W} = \begin{cases} s & \text{if } \xi < \beta, \\ s^\frown\langle F\rangle & \text{if } \xi = \beta. \end{cases}$$

If $\mathrm{dom}(u) > \beta+1$, then i_s exists precisely when $s = e^{\mathcal{T}}_{\xi}$ for some $\xi \geq \beta+1$, and

$$e^{\mathcal{W}}_{u(\xi)} = \begin{cases} s & \text{if } \xi < \beta, \\ s^\frown\langle F\rangle & \text{if } \xi = \beta, \\ s\restriction i_s{}^\frown\langle F\rangle^\frown\langle p^{\mathcal{T},F}(s(i)) \mid i \geq i_s\rangle & \text{if } \mathrm{crit}(F) \leq \mathrm{crit}(s(i_s)), \\ s\restriction i_s{}^\frown\langle p^{\mathcal{T},F}(s(i)) \mid i \geq i_s\rangle & \text{if } \mathrm{crit}(s(i_s)) < \mathrm{crit}(F). \end{cases}$$

So if E is used before H in $e^{\mathcal{T}}_{\xi}$, then $p_{\mathcal{T},F}(E)$ is used before $p_{\mathcal{T},F}(H)$ in $e^{\mathcal{W}}_{u(\xi)}$.

DEFINITION 6.3.2. Let $\mathcal{W} = W(\mathcal{T},F)$, and suppose $s \in \mathcal{T}^{\mathrm{ext}}$ is such that $\forall \mu \in \mathrm{dom}(s)$, $p_{\mathcal{T},F}(s(\mu))\downarrow$; then

$$\hat{p}_{\mathcal{T},F}(s) = \text{unique shortest } t \in \mathcal{W}^{\mathrm{ext}} \text{ such that}$$
$$\forall \mu \in \mathrm{dom}(s),\ p_{\mathcal{T},F}(s(\mu)) \in \mathrm{ran}(t).$$

For $\hat{p} = \hat{p}_{\mathcal{T},F}$, we have that $\hat{p}(e^{\mathcal{T}}_{\xi}) = e^{\mathcal{W}}_{u(\xi)}$, except when $\xi = \beta$. At β, we have $e^{\mathcal{W}}_{u(\beta)} = \hat{p}(e^{\mathcal{T}}_{\beta})^\frown\langle F\rangle$. The map $\hat{p}\colon \mathcal{T}^{\mathrm{ext}} \to W(\mathcal{T},F)^{\mathrm{ext}}$ does preserve $\subseteq$.

PROPOSITION 6.3.3. *Let $s,t \in \mathrm{dom}(\hat{p}^{\mathcal{T},F})$; then*

(1) $s \subseteq t \implies \hat{p}(s) \subseteq \hat{p}(t)$, *and*
(2) $s \perp t \implies \hat{p}(s) \perp \hat{p}(t)$.

6.4. Tree embeddings

An iteration strategy Σ for M *condenses well* iff whenever $\mathcal{U}$ is by Σ, and π is a sufficiently elementary embedding from $\mathcal{T}$ into $\mathcal{U}$ such that $\pi\restriction(M \cup \{M\})$ is the identity, then $\mathcal{T}$ is by Σ. By weakening the elementarity required of π, we obtain stronger condensation properties.

In the Hull Condensation property of [42], one is given a map $\sigma : \mathrm{lh}(\mathcal{T}) \to \mathrm{lh}(\mathcal{U})$, and embeddings $\tau_\alpha : \mathcal{M}^{\mathcal{T}}_{\alpha} \to \mathcal{M}^{\mathcal{U}}_{\sigma(\alpha)}$ for $\alpha < \mathrm{lh}(\mathcal{T})$. σ preserves tree order and tree predecessor. The τ_α's have the agreement one would get from a copying construction, and they commute with the branch embeddings of $\mathcal{T}$ and $\mathcal{U}$. Moreover, $\tau_\alpha(E^{\mathcal{T}}_{\alpha}) = E^{\mathcal{U}}_{\sigma(\alpha)}$. A simple example is the way $\mathcal{T} = \pi\mathcal{W}$ sits inside $\mathcal{U} = \pi(\mathcal{W})$, in the case $\pi\colon H \to V$ is elementary and $\pi\restriction(M \cup \{M\}) = \mathrm{id}$.

A hull embedding $(\sigma, \vec{\tau})$ as above induces a map $p : \mathrm{Ext}(\mathcal{T}) \to \mathrm{Ext}(\mathcal{U})$ by

$$p(E^{\mathcal{T}}_{\alpha}) = \tau_\alpha(E^{\mathcal{T}}_{\alpha}).$$

We then get $\hat{p}\colon \mathcal{T}^{\mathrm{ext}} \to \mathcal{U}^{\mathrm{ext}}$ from p as in Definition 6.3.2. $\hat{p}$ preserves $\subseteq$ and incompatibility in the extender trees. $\hat{p}$ is related to σ by

$$\hat{p}(e^{\mathcal{T}}_{\alpha+1}) = e^{\mathcal{U}}_{\sigma(\alpha+1)}.$$

But for λ a limit, $\hat{p}(e^{\mathcal{T}}_{\lambda})$ may be a proper initial segment of $e^{\mathcal{U}}_{\sigma(\lambda)}$.

We now define the notion of a tree embedding from $\mathcal{T}$ into $\mathcal{U}$. This will be a tuple with most of the properties of σ, $\vec{\tau}$, ψ above. The pair $(\sigma, \vec{\tau})$ is resolved into two pairs: the pair $(v, \vec{s})$, which embeds the models of $\mathcal{T}$ into models of $\mathcal{U}$ in a minimal way, and the pair $(u, \vec{t})$, which connects the exit extenders of $\mathcal{T}$ to exit extenders in $\mathcal{U}$. The requirement that σ preserves tree predecessors is relaxed to the requirement that if $\beta = T\text{-pred}(\gamma+1)$, then $U\text{-pred}(u(\gamma)+1) \in [v(\beta), u(\beta)]_U$. We shall also allow the t_α's to be partial, in a controlled way. Recall here the partial branch embeddings $\hat{\imath}^{\mathcal{U}}_{\alpha,\beta}$. Recall also that $\varepsilon(E) = \text{lh}(E)$ if E has plus type, and $\varepsilon(E) = \lambda(E)$ otherwise.

DEFINITION 6.4.1. Let $\mathcal{T}$ and $\mathcal{U}$ be plus trees on a premouse M, with $\text{lh}(\mathcal{T}) > 1$. A *tree embedding* of $\mathcal{T}$ into $\mathcal{U}$ is a system

$$\langle u, v, \langle s_\beta \mid \beta < \text{lh}(\mathcal{T})\rangle, \langle t_\beta \mid \beta+1 < \text{lh}(\mathcal{T})\rangle\rangle$$

such that

(a) $u : \{\alpha \mid \alpha+1 < \text{lh}(\mathcal{T})\} \to \{\alpha \mid \alpha+1 < \text{lh}(\mathcal{U})\}$, and $\alpha < \beta \implies u(\alpha) < u(\beta)$.
(b) v: $\text{lh}(\mathcal{T}) \to \text{lh}(\mathcal{U})$, v preserves tree order and is continuous at limit ordinals, $v(0) = 0$, and $v(\alpha+1) = u(\alpha)+1$.
(c) $s_\beta : \mathcal{M}^{\mathcal{T}}_\beta \to \mathcal{M}^{\mathcal{U}}_{v(\beta)}$ is elementary, and $s_0 = \text{id}$; moreover for $\alpha <_T \beta$,

$$s_\beta \circ \hat{\imath}^{\mathcal{T}}_{\alpha,\beta} = \hat{\imath}^{\mathcal{U}}_{v(\alpha),v(\beta)} \circ s_\alpha.$$

In particular, the two sides have the same domain.
(d) For $\alpha+1 < \text{lh}(\mathcal{T})$, $v(\alpha) \leq_U u(\alpha)$, and

$$t_\alpha = \hat{\imath}^{\mathcal{U}}_{v(\alpha),u(\alpha)} \circ s_\alpha.$$

Moreover, if $E^{\mathcal{T}}_\alpha$ is of plus type, then

$$E^{\mathcal{U}}_{u(\alpha)} = t_\alpha(E^{\mathcal{T}}_\alpha),$$

and if $E^{\mathcal{T}}_\alpha$ is not of plus type, then

$$E^{\mathcal{U}}_{u(\alpha)} \in \{t_\alpha(E^{\mathcal{T}}_\alpha), t_\alpha(E^{\mathcal{T}}_\alpha)^+\}.$$

(e) For $\alpha < \beta < \text{lh}(\mathcal{T})$,

$$s_\beta \restriction \varepsilon(E^{\mathcal{T}}_\alpha) = t_\alpha \restriction \varepsilon(E^{\mathcal{T}}_\alpha).$$

(f) If $\beta = T\text{-pred}(\alpha+1)$, then $U\text{-pred}(u(\alpha)+1) \in [v(\beta), u(\beta)]_U$, and setting $\beta^* = U\text{-pred}(u(\alpha)+1)$, $P = \mathcal{M}^{*,\mathcal{T}}_{\alpha+1}$, and $Q = \mathcal{M}^{*,\mathcal{U}}_{u(\alpha)+1}$,

$$s_{\alpha+1}([a,f]^P_{E^{\mathcal{T}}_\alpha}) = [t_\alpha(a), \hat{\imath}^{\mathcal{U}}_{v(\beta),\beta^*} \circ s_\beta(f)]^Q_{E^{\mathcal{U}}_{u(\alpha)}}.$$

The map $s_{\alpha+1}$ in clause (f) is essentially the copy map associated to $(t_\alpha, \hat{\imath}^{\mathcal{U}}_{v(\beta),\beta^*} \circ s_\beta, E^{\mathcal{T}}_\alpha)$. (It is not literally that if $E^{\mathcal{U}}_\alpha$ is of plus type but $E^{\mathcal{T}}_\alpha$ is not.) We shall show that there is always enough agreement between t_α and $\hat{\imath}^{\mathcal{U}}_{v(\beta),\beta^*} \circ s_\beta$ that the Shift Lemma applies.[10]

Figure 6.4.1 is a diagram to go with (f) of Definition 6.4.1, in the non-dropping case.

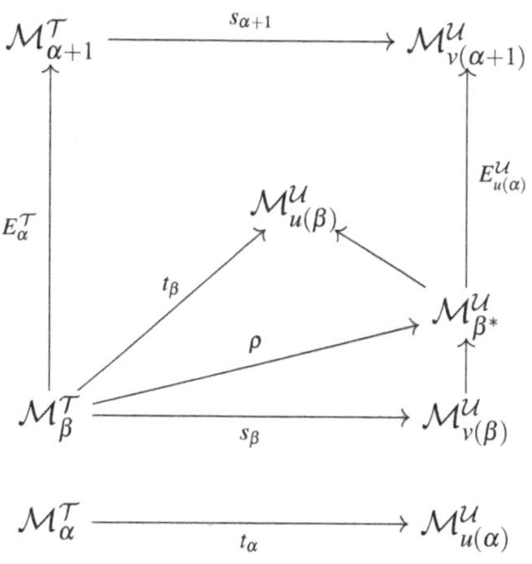

FIGURE 6.4.1. Extending a tree embedding at a successor step.

DEFINITION 6.4.2. For plus trees $\mathcal{T}$ and $\mathcal{U}$,
(a) $\Phi\colon \mathcal{T} \to \mathcal{U}$ iff Φ is a tree embedding of $\mathcal{T}$ into $\mathcal{U}$,
(b) if $\Phi\colon \mathcal{T} \to \mathcal{U}$, then $u^\Phi, v^\Phi, s^\Phi_\alpha$, and t^Φ_α are the component maps of Φ, and
(c) $\mathcal{T}$ is a *pseudo-hull* of $\mathcal{U}$ iff there is a tree embedding of $\mathcal{T}$ into $\mathcal{U}$.

Remark 6.4.3. It is easy to see that $\Phi\colon \mathcal{T} \to \mathcal{U}$ if and only if $\Phi\colon \mathcal{T} \to \mathcal{U} \restriction \gamma$, where $\gamma = \sup(\{v^\Phi(\alpha)+1 \mid \alpha < \mathrm{lh}(\mathcal{T})\})$.

DEFINITION 6.4.4. A tree embedding $\Phi\colon \mathcal{T} \to \mathcal{U}$ is *cofinal* iff

$$\mathrm{lh}(\mathcal{U}) = \sup(\{v^\Phi(\alpha)+1 \mid \alpha < \mathrm{lh}(\mathcal{T})\}).$$

Remark 6.4.5. $v(0) = 0$, but it is possible that $u(0) > 0$. The map u may not preserve tree order.

If $\Phi\colon \mathcal{T} \to \mathcal{U}$ is a tree embedding, then $\mathcal{T}$ and $\mathcal{U}$ have the same base model, and s^Φ_0 is the identity map. One might ask whether there is a natural more general concept, one that allows $\mathcal{M}^{\mathcal{T}}_0 \neq \mathcal{M}^{\mathcal{U}}_0$. Indeed there is, but it reduces to the notion above. Namely, one can have an elementary $\pi\colon \mathcal{M}^{\mathcal{T}}_0 \to \mathcal{M}^{\mathcal{U}}_0$, together with a tree

[10] One can show that, in the notation of clause (f), $\langle \hat{\imath}^{\mathcal{U}}_{v(\beta),\beta^*} \circ s_\beta, t_\alpha\rangle \colon (P, E^{\mathcal{T}}_\alpha) \stackrel{*}{\to} (Q, E^{\mathcal{U}}_{u(\alpha)})$. This is what we need in order to see that $s_{\alpha+1}$ is elementary. The proof is similar to the proof of the corresponding fact in the Copy Lemma. See Lemma 8.2.3.

embedding from the copied tree $\pi\mathcal{T}$ into $\mathcal{U}$. This seems to be the natural way to relate trees on different base models.

Any tree embedding $\Phi\colon \mathcal{T} \to \mathcal{U}$ induces an embedding of extender trees. Namely, let $p : \mathrm{Ext}(\mathcal{T}) \to \mathrm{Ext}(\mathcal{U})$ be given by

$$p(E_\alpha^\mathcal{T}) = E_{u(\alpha)}^\mathcal{U}.$$

We write $p = p^\Phi$. It is easy to see that E is used before F on the same branch of $\mathcal{T}$ iff $p(E)$ is used before $p(F)$ on the same branch of $\mathcal{U}$, so that p induces $\hat{p} : \mathcal{T}^{\mathrm{ext}} \to \mathcal{U}^{\mathrm{ext}}$ as in Definition 6.3.2. The map v on model indices corresponds to the map $\hat{p}$ on extender trees via

$$e_{v(\beta)}^\mathcal{U} = \hat{p}(e_\beta^\mathcal{T}).$$

PROPOSITION 6.4.6. *Let $\Phi\colon \mathcal{T} \to \mathcal{U}$ be a tree embedding, let $p = p^\Phi$, and let $\hat{p}\colon \mathcal{T}^{\mathrm{ext}} \to \mathcal{U}^{\mathrm{ext}}$ be the induced map on extender trees; then Let $s,t \in \mathrm{dom}(\hat{p}^{\mathcal{T},F})$; then*

(1) $s \subseteq t \implies \hat{p}(s) \subseteq \hat{p}(t)$, and
(2) $s \perp t \implies \hat{p}(s) \perp \hat{p}(t)$.

Remark 6.4.7. Given $u(\alpha)$ and t_α, we can characterize $v(\alpha)$ as the least $\xi \le_U u(\alpha)$ such that $\mathrm{ran}(t_\alpha) \subseteq \mathrm{ran}(\hat{\imath}_{\xi,u(\alpha)}^\mathcal{U})$.

Let us record the agreement properties of the maps in a tree embedding. In the context of pfs premice, embeddings that agree on $\mathrm{lh}(E)$ will generally be forced to agree on $\mathrm{lh}(E)+1$. For example, in clause (e) of Definition 6.4.1, $s_{\alpha+1}$ agrees with t_α on $\mathrm{lh}(E_\alpha^\mathcal{T})+1$, because the Shift Lemma produces this kind of agreement. One does encounter embeddings that agree on λ_E, but not on λ_E+1. With this in mind, we see that

LEMMA 6.4.8. *Let $\langle u, v, \langle s_\beta \mid \beta < \mathrm{lh}\,\mathcal{T}\rangle, \langle t_\beta \mid \beta+1 < \mathrm{lh}\,\mathcal{T}\rangle\rangle$ be a tree embedding of $\mathcal{T}$ into $\mathcal{U}$; then*

(a) if $\alpha+1 < \mathrm{lh}(\mathcal{T})$, then t_α agrees with s_α on $\varepsilon_\alpha^\mathcal{T}$,
(b) if $\beta < \alpha < \mathrm{lh}(\mathcal{T})$, then s_α agrees with t_β on $\varepsilon(E_\beta^\mathcal{T})$, and
(c) if $\beta < \alpha < \mathrm{lh}(\mathcal{T})$, then s_α agrees with s_β on $\varepsilon_\beta^\mathcal{T}$.

PROOF. For (a), notice that if F is used in $e_\alpha^\mathcal{T}$, then $p(F)$ is used in $e_{v(\alpha)}^\mathcal{U}$, and so $\varepsilon(p(F)) \le \mathrm{crit}(\hat{\imath}_{v(\alpha),u(\alpha)}^\mathcal{U})$. Thus $\sup s_\alpha``\varepsilon_\alpha^\mathcal{T} \le \mathrm{crit}(\hat{\imath}_{v(\alpha),u(\alpha)}^\mathcal{U})$. But $t_\alpha = \hat{\imath}_{v(\alpha),u(\alpha)}^\mathcal{U} \circ s_\alpha$, so we have (a).

Part (b) is just a clause in the definition. Part (c) follows at once from (a) and (b). ⊣

One could not replace $\varepsilon_\alpha^\mathcal{T}$ by $\sup\{\mathrm{lh}(F) \mid F \in \mathrm{ran}(e_\alpha^\mathcal{T})\}$ in the lemma above, even if $\mathcal{T}$ and $\mathcal{U}$ are assumed to be normal. The reason is that there could be a last extender F used in $e_\alpha^\mathcal{T}$. (So $F = E_\beta^\mathcal{T}$ where $\alpha = \beta+1$.) Then $p(F)$ is the last extender used in $e_{v(\alpha)}^\mathcal{U}$. It could be that $\mathrm{crit}(\hat{\imath}_{v(\alpha),u(\alpha)}^\mathcal{U}) = \lambda_{p(F)}$, and thus t_α and

$s_{\gamma+1}$ both disagree with s_α at λ_F. This is the only way the stronger agreement lemma can fail in the case of normal trees.

Remark 6.4.9. The proof of Lemma 8.2.3 in Chapter 8 gives a formula for the point of application of $E^{\mathcal{U}}_{u(\alpha)}$ under a tree embedding of $\mathcal{T}$ into $\mathcal{U}$, namely

$$U\text{-pred}(u(\alpha)+1) = \text{least } \eta \in [v(\beta), u(\beta)]_U \text{ such that}$$
$$\text{crit}\, \hat{\imath}^{\mathcal{U}}_{\eta,u(\beta)} > \hat{\imath}^{\mathcal{U}}_{v(\beta),\eta} \circ s_\beta(\mu),$$

where $\beta = T\text{-pred}(\alpha+1)$ and $\mu = \text{crit}(E^{\mathcal{T}}_\alpha)$.

Remark 6.4.10. It is easy to see that $\mathcal{T}, \mathcal{U}$, and u determine the rest of the tree embedding. For p is given by $p(E^{\mathcal{T}}_\alpha) = E^{\mathcal{U}}_{u(\alpha)}$, and p determines $\hat{p}$ and v. We then determine the copy maps s_α and t_α by induction on α. t_α is determined from s_α by $t_\alpha = \hat{\imath}^{\mathcal{U}}_{v(\alpha),u(\alpha)} \circ s_\alpha$. If α is a limit, we easily get s_α from $v(\alpha)$ and the fact that $s_\alpha \circ \hat{\imath}^{\mathcal{T}}_{\beta,\alpha} = \hat{\imath}^{\mathcal{U}}_{v(\beta),v(\alpha)} \circ s_\beta$ holds whenever $\beta <_T \alpha$. Clause (e) determines $s_{\alpha+1}$ from earlier s and t values.

p determines u, hence p determines the whole of the tree embedding as well. In other words, a tree embedding from $\mathcal{T}$ into $\mathcal{U}$ is an appropriately elementary way of connecting the exit extenders of $\mathcal{T}$ to exit extenders of $\mathcal{U}$.

Remark 6.4.11. Suppose that $\text{lh}(\mathcal{T}) = \alpha+1$ and $\Phi\colon \mathcal{T} \to \mathcal{U}$ is a tree embedding. Let $s = s^\Phi$, $u = u^\Phi$, etc., so that $s_\alpha\colon \mathcal{M}^{\mathcal{T}}_\alpha \to \mathcal{M}^{\mathcal{U}}_{v(\alpha)}$ is our enlargement of the last model of $\mathcal{T}$. Then for all $\beta < \alpha$,

$$s_\alpha(\text{lh}(E^{\mathcal{T}}_\beta)) = \text{lh}(E^{\mathcal{U}}_{u(\beta)}),$$

by Lemma 6.4.8. Thus $s_\alpha, \mathcal{T}$, and $\mathcal{U} \restriction v(\alpha)+1$ determine u, and hence the whole of Φ. As far as Φ is concerned, $M^{\mathcal{U}}_{v(\alpha)}$ is the last relevant model of $\mathcal{U}$. So we can say that if $\mathcal{T}$ has successor length, then a tree embedding from $\mathcal{T}$ to $\mathcal{U}$ is just a map from the last model of $\mathcal{T}$ into some model of $\mathcal{U}$ that is elementary in a certain strong sense.

The reader might wonder why the u-map and t-maps of $\Phi\colon \mathcal{T} \to \mathcal{U}$ are undefined at α, where $\alpha+1 = \text{lh}(\mathcal{T})$. In general, forcing Φ to include a value for $u(\alpha)$ is wrong, because u is being used to connect exit extenders, and $\mathcal{T}$ has not yet chosen an exit extender at α. If we demand Φ include a value for $u(\alpha)$, then what we would like to call extensions of Φ may have to revise this value. That is awkward. (See Lemma 8.2.3 for a characterization of when it is possible to extend $\Phi\colon \mathcal{T} \to \mathcal{U}$ to $\Psi\colon \mathcal{T}^\frown\langle F\rangle \to \mathcal{U}$.)

In the case $\mathcal{U} = W(\mathcal{T}, F)$, there is a natural way to define u and $\vec{t}$ at $\alpha = \text{lh}(\mathcal{T}) - 1$, namely, $u(\alpha) = \text{lh}(\mathcal{U}) - 1$, and $t_\alpha = \hat{\imath}^{\mathcal{U}}_{v(\alpha),u(\alpha)} \circ s_\alpha$. It helps to make a definition here.

DEFINITION 6.4.12. Let $\mathcal{T}$ and $\mathcal{U}$ be normal iteration trees of lengths $\alpha+1>1$ and $\beta+1$, and let $\Phi\colon \mathcal{T}\to\mathcal{U}$ be a tree embedding, with

$$\Phi=\langle u,v,\langle s_\xi \mid \xi\leq\alpha\rangle,\langle t_\xi\mid\xi<\alpha\rangle\rangle.$$

Suppose that $v(\alpha)\leq_U \beta$; then we define

$$\Psi(\Phi,\mathcal{U})=\langle u\cup\{\langle\alpha,\beta\rangle\},v,\langle s_\xi\mid\xi\leq\alpha\rangle,\langle t_\xi\mid\xi<\alpha\rangle^\frown\langle \hat{\imath}^{\mathcal{U}}_{v(\alpha),\beta}\circ s_\alpha\rangle\rangle.$$

We say that Ψ is an *extended tree embedding* iff $\Psi=\Psi(\Phi,\mathcal{U})$ for some Φ and $\mathcal{U}$, and write $\Phi=c(\Psi)$ and $\mathcal{U}=r(\Psi)$ for the unique such Φ and $\mathcal{U}$.

Notice that in Definition 6.4.12 the interval $(v(\alpha),\beta]$ may drop in $\mathcal{U}$, and consequently the last t-map t_α may be only defined on a proper initial segment of $\mathcal{M}^{\mathcal{T}}_\alpha$. Of course, the same was true for the t_ξ such that $\xi<\alpha$.

Extended tree embeddings are not tree embeddings, they are tree embeddings that have been extended in a small way. If $\Phi\colon\mathcal{T}\to\mathcal{U}$ is a cofinal tree embedding, then its extension $\Psi(\Phi,\mathcal{U})$ is completely trivial. In general, an extended tree embedding from $\mathcal{T}$ into $\mathcal{U}$ is completely determined by $\mathcal{T},\mathcal{U}$, and its last s-map.

Remark 6.4.13. $\mathcal{T}$ is a pseudo-hull of $W(\mathcal{T},F)$, and in fact, there is an extended tree embedding $\Psi=\langle u,v,\vec{s},\vec{t},\rangle$ from $\mathcal{T}$ into $W(\mathcal{T},F)$. In our embedding normalization notation, $u=u^{\mathcal{T},F}$, $t_\beta=t^{\mathcal{T},F}_\beta$, and $p(E^{\mathcal{T}}_\xi)=E^{W(\mathcal{T},F)}_{u(\xi)}$ for $\xi+1<\mathrm{lh}(\mathcal{T})$. This determines $\hat{p}$ and v. u agrees with v except at $\beta=\beta^{\mathcal{T},F}$, where we have $v(\beta)=\beta$ and $u(\beta)=\alpha^{\mathcal{T},F}+1$.

Letting $\Phi=c(\Psi)$ be the associated tree embedding, it is easy to see that Φ is cofinal iff $\mathcal{T}^\frown\langle F\rangle$ is not normal.

DEFINITION 6.4.14. Let Φ be a tree embedding from $\mathcal{T}$ into $\mathcal{U}$, and Ψ be a tree embedding from $\mathcal{U}$ into $\mathcal{V}$; then $\Psi\circ\Phi$ is the tree embedding from $\mathcal{T}$ into $\mathcal{V}$ obtained by composing the corresponding component maps of Φ and Ψ. Similarly, if Φ and Ψ are extended tree embeddings, then $\Psi\circ\Phi$ is the extended tree embedding obtained by composing corresponding maps.

It is easy to check that composing corresponding maps does indeed produce a tree embedding or extended tree embedding, as the case may be.

One can extend Definition 6.4.1 in a natural way by allowing s_α to be only nearly elementary, and to map $\mathcal{M}^{\mathcal{T}}_\alpha$ into a proper initial segment of $\mathcal{M}^{\mathcal{U}}_{v(\alpha)}$. One can think of the natural embedding of $\mathcal{T}$ into $\mathcal{T}^+$ as a tree embedding in this sense, with $u=v=\mathrm{id}$. The more general notion of tree embedding leads to a strengthening of strong hull condensation that subsumes internal lift consistency.

6.5. Normalizing $\mathcal{T}^\frown\mathcal{U}$

In this section we define the embedding normalization $W(\mathcal{T},\mathcal{U})$ of a maximal M-stack $\langle\mathcal{T},\mathcal{U}\rangle$ of length 2. It is not hard to extend our definitions so that they apply to arbitrary M-stacks of length 2, but the additional notation introduced by gratuitous dropping would be a burden. We do not need to deal directly with arbitrary finite stacks because, in our context of interest, they can be reduced to maximal stacks. (See Section 5.4.)

To begin with, note that $W(\mathcal{T},F)$ makes sense in somewhat greater generality. Let $\mathcal{T}$ be a normal tree on the premouse M.[11] Let $\mathcal{S}$ be another normal tree on M, and F be on the sequence of the last model of $\mathcal{S}$. Let α be least such that F is on the sequence of $\mathcal{M}_\alpha^\mathcal{S}$, so that $\mathcal{S}\restriction(\alpha+1) = \mathcal{S}^{<\mathrm{lh}(F)}$. Let β be such that $\beta = \mathcal{S}\text{-pred}(\alpha+1)$ would hold in any normal $\mathcal{S}'$ extending $\mathcal{S}\restriction(\alpha+1)$ such that $F = E_\alpha^{\mathcal{S}'}$. That is, $\mathcal{S}\restriction\beta+1 = \mathcal{S}^{<\mathrm{crit}(F)}$. Suppose that

$$\mathcal{T}\restriction\beta+1 = \mathcal{S}\restriction\beta+1.$$

Suppose also that if $\beta+1 < \mathrm{lh}(\mathcal{T})$, then $\mathrm{dom}(F) = \mathcal{M}_\beta^\mathcal{T}|\eta$ for some $\eta < \mathrm{lh}(E_\beta^\mathcal{T})$, that is, assume that

$$\mathcal{T}\restriction\beta+1 = \mathcal{T}^{<\mathrm{crit}(F)}.$$

We define a normal tree $W(\mathcal{T},\mathcal{S},F)$.

Remark 6.5.1. The last supposition holds if either $\alpha=\beta$ and $\mathrm{lh}(F) < \mathrm{lh}(E_\beta^\mathcal{T})$, or $\alpha > \beta$, and $\mathrm{lh}(E_\beta^\mathcal{S}) \le \mathrm{lh}(E_\beta^\mathcal{T})$. This will be the case when we use $W(\mathcal{T},\mathcal{S},F)$ to define $W(\mathcal{T},\mathcal{U})$.

Let $Q \trianglelefteq N = \mathcal{M}_\theta^\mathcal{T}$, where $\theta+1 = \mathrm{lh}(\mathcal{T})$, and let

$$\mu = \mathrm{crit}(F).$$

Suppose that $\mathrm{Ult}(Q,F)$ makes sense, that is, $\mathrm{dom}(F) \le \rho_{k(Q)}(Q)$. Suppose also that Q is the longest initial segment of N to which F applies, that is, either $Q=N$, or $\rho(Q) \le \mu < \rho_{k(Q)}(Q)$. We want to define $W(\mathcal{T},\mathcal{S},F)$ so that $\mathrm{Ult}(Q,F)$ embeds into the last model of $W(\mathcal{T},\mathcal{S},F)$ via a nearly elementary map.

There are three cases.

Case 1. $Q \neq N$.
In this case Q is a proper initial segment of $\mathcal{M}_\beta^\mathcal{T}|\mathrm{lh}(E_\beta^\mathcal{T})$, by the argument given in the dropping case of the definition of $W(\mathcal{T},F)$.

$$W(\mathcal{T},\mathcal{S},F) = \mathcal{S}\restriction(\alpha+1)^\frown\langle F\rangle$$

is the unique normal continuation $\mathcal{W}$ of $\mathcal{S}\restriction(\alpha+1)$ of length $\alpha+2$ such that $E_\alpha^\mathcal{W} = F$. Note here that $\mathcal{M}_\beta^\mathcal{T} = \mathcal{M}_\beta^\mathcal{S}$, and Q is what F would be applied to in a normal continuation of $\mathcal{S}\restriction\alpha+1$. (Unlike the case $\mathcal{T}=\mathcal{S}$ we discussed before, it is possible that $Q \neq N$ and $\alpha > \beta$.) In this dropping case, the last model of $W(\mathcal{T},\mathcal{S},F)$ is equal to $\mathrm{Ult}(Q,F)$, and does not just embed it.

Case 2. $Q=N$, and $\mathrm{lh}(\mathcal{T}) = \beta+1$.

[11] Normal trees may use extenders of plus type.

Again

$$W(\mathcal{T},\mathcal{S},F) = \mathcal{S}\restriction(\alpha+1)^\frown\langle F\rangle$$

is the unique normal $\mathcal{S}'$ of length $\alpha+2$ extending $\mathcal{S}$ such that $E_\alpha^{\mathcal{S}'} = F$. $Q = N = \mathcal{M}_\beta^{\mathcal{T}}$, and so $\mathrm{Ult}(Q,F)$ is equal to the last model of $W(\mathcal{T},\mathcal{S},F)$.

Case 3. $\mathrm{lh}(\mathcal{T}) > \beta+1$, and $Q = N$.
In this case, we construct $\mathcal{W} = W(\mathcal{T},\mathcal{S},F)$ just as before. We set

$$\mathcal{W}\restriction(\alpha+1) = \mathcal{S}\restriction(\alpha+1),$$

and

$$\mathcal{M}_{\alpha+1}^{\mathcal{W}} = \mathrm{Ult}(\mathcal{M}_\beta^{\mathcal{T}}|\langle\gamma,k\rangle,F),$$

where k,γ are appropriate for normality. (Note $\mathcal{M}_\beta^{\mathcal{T}} = \mathcal{M}_\beta^{\mathcal{S}} = \mathcal{M}_\beta^{\mathcal{W}}$.) Let $u(\xi) = \xi$ for $\xi < \beta$, and $u(\xi) = (\alpha+1)+(\xi-\beta)$ for $\xi \geq \beta$. Let $t_\xi = \mathrm{id}$ for $\xi < \beta$, and $t_\beta : \mathcal{M}_\beta^{\mathcal{T}}|\langle\gamma,k\rangle \to \mathcal{M}_{\alpha+1}^{\mathcal{W}}$ be the canonical embedding. Note that by our case hypothesis, F applies to $\mathcal{M}_\theta^{\mathcal{T}}$, and hence to $\mathcal{M}_\beta^{\mathcal{T}}|\mathrm{lh}(E_\beta^{\mathcal{T}})$, so $\langle\mathrm{lh}(E_\beta^{\mathcal{T}}),0\rangle \leq \langle\gamma,k\rangle$. Thus t_β moves $E_\beta^{\mathcal{T}}$. So we can use the Shift Lemma to lift the rest of $\mathcal{T}$, defining an elementary

$$t_\xi : \mathcal{M}_\xi^{\mathcal{T}} \to \mathcal{M}_{u(\xi)}^{\mathcal{W}}$$

for $\xi > \beta$, by induction on ξ. If $\sigma = T\text{-pred}(\xi)$, then $u(\sigma) = W\text{-pred}(u(\xi))$, unless $\sigma = \beta$ and $\mathrm{crit}(E_{\xi-1}^{\mathcal{T}}) < \mu$. In this case, $\mathrm{crit}(E_{u(\xi)-1}^{\mathcal{W}}) = \mathrm{crit}(E_{\xi-1}^{\mathcal{T}}) < \mu$, so $W\text{-pred}(u(\xi)) = \beta$, rather than $u(\beta)$. We write

$$W(\mathcal{T},\mathcal{S},F) = \mathcal{S}^{<\mathrm{lh}(F)\frown}\langle F\rangle^\frown i_F\text{``}\mathcal{T}^{>\mathrm{crit}(F)}$$

in this case.

DEFINITION 6.5.2. For $\mathcal{T},\mathcal{S},F$ as above, $\alpha^{\mathcal{T},\mathcal{S},F} = \alpha(\mathcal{S},F)$ is the least γ such that F^- is on the $\mathcal{M}_\gamma^{\mathcal{S}}$-sequence, and $\beta^{\mathcal{T},\mathcal{S},F} = \beta(\mathcal{S},F)$ is the least γ such that $\mathrm{crit}(F) < \hat{\lambda}(E_\gamma^{\mathcal{S}})$ or $\gamma+1 = \mathrm{lh}(\mathcal{S})$. In Case 3, $u^{\mathcal{T},\mathcal{S},F}$ and $t_\xi^{\mathcal{T},\mathcal{S},F}$, for $\xi < \mathrm{lh}\,\mathcal{T}$, are the maps u and t_ξ described above. In Cases 1 and 2, let $\mathrm{dom}(u^{\mathcal{T},\mathcal{S},F}) = \beta+1$, with $u^{\mathcal{T},\mathcal{S},F}(\xi) = \xi$ if $\xi < \beta$, and $u^{\mathcal{T},\mathcal{S},F}(\beta) = \alpha+1$. (Where $\alpha = \alpha^{\mathcal{T},\mathcal{S},F}$ and $\beta = \beta^{\mathcal{T},\mathcal{S},F}$.) Let $t_\xi^{\mathcal{T},\mathcal{S},F} = \mathrm{id}$ if $\xi < \beta$, and $t_\beta^{\mathcal{T},\mathcal{S},F} : \mathcal{M}_{\alpha+1}^{*,\mathcal{W}} = \mathcal{M}_\beta^{\mathcal{T}}|\xi \to \mathcal{M}_{\alpha+1}^{\mathcal{W}}$ be the canonical embedding in those cases.

In cases 2 and 3, we have an extended tree embedding

$$\Phi_{\mathcal{T},\mathcal{S},F} = \langle u,v,\langle s_\xi \mid \xi < \mathrm{lh}\,\mathcal{T}\rangle,\langle t_\xi \mid \xi+1 < \mathrm{lh}(\mathcal{T})\rangle\rangle$$

from $\mathcal{T}$ into $W(\mathcal{T},\mathcal{S},F)$. It is determined by setting

$$u = u^{\mathcal{T},\mathcal{S},F}.$$

Some of its other maps are given by

$$t_\xi = t_\xi^{\mathcal{T},\mathcal{S},F}$$

and

$$p(E_\xi^{\mathcal{T}}) = t_\xi^{\mathcal{T},\mathcal{S},F}(E_\xi^{\mathcal{T}}).$$

In case 1, these objects determine a partial extended tree embedding from $\mathcal{T} \upharpoonright \beta + 1$ into $\mathcal{W}(\mathcal{T},\mathcal{S},F)$. This is a system with all the properties of an extended tree embedding, except that its last map t_β may only be defined on some $Q \trianglelefteq \mathcal{M}_\beta^{\mathcal{T}}$. We call it $\Phi_{\mathcal{T},\mathcal{S},F}$ as well.

The illustrations associated to $W(\mathcal{T},\mathcal{S},F)$ are pretty much the same as before, allowing for the possibility that $\mathcal{S} \neq \mathcal{T}$. In particular, if $\xi \geq \beta^{\mathcal{T},\mathcal{S},F}$, then F either appears directly as one of the extenders used in $[0,u(\xi))_W$, or appears indirectly via some extender $F(G)$ used in $[0,u(\xi))_W$, where $\mathrm{crit}(G) < \mu < \lambda(G)$ and G is used in $[0,\xi)_T$.

Now let $\mathcal{T}$ be a normal tree on a premouse M, with last model Q, and let $\mathcal{U}$ be a normal tree on Q. We do not assume that $\mathcal{U}$ has a last model. We shall define $W(\mathcal{T},\mathcal{U}) = \mathcal{W}$, the embedding normalization of $\mathcal{T}^\frown\mathcal{U}$. For this, we define

$$\mathcal{W}_\gamma = W(\mathcal{T},\mathcal{U}|(\gamma+1)),$$

the embedding normalization of $\mathcal{T}^\frown\mathcal{U}|(\gamma+1)$, by induction on γ. Let us write

$$Q_\gamma = \mathcal{M}_\gamma^{\mathcal{U}} = \text{last model of } \mathcal{U}|(\gamma+1).$$

We shall maintain that each $\mathcal{W}_\gamma$ successor length $z(\gamma)+1$, with last model

$$\begin{aligned} R_\gamma &= \text{last model of } \mathcal{W}_\gamma \\ &= \mathcal{M}_{z(\gamma)}^{\mathcal{W}_\gamma}, \end{aligned}$$

and that there is a nearly elementary embedding

$$\sigma_\gamma : Q_\gamma \to R_\gamma.$$

As we go we construct extended tree embeddings $\Phi_{\eta,\gamma}$, for $\eta <_U \gamma$, from an appropriate initial segment of $\mathcal{W}_\eta$ to $\mathcal{W}_\gamma$.[12] $\Phi_{\eta,\gamma}$ is determined by its u-map $u_{\eta,\gamma}$ acting on an initial segment of $\mathrm{lh}(\mathcal{W}_\eta)$, and its t-maps we call

$$t_\tau^{\eta,\gamma} : \mathcal{M}_\tau^{\mathcal{W}_\eta} \to \mathcal{M}_{u_{\eta,\gamma}(\tau)}^{\mathcal{W}_\gamma},$$

defined when $\tau \in \mathrm{dom}(u_{\eta,\gamma})$. (There is the possibility that $t_\tau^{\eta,\gamma}$ acts only on some proper initial segment of $\mathcal{M}_\tau^{\mathcal{W}_\eta}$. That happens iff $(\eta,\gamma]_U$ has a drop.) Roughly, the system

$$(\langle \mathcal{W}_\gamma \mid \gamma < \mathrm{lh}(\mathcal{U})\rangle, \langle \Phi_{\eta,\gamma} \mid \eta <_U \gamma \rangle)$$

[12] The s and t maps of $\Phi_{\eta,\gamma}$ are elementary, as required by the definition of tree embeddings. σ_γ is a factor map, and so may be only nearly elementary.

is an iteration tree of iteration trees[13], whose base node is $\mathcal{W}_0 = \mathcal{T}$, and whose overall structure is induced by $\mathcal{U}$. The $\Phi_{\eta,\gamma}$ are the branch embeddings of this tree.

We set

$$\mathcal{W}_0 = \mathcal{T},$$

and let σ_0 be the identity. Now suppose everything is given up to γ. We let

$$F_\gamma = \sigma_\gamma(E^{\mathcal{U}}_\gamma),$$

and

$$\alpha_\gamma = \text{least } \xi \text{ such that } F_\gamma \text{ is on the sequence of } \mathcal{M}^{\mathcal{W}_\gamma}_\xi.$$

So F_γ is on the extended sequence of $\mathcal{M}^{\mathcal{W}_\gamma}_\xi$ for all ξ such that $\alpha_\gamma \leq \xi \leq z(\gamma)$. We assume the following agreement hypotheses:

$(*)_\gamma$ (i) For $\eta \leq \xi \leq \gamma$, $\sigma_\eta \restriction (\text{lh}(E^{\mathcal{U}}_\eta)+1) = \sigma_\xi \restriction (\text{lh}(E^{\mathcal{U}}_\eta)+1)$.

(ii) For $\eta < \xi < \gamma$, $\alpha_\eta < \alpha_\xi$ and $\text{lh}(F_\eta) < \text{lh}(F_\xi)$.

(iii) For $\eta < \xi \leq \gamma$, R_η agrees with R_ξ up to $\text{lh}(F_\eta)$, but $\text{lh}(F_\eta)$ is a cardinal of R_ξ, so they disagree at $\text{lh}(F_\eta)$.

(iv) For $\eta < \xi \leq \gamma$, $\mathcal{W}_\eta \restriction (\alpha_\eta + 1) = \mathcal{W}_\xi \restriction (\alpha_\eta + 1)$, and $E^{\mathcal{W}_\xi}_{\alpha_\eta} = F_\eta$.

(v) For $\eta < \gamma$,

(a) for all $\xi < \alpha_\eta$, $\text{lh}(E^{\mathcal{W}_\eta}_\xi) < \text{lh}(F_\eta)$, and

(b) if $\alpha_\eta < z(\eta)$, then $\text{lh}(F_\eta) < \text{lh}(E^{\mathcal{W}_\eta}_{\alpha_\eta})$.

CLAIM 6.5.3. *(ii) and (v) of $(*)_{\gamma+1}$ hold.*

PROOF. For (ii), if $\eta < \gamma$, then $\text{lh}(E^{\mathcal{U}}_\eta) < \text{lh}(E^{\mathcal{U}}_\gamma)$, so $\text{lh}(F_\eta) < \text{lh}(F_\gamma)$ by (i) at γ. Moreover, if $\alpha_\gamma \leq \alpha_\eta$, then by (iv), F^-_γ is on the sequence of $\mathcal{M}^{\mathcal{W}_\gamma}_{\alpha_\eta} = \mathcal{M}^{\mathcal{W}_\eta}_{\alpha_\eta}$.[14] But F^-_η is also on the $\mathcal{M}^{\mathcal{W}_\gamma}_{\alpha_\eta}$ sequence, by (iv). Since $\text{lh}(F_\eta) < \text{lh}(F_\gamma)$ and F^-_γ is on the R_γ sequence, we get that F^-_η is on the R_γ sequence. However, F_η is used in $\mathcal{W}_\gamma$ by (iv) at γ, and thus F^-_η is not on the R_γ sequence.

(v)(a) holds because otherwise F^-_γ would be on the sequence of some $\mathcal{M}^{\mathcal{W}_\gamma}_\xi$ for $\xi < \alpha_\gamma$. For (v)(b), suppose $\alpha_\gamma < z(\gamma)$. Since F^-_γ is on the sequences of $\mathcal{M}^{\mathcal{W}_\gamma}_{\alpha_\gamma}$ and of $\mathcal{M}^{\mathcal{W}_\gamma}_{\alpha_\gamma+1}$, we must have $\text{lh}(F_\gamma) < \text{lh}(E^{\mathcal{W}_\gamma}_{\alpha_\gamma})$. ⊣

Now suppose $\eta = U\text{-pred}(\gamma+1)$. We set

$$\mathcal{W}_{\gamma+1} = W(\mathcal{W}_\eta, \mathcal{W}_\gamma, F_\gamma).$$

Let us check that this makes sense. Let us write $F = F_\gamma$ and $\alpha = \alpha_\gamma$. Clearly $\alpha = \alpha^{\mathcal{W}_\eta, \mathcal{W}_\gamma, F}$. Let

$$\bar{\mu} = \text{crit}(E^{\mathcal{U}}_\gamma),$$

[13] A *meta-iteration tree.*

[14] Equivalently, F_γ is on the extended sequence.

and

$$\mu = \sigma_\gamma(\bar{\mu}) = \text{crit}(F).$$

Let

$$\begin{aligned}\beta &= \beta^{\mathcal{W}_\eta,\mathcal{W}_\gamma,F}\\ &= \text{least } \xi \text{ such that } \mu < \hat{\lambda}(E_\xi^{\mathcal{W}_\gamma}) \text{ or } \xi = z(\gamma)\end{aligned}$$

be the tree predecessor of $\alpha+1$ in any normal continuation $\mathcal{S}$ of $\mathcal{W}_\gamma\restriction(\alpha+1)$ that uses F. Since η is the least ξ such that $\bar{\mu} < \hat{\lambda}(E_\xi^{\mathcal{U}})$, we have by (i) of $(*)_\gamma$ that

$$\eta = \text{the least } \xi \text{ such that } \mu < \hat{\lambda}(F_\xi).$$

But $\mathcal{W}_\eta\restriction(\alpha_\eta+1) = \mathcal{W}_\gamma\restriction(\alpha_\eta+1)$, and $E_{\alpha_\eta}^{\mathcal{W}_\gamma} = F_\eta$ or else $\eta = \gamma$. In either case, $\beta \le \alpha_\eta$, so

$$\mathcal{W}_\eta\restriction(\beta+1) = \mathcal{W}_\gamma\restriction(\beta+1).$$

Moreover, since $\beta \le \alpha_\eta$, if $\beta < z(\eta)$ then

$$\text{lh}(E_\beta^{\mathcal{W}_\gamma}) \le \text{lh}(E_\beta^{\mathcal{W}_\eta}),$$

with equality holding iff $\beta < \alpha_\eta$. These are the conditions we needed to check, so $W(\mathcal{N}_\eta,\mathcal{W}_\gamma,F)$ makes sense.

Let $\Phi_{\eta,\gamma+1}$ be the (possibly partial) extended tree embedding $\Phi_{\mathcal{W}_\eta,\mathcal{W}_\gamma,F}$. Its u-map is

$$u_{\eta,\gamma+1} = u^{\mathcal{W}_\eta,\mathcal{W}_\gamma,F},$$

and its t maps are

$$t_\tau^{\eta,\gamma+1} = t_\tau^{\mathcal{W}_\eta,\mathcal{W}_\gamma,F}.$$

For $\delta <_U \eta$,

$$\Phi_{\delta,\gamma+1} = \Phi_{\eta,\gamma+1}\circ\Phi_{\delta,\eta}.$$

This of course means that $u_{\delta,\gamma+1} = u_{\eta,\gamma+1}\circ u_{\delta,\eta}$, and $t_\tau^{\delta,\gamma+1} = t_{u_{\delta,\eta}(\tau)}^{\eta,\gamma+1}\circ t_\tau^{\delta,\eta}$. Here the compositions are considered as defined wherever they make sense.

Note that $\Phi_{\eta,\gamma+1}$ is partial iff $\gamma+1 \in D^{\mathcal{U}}$. If $\gamma+1 \in D^{\mathcal{U}}$, then $\text{dom}(u_{\eta,\gamma+1}) = \beta+1$, and $t_\beta^{\eta,\gamma+1}$ acts on a proper initial segment of $\mathcal{M}_\beta^{\mathcal{W}_\eta}$.

$\sigma_{\gamma+1}$ is determined as follows. Let

$$Q_{\gamma+1} = \text{Ult}(Q^*, E_\gamma^{\mathcal{U}}),$$

where $Q^* \trianglelefteq Q_\eta$.

Let $R^* = R_\eta$ if $Q^* = Q_\eta$, and $R^* = \sigma_\eta(Q^*)$ otherwise. $\sigma_\eta\restriction Q^*$ is elementary from Q^* to R^*.

Suppose first that we drop in $\mathcal{U}$, i.e., $Q^* \neq Q_\eta$. Then $\rho(Q^*) \leq \bar{\mu}$, and σ_η is a near $k(Q^*)+1$ embedding, so

$$\mu = \sigma_\gamma(\bar{\mu}) = \sigma_\eta(\bar{\mu}) \leq \rho(R^*),$$

while $\rho_{k(R^*)}(R^*) = \sigma_\eta(\rho_{k(Q)}(Q)) > \mu$. So R^* is what we would apply F to in a normal continuation of $\mathcal{W}_\gamma \restriction (\alpha+1)$. Moreover,

$$\mathcal{W}_{\gamma+1} = \mathcal{W}_\gamma^{<\mathrm{lh}(F)} {}^\frown \langle F \rangle {}^\frown \mathrm{Ult}(R^*, F)$$

because we are in case 1 of the definition of $W(\mathcal{W}_\eta, \mathcal{W}_\gamma, F)$. So $R_{\gamma+1} = \mathrm{Ult}(R^*, F)$, and we can take $\sigma_{\gamma+1}$ to be the Shift Lemma map.

Suppose next that $Q^* = Q_\eta$, so that we are in case 2 or case 3, and

$$\mathcal{W}_{\gamma+1} = \mathcal{W}_\gamma^{<\mathrm{lh}(F)} {}^\frown \langle F \rangle {}^\frown i_F \text{“} \mathcal{W}_\eta^{>\mathrm{crit}(F)}.$$

For $\tau \leq z(\eta)$, we have an elementary $t_\tau^{\eta,\gamma+1} : \mathcal{M}_\tau^{\mathcal{W}_\eta} \to \mathcal{M}_{u_{\eta,\gamma+1}(\tau)}^{\mathcal{W}_{\gamma+1}}$. Since we are not dropping in $\mathcal{U}$,

$$Q_{\gamma+1}^{\mathcal{U}} = \mathrm{Ult}(Q_\eta^{\mathcal{U}}, E_\gamma^{\mathcal{U}}).$$

and

$$u_{\eta,\gamma+1}(z(\eta)) = z(\gamma+1).$$

We have then the diagram

$$\begin{array}{ccccc}
Q_{\gamma+1} & \xrightarrow{\theta} & \mathrm{Ult}(R_\eta, F) & \xrightarrow{\psi} & R_{\gamma+1} = \mathcal{M}_{z(\gamma+1)}^{\mathcal{W}_{\gamma+1}} \\
\uparrow {\scriptstyle i_{\eta,\gamma+1}^{\mathcal{U}}} & & \uparrow & \nearrow {\scriptstyle t_{z(\eta)}^{\eta,\gamma+1}} & \\
Q_\eta & \xrightarrow[\sigma_\eta]{} & R_\eta = \mathcal{M}_{z(\eta)}^{\mathcal{W}_\eta} & &
\end{array}$$

Here θ is given by the Shift Lemma, and ψ comes from the fact that F is an initial segment of the extender of $t_{z(\eta)}^{\eta,\gamma+1}$, as we remarked before. (So $\psi \restriction \mathrm{lh}(F) = \mathrm{id}$.) We then set

$$\sigma_{\gamma+1} = \psi \circ \theta.$$

So when $\gamma+1 \notin D^{\mathcal{U}}$, we have the diagram

$$\begin{array}{ccc}
\mathcal{M}_{\gamma+1}^{\mathcal{U}} & \xrightarrow{\sigma_{\gamma+1}} & R_{\gamma+1} \\
\uparrow {\scriptstyle i_{\eta,\gamma+1}^{\mathcal{U}}} & & \uparrow {\scriptstyle t_{z(\eta)}^{\eta,\gamma+1}} \\
\mathcal{M}_\eta^{\mathcal{U}} & \xrightarrow[\sigma_\eta]{} & R_\eta
\end{array}$$

When $\gamma+1 \in D^{\mathcal{U}}$, we have the diagram

$$\begin{array}{ccc}
\mathcal{M}^{\mathcal{U}}_{\gamma+1} & \xrightarrow{\sigma_{\gamma+1}} & R_{\gamma+1} \\
\uparrow{\scriptstyle i^{*,\mathcal{U}}_{\gamma+1}} & & \uparrow{\scriptstyle i^{\eta,\gamma+1}_{\beta}} \\
\mathcal{M}^{*,\mathcal{U}}_{\gamma+1} & \xrightarrow[\sigma_\eta]{} & \sigma_\eta(\mathcal{M}^{*,\mathcal{U}}_{\gamma+1})
\end{array}$$

where $\beta = \beta^{W_\eta, \mathcal{W}_\gamma, F}$.

CLAIM 6.5.4. $(*)_{\gamma+1}$ *holds.*

PROOF. Left to the reader. ⊣

We have completed the definition of $\mathcal{W}_{\gamma+1}$.

If $\lambda < \mathrm{lh}(\mathcal{U})$ is a limit ordinal, then

$$\mathcal{W}_\lambda = \lim_{\alpha <_U \lambda} \mathcal{W}_\alpha,$$

where we make sense of the direct limit using the tree embeddings $\Phi_{\eta,\gamma}$ for $\eta <_U \gamma <_U \lambda$. We give a little more detail on this below.

In our context of interest, $\langle \mathcal{T}, \mathcal{U} \rangle$ is played by a background induced iteration strategy Σ for M, and we shall show that all $\mathcal{W}_\alpha$ are by Σ. So in our context of interest, all models above are wellfounded.

Here are a couple illustrations that the reader may or may not find helpful. Let $\gamma_0\, U\, \gamma_1\, U\, \gamma_2\, U\, \gamma_3$ be successive elements of a branch of U. Write $u_i = u_{\gamma_i,\gamma_{i+1}}$. Let $\beta_i = \beta^{W_{\gamma_i}, \mathcal{W}_{\tau_i}, F_i}$, where $\tau_i = \gamma_{i+1} - 1$ and $F_i = \sigma_{\tau_i}(E^{\mathcal{U}}_{\tau_i})$.[15] Thus $\mathcal{W}_{\gamma_{i+1}} = W(\mathcal{W}_{\gamma_i}, \mathcal{W}_{\tau_i}, F_i)$, and $\beta_i = \mathrm{crit}(u_i)$. The u_i might look like:

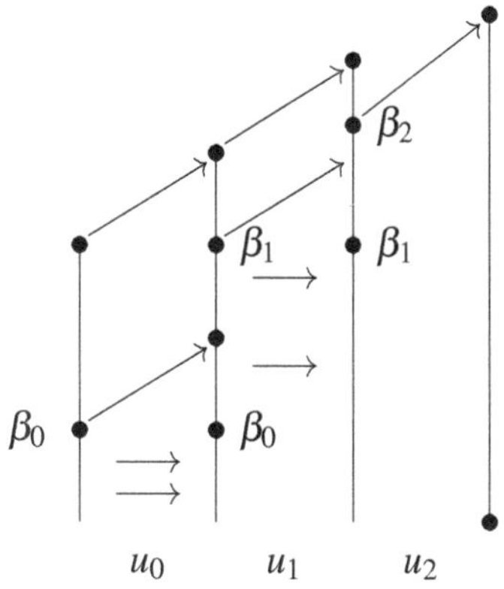

The last step pictured involves a drop. Notice that $\beta_{i+1} \geq u_i(\beta_i)$. (Equality is possible.) This is because $\mathcal{U}$ is normal. In $\mathcal{W}_{\gamma_{i+1}}$, $\mathcal{M}^{\mathcal{W}_{\gamma_{i+1}}}_{u_i(\beta_i)}$ is immediately above $\mathcal{M}^{\mathcal{W}_{\gamma_{i+1}}}_{\beta_i}$ via an F_i-ultrapower. Moreover, $\mathcal{W}_{\gamma_{i+1}} \upharpoonright (\alpha+1) = \mathcal{W}_{\tau_i} \upharpoonright (\alpha+1)$, where

[15] F_i is what we called F_{τ_i} in the general definition.

$\alpha+1=u_i(\beta_i)$. By our choice of α, $\hat{\lambda}(E_\xi^{\mathcal{W}_{\tau_i}})\le\hat{\lambda}(F_i)$ for all $\xi<\alpha$. But $\hat{\lambda}(F_i)\le \mathrm{crit}(F_{i+1})$, since $\mathcal{U}$ is normal, so F_{i+1} cannot be applied to any $\mathcal{M}_\xi^{\mathcal{W}_{\gamma_{i+1}}}$ for $\xi< u_i(\beta_i)$.

Because $\beta_{i+1}\ge u_i(\beta_i)$, and above $u_i(\beta_i)$, $\mathrm{ran}(u_i)$ is an initial segment of $\mathrm{ORD}-u(\beta_i)$, we see that along any branch b of $\mathcal{U}$, the direct limit of the $u_{\gamma,\eta}$ for $\gamma,\eta\in b$ is wellfounded.

In fact, the direct limit has order type $\lambda+\theta$, where $\lambda=\sup_{\eta\in b}\mathrm{crit}(u_{\eta,b})$, and $\theta=\mathrm{lh}\,\mathcal{T}-\beta$, where β is least such that $u_{0,b}(\beta)\ge\lambda$.

In addition to the u-maps on indices of models, we have the t-maps on the models. Let $\mu_i=\mathrm{crit}(F_i)$, and let $\mathrm{lh}(\mathcal{W}_{\gamma_1})=\theta+1$. Let η be the level of R_{γ_2}, or equivalently $\mathcal{M}_{\beta_2}^{\mathcal{W}_{\gamma_2}}$, that we drop to when we apply F_2. The picture is

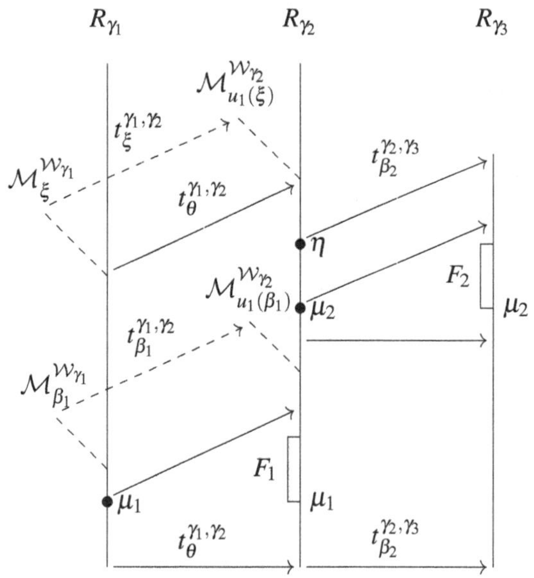

One can look at $\Phi_{\eta,\gamma}$, for $\eta<_U\gamma$, as a map on the extender trees. Let $p_{\eta,\gamma}$ be the p-map of $\Phi_{\eta,\gamma}$, that is

$$p_{\eta,\gamma}:\mathrm{Ext}(\mathcal{W}_\eta)\to\mathrm{Ext}(\mathcal{W}_\gamma)$$

and

$$p_{\eta,\gamma}(E_\xi^{\mathcal{W}_\eta})=t_\xi^{\eta,\gamma}(E_\xi^{\mathcal{W}_\eta})=E_{u_{\eta,\gamma}(\xi)}^{\mathcal{W}_\gamma}.$$

So $p_{\eta,\gamma}(E_\xi^{\mathcal{W}_\eta})\downarrow$ iff $\xi\in\mathrm{dom}\,u_{\eta,\gamma}$. Let

$$\hat{p}(s)=\text{least } t\in\mathcal{W}_\gamma^{\mathrm{ext}} \text{ such that } p\text{``}\mathrm{ran}(s)\subseteq\mathrm{ran}(t).$$

By Proposition 6.4.6, $\hat{p}_{\eta,\gamma}$ preserves extender tree order and incompatibility; that is $s\subseteq t\implies\hat{p}_{\eta,\gamma}(s)\subseteq\hat{p}_{\eta,\gamma}(t)$, and $s\perp t\implies\hat{p}_{\eta,\gamma}(s)\perp\hat{p}_{\eta,\gamma}(t)$. Moreover

PROPOSITION 6.5.5. *Let $\eta <_U \gamma$ and $u_{\eta,\gamma}(\alpha)\downarrow$, and suppose whenever $\eta \leq_U \xi <_U \gamma$, then $u_{\eta,\xi}(\alpha) \geq \mathrm{crit}(u_{\xi,\gamma})$. Then for $s = e_\alpha^{\mathcal{W}_\eta}$,*

$$e_{u_{\eta,\gamma}(\alpha)}^{\mathcal{W}_\gamma} = \hat{p}_{\eta,\gamma}(s)^\frown\langle F_\tau \mid \tau+1 \leq_U \gamma \text{ and for all } i \in \mathrm{dom}\,\hat{p}_{\eta,\gamma}(s),$$
$$\hat{\lambda}(\hat{p}_{\eta,\gamma}(s)(i)) \leq \mathrm{crit}(F_\tau)\rangle.$$

We omit the simple proof. The proposition says that the branch extender to $\mathcal{M}_{u_{\eta,\gamma}(\alpha)}^{\mathcal{W}_\gamma}$ consists of blow-ups by $p_{\eta,\gamma}$ of extenders used in the branch to $\mathcal{M}_\alpha^{\mathcal{W}_\eta}$, together with certain F_τ's used in $\mathcal{U}$ from η to γ. It generalizes our pictures on page 240 and before.

Suppose now that $\lambda \leq \mathrm{lh}(\mathcal{U})$ is a limit ordinal, and we have defined $\mathcal{W}_\gamma, \sigma_\gamma$, and the $\Phi_{\eta,\gamma}$ for $\eta, \gamma < \lambda$. We let $W(\mathcal{T},\mathcal{U} \upharpoonright \lambda)$ be the lim inf of the $\mathcal{W}_\gamma$ for $\gamma < \lambda$. More precisely, let

$$F_\gamma = \sigma_\gamma(E_\gamma^{\mathcal{U}})$$

and

$$\begin{aligned}\alpha_\gamma &= \text{least } \alpha \text{ such that } F_\gamma^- \text{ is on the sequence of } \mathcal{M}_\alpha^{\mathcal{W}_\gamma}\\ &= \text{largest } \alpha \text{ such that } \mathcal{W}_{\gamma+1} \upharpoonright(\alpha+1) = \mathcal{W}_\gamma \upharpoonright(\alpha+1).\end{aligned}$$

We put

$$W(\mathcal{T},\mathcal{U} \upharpoonright \lambda) = \bigcup_{\gamma<\lambda} \mathcal{W}_\gamma \upharpoonright(\alpha_\gamma+1).$$

Since $\gamma < \eta \implies \alpha_\gamma < \alpha_\eta$, $W(\mathcal{T},\mathcal{U} \upharpoonright \lambda)$ has limit length. There are no new σ's or Φ's to be defined at this stage.

Now let b be a cofinal branch of $\mathcal{U} \upharpoonright \lambda$ (not necessarily a wellfounded one). We define the embedding normalization

$$\mathcal{W}_b = W(\mathcal{T},\mathcal{U} \upharpoonright \lambda^\frown b)$$

by forming the direct limit of the $\mathcal{W}_\gamma$, for $\gamma \in b$, under the $\Phi_{\eta,\gamma}$ for $\eta <_U \gamma$ in b.

We begin with $\mathrm{lh}(\mathcal{W}_b)$. Let us put

$$\langle \eta, \xi \rangle \in I \text{ iff } \eta \in b, \text{ and for all sufficiently large } \gamma \in b,\ u_{\eta,\gamma}(\xi)\downarrow.$$

Put

$$\langle \eta, \xi \rangle \leq_I \langle \delta, \theta \rangle \text{ iff for all sufficiently large } \gamma \in b,\ u_{\eta,\gamma}(\xi) \leq u_{\delta,\gamma}(\theta).$$

It is easy to see that $\leq_I$ is a prewellorder (even if b is illfounded, or drops infinitely often). We set

$$\mathrm{lh}(\mathcal{W}_b) = \mathrm{otp}(I, \leq_I).$$

For $\eta \in b$, we let $u_{\eta,b}(\xi)\downarrow$ iff $\langle \eta, \xi \rangle \in I$, and in that case, set

$$u_{\eta,b}(\xi) = \text{rank of } \langle \eta, \xi \rangle \text{ in } (I, \leq_I).$$

We define the tree order $\leq_{W_b}$ by: given $\langle \eta, \xi \rangle$ and $\langle \delta, \theta \rangle \in I$,

$$u_{\eta,b}(\xi) \leq_{W_b} u_{\delta,b}(\theta) \text{ iff for all sufficiently large } \gamma \in b,\ u_{\eta,\gamma}(\xi) \leq_{W_\gamma} u_{\delta,\gamma}(\theta).$$

Although the $u_{\eta,\gamma}$ do not completely preserve tree order, they almost do so. See clause (4) in the list following Remark 6.2.2, and the illustration on p.238. Using this, we can show $\leq_{W_b}$ is a tree order. $u_{\eta,b}$ may fail to preserve tree order, but again, this can only happen in a way similar to the possible failure described after Remark 6.2.2. We record this in a proposition.

PROPOSITION 6.5.6. *Let $\langle \eta, \xi \rangle$, $\langle \eta, \delta \rangle \in I$, and suppose that $\xi \leq_{W_\eta} \delta$ but $u_{\eta,b}(\xi) \not\leq_{W_b} u_{\eta,b}(\delta)$. Then there is a unique $\gamma \geq \eta$ in b such that letting U-pred$(\theta+1) = \gamma$ with $\theta + 1 \in b$, $F = F_\theta$, and $\beta = \beta^{\mathcal{W}_\gamma, \mathcal{W}_\theta, F}$, we have*

1. *$\beta = u_{\eta,\gamma}(\xi) \leq_{W_\gamma} u_{\eta,\gamma}(\delta)$, and*
2. *letting G be the first extender used in $[0, u_{\eta,\gamma}(\delta))$ such that $\hat{\lambda}(G) \geq \hat{\lambda}(E_\beta^{\mathcal{W}_\gamma})$, we have $\mathrm{crit}(G) < \mathrm{crit}(F) < \hat{\lambda}(G)$.*

Moreover, in this case, if $\xi = W_\eta$-pred(δ), $\beta = u_{\eta,\gamma}(\xi) = W_\gamma$-pred$(u_{\eta,\gamma}(\delta))$, and

$$W_{\theta+1}\text{-pred}(u_{\eta,\theta+1}(\delta)) = \beta = W_{\theta+1}\text{-pred}(u_{\eta,\theta+1}(\xi)).$$

We omit the easy proof. Using such arguments, we can show $\leq_{W_b}$ is a tree order, and

PROPOSITION 6.5.7. *Let $\langle \eta, \xi \rangle$ and $\langle \delta, \theta \rangle \in I$. Then*

$$u_{\eta,b}(\xi) = W_b\text{-pred}(u_{\delta,b}(\theta))$$

iff for all sufficiently large $\gamma \in b$, $u_{\eta,\gamma}(\xi) = W_\gamma$-pred$(u_{\delta,\gamma}(\theta))$.

Here is a more concrete description of $\mathrm{lh}(\mathcal{W}_b)$ and $u_{\eta,b}$. Let

$$\begin{aligned} \delta &= \mathrm{lh}(W(\mathcal{T}, \mathcal{U} \restriction \lambda)) \\ &= \sup\{\alpha_\gamma \mid \gamma < \lambda\} \\ &= \sup\{\mathrm{crit}(u_{\eta,\gamma}) \mid \eta <_U \gamma \wedge \gamma \in b\}. \end{aligned}$$

(The last equality holds because if $\eta = U$-pred$(\gamma+1)$ and $\gamma + 1 \leq_U \tau$ where $\tau \in b$, then $\mathrm{crit}(u_{\eta,\gamma+1}) \leq \alpha_\gamma < \mathrm{crit}(u_{\gamma+1,\tau})$.)

Case 1. b drops somewhere.

Let $\gamma+1$ be least in $b \cap D^{\mathcal{U}}$, and $\eta = U$-pred$(\gamma+1)$, and $\beta = \beta^{\mathcal{W}_\eta, \mathcal{W}_\gamma, F_\gamma} = \mathrm{crit}(u_{\eta,\gamma+1})$. Let $\beta = u_{0,\eta}(\tau)$. Then for all $\gamma + 1 \leq_U \theta <_U \rho$, with $\rho \in b$,

$$\begin{aligned} \mathrm{crit}(u_{\theta,\rho}) &= u_{\eta,\theta}(\beta) \\ &= \mathrm{lh}(\mathcal{W}_\theta) - 1. \end{aligned}$$

(Further dropping cuts down on the domains of the t-maps, not on that of the u-maps.) Thus

$$\begin{aligned} \mathrm{lh}(\mathcal{W}_b) &= \delta + 1 \\ &= u_{\eta,b}(\beta) + 1 = u_{0,b}(\tau) + 1. \end{aligned}$$

Case 2. b does not drop.

Let

$$\tau = \tau_b = \text{least } \alpha < \text{lh}(\mathcal{T}) \text{ such that for all } \gamma <_U \xi$$
$$\text{with } \xi \in b,\ u_{0,\gamma}(\alpha) \geq \text{crit}(u_{\gamma,\xi}).$$

Then

$$u_{0,b}(\tau) = \delta,$$
$$\text{lh}(\mathcal{W}_b) = \delta + (\text{lh}(\mathcal{T}) - \tau),$$

and for $\xi \geq \tau$ with $\xi < \text{lh}(\mathcal{T})$,

$$u_{0,b}(\xi) = \delta + (\xi - \tau).$$

This case can happen in two ways: it can be that $u_{0,\eta}(\tau) = \text{crit}(u_{\eta,\gamma})$ for some $\eta <_U \gamma$ with $\gamma \in b$, in which case that is true for all sufficiently large such η, γ. Or it can happen that $u_{0,\eta}(\tau) > \text{crit}(u_{\eta,\gamma})$, for all $\eta <_U \gamma$ with $\gamma \in b$. In the latter case, τ is a limit ordinal, and the extenders in b are being inserted cofinally into the branch extender of $[0,\tau)_T$.

It can happen in Case 2 that τ is a limit ordinal, but some $u_{0,\eta}(\tau)$ and its images are in the "eventual critical points" along b. In that case, some tail of the extenders used in b are being inserted after the blow-ups of all those in $[0,\tau)_T$.

Now we define the models and extenders of $\mathcal{W}_b$. Suppose $\alpha = u_{\eta,b}(\gamma) < \text{lh}(\mathcal{W}_b)$. Suppose $\eta \leq \xi < \delta \in b$. Then we have the map $t^{\xi,\delta}_{u_{\eta,\xi}(\gamma)}$ acting on either $\mathcal{M}^{\mathcal{W}_\xi}_{u_{\eta,\xi}(\gamma)}$ or an initial segment thereof. We let

$$\mathcal{M}^{\mathcal{W}_\gamma}_\alpha = \text{dir lim of the } \mathcal{M}^{\mathcal{W}_\xi}_{u_{\eta,\xi}(\gamma)} \text{ under the } t^{\xi,\delta}_{u_{\eta,\xi}(\gamma)}\text{'s.}$$

If b does not drop after η, then we have

$$t^{\eta,b}_\gamma : \mathcal{M}^{\mathcal{W}_\eta}_\gamma \to \mathcal{M}^{\mathcal{W}_b}_{u_{\eta,b}(\gamma)}$$

as the direct limit map. Otherwise $t^{\eta,b}_\gamma$ may (or may not) act on a proper initial segment of $\mathcal{M}^{\mathcal{W}_\eta}_\gamma$.

Finally, if $\alpha = u_{\eta,b}(\gamma) < \text{lh}(\mathcal{W}_b)$ and $\alpha + 1 < \text{lh}(\mathcal{W}_\gamma)$, then

$$E^{\mathcal{W}_b}_\alpha = t^{\eta,b}_\gamma(E^{\mathcal{W}_\eta}_\gamma).$$

One can check that with this choice of extenders, $\mathcal{W}_b$ is a normal iteration tree on M. For example, suppose that $\eta \in b$ and that for all $\xi \geq \eta$ in b, $W_\xi\text{-pred}(u_{\eta,\xi}(\gamma+1)) = u_{\eta,\xi}(\theta)$, and we are not dropping, so

$$\mathcal{M}^{\mathcal{W}_\xi}_{u_{\eta,\xi}(\gamma+1)} = \text{Ult}(\mathcal{M}^{\mathcal{W}_\xi}_{u_{\eta,\xi}(\theta)}, E^{\mathcal{W}_\xi}_{u_{\eta,\xi}(\gamma)}).$$

Then

$$\mathcal{M}^{\mathcal{W}_b}_{u_{\eta,b}(\gamma+1)} = \text{Ult}(\mathcal{M}^{\mathcal{W}_b}_{u_{\eta,b}(\theta)}, E^{\mathcal{W}_b}_{u_{\eta,b}(\gamma)}).$$

because each of the three objects in this equation is a direct limit of its ξ-approximations, for $\xi \in b$, and the maps commute appropriately. We omit further detail.

Now we also have the natural map

$$\sigma_b \colon \mathcal{M}_b^{\mathcal{U}} \to R_b,$$

where R_b is the last model of $\mathcal{W}_b$, given by

$$\sigma_b(i_{\gamma,b}^{\mathcal{U}}(x)) = t_{z(\gamma)}^{\gamma,b}(\sigma_\gamma(x)).$$

In the abstract, it may happen that not all models of $\mathcal{W}_b$ are wellfounded. In our context of interest, $\langle \mathcal{T}, \mathcal{U}^\frown b\rangle$ is played according to an iteration strategy Σ for M, and we show that Σ is sufficiently good that $\mathcal{W}_b$ is also played by Σ.

Now suppose $\lambda < \mathrm{lh}(\mathcal{U})$ and $b = [0,\lambda)_U$, and all models of $\mathcal{W}_b$ are wellfounded. Then we set

$$\begin{aligned} \mathcal{W}_\lambda &= \mathcal{W}_b, \\ u_{\eta,\lambda} &= u_{\eta,b}, \\ t_\gamma^{\eta,\lambda} &= t_\gamma^{\eta,b}, \\ \sigma_\lambda &= \sigma_b, \end{aligned}$$

and continue with the inductive construction of $W(\mathcal{T},\mathcal{U})$. If some model of $\mathcal{W}_b$ is illfounded, we stop the construction, and say that $W(\mathcal{T},\mathcal{U})$ is undefined.

Finally, if $\mathcal{U}$ has a last model, we set $W(\mathcal{T},\mathcal{U}) = \mathcal{W}_\gamma$, where $\mathrm{lh}(\mathcal{U}) = \gamma+1$. If $\mathcal{U}$ has limit length λ, then $W(\mathcal{T},\mathcal{U}) = W(\mathcal{T},\mathcal{U}\restriction\lambda)$ has already been defined.

To summarize our notation associated to $W(\mathcal{T},\mathcal{U})$: for $\gamma < \mathrm{lh}(\mathcal{U})$,

$$F_\gamma = \sigma_\gamma(E_\gamma^{\mathcal{U}})$$

where $\sigma_\gamma : \mathcal{M}_\gamma^{\mathcal{U}} \to R_\gamma = \mathcal{M}_{z(\gamma)}^{\mathcal{W}_\gamma}$ is the last model of $\mathcal{W}_\gamma$, and

$$\mathcal{W}_{\gamma+1} = W(\mathcal{W}_\eta, \mathcal{W}_\gamma, F_\gamma)$$

where $\eta = U\text{-pred}(\gamma+1)$.

If $\mathcal{T}$ and $\mathcal{U}$ are λ-separated, then so is $W(\mathcal{T},\mathcal{U})$. Similarly, if both $\mathcal{T}$ and $\mathcal{U}$ are λ-tight, then so is $W(\mathcal{T},\mathcal{U})$. In these cases, granted that all $\mathcal{W}_\gamma$ are played by the same iteration strategy, R_γ and $\mathcal{W}_\gamma$ determine each other, while F_γ and $\mathcal{W}_\gamma\restriction(\alpha_\gamma+1)$ determine each other, modulo the λ-separation or λ-tightness of $\mathcal{W}_\gamma$. The case that all trees are λ-separated is the most important one in this book.

The R_γ's are not the models of a single iteration tree; they constitute an enlargement of $\mathcal{U}$, with accompanying maps $\sigma_\gamma \colon \mathcal{M}_\gamma^{\mathcal{U}} \to R_\gamma$. We proved the basic facts about agreement of models and maps in this enlargement in $(*)_\gamma$ above; we list some of them again here for reference.

PROPOSITION 6.5.8. *Let* $\gamma < \eta < \mathrm{lh}(\mathcal{U})$. *Then*

(*a*) R_γ *agrees with* R_η *below* $\mathrm{lh}(F_\gamma)$,

(b) $\sigma_\eta \restriction (\mathrm{lh}(E^{\mathcal{U}}_\eta)+1) = \sigma_\gamma \restriction (\mathrm{lh}(E^{\mathcal{U}}_\eta)+1)$, *and*

(c) F^-_γ *is on the sequence of* R_γ, *but not that of* R_η. *In fact,* $\mathrm{lh}(F_\gamma)$ *is a cardinal of* R_η.

The following diagram summarizes the situation. We draw the diagram as if the maps in question exist, although sometimes they may not, because of dropping. Let $z(\eta)+1 = \mathrm{lh}(\mathcal{W}_\eta)$, and let $i^{\mathcal{W}_\eta} : M \to R_\eta$ be the canonical embedding (assuming M-to-R_η does not drop).

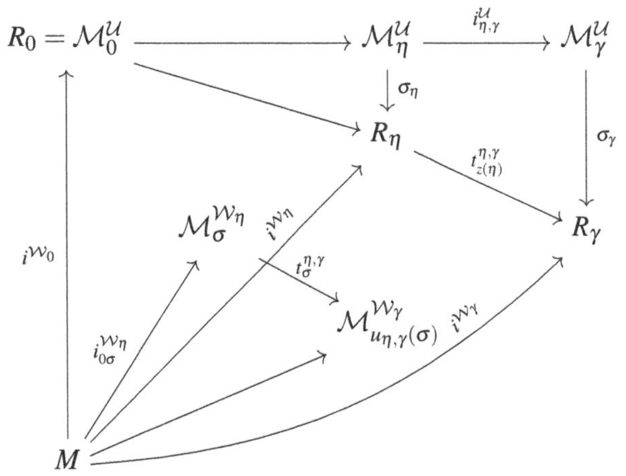

The various embeddings all commute:

(i) $i^{\mathcal{W}_\gamma} = t^{\eta,\gamma}_{z(\eta)} \circ i^{\mathcal{W}_\eta}$.

(ii) $t^{\eta,\gamma}_\sigma \circ i^{\mathcal{W}_\eta}_{\xi,\sigma} = i^{\mathcal{W}_\gamma}_{u_{\eta,\gamma}(\xi),u_{\eta,\gamma}(\sigma)} \circ t^{\eta,\gamma}_\xi$. (General version of (i).)

(iii) $\sigma_\gamma \circ i^{\mathcal{U}}_{\eta,\gamma} = t^{\eta,\gamma}_{z(\eta)} \circ \sigma_\eta$.

Remark 6.5.9. One can regard the sequence of iteration trees $\langle \mathcal{W}_\gamma \mid \gamma < \mathrm{lh}(\mathcal{U})\rangle$ that occurs in the definition of $W(\mathcal{T},\mathcal{U})$ as an iteration tree of iteration trees. One might call such a system a *meta-iteration tree*, or *meta-tree*. The nodes in the meta-tree are iteration trees, with $\mathcal{T}$ being the base node. The F_γ are used to extend the meta-tree at successor steps, via the W-operation. We have tree embeddings from one node to the later ones along branches of our meta-tree.

The meta-tree associated to $W(\mathcal{T},\mathcal{U})$ is not the general case, however, because there is in general no need to require that the F_γ be obtained by lifting extenders used in some tree $\mathcal{U}$ on the last model of $\mathcal{T}$. This was first realized by Schlutzenberg, who defined the general notion of "meta-iterate of $\mathcal{T}$". (Schlutzenberg's term is "inflation of $\mathcal{T}$".) Schlutzenberg also showed that if $\mathcal{T}$ is played by a strategy Σ with the weak Dodd-Jensen property, then Σ induces a meta-iteration strategy for $\mathcal{T}$. See [57]. Schlutzenberg's work was streamlined and re-written by Jensen, who introduced the general notion of meta-tree. See [21]. Further general results on meta-iteration trees and strategies can be found in [64], along with a more detailed discussion of the evolution of the idea.

Coarse embedding normalization. We must also define $W(\mathcal{T},\mathcal{U})$ in the coarse case. Suppose that M is a transitive model of ZFC, that $\mathcal{T}$ is a nice, normal tree on M with last model P, and that $\mathcal{U}$ is a nice, normal tree on P with last model Q. We define $W(\mathcal{T},\mathcal{U})$ as above:

$$\begin{aligned}\mathcal{W}_\gamma &= W(\mathcal{T},\mathcal{U}\restriction\gamma+1),\\ \sigma_\gamma &: \mathcal{M}^{\mathcal{U}}_\gamma \to \mathcal{M}^{\mathcal{W}_\gamma}_{z(\gamma)},\\ F_\gamma &= \sigma_\gamma(E^{\mathcal{U}}_\gamma),\end{aligned}$$

and

$$\begin{aligned}\mathcal{W}_{\gamma+1} &= W(\mathcal{W}_\eta,\mathcal{W}_\gamma,F_\gamma)\\ &= \mathcal{W}_\gamma\restriction(\alpha_\gamma+1)^\frown\langle F_\gamma\rangle^\frown i_{F_\gamma}\text{``}\mathcal{W}_\nu^{>\operatorname{crit}(F_\gamma)}.\end{aligned}$$

Here

$$\alpha_\gamma = \begin{cases}\text{least } \alpha \text{ such that } \operatorname{lh}(F_\gamma) \le \operatorname{lh}(E^{\mathcal{W}_\gamma}_\alpha) & \text{if one exists,}\\ \operatorname{lh}(\mathcal{W}_\gamma)-1 & \text{otherwise;}\end{cases}$$

$$\eta = \text{least } \xi \text{ such that } \operatorname{crit}(F_\gamma) < \operatorname{lh}(F_\xi).$$

In this coarse case we shall have

$$\mathcal{M}^{\mathcal{U}}_\gamma = R_\gamma$$

and

$$\sigma_\gamma = \text{identity},$$

so $F_\gamma = E^{\mathcal{U}}_\gamma$, for all γ. This we prove by induction, the successor step being essentially the same as the proof of Proposition 6.2.9. So in the coarse case, embedding normalization coincides with full normalization.

One can also characterize α_γ as the least η such that for $\xi = \operatorname{lh}(F_\gamma)$, $V_\xi^{\mathcal{M}^{\mathcal{W}_\gamma}_\eta} = V_\xi^{R_\gamma}$. α_γ may not be the least η such that $F_\gamma \in \mathcal{M}^{\mathcal{W}_\gamma}_\eta$, but it is the least η such that $\mathcal{M}^{\mathcal{W}_\gamma}_\eta \models$ "F_γ is nice".

We might also be normalizing a stack of coarse $\mathcal{F}$-trees, for some collection $\mathcal{F}$ of nice extenders in the base model M. In that case, α_γ should be the least η such that $F_\gamma \in i^{\mathcal{W}_\gamma}_{0,\eta}(\mathcal{F})$. It is easy to see that if $\mathcal{F}$ is part of a coherent pair $(w,\mathcal{F})$ in M, then this is equivalent to the definition of α_γ given above. In practice, when we normalize $\mathcal{F}$-trees, $\mathcal{F}$ will be part of such a coherent pair.

We are not assuming the extenders used in $\mathcal{T}$ and $\mathcal{U}$ come from a coherent sequence, but it is not too hard to show that $W(\mathcal{T},\mathcal{U})$ is normal, provided its models are wellfounded.

We have seen that conversion systems can produce non-normal trees on the background universe when applied to a λ-separated tree on some premouse. In proving that background induced strategies normalize well, we shall therefore look at "quasi-normalizations" of stacks of quasi-normal trees on the background universe. We do this in Section 6.7.

6.6. The branches of $W(\mathcal{T},\mathcal{U})$

Let M be a pfs premouse, $\mathcal{T}$ a normal plus tree on M, and $\mathcal{U}$ a normal plus tree on the last model of $\mathcal{T}$. Let us adopt our standard notation, so that we have

(a) $\mathcal{W}_\gamma = W(\mathcal{T},\mathcal{U}\upharpoonright\gamma+1)$,
(b) $\sigma_\gamma \colon \mathcal{M}^{\mathcal{U}}_\gamma \to \mathcal{M}^{\mathcal{W}_\gamma}_{z(\gamma)}$,
(c) $F_\gamma = \sigma_\gamma(E^{\mathcal{U}}_\gamma)$,

and when $\eta = U\text{-pred}(\gamma+1)$,

(d) $\alpha_\gamma = \alpha(\mathcal{W}_\eta,\mathcal{W}_\gamma,F_\gamma)$ and $\beta_\gamma = \beta(\mathcal{W}_\eta,\mathcal{W}_\gamma,F_\gamma)$, and
(e) $\Phi_{\eta,\gamma}\colon \mathcal{W}_\eta \to \mathcal{W}_\gamma$ is the associated extended tree embedding, with u-maps $u_{\eta,\gamma}$, v-maps $v_{\eta,\gamma}$, t-maps $t^{\eta,\gamma}_\tau$, and s-maps $s^{\eta,\gamma}_\tau$.

Suppose $\mathrm{lh}(\mathcal{U})$ is a limit ordinal θ, and let

$$\lambda = \mathrm{lh}(W(\mathcal{T},\mathcal{U})) = \sup_{\gamma<\theta}\alpha_\gamma.$$

Here we assume $W(\mathcal{T},\mathcal{U})$ exists, i.e., embedding normalization has so far produced only wellfounded models. Let b be a cofinal branch of $\mathcal{U}$. We do not assume $\mathcal{M}^{\mathcal{U}}_b$ is wellfounded. Note that $\mathcal{W}_b$ still makes sense, as defined above.

PROPOSITION 6.6.1. $\lambda = u_{0,b}(\tau)$, *where* τ *is least such that whenever* $\eta,\gamma\in b$ *and* $\eta <_U \gamma$, *then* $\mathrm{crit}(u_{\eta,\gamma}) \le u_{0,\eta}(\tau)$.

PROOF. Let $\eta+1\in b$, and $\sigma = U\text{-pred}(\eta+1)$. Then $u_{\sigma,\eta+1}(\mathrm{crit}(u_{\sigma,\eta+1})) = \alpha_\eta+1$, so $\alpha_\eta+1 \le \mathrm{crit}(u_{\eta+1,\xi})$ for all $\xi\in b$. It follows that $u_{0,b}(\tau)\ge\lambda$. But if $\sigma<\tau$, we can find $\gamma+1\in b$ with $\eta = U\text{-pred}(\gamma+1)$ such that $u_{0,\eta}(\sigma) < \mathrm{crit}(u_{\eta,\gamma+1})$. Then $u_{0,b}(\sigma) = u_{0,\eta}(\sigma) < \alpha_\gamma < \lambda$. Finally, $\lambda\in\mathrm{ran}\,u_{0,b}$ (because any $\xi<\mathrm{lh}(\mathcal{W}_\gamma)$ not in $\mathrm{ran}\,u_{0,\gamma}$ is fixed by $u_{\gamma,b}$), so $\lambda = u_{0,b}(\tau)$. ⊣

PROPOSITION 6.6.2. *Let* $a = [0,\lambda)_{W_b}$ *and* $\lambda = u_{0,b}(\tau)$; *then*

$$\xi\in a \text{ iff } \exists\eta\in b(\xi\le\mathrm{crit}(u_{\eta,b})\wedge\xi\le_{W_\eta}u_{0,\eta}(\tau)).$$

We omit the easy proof.

Remark 6.6.3. We do not get a "continuously" from b. If τ is fixed in advance, then continuously in those b such that $\tau=\tau_b$, we can produce the corresponding a's.

DEFINITION 6.6.4. In the situation above, we write

$$a = \mathrm{br}(b, \mathcal{T}, \mathcal{U})$$

and

$$\tau = m(b, \mathcal{T}, \mathcal{U})$$

for the branch of $W(\mathcal{T},\mathcal{U})$ and model of $\mathcal{T}$ determined by b.

Remark 6.6.5. Let E_b be the extender of $i^{\mathcal{U}}_b$. It is an extender over the model $\mathcal{M}^{\mathcal{T}}_\xi$, where $\xi + 1 = \mathrm{lh}(\mathcal{T})$. One can show that τ is the least α such that either E_b is an extender over $\mathcal{M}^{\mathcal{T}}_\alpha | \mathrm{lh}(E^{\mathcal{T}}_\alpha)$ (that is, $\mathrm{dom}(E_b) \subseteq \mathcal{M}^{\mathcal{T}}_\alpha | \mathrm{lh}(E^{\mathcal{T}}_\alpha)$), or $\alpha = \xi$.

The branch extender of a is given by

PROPOSITION 6.6.6. *Let $a = \mathrm{br}(b, \mathcal{T}, \mathcal{U})$ and $\tau = m(b, \mathcal{T}, \mathcal{U})$ be as above; then*

$$e_a^{W(\mathcal{T},\mathcal{U})} =$$

$$\hat{p}_{0,b}(e^{\mathcal{T}}_\tau)^\frown \langle F_\sigma \mid \sigma + 1 \in b \wedge \forall i \in \mathrm{dom}(\hat{p}_{0,b}(e^{\mathcal{T}}_\tau))\varepsilon(\hat{p}_{0,b}(e^{\mathcal{T}}_\tau)(i)) \leq \mathrm{crit}(F_\sigma)\rangle.$$

Here we are writing $e_a^{W(\mathcal{T},\mathcal{U})}$ for $e_\lambda^{\mathcal{W}_b}$, because $e_a^{W(\mathcal{T},\mathcal{U})}$ really only depends on a and $W(\mathcal{T},\mathcal{U})$. We omit the proof of Proposition 6.6.6. For what it is worth, here is a picture

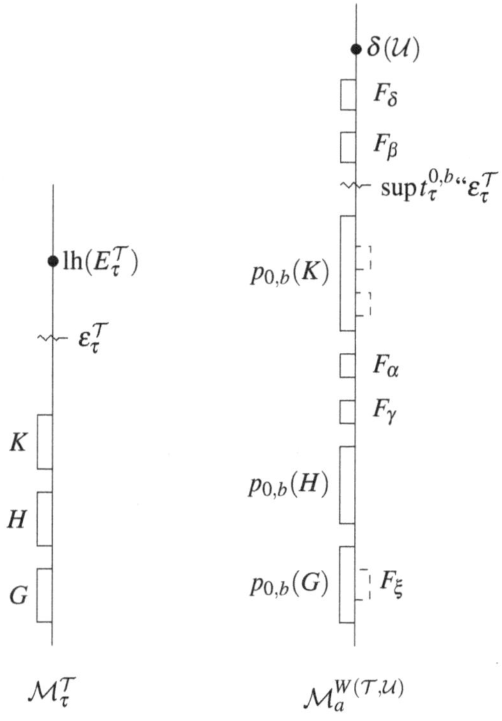

Note $\delta(\mathcal{U}) = \delta(W(\mathcal{T},\mathcal{U}))$. The F's in the picture were all used in b. Some got put directly into $e_a^{W(\mathcal{T},\mathcal{U})}$, others indirectly via some $p_{0,b}(G)$. $\varepsilon_\tau^{\mathcal{T}}$ is the sup of the generators of extenders used to get to $\mathcal{M}_\tau^{\mathcal{T}}$. (In general, $\varepsilon_\tau^{\mathcal{T}} < \hat{\lambda}(E_\tau^{\mathcal{T}})$.) The extenders in $e_a^{W(\mathcal{T},\mathcal{U})}$ with generators beyond $\sup t_\tau^{0,b}$ "$\varepsilon_\tau^{\mathcal{T}}$ are all directly inserted F's.

Branches of $W(\mathcal{T},\mathcal{U})$ of the form $\mathrm{br}(b,\mathcal{T},\mathcal{U})$ come from cofinal branches of $\mathcal{U}$ and *models* of $\mathcal{T}$. There may also be cofinal branches of $W(\mathcal{T},\mathcal{U})$ coming from cofinal branches of $\mathcal{U}$ and *maximal* (perhaps not cofinal) *branches* of $\mathcal{T}$. So we extend our definitions.

DEFINITION 6.6.7. Let $\mathcal{T}$ be a normal on M and $\mathcal{U}$ a normal tree on the last model of $\mathcal{T}$. For $\alpha \leq_U \beta$, let $u_{\alpha,\beta} = u^{\Phi_{\alpha,\beta}}$ and $v_{\alpha,\beta} = v^{\Phi_{\alpha,\beta}}$, where $\Phi_{\alpha,\beta} : \mathcal{W}_\alpha \to \mathcal{W}_\beta$ is the tree embedding of the meta-tree asssociated to $W(\mathcal{T},\mathcal{U})$. Let $\xi < \mathrm{lh}(\mathcal{T})$, $\gamma+1 < \mathrm{lh}(\mathcal{U})$, and $\eta = U\text{-pred}(\gamma+1)$; then

$$\mathrm{nd}(\xi,\gamma+1) \text{ is defined iff } v_{0,\eta}(\xi) \leq \mathrm{crit}(u_{\eta,\gamma+1}),$$

and if it is defined, then

$$\mathrm{nd}(\xi,\gamma+1) = \begin{cases} u_{0,\eta}(\xi) & \text{if } u_{0,\eta}(\xi)\downarrow \text{ and } u_{0,\eta}(\xi) = \mathrm{crit}(u_{\eta,\gamma+1}); \\ v_{0,\eta}(\xi) & \text{otherwise.} \end{cases}$$

We should write $\mathrm{nd}_{\mathcal{T},\mathcal{U}}(\xi,\gamma+1)$, but will usually drop the subscripts. We think of $\mathrm{nd}(\xi,\gamma+1)$ as the node in $\mathcal{W}$ obtained by shuffling into $e_\xi^{\mathcal{T}}$ an initial segment of the extenders F_τ, for $\tau+1 \leq_U \gamma+1$. The first case in its definition corresponds to the case when F_γ is being added as itself at the end, past all the images of extenders in $e_\xi^{\mathcal{T}}$.

Here are some observations about the u and v maps above that we shall use:

(i) $v_{0,\eta}(\xi) \leq \mathrm{crit}(u_{\eta,\mu}) \Longrightarrow v_{0,\eta}(\xi) = v_{0,\mu}(\xi)$.
(ii) $\mathrm{crit}(u_{\eta,\mu}) \leq_{W_\mu} u_{\eta,\mu}(\mathrm{crit}(u_{\eta,\mu}))$.
(iii) $v_{0,\eta}(\xi) \leq_{W_\eta} u_{0,\eta}(\xi)$.
(iv) If $\mathrm{crit}(u_{\theta,\mu}) < v_{0,\theta}(\xi)$ whenever $\theta <_U \mu <_U \eta$, then $v_{0,\eta}(\xi) = u_{0,\eta}(\xi)$.

Here are some simple facts about the node function:

PROPOSITION 6.6.8. *Let $\mathcal{W} = W(\mathcal{T},\mathcal{U})$ and $\eta = U\text{-pred}(\gamma+1)$; then*

(1) $\mathrm{nd}(\xi,\gamma+1)\downarrow$ *iff* $v_{0,\eta}(\xi) \leq \mathrm{crit}(u_{\eta,\gamma+1})$,
(2) if $\sigma = \mathrm{nd}(\xi,\gamma+1)$, *then* $\mathcal{W} \restriction \sigma+1 = \mathcal{W}_\eta \restriction \sigma+1$,
(3) if $\gamma+1 \leq_U \delta+1$ *and* $\theta = U\text{-pred}(\delta+1)$, *then either*
  (a) $\mathrm{nd}(\xi,\gamma+1) \leq_W \mathrm{nd}(\xi,\delta+1)$, *or*
  (b) $\mathrm{nd}(\xi,\gamma+1) = \mathrm{crit}(u_{\eta,\gamma+1}) = u_{0,\eta}(\xi)$, *and* $\mathrm{nd}(\xi,\delta+1) = v_{0,\eta}(\xi) = v_{0,\theta}(\xi)$,
(4) if $\xi <_T \theta$ *and* $\mathrm{nd}(\theta,\gamma+1)\downarrow$, *then* $\mathrm{nd}(\xi,\gamma+1)\downarrow$*; moreover, for all* $\beta \geq_U \gamma+1$,
  (i) $\mathrm{nd}(\xi,\gamma+1) = v_{0,\eta}(\xi) = v_{0,\beta}(\xi) = \mathrm{nd}(\xi,\beta)$, *and*
  (ii) $\mathrm{nd}(\xi,\beta) <_W \mathrm{nd}(\theta,\beta)$.

PROOF. (1) is part of the definition, and (2) holds because $\mathcal{W}_\eta$ agrees with $\mathcal{W}$ up to $\alpha_\eta + 1$, and $\text{crit}(u_{\eta,\gamma}) \leq \alpha_\eta$.

For (3), let $\theta = U\text{-pred}(\delta+1)$ and $\sigma = \text{nd}(\xi, \gamma+1)$. Suppose first that $\sigma = v_{0,\eta}(\xi)$; then by observation (i) above, $\sigma = v_{0,\theta}(\xi)$, and

$$\sigma \leq_{W_\theta} u_{\theta,\delta+1}(\text{crit}(u_{\theta,\delta+1})).$$

But $\text{nd}(\xi, \delta+1)$ is either $v_{0,\theta}(\xi)$ or $u_{0,\theta}(\xi)$, and by (iii), $\sigma \leq_W \text{nd}(\xi, \delta+1)$ in either case, so (3)(a) holds.

Next, suppose $\sigma = u_{0,\eta}(\xi) = \text{crit}(u_{\eta,\gamma+1})$. By (ii), $\sigma \leq_{W_\theta} u_{\eta,\theta}(\sigma) = u_{0,\theta}(\xi)$. So if $\text{nd}(\xi, \delta+1) = u_{0,\theta}(\xi)$ then (3)(a) holds, and we are done. If not, then $\text{nd}(\xi, \delta+1) = v_{0,\theta}(\xi)$. But $v_{0,\eta}(\xi) \leq \text{crit}(u_{\eta,\theta})$, so $v_{0,\eta}(\xi) = v_{0,\theta}(\xi)$ by (i), so then (3)(b) holds. This proves (3).

For (4), $\text{nd}(\xi, \gamma+1)\downarrow$ since $v_{0,\eta}(\xi) < v_{0,\eta}(\theta)$. Let $\sigma = \text{nd}(\xi, \gamma+1)$ and $\tau = \text{nd}(\theta, \gamma+1)$. We claim first that it is not the case that $\sigma = u_{0,\eta}(\xi) = \text{crit}(u_{\eta,\gamma+1})$. For then

$$\text{crit}(u_{\mu,\beta}) < u_{0,\beta}(\theta)$$

whenever $\mu <_U \beta <_U \eta$, so using (iv),

$$v_{0,\eta}(\theta) = u_{0,\eta}(\theta) = \tau.$$

But $u_{0,\eta}(\xi) < u_{0,\eta}(\theta)$ (as ordinals), which gives us $\text{crit}(u_{\eta,\gamma+1}) < v_{0,\eta}(\theta)$, so that $\text{nd}(\theta, \gamma+1)$ is not defined after all.

Thus $\sigma = v_{0,\eta}(\xi) < \text{crit}(u_{\eta,\gamma+1})$. This implies $\sigma <_{W_\eta} v_{0,\eta}(\theta) \leq_{W_\eta} u_{0,\eta}(\theta)$, so $\sigma <_{W_\eta} \tau$, so $\sigma <_W \tau$. This verifies (4) at $\beta = \gamma+1$. We leave the case $\gamma+1 \leq_U \beta$ to the reader. ⊣

Now let b be a cofinal branch of $\mathcal{U}$, and let c be a branch of $\mathcal{T}$. We allow c to be cofinal, or maximal and not cofinal, or to have a largest element. We may be able to use the node function to generate from b and c a cofinal branch $\text{br}(c,b)$ of $W(\mathcal{T},\mathcal{U})$.

DEFINITION 6.6.9. Let c be a branch of $\mathcal{T}$ and b be a cofinal branch of $\mathcal{U}$. We say that $\text{br}(c,b)$ *is defined* (or $\text{br}(c,b)\downarrow$) iff either

(1) c has a largest element ξ, and for all sufficiently large $\gamma+1 \in b$, $\text{nd}(\xi,\gamma+1)\downarrow$, and $\text{nd}(\xi,\gamma+1) = \text{crit}(u_{\eta,\gamma+1})$, where $\eta = U\text{-pred}(\gamma+1)$, or
(2) c has no largest element, and for all $\xi \in c$, there is a $\gamma+1 \in b$ such that $\text{nd}(\xi,\gamma+1)\downarrow$.

In case (1), we set

$$\text{br}(c,b) = \{\tau \mid \tau \leq_W \text{nd}(\xi,\gamma+1) \text{ for all sufficiently large } \gamma+1 \in b\},$$

where ξ is the largest element of c. In case (2), we set

$$\begin{aligned}\text{br}(c,b) = \{\tau \mid \exists\xi \in c \exists\gamma+1 \in b(&\text{nd}(\xi,\gamma+1)\downarrow \text{ and} \\ &\tau \leq_{W_\eta} v_{0,\eta}(\xi), \text{where } \eta = U\text{-pred}(\gamma+1))\}.\end{aligned}$$

In case (2), $v_{0,\eta}(\xi)$ is just the common value of $\mathrm{nd}(\xi,\gamma+1)$ for all sufficiently large $\gamma+1 \in b$. This follows from part (4) of Proposition 6.6.8.

DEFINITION 6.6.10. Suppose that $\mathrm{br}(c,b)\downarrow$; then we say that c is *b-cofinal* iff c has a largest element, or c has no largest element, and for all $\gamma+1 \in b$ there is a $\xi \in c$ such that $\mathrm{nd}(\xi,\gamma+1)$ is not defined.

DEFINITION 6.6.11. We say that $\gamma+1$ is (ξ,σ)-*minimal* iff $\mathrm{nd}(\xi,\gamma+1)=\sigma$, and whenever $\delta+1 <_U \gamma+1$, $\mathrm{nd}(\xi,\delta+1)\neq\sigma$.

LEMMA 6.6.12. *Suppose that* $\mathrm{br}(c,b)\downarrow$ *and c is b-cofinal; then* $\mathrm{br}(c,b)$ *is a cofinal branch of* $\mathcal{W}$*. Moreover, there are cofinally many* $\sigma \in \mathrm{br}(c,b)$ *such that for some* $\xi \in c$ *and* $\gamma+1 \in b$*,* $\gamma+1$ *is* (ξ,σ)*-minimal.*

PROOF. Suppose first that c has largest element ξ. For $\gamma+1 <_U \delta+1 \in b$ sufficiently large, letting $\eta = U\text{-pred}(\gamma+1)$, we have that $u_{0,\eta}(\xi)=\mathrm{crit}(u_{\eta,\gamma+1})$. Letting $\mathcal{W}_b = W(\mathcal{T},\mathcal{U}^\frown b)$, this easily implies that $\mathrm{br}(c,b)=[0,\lambda)_{W_b}$, so that it is a cofinal branch of $\mathcal{W}$. Moreover, all sufficiently large $\sigma \in \mathrm{br}(c,b)$ are of the form $\mathrm{nd}(\xi,\gamma+1)$, where $\gamma+1 \in b$ and $\gamma+1$ is (ξ,σ)-minimal.

Suppose next that c has no largest element. Proposition 6.6.8 part (4) implies that $\mathrm{br}(c,b)$ is a branch of $\mathcal{W}$. To see that it is cofinal, let $\mu < \mathrm{lh}(\mathcal{W})$, and pick $\gamma+1 \in b$ such that $\mu < \alpha_\gamma$. Since c is b-cofinal, we have $\xi \in c$ such that $\mathrm{nd}(\xi,\gamma+1)$ is not defined. Let $\delta+1$ be least such that $\gamma+1 <_U \delta+1 \in b$ and $\mathrm{nd}(\xi,\delta+1)$ is defined. Let $\sigma = \mathrm{nd}(\xi,\delta+1)$, so that $\delta+1$ is (ξ,σ)-minimal. We shall show that $\sigma \in \mathrm{br}(c,b)$, and $\mu < \sigma$.

Let η and θ be the U-predecessors of $\gamma+1$ and $\delta+1$. By the minimality of $\delta+1$, we have that

$$\mathrm{crit}(u_{\nu,\theta}) < v_{0,\nu}(\xi) \text{ for all } \nu <_U \theta,$$

and thus

$$u_{0,\nu}(\xi) = v_{0,\nu}(\xi) \text{ for all } \nu \leq_U \theta,$$

by observation (iv). Thus $\sigma = v_{0,\theta}(\xi) = u_{0,\theta}(\xi)$, so $\sigma \in \mathrm{br}(c,b)$. Also,

$$\begin{aligned}\mu < \alpha_\gamma + 1 &= u_{\eta,\gamma+1}(\mathrm{crit}(u_{\eta,\gamma+1}))\\ &< u_{\eta,\gamma+1}(u_{0,\eta}(\xi)) = u_{0,\gamma+1}(\xi)\\ &= v_{0,\gamma+1}(\xi) \leq v_{0,\theta}(\xi) = \sigma.\end{aligned}$$ ⊣

We shall show that if a is a cofinal branch of $W(\mathcal{T},\mathcal{U})$, then $a = \mathrm{br}(c,b)$ for some cofinal branch b of $\mathcal{U}$ and some c; moreover, there is a unique such b, and a unique such b-cofinal c. First, let us recall some simple facts about the agreement between the $\mathcal{W}_\gamma$'s. Let R_γ be the last model of $\mathcal{W}_\gamma$.

LEMMA 6.6.13. *Let* $\gamma < \delta < \mathrm{lh}(\mathcal{U})$*; then*

(a) $R_\gamma || \mathrm{lh}(F_\gamma) = R_\delta || \mathrm{lh}(F_\gamma)$,

(b) F_γ *is on the sequence of* R_γ*, and not on the sequence of* R_δ*,*

(c) for all $\xi \geq \alpha_\gamma$, *and all* ν, $M_\xi^{\mathcal{W}_\gamma}|\mathrm{lh}(F_\gamma)$ *is not an initial segment of* $M_\nu^{\mathcal{W}_\delta}$, *and*
(d) if $s^\frown\langle H\rangle \in \mathcal{W}_\gamma^{\mathrm{ext}} \cap \mathcal{W}_\delta^{\mathrm{ext}}$, *then* $\mathrm{lh}(H) < \mathrm{lh}(F_\gamma)$.

PROOF. We have already proved (a)-(c), and part (d) is an immediate consequence of (c). ⊣

The following is the key lemma.

LEMMA 6.6.14. *Let* $\mathcal{T}, \mathcal{U}$ *be as above. Let* γ *and* δ *be* $\leq_U$*-incomparable, and let* η *be largest such that* $\eta <_U \gamma$ *and* $\eta <_U \delta$. *Let* $\alpha = u_{\eta,\gamma}(\bar{\alpha})$ *and* $\varepsilon = u_{\eta,\delta}(\bar{\varepsilon})$, *where* $\bar{\alpha} \geq \mathrm{crit}(u_{\eta,\gamma})$ *and* $\bar{\varepsilon} \geq \mathrm{crit}(u_{\eta,\delta})$*; then* $e_\alpha^{\mathcal{W}_\gamma}$ *is incompatible with* $e_\varepsilon^{\mathcal{W}_\delta}$.

PROOF. Let $a = e_\alpha^{\mathcal{W}_\gamma}$, $\bar{a} = e_{\bar{\alpha}}^{\mathcal{W}_\eta}$, $e = e_\varepsilon^{\mathcal{W}_\delta}$ and $\bar{e} = e_{\bar{\varepsilon}}^{\mathcal{W}_\eta}$. Assume toward contradiction that either $a \subseteq e$, or $e \subseteq a$.

Let

$$\gamma_0 + 1 = \text{least } \xi \in (\eta, \gamma]_U,$$
$$\delta_0 + 1 = \text{least } \xi \in (\eta, \delta]_U,$$

so that $E_{\gamma_0}^{\mathcal{U}}$ and $E_{\delta_0}^{\mathcal{U}}$ are the extenders used in $\mathcal{U}$ along the two branches of $\mathcal{U}$ at the point where they diverge, and F_{γ_0} and F_{δ_0} stretch $\mathcal{W}_\eta$ into $\mathcal{W}_{\gamma_0+1}$ and $\mathcal{W}_{\delta_0+1}$. Let

$$k(\bar{a}) = \begin{cases} \text{least } i \text{ such that } \mathrm{crit}(F_{\gamma_0}) < \hat{\lambda}(\bar{a}(i)) & \text{if this exists,} \\ \mathrm{dom}(\bar{a}) & \text{otherwise,} \end{cases}$$

and

$$k(\bar{e}) = \begin{cases} \text{least } i \text{ such that } \mathrm{crit}(F_{\delta_0}) < \hat{\lambda}(\bar{e}(i)) & \text{if this exists,} \\ \mathrm{dom}(\bar{e}) & \text{otherwise.} \end{cases}$$

CLAIM 6.6.15. $k(\bar{a}) = k(\bar{e})$, *and for* $k = k(\bar{a})$, $\bar{a}\restriction k = \bar{e}\restriction k = a\restriction k = e\restriction k$.

PROOF. Let $k = k(\bar{a})$. If $k < k(\bar{e})$, then $e(k) = \bar{e}(k)$, so $\hat{\lambda}(e(k)) \leq \mathrm{crit}(F_{\delta_0})$. But $\hat{\lambda}(a(k)) \geq \hat{\lambda}(F_{\gamma_0})$. [$e_{u_{\eta,\gamma_0+1}(\bar{\alpha})}^{W_{\gamma_0+1}}(k) = H$ is defined because $\bar{\alpha} \geq \mathrm{crit}(u_{\eta,\gamma_0+1})$. H is either F_{γ_0} or the stretch by F_{γ_0} of some G such that $\mathrm{crit}(G) < \mathrm{crit}(F_{\gamma_0})$. In either case, $\hat{\lambda}(H) \geq \hat{\lambda}(F_{\gamma_0})$. $a(k) = p_{\gamma_0+1,\gamma}(H)$, so $\hat{\lambda}(a(k)) \geq \hat{\lambda}(H)$.] Since $a(k) = e(k)$, we have $\hat{\lambda}(F_{\gamma_0}) \leq \mathrm{crit}(F_{\delta_0})$, so F_{γ_0} and F_{δ_0} do not overlap, contradiction. $k(\bar{e}) < k(\bar{a})$ leads to a parallel contradiction. So we have $k(\bar{a}) = k(\bar{e}) = k$.

For $i < k$, $a(i) = \bar{a}(i)$ and $e(i) = \bar{e}(i)$. So $\bar{a}\restriction k = \bar{e}\restriction k = a\restriction k = e\restriction k$. ⊣

Fix $k = k(\bar{a})$. We may assume by symmetry that $\gamma_0 < \delta_0$.

CLAIM 6.6.16. $k \in \mathrm{dom}(\bar{a})$, *and moreover,* $\mathrm{crit}(\bar{a}(k)) < \mathrm{crit}(F_{\gamma_0})$.

PROOF. If either statement fails, then

$$e_{u_{\eta,\gamma_0+1}(\bar{\alpha})}^{\mathcal{W}_{\gamma_0+1}}(k) = F_{\gamma_0}.$$

Since the extenders used in $(\gamma_0+1,\gamma]_U$ have critical point at least $\hat{\lambda}(E^{\mathcal{U}}_{\gamma_0})$, we get

$$p_{\gamma_0+1,\gamma}(F_{\gamma_0}) = F_{\gamma_0}.$$

(In fact, $u_{\gamma_0+1,\gamma}\restriction(\gamma_0+1) =$ identity, and $t^{\gamma_0+1,\gamma}_{\gamma_0} =$ identity.) So

$$a(k) = F_{\gamma_0}.$$

But $k = k(\bar{e})$, and from this we get

$$\hat{\lambda}(F_{\delta_0}) \leq \hat{\lambda}(e(k))$$

as in Claim 6.6.15. Since $\hat{\lambda}(F_{\gamma_0}) < \hat{\lambda}(F_{\delta_0})$, we have a contradiction. ⊣

Let $G = \bar{a}(k)$ and $H = a(k)$. By Claim 6.6.16, along the branch from η to γ, G is being stretched above its critical point into H, by the copy maps corresponding to the F_τ for $\tau+1 \leq_U \gamma$ and $\eta \leq \tau$. Let $\gamma_1 \leq \gamma$ be least such that the stretching is finished at γ_1. That is, setting

$$\begin{aligned} G &= E^{\mathcal{W}_\eta}_\xi \\ \gamma_1 &= \text{least } \tau \leq \gamma \text{ such that } \operatorname{crit}(u_{\tau,\gamma}) > u_{\eta,\tau}(\xi) \\ &= \text{least } \tau \leq \gamma \text{ such that } t^{\eta,\tau}_\xi(G) = H. \end{aligned}$$

If $\eta <_U \tau+1 \leq_U \gamma_1$, so that F_τ was used in producing $\mathcal{W}_{\gamma_1}$ from $\mathcal{W}_\eta$, then F_τ is an initial segment of all the extenders of copy maps $t^{\mu,\tau+1}_\rho$, where $\mu = U\text{-pred}(\tau+1)$, and $\rho \geq \operatorname{crit}(u_{\mu,\tau+1})$.

From this we get

CLAIM 6.6.17. *For $\eta <_U \tau+1 \leq_U \gamma_1$, $\operatorname{lh}(F_\tau) < \operatorname{lh}(H)$.*

PROOF. $F_\tau = E^{\mathcal{W}_{\tau+1}}_{\alpha_\tau}$, so for $\eta <_U \tau+1 \leq \gamma_1$,

$$\operatorname{lh}(F_\tau) < \operatorname{lh}(E^{\mathcal{W}_{\tau+1}}_{u_{\eta,\tau+1}(\xi)}) \leq \operatorname{lh}(E^{\mathcal{W}_{\gamma_1}}_{u_{\eta,\gamma_1}(\xi)}) = \operatorname{lh}(H).$$ ⊣

CLAIM 6.6.18. $H \neq F_{\delta_0}$.

PROOF. Suppose $H = F_{\delta_0}$. By Claim 6.6.17, $\operatorname{lh}(F_\tau) < \operatorname{lh}(F_{\delta_0})$ whenever $\tau+1 \leq_U \gamma_1$. Since $\mathcal{U}$ is normal, this implies $\gamma_1 \leq \delta_0$. But γ_1 and δ_0 are incomparable in $\mathcal{U}$, so $\gamma_1 < \delta_0$.

But then $a\restriction k^\frown\langle H\rangle \in \mathcal{W}^{\text{ext}}_{\gamma_1} \cap \mathcal{W}^{\text{ext}}_{\delta_0}$, and $\operatorname{lh}(F_{\gamma_1}) < \operatorname{lh}(H) = \operatorname{lh}(F_{\delta_0})$. This contradicts part (d) of Lemma 6.6.13. ⊣

By Claim 6.6.18, $k \in \operatorname{dom}(\bar{e})$, and letting $L = \bar{e}(k)$, $\operatorname{crit}(L) < \operatorname{crit}(F_{\delta_0})$. So L is being stretched above its critical point into H along the branch from η to δ. Let $\delta_1 \leq \delta$ be least such that the stretching is over with at δ_1; that is, setting

$$\begin{aligned} L &= E^{\mathcal{W}_\eta}_\mu \\ \delta_1 &= \text{least } \tau \leq_U \delta \text{ such that} \operatorname{crit}(u_{\tau,\delta}) > u_{\eta,\tau}(\mu) \\ &= \text{least } \tau \leq_U \delta \text{ such that } t^{\eta,\tau}_\mu(L) = H. \end{aligned}$$

We have that $\gamma_1 \neq \delta_1$. Suppose that $\gamma_1 < \delta_1$; it no longer matters whether $\gamma_0 < \delta_0$, so this is not a loss of generality. Since $\mathcal{U}$ is normal, we have a $\tau + 1 \leq_U \delta_1$ such that $\gamma_1 \leq \tau$. By the proof of Claim 6.6.17,

$$\mathrm{lh}(F_{\gamma_1}) \leq \mathrm{lh}(F_\tau) < \mathrm{lh}(H).$$

But $a \restriction k^\frown \langle H \rangle$ is in the extender trees of both $\mathcal{W}_{\gamma_1}$ and $\mathcal{W}_{\delta_1}$, so by Lemma 6.6.13(d), $\delta_1 \leq \gamma_1$. This contradiction completes the proof of Lemma 6.6.14. ⊣

COROLLARY 6.6.19. *Let* $\sigma = \mathrm{nd}(\xi, \gamma_0 + 1)$ *and* $\tau = \mathrm{nd}(\rho, \gamma_1 + 1)$, *where* $\gamma_0 + 1$ *is* (ξ, σ)*-minimal and* γ_1 *is* (ρ, τ)*-minimal. Suppose that* U-pred$(\gamma_0 + 1)$ *is* $\leq_U$*-incomparable with* U-pred$(\gamma_1 + 1)$*; then* σ *and* τ *are* $\leq_{W(\mathcal{T},\mathcal{U})}$*-incomparable.*

PROOF. Let η be largest such that $\eta <_U \gamma_0 + 1$ and $\eta <_U \gamma_1 + 1$. Let $\eta = U\text{-pred}(\eta_0 + 1) = U\text{-pred}(\eta_1 + 1)$, where $\eta_0 + 1 \leq_U \gamma_0 + 1$ and $\eta_1 + 1 \leq_U \gamma_1 + 1$. By the minimality of γ_0 and γ_1,

$$\mathrm{crit}(u_{\eta,\eta_0+1}) \leq u_{0,\eta}(\xi)$$

and

$$\mathrm{crit}(u_{\eta,\eta_1+1}) \leq u_{0,\eta}(\rho).$$

To see this, suppose $u_{0,\eta}(\xi) < \mathrm{crit}(u_{\eta,\eta_0+1})$, and let $\theta = U\text{-pred}(\gamma_0 + 1)$. Recall that the u maps along a branch of $\mathcal{U}$ have increasing critical points. Thus

$$\sigma \leq u_{0,\theta}(\xi) < \mathrm{crit}(u_{\theta,\gamma_0+1}),$$

so

$$\sigma = v_{0,\theta}(\xi) = v_{0,\eta}(\xi) \leq u_{0,\eta}(\xi) < \mathrm{crit}(u_{\eta,\eta_0+1}).$$

Thus $\sigma = \mathrm{nd}(\xi, \eta_0 + 1)$, contrary to the minimality of $\gamma_0 + 1$.

So if $\mathrm{crit}(u_{\eta,\eta_0+1}) > u_{0,\eta}(\xi)$, then $\sigma = u_{0,\eta}(\xi)$, so $\sigma = \mathrm{nd}(\xi, \eta_0 + 1)$. The proof that $\mathrm{crit}(u_{\eta,\eta_1+1}) \leq u_{0,\eta}(\rho)$ is the same. But then

$$\mathrm{crit}(u_{\eta,\gamma_0+1}) \leq u_{0,\eta}(\xi)$$

and

$$\mathrm{crit}(u_{\eta,\gamma_1+1}) \leq u_{0,\eta}(\rho).$$

By Lemma 6.6.14,

$$e_\sigma^{\mathcal{W}_{\gamma_0+1}} \perp e_\tau^{\mathcal{W}_{\gamma_1+1}}.$$

But $\sigma \leq \beta_{\gamma_0}$ by the definition of $\mathrm{nd}(\xi, \gamma_0 + 1)$, so $\sigma \leq \alpha_{\gamma_0}$, so $e_\sigma^{\mathcal{W}_{\gamma_0+1}} = e_\sigma^{W(\mathcal{T},\mathcal{U})}$. Similarly, $e_\tau^{\mathcal{W}_{\gamma_1+1}} = e_\tau^{W(\mathcal{T},\mathcal{U})}$, so we are done. ⊣

COROLLARY 6.6.20. *Suppose that* $a = \mathrm{br}(c_0, b_0) = \mathrm{br}(c_1, b_1)$, *where* c_i *is* b_i*-cofinal for* $i = 0, 1$*; then* $b_0 = b_1$ *and* $c_0 = c_1$.

PROOF. Let $a = \mathrm{br}(c_0, b_0) = \mathrm{br}(c_1, b_1)$. Suppose toward contradiction that $b_0 \neq b_1$. Let $\eta_0 \in b_0$ and $\eta_1 \in b_1$ be $\leq_U$-incomparable. By Lemma 6.6.12 we can find $\xi_i \in c_i$ and $\gamma_i + 1 \in b_i$, for $i = 0, 1$, such that letting $\sigma_i = \mathrm{nd}(\xi_i, \gamma_i + 1)$:

(a) $\sigma_i \in \text{br}(c_i, b_i)$,
(b) $\eta_i <_U \gamma_i + 1$, and
(c) γ_i is (ξ_i, σ_i)-minimal.

But then $\gamma_0 + 1$ is $\mathcal{U}$-incomparable with $\gamma_1 + 1$, so σ_0 is $\mathcal{W}$-incomparable with σ_1 by Corollary 6.6.19. Since σ_0 and σ_1 are in a, we have a contradiction.

So let $b = b_0 = b_1$, and suppose that $c_0 \neq c_1$. Since they are b-cofinal, we have $\xi \in c_0$ and $\nu \in c_1$ such that ξ and ν are $\mathcal{T}$-incomparable. For all sufficiently large $\eta \in b$, $v_{0,\eta}(\xi) \in a$ and $v_{0,\eta}(\nu) \in a$, so there is an η such that $v_{0,\eta}(\xi)$ and $v_{0,\eta}(\nu)$ are defined and comparable in $\mathcal{W}_\eta$. Since $v_{0,\eta}$ preserves incompatibility, we have a contradiction. ⊣

Finally, we show that our branch-merging function is surjective.

LEMMA 6.6.21. *For any cofinal branch a of $W(\mathcal{T},\mathcal{U})$, there is a cofinal branch b of $\mathcal{U}$ and a branch c of $\mathcal{T}$ such that* $\text{br}(c,b) = a$.

PROOF. We begin by decoding nodes of $\mathcal{U}$ from nodes of $W(\mathcal{T},\mathcal{U})$. For $\xi < \text{lh}(W(\mathcal{T},\mathcal{U}))$, set

$$d(\xi) = \text{least } \gamma \text{ such that } \xi \leq \alpha_\gamma.$$

CLAIM 1.

$$\begin{aligned} d(\xi) &= \textit{least } \gamma \textit{ such that } e_\xi^{\mathcal{W}_\gamma} = e_\xi^{W(\mathcal{T},\mathcal{U})} \\ &= \textit{least } \gamma \textit{ such that } \mathcal{M}_\xi^{\mathcal{W}_\gamma} = \mathcal{M}_\xi^{W(\mathcal{T},\mathcal{U})}. \end{aligned}$$

PROOF. If $\xi \leq \alpha_\gamma$, then $\mathcal{W}_\gamma \restriction \xi + 1 = W(\mathcal{T},\mathcal{U}) \restriction \xi + 1$, so $e_\xi^{\mathcal{W}_\gamma} = e_\xi^{W(\mathcal{T},\mathcal{U})}$ and $\mathcal{M}_\xi^{\mathcal{W}_\gamma} = \mathcal{M}_\xi^{W(\mathcal{T},\mathcal{U})}$. On the other hand, F_γ is used in $W(\mathcal{T},\mathcal{U})$ but not in $\mathcal{W}_\gamma$, so if $\xi > \alpha_\gamma$, then F_γ is on the sequence of $\mathcal{M}_\xi^{\mathcal{W}_\gamma}$ but not that of $\mathcal{M}_\xi^{W(\mathcal{T},\mathcal{U})}$. So $\mathcal{M}_\xi^{\mathcal{W}_\gamma} \neq \mathcal{M}_\xi^{W(\mathcal{T},\mathcal{U})}$, and hence $e_\xi^{\mathcal{W}_\gamma} \neq e_\xi^{W(\mathcal{T},\mathcal{U})}$. ⊣

CLAIM 2. $\xi_0 \leq_{W(\mathcal{T},\mathcal{U})} \xi_1 \implies d(\xi_0) \leq_U d(\xi_1)$.

PROOF. Let $\gamma_0 = d(\xi_0)$ and $\gamma_1 = d(\xi_1)$. We claim that $\xi_0 \in \text{ran}\, u_{0,\gamma_0}$. For let τ be least such that $u_{0,\gamma_0}(\tau) \geq \xi_0$. If $u_{0,\gamma_0}(\tau) \neq \xi_0$, then there must be $0 \leq_U \eta <_U \sigma + 1 \leq_U \gamma_0$ such that

$$\text{crit}(u_{\eta,\sigma+1}) \leq \xi_0 < u_{\eta,\sigma+1}(\text{crit}(u_{\eta,\sigma+1}))$$

and $\eta = U\text{-pred}(\sigma+1)$. (All discontinuities in u_{0,γ_0} arise this way.) But then $\xi_0 < \alpha_\sigma + 1$, so $\xi_0 \leq \alpha_\sigma$, and $\sigma < \gamma_0$, contradiction.

Similarly, $\xi_1 \in \text{ran}\, u_{0,\gamma_1}$.

We claim that γ_0 and γ_1 are comparable in $\mathcal{U}$. Suppose not, and let η be largest such that $\eta <_U \gamma_0$ and $\eta <_U \gamma_1$. Let

$$\xi_0 = u_{\eta,\gamma_0}(\bar{\xi}_0) \quad \text{and} \quad \xi_1 = u_{\eta,\gamma_1}(\bar{\xi}_1).$$

The hypotheses of Lemma 6.6.14 are satisfied, noting that $\bar{\xi}_0 \geq \text{crit}(u_{\eta,\gamma_0})$ because otherwise $e_{\xi_0}^{\mathcal{W}_{\gamma_0}} = e_{\xi_0}^{\mathcal{W}_\eta}$, whilst γ_0 was least such that $e_{\xi_0}^{\mathcal{W}_{\gamma_0}}$ appears as a branch extender. Similarly, $\bar{\xi}_1 \geq \text{crit}(u_{\eta,\gamma_1})$. The other hypotheses of Lemma 6.6.14 hold, so we conclude $e_{\xi_0}^{\mathcal{W}_{\gamma_0}}$ is incompatible with $e_{\xi_1}^{\mathcal{W}_{\gamma_1}}$. This implies ξ_0 and ξ_1 are incomparable in $W(\mathcal{T},\mathcal{U})$.

Finally, $\xi_0 \leq_{W(\mathcal{T},\mathcal{U})} \xi_1 \Longrightarrow \xi_0 \leq \xi_1$, and trivially $\xi_0 \leq \xi_1 \Longrightarrow d(\xi_0) \leq d(\xi_1)$. Since $d(\xi_0)$ and $d(\xi_1)$ are $\leq_U$-comparable, $d(\xi_0) \leq_U d(\xi_1)$, as desired. ⊣

CLAIM 3. $d : \text{lh}(W(\mathcal{T},\mathcal{U})) \to \text{lh}(\mathcal{U})$ *is an order-homomorphism, and* $\text{ran}(d)$ *is cofinal in* $\text{lh}(\mathcal{U})$.

PROOF. As we remarked, $\xi_0 \leq \xi_1 \Longrightarrow d(\xi_0) \leq d(\xi_1)$ is trivial. Pick any $\gamma < \text{lh}(\mathcal{U})$, and $\xi < \text{lh}(W(\mathcal{T},\mathcal{U}))$ with $\xi > \alpha_\gamma$. (The α_γ's are strictly increasing.) Then $d(\xi) > \gamma$. ⊣

It follows that for any branch a of $W(\mathcal{T},\mathcal{U})$, we can set

$$d(a) = \{\gamma \mid \exists \xi \in a(\gamma \leq_U d(\xi))\},$$

and $d(a)$ is a branch of $\mathcal{U}$. If a is cofinal in $W(\mathcal{T},\mathcal{U})$, then $d(a)$ is cofinal in $\mathcal{U}$.

Next we decode nodes of $\mathcal{T}$. For any $\xi < \text{lh}(W(\mathcal{T},\mathcal{U}))$, set

$$e(\xi) = \text{unique } \alpha < \text{lh}\,\mathcal{T} \text{ such that } u_{0,d(\xi)}(\alpha) = \xi.$$

We showed in the proof of Claim 2 that $\xi \in \text{ran}(u_{0,d(\xi)})$.

CLAIM 4. $\xi_0 \leq_{W(\mathcal{T},\mathcal{U})} \xi_1 \Longrightarrow e(\xi_0) \leq_T e(\xi_1)$.

PROOF. Let $\gamma_i = d(\xi_i)$ and $\bar{\xi}_i = e(\xi_i)$. As we noted above, the u maps do not introduce new tree-order relationships in $\text{ran}\,u$.

SUBCLAIM A. *If* $u_{\eta,\gamma}(\mu) \leq_{W_\gamma} u_{\eta,\gamma}(\nu)$, *then* $\mu \leq_{W_\eta} \nu$.

PROOF. Easy induction on γ. ⊣

So if $\bar{\xi}_0 \not\leq_T \bar{\xi}_1$, then $u_{0,\gamma_0}(\bar{\xi}_0) \not\leq_{W_{\gamma_0}} u_{0,\gamma_0}(\bar{\xi}_1)$. That is, $\xi_0 \not\leq_{W_{\gamma_0}} u_{0,\gamma_0}(\bar{\xi}_1)$. If $\text{crit}(u_{\gamma_0,\gamma_1}) > \xi_0$, then we get $\xi_0 \not\leq_{W_{\gamma_1}} \xi_1$, and since $\xi_1 \leq \alpha_{\gamma_1}$, $\xi_0 \not\leq_{W(\mathcal{T},\mathcal{U})} \xi_1$, as desired. So assume $\xi_0 \geq \text{crit}(u_{\gamma_0,\gamma_1})$.

If $\xi_0 = \text{crit}(u_{\gamma_0,\gamma_1})$, then $\xi_0 \leq_{W_{\gamma_1}} u_{\gamma_0,\gamma_1}(\sigma)$ iff $\xi_0 \leq_{W_{\gamma_0}} \sigma$ for all σ. Since $\xi_0 \not\leq_{W_{\gamma_0}} u_{0,\gamma_0}(\bar{\xi}_1)$, this yields $\xi_0 \not\leq_{W_{\gamma_1}} \xi_1$, so $\xi_0 \not\leq_{W(\mathcal{T},\mathcal{U})} \xi_1$, as desired.

Finally, suppose $\xi_0 > \text{crit}(u_{\gamma_0,\gamma_1})$. So letting $\tau + 1 \leq_U \gamma_1$ be least such that $\gamma_0 <_U \tau + 1$, and

$$\beta = \beta(\mathcal{W}_{\gamma_0}, \mathcal{W}_\tau, F_\tau),$$

we have

$$\beta < \xi_0 \leq \alpha_{\gamma_0} < \alpha_\tau.$$

No extender of the form $E^{\mathcal{W}_{\gamma_1}}_{u_{\gamma_0,\gamma_1}(\sigma)}$ can have critical point in the interval

$$[\mathrm{crit}(F_\tau), \hat{\lambda}(F_\tau)).$$

This implies that if $\tau+1 \leq_U \gamma$ and $\beta < \xi \leq \alpha_\tau$, then for all $\sigma \in \mathrm{dom}\, u_{\gamma_0,\gamma}$, $\xi \not\leq_{W_\gamma} u_{\gamma_0,\gamma}(\sigma)$. In particular, $\xi_0 \not\leq_{W_{\gamma_1}} \xi_1$, so $\xi_0 \not\leq_{W(\mathcal{T},\mathcal{U})} \xi_1$, as desired. ⊣

For a branch a of $W(\mathcal{T},\mathcal{U})$, we set

$$e(a) = \{\beta \mid \exists \xi \in a(\beta \leq_T e(\xi))\}.$$

So $e(a)$ is a branch of $\mathcal{T}$. Even if a is cofinal in $W(\mathcal{T},\mathcal{U})$, $e(a)$ may not be cofinal in $\mathcal{T}$. $e(a)$ may have a largest element, or be a maximal branch of $\mathcal{T}$ not chosen by $\mathcal{T}$.

CLAIM 5. *Let a be cofinal in $W(\mathcal{T},\mathcal{U})$. Then $a = \mathrm{br}_{\mathcal{W}}(e(a),d(a))$, and $e(a)$ is $d(a)$-cofinal.*

PROOF. Let $b = d(a)$ and $c = e(a)$. Suppose first that c has largest element ξ. So for $\sigma \in a$ sufficiently large, $e(\sigma) = \xi$, that is, $\sigma = u_{0,\gamma+1}(\xi)$ for $\gamma+1 \in b$ least such that $\sigma \leq \alpha_\gamma$. This implies that $u_{0,\eta}(\xi) = \mathrm{crit}(u_{\eta,\gamma+1})$ for $\gamma+1 \in b$ sufficiently large and $\eta = U\text{-pred}(\gamma+1)$. So for $\sigma \in a$ sufficiently large, $\sigma \in \mathrm{br}(c,b)$, as witnessed by ξ and some (unique) $\gamma+1 \in b$. Thus $a = \mathrm{br}(c,b)$, and c is b-cofinal.

Now suppose c has no largest element. To see that $\mathrm{br}(c,b) = a$, it suffices to show that cofinally many points in $\mathrm{br}(c,b)$ are in a. So let $\mu < \sup(c)$ and $\nu < \sup(b)$. We must find $\xi \in c - \mu$ and $\gamma+1 \in b - \nu$ such that for $\eta = U\text{-pred}(\gamma+1)$, $v_{0,\eta}(\xi) \leq \mathrm{crit}(u_{\eta,\gamma+1})$ and $v_{0,\eta}(\xi) \in a$. But let $\sigma \in a$ be such that $e(\sigma) \geq \mu$ and $\alpha_\nu < \sigma$. Let $\xi = e(\sigma)$ and $\eta = d(\sigma)$, and let $\gamma+1 \in b$ be least such that $\eta <_U \gamma+1$. We have that

$$v_{0,\eta}(\xi) \leq_{W_\eta} u_{0,\eta}(\xi) = \sigma \leq \alpha_\eta < \mathrm{crit}(u_{\eta,\gamma+1}).$$

Since $\mathcal{W}_\eta \restriction \alpha_\eta + 1 = \mathcal{W} \restriction \alpha_\eta + 1$, we have $v_{0,\eta}(\xi) \leq_W \sigma$, so $v_{0,\eta}(\xi) \in a$.

Lastly, we must see that c is b-cofinal. Let $\gamma+1 \in b$; we seek $\xi \in c$ such that $v_{0,\eta}(\xi) > \mathrm{crit}(u_{\eta,\gamma+1})$, where $\eta = U\text{-pred}(\gamma+1)$. But pick any $\sigma \in a$ such that $\delta = d(\sigma) \geq \gamma+1$, and let $\xi \in c$ with $\xi > e(\sigma)$. We get that $v_{0,\delta}(\xi) > \sigma \geq \mathrm{crit}(u_{\eta,\gamma+1})$, which implies $v_{0,\eta}(\xi) > \sigma \geq \mathrm{crit}(u_{\eta,\gamma+1})$, as desired. ⊣

This completes the proof of Lemma 6.6.21. ⊣

DEFINITION 6.6.22. Given $\mathcal{T}$ normal on M, and $\mathcal{U}$ normal on the last model of $\mathcal{T}$, we write $\mathrm{br}_{\mathcal{W}}(\mathcal{T},\mathcal{U})$ for the function $\mathrm{br}_{\mathcal{W}}$ (defined on pairs of nodes and pairs of branches) defined above. We write $\mathrm{br}^{\mathcal{W}}_{\mathcal{U}}$ for the function d and $\mathrm{br}^{\mathcal{W}}_{\mathcal{T}}$ for the function e defined above.

Notation 6.6.22.1. To reconcile with our previous notation: if b is cofinal in $\mathcal{U}$, there is exactly one branch c of $\mathcal{T}$ such that

(*i*) c is chosen by $\mathcal{T}$, in that $c = [0,\tau]_T$ or $c = [0,\tau)_T$ for some $\tau < \mathrm{lh}(\mathcal{T})$, and
(*ii*) $\mathrm{br}_{\mathcal{W}}(c,b)$ is cofinal in $W(\mathcal{T},\mathcal{U})$.

This uses that $\mathcal{T}$ has a last model. We defined $\mathrm{br}(b,\mathcal{T},\mathcal{U})$ to be $\mathrm{br}_{\mathcal{W}}(c,b)$, for the unique such c. We defined $m(b,\mathcal{T},\mathcal{U})$ to be the unique τ as in (i). We shall not use this earlier notation much.

For τ in (i) a limit ordinal, the earlier notation does not distinguish between $c=[0,\tau)_T$ and $c=[0,\tau]_T$, whereas the current one does. $c=[0,\tau)_T$ is the case where, roughly speaking, the measures in E_b concentrate on proper initial segments of $\mathcal{M}^{\mathcal{T}}_c|\delta(\mathcal{T}\restriction\sup c)=\mathcal{M}^{\mathcal{T}}_\tau|\lambda^{\mathcal{T}}_\tau$. In the case $c=[0,\tau]_T$, some tail end of the extenders used in b are being added "as themselves" to the inflation of $e^{\mathcal{T}}_\tau$ by the earlier extenders used in b.

Remark 6.6.23. We assumed $\mathcal{T}$ has a last model, but one could generalize some of this by dropping that, and assuming that $\mathcal{U}$ is on $\mathcal{M}(\mathcal{T})$.

Remark 6.6.24. There are two special cases worth mentioning.

(a) $\mathcal{T}^\frown\mathcal{U}$ is already normal. Then $W(\mathcal{T}^\frown\mathcal{U})=\mathcal{T}^\frown\mathcal{U}$, and $\mathrm{br}_W(c,b)=c^\frown b$.

(b) $\mathcal{U}$ is a tree on $M|\kappa$, where $\kappa=\inf\{\mathrm{crit}(E^{\mathcal{T}}_\eta)\mid\eta+1<\mathrm{lh}\,\mathcal{T}\}$. Then if $\mathcal{U}$ has limit length, then $W(\mathcal{T},\mathcal{U})=\mathcal{U}$-on-$M$, i.e., $\mathcal{U}$ regarded as a tree on M. For b a cofinal branch of $\mathcal{U}$, $\mathcal{W}_b=W(\mathcal{T},\mathcal{U}^\frown b)=\mathcal{U}^\frown b^\frown(i^{\mathcal{U}}_b)\mathcal{T}$, and $\mathrm{br}_W(c,b)=b^\frown u``c$, where $u(\eta)=\mathrm{lh}(\mathcal{U})+\eta$.

In our application, however, $\mathcal{T}$ and $\mathcal{U}$ will definitely not be separated this way.

The coarse case. The results and proofs of this section go over to the coarse case in a straightforward way. Suppose that M is a transitive model of ZFC, and $\langle\mathcal{T},\mathcal{U}\rangle$ is a stack of nice, normal trees on M, and that $\mathrm{lh}(\mathcal{U})$ is a limit ordinal. Let $\mathcal{W}=W(\mathcal{T},\mathcal{U})$. We defined the node merging function $\mathrm{nd}(\xi,\gamma+1)$, the branch merging function $\mathrm{br}_{\mathcal{W}}(c,b)$, and the branch decoding functions $\mathrm{br}^{\mathcal{W}}_{\mathcal{T}}$ and $\mathrm{br}^{\mathcal{W}}_{\mathcal{U}}$ from the meta-tree structure of the $\mathcal{W}_\gamma$'s. The definitions made no reference to the intrinsic structure of M.

Lemma 6.6.13 on the agreement of the $\mathcal{W}_\gamma$'s did make use of the fact that M was a premouse. In the coarse case, the analog of "$Q|\mathrm{lh}(F)$" is "$(V^Q_{\mathrm{lh}(F)},F)$". The analog of "$Q|\mathrm{lh}(F)\trianglelefteq R$" is "$V^Q_{\mathrm{lh}(F)}=V^R_{\mathrm{lh}(F)}$ and $F\in R$", or equivalently, "$R\models F$ is nice". The counterpart to Lemma 6.6.13 is

LEMMA 6.6.25. *Let $\gamma<\delta<\mathrm{lh}(\mathcal{U})$ and $\eta=\mathrm{lh}(F_\gamma)$; then*

(a) $V^{R_\gamma}_\eta=V^{R_\delta}_\eta$,

(b) $\mathcal{M}^{\mathcal{W}_\gamma}_\xi\models$ *"F_γ is nice" iff $\xi\geq\alpha_\gamma$,*

(c) for all $\xi\geq\alpha_\gamma$, $F_\gamma\notin\mathcal{M}^{\mathcal{W}_\delta}_\xi$, and

(d) if $s^\frown\langle H\rangle\in\mathcal{W}^{\mathrm{ext}}_\gamma\cap\mathcal{W}^{\mathrm{ext}}_\delta$, then $\mathrm{lh}(H)<\mathrm{lh}(F_\gamma)$.

Part (d) of Lemma 6.6.25 played a role in our proof that the branch merging function $\mathrm{br}_{\mathcal{W}}$ is injective and surjective. In particular, the key lemma, Lemma 6.6.14, made use of it.

6.7. Quasi-normalizing stacks of plus trees

We shall show in the next chapter that background-induced strategies normalize well for stacks of the form $\langle \mathcal{T}, \mathcal{U} \rangle$, where $\mathcal{T}$ is λ-separated. More precisely, if Σ^* is a strongly unique iteration strategy for V, and $\langle \mathcal{T}, \mathcal{U} \rangle$ is a normal stack by $\Omega(\mathbb{C}, M, \Sigma^*)$ such that $\mathcal{T}$ is λ-separated, then $W(\mathcal{T}, \mathcal{U})$ is by $\Omega(\mathbb{C}, M, \Sigma^*)$. What the proof gives when $\mathcal{T}$ is not λ-separated is the same conclusion, but with $W(\mathcal{T}, \mathcal{U})$ replaced by a "quasi-normalization" of $\langle \mathcal{T}, \mathcal{U} \rangle$ that we call $V(\mathcal{T}, \mathcal{U})$. If $\mathcal{T}$ is λ-separated, then $W(\mathcal{T}, \mathcal{U}) = V(\mathcal{T}, \mathcal{U})$, but in general they can be different, even when both $\mathcal{T}$ and $\mathcal{U}$ are ordinary normal trees. In general, if $\langle \mathcal{T}, \mathcal{U} \rangle$ is a normal stack, then $W(\mathcal{T}, \mathcal{U})$ is the normal companion of $V(\mathcal{T}, \mathcal{U})$.

Quasi-normalization is a small variant on embedding normalization, so we shall not describe it in the detail we gave for $W(\mathcal{T}, \mathcal{U})$. We shall instead just say enough that the reader can see how similar the two normalization methods are, and where the difference lies.

To see the difference between normalizing and quasi-normalizing, suppose $\mathcal{T}$ is normal, F^- is on the sequence of its last model, and $\alpha = \alpha(\mathcal{T}, F)$ is least such that F^- is on the $\mathcal{M}_\alpha^\mathcal{T}$ sequence. We shall have $V(\mathcal{T}, F) = W(\mathcal{T}, F)$ unless $\alpha + 1 < \text{lh}(\mathcal{T})$, $E_\alpha^\mathcal{T}$ is not of plus type,[16] and

$$\lambda(E_\alpha^\mathcal{T}) \leq \hat{\lambda}(F) < \text{lh}(F) < \text{lh}(E_\alpha^\mathcal{T}).$$

(Note $\text{lh}(F) < \text{lh}(E_\alpha^\mathcal{T})$ by the definition of α.) In this case, $V(\mathcal{T}, F)$ does not replace $E_\alpha^\mathcal{T}$ with F the way $W(\mathcal{T}, F)$ did; instead, $\mathcal{V} \restriction \alpha + 2 = \mathcal{T} \restriction \alpha + 2$, and $E_{\alpha+1}^\mathcal{V} = F$. The rest of $\mathcal{V}$ is i_F"$\mathcal{T}^{>\text{crit}(F)}$, as before, so

$$\mathcal{M}_{\alpha+2+\xi}^\mathcal{V} = \mathcal{M}_{\alpha+1+\xi}^\mathcal{W}$$

for all $\xi \in [\beta(\mathcal{T}, F), \text{lh}(\mathcal{T}))$. There is one nontrivial delay interval in $\mathcal{V}$, namely $[\alpha, \alpha + 1]$, and $\mathcal{W}$ is the normal companion of $\mathcal{V}$.

More generally, if $\mathcal{T}$ is merely quasi-normal, then $V(\mathcal{T}, F)$ keeps all $E_\xi^\mathcal{T}$ such that $\xi \geq \alpha$, $E_\xi^\mathcal{T}$ is not of plus type, and $\lambda(E_\xi^\mathcal{T}) \leq \hat{\lambda}(F)$. This is a (perhaps empty) initial segment of the delay interval in $\mathcal{T}$ that starts at α. Then $V(\mathcal{T}, F)$ inserts F, and proceeds with copying $\mathcal{T}^{>\text{crit}(F)}$.

DEFINITION 6.7.1. Let $\mathcal{T}$ be a plus tree on M, and suppose F^- is on the sequence of its last model; then $\alpha_0(\mathcal{T}, F)$ is the least ξ such that

(a) $\alpha(\mathcal{T}, F) \leq \xi$, and
(b) $\text{lh}(F) < \hat{\lambda}(E_\xi^\mathcal{T})$, or $E_\xi^\mathcal{T}$ is of plus type, or $\xi + 1 = \text{lh}(\mathcal{T})$.

LEMMA 6.7.2. *Let $\mathcal{T}$ be a quasi-normal plus tree on M and F be on the extended sequence of its last model; then $\alpha(\mathcal{T}, F)$ begins a delay interval in $\mathcal{T}$,*

[16] Equivalently, $\lambda(E_\alpha^\mathcal{T}) = \hat{\lambda}(E_\alpha^\mathcal{T})$.

and $\{\xi \mid \alpha(\mathcal{T},F) \le \xi < \alpha_0(\mathcal{T},F)\}$ *is a (perhaps empty) initial segment of that interval.*

PROOF. Let $E_\xi = E^{\mathcal{T}}_\xi$ and $M_\xi = \mathcal{M}^{\mathcal{T}}_\xi$. Let $\alpha = \alpha(\mathcal{T},F)$ and $\alpha_0 = \alpha_0(\mathcal{T},F)$.

Since F^- is on the M_α-sequence and the $M_{\mathrm{lh}(\mathcal{T})-1}$ sequence, $M_\alpha|\mathrm{lh}(F) = M_{\mathrm{lh}(\mathcal{T})-1}|\mathrm{lh}(F)$, so for all $\theta \ge \alpha$, $\mathrm{lh}(E^{\mathcal{T}}_\theta) > \mathrm{lh}(F)$ by coherence. But if $\delta < \alpha$ and $\forall\theta \ge \delta(\mathrm{lh}(F) < \mathrm{lh}(E^{\mathcal{T}}_\theta))$, then $\mathcal{M}_\delta|\mathrm{lh}(F) = \mathcal{M}_\alpha|\mathrm{lh}(F)$, so F is on the $\mathcal{M}^{\mathcal{T}}_\delta$-sequence, contradiction. Thus

$$\alpha = \text{least } \delta \text{ such that } \forall\theta \ge \delta(\mathrm{lh}(F) < \mathrm{lh}(E^{\mathcal{T}}_\theta)).$$

So if $\delta < \alpha$, $\mathrm{lh}(E_\delta) < \mathrm{lh}(E_\alpha)$, and since the lengths of exit extenders decrease within a delay interval, α and δ are in different delay intervals. Thus α begins a (perhaps trivial) delay interval in $\mathcal{T}$.

If $\alpha \le \xi < \alpha_0$, then $\hat{\lambda}(E_\xi) < \mathrm{lh}(F)$, so $\mathrm{lh}(E_\xi) < \mathrm{lh}(E_\alpha)$. Also E_ξ is not of plus type. These two facts imply that ξ is in the delay interval of $\mathcal{T}$ that begins with α. ⊣

The maximal delay interval in $\mathcal{T}$ that starts at $\alpha(\mathcal{T},F)$ may or may not have $\alpha_0(\mathcal{T},F)$ in it, and may or may not continue beyond $\alpha_0(\mathcal{T},F)$. While the lengths of the exit extenders in this interval are strictly descreasing, and all $> \mathrm{lh}(F)$, their $\hat{\lambda}$'s may strictly increase, and one of those may exceed $\mathrm{lh}(F)$.

In $V(\mathcal{T},F)$, we replace $E^{\mathcal{T}}_{\alpha_0}$ with F. More precisely, let $\mathcal{T}$ be a plus tree and F be an extender such that F^- is on the sequence of last model of $\mathcal{T}$.[17] Let

$$\alpha_0 = \alpha_0(\mathcal{T},F).$$

We define the *quasi-normalization* $\mathcal{V} = V(\mathcal{T},F)$ by

$$\mathcal{V} \restriction \alpha_0 + 1 = \mathcal{T} \restriction \alpha_0 + 1,$$

and

$$\mathcal{M}^{\mathcal{V}}_{\alpha_0+2} = \mathrm{Ult}(P,F),$$

where for $\beta = \beta(\mathcal{T},F)$, P is the longest $Q \trianglelefteq \mathcal{M}^{\mathcal{T}}_\beta$ such that $o(Q) \le \mathrm{lh}(E^{\mathcal{T}}_\beta)$ and $\rho_{k(Q)}(Q) > \mathrm{crit}(F)$, and defining $\mathcal{M}^{\mathcal{V}}_{\alpha_0+1+\xi}$ for $\xi > \beta$ by copying, just as we did in the W-case. Heuristically,

$$V(\mathcal{T},F) = \mathcal{T} \restriction (\alpha_0+1)^\frown\langle F\rangle^\frown i_F\text{``}\mathcal{T}^{>\mathrm{crit}(F)}.$$

More generally, suppose $\mathcal{S}$ and $\mathcal{T}$ are maximal plus trees on M, and F is an extender such that F^- is on the sequence of the last model of $\mathcal{S}$. Let $\alpha_0 = \alpha_0(\mathcal{S},F^-)$, $\beta = \beta(\mathcal{S},F)$. Suppose $\mathcal{T} \restriction \beta+1 = \mathcal{S} \restriction \beta+1$, and if $\beta < \mathrm{lh}(\mathcal{T})$, then $\mathrm{dom}(F) = \mathcal{M}^{\mathcal{T}}_\beta|\eta$ for some $\eta < \mathrm{lh}(E^{\mathcal{T}}_\beta)$. We define

$$V(\mathcal{T},\mathcal{S},F) = \mathcal{S} \restriction (\alpha_0+1)^\frown\langle F\rangle^\frown i_F\text{``}\mathcal{T}^{>\mathrm{crit}(F)}.$$

[17] We are allowing the possibility that $F = F^-$.

Again, this is the same formula that defined $W(\mathcal{T},\mathcal{S},F)$, with $\alpha_0(\mathcal{S},F)$ replacing $\alpha(\mathcal{S},F)$. There is a natural tree embedding Φ of $\mathcal{T}$ into $V(\mathcal{T},\mathcal{S},F)$. If $\mathcal{T}$ is normal and $\Phi^*: \mathcal{T} \to W(\mathcal{T},\mathcal{S},F)$ is the natural tree embedding, then Φ is essentially the same as Φ^*, modulo the fact that $u(\xi) = n + u^*(\xi)$ when $\xi \geq \beta$ and $\alpha_0(\mathcal{S},F) = \alpha(\mathcal{S},F) + n$.

Note that $V(\mathcal{T},\mathcal{S},F)$ is maximal. The delay intervals in $V(\mathcal{T},\mathcal{S},F)$ are described by

LEMMA 6.7.3. *Let $\Phi\colon \mathcal{T} \to V(\mathcal{T},\mathcal{S},F)$ be the natural tree embedding; say $\Phi = \langle u, v, \vec{s}, \vec{t}\rangle$. Equivalent are*

(1) *I is a maximal delay interval in $V(\mathcal{T},\mathcal{S},F)$,*
(2) *Either*
   (a) *I is a maximal delay interval in $\mathcal{S} \restriction \alpha(\mathcal{T},F)$, or*
   (b) *$I = [\alpha(\mathcal{S},F), \alpha_0(\mathcal{S},F)]$, or*
   (c) *$I = [u(\xi), u(\gamma)]$, where $[\xi,\gamma]$ is a maximal delay interval in $\mathcal{T}$ and $\beta(\mathcal{S},F) \leq \xi$.*

We omit the simple proof.

DEFINITION 6.7.4. Let $\mathcal{S}$ be a plus tree; then

(1) $\mathcal{S}$ is *length-increasing above α* iff whenever $\alpha \leq \beta < \gamma < \mathrm{lh}(\mathcal{S}) - 1$, then $\mathrm{lh}(E_\beta^{\mathcal{S}}) < \mathrm{lh}(E_\gamma^{\mathcal{S}})$.
(2) $\mathcal{S}$ is *λ-separated above α* iff whenever $\alpha \leq \beta < \mathrm{lh}(\mathcal{S}) - 1$, then $E_\beta^{\mathcal{S}}$ is of plus type.

Being λ-separated above α implies being length-increasing above α.

LEMMA 6.7.5. *Let $\Phi\colon \mathcal{T} \to V(\mathcal{T},\mathcal{S},F)$ be the natural tree embedding; say $\Phi = \langle u, v, \vec{s}, \vec{t}\rangle$. Suppose that $\mathcal{S}$ is length-increasing above $\alpha(\mathcal{S},F)$ and $\mathcal{T}$ is length-increasing above $\beta(\mathcal{S},F)$; then*

(1) *$V(\mathcal{T},\mathcal{S},F)$ is length-increasing above $\alpha_0(\mathcal{S},F)$,*
(2) *$\alpha_0(\mathcal{S},F) \leq \alpha(\mathcal{S},F) + 1$,*
(3) *if $E_\alpha^{\mathcal{S}}$ is of plus type, then $\alpha_0(\mathcal{S},F) = \alpha(\mathcal{S},F)$, and*
(3) *the nontrivial delay intervals of $V(\mathcal{T},\mathcal{S},F)$ are those of $\mathcal{S} \restriction \alpha(\mathcal{S},F)$, together perhaps with $[\alpha(\mathcal{S},F), \alpha_0(\mathcal{S},F)]$; moreover*
(5) *if $\mathcal{T}$ is λ-separated above $\beta(\mathcal{S},F)$, then $V(\mathcal{T},\mathcal{S},F)$ is λ-separated above $\alpha_0(\mathcal{S},F)$.*

Again, we omit the easy proof. The lemma implies that if $\mathcal{T}$ and $\mathcal{S}$ are normal, then $W(\mathcal{T},\mathcal{S},F)$ is the normal companion of $V(\mathcal{T},\mathcal{S},F)$, with $[\alpha(\mathcal{S},F), \alpha(\mathcal{S},F) + 1]$ being the only possiblity for a nontrivial delay interval.

Now suppose that $\langle \mathcal{T}, \mathcal{U}\rangle$ is a maximal M-stack. We define $V(\mathcal{T},\mathcal{U})$ by induction on $\mathrm{lh}(\mathcal{U})$, just as in the W-case. Setting $\mathcal{V}_\xi = V(\mathcal{T}, \mathcal{U} \restriction \xi + 1)$, we have

$$\mathcal{V}_{\gamma+1} = V(\mathcal{V}_\nu, \mathcal{V}_\gamma, F_\gamma),$$

where $\nu = U\text{-pred}(\gamma+1)$ and

$$F_\gamma = \sigma_\gamma(E_\gamma^{\mathcal{U}}),$$

for $\sigma_\gamma \colon \mathcal{M}_\gamma^{\mathcal{U}} \to \mathcal{M}_{z(\gamma)}^{\mathcal{V}_\gamma}$ the natural map. We have a tree embedding $\Phi_{\nu,\gamma+1} \colon \mathcal{V}_\nu \to \mathcal{V}_{\gamma+1}$. For λ a limit,

$$\mathcal{V}_\lambda = \lim_{\xi <_U \lambda} \mathcal{V}_\xi,$$

under the $\Phi_{\xi,\gamma}$ for $\xi <_U \gamma <_U \lambda$. One can think of the $\mathcal{V}_\xi$ as the nodes in a meta-iteration tree.

DEFINITION 6.7.6. Let $\langle \mathcal{T}, \mathcal{U} \rangle$ be a maximal M-stack; then $V(\mathcal{T},\mathcal{U})$ is the *quasi-normalization* of $\langle \mathcal{T}, \mathcal{U} \rangle$. For longer stacks s, the quasi-normalization $V(s)$ is defined "bottom up": $V(s^\frown\langle \mathcal{U} \rangle) = V(V(s), \pi\mathcal{U})$, for π the t-map on last models, with direct limits under the associated tree embeddings for s of limit length.

We have no use for $V(\mathcal{T},\mathcal{U})$ when $\mathcal{U}$ is not normal, and the basic agreement facts about the meta-tree structure producing it are a bit easier to state if $\mathcal{U}$ is normal. In that case, the nontrivial delay intervals in $V(\mathcal{T},\mathcal{U})$ are all either blowups of delay intervals in $\mathcal{T}$, or of the form $[\alpha, \alpha+1]$, where $E_{\alpha+1}^{\mathcal{V}} = F_\gamma = \sigma_\gamma(E_\gamma^{\mathcal{U}})$ is one of the inserted extenders. The following is the counterpart of Proposition 6.5.8.

PROPOSITION 6.7.7. *Let $\langle \mathcal{T}, \mathcal{U} \rangle$ be a maximal stack, and suppose that $\mathcal{U}$ is normal. Let $\mathcal{V}_\gamma = V(\mathcal{T}, \mathcal{U} \restriction \gamma+1)$, $R_\gamma = \mathcal{M}_{z(\gamma)}^{\mathcal{V}_\gamma}$ be its last model, $\sigma_\gamma \colon \mathcal{M}_\gamma^{\mathcal{U}} \to R_\gamma$ the natural map, and $F_\gamma = \sigma_\gamma(E_\gamma^{\mathcal{U}})$. If $\gamma < \eta < \text{lh}(\mathcal{U})$, then*

(a) letting $\alpha = \alpha_0(\mathcal{V}_\gamma, F_\gamma)$, $F_\gamma = E_\alpha^{\mathcal{V}_\eta}$ and $\text{lh}(F_\gamma) < \text{lh}(E_\xi^{\mathcal{V}_\eta})$ for all $\xi > \alpha$,
(b) if $\eta+1 < \text{lh}(\mathcal{U})$, then $\text{lh}(F_\gamma) < \text{lh}(F_\eta)$,
(c) R_γ agrees with R_η below $\text{lh}(F_\gamma)$,
(d) $\sigma_\eta \restriction (\text{lh}(E_\gamma^{\mathcal{U}})+1) = \sigma_\gamma \restriction (\text{lh}(E_\gamma^{\mathcal{U}})+1)$, and
(e) F_γ is on the sequence of R_γ, but not that of R_η. In fact, $\text{lh}(F_\gamma)$ is a cardinal of R_η.

We are mainly interested in normal M-stacks.

LEMMA 6.7.8. *Let $\langle \mathcal{T}, \mathcal{U} \rangle$ be a normal M-stack; then*

(1) $W(\mathcal{T},\mathcal{U})$ is the normal companion of $V(\mathcal{T},\mathcal{U})$,
(2) if $\mathcal{T}$ is λ-separated, then $W(\mathcal{T},\mathcal{U}) = V(\mathcal{T},\mathcal{U})$, and
(3) if both $\mathcal{T}$ and $\mathcal{U}$ are λ-separated, then $V(\mathcal{T},\mathcal{U})$ is λ-separated.

PROOF. The meta-tree leading to $V(\mathcal{T},\mathcal{U})$ has nodes $\mathcal{V}_\xi = V(\mathcal{T}, \mathcal{U} \restriction \xi+1)$, where $\mathcal{V}_{\gamma+1} = V(\mathcal{V}_\nu, \mathcal{V}_\gamma, F_\gamma)$ for $\nu = U\text{-pred}(\gamma+1)$, and $F_\gamma = \sigma_\gamma(E_\gamma^{\mathcal{U}})$. Here σ_γ is the natural map from $\mathcal{M}_\gamma^{\mathcal{U}}$ to $\mathcal{M}_{z(\gamma)}^{\mathcal{V}_\gamma}$. Similarly, let $\mathcal{W}_\xi = W(\mathcal{T}, \mathcal{U} \restriction \xi+1)$ be the ξ-th tree in the meta-tree leading to $W(\mathcal{T},\mathcal{U})$, and $\tau_\xi \colon \mathcal{M}_\xi^{\mathcal{U}} \to \mathcal{M}_{z(\xi)}^{\mathcal{W}_\xi}$ the natural map. Let also

$$\alpha_\gamma = \alpha(\mathcal{V}_\gamma, F_\gamma)$$

and

$$\alpha_\gamma^0 = \alpha_0(\mathcal{V}_\gamma, F_\gamma).$$

As in the case of embedding normalization, the $\mathrm{lh}(F_\gamma)$ are strictly increasing with γ, and

$$\gamma < \eta \Longrightarrow \alpha_\gamma \leq \alpha_\gamma^0 < \alpha_\eta,$$

basically because $\mathcal{U}$ is normal. $F_\gamma = E_{\alpha_\gamma^0}^{\mathcal{V}_{\gamma+1}}$, and $\mathcal{V}_{\gamma+1} \restriction \alpha_\gamma^0 + 2 = \mathcal{V}_\eta \restriction \alpha_\gamma^0 + 2$ for all $\eta \geq \gamma$.

CLAIM 6.7.9. *(i)* $\mathcal{V}_\gamma$ *is length-increasing above* $\sup\{\alpha_\xi^0 \mid \xi < \gamma\}$.

(ii) If $\mathcal{T}$ *is* λ*-separated, then* $\mathcal{V}_\gamma$ *is* λ*-separated above* $\sup\{\alpha_\xi^0 \mid \xi < \gamma\}$.

PROOF. By induction on γ. Suppose it is true at all $\eta \leq \gamma$, and let $\nu = \mathcal{U}\text{-pred}(\gamma+1)$, so that $\mathcal{V}_{\gamma+1} = V(\mathcal{V}_\nu, \mathcal{V}_\gamma, F_\gamma)$. Letting $\beta = \beta(\mathcal{V}_\gamma, F_\gamma)$, we have $\alpha_\xi^0 \leq \beta$ for all $\xi < \beta$. (Because $F_\xi = E_{\alpha_\xi^0}^{\mathcal{V}_\gamma}$ and $\hat{\lambda}(F_\xi) \leq \mathrm{crit}(F_\gamma)$ for $\xi < \nu$.) Thus $\mathcal{V}_\nu$ is length-increasing above β, and λ-separated above β if $\mathcal{T}$ is λ-separated.

Similarly, $\mathcal{V}_\gamma$ is length-increasing above α_γ, and λ-separated above α_γ if $\mathcal{T}$ is λ-separated.

We now get (i) at $\gamma+1$ from Lemma 6.7.5(1). That is, $\mathcal{V}_{\gamma+1}$ is length-increasing above α_γ^0. If $\mathcal{T}$ is λ-separated, then $\mathcal{V}_{\gamma+1}$ is λ-separated above α_γ^0 by Lemma 6.7.5(5), as required for (ii).

We leave the limit case of the induction to the reader. ⊣

Let us now prove item (2) of Lemma 6.7.8. Suppose $\mathcal{T}$ is λ-separated. Since $\alpha_\xi^0 < \alpha_\gamma$ for all $\xi < \gamma$, we have that $\mathcal{V}_\gamma$ is λ-separated above α_γ. By Lemma 6.7.5(3), $\alpha_\gamma^0 = \alpha_\gamma$. Since this holds at all γ, $V(\mathcal{T}, \mathcal{U}) = W(\mathcal{T}, \mathcal{U})$.

For part (3) of Lemma 6.7.8, suppose both $\mathcal{T}$ and $\mathcal{U}$ are λ-separated. We show by induction that each $\mathcal{V}_\gamma$ is λ-separated. Consider the successor step: since $\alpha_\gamma = \alpha_\gamma^0$,

$$\mathcal{V}_{\gamma+1} = \mathcal{V}_\gamma \restriction \alpha_\gamma + 1 ^\frown \langle F_\gamma \rangle ^\frown i_{F_\gamma} \text{``} \mathcal{V}_\nu^{>\mathrm{crit}(F_\gamma)}.$$

By induction, all extenders used here have plus type, except perhaps F_γ. But $F_\gamma = \sigma_\gamma(E_\gamma^{\mathcal{U}})$, and $E_\gamma^{\mathcal{U}}$ has plus type, so F_γ has plus type too.

To prove (1), let $\mathcal{W}_\xi = W(\mathcal{T}, \mathcal{U} \restriction \xi+1)$ be the ξ-th tree in the meta-tree leading to $W(\mathcal{T}, \mathcal{U})$, and $\tau_\xi : \mathcal{M}_\xi^{\mathcal{U}} \to \mathcal{M}_{z(\xi)}^{\mathcal{W}_\xi}$ the natural map. One can prove by induction that $\mathcal{W}_\xi$ is the normal companion of $\mathcal{V}_\xi$, that $\tau_\xi = \sigma_\xi$, and that the tree embeddings of the W-system are restrictions of the corresponding tree embeddings in the V-system. We omit the completely routine details. ⊣

The most important case for strategy comparison is covered in part (3) of the lemma.

Coarse quasi-normalization. Let M be a transitive model of ZFC, $\mathcal{T}$ a nice tree on M, and $\mathcal{U}$ a nice tree on the last model of $\mathcal{T}$. In the coarse case, we generally deal with nice trees that have unique cofinal wellfounded branches, in which case we could just replace all trees by their normal companions. But this is inconvenient. In one important context, we have $\langle\mathcal{T},\mathcal{U}\rangle = \text{lift}(\bar{\mathcal{T}},\bar{\mathcal{U}})_0$, and must show that $\text{lift}(V(\bar{\mathcal{T}},\bar{\mathcal{U}}))_0$ is a quasi-normalization of $\langle\mathcal{T},\mathcal{U}\rangle$. Replacing the nice trees on the background universe with their normal companions will cause a notational mess.

At the same time, we do not want to define a *unique* quasi-nomalization of $\langle\mathcal{T},\mathcal{U}\rangle$ in the coarse case, because the lift of $V(\langle\bar{\mathcal{T}},\bar{\mathcal{U}}\rangle)$ is not a function of $\text{lift}(\langle\bar{\mathcal{T}},\bar{\mathcal{U}}\rangle)_0$. The freedom here amounts to the freedom to choose any α in the delay interval beginning with $\alpha(\mathcal{V}_\gamma,F_\gamma)$ to serve as α_γ.

So in this coarse case, by a quasi-normalization of $\langle\mathcal{T},\mathcal{U}\rangle$ we mean any system ("meta-tree")

$$\langle\langle\mathcal{V}_\gamma \mid \gamma < \text{lh}(\mathcal{U})\rangle, \langle F_\gamma, \alpha_\gamma \mid \gamma+1 < \text{lh}(\mathcal{U})\rangle, \langle\Phi_{\eta,\xi} \mid \eta <_U \xi\rangle\rangle$$

such that $\mathcal{V}_0 = \mathcal{T}$, each $\mathcal{V}_\gamma$ is a nice, quasi-normal tree on M with last model $\mathcal{M}^{\mathcal{U}}_\gamma$, $F_\gamma = E^{\mathcal{U}}_\gamma$, and when $\eta = U\text{-pred}(\gamma+1)$,

$$\mathcal{V}_{\gamma+1} = \mathcal{V}_\gamma \restriction \alpha_\gamma + 1)^\frown \langle F_\gamma\rangle^\frown i_{F_\gamma} \text{``}\mathcal{V}_\eta^{>\text{crit}(F_\gamma)},$$

where α_γ is in the delay interval of $\mathcal{V}_\gamma$ that begins with $\alpha(\mathcal{V}_\gamma,F_\gamma)$. As in the fine case, $\alpha(\mathcal{V},F_\gamma)$ indexes the first model in $\mathcal{V}_\gamma$ to which F_γ belongs. The maps $\Phi_{\eta,\delta}\colon \mathcal{V}_\eta \to \mathcal{V}_\delta$ are the coarse tree embeddings associated to the system.

These conventions make use of the fact that in the coarse case, embedding normalization and full normalization coincide, so $\mathcal{M}^{\mathcal{U}}_\gamma$ is the last model of $\mathcal{V}_\gamma$ and $E^{\mathcal{U}}_\gamma = F_\gamma$. Also, in the coarse case we shall never need to let $\alpha_\gamma - 1$ end the delay interval starting at $\alpha(\mathcal{V}_\gamma,F_\gamma)$.

Once again, in the coarse case, these quasi-normalization meta-trees are just a way of keeping the books efficiently in certain lifting constructions.

6.8. Copying commutes with normalization

We prove that both kinds of normalization commute with copying. The proof is completely straightforward, but takes a while to put on paper, because of the many embeddings involved. We shall use this fact to show that normalizing well, in either sense, passes from a strategy to its pullbacks. The proof also serves as an introduction to our proof that quasi-normalization commutes with lifting to a background universe. That in turn is used in the proof that if a strategy for the background universe quasi-normalizes well, then so do the strategies on premice that it induces. (See Theorem 7.4.1.)

THEOREM 6.8.1. *Let $\langle \mathcal{T}, \mathcal{U} \rangle$ be an M-stack, and let $\psi \colon M \to N$ be nearly elementary. Let $\langle \mathcal{T}^*, \mathcal{U}^* \rangle = \psi \langle \mathcal{T}, \mathcal{U} \rangle$ be the stack on N obtained by copying.*

(A) Suppose that $\langle \mathcal{T}, \mathcal{U} \rangle$ is maximal, $\mathcal{U}$ is normal, and $V(\mathcal{T}^*, \mathcal{U}^*)$ exists; then
  (1) $V(\mathcal{T}, \mathcal{U})$ exists, and $\psi V(\mathcal{T}, \mathcal{U}) = V(\mathcal{T}^*, \mathcal{U}^*)$, and
  (2) let $\mathcal{U}$ and $\mathcal{U}^*$ have last models Q and Q^* respectively, and $V(\mathcal{T}, \mathcal{U})$ and $V(\mathcal{T}^*, \mathcal{U}^*)$ have last model R and R^* respectively, and let
    (i) $\rho \colon Q \to Q^*$ be the map from copying $\langle \mathcal{T}, \mathcal{U} \rangle$ to $\langle \mathcal{T}^*, \mathcal{U}^* \rangle$,
    (ii) $\sigma \colon Q \to R$ be the normalization map associated to $V(\mathcal{T}, \mathcal{U})$,
    (iii) $\theta \colon R \to R^*$ be the map from copying $V(\mathcal{T}, \mathcal{U})$ to $V(\mathcal{T}^*, \mathcal{U}^*)$, and
    (iv) $\sigma^* \colon Q^* \to R^*$ be the normalization map associated to $V(\mathcal{T}^*, \mathcal{U}^*)$;
    then $\theta \circ \sigma = \sigma^* \circ \rho$.

(B) Suppose that $\langle \mathcal{T}, \mathcal{U} \rangle$ is normal, and $W(\mathcal{T}^*, \mathcal{U}^*)$ exists; then the conclusions of part (A) hold, with "W" replacing "V" everywhere.

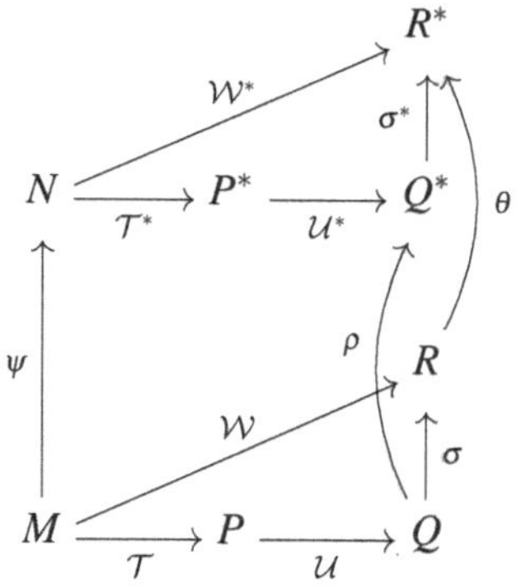

PROOF. We prove (A). The proof of (B) is nearly the same.

The quasi-normalization $V(\mathcal{T}, \mathcal{U})$ has associated to it quasi-normal trees

$$\mathcal{W}_\gamma = V(\mathcal{T}, \mathcal{U} \upharpoonright \gamma + 1)$$

on M, for $\gamma < \mathrm{lh}(\mathcal{U})$. (We called the nodes of the quasi-normalization meta-tree $\mathcal{V}_\gamma$ before, but $\mathcal{W}$ is easier to read in various places, so let us switch.) We also have extended tree embeddings

$$\Phi_{\eta,\gamma} \colon W_\eta \to W_\gamma,$$

defined for $\eta \leq_U \gamma$. For $\eta \leq_U \gamma$, we set

$$\phi_{\eta,\gamma} = u^{\Phi_{\eta,\gamma}},$$

so that $\phi_{\eta,\gamma} \colon \mathrm{lh}(\mathcal{W}_\eta) \to \mathrm{lh}(\mathcal{W}_\gamma)$, and for $\tau \in \mathrm{dom}\, \phi_{\eta,\gamma}$,

$$\pi_\tau^{\eta,\gamma} = t_\tau^{\Phi_{\eta,\gamma}},$$

so that $\pi_\tau^{\eta,\gamma} : \mathcal{M}_\tau^{\mathcal{W}_\eta} \to \mathcal{M}_{\phi_{\eta,\gamma}(\tau)}^{\mathcal{W}_\gamma}$. Let R_γ be the last model of $\mathcal{W}_\gamma$, $\sigma_\gamma : \mathcal{M}_\gamma^{\mathcal{U}} \to R_\gamma$ as before, and $F_\gamma = \sigma_\gamma(E_\gamma^{\mathcal{U}})$. So

$$\mathcal{W}_{\gamma+1} = V(\mathcal{W}_\eta, F_\gamma)$$

when $\eta = U\text{-pred}(\gamma+1)$.

Similarly, $V(\mathcal{T}^*, \mathcal{U}^*)$ has associated trees

$$\mathcal{W}_\gamma^* = V(\mathcal{T}^*, \mathcal{U}^* \restriction \gamma+1)$$

on N for $\gamma < \text{lh}(\mathcal{U}^*) = \text{lh}\mathcal{U}$, together tree embeddings

$$\Phi_{\eta,\gamma}^* \colon \mathcal{W}_\eta^* \to \mathcal{W}_\gamma^*$$

defined when $\eta \leq_U \gamma$. We call the u maps of these tree embeddings $\phi_{\eta,\gamma}^* : \text{lh}(\mathcal{W}_\eta^*) \to \text{lh}(\mathcal{W}_\gamma^*)$, and for $\tau \in \text{dom}\,\phi_{\eta,\gamma}^*$, the t map is $\overset{*}{\pi}{}_\tau^{\eta,\gamma}$. We let R_γ^* be the last model of $\mathcal{W}_\gamma^*$, $\sigma_\gamma^* : \mathcal{M}_\gamma^{\mathcal{U}^*} \to R_\gamma^*$, and $F_\gamma^* = \sigma_\gamma^*(E_\gamma^{\mathcal{U}^*})$. We have that $\mathcal{W}_{\gamma+1}^* = V(\mathcal{W}_\eta^*, F_\gamma^*)$ when $\eta = U^*\text{-pred}(\gamma+1)$ (equivalently, $\eta = U\text{-pred}(\gamma+1)$).

We shall prove that for all γ,

$$\psi\mathcal{W}_\gamma = \mathcal{W}_\gamma^*.$$

The proof is by induction on γ, with a subinduction on initial segments of $\mathcal{W}_\gamma$. Given that we know this holds for $\mathcal{W}_\gamma \restriction \eta$, we have copy maps

$$\psi_\tau^\gamma \colon \mathcal{M}_\tau^{\mathcal{W}_\gamma} \to \mathcal{M}_\tau^{\mathcal{W}_\gamma^*}$$

defined for all $\tau < \eta$. $\psi_0^\gamma = \psi$ for all γ.

For $\gamma < \text{lh}(\mathcal{U})$, let

$$\psi_\gamma^{\mathcal{U}} \colon \mathcal{M}_\gamma^{\mathcal{U}} \to \mathcal{M}_\gamma^{\mathcal{U}^*}$$

be the copy map. So $\psi_0^{\mathcal{U}}$ is the copy map given by the fact that $\mathcal{T}^* = \psi\mathcal{T}$, and the remaining $\psi_\gamma^{\mathcal{U}}$ come from the fact that $\mathcal{U}^* = (\psi_0^{\mathcal{U}})\mathcal{U}$.

We write $z(\nu)$ for $\text{lh}(\mathcal{W}_\nu) - 1$ and $z^*(\nu)$ for $\text{lh}(\mathcal{W}_\nu^*) - 1$. (Once we have shown that $\psi\mathcal{W}_\nu = \mathcal{W}_\nu^*$, we get $z(\nu) = z^*(\nu)$, of course.) We may use ∞ for $z(\nu)$ or $z^*(\nu)$ when context permits. So $R_\nu = \mathcal{M}_{z(\nu)}^{\mathcal{W}_\nu} = \mathcal{M}_\infty^{\mathcal{W}_\nu}$. If $(\nu, \gamma]_U$ does not drop, then $\phi_{\nu,\gamma}(z(\nu)) = z(\gamma)$, and $\pi_{z(\nu)}^{\nu,\gamma} = \pi_\infty^{\nu,\gamma} \colon R_\nu \to R_\gamma$.

LEMMA 6.8.2. *Let $\gamma < \text{lh}(\mathcal{U})$. Then*

(1) $\mathcal{W}_\gamma^* = \psi\mathcal{W}_\gamma$.
(2) $\phi_{\eta,\nu} = \phi_{\eta,\nu}^*$, *if* $\eta, \nu \leq \gamma$ *and* $\eta \leq_U \nu$.
(3) *Whenever* $\nu <_U \gamma$ *and* $(\nu, \gamma]_U$ *does not drop in model or degree, then for all* $\tau < \text{lh}(\mathcal{W}_\nu)$, $\psi_{\phi_{\nu,\gamma}(\tau)}^\gamma \circ \pi_\tau^{\nu,\gamma} = \overset{*}{\pi}{}_\tau^{\nu,\gamma} \circ \psi_\tau^\nu$.
(4) $\psi_{z(\gamma)}^\gamma \circ \sigma_\gamma = \sigma_\gamma^* \circ \psi_\gamma^{\mathcal{U}}$.

Letting Ω_η be the system of all copy maps from W_η to W_η^*, item (3) is keeping track of the sense in which $\Omega_\gamma \circ \Phi_{\nu,\gamma} = \Phi_{\nu,\gamma}^* \circ \Omega_\nu$. Here is a diagram of (3):

$$\begin{array}{ccc}
\mathcal{M}^{\mathcal{W}_\gamma}_{\phi_{\nu,\gamma}(\tau)} & \xrightarrow{\psi^{\gamma}_{\phi_{\nu,\gamma}(\tau)}} & \mathcal{M}^{\mathcal{W}^*_\gamma}_{\phi^*_{\nu,\gamma}(\tau)} \\
\uparrow{\scriptstyle \pi^{\nu,\gamma}_\tau} & & \uparrow{\scriptstyle \overset{*}{\pi}^{\nu,\gamma}_\tau} \\
\mathcal{M}^{\mathcal{W}_\nu}_\tau & \xrightarrow{\psi^{\nu}_{\tau}} & \mathcal{M}^{\mathcal{W}^*_\nu}_\tau
\end{array}$$

There is a diagram related to (4) and the case $\tau = z(\nu)$ of (3) near the end of the proof.

PROOF. We prove Lemma 6.8.2 by induction. Suppose that it is true at all $\nu \leq \gamma$. We show it at $\gamma + 1$. Let $\nu = U\text{-pred}(\gamma+1)$, and

$$F = F_\gamma = \sigma_\gamma(E^{\mathcal{U}}_\gamma),$$

and

$$\alpha = \alpha_0(\mathcal{W}_\nu, \mathcal{W}_\gamma, F).$$

That is, α is the least ξ such that F^- is on the $\mathcal{M}^{\mathcal{W}_\gamma}_\xi$-sequence, and either $\mathrm{lh}(F) < \hat{\lambda}(E^{\mathcal{W}_\gamma}_\xi)$, or $E^{\mathcal{W}_\gamma}_\xi$ is of plus type, or $\xi + 1 = \mathrm{lh}(\mathcal{W}_\gamma)$. So

$$\begin{aligned}
\mathcal{W}_{\gamma+1} &= V(\mathcal{W}_\nu, \mathcal{W}_\gamma, F) \\
&= \mathcal{W}_\gamma \restriction (\alpha+1)^\frown \langle F \rangle^\frown i_F\text{“}\mathcal{W}_\nu^{>\mathrm{crit}(F)}.
\end{aligned}$$

Let also

$$F^* = F^*_\gamma = \sigma^*_\gamma(E^{\mathcal{U}^*}_\gamma).$$

Since $\mathcal{U}^*$ is a copy of $\mathcal{U}$, $\nu = U^*\text{-pred}(\gamma+1)$, so

$$\mathcal{W}^*_{\gamma+1} = V(\mathcal{W}^*_\nu, \mathcal{W}^*_\gamma, F^*).$$

CLAIM 6.8.3. *(1)* $\psi^\gamma_{z(\gamma)}(F) = F^*$,
(2) $\alpha = \alpha_0(\mathcal{W}^*_\gamma, \mathcal{W}^*_\nu, F^*)$*, and*
(3) $\beta(\mathcal{W}_\nu, \mathcal{W}_\gamma, F) = \beta(\mathcal{W}^*_\nu, \mathcal{W}^*_\gamma, F^*)$.

PROOF. For (1), we have

$$\begin{aligned}
\psi^\gamma_{z(\gamma)}(F) &= \psi^\gamma_{z(\gamma)} \circ \sigma_\gamma(E^{\mathcal{U}}_\gamma) \\
&= \sigma^*_\gamma \circ \psi^{\mathcal{U}}_\gamma(E^{\mathcal{U}}_\gamma) \\
&= \sigma^*_\gamma(E^{\mathcal{U}^*}_\gamma) \\
&= F^*.
\end{aligned}$$

For (2), we show first that $\beta = \beta^*$, where $\beta = \alpha(\mathcal{W}_\gamma, F)$ and $\beta^* = \alpha(\mathcal{W}^*_\gamma, F^*)$. For all $\tau \geq \beta$, $\mathrm{lh}(E^{\mathcal{W}_\gamma}_\tau) > \mathrm{lh}(F)$, so for all $\tau \geq \beta$, ψ^γ_τ agrees with $\psi^\gamma_{z(\gamma)}$ on $\mathrm{lh}(F)+1$ by our lemma on the agreement of copy maps. So for $\tau \geq \beta$,

$$\mathrm{lh}(F^*) = \psi^\gamma_{z(\gamma)}(\mathrm{lh}(F)) = \psi^\gamma_\tau(\mathrm{lh}(F)) < \psi^\gamma_\tau(\mathrm{lh}(E^{\mathcal{W}_\gamma}_\tau)) = \mathrm{lh}(E^{\mathcal{W}^*_\gamma}_\tau).$$

This implies F^* is on the $\mathcal{M}_\beta^{\mathcal{W}_\gamma^*}$-sequence, so $\beta \le \beta^*$. On the other hand, if $\xi < \beta$, and F^* is on the $\mathcal{M}_\xi^{\mathcal{W}_\gamma^*}$-sequence, then for all $\tau \ge \xi$, F^* is on the $\mathcal{M}_\tau^{\mathcal{W}_\gamma^*}$-sequence. But let $\xi < \tau < \alpha$ be such that τ ends a delay interval. (Such a τ exists because α begins a delay interval.) Then $\mathrm{lh}(E_\tau^{\mathcal{W}_\gamma}) < \mathrm{lh}(F)$ but $\mathrm{lh}(E_\tau^{\mathcal{W}_\gamma^*}) > \mathrm{lh}(F^*)$. This contradicts the fact that ψ_τ^γ agrees with $\psi_{z(\gamma)}^\gamma$ on $\mathrm{lh}(E_\tau^{\mathcal{W}_\gamma}) + 1$.

But then $\alpha = \beta + n$, for some n, and is the least $\xi \ge \beta$ such that either $\mathrm{lh}(F) < \hat{\lambda}(E_\xi^{\mathcal{W}_\gamma})$ or $E_\xi^{\mathcal{W}_\gamma}$ is of plus type or $\xi + 1 = \mathrm{lh}(\mathcal{W}_\gamma)$. $\mathcal{W}_\gamma$ and $\mathcal{W}_\gamma^*$ have the same length, and agree on whether extenders have plus type. Moreover, ψ_τ^γ agrees with $\psi_{z(\gamma)}^\gamma$ on $\hat{\lambda}(E_\tau^{\mathcal{W}_\gamma}) + 1$, for all τ. Thus $\beta = \alpha(\mathcal{W}_\gamma^*, F^*)$, as desired.

Remark 6.8.4. In the proof of (B), we set $\alpha = \alpha(\mathcal{W}_\gamma, F)$, and it is just the first part of the proof of (2) that applies. Otherwise, the two arguments are the same.

For (3), we must show that $\mathrm{crit}(F) < \hat{\lambda}(E_\tau^{\mathcal{W}_\gamma})$ if and only if $\mathrm{crit}(F^*) < \hat{\lambda}(E_\tau^{\mathcal{W}_\gamma^*})$. But this follows from the fact that $\psi_{z(\gamma)}^\gamma$ agrees with ψ_τ^γ on $\hat{\lambda}(E_\tau^{\mathcal{W}_\gamma}) + 1$ $\dashv$

The claim easily implies that $\phi_{\nu,\gamma+1} = \phi_{\nu,\gamma+1}^*$, which then gives us (2) of Lemma 6.8.2 at $\gamma + 1$.

We now define the copy maps $\psi_\tau^{\gamma+1} : \mathcal{M}_\tau^{\mathcal{W}_{\gamma+1}} \to \mathcal{M}_\tau^{\mathcal{W}_{\gamma+1}^*}$ that witness $\mathcal{W}_{\gamma+1}^* = \psi\mathcal{W}_{\gamma+1}$. As we do so, we show that (3) of Lemma 6.8.2 holds, that is, the ψ^ν and $\psi^{\gamma+1}$ maps commute with the quasi-normalization maps of models of $\mathcal{W}_\nu$ into models of $\mathcal{W}_{\gamma+1}$ and models of $\mathcal{W}_\nu^*$ into models of $\mathcal{W}_{\gamma+1}^*$.

We have $\mathcal{W}_{\gamma+1} \restriction (\alpha+1) = \mathcal{W}_\gamma \restriction (\alpha+1)$ and $\mathcal{W}_{\gamma+1}^* \restriction (\alpha+1) = \mathcal{W}_\gamma^* \restriction (\alpha+1)$, so we can set

$$\psi_\tau^{\gamma+1} = \psi_\tau^\gamma, \text{ for all } \tau \le \alpha.$$

Now $F = E_\alpha^{\mathcal{W}_{\gamma+1}}$ and $F^* = E_\alpha^{\mathcal{W}_{\gamma+1}^*}$, moreover $\psi_\alpha^\gamma(F) = \psi_{z(\gamma)}^\gamma(F) = F^*$ because $\mathrm{lh}(F) < \mathrm{lh}(E_\tau^{\mathcal{W}_\gamma})$ for all $\tau \in [\alpha, z(\gamma))$. Letting $P = \mathcal{M}_\beta^{\mathcal{W}_\nu} | \langle \eta, k \rangle$ be such that

$$\mathcal{M}_{\alpha+1}^{\mathcal{W}_{\gamma+1}} = \mathrm{Ult}(P, F),$$

we have

$$\mathcal{M}_{\alpha+1}^{\mathcal{W}_{\gamma+1}^*} = \mathrm{Ult}(P^*, F^*),$$

where $P^* = \mathcal{M}_\beta^{\mathcal{W}_\nu^*} | \langle \psi_\beta^\nu(\eta), k \rangle$. (Here we make the usual convention if $\eta = o(\mathcal{M}_\beta^{\mathcal{W}_\nu})$.) This is because $\mathcal{W}_\nu \restriction (\beta+1) = \mathcal{W}_\gamma \restriction (\beta+1)$, and similarly at the (*) level, by the properties of quasi-normalization. So $\psi_\beta^\nu = \psi_\beta^\gamma$, and thus agrees with $\psi_{z(\gamma)}^\gamma$ up to $\hat{\lambda}(E_\beta^{\mathcal{W}_\gamma})$, hence past $\mathrm{crit}(F)$. So we can let

$$\psi_{\alpha+1}^{\gamma+1}([a, f]_F^P) = [\psi_\alpha^{\gamma+1}(a), \psi_\beta^{\gamma+1}(f)]_{F^*}^{P^*},$$

by the Shift Lemma, and we have $\psi\mathcal{W}_{\gamma+1}\restriction(\alpha+2) = \mathcal{W}^*_{\gamma+1}\restriction(\alpha+2)$. Note that $\alpha+1 = \phi_{\nu,\gamma+1}(\beta)$, so $\psi^{\gamma+1}_{\phi_{\nu,\gamma+1}(\beta)} \circ \pi^{\nu,\gamma+1}_{\beta} = \overset{*}{\pi}{}^{\nu,\gamma+1}_{\beta} \circ \psi^{\nu}_{\beta}$ by the Shift Lemma, and this gives us the new instance of (3) of Lemma 6.8.2.

The general successor case above $\alpha+1$ is similar. Suppose we have

$$\psi\mathcal{W}_{\gamma+1}\restriction(\eta+1) = \mathcal{W}^*_{\gamma+1}\restriction(\eta+1)$$

as witnessed by $\psi^{\gamma+1}_{\tau}$ for $\tau \leq \eta$. Suppose $\eta > \alpha$. Let

$$\eta = \phi_{\nu,\gamma+1}(\xi) = \phi^*_{\nu,\gamma+1}(\xi), \quad G = E^{\mathcal{W}_{\gamma+1}}_{\eta}, \quad \text{and} \quad G^* = E^{\mathcal{W}^*_{\gamma+1}}_{\eta}.$$

Then

$$\begin{aligned}
\psi^{\gamma+1}_{\eta}(G) &= \psi^{\gamma+1}_{\phi_{\nu,\gamma+1}(\xi)}(\pi^{\nu,\gamma+1}_{\xi}(E^{\mathcal{W}_\nu}_{\xi})) \\
&= \overset{*}{\pi}{}^{\nu,\gamma+1}_{\xi}(\psi^{\nu}_{\xi}(E^{\mathcal{W}_\nu}_{\xi})) \\
&= \overset{*}{\pi}{}^{\nu,\gamma+1}_{\xi}(E^{\mathcal{W}^*_\nu}_{\xi}) \\
&= E^{\mathcal{W}^*_{\gamma+1}}_{\eta} = G^*.
\end{aligned}$$

The Shift Lemma now gives us $\psi^{\gamma+1}_{\eta+1}$ as above, and we have $\psi\mathcal{W}_{\gamma+1}\restriction(\eta+2) = \mathcal{W}^*_{\gamma+1}\restriction(\eta+2)$.

We leave the limit case of the subinduction to the reader. This finishes the subinduction proving (1), (2), and (3) of Lemma 6.8.2 at step $\gamma+1$. For (4), let us set $\tau = \gamma+1$. To simplify things, let us assume that $(\nu,\gamma+1]_U$ is not a drop. Consider the diagram

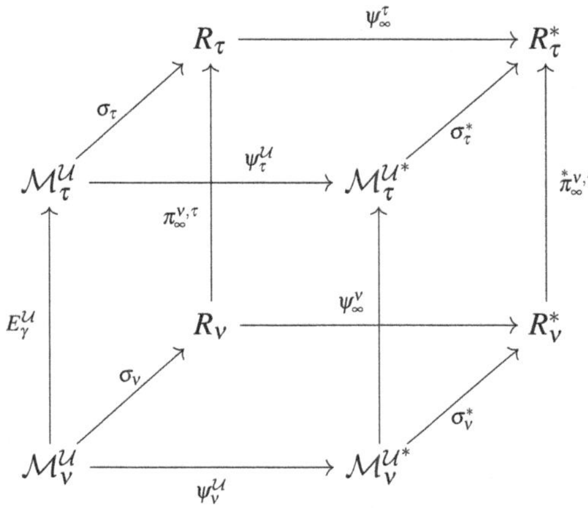

We are asked to show that $\sigma^*_\tau \circ \psi^{\mathcal{U}}_\tau = \psi^\tau_\infty \circ \sigma_\tau$, in other words, that the square on the top face of the cube commutes. The square on the bottom commutes by our induction hypothesis. The square in front commutes because $\mathcal{U}^*$ is a copy of $\mathcal{U}$. That the square in back commutes is clause (3) of our lemma at $\gamma+1$, which we just proved. The squares on the left and right faces commute by the properties of embedding normalization.

It is clear from these facts that the top square commutes on $\mathrm{ran}(i^{\mathcal{U}}_{\nu,\tau})$. Since $\mathcal{M}^{\mathcal{U}}_{\tau}$ is generated by $\mathrm{ran}(i^{\mathcal{U}}_{\nu,\tau}) \cup (\hat{\lambda}(E^{\mathcal{U}}_{\gamma})+1)$, it is enough to see that the top square commutes on $\hat{\lambda}(E^{\mathcal{U}}_{\gamma})+1$.

Let $a \in [\hat{\lambda}(E^{\mathcal{U}}_{\gamma})+1]^{<\omega}$. So $\sigma_{\gamma}(a) \in [\hat{\lambda}(F)+1]^{<\omega}$, and $\sigma_{\tau}(a) = \sigma_{\gamma}(a)$ by Proposition 6.7.7 on the agreement properties of quasi-normalization maps.[18] Thus

$$\begin{aligned}\psi^{\tau}_{\infty}(\sigma_{\tau}(a)) &= \psi^{\tau}_{\infty}(\sigma_{\gamma}(a)) \\ &= \psi^{\gamma}_{\infty}(\sigma_{\gamma}(a)),\end{aligned}$$

using that the copy maps ψ^{τ}_{∞} and ψ^{γ}_{∞} both agree with ψ^{γ}_{α} on $\hat{\lambda}(F)+1$. On the other hand, $\psi^{\mathcal{U}}_{\tau}(a) \in [\lambda(E^{\mathcal{U}^*}_{\tau})]^{<\omega}$, so

$$\begin{aligned}\sigma^*_{\tau}(\psi^{\mathcal{U}}_{\tau}(a)) &= \sigma^*_{\gamma}(\psi^{\mathcal{U}}_{\tau}(a)) \\ &= \sigma^*_{\gamma}(\psi^{\mathcal{U}}_{\gamma}(a)),\end{aligned}$$

by the agreement in normalization maps on the $\mathcal{W}^*$ side. But $\psi^{\gamma}_{\infty} \circ \sigma_{\gamma} = \sigma^*_{\gamma} \circ \psi^{\mathcal{U}}_{\gamma}$ by induction, so

$$\begin{aligned}\psi^{\tau}_{\infty} \circ \sigma_{\tau}(a) &= \psi^{\gamma}_{\infty} \circ \sigma_{\gamma}(a) \\ &= \sigma^*_{\gamma} \circ \psi^{\mathcal{U}}_{\gamma}(a) \\ &= \sigma^*_{\tau} \circ \psi^{\mathcal{U}}_{\tau}(a),\end{aligned}$$

as desired.

This finishes the step from γ to $\gamma+1$ in the inductive proof of Lemma 6.8.2. We leave the limit step to the reader. ⊣

It is easy to see that Theorem 6.8.1 follows from Lemma 6.8.2. ⊣

6.9. Normalizing longer stacks

There seem to be in the abstract many different ways to normalize a stack $\langle \mathcal{U}_1, \dots, \mathcal{U}_n \rangle$, one for each way of associating the $\mathcal{U}_i$. If we are in the case that embedding normalization coincides with full normalization, and there is a fixed strategy Σ for M according to which all these normalizations are played, such that for any N there is at most one normal Σ-iteration from M, then clearly all these normalizations are the same. They are just the unique normal tree by Σ from M to the last model of $\vec{\mathcal{U}}$. We shall be in that situation below when we deal with coarse iterations of a background universe. But in general, it seems that the various normalizations of $\vec{\mathcal{U}}$ might all be different from one another.

[18] Here we use the hypothesis that $\mathcal{U}$ is normal. With more work, one could probably avoid it.

We shall define Σ *normalizes well* by demanding that whenever $\vec{\mathcal{U}}$ is a finite stack by Σ, then all normalizations of $\vec{\mathcal{U}}$ are by Σ. In addition, we demand that Σ pull back to itself under normalization maps.

DEFINITION 6.9.1. Let $\vec{\mathcal{U}} = \langle \mathcal{U}_1, \dots, \mathcal{U}_n \rangle$ be a finite stack of normal trees on M, where $n > 1$. Let $M_0 = M$, and M_i be the last model of $\mathcal{U}_i$ for $1 \leq i \leq n$. A *1-step normalization* of $\vec{\mathcal{U}}$ is a triple $\langle k, \vec{\mathcal{V}}, \vec{\pi} \rangle$ such that $\vec{\mathcal{V}}$ is a stack of length $n-1$ on $M = M_0$, and

(1) $1 \leq k < n$,
(2) $\mathcal{V}_m = \mathcal{U}_m$ for all $m < k$, and $\mathcal{V}_k = W(\mathcal{U}_k, \mathcal{U}_{k+1})$,
(3) Letting $N_0 = M$ and N_i be the last model of $\mathcal{V}_i$ for $i < n$, we have that
  (a) $\pi_i \colon M_i \to N_i$ is the identity for $i < k$,
  (b) $\pi_k \colon M_{k+1} \to N_k$ is the σ-map given by embedding normalization, and
  (c) for $k < i < n$, $\mathcal{V}_i = \pi_{i-1}\mathcal{U}_{i+1}$, and $\pi_i \colon M_{i+1} \to N_i$ is the copy map.

Clearly, $\vec{\mathcal{U}}$ and k determine the rest of the normalization.

DEFINITION 6.9.2. Let $\vec{\mathcal{U}} = \langle \mathcal{U}_1, \dots, \mathcal{U}_n \rangle$ be a finite, maximal M-stack, with $n > 1$. Let $1 \leq t < n$; then a *t-step normalization* of $\vec{\mathcal{U}}$ is a sequence s with domain $t+1$ such that $s(0) = (0, \vec{\mathcal{U}}, \emptyset)$, and whenever $0 \leq i < t$, $s(i+1)$ is a 1-step normalization of $\vec{\mathcal{V}}$, where $\vec{\mathcal{V}}$ is the second coordinate of $s(i)$.

A *complete normalization* of $\langle \mathcal{U}_1, \dots, \mathcal{U}_n \rangle$ is an $n-1$ step normalization of $\langle \mathcal{U}_1, \dots, \mathcal{U}_n \rangle$. We shall sometimes identify a t-step normalization s of $\vec{\mathcal{U}}$ with the stack of trees in the second coordinate of $s(t)$. If $t = n-1$, then this is a single normal tree on M.

Remark 6.9.3. Benjamin Siskind has shown in [63, Theorem 3.2.16] that the normalization operation is associative, in that if $\vec{\mathcal{U}}$ is a finite stack of normal trees on a premouse M, then all complete normalizations of s produce the same normal tree on M. This is not at all obvious, even in the case that $\mathrm{lh}(\vec{\mathcal{U}}) = 3$, where there are only two possible ways to normalize $\vec{\mathcal{U}}$.

For $m \geq 1$, and $i \geq 0$, let us write $\mathcal{V}_m^{s(i)}$ for the m-th tree in the second coordinate of $s(i)$ (or in its third coordinate, if $i > 0$), and $N_m^{s(i)}$ for the last model of $\mathcal{V}_m^{s(i)}$. Let $N_0^{s(i)} = M$, for all i. For any $e < i < n$, and any m such that $N_m^{s(i)}$ exists, there is a unique l such that $N_m^{s(i)}$ comes from $N_l^{s(e)}$, in the sense that $s(e) \restriction (l+1)$ is normalized to $s(i) \restriction (m+1)$ by $s \restriction (e, i]$. Let us write

$$l = o^{s,i,e}(m)$$

in this case. Composing normalization maps and copy maps given by $s \restriction (e, i]$ yields a canonical

$$\pi_{l,m}^{s,i,e} : N_l^{s(e)} \to N_m^{s(i)},$$

where $l = o^{s,i,e}(m)$. So if s is a normalization of $\langle \mathcal{U}_1, \ldots, \mathcal{U}_n \rangle$ with $\mathrm{dom}(s) = i+1$, then the stack $\vec{\mathcal{V}}^{s(i)}$ has last model $N_m^{s(i)}$, where $m = n - i$, and $n = o^{s,i,0}(m)$, and $\pi_{n,m}^{s,i,0}$ is the natural map from the last model of $\vec{\mathcal{U}}$ to the last model of $\vec{\mathcal{V}}$. Let us write

$$\pi^s = \pi_{n,m}^{s,i,0}$$

in this case. So π^s is the natural map from the last model of $s(0)$ to the last model of the stack in $s(\mathrm{dom}(s)-1)$ that is given by s. All $\pi_{l,m}^{s,i,e}$ have the form π^u, for u obtained from s in a simple way.

Probably the most natural order in which to normalize a stack is bottom-up.

DEFINITION 6.9.4. Let $\vec{\mathcal{U}} = \langle \mathcal{U}_1, \ldots, \mathcal{U}_n \rangle$ be a finite, maximal stack of normal trees on M; then the *bottom-up normalization* of $\vec{\mathcal{U}}$ is the complete normalization s of $\vec{\mathcal{U}}$ such that for each $i \geq 1$ in $\mathrm{dom}(s)$, $s(i)$ has first coordinate 1. We write $W(\vec{\mathcal{U}})$ for the normal tree on M in the second coordinate of $s(\mathrm{dom}(s)-1)$, and also call $W(\vec{\mathcal{U}})$ the bottom-up normalization of $\vec{\mathcal{U}}$.

The definitions above extend to stacks $\vec{\mathcal{U}}$ on M of infinite length. Again, it seems to makes sense to normalize in any order, but the most natural way is bottom-up. Suppose for example that $\vec{\mathcal{U}} = \langle \mathcal{U}_n \mid n < \omega \rangle$. Let $\mathcal{W}_0 = \mathcal{U}_0$, and for $n \geq 1$ let

$$\mathcal{W}_n = W(\langle \mathcal{U}_i \mid i \leq n \rangle).$$

For $n \geq 0$, let

$$\Phi_n \colon \mathcal{W}_n \to \mathcal{W}_{n+1}$$

be the tree embedding given by the fact that $\mathcal{W}_{n+1} = W(\mathcal{W}_n, \pi\mathcal{U}_{n+1})$ for the appropriate π. (Φ_n is partial iff $\mathcal{U}_{n+1}$ drops along its main branch.) Then we set

$$W(\vec{\mathcal{U}}) = \lim_n \mathcal{W}_n,$$

where the limit is taken using the Φ_n. It is clear how to define this limit as an algebraic structure, but not at all clear that it is an iteration tree. Its length may be illfounded, and the models occurring in it may be illfounded. As in the case of finite stacks, what we need is that $\vec{\mathcal{U}}$ has been played according to a sufficiently good iteration strategy. The optimal result in this direction is due to Schlutzenberg; see [57]. We discuss this matter further in the next chapter.

One can continue further into the transfinite. $W(\vec{\mathcal{U}})$ makes sense as an algebraic structure for stacks $\vec{\mathcal{U}}$ of normal trees of any length, and under appropriate iterability hypotheses it is an iteration tree. In fact, one could go beyond linear stacks of normal trees, and consider normalizing arbitrary trees on M. There is as of now no good theory at this level of generality.

In this book we shall not need more than normalization for finite stacks of normal trees.

Chapter 7

STRATEGIES THAT CONDENSE AND NORMALIZE WELL

In this chapter we define what it is for an iteration strategy to normalize well and to have strong hull condensation. We prove some elementary facts related to these two properties, and we show that in the coarse case, they follow from strong unique iterability. Moreover, unlike strong unique iterability, they pass from an iteration strategy for a coarse premouse to the iteration strategies for fine premice that it induces via a background construction. Countable mice with Woodin cardinals do not have strongly unique iteration strategies; on the other hand, we shall see in Definition 7.6.1 that every iterable pure extender premouse has an iteration strategy that normalizes well and has strong hull condensation.[1]

Assuming AD^+, one can obtain strongly uniquely iterable coarse premice having Woodin cardinals via the Γ-Woodin construction. We discuss this in Section 7.2. In Section 7.3, we show that UBH together with the existence of a Woodin cardinal above a supercompact cardinal implies the existence of strongly uniquely iterable coarse premice with Woodin cardinals. These are our main existence theorems for coarse premice with strongly unique iteration strategies.

In Section 7.4, we show that if $\mathbb{C}$ is a background construction done inside a coarse premouse N^* with an iteration strategy Σ^* that normalizes well, then for any model M of $\mathbb{C}$, the induced strategy $\Omega(\mathbb{C}, M, \Sigma^*)$ for M normalizes well. In Section 7.5 we show that strong hull condensation is similarly preserved. In particular, if Σ^* is a strongly unique strategy for N^*, then the background-induced strategies $\Omega(\mathbb{C}, M, \Sigma^*)$ all normalize well and have strong hull condensation. This (together with its counterpart later for strategy mice) is our main existence theorem for fine structural mice with strategies that normalize well and have strong hull condensation.

In Section 7.6 we collect the key regularity properties of iteration strategies for pfs premice in the notion of a *pure extender pair*. Pure extender pairs are one of the types of mouse pair to which our comparison theorem applies.

[1] Let $M \models$ "δ is Woodin", and suppose M is coded by the real x. Let $i\colon M \to N$ be a genericity iteration such that $x \in N[g]$ for g being $\mathrm{Col}(\omega, i(\delta))$-generic over N. If the iteration tree producing i picks unique wellfounded branches at limit steps, then $i \restriction \delta^{+,M} \in N[g]$. But i is continuous at $\delta^{+,M}$, so then $i(\delta)^{+,N}$ is not a cardinal in $N[g]$, a contradiction. This argument has many refinements; see [51] and [82, §6.2], for example.

7.1. The definitions

The definitions in this section apply to both fine-structural premice and coarse premice.

DEFINITION 7.1.1. Let Σ be a complete (λ,θ)-iteration strategy for a pfs premouse M, where $\lambda > 1$. We say that Σ *quasi-normalizes well* iff whenever s is an M-stack by Σ, and $\langle \mathcal{T},\mathcal{U}\rangle$ is a maximal 2-stack by Σ_s such that $\mathcal{U}$ is normal, then

(a) $V(\mathcal{T},\mathcal{U})$ is by Σ_s, and
(b) letting $\mathcal{V} = V(\mathcal{T},\mathcal{U})$ and $\pi\colon \mathcal{M}^{\mathcal{U}}_\infty \to \mathcal{M}^{\mathcal{V}}_\infty$ be the last σ-map of the quasi-normalization[2], we have that $\Sigma_{s^\frown\langle\mathcal{T},\mathcal{U}\rangle} = (\Sigma_{s^\frown\vec{\mathcal{V}}})^\pi$.

In clause (b), the map π may be only nearly elementary, but that is sufficient to pull back an iteration strategy. In the coarse case, the last σ-map of a quasi-normalization is the identity, so the counterpart to Definition 7.1.1 is

DEFINITION 7.1.2. Let M be a transitive model of ZFC, and Σ be a $(\lambda,\theta,\mathcal{F})$ iteration strategy for M, where $\lambda > 1$. We say that Σ *quasi-normalizes well* iff whenever s is an M-stack by Σ, and $\langle\mathcal{T},\mathcal{U}\rangle$ is a maximal 2-stack by Σ_s such that $\mathcal{U}$ is normal, and $\mathcal{V}$ is a quasi-normalization of $\langle\mathcal{T},\mathcal{U}\rangle$, then

(a) $\mathcal{V}$ is by Σ_s, and
(b) $\Sigma_{s^\frown\langle\mathcal{T},\mathcal{U}\rangle} = \Sigma_{s^\frown\mathcal{V}}$.

DEFINITION 7.1.3. Let Σ be a complete (λ,θ)-iteration strategy for M, where $\lambda > 1$; then Σ *normalizes well* iff

(a) Σ quasi-normalizes well, and
(b) whenever s is an M-stack by Σ, and $\mathcal{T}$ is a plus tree by Σ_s, and $\mathcal{U}$ is the normal companion of $\mathcal{T}$, then $\mathcal{U}$ is by Σ_s.

Clearly, if Σ normalizes or quasi-normalizes well, then so do all its tail strategies. Recall that if $\langle\mathcal{T},\mathcal{U}\rangle$ is maximal and $\mathcal{U}$ is normal, then $W(\mathcal{T},\mathcal{U})$ is the normal companion of $V(\mathcal{T},\mathcal{U})$. Thus if Σ normalizes well, then conclusions (a) and (b) of Definition 7.1.1 hold with embedding normalization replacing quasi-normalization.[3]

We shall see that coarse strategies that are strongly unique normalize well. This is the important case for our main results. In the coarse case, beyond its bookkeeping value, we have no reason to distinguish between normalizing well and quasi-normalizing well.[4]

In the rest of this section, we shall focus on normalization in the fine-structural case. The adaptations needed in the coarse structural case are simple, and usually

[2] In the notation of Proposition 6.5.8, $\pi = \sigma_\gamma$, where $\mathrm{lh}(\mathcal{U}) = \gamma + 1$.

[3] For Definition 7.1.1(b), note that $V(\mathcal{T},\mathcal{U})$ and $W(\mathcal{T},\mathcal{U})$ have the same last model R, and both systems generate the same map from $\mathcal{M}^{\mathcal{U}}_\infty$ to R.

[4] See the remarks at the end of Section 6.7.

obvious. In the coarse case, the fundamental regularity property of iteration strategies we assume is strong uniqueness. Normalizing well and the other regularity properties we shall consider follow easily from this.

In Definition 7.1.1(1), we restrict ourselves to maximal stacks $\langle \mathcal{T}, \mathcal{U} \rangle$ because we have only defined $V(\mathcal{T}, \mathcal{U})$ in this case. We restricted ourselves to normal $\mathcal{U}$ because we have only proved some of the basic properties of the $V(\mathcal{T}, \mathcal{U})$ meta-tree in that case. One can probably extend Definition 7.1.1 to arbitrary stacks $\langle \mathcal{T}, \mathcal{U} \rangle$, and prove Theorems 7.2.9 and 7.4.1 in this greater generality. Since we do not need this generality, and it complicates the notation, we have not done that.

We defined normalizing well using stacks $\langle \mathcal{T}, \mathcal{U} \rangle$ of length 2, but this implies the corresponding behavior with respect to bottom-up normalizations of arbitrary finite, normal stacks.

LEMMA 7.1.4. *Suppose that Σ quasi-normalizes well, let s be an M-stack by Σ, and let t be a finite normal stack by Σ_s; then*

(a) *$V(t)$ is by Σ_s, and*

(b) *letting $\pi\colon \mathcal{M}_\infty(t) \to \mathcal{M}_\infty^{V(t)}$ be the last σ-map of the quasi-normalization, we have that $\Sigma_{s^\frown t} = (\Sigma_{s^\frown V(t)})^\pi$.*

If Σ normalizes well, then the same holds true with "W" replacing "V" everywhere.

PROOF. The easy proof is by induction on the length of t. It is essentially the same as the proof of Proposition 7.1.5 to follow, so we omit further detail. $\dashv$

In the case of embedding normalization, we looked in Section 6.9 at normalizing finite normal stacks in an arbitrary order, not just bottom-up.[5] We now show that if Σ normalizes well, then it behaves well with respect to all these normalizations.

PROPOSITION 7.1.5. *Let Σ be an complete (λ, θ)-iteration strategy for M that normalizes well, and let r be a stack of length $< \lambda$ by Σ. Suppose $\vec{\mathcal{U}}$ is a finite maximal stack by Σ_r, and s is a t-step normalization of $\vec{\mathcal{U}}$, and $\vec{\mathcal{V}} = \vec{\mathcal{V}}^{s(t)}$ is the stack in $s(t)$, then*

(1) *$\vec{\mathcal{V}}$ is by Σ_r, and*

(2) *if $\pi = \pi^s$ is the natural map from the last model Q of $\vec{\mathcal{U}}$ to the last model R of $\vec{\mathcal{V}}$, then $\Sigma_{r^\frown \vec{\mathcal{U}}, Q} = (\Sigma_{r^\frown \vec{\mathcal{V}}, R})^\pi$.*

PROOF. The proof is by induction on $\mathrm{lh}(s)$. We give it for Σ, but the same proof works for the tails Σ_r of Σ.

For $\mathrm{lh}(s) = 2$ this is true by hypothesis. Let $\vec{\mathcal{T}}^\frown \langle \mathcal{U}_1, \mathcal{U}_2 \rangle^\frown \vec{\mathcal{S}}$ be a stack of length $n+1$ by Σ. We want to see that the 1-step normalization obtained by replacing $\langle \mathcal{U}_1, \mathcal{U}_2 \rangle$ by $W(\mathcal{U}_1, \mathcal{U}_2)$, and $\vec{\mathcal{S}}$ by $\pi\vec{\mathcal{S}}$ for π the normalization map, behaves well. It is clear that this implies t-step normalizations behave well, for all t.

[5]Doing this for quasi-normalization would involve defining $V(\mathcal{T}, \mathcal{U})$ when $\mathcal{U}$ is not normal, and we have decided to avoid that.

Let $\mathcal{V}$ be a complete normalization of $\vec{\mathcal{T}}$, with θ the normalization map from $N = \mathcal{M}_\infty^{\vec{\mathcal{T}}}$ to $N^* = \mathcal{M}_\infty^{\mathcal{V}}$. θ lifts $\mathcal{U}_1$ to $\theta\mathcal{U}_1$; let $\rho\colon \mathcal{M}_\infty^{\mathcal{U}_1} \to \mathcal{M}_\infty^{\theta\mathcal{U}_1}$ be the copy map. Note that $\langle \mathcal{V}, \theta\mathcal{U}_1, \rho\mathcal{U}_2 \rangle$ is a stack by Σ, because $\Sigma_{\mathcal{V},N^*}$ pulls back under θ to $\Sigma_{\vec{\mathcal{T}},N}$ by our induction hypothesis. Let Q^* be its last model. Let

$$\mathcal{W}^* = W(\theta\mathcal{U}_1, \rho\mathcal{U}_2),$$

and let R^* be the last model of $\mathcal{W}^*$, and $\sigma^*\colon Q^* \to R^*$ the normalization map. The hypothesis of our proposition tells us that $\langle \mathcal{V}, \mathcal{W}^* \rangle$ is by Σ, and that

$$\Sigma_{\langle \mathcal{V}, \theta\mathcal{U}_1, \rho\mathcal{U}_2 \rangle, Q^*} = (\Sigma_{\langle \mathcal{V}, \mathcal{W}^* \rangle, R^*})^{\sigma^*}.$$

Let Q be the last model of $\vec{\mathcal{T}}^\frown \langle \mathcal{U}_1, \mathcal{U}_2 \rangle$, let

$$\mathcal{W} = W(\mathcal{U}_1, \mathcal{U}_2),$$

and let R be the last model of $\mathcal{W}$. Let $\sigma\colon Q \to R$ be the normalization map. The situation can be encapsulated in the following diagram.

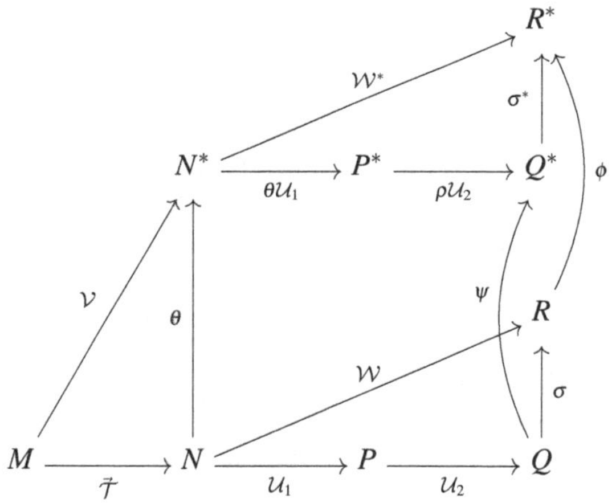

Here $P = \mathcal{M}_\infty^{\mathcal{U}_1}$, and $P^* = \mathcal{M}_\infty^{\theta\mathcal{U}_1}$, and $\rho\colon P \to P^*$ is the copy map. The maps $\psi\colon Q \to Q^*$ and $\phi\colon R \to R^*$ are copy maps. We get ϕ from Theorem 6.8.1; in this case, copying $\langle \mathcal{U}_1, \mathcal{U}_2 \rangle$ via θ commutes with normalizing $\langle \mathcal{U}_1, \mathcal{U}_2 \rangle$. We have

$$\phi \circ \sigma = \sigma^* \circ \psi$$

from Theorem 6.8.1.

Since $\theta\mathcal{W} = \mathcal{W}^*$, and Σ pulls back to itself under θ by induction, we have that $\vec{\mathcal{T}}^\frown\langle \mathcal{W} \rangle$ is by Σ, and $\Sigma_{\vec{\mathcal{T}}^\frown\langle \mathcal{W}\rangle, R} = (\Sigma_{\langle \mathcal{V}, \mathcal{W}^* \rangle, R^*})^{\phi}$. It follows that

$$\begin{aligned}(\Sigma_{\vec{\mathcal{T}}^\frown\langle \mathcal{W}\rangle, R})^{\sigma} &= (\Sigma_{\langle \mathcal{V}, \mathcal{W}^* \rangle, R^*})^{\phi\circ\sigma}\\ &= (\Sigma_{\langle \mathcal{V}, \mathcal{W}^* \rangle, R^*})^{\sigma^*\circ\psi}\\ &= ((\Sigma_{\langle \mathcal{V}, \mathcal{W}^* \rangle, R^*})^{\sigma^*})^{\psi}\end{aligned}$$

$$= (\Sigma_{\langle \mathcal{V}, \theta\mathcal{U}_1, \rho\mathcal{U}_2\rangle, Q^*})^{\psi}$$
$$= \Sigma_{\vec{\mathcal{T}}^\frown\langle \mathcal{U}_1,\mathcal{U}_2\rangle, Q}.$$

Line 1 holds because Σ normalizes well for $\vec{\mathcal{T}}$, line 2 comes from Theorem 6.8.1, line 4 holds because $\Sigma_{\mathcal{V},N^*}$ 2-normalizes well, and line 5 holds because Σ normalizes well for $\vec{\mathcal{T}}$.

This takes care of the case $\vec{\mathcal{S}} = \emptyset$. The general case follows easily. Since $(\Sigma_{\vec{\mathcal{T}}^\frown\langle \mathcal{W}\rangle,R})^\sigma = \Sigma_{\vec{\mathcal{T}}^\frown\langle \mathcal{U}_1,\mathcal{U}_2\rangle,Q}$ and $\vec{\mathcal{S}}$ is by $\Sigma_{\vec{\mathcal{T}}^\frown\langle \mathcal{U}_1,\mathcal{U}_2\rangle,Q}$, we have that $\sigma\vec{\mathcal{S}}$ is by $\Sigma_{\vec{\mathcal{T}}^\frown\langle \mathcal{W}\rangle,R}$, and moreover the $\vec{\mathcal{T}}^\frown\langle \mathcal{W}\rangle^\frown\sigma\vec{\mathcal{S}}$-tail of Σ pulls back under the relevant copy map to the $\vec{\mathcal{T}}^\frown\langle \mathcal{U}_1,\mathcal{U}_2\rangle^\frown\vec{\mathcal{S}}$-tail of Σ. ⊣

A very similar argument shows that the property of normalizing well passes to pullback strategies.

THEOREM 7.1.6. *Let Σ be an iteration strategy for N, and let $\pi\colon M \to N$ be nearly elementary; then*

(a) *if Σ quasi-normalizes well, then Σ^π quasi-normalizes well, and*
(b) *if Σ normalizes well, then Σ^π normalizes well.*

PROOF. We start with (a). Let $\langle \mathcal{V},\mathcal{U}_1,\mathcal{U}_2\rangle$ be a stack by Σ^π, with last model Q. Let $\mathcal{W} = V(\mathcal{U}_1,\mathcal{U}_2)$ have last model R, and $\sigma\colon Q \to R$ be the quasi-normalization map. We want to see that $\langle \mathcal{V},\mathcal{W}\rangle$ is by Σ^π, and that the $\langle \mathcal{V},\mathcal{W}\rangle$-tail of Σ^π pulls back under σ to the $\langle \mathcal{V},\mathcal{U}_1,\mathcal{U}_2\rangle$-tail of Σ^π.

We have the diagram

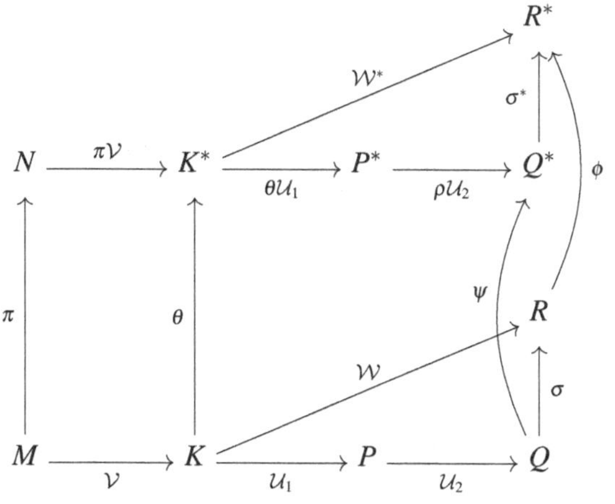

Here $\theta\colon K \to K^*$ and $\rho\colon P \to P^*$ are copy maps generated by π, and $\mathcal{W}^*$ is the normalization of $\langle \theta\mathcal{U}_1, \rho\mathcal{U}_2\rangle$. σ^* is the associated normalization map. ψ and ϕ are copy maps, which we have because copying commutes with normalization. $\phi \circ \sigma = \sigma^* \circ \psi$ by Theorem 6.8.1.

The copy map ϕ tells us that $\langle \mathcal{V}, \mathcal{W} \rangle$ is by Σ^π. The rest is given by

$$
\begin{aligned}
(\Sigma^\pi_{\langle \mathcal{V},\mathcal{W}\rangle,R})^\sigma &= (\Sigma_{\langle \pi\mathcal{V},\mathcal{W}^*\rangle,R^*})^{\phi\circ\sigma} \\
&= (\Sigma_{\langle \pi\mathcal{V},\mathcal{W}^*\rangle,R^*})^{\sigma^*\circ\psi} \\
&= ((\Sigma_{\langle \pi\mathcal{V},\mathcal{W}^*\rangle,R^*})^{\sigma^*})^{\psi} \\
&= (\Sigma_{\langle \pi\mathcal{V},\theta\mathcal{U}_1,\rho\mathcal{U}_2\rangle,Q^*})^{\psi} \\
&= \Sigma^\pi_{\mathcal{T}^\frown\langle \mathcal{U}_1,\mathcal{U}_2\rangle,Q}.
\end{aligned}
$$

This is what we want.

For (b), suppose that $\mathcal{T}$ is a plus tree on M by Σ^π, and let $\psi_\alpha \colon \mathcal{M}^{\mathcal{T}}_\alpha \to \mathcal{M}^{\pi\mathcal{T}}_\alpha$ be the copy map. Since ψ_α agrees with $\psi_{\alpha+1}$ on $\mathrm{lh}(E^{\mathcal{T}}_\alpha)$, $\mathrm{lh}(E^{\mathcal{T}}_\alpha) > \mathrm{lh}(E^{\mathcal{T}}_{\alpha+1})$ iff $\mathrm{lh}(E^{\pi\mathcal{T}}_\alpha) > \mathrm{lh}(E^{\pi\mathcal{T}}_{\alpha+1})$. Thus $\mathcal{T}$ and $\pi\mathcal{T}$ have the same maximal delay intervals, and $(\pi\mathcal{T})^{\mathrm{nrm}} = \pi\mathcal{T}^{\mathrm{nrm}}$. But $(\pi\mathcal{T})^{\mathrm{nrm}}$ is by Σ, so $\mathcal{T}^{\mathrm{nrm}}$ is by Σ^π. The same proof works for the tails of Σ, so we have (b). ⊣

We conclude this elementary discussion by showing that a strategy that normalizes well is determined by its action on normal trees.

Suppose that Σ normalizes well, and $\mathcal{T}$ is a normal tree on M with last model Q that is according to Σ. Let $\mathcal{U}$ on Q be normal and by $\Sigma_{\mathcal{T},Q}$ and of limit length, and let

$$b = \Sigma_{\mathcal{T},Q}(\mathcal{U}) = \Sigma(\langle \mathcal{T},\mathcal{U}\rangle),$$

and

$$a = \Sigma(W(\mathcal{T},\mathcal{U})).$$

Then

$$a = \mathrm{br}^{\mathcal{T},\mathcal{U}}_W(c,b),$$

where c is some branch $[0,\tau)_{\mathcal{T}}$ or $[0,\tau]_{\mathcal{T}}$ of $\mathcal{T}$ that is chosen by Σ. Moreover,

$$b = \mathrm{br}^{\mathcal{T},\mathcal{U}}_{\mathcal{U}}(a).$$

In other words, $\Sigma(\langle \mathcal{T},\mathcal{U}\rangle)$ and $\Sigma(W(\mathcal{T},\mathcal{U}))$ determine each other, modulo $\mathcal{T}$.

PROPOSITION 7.1.7. *Let Σ and Ψ be complete strategies for M with scope H_δ that normalize well, and suppose that Σ agrees with Ψ on normal trees; then Σ agrees with Ψ on finite, maximal M-stacks.*

PROOF. We just gave the proof for stacks of length 2. Let s be finite, maximal stack by both strategies such that $\Sigma_s = \Psi_s$, and $\mathcal{T}$ be a normal tree on $\mathcal{M}_\infty(s)$ with last model Q. We want to see that $\Sigma_{s^\frown\langle \mathcal{T}\rangle,Q} = \Psi_{s^\frown\langle \mathcal{T}\rangle,Q}$, so let $\mathcal{U}$ be a normal tree of limit length by both strategies. Let $b = \Sigma_{s^\frown\langle \mathcal{T}\rangle,Q}(\mathcal{U}) = \Sigma_s(\langle \mathcal{T},\mathcal{U}\rangle)$ and $\mathcal{W}_b = W(\mathcal{T},\mathcal{U}^\frown b)$. Since Σ_s normalizes well, $\mathcal{W}_b$ is by Σ_s. But then $\mathcal{W}_b$ is by Ψ_s, and since $\mathcal{W}_b$ determines b modulo $\langle \mathcal{T},\mathcal{U}\rangle$, $b = \Psi_s(\langle \mathcal{T},\mathcal{U}\rangle)$, as desired. ⊣

Strong hull condensation. We turn to strong hull condensation. The following convention will smooth our terminology. Let us regard the empty tree on M as a pseudo-hull of every plus tree on M, under the empty tree embedding.

DEFINITION 7.1.8. Let $\mathcal{U}$ be a plus tree on M of length $\beta+1$; then the empty tree on M is a pseudo-hull of $\mathcal{U}$, and $\hat{\imath}^{\mathcal{U}}_{0,\beta}$ is the t-map of an extended tree embedding of the empty tree into $\mathcal{U}$.

The point of this terminology is to streamline the following definition.

DEFINITION 7.1.9. Let Σ be a complete (λ,θ) iteration strategy for a pfs premouse M; then Σ has *strong hull condensation* iff whenever s is a stack of plus trees by Σ and $N \trianglelefteq \mathcal{M}_\infty(s)$, and $\mathcal{U}$ is a plus tree on N by $\Sigma_{s,N}$, then for any plus tree $\mathcal{T}$ on N,

(a) if $\mathcal{T}$ is a pseudo-hull of $\mathcal{U}$, then $\mathcal{T}$ is by $\Sigma_{s,N}$, and
(b) if $\Phi\colon \mathcal{T} \to \mathcal{U}$ is an extended tree embedding, with last t-map π, and $Q \trianglelefteq \operatorname{dom}(\pi)$, then $\Sigma_{s^\frown\langle\mathcal{T}\rangle,Q} = (\Sigma_{s^\frown\langle\mathcal{U}\rangle,\pi(Q)})^\pi$.

Because less is required of a tree embedding than is required of a hull embedding in [42], the property is stronger than the property called Hull Condensation in [42]. Hence its name.

Clause (b) was not part of our original definition of strong hull condensation. Benjamin Siskind then showed that (b) follows abstractly from (a) and normalizing well (see [64]), via a strategy-comparison argument. We have made clause (b) part of the definition here because it is useful, and one can obtain it directly for background-induced strategies.

Because we have included clause (b) in the definition of strong hull condensation, it implies pullback consistency. Recall that a pullback consistent strategy is one that pulls back to itself under its own iteration maps. (See Definition 5.3.1.) It is important here that in clause (b) of Definition 7.1.9 we have allowed Q to be a proper initial segment of $\operatorname{dom}(\pi)$. This leads to pullback consistency for partial iteration maps, and thus the very mild form of positionality described in Proposition 5.2.2.

LEMMA 7.1.10. *Let Σ be a complete strategy for M that has strong hull condensation; then*

(a) Σ is pullback consistent, and
(b) Σ is mildly positional.

PROOF. For pullback consistency: suppose that $\mathcal{U}$ is a plus tree on $M_\infty(s)$ by Σ_s of length $\beta+1$, and that $\alpha \leq_U \beta$. Let $\mathcal{T} = \mathcal{U} \upharpoonright \alpha+1$ and $\pi = \hat{\imath}^{\mathcal{U}}_{\alpha,\beta}$; then π is the last t-map of an extended tree embedding from $\mathcal{T}$ into $\mathcal{U}$. (If $\alpha > 0$, its associated tree embedding is just the identity on $\mathcal{T} \upharpoonright \alpha+1$, and if $\alpha = 0$, we appeal to Definition 7.1.8.) Suppose that $Q \trianglelefteq \mathcal{M}^{\mathcal{T}}_\alpha$ and $Q \subseteq \operatorname{dom}(\pi)$. By part (b) of Definition 7.1.9, $\Sigma_{s^\frown\langle\mathcal{T}\rangle,Q} = (\Sigma_{s^\frown\langle\mathcal{U}\rangle,\pi(Q)})^\pi$, which is what we need. It is routine the extend this argument to finite stacks by Σ_s, by pulling back under the branch embeddings of the constituent normal trees, one at a time.

Part (b) follows from pullback consistency. We must see that if s is a stack by Σ and $P \trianglelefteq N \trianglelefteq \mathcal{M}_\infty(s)$, then $(\Sigma_{s,N})_P = \Sigma_{s,P}$. Let us assume $s = \emptyset$ for simplicity. $\Sigma_N = \Sigma_t$, where $t = \langle \mathcal{T}, N \rangle$ is the empty tree $\mathcal{T}$ on M followed by a gratuitous drop to N. $(\Sigma_N)_P = \Sigma_u$, where $u = \langle \mathcal{T}, N, \mathcal{U}, P \rangle$, for $\mathcal{U}$ the empty tree on N. Letting π be the identity map on N, π is the main branch embedding of $\mathcal{U}$, and $\pi(P) = P$. So we can pull back by π, and we get $\Sigma_P = \Sigma_{\mathcal{T},P} = \Sigma_u^\pi = \Sigma_u = (\Sigma_N)_P$, as desired. ⊣

Strong hull condensation is preserved by pullbacks:

PROPOSITION 7.1.11. *Let $\pi\colon M \to N$ be nearly elementary, and let Σ be a strategy for N having strong hull condensation; then Σ^π has strong hull condensation.*

PROOF SKETCH. There is a relevant diagram below. Let s be a stack on M with last model K, and let K^* be the last model of πs, with $\theta\colon K \to K^*$ the copy map. Let $\mathcal{U}$ be on K and by $(\Sigma^\pi)_s$, and let $\mathcal{T}$ be a pseudo-hull of $\mathcal{U}$. It is not hard to see that $\theta\mathcal{T}$ is a pseudo-hull of $\theta\mathcal{U}$. Since $\theta\mathcal{U}$ is by $\Sigma_{\pi s,K^*}$, $\theta\mathcal{T}$ is by $\Sigma_{\pi s,K^*}$, so $\mathcal{T}$ is by $(\Sigma^\pi)_s$, as desired for part (a).

For part (b), let $\Phi\colon \mathcal{T} \to \mathcal{U}$ be an extended tree embedding with last t-map $\sigma\colon Q \to R$. By the (suppressed) construction of the first part, we have an extended tree embedding $\Psi\colon \theta\mathcal{T} \to \theta\mathcal{U}$. Let $\sigma^*\colon Q^* \to R^*$ be the last t-map of Ψ. Let $\psi\colon Q \to Q^*$ come from the copying of $\mathcal{T}$ to $\theta\mathcal{T}$, and $\phi\colon R \to R^*$ come from copying $\mathcal{U}$ to $\theta\mathcal{U}$. We have the diagram

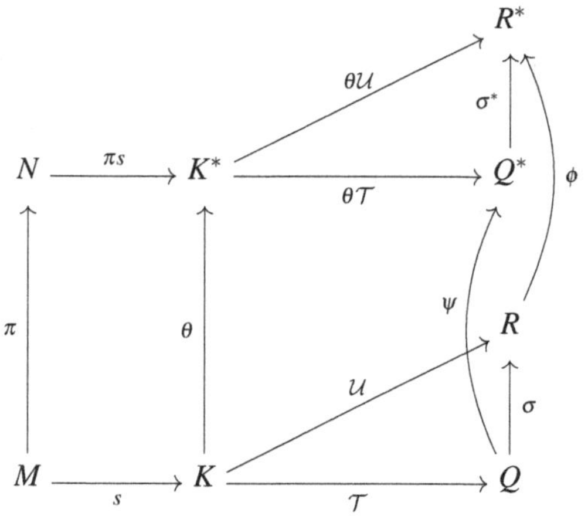

This is quite similar to the diagram in Theorem 7.1.6, because the situations are quite similar. Again, we calculate

$$
\begin{aligned}
(\Sigma_{\langle s^\frown \langle \mathcal{U} \rangle, R \rangle})^\sigma &= (\Sigma_{\langle \pi s, \theta\mathcal{U} \rangle, R^*})^{\phi \circ \sigma} \\
&= (\Sigma_{\langle \pi s, \theta\mathcal{U} \rangle, R^*})^{\sigma^* \circ \psi}
\end{aligned}
$$

$$= ((\Sigma_{\langle \pi s, \theta \mathcal{U}\rangle, R^*})^{\sigma^*})^{\psi}$$
$$= (\Sigma_{\langle \pi s, \theta \mathcal{T}\rangle, Q^*})^{\psi}$$
$$= (\Sigma_{s^\frown\langle \mathcal{T}\rangle, Q})^{\psi}. \quad \dashv$$

We built pullback consistency into strong hull condensation. Internal lift consistency is a form of strong hull condensation for iteration strategies that are defined on non-maximal trees, but since we are avoiding non-maximal trees, internal lift consistency will remain an independent regularity property.

All the regularity properties of iteration strategies we have encountered so far are implied by strong hull condensation, internal lift consistency, and quasi-normalizing well. We showed in Section 5.4 that if Σ^* is a strongly unique iteration strategy for V, then the iteration strategies it induces via PFS constructions are internally lift consistent. In this chapter, we shall show that they quasi-normalize well and have strong hull condensation.

We believe that a complete strategy with strong hull condensation need not normalize well, although we have no example at the moment. However, any complete strategy for normal trees that has strong hull condensation can be extended in a unique way to a complete strategy for finite stacks of normal trees that has strong hull condensation and normalizes well. This is a result of Schlutzenberg and the author. Schlutzenberg also proved a stronger version of the theorem in which the extended strategy can act on infinite stacks. See [57] and [64], and Theorem 7.3.11 in the next section.

Remark 7.1.12. The papers [77] and [64] introduce a still weaker sort of embedding of iteration trees, and make use of the resulting "very strong hull condensation". It turns out that strategies for premice that have strong hull condensation also have very strong hull condensation, and this implies that they fully normalize well. However, the proof of this requires a strategy-comparison argument. Strong hull condensation has the virtue that we can verify it directly for background-induced strategies, so we can use it in proving a comparison theorem.

The following elementary lemma on extending tree embeddings at limit steps will be useful.

LEMMA 7.1.13. *Let Σ be a strategy for the premouse M having strong hull condensation, and let $\mathcal{T}$ and $\mathcal{U}$ be trees of limit length by Σ. Let $\Phi\colon \mathcal{T} \to \mathcal{U}$ be a tree embedding such that*

$$\exists \alpha < \mathrm{lh}(\mathcal{U}) \forall \beta (\alpha < \beta < \mathrm{lh}(\mathcal{U}) \Rightarrow \beta \in \mathrm{ran}(u^{\Phi})).$$

Let $b = \Sigma(\mathcal{T})$ and $c = \Sigma(\mathcal{U})$; then there is a unique tree embedding $\Psi\colon \mathcal{T}^\frown b \to \mathcal{U}^\frown c$ such that $\Phi \subseteq \Psi$.

PROOF. Let $u = u^{\Phi}$, and $d = u^{-1}$"c. By our hypothesis that $\mathrm{ran}(u)$ contains a final segment of the ordinals below $\mathrm{lh}(\mathcal{U})$, we see that d is cofinal in $\mathrm{lh}(\mathcal{T})$.

Moreover, Φ extends to a tree embedding of $\mathcal{T}^\frown d$ into $\mathcal{U}^\frown c$. By strong hull condensation, $d = \Sigma(\mathcal{T}) = b$, so we are done. ⊣

If one weakens the hypothesis of Lemma 7.1.13 by requiring only that $\mathrm{ran}(u^\Phi)$ be cofinal in $\mathrm{lh}(\mathcal{U})$, then the conclusion may not hold.[6]

7.2. Coarse Γ-Woodins and Γ-universality

Of course, one cannot prove that there are any nontrivial iteration strategies without making assumptions that go beyond ZF. Determinacy assumptions are particularly useful in this regard. Under AD^+, every Suslin-co-Suslin set is Wadge reducible to an iteration strategy; in fact, there are countable iterable structures at every Suslin-co-Suslin degree of correctness. More precisely

DEFINITION 7.2.1. Let $A \subseteq \mathbb{R}$. We say that $(M, \delta, \tau, \Sigma)$ *captures* A iff

(a) $M \models \mathsf{ZFC} +$ "δ is Woodin",
(b) δ is countable, and Σ is a complete strategy with scope HC for $V^M_{\delta+1}$, and
(c) $\tau \in M$ is a $\mathrm{Col}(\omega, \delta)$-term for a set of reals, and
(d) whenever $i \colon M \to N$ is by Σ and g is $\mathrm{Col}(\omega, i(\delta))$-generic over N, then $i(\tau)_g = A \cap N[g]$.

Notice here that $(M, \delta, \tau, \Sigma)$ does indeed determine A, because for every real x there are N and g as in (d) such that $x \in N[g]$.

The following came out of Woodin's work in the late 1980s on large cardinals in HOD under determinacy hypotheses. See [26] and [71].

THEOREM 7.2.2 (Woodin). *Assume* AD; *then for any Suslin and co-Suslin set* A*, there is a tuple* $(M, \delta, \tau, \Sigma)$ *that captures* A.

Unfortunately, the models M produced by the proof of 7.2.2 are not given as fine structural. However, one can use M as a background universe for the construction of some fine structural premouse N, and hope to show that N and its induced strategy capture some set close to A. This is the basic plan behind the proofs we currently have for fragments of LEC and HPC, and it is therefore the main source for the iteration strategies to which the theorems of this book apply.

In this context, it helps to be working with a background universe M having more structure than is recorded in Definition 7.2.1. We shall call the resulting pairs *coarse* Γ*-Woodin pairs*.

Assume AD^+, and let Γ, Γ_1 be good (i.e., closed under $\exists^{\mathbb{R}}$) lightface pointclasses with the scale property such that $\Gamma \subseteq \Delta_1$. Let A be a universal Γ_1 set, and let $U \subset \mathbb{R}$ code $\{\langle \varphi, x\rangle \mid (V_{\omega+1}, \in, A) \models \varphi[x]\}$. Let S and T be trees on some $\omega \times \kappa$ that project to U and $\neg U$. Using his work in [26], Woodin has shown ([71, Lemma

[6]There is a counterexample in [64], just after Definition 1.3.

3.13]) that there is a countable transitive $N^* \in \mathrm{HC}$, a wellorder w of N^*, and an iteration strategy Σ such that for $\delta = o(N^*)$,

(a) (fullness) $N^* = V_\delta^{L(N^* \cup \{S,T,w\})}$,
(b) N^* is f-Woodin, for all $f\colon \delta \to \delta$ such that $f \in C_\Gamma(N^*, w)$,[7]
(c) for all $\eta \leq \delta$, there is an $f\colon \eta \to \eta$ such that $f \in C_{\Gamma_1}(V_\eta^{N^*}, w \cap V_\eta^{N^*})$ and $V_\eta^{N^*}$ is not f-Woodin, and
(d) Σ is an (ω_1, ω_1)-iteration strategy for $L(N^*, S, T, \lhd)$, with respect to nice trees based on N^*.

Concerning item (d), recall that ω_1-iterability implies $\omega_1 + 1$-iterability, granted AD.

DEFINITION 7.2.3. Assume AD^+, and let Γ be a good pointclass with the scale property, and let $N^*, \delta, S, T, \lhd$, and Σ be as in (a)-(d); then

(1) we call $\langle N^*, \delta, S, T, w, \Sigma\rangle$ a *coarse* Γ-*Woodin tuple*, and
(2) letting $M = (L[N^*, S, T, w], \in, S, T)$, we call (M, Σ) a *coarse* Γ-*Woodin pair.*

Of course, S and T determine U, and hence A and Γ_1. U is self-dual, so S is only there for notational convenience. We write $A = A_T$. If (M, Σ) is a coarse Γ-Woodin pair, then we write δ^M, w^M, T^M, and S^M for the associated objects.

From [26] (see also [71, Lemma 3.13]), we have

THEOREM 7.2.4 (Woodin). *Let Γ be a good lightface pointclass with the scale property, and assume that all sets in $\check{\Gamma}$ are Suslin; then for any real x there is a coarse Γ-Woodin pair (M, Σ) such that $x \in M$.*

LEMMA 7.2.5. *Let (M, Σ) be a coarse Γ-Woodin pair, $\delta = \delta^M$, $T = T^M$, and $S = S^M$. Let s be a M-stack all of whose models are wellfounded, with iteration map $i\colon M \to Q$; then*

(i) $p[i(T)] = p[T]$ and $p[i(S)] = p[S]$, and
(ii) if g is $\mathrm{Col}(\omega, i(\delta))$-generic over Q, then for $A = A_T$, $(V_{\omega+1}^{Q[g]}, \in, A \cap Q[g]) \prec (V_{\omega+1}, \in, A)$.

PROOF. As usual: $p[T] \subseteq p[i(T)]$ and $p[S] \subseteq p[i(S)]$, while $p[i(T)] \cap p[i(S)] = \emptyset$ because Q is wellfounded, and wellfoundedness is absolute to wellfounded models. This gives us (i). For (ii), we use the Tarski-Vaught criterion. Suppose $x \in N[g]$ and $(V_{\omega+1}, \in, A) \models \exists y \in \mathbb{R}\varphi[y, x]$. There is then a branch of T of the form $(\varphi, \langle y, x\rangle, f)$. But then $(\varphi, \langle y, x\rangle, i(f))$ is a branch of $i(T)$, so there is a branch $(\varphi, \langle y, x\rangle, h)$ of $i(T)$ such that $y \in N[g]$, as desired. ⊣

[7] For a countable and transitive, $C_\Gamma(a)$ is the largest countable $\Gamma(a \cup \{a\})$ subset of $P(a)$. Its theory (under determinacy hypotheses) was first developed by Kechris and Moschovakis. See [23], [24], and the survey [72]. Harrington and Kechris showed in [15] that $C_\Gamma(a) = P(a) \cap L[T, a]$, for any tree T of a Γ scale on a universal Γ set. This is probably the most useful characterization of $C_\Gamma(a)$ in our context.

Note that we did not assume in the lemma that s was by Σ. We shall show in a moment that this follows, that is, that Σ witnesses strong unique iterability.

If we drop down from M to $L(N^*, W, w)$, where W is the tree of a Γ-scale on a universal Γ set, then δ becomes Woodin, and Lemma 7.2.5 yields a pair capturing Γ in the sense of Definition 7.2.1.

COROLLARY 7.2.6. *Let (M,Σ) be a coarse Γ-Woodin pair, and $\delta = \delta^M$. Let W be the tree of a scale on a universal Γ set, and let τ be the natural term for $p[W]$; then $(L[V_\delta^M, w^M, W], \delta, \tau, \Sigma)$ captures $p[W]$.*

Let $M = L[N^*, S, T, w]$, where $(N^*, \delta, S, T, w, \Sigma)$ is a coarse Γ-Woodin tuple. Let $A = A_T$, and let Γ_1 be the good pointclass whose universal set is A. If P is a wellfounded iterate of M, and g is P-generic over $\mathrm{Col}(\omega, i(\delta))$, then $P[g]$ is projectively-in-A correct. Thus the C_Γ and C_{Γ_1} operators are correctly defined over $P[g]$. It follows that M and its iterates are C_{Γ_1}-full, and Σ is guided at $\mathcal{T}$ by a Q-structure in $C_{\Gamma_1}(\mathcal{M}(\mathcal{T}))$, where

$$\mathcal{M}(\mathcal{T}) = (\bigcup_{\alpha<lh(\mathcal{T})} V^{M_\alpha}_{\mathrm{lh}(E_\alpha)}, \bigcup_{\alpha<\mathrm{lh}(\mathcal{T})} i_{0,\alpha}(w) \cap V^{M_\alpha}_{\mathrm{lh}(E_\alpha)}).$$

(We have omitted some superscript $\mathcal{T}$'s here.) That is,

LEMMA 7.2.7. *Assume AD^+, and let (M,Σ) be a coarse Γ-Woodin pair. Let $\vec{\mathcal{T}}, \mathcal{U}$ be a stack of nice trees played by Σ; then the following are equivalent*

(1) $\Sigma_{\vec{\mathcal{T}}}(\mathcal{U}) = b$,
(2) $C_{\Gamma_1}(\mathcal{M}(\mathcal{U})) \subseteq \mathcal{M}^{\mathcal{U}}_b$,
(3) $\mathcal{M}^{\mathcal{U}}_b$ is wellfounded.

PROOF. Just outlined. ⊣

It follows that if (M,Σ) is a coarse Γ-Woodin pair, then all its iterates are coarse Γ-Woodin pairs, and Σ is positional, that is, $\Sigma_{s,Q}$ depends only on Q. (Cf. Lemma 9.3.9.) Moreover, if Q is an iterate of M via the stack s, then for $\theta = \omega_1^V$,

(i) Q is strongly uniquely (θ,θ)-iterable, and
(ii) $Q \models$ "I am strongly uniquely (θ,θ)-iterable".

The strategy witnessing (i) is Σ_Q, and the strategy witnessing (ii) is $\Sigma_Q \restriction Q$. Moreover, Σ_Q is definable over $(V_{\omega+1}, \in, A)$ from the parameter $(V^Q_{\delta^Q}, w^Q)$, uniformly in Q, and Q and its generic extensions are correct for the theory of $(V_{\omega+1}, \in, A)$. So we have

COROLLARY 7.2.8. *Assume AD^+, and let (M,Σ) be a coarse Γ-Woodin pair; then M is strongly uniquely iterable for countable stacks of countable normal trees. Moreover, for $\kappa = \omega_1^V$,*

$$M \models \text{"I am strongly uniquely } (\kappa,\kappa)\text{-iterable"}.$$

If (M,Σ) is a coarse Γ-Woodin pair, and $\mathbb{C}$ is any PFS-construction in the sense of M, then $\mathbb{C}$ is good (never breaks down), because all its levels have iteration strategies induced by Σ. If there are enough extenders in $\mathcal{F}^{\mathbb{C}}$ to witness that δ^M is Γ-Woodin in M, then $\mathbb{C}$ is Γ-universal, in the sense that every pfs premouse in V_δ^M that has a Γ iteration strategy iterates into some level of $\mathbb{C}$. We shall prove this in Section 8.1.

It is easy to see that a strongly unique strategy has strong hull condensation and normalizes well.

THEOREM 7.2.9. *Let $(M, \in, w, \mathcal{F})$ be a coarse extender premouse, and let Σ witness that M is is strongly uniquely $(\eta, \theta, \mathcal{F})$-iterable; then Σ has strong hull condensation and normalizes well.*

PROOF. Strong hull condensation is immediate. For if $\mathcal{U}$ is by Σ_s and $\mathcal{T}$ is a pseudo-hull of $\mathcal{U}$, then all models of $\mathcal{T}$ are wellfounded, so $\mathcal{T}$ is by Σ_s. Further, if π is the map on last models, then $\Sigma^\pi_{s,\mathcal{U}} = \Sigma_{s,\mathcal{T}}$ because $\Sigma^\pi_{s,\mathcal{U}}$ chooses wellfounded branches, and $\Sigma_{s,\mathcal{T}}$ chooses unique wellfounded branches.

We show now that the complete strategy induced by Σ normalizes well. So let s be by Σ and $\langle \mathcal{T}, \mathcal{U} \rangle$ by Σ_s; we must see that $W(\mathcal{T}, \mathcal{U})$ is by Σ_s. Since Σ is strongly unique, this implies that $V(\mathcal{T}, \mathcal{U})$ is by Σ.

Let $\mathrm{lh}(\mathcal{U}) = \mu + 1$, and for $\gamma \le \mu$ set

$$\mathcal{W}_\gamma = W(\mathcal{T}, \mathcal{U} \restriction \gamma + 1).$$

We show by induction on γ that $\mathcal{W}_\gamma$ is by Σ_s.

$\mathcal{W}_0 = \mathcal{T}$ is by Σ_s. Suppose now that $\mathcal{W}_\gamma$ is by Σ_s, and let

$$\mathcal{W}_{\gamma+1} = W(\mathcal{W}_\nu, \mathcal{W}_\gamma, F),$$

where $F = \sigma_\gamma(E^{\mathcal{U}}_\gamma)$. Since we are in the coarse case, full normalization coincides with embedding normalization, and σ_γ is the identity, but we do not need this. Let $\alpha = \alpha(\mathcal{W}_\nu, \mathcal{W}_\gamma, F)$ an $\beta = \beta(\mathcal{W}_\nu, \mathcal{W}_\gamma, F)$. We assumed that $(w, \mathcal{F})$ is a coherent pair, so α is the least η such that $\mathrm{lh}(E^{\mathcal{W}_\gamma}_\eta) \ge \mathrm{lh}(F)$. We have that $\mathcal{W}_{\gamma+1} \restriction \alpha + 1 = \mathcal{W}_\gamma \restriction \alpha + 1$ is by Σ_s. So it is enough to show by induction that $\mathcal{W}_{\gamma+1} \restriction \alpha + \lambda + 1$ is by Σ_s for all $\lambda < \mathrm{lh}(\mathcal{W}_\eta)$. Clearly, we may assume that λ is a limit ordinal.

The construction of $W(\mathcal{W}_\nu, \mathcal{W}_\gamma, F)$ gives us a tree embedding Φ from $\mathcal{W}_\eta \restriction \beta + \lambda$ into $\mathcal{W}_{\gamma+1} \restriction \alpha + \lambda$ whose u-map satisfies $u(\beta + \xi) = \alpha + 1 + \xi$ for all $\xi < \lambda$. We can use Lemma 7.1.13 to extend Φ. If

$$c = \Sigma_s(\mathcal{W}_{\gamma+1} \restriction \alpha + \lambda),$$

then letting $b = u^{-1}$"c, we can extend Φ to a tree embedding of $(\mathcal{W}_\nu \restriction \beta + \lambda)^\frown b$ to $(\mathcal{W}_{\gamma+1} \restriction \alpha + \lambda)^\frown c$, and since pseudo-hulls of normal trees by Σ are by Σ,

$$b = \Sigma_s(\mathcal{W}_\nu \restriction \beta + \lambda).$$

So $b = [0, \beta + \lambda]_{W_\nu}$, so $c = [0, \alpha + \lambda]_{W_{\gamma+1}}$, as desired.

Now suppose λ is a limit ordinal. We want to see $\mathcal{W}_\lambda$ is by Σ_s. Let $\mathcal{W} = W(\mathcal{T}, \mathcal{U} \restriction \lambda)$ and let $a = \Sigma_s(\mathcal{W})$. The results of Section 6.6 go through in the coarse case, as we explained at the end of that section. Adopting the notation of Section 6.6, let

$$b = \mathrm{br}_{\mathcal{U}}^{\mathcal{W}}(a)$$

be the cofinal branch of $\mathcal{U} \restriction \lambda$ determined by a. So $W(\mathcal{T},\mathcal{U})^\frown a$ is an initial segment of $\mathcal{W}_b$, and is by Σ_s.

We show by induction on ξ that $\mathcal{W}_b \restriction \xi+1$ is by Σ_s, the proof being like the one in the successor case above. Let $\eta = \mathrm{lh}(W(\mathcal{T},\mathcal{U} \restriction \lambda))$. Let

$$\Phi = \Phi_{0,b} \colon \mathcal{T} \to \mathcal{W}_b$$

be the "putative tree embedding" we get from the construction of $\mathcal{W}_b$. (We do not know yet that the models of $\mathcal{W}_b$ are wellfounded, so Φ may not be a true tree embedding.) Let $u = u^\Phi$, and let τ be such that

$$\eta = \sup_{\gamma<\lambda} \alpha_\gamma = u(\tau),$$

so that $\tau < \mathrm{lh}(\mathcal{T})$, and $\tau = m(b, \mathcal{T}, \mathcal{U} \restriction \lambda)$. We show by induction on ξ that if $\eta \leq \xi < \mathrm{lh}(\mathcal{W}_b)$, then $\mathcal{W}_b \restriction (\xi+1)$ is by Σ_s. This is trivial if ξ is a successor ordinal, because Σ_s cannot lose at a successor step. But if ξ is a limit, then we have

$$\xi = u(\bar{\xi})$$

for some limit ordinal $\bar{\xi} < \mathrm{lh}(\mathcal{T})$. Moreover, $\xi - \eta$ is contained in $\mathrm{ran}(u)$. Thus by Lemma 7.1.13, letting $c = \Sigma_s(\mathcal{W}_b \restriction \xi)$ and $d = [0, \bar{\xi})_T = \Sigma_s(\mathcal{T} \restriction \bar{\xi})$, we have $u``d \subseteq c$. It follows that $c = [0, \xi)_{W_b}$, so that $\mathcal{W}_b \restriction \xi+1$ is by Σ_s, as desired.

So $\mathcal{W}_b$ is by Σ_s. But there is an embedding of $\mathcal{M}_b^{\mathcal{U}}$ into the last model of $\mathcal{W}_b$, so $\mathcal{M}_b^{\mathcal{U}}$ is wellfounded, so $b = \Sigma_s(\langle \mathcal{T}, \mathcal{U} \restriction \lambda \rangle)$, that is $b = [0, \lambda)_U$, and $\mathcal{W}_\lambda = \mathcal{W}_b$ is by Σ_s, as desired.

This shows that $W(\mathcal{T},\mathcal{U})$ is by Σ_s. Let π be the embedding normalization map from the last model of $\mathcal{U}$ to the last model of $W(\mathcal{T},\mathcal{U})$. ($\pi$ is the identity in this coarse case, but we do not need that.) Then $\Sigma^\pi_{s^\frown \langle W(\mathcal{T},\mathcal{U}) \rangle} = \Sigma_{s^\frown \langle \mathcal{T},\mathcal{U} \rangle}$ because the π-pullback strategy picks wellfounded branches, and these are unique. ⊣

Let us assume AD^+ for a while. Let (M, Σ) be a coarse Γ-Woodin pair. M is uncountable, because it incorporates the trees S and T. Σ acts on countable iteration trees based on V_δ^M, which is countable, but if we think of Σ as moving only V_α^M for some $\alpha < \omega_1^V$, then there will no longer be unique wellfounded branches, just unique C_{Γ_1}-full branches. To get equivalent (3) of Lemma 7.2.7, we really needed to let $i_b^{\mathcal{U}}$ act on S and T. This showed up in the proof of Lemma 7.2.5.

In the AD^+ context it is natural to be working with countable base models. This leads us to

DEFINITION 7.2.10. A *coarse extender pair* is a pair $((N, w, \mathcal{F}), \Sigma)$ such that $(N, w, \mathcal{F})$ is a coarse extender premouse, and for some $\theta \geq \omega_1, \Sigma$ is a (θ, θ)-iteration strategy for $(N, w, \mathcal{F})$ that normalizes well and has strong hull condensation.

We can reformulate some of the results of this section as follows:

THEOREM 7.2.11. *Assume* AD^+, *and let* $A \subseteq \mathbb{R}$ *be Suslin and co-Suslin; then there is a coarse extender pair* $((M, \in, w, \mathcal{F}), \Sigma)$ *such that*

(a) M *is countable,*
(b) Σ *is an* (ω_1, ω_1) *iteration strategy for* $(M, \in, w, \mathcal{F})$,
(c) $\delta(w)$ *is* $\mathcal{F}$*-Woodin in* M, *and*
(d) $((M, \in, w, \mathcal{F}), \Sigma)$ *captures* A.

PROOF. Let (N, Σ) be a coarse Γ-Woodin pair, where $A \in \Gamma \cap \check{\Gamma}$. Let $\delta = \delta^N$, $w = w^M$, and let $\mathcal{F}$ be the set of all nice extenders $E \in V_\delta^N$ such that for $\eta = \mathrm{lh}(E)$, $i_E(w) \cap V_{\eta+1}^{\mathrm{Ult}(N,E)} = w \cap V_{\eta+1}^{\mathrm{Ult}(N,E)}$. Let $\delta < \alpha < \omega_1^V$ be such that $V_\alpha^N \models \mathsf{ZFC}$. The results above yield at once that $((V_\alpha^N, \in, w, \mathcal{F}), \Sigma)$ is a coarse extender pair satisfying (a)-(d). ⊣

In Chapter 9 we shall introduce *coarse strategy pairs*. See Definition 9.4.14. These are the appropriate background universes for a strategy mouse construction. The analog of Theorem 7.2.11 is Theorem 9.4.16, according to which AD^+ implies that every Suslin-co-Suslin set is captured by a coarse strategy pair.

7.3. Strong unique iterability from UBH

We now look at consequences of the Unique Branches Hypothesis for for the existence of iteration strategies. The value of these iterability proofs that assume UBH is an open question. Perhaps they will play an important role in the ultimate construction of iteration strategies for mice with very large cardinals, perhaps not. Perhaps in the end UBH will be simply be a corollary of strategy-existence theorems that are proved without assuming it. This is closer to the way inner model theory has developed so far. In any case, we devote this section to describing some consequences of UBH for iterability.

DEFINITION 7.3.1. Let $\mathcal{F}$ be a set or class of extenders; then $\mathcal{F} - \mathsf{UBH}$ holds iff whenever $\mathcal{T}$ is a normal $\mathcal{F}$-tree on V, then $\mathcal{T}$ has at most one cofinal, wellfounded branch.

In particular, nice-UBH is the restriction of UBH to nice, normal trees. Every nice tree of limit length has the same cofinal branches as its normal companion, so nice-UBH is equivalent to UBH for arbitrary nice trees.

Woodin has observed that a Löwenheim-Skolem argument shows that $\mathcal{F}$-UBH follows from $\mathcal{F}$-UBH for countable trees.

Although $\mathcal{F}$-UBH involves only normal trees, we can show

LEMMA 7.3.2. *Let* $\mathcal{F}$ *be a class of nice extenders, and suppose that* $\mathcal{F}$-UBH *holds; then whenever* s *is a stack of* $\mathcal{F}$*-trees with last tree* $\mathcal{U}$, *then* $\mathcal{U}$ *has at most one cofinal, wellfounded branch.*

PROOF. Suppose first that we have a stack $s = \langle \vec{\mathcal{T}}, \mathcal{U} \rangle$ of length two. Let b and c be cofinal, wellfounded branches of $\mathcal{U}$. Let $\mathcal{W} = W(\mathcal{T}, \mathcal{U})$, and let

$$a = \mathrm{br}(b, \mathcal{T}, \mathcal{U})$$

and

$$d = \mathrm{br}(c, \mathcal{T}, \mathcal{U}).$$

It will be enough to show that $a = d$, for then $b = c$ by the results of Section 6.6. We have assumed $\mathcal{F}$-UBH for normal trees, so it is enough to show that $\mathcal{M}_a^{\mathcal{W}}$ and $\mathcal{M}_d^{\mathcal{W}}$ are wellfounded. The situation is symmetric, so it is enough to show $\mathcal{M}_a^{\mathcal{W}}$ is wellfounded. So suppose toward contradiction that

$$\mathcal{M}_a^{\mathcal{W}} \text{ is illfounded.}$$

Adopting our usual notation for embedding normalization, let τ be such that $u_{0,b}(\tau) = \mathrm{lh}(W(\mathcal{T}, \mathcal{U}))$. We have then that

$$\mathcal{M}_a^{\mathcal{W}} = \mathrm{Ult}(\mathcal{M}_\tau^{\mathcal{T}}, E_b),$$

where E_b is the extender of b.

We need some elementary covering properties of the models in $\mathcal{T}$. For $\eta < \mathrm{lh}(\mathcal{T})$, let

$$\nu_\eta = \sup(\{\mathrm{lh}(G) \mid G \text{ is used in } [0, \eta)_T\}).$$

It is clear that ν_η is either inaccessible or a limit of inaccessibles in $\mathcal{M}_\eta^{\mathcal{T}}$.

CLAIM 7.3.3. *Let $X \subseteq \mathcal{M}_\eta^{\mathcal{T}}$ be countable in V; then there is a $Y \supseteq X$ such that $Y \in \mathcal{M}_\eta^{\mathcal{T}}$ and $\mathcal{M}_\eta^{\mathcal{T}} \models |Y| \leq \nu_\eta$.*

PROOF. There are $f_n \in V$, for $n < \omega$, such that every $x \in X$ is of the form $i_{0,\eta}(f_n)(a)$, for some $a \in [\nu_\eta]^{<\omega}$. So we can take $Y = \{i_{0,\eta}(f_n)(a) \mid n < \omega \text{ and } a \in [\nu_\eta]^{<\omega}\}$. ⊣

CLAIM 7.3.4. *Suppose $\mathcal{M}_\eta \models$ "θ is regular but not measurable"; then θ has uncountable cofinality in V.*

PROOF. We prove this by induction on η. It is trivial for $\eta = 0$. Suppose we have it for $\eta < \lambda$, where λ is a limit ordinal. Let θ be regular but not measurable in $\mathcal{M}_\lambda$, and let $\theta = i_{\alpha,\lambda}(\beta)$. By induction, $\mathrm{cof}^V(\beta) > \omega$. But $i_{\alpha,\lambda}$ is continuous at β, because β is regular but not measurable in $\mathcal{M}_\alpha$. Thus $\mathrm{cof}^V(\theta) > \omega$.

Finally, suppose the claim holds at η, and let θ be regular but not measurable in $\mathcal{M}_{\eta+1}$. Let $\nu = \mathrm{lh}(E_\eta^{\mathcal{T}}) = \nu_{\eta+1}$. If $\theta < \nu$, then the agreement between $\mathcal{M}_\eta$ and $\mathcal{M}_{\eta+1}$ implies θ is regular but not measurable in $\mathcal{M}_\eta$, so $\mathrm{cof}^V(\theta) > \omega$ by induction. If $\theta = \nu$, then θ is regular but not measurable in $\mathcal{M}_\eta$ by our hypothesis on the extenders in $\vec{F}$, so again $\mathrm{cof}^V(\theta) > \omega$. Finally, if $\theta > \nu$ and $\mathrm{cof}^V(\theta) = \omega$, then θ is singular in $\mathcal{M}_{\eta+1}$ by Claim 7.3.3, contradiction. ⊣

Now let $\nu = \nu_{\tau+1} = \mathrm{lh}(E_\tau^{\mathcal{T}})$. We have that $i_b^{\mathcal{U}}(\nu) \geq \delta(\mathcal{U})$, for if not, then $u_{0,b}(\tau) < \lambda$. (See Proposition 6.6.1, and the discussion near it.) But ν is regular and not measurable in $\mathcal{M}_0^{\mathcal{U}} = \mathcal{M}_\infty^{\mathcal{T}}$, so $i_b^{\mathcal{U}}$ is continuous at ν. Moreover, $\mathrm{cof}^V(\nu) > \omega$, while $\mathrm{cof}^V(\delta(U)) = \omega$ because b is not the only cofinal branch of $\mathcal{U}$. Thus we can fix ρ such that

$$\rho < \nu \text{ and } i_b^{\mathcal{U}}(\rho) > \delta(\mathcal{U}).$$

Since the measures in E_b all concentrate on bounded subsets of ρ, we also have

$$\nu_\tau \leq \rho.$$

Let us fix a witness to the illfoundedness of $\mathrm{Ult}(\mathcal{M}_\tau^{\mathcal{T}}, E_b)$, namely $f_n \in \mathcal{M}_\tau$ and $a_n \in [\delta(\mathcal{U})]^{<\omega}$ such that $\pi(f_{n+1})(a_{n+1}) \in \pi(f_n)(a_n)$ for all n, where

$$\pi\colon \mathcal{M}_\tau \to \mathrm{Ult}(\mathcal{M}_\tau^{\mathcal{T}}, E_b)$$

is the canonical embedding. By Claim 7.3.3, we can cover $\{f_n \mid n < \omega\}$ by a set $Y \in \mathcal{M}_\tau^{\mathcal{T}}$ such that $|Y| \leq \rho$ in $\mathcal{M}_\tau^{\mathcal{T}}$. Let $Y \subseteq N$, where N is a rank initial segment of $\mathcal{M}_\tau^{\mathcal{T}}$, and let P be the transitive collapse of $\mathrm{Hull}^N(Y \cup \rho)$. Letting g_n be the collapse of f_n, we see that

$$\mathrm{Ult}(P, E_b) \text{ is illfounded},$$

as witnessed by the g_n's and a_n's. But $\mathcal{M}_0^{\mathcal{U}}$ agrees with $\mathcal{M}_\tau^{\mathcal{T}}$ up to ν, so

$$P \in \mathcal{M}_0^{\mathcal{U}}.$$

Further, $\mathrm{Ult}(P, E_b)$ embeds into $i_b^{\mathcal{U}}(P)$, so $i_b^{\mathcal{U}}(P)$ is illfounded. But $i_b^{\mathcal{U}}(P)$ is wellfounded in $\mathcal{M}_b^{\mathcal{U}}$, so $\mathcal{M}_b^{\mathcal{U}}$ is illfounded, contradiction.

This takes care of the case that s has length two. Given an arbitrary finite stack $s = t^\frown\mathcal{U}$, with t having last model N, set $\mathcal{T} = W(t)$. Because we are in the coarse case, $\mathcal{T}$ has last model N. But $\mathcal{T}$ is normal, so the proof above shows that $\mathcal{U}$ has at most one cofinal, wellfounded branch.

One can prove the full lemma for arbitrary stacks using the normalizability of such stacks. This is shown in [57]. ⊣

We shall see below that one cannot drop the niceness hypothesis in Lemma 7.3.2 completely.

We turn to branch existence. The main results here come from [30]. That paper shows that nice-UBH implies that every countable, normal tree on V has a cofinal wellfounded branch. Combining it with Lemma 7.3.2, we get

LEMMA 7.3.5. *Let $\mathcal{F}$ be a class of nice extenders, and suppose that $\mathcal{F}$-UBH holds; then V is strongly uniquely $(\omega_1, \omega_1, \vec{F})$- iterable.*

For iterations of uncountable length, we need UBH in the appropriate collapse extension.

THEOREM 7.3.6 (Folklore). *Let $\mathcal{F}$ be a class of nice extenders such that $\theta <$ crit(G) for all $G \in \mathcal{F}$. Suppose that $\mathcal{F}$-UBH holds in $V[G]$, where G is $\mathrm{Col}(\omega, \theta)$ generic over V; then V is strongly uniquely $(\theta^+, \theta^+, \mathcal{F})$-iterable.*

PROOF SKETCH. Given $\mathcal{T}$ in V of limit length $< \theta^+$, we can regard $\mathcal{T}$ as a tree on $V[G]$ because $\theta < \kappa$. In $V[G]$, $\mathcal{T}$ is countable, so by UBH in $V[G]$ and [30] in $V[G]$, it has a unique cofinal, wellfounded branch. Because the collapse is homogeneous, this branch is in V. ⊣

In one situation, UBH in V implies instances of UBH in $V[G]$:

THEOREM 7.3.7 (Woodin). *Let δ be Woodin, and assume that $\mathcal{F}$-UBH holds, where $\mathcal{F}$ is a class of extenders with all critical points $> \delta$. Let $\mathcal{T}$ be a normal $\mathcal{F}$-tree, with $|\mathcal{T}| < \delta$, and let G be V-generic for a poset of size $< \delta$; then $V[G] \models$ "$\mathcal{T}$ has at most one cofinal, wellfounded branch".*

PROOF SKETCH. We may assume G is countable in $V[H]$, where H is V-generic for the countable stationary tower $\mathbb{Q}_{<\delta}$. Suppose toward contradiction that b and c are distinct cofinal branches of $\mathcal{T}$ in $V[G]$. $\mathcal{T}$ can be regarded as a tree on $V[H]$, and b and c are still wellfounded when it is regarded this way.

But let $\pi \colon V \to M = \mathrm{Ult}(V, H)$ be the generic elementary embedding. Since M is closed under countable sequences in $V[H]$, $\pi\mathcal{T} \in M$, and one can check that b and c are wellfounded as branches of $\pi\mathcal{T}$. (Essentially the same functions into the ordinals are used in forming $\mathcal{M}_b^{\mathcal{T}}$ and $\mathcal{M}_b^{\pi\mathcal{T}}$, for example.) One can also check that in M, $\pi\mathcal{T}$ is a $\pi(\mathcal{F})$-tree. Thus $\pi(\mathcal{F})$-UBH fails in M, contrary to the elementarity of π. ⊣

At supercompacts, we catch our tail:

THEOREM 7.3.8 (Woodin). *Suppose that κ is supercompact, $\mathcal{F}$ is a class of nice extenders such that* crit$(G) > \kappa$ *for all $G \in \mathcal{F}$, and $\mathcal{F}$-UBH holds; then for all θ, V is strongly uniquely $(\theta, \theta, \mathcal{F})$-iterable.*

PROOF. Given s an $\mathcal{F}$-stack on V with last normal tree $\mathcal{T}$, with $s \in V_\theta$, let $j : V \to M$, $\mathrm{crit}(j) = \kappa$, $j \upharpoonright V_\theta \in M$. In M, the lifted stack js has size $< j(\kappa)$, and all its critical points are above $j(\kappa)$. So by Theorems 7.3.6 and 7.3.7, $j\mathcal{T}$ has a cofinal wellfounded branch b in M. (Note $j(\kappa)$ is a limit of Woodin cardinals in M.) The copy map $\sigma : M_b^{\mathcal{T}} \to M_b^{j\mathcal{T}}$ witnesses that b is wellfounded branch of $\mathcal{T}$. ⊣

In the theory of strategy mice, it is important that strategies be moved to their tails by their own iteration maps. We call this property *pushforward consistency*. More precisely, we would like to know that if $i \colon M \to N$ comes from a stack of trees $\vec{\mathcal{T}}$ by Σ, then $i(\Sigma \cap M) = \Sigma_{\vec{\mathcal{T}},N} \cap N$. We shall obtain this from the corresponding property of coarse strategies Σ such that Σ witnesses that V is strongly uniquely $(\lambda, \theta, \mathcal{F})$-iterable.

LEMMA 7.3.9. *Let $\mathcal{F}$ be a class of nice extenders, and let Σ witness that V is strongly uniquely $(\lambda, \theta, \mathcal{F})$-iterable. Suppose that $i \colon V \to N$ comes from a stack of trees $\mathcal{T}$ by Σ; then $i(\Sigma) = \Sigma_{\mathcal{T},N} \cap N$.*

PROOF. Both $i(\Sigma)$ and $\Sigma_{\mathcal{T},N}$ choose wellfounded branches. Since these are unique (in V!), the two strategies cannot disagree. ⊣

Some failures of UBH. The remainder of this section contains examples and results related to unique iterability that are somewhat removed from the main line of this book.

First, there are some counterexamples to forms of UBH to keep in mind when considering strong unique iterability for stacks on V. The counterexamples involve extenders overlapping Woodin cardinals, and thus do not apply to the Γ-Woodin models of Definition 7.2.3, which have no such extenders. They involve stacks of trees that are not nice.

If we allow our trees to use extenders that do not have ω-closed ultrapowers in the models where they appear, then Woodin has shown in [84] that there are in fact normal trees of length ω on V having distinct wellfounded branches. (His construction requires a supercompact cardinal.) The construction relies heavily on the non-ω-closure, and it is quite plausible to the author that normal trees on V using only extenders that are ω-closed in the models they are taken from can have at most one cofinal wellfounded branch.

When one moves to stacks of normal trees, ω-closure is no longer enough to avoid counterexamples, as Woodin has shown. His example builds on one due to Neeman and the author. In [40], they construct a stack $\vec{\mathcal{U}} = \langle \mathcal{U}_0, \mathcal{U}_1 \rangle$ of normal iteration trees on V such that for some strong limit cardinal δ of cofinality ω,

(i) $\mathcal{U}_0 = \langle F \rangle$, where $\operatorname{lh} F = \operatorname{strength}(F) = \delta$,
(ii) $\mathcal{U}_1$ is an alternating chain on $V_\delta = V_\delta^{\mathrm{Ult}(V,F)}$, with distinct branches b and c, and
(iii) both $\mathcal{M}_b^{\mathcal{U}_1}$ and $\mathcal{M}_c^{\mathcal{U}_1}$ are wellfounded.

The key here is that because $V_\delta = V_\delta^{\mathrm{Ult}(V,F)}$, both $i_b^{\mathcal{U}_1}$ and $i_c^{\mathcal{U}_1}$ can be extended so as to act on V, and the construction arranges that $i_b(F) = i_c(F)$. But then $\mathcal{M}_b^{\mathcal{U}_1} = \mathrm{Ult}(V, i_b(F)) = \mathrm{Ult}(V, i_c(F)) = \mathcal{M}_c^{\mathcal{U}_1}$. So not only are b and c both wellfounded as branches of $\vec{\mathcal{U}}$, in fact $\mathcal{M}_b^{\mathcal{U}_1} = \mathcal{M}_c^{\mathcal{U}_1}$!

In the example above, $\mathrm{Ult}(V,F)$ is not closed under ω-sequences. However, Woodin showed that under stronger large cardinal assumptions, we can modify the example so as to get a stack of length 2 of "almost nice" trees on V. Namely, suppose we start with μ a normal measure on δ_0, where δ_0 is Woodin, and F_0 an extender with length = strength equal to δ_0. Let $\mathcal{I}$ be a linear iteration of μ of length ω, with direct limit model N. Let F and δ be the images in N of F_0 and δ_0. Then let $\mathcal{U}_0$ be the normal tree determined by $\mathcal{I}^\frown\langle F \rangle$, so that the last model of $\mathcal{U}_0$ is $M = \mathrm{Ult}(V,F)$. and let $\mathcal{U}_1$ be an alternating chain on M with branches b and c which, when acting on N, satisfy $i_b(F) = i_c(F)$. The construction of [40] gives us this $\mathcal{U}_2$; we only need $\mathrm{cof}(\delta) = \omega$ to hold in V, it need not hold in N. Again we have $\mathcal{M}_b^{\vec{\mathcal{U}}} = \mathcal{M}_c^{\vec{\mathcal{U}}}$, so both branches are wellfounded. But now $\vec{\mathcal{U}}$ is satisfies all the requirements of niceness, with the exception that $\mathrm{lh}(F_0)$ is measurable in M.

Remark 7.3.10. We saw in Lemma 7.3.2 that this apparently small departure from niceness is essential.

In both examples, the branches b and c are not equally good. For example, consider the first example. Let E_b and E_c be the two branch extenders. Since our chain was constructed by the one-step method, exactly one of $\mathrm{Ult}(V, E_b)$ and $\mathrm{Ult}(V, E_c)$ is wellfounded. But in $\langle \mathcal{U}_0, \mathcal{U}_1{}^\frown b\rangle$ and $\langle \mathcal{U}_0, \mathcal{U}_1{}^\frown c\rangle$, these branch extenders are applied to $\mathrm{Ult}(V, F)$ rather than V. We have taken advantage of non-normality to hide the difference between b and c. If we normalize, the difference shows up:

$$W(\mathcal{U}_0, \mathcal{U}_1{}^\frown b) = \mathcal{U}_1{}^\frown b^\frown i_b^{\mathcal{U}_1}(F)$$

and

$$W(\mathcal{U}_0, \mathcal{U}_1{}^\frown c) = \mathcal{U}_1{}^\frown c^\frown i_c^{\mathcal{U}_1}(F).$$

Here $\mathcal{U}_1{}^\frown b$ and $\mathcal{U}_1{}^\frown c$ are acting on V, where only one of the two is actually an iteration tree, in that all its models are wellfounded.

Strategy extension. Our analysis of the counterexample above suggests that we might iterate V for finite stacks by simply choosing branches that are consistent with the iteration tree we get by normalizing. We shall show now that in fact *any* iteration strategy with strong hull condensation that acts on normal trees can be extended in this way.

In the fine-structural context, this was first proved independently by Schlutzenberg and the author. Schlutzenberg went on to prove a stronger form of the theorem, in which the extended strategy acts on infinite stacks. (See [57].) The proof of Schlutzenberg's stronger form requires significant new ideas. The construction in the finite-stack case is at bottom the same as the one we are about to give in a coarse setting. The details are simpler in the coarse case, however, because our assumptions will imply embedding normalization coincides with full normalization, and hence various maps are the identity that would not otherwise be.

We shall not actually use Theorem 7.3.11 anywhere later in the book. Instead Theorem 7.2.9 will be our source for coarse iteration strategies that normalize well and have strong hull condensation.

THEOREM 7.3.11. *Let $M \models \mathsf{ZFC} +$ "$\vec{F}$ is coarsely coherent," and let Σ be a $(1, \theta, \vec{F})$ iteration strategy for M. Suppose that Σ has strong hull condensation; then there is a unique $(\omega, \theta, \vec{F})$ strategy Σ^* such that*

(a) $\Sigma \subseteq \Sigma^*$, and

(b) Σ^* normalizes well, and has strong hull condensation.

Remark 7.3.12. Let s be a stack of length ω all of whose finite initial segments are by Σ^*. We do not demand that the direct limit along s be wellfounded, as would be required if Σ^* were to be a complete strategy. Adding this demand would take us into the difficulties that Schlutzenberg overcame in the fine-structural case.

Remark 7.3.13. We do not assume in Theorem 7.3.11 that Σ witnesses strong unique iterability. Coarse coherence simplifies a few things, but could probably be weakened or avoided altogether.

PROOF. Since $\vec{F}$ is coarsely coherent, quasi-normalization, embedding normalization, and full normalization coincide. In particular, if $\langle \mathcal{T}, \mathcal{U} \rangle$ is an $\vec{F}$-stack on M, with Q being the last model of $\mathcal{T}$ and N the last model of $\mathcal{U}$, and $W(\mathcal{T}, \mathcal{U})$ exists, then $W(\mathcal{T}, \mathcal{U})$ also has last model N. The embedding normalization map $\sigma \colon N \to N$ is the identity, and the last t-map of the extended tree embedding from $\mathcal{T}$ into $\mathcal{U}$ is equal to the main branch embedding $i^{\mathcal{U}} \colon Q \to N$.

We begin by extending Σ to Σ_2, acting on stacks of length ≤ 2. Let $\langle \mathcal{T}, \mathcal{U} \rangle$ be a 2-stack of $\vec{F}$-trees, with $\mathcal{T}$ by Σ. We define $\Sigma_2(\langle \mathcal{T}, \mathcal{U} \rangle)$ by induction on $\mathrm{lh}(\mathcal{U})$, maintaining by induction that $W(\mathcal{T}, \mathcal{U})$ is by Σ. Let us write

$$\mathcal{W}_\gamma = W(\mathcal{T}, \mathcal{U} \restriction \gamma + 1)$$

as before.

Suppose that $\mathcal{W}_\gamma$ is by Σ; we wish to show that $\mathcal{W}_{\gamma+1}$ is by Σ. For let η be such that

$$\mathcal{W}_{\gamma+1} = W(\mathcal{W}_\eta, F),$$

where $F = E^{\mathcal{U}}_\gamma$. Let $\alpha = \alpha(\mathcal{W}_\eta, \mathcal{W}_\gamma, F)$ an $\beta = \beta(\mathcal{W}_\eta, \mathcal{W}_\gamma, F)$. We have that $\mathcal{W}_{\gamma+1} \restriction \alpha + 1 = \mathcal{W}_\gamma \restriction \alpha + 1$ is by Σ. So it is enough to show by induction that $\mathcal{W}_{\gamma+1} \restriction \alpha + \lambda + 1$ is by Σ for all $\lambda < \mathrm{lh}(\mathcal{W}_\eta)$. Clearly, we may assume that λ is a limit ordinal.

But now the construction of $W(\mathcal{W}_\eta, \mathcal{W}_\gamma, F)$ gives us a tree embedding Φ from $\mathcal{W}_\eta \restriction \beta + \lambda$ into $\mathcal{W}_{\gamma+1} \restriction \alpha + \lambda$ whose u-map satisfies $u(\beta + \xi) = \alpha + 1 + \xi$. We can use Lemma 7.1.13 to extend Φ. To repeat its proof: if

$$c = \Sigma(\mathcal{W}_{\gamma+1} \restriction \alpha + \lambda),$$

then letting $b = u^{-1}``c$, we can extend Φ to a tree embedding of $(\mathcal{W}_\eta \restriction \beta + \lambda)^\frown b$ to $(\mathcal{W}_{\gamma+1} \restriction \alpha + \lambda)^\frown c$, and since pseudo-hulls of normal trees by Σ are by Σ,

$$b = \Sigma(\mathcal{W}_\eta \restriction \beta + \lambda).$$

So $b = [0, \beta + \lambda]_{W_\eta}$, so $c = [0, \alpha + \lambda]_{W_{\gamma+1}}$, as desired.

Now suppose $\mathcal{U}$ of limit length λ. It is enough show that there is a unique cofinal branch b of $\mathcal{U}$ such that setting

$$\mathcal{W}_b = W(\mathcal{T}, \mathcal{U}^\frown b),$$

$\mathcal{W}_b$ is by Σ. For then we can set

$$\Sigma_2(\langle \mathcal{T}, \mathcal{U} \rangle) = b,$$

and our induction hypothesis remains true at $\lambda + 1$. To show this, let $\mathcal{W} = W(\mathcal{T}, \mathcal{U})$ and let $a = \Sigma(\mathcal{W})$. Adopting the notation of Section 6.6, let

$$b = \mathrm{br}^{\mathcal{W}}_{\mathcal{U}}(a)$$

be the cofinal branch of $\mathcal{U}$ determined by a. So $W(\mathcal{T},\mathcal{U})^\frown a$ is an initial segment of $\mathcal{W}_b$, and is by Σ. One can show by induction on ξ that $\mathcal{W}_b \restriction \xi+1$ is by Σ. The proof is identical to the corresponding argument in the proof of Lemma 7.2.9, so we omit it.

This completes the definition of Σ_2 on stacks of length ≤ 2. Clearly, normalizations of stacks by Σ_2 are by Σ. Suppose now we have Σ_n where $n \geq 2$, and

$(*)_n$ whenever $\vec{\mathcal{T}}$ is an $\vec{F}$-stack of length $\leq n$ played by Σ_n, and having last model R, then there is a normal $\mathcal{F}$-iteration tree on V with last model R.

There is then exactly one such $\mathcal{T}$ by Lemma 2.9.12, and we write

$$\mathcal{T} = W(\vec{\mathcal{T}}).$$

We define Σ_{n+1} as follows: if $\vec{\mathcal{T}}^\frown\langle\mathcal{U}\rangle$ is a stack of length $\leq n+1$ played by Σ_{n+1},

$$\Sigma_{n+1}(\vec{\mathcal{T}}^\frown\langle\mathcal{U}\rangle) = \Sigma_2(\langle W(\vec{\mathcal{T}}),\mathcal{U}\rangle).$$

Clearly, Σ_{n+1} is an $\vec{F}$-iteration strategy defined on stacks of length at most $n+1$, extending Σ_n. If $\vec{\mathcal{T}}^\frown\langle U\rangle$ is a stack on V by Σ_{n+1} with last model R, then $\langle W(\vec{\mathcal{T}}),\mathcal{U}\rangle$ is a 2-stack by Σ_2 with last model R, so $W(W(\vec{\mathcal{T}}),\mathcal{U})$ is a normal tree with last model R. Thus $(*)_{n+1}$ holds, and we can go on.

Let

$$\Sigma^* = \bigcup_n \Sigma_n.$$

We now show that Σ normalizes well. For this, the following definition is useful.

DEFINITION 7.3.14. (1) Let $\mathcal{W}$ be a normal iteration tree, and δ a limit ordinal. We say that b is a *δ-branch* of $\mathcal{W}$ iff $\delta = \sup\{lh(E^{\mathcal{W}}_\alpha) \mid \alpha+1 \in b\}$.

(2) Let $\mathcal{W}$ and $\mathcal{U}$ be normal iteration trees, let b be a branch of $\mathcal{U}$ of limit order type (perhaps maximal), and let c be a branch of $\mathcal{W}$ (perhaps maximal). We say that *b fits into c* iff for any extender F used in b, there is an extender G used in c such that $\operatorname{crit}(G) \leq \operatorname{crit}(F) \leq \operatorname{lh}(F) \leq \operatorname{lh}(G)$.

LEMMA 7.3.15. *Let $\mathcal{W}$ and $\mathcal{U}$ be normal iteration trees, and let δ be a limit ordinal; then for any δ-branch c of $\mathcal{W}$, there is at most one δ-branch b of $\mathcal{U}$ such that b fits into c.*

PROOF. Suppose a and b fit into c, where $a \neq b$. We get the zipper pattern, that is F_n's used in a and G_n's used in b such that $\operatorname{crit}(F_n) \leq \operatorname{crit}(G_n) < \nu(F_n) < \operatorname{crit}(F_{n+1}) < \nu(G_n)$. If H is used in c and F_0 fits into H, then G_0 must also fit into H, since it does not fit anywhere else in c. By induction, all the F_n and G_n fit into H. But then $\delta \leq \nu(H)$, contradiction. ⊣

LEMMA 7.3.16. *Let $\langle\mathcal{T},\mathcal{U}\rangle$ be a stack of nice iteration trees on M, and b be a cofinal branch of $\mathcal{U}$; then b fits into* $\operatorname{br}(b,\mathcal{T},\mathcal{U})$.

PROOF. This is clear from the construction, and the fact that the σ-maps of embedding normalization are the identity in this coarse case. See the earlier diagrams of the extender tree of $W(\mathcal{T},\mathcal{U})$. $\dashv$

We show now that all tails of Σ 2-normalize well. So let $\vec{\mathcal{S}}$ be a stack by Σ with last model Q, and let $\langle\mathcal{T},\mathcal{U}\rangle$ be by $\Sigma_{\vec{\mathcal{S}},Q}$ with last model R. We must see that $W(\mathcal{T},\mathcal{U})$ is by $\Sigma_{\vec{\mathcal{S}},Q}$, and that $\Sigma_{\vec{\mathcal{S}}^\frown\langle\mathcal{T},\mathcal{U}\rangle,R} = \Sigma_{\vec{\mathcal{S}}^\frown\langle W(\mathcal{T},\mathcal{U})\rangle,R}$. Here we are making use of the fact that the σ-maps in this coarse case are all the identity.

The proof is by induction on $\mathrm{lh}(\mathcal{U})$, and the harder case is $\mathrm{lh}(\mathcal{U}) = \lambda+1$ for some limit ordinal λ, so let us just handle that case. Let $b = [0,\lambda)_U$, and $\delta = \delta(\mathcal{U})$. Since $\vec{\mathcal{S}}^\frown\langle\mathcal{T},\mathcal{U}\rangle$ is by Σ, we see from the definition of Σ that

$$\mathcal{W}_0 = W(W(\vec{\mathcal{S}}^\frown\langle\mathcal{T}\rangle),\mathcal{U})$$

is the unique normal $\vec{F}$-tree on V with last model $R = \mathcal{M}^{\mathcal{U}}_\lambda$. Moreover $\mathcal{W}_0$ chooses the δ-branch

$$a = \mathrm{br}(b,\mathcal{V},\mathcal{U}) = \Sigma(\mathcal{W}_0 \restriction \eta),$$

where we have set $W(\vec{\mathcal{S}}^\frown\langle\mathcal{T}\rangle) = \mathcal{V}$. Letting

$$c = \mathrm{br}(b,\mathcal{T},\mathcal{U}),$$

and

$$c_1 = \Sigma_2(\langle W(\vec{\mathcal{S}}), W(\mathcal{T},\mathcal{U}\restriction\lambda)\rangle),$$

we must show that $c = c_1$. Setting

$$\mathcal{W}_1 = W(W(\vec{\mathcal{S}}), W(\mathcal{T},\mathcal{U}\restriction\lambda)),$$

we have by our induction hypothesis that $\mathcal{W}_1$ is according to Σ. Because the embedding normalization σ-maps are the identity, the common part model $\mathcal{M}(\mathcal{W}_1) = V^{W(\mathcal{T},\mathcal{U}\restriction\lambda)}_\delta = V^R_\delta$. By our uniqueness lemma for normal $\mathcal{F}$-iterations,

$$\mathcal{W}_1 = \mathcal{W}_0 \restriction \eta,$$

so c_1 fits into $\Sigma(\mathcal{W}_1) = a$. Thus it is enough to see that c also fits into a.

Let $\tau = m(b,\mathcal{T},\mathcal{U})$, and

$$p\colon \mathrm{Ext}(\mathcal{T}) \to \mathrm{Ext}(W(\mathcal{T},\mathcal{U}))$$

be the map on extenders induced by the tree embedding Φ of $\mathcal{T}$ into $W(\mathcal{T},\mathcal{U})$. Suppose F is used in c; we must see that F fits into some H used in a. This is true if F is used in b, since b fits into a. The other possibility is that $F = p(G)$, where $G \in \mathrm{ran}(s^{\mathcal{T}}_\tau)$, so assume that. Let

$$q\colon \mathrm{Ext}(\mathcal{V}) \to \mathrm{Ext}(\mathcal{W}_0)$$

be induced by the tree embedding Ψ of $\mathcal{V}$ into $W(\mathcal{V},\mathcal{U})$, and let $\rho = m(b,\mathcal{V},\mathcal{U})$. Letting E_b be the extender of $i^{\mathcal{U}}_b$, we have that τ is least such that E_b is an extender over $\mathcal{M}^{\mathcal{T}}_\tau$, and ρ is least such that E_b is an extender over $\mathcal{M}^{\mathcal{V}}_\rho$, so that ρ is least such that $\mathcal{M}^{\mathcal{V}}_\rho$ agrees with $\mathcal{M}^{\mathcal{T}}_\tau$ through $\mathrm{dom}(E_b)$. It follows that

$br([0,\tau)_T, W(\vec{\mathcal{S}}), \mathcal{T}\restriction \tau+1) = [0,\rho)_V$, and thus G fits into some K that is used in $[0,\rho)_V$. But then $F = p(G)$ fits into $q(K)$, because t_τ^Φ and t_ρ^Ψ are both E_b-ultrapower maps, so agree with one another on $\mathrm{lh}(K)+1$. (Letting N be the last model of $\mathcal{T}$ and $i^{\mathcal{U}}: N \to R$ the canonical embedding, t_τ^Φ and t_ρ^Ψ agree with the common last t-map $i^{\mathcal{U}}$ of Φ and Ψ this far.) Since $q(K)$ is used in a, we are done.

We shall not give a full proof that Σ has strong hull condensation. To see how it goes, suppose $\Phi\colon \mathcal{T} \to \mathcal{U}$ is an extended tree embedding, where $\mathcal{U}$ is by Σ. Let $\pi\colon N \to P$ be its last t-map, where these are the last models of $\mathcal{T}$ and $\mathcal{U}$. We must see that $\Sigma_{\mathcal{T},N} = \Sigma^\pi_{\mathcal{U},P}$. Let $\mathcal{V}$ be of limit length and by both strategies. Now $\Sigma_{\mathcal{T},N}(\mathcal{V})$ is determined by $\Sigma(W(\mathcal{T},\mathcal{V}))$, and $\Sigma_{\mathcal{U},P}(\pi\mathcal{V})$ is determined by $\Sigma(W(\mathcal{U},\pi\mathcal{V}))$. Using Φ, we can obtain a tree embedding from $W(\mathcal{T},\mathcal{V})$ into $W(\mathcal{U},\pi\mathcal{V})$. We can then use the fact that Σ condenses well on normal trees to show that $\Sigma_{\mathcal{T},N}(\mathcal{V}) = \Sigma_{\mathcal{U},P}(\pi\mathcal{V})$. ⊣

This gives us a result on strong unique iterability that does not require a supercompact.

THEOREM 7.3.17. *Let $\vec{F}$ be coarsely coherent, and suppose that V is strongly uniquely $(1,\theta,\vec{F})$- iterable; then V is strongly uniquely $(\omega,\theta,\vec{F})$-iterable. Moreover, letting Σ be the complete strategy that witnesses this,*

(a) Σ normalizes well and has strong hull condensation, and

(b) if s is a stack of length ω of countable normal trees on V with last models $M_i(s)$, then the direct limit of the $M_i(s)$ under the iteration maps of s is wellfounded.

PROOF. By the first part of the proof of Theorem 7.3.11, we have a strategy Σ witnessing that V is $(\omega,\theta,\vec{F})$-iterable. Our hypothesis implies $\vec{F}$-UBH, so by Lemma 7.3.2, Σ witnesses strong uniqueness.

That Σ normalizes well and has strong hull condensation follows from Theorem 7.2.9. Item (b) in the conclusion comes from the branch existence arguments of [30]. Note for example that each $\mathcal{T}_i(s)$ is continuously illfounded off the branches it chooses. ⊣

7.4. Fine strategies that normalize well

Next, we show that if Σ^* is an iteration strategy for a coarse N^* that quasi-normalizes well, then the strategies for premice induced by Σ^* via a full background extender construction also quasi-normalize well. The reason is simply that quasi-normalization commutes with our conversion method. The proof of that is like the proof that quasi-normalization commutes with copying given in Theorem 6.8.1, but there is more to it because in addition to copying, we are passing to resurrected background extenders. Indeed, this gap led to the resurrection consistency problem, which we have solved by moving to pfs premice, and to the background coherence problem, which we have solved by moving to plus trees and quasi-normalization. With these changes to the basic definitions, the relevant diagrams now commute as they should, and our job here is just to verify that.

THEOREM 7.4.1. *Let N^* be a coarse premouse, Σ^* be a $\langle\lambda,\theta\rangle$ iteration strategy for N^* that quasi-normalizes well, and $c = \langle M,\pi,P,\mathbb{C},N^*\rangle$ be a conversion stage; then the induced strategy $\Omega(c,\Sigma^*)$ for M quasi-normalizes well.*

Remark 7.4.2. We believe that the proof of Theorem 7.4.1 works even if the construction $\mathbb{C}$ is allowed to use extenders that are not nice, so that embedding normalization does not coincide with full normalization at the background level. This just means that certain embeddings are no longer the identity, and hence must be given names in the proof to follow.

PROOF. We must show that all tails Σ_s of Σ 2-normalize well. This reduces at once to the case that s is empty, so we assume that.

Let $\langle\mathcal{T},\mathcal{U}\rangle$ be a maximal stack of plus trees on M of length two, and let

$$\langle\mathcal{T}^*,\mathcal{U}^*\rangle = \text{lift}(\langle\mathcal{T},\mathcal{U}\rangle,c)_0$$

be the converted stack on N^*. It suffices to show that $V(\mathcal{T},\mathcal{U})$ lifts to an initial segment of a quasi-normalization $\mathcal{V}^*$ of $\langle\mathcal{T}^*,\mathcal{U}^*\rangle$.[8]

The quasi-normalization $V(\mathcal{T},\mathcal{U})$ has associated to it plus trees $\mathcal{W}_\gamma$ on M[9] , for $\gamma < \text{lh}(\mathcal{U})$, and extended tree embeddings

$$\Phi_{\eta,\gamma}\colon \mathcal{W}_\eta \to \mathcal{W}_\gamma$$

defined when $\eta <_U \gamma$. The components of $\Phi_{\eta,\gamma}$ are

$$\Phi_{\eta,\gamma} = \langle u_{\eta,\gamma}, v_{\eta,\gamma}, \langle s_\nu^{\eta,\gamma} \mid \nu\in\text{dom}(u_{\eta,\gamma})\rangle, \langle t_\nu^{\eta,\gamma} \mid \nu\in\text{dom}(u_{\eta,\gamma})\rangle\rangle.$$

So $u_{\eta,\gamma}$ maps an initial segment of $\text{lh}(\mathcal{W}_\eta)$ to $\text{lh}(\mathcal{W}_\gamma)$, and $t_\nu^{\eta,\gamma}$ is a perhaps partial map from $\mathcal{M}_\nu^{\mathcal{W}_\eta}$ to $\mathcal{M}_{u_{\eta,\gamma}(\nu)}^{\mathcal{W}_\gamma}$. We have also

$$R_\gamma = \mathcal{M}_{z(\gamma)}^{\mathcal{W}_\gamma} = \text{last model of } \mathcal{W}_\gamma,$$

and $\sigma_\gamma : \mathcal{M}_\gamma^{\mathcal{U}} \to R_\gamma$, and $F_\gamma = \sigma_\gamma(E_\gamma^{\mathcal{U}})$, so that

$$\mathcal{W}_{\gamma+1} = V(\mathcal{W}_\eta,\mathcal{W}_\gamma,F_\gamma) = \mathcal{W}_\gamma\restriction(\alpha_\gamma+1)^\frown\langle F_\gamma\rangle^\frown i_{F_\gamma}\text{“}\mathcal{W}_\eta^{>\text{crit}(F_\gamma)},$$

where $\eta = U\text{-pred}(\gamma+1)$ and $\alpha_\gamma = \alpha_0(\mathcal{W}_\gamma,F_\gamma)$.

$\mathcal{W}_\gamma$ is a tree on M, so it can be converted to a tree on N^* using c. Let

$$\text{lift}(\mathcal{W}_\gamma,c) = \langle\mathcal{W}_\gamma^*,\langle c_\xi^\gamma \mid \xi\le z(\gamma)\rangle\rangle$$

where

$$c_\xi^\gamma = \langle\mathcal{M}_\xi^{\mathcal{W}_\gamma},\pi_\xi^\gamma,P_\xi^\gamma,\mathbb{C}_\xi^\gamma,\mathcal{M}_\xi^{\mathcal{W}_\gamma^*}\rangle.$$

[8] If $\mathcal{U}$ has successor length, then dropping along the main branch of $\mathcal{U}$ can cause $V(\mathcal{T},\mathcal{U})$ to lift to a proper initial segment of our $\mathcal{V}^*$.

[9] We called them $\mathcal{V}_\gamma$ in Section 6.7. They may not be normal.

We have $\mathcal{W}_0 = \mathcal{T}$ and $\mathcal{W}_0^* = \mathcal{T}^*$. For all γ, $c_0^\gamma = c$. The conversion system that lifts $\mathcal{U}$ is

$$\mathrm{lift}(\mathcal{U}, c^0_{z(0)}) = \langle \mathcal{U}^*, \langle d_\xi \mid \xi < \mathrm{lh}(\mathcal{U})\rangle\rangle,$$

where

$$d_\xi = \langle \mathcal{M}^{\mathcal{U}}_\xi, \psi_\xi, Q_\xi, \mathbb{D}_\xi, \mathcal{M}^{\mathcal{U}^*}_\xi\rangle.$$

Thus $\psi_0 = \pi^0_{z(0)}$, $Q_0 = P^0_{z(0)}$, and $\mathbb{D}_0 = \mathbb{C}^0_{z(0)}$.

Finally, the quasi-normalization $\mathcal{V}^*$ has associated trees $\mathcal{V}^*_\gamma$ on N^* for $\gamma < \mathrm{lh}(\mathcal{U}^*) = \mathrm{lh}(\mathcal{U})$, together with tree embeddings

$$\Phi^*_{\eta,\gamma} \colon \mathcal{V}^*_\eta \to \mathcal{V}^*_\gamma$$

defined when $\eta <_{U^*} \gamma$, or equivalently, $\eta <_U \gamma$. $\Phi^*_{\eta,\gamma}$ determines a u-map $u^*_{\eta,\gamma}\colon \mathrm{lh}(\mathcal{V}^*_\eta) \to \mathrm{lh}(\mathcal{V}^*_\gamma)$, and for $\nu \in \mathrm{dom}(u^*_{\eta,\gamma})$, a t-map $\overset{*}{t}{}^{\eta,\gamma}_{\gamma}$. Since Σ^* normalizes well, the $\mathcal{V}^*_\gamma$ are by Σ^*; moreover, by Proposition 6.2.9, the last model of $\mathcal{V}^*_\gamma$ is

$$\mathcal{M}^{\mathcal{V}^*_\gamma}_{z^*(\gamma)} = \mathcal{M}^{\mathcal{U}^*}_\gamma.$$

When $\eta = U^*\text{-pred}(\gamma+1)$ (equivalently, $\eta = U\text{-pred}(\gamma+1)$), we shall have that

$$\mathcal{V}^*_{\gamma+1} = \mathcal{V}^*_\gamma \restriction (\alpha_\gamma + 1)^\frown \langle G^*_\gamma\rangle^\frown i_{G^*_\gamma}``(\mathcal{V}^*_\eta)^{>\mathrm{crit}(G^*_\gamma)}$$

where

$$G^*_\gamma = E^{\mathcal{U}^*}_\gamma.$$

We shall prove that $\mathcal{W}^*_\gamma = \mathcal{V}^*_\gamma \restriction z(\gamma) + 1$ for all γ. Since $\mathcal{V}^*_\gamma$ is by Σ^*, we get that $\mathcal{W}_\gamma$ is by $\Omega(c, \Sigma^*)$, as desired. The proof is by induction on γ, with a subinduction on initial segments of $\mathcal{W}_\gamma$. Our overall plan is summarized in the diagram:

$$\begin{array}{ccc}
\mathcal{W}_\gamma & \xrightarrow{\ \text{lift}\ } & \mathcal{W}^*_\gamma \trianglelefteq \mathcal{V}^*_\gamma \\
\uparrow {\scriptstyle \Phi_{\nu,\gamma}} & & \uparrow {\scriptstyle \Phi^*_{\nu,\gamma}} \\
\mathcal{W}_\nu & \xrightarrow{\ \text{lift}\ } & \mathcal{W}^*_\nu \trianglelefteq \mathcal{V}^*_\nu
\end{array}$$

LEMMA 7.4.3. *Let* $\gamma < \mathrm{lh}(\mathcal{U})$. *Then*

(1) $\mathcal{W}^*_\gamma = \mathcal{V}^*_\gamma \restriction z(\gamma) + 1$.

(2) *Whenever* $\nu <_U \gamma$ *and* $(\nu, \gamma]_U$ *does not drop in model or degree, then for all* $\xi < \mathrm{lh}(\mathcal{W}_\nu)$,

  (*i*) $P^\gamma_{u_{\nu,\gamma}(\xi)} = \overset{*}{t}{}^{\nu,\gamma}_\xi(P^\nu_\xi)$, *and*

  (*ii*) $\pi^\gamma_{u_{\nu,\gamma}(\xi)} \circ t^{\nu,\gamma}_\xi = \overset{*}{t}{}^{\nu,\gamma}_\xi \circ \pi^\nu_\xi$.

(3) $u_{\eta,\nu} \subseteq u^*_{\eta,\nu}$, *if* $\eta, \nu \le \gamma$ *and* $\eta \le_U \nu$.

(4) (*i*) $P^{\gamma}_{z(\gamma)} = Q_\gamma$, *and there is an* η *such that* $Q_\gamma \in \mathrm{lev}(\mathbb{D}_\gamma \upharpoonright \eta)$ *and* $\mathbb{C}^{\gamma}_{z(\gamma)} \upharpoonright \eta \doteq \mathbb{D}_\gamma \upharpoonright \eta$.

(*ii*) *If* $[0,\gamma]_U \cap D^{\mathcal{U}} = \emptyset$, *then* $z(\gamma) = z^*(\gamma)$, $\mathcal{W}^*_\gamma = \mathcal{V}^*_\gamma$ *and* $\mathbb{C}^{\gamma}_{z(\gamma)} = \mathbb{D}_\gamma$.[10]

(*iii*) $\pi^{\gamma}_{z(\gamma)} \circ \sigma_\gamma = \psi_\gamma$.

PROOF. Here is a diagram related to Lemma 7.4.3:

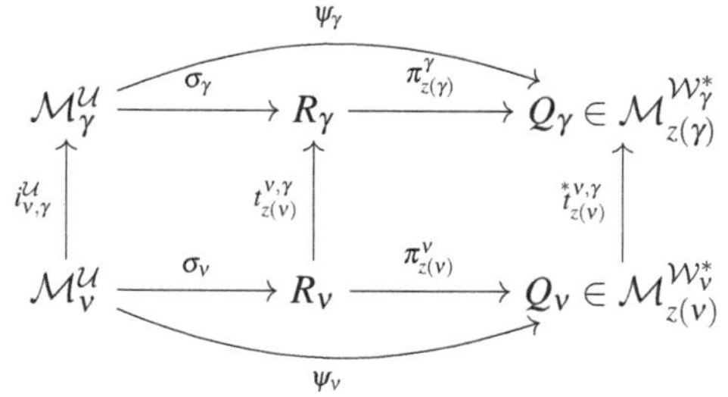

The fact that $\pi^{\nu}_{z(\nu)}$ and $\pi^{\gamma}_{z(\gamma)}$ map to Q_ν and Q_γ is (4)(*i*). The fact that the triangle on the top commutes is (4)(*iii*). That the square on the right commutes is (2), in the case $\xi = z(\nu)$. We of course need (2) at other τ as well. That square on the left commutes is a basic fact about quasi-normalization.

The reader might look back at the diagram near the end of the proof of Lemma 6.8.2. $\mathcal{M}^{\mathcal{U}^*}_\nu$ in that diagram corresponds to Q_ν in the present one. We can take R^*_ν of that diagram to also be Q_ν in the present one, because our tree on the background universe is nice. We do not actually need that; if the background extenders were not nice, then in the present case we would be introducing some $\sigma^*_\nu \colon Q_\nu \to R^*_\nu$ via a quasi-normalization of $\langle \mathcal{T}^*, \mathcal{U}^* \rangle$. $\pi^{\nu}_{z(\nu)}$ would map into R^*_ν, rather than Q_ν, and the present diagram would transform into the previous one. (See Remark 7.4.2 above.)

We prove Lemma 7.4.3 by induction on γ. For $\gamma = 0$, $\mathcal{W}_0 = \mathcal{T}$ and $\mathcal{V}^*_0 = \mathcal{T}^*$, so (1) holds. Since $P^{0}_{z(0)} = Q_0$, $\psi_0 = \pi^{0}_{z(0)}$, $\sigma_0 = \mathrm{id}$, and $\mathbb{D}_0 = \mathbb{C}^{0}_{z(0)}$, (4) holds. (2) and (3) are vacuous.

Now suppose Lemma 7.4.3 is true at all $\nu \leq \gamma$. We show it at $\gamma + 1$. Let $\nu = U\text{-pred}(\gamma+1)$, and

$$H = \psi_\gamma(E^{\mathcal{U}}_\gamma),$$

$$G = \sigma_{Q_\gamma}[Q_\gamma | \mathrm{lh}(H)]^{\mathbb{D}_\gamma}(H),$$

and

$$G^* = B^{\mathbb{D}_\gamma}(G),$$

so that

$$G^* = E^{\mathcal{U}^*}_\gamma.$$

[10]This item is not needed to carry through the induction, it is just a simpler case to keep in mind. The meta-tree determined by the $\mathcal{W}_\gamma$'s drops whenever $\mathcal{U}$ drops. The meta-tree determined by the $\mathcal{V}^*_\gamma$'s is coarse, so does not drop. Thus a drop in $[0,\gamma]_U$ can lead to $z(\gamma) < z^*(\gamma)$.

Let also

$$F = F_\gamma = \sigma_\gamma(E_\gamma^{\mathcal{U}}),$$
$$\alpha = \alpha_0(\mathcal{W}_\gamma, F),$$
$$K = \pi_\alpha^\gamma(F).$$

F is on the extended sequence of $\mathcal{M}_\alpha^{\mathcal{W}_\gamma}$, and we can lift it by π_α^γ to K on the extended sequence of $\mathcal{M}_\alpha^{\mathcal{W}_\gamma^*}$. Moreover,

$$H = \pi_{z(\gamma)}^\gamma(F)$$

because $\psi_\gamma = \pi_{z(\gamma)}^\gamma \circ \sigma_\gamma$.

The following is the main claim.

CLAIM 7.4.4. *(a)* $G = \sigma_{\mathrm{P}_\alpha^\gamma}[P_\alpha^\gamma | \mathrm{lh}(K)]^{\mathbb{C}_\alpha^\gamma}(K)$.

(b) $G^* = B^{\mathbb{C}_\alpha^\gamma}(G)$.

(c) If $\xi < \alpha$, *then* $\mathrm{lh}(E_\xi^{\mathcal{W}_\gamma^*}) \leq \mathrm{lh}(G^*)$, *and if* $\alpha < z(\gamma)$, *then* $\mathrm{lh}(G^*) \leq \mathrm{lh}(E_\alpha^{\mathcal{W}_\gamma^*})$.

PROOF. We assume that $\alpha < z(\gamma)$, and leave the other case to the reader. Let $E = E_\alpha^{\mathcal{W}_\gamma}$. Since $\alpha = \alpha_0(\mathcal{W}_\gamma, F)$, either $\mathrm{lh}(F) < \hat{\lambda}(E)$, or E is of plus type and $\mathrm{lh}(F) < \mathrm{lh}(E)$. In either case, the agreement properties of PFS conversion systems imply that[11]

$$\pi_{z(\gamma)}^\gamma \restriction \mathrm{lh}(F) + 1 = \mathrm{res}_\alpha^\gamma \circ \pi_\alpha^\gamma \restriction \mathrm{lh}(F) + 1.$$

Here $\mathrm{res}_\alpha^\gamma$ is the α-th generator map of lift$(\mathcal{W}_\gamma, c)$, that is, setting $E_1 = \psi_\alpha(E)$ and $X = P_\alpha^\gamma | \mathrm{lh}(E_1)$,

$$\mathrm{res}_\alpha^\gamma = \sigma_{P_\alpha^\gamma}^{\mathbb{C}_\alpha^\gamma}[X].$$

Thus

$$\begin{aligned} H &= \pi_{z(\gamma)}^\gamma(F) \\ &= \mathrm{res}_\alpha^\gamma \circ \pi_\alpha^\gamma(F) \\ &= \mathrm{res}_\alpha^\gamma(K). \end{aligned}$$

Let $Y = \mathrm{Res}_{\mathrm{P}_\alpha^\gamma}^{\mathbb{C}_\alpha^\gamma}[X]$, and $E_2 = \mathrm{res}_\alpha^\gamma(E_1)$. The last extender of Y is E_2^-. To prove (a) of the claim, we must show that

$$\sigma_{\mathrm{Y}}^{\mathbb{C}_\alpha^\gamma}[Y | \mathrm{lh}(H)](H) = G.$$

[11]See induction hypothesis (3)(a) in Section 4.8. This is where the fact that we are only quasi-normalizing comes into play. If $E_\alpha^{\mathcal{W}_\gamma}$ is not of plus type, and $\lambda(E_\alpha^{\mathcal{W}_\gamma}) < \mathrm{lh}(F) < \mathrm{lh}(E_\alpha^{\mathcal{W}_\gamma})$, then we do not have enough agreement between π_α^γ and $\pi_{z(\gamma)}^\gamma$ to go on.

For then

$$G = \sigma_Y[Y|\mathrm{lh}(H)] \circ \sigma_{P^\gamma_\alpha}[X](K)$$
$$= \sigma_{P^\gamma_\alpha}[P^\gamma_\alpha|\mathrm{lh}(K)](K),$$

as desired. The last step uses resurrection consistency in $\mathbb{C}^\gamma_\alpha$, and is the reason we moved to PFS constructions.

We have that $E_\alpha^{\mathcal{W}^*_\gamma} = B^{\mathbb{C}^\gamma_\alpha}(E_2^-)$. Let ξ be such that $Y||o(Y)$ is the last model of $\mathbb{C}^\gamma_\alpha \restriction \xi$. By our coherence lemma 4.7.7 for constructions,

$$\mathbb{C}^\gamma_\alpha \restriction \xi = \mathbb{C}^\gamma_{z(\gamma)} \restriction \xi.$$

But $\mathrm{lh}(H) < o(Y)$, so

$$\sigma_Y[Y|\mathrm{lh}(H)]^{\mathbb{C}^\gamma_\alpha}(H) = \sigma_{Y||o(Y)}[Y|\mathrm{lh}(H)]^{\mathbb{C}^\gamma_{z(\gamma)}}(H)$$
$$= \sigma_{P^\gamma_{z(\gamma)}}[P^\gamma_{z(\gamma)}|\mathrm{lh}(H)]^{\mathbb{C}^\gamma_{z(\gamma)}}(H)$$
$$= G,$$

as desired. Here we use that $o(Y)$ is a cardinal of $P^\gamma_{z(\gamma)}$ and $o(Y) \leq \rho^-(P^\gamma_{z(\gamma)})$.

This proves part (a) of Claim 7.4.4. Part (b) follows from (a) and the fact that $\mathbb{C}^\gamma_\alpha \restriction \xi = \mathbb{C}^\gamma_{z(\gamma)} \restriction \xi$, where ξ is as above.

For (c), $\mathrm{lh}(G^*) < \mathrm{lh}(E_\alpha^{\mathcal{W}^*_\gamma})$ because, in the notation above, $E_\alpha^{\mathcal{W}^*_\gamma} = B^{\mathbb{C}^\gamma_\alpha}(E_2^-)$, and E_2^- is the last extender of Y, and G is the last extender of some Z such that $Z <_{\mathbb{C}^\gamma_\alpha} Y$. On the other hand,

$$(\mathcal{W}^*_\gamma \restriction \alpha+1)^\frown\langle G^*\rangle = \mathrm{lift}((\mathcal{W}_\gamma \restriction \alpha+1)^\frown\langle F\rangle, c)_0,$$

and the lift of a quasi-normal tree is quasi-normal. Thus $\mathrm{lh}(E_\xi^{\mathcal{W}^*_\gamma}) \leq \mathrm{lh}(G^*)$ for all $\xi < \alpha$. ⊣

Part (c) tells us that setting $\mathcal{V}^*_{\gamma+1} = \mathcal{V}^*_\gamma \restriction (\alpha_\gamma+1)^\frown\langle G^*_\gamma\rangle^\frown i_{G^*_\gamma}``(\mathcal{V}^*_\eta)^{>\mathrm{crit}(G^*_\gamma)}$, as we have done, is indeed a legitimate step of quasi-normalization at the background level.

By definition, $\mathcal{W}_{\gamma+1} \restriction \alpha+2 = \mathcal{W}_\gamma \restriction (\alpha+1)^\frown\langle F\rangle$. So at the background level, we have

CLAIM 7.4.5. *(a)* $\mathcal{W}^*_{\gamma+1} \restriction \alpha+2 = \mathcal{W}^*_\gamma \restriction (\alpha+1)^\frown\langle G^*\rangle = \mathcal{V}^*_{\gamma+1} \restriction \alpha+2$.
(b) $\beta(\mathcal{W}_\gamma, F) = \beta(\mathcal{V}^*_\gamma, G^*)$.

PROOF. That $\mathcal{W}^*_{\gamma+1} \restriction \alpha+2 = \mathcal{W}^*_\gamma \restriction (\alpha+1)^\frown\langle G^*\rangle$ is just Claim 7.4.4 restated. By definition, $\mathcal{V}^*_{\gamma+1} \restriction \alpha+2 = \mathcal{V}^*_\gamma \restriction (\alpha+1)^\frown\langle G^*\rangle$.

Part (b) follows at once from the agreement between $\mathcal{W}^*_\gamma$ and $\mathcal{V}^*_\gamma$, together with the fact that $\mathcal{W}^*_{\gamma+1}$ is maximal. ⊣

Let $\beta = \beta(\mathcal{W}_\gamma, F)$. Since the quasi-normalizations producing $\mathcal{W}_{\gamma+1}$ and $\mathcal{V}^*_{\gamma+1}$ have the same α and β, we have $u_{\nu,\gamma+1} \subseteq u^*_{\nu,\gamma+1}$. Moreover, if $[0,\gamma+1]_U \cap D^{\mathcal{U}} = \emptyset$, then $z(\nu) = z^*(\nu)$ by (ii), so $z(\gamma+1) = (\alpha+1)+(z(\nu)-\beta) = (\alpha+1)+(z^*(\nu)-\beta) = z^*(\gamma+1)$. We have $u_{\nu,\gamma+1} = u^*_{\nu,\gamma+1}$ in this case.

Remark 7.4.6. If $D^{\mathcal{U}} \cap [0,\gamma+1]_U = \varnothing$, then $\mathrm{lh}(\mathcal{W}_{\gamma+1}) = \mathrm{lh}(W^*_{\gamma+1})$, and $u_{\nu,\gamma+1} = u^*_{\nu,\gamma+1}$.

We now show that (1) and (2) of Lemma 7.4.3 hold at $\gamma+1$. For this, we show by induction on ξ:

Induction Hypothesis $(\dagger)_\xi$*:*

(1) $\mathcal{W}^*_{\gamma+1} \restriction \xi = \mathcal{V}^*_{\gamma+1} \restriction \xi$.
(2) If $(\nu,\gamma+1]_U$ does not drop in model or degree, and $u_{0,\gamma+1}(\tau) < \xi$, then
 (a) $P^{\gamma+1}_{u_{\nu,\gamma+1}(\tau)} = t^{*\nu,\gamma+1}_\tau(P^\nu_\tau)$, and
 (b) $\pi^{\gamma+1}_{u_{\nu,\gamma+1}(\tau)} \circ t^{\nu,\gamma+1}_\gamma = t^{*\nu,\gamma+1}_\tau \circ \pi^\nu_\tau$.

Note that the limit step in the inductive proof of $(\dagger)_\xi$ is trivial.

Base Case 1. $\xi = \alpha+1$.

We have $\mathcal{W}_{\gamma+1} \restriction (\alpha+1) = \mathcal{W}_\gamma \restriction (\alpha+1)$ and $\mathcal{W}^*_{\gamma+1} \restriction (\alpha+1) = \mathcal{W}^*_\gamma \restriction (\alpha+1)$. Since Lemma 7.4.3 holds at γ, we get $(\dagger)_\xi(1)$. For $(\dagger)_\xi(2)$, let $u_{\nu,\gamma+1}(\tau) < \alpha+1$. Then $\tau < \beta$ and $u_{\nu,\gamma+1}(\tau) = \tau$. Moreover $t^{\nu,\gamma+1}_\tau$ and $t^{*\nu,\gamma+1}_\tau$ are the identity. So $(\dagger)_\xi(2)$ boils down to $P^{\gamma+1}_\tau = P^\nu_\tau$, and $\pi^{\gamma+1}_\tau = \pi^\nu_\tau$. This holds because $\mathcal{W}_\nu \restriction (\tau+1) = \mathcal{W}_{\gamma+1} \restriction (\tau+1)$, so their lifts are equal.

Base Case 2. $\xi = \alpha+2$.

We proved $(\dagger)_\xi(1)$ in Claim 7.4.5.

For $(\dagger)_\xi(2)$, the new case to consider is $\tau = \beta$. We have

$$\pi^\nu_\beta = \pi^{\gamma+1}_\beta,$$

$$t^{\nu,\gamma+1}_\beta = i^{\mathcal{W}_{\gamma+1}}_{\beta,\alpha+1}$$

and

$$t^{*\nu,\gamma+1}_\beta = i^{\mathcal{W}^*_{\gamma+1}}_{\beta,\alpha+1}.$$

The first because $\mathcal{W}_{\gamma+1} \restriction (\beta+1) = \mathcal{W}_\nu \restriction (\beta+1)$, and the second two by our definition of quasi- normalization. (Note we are in the case that $(\beta,\alpha+1]_{\mathcal{W}_{\gamma+1}}$ is not a drop in model or degree.) But

$$\pi^{\gamma+1}_{\alpha+1} \circ i^{\mathcal{W}_{\gamma+1}}_{\beta,\alpha+1} = i^{\mathcal{W}^*_{\gamma+1}}_{\beta,\alpha+1} \circ \pi^{\gamma+1}_\beta$$

holds because lifting maps commute with the branch embeddings in a conversion system. This gives

$$\pi_{\alpha+1}^{\gamma+1} \circ t_\beta^{\nu,\gamma+1} = \overset{*}{t}{}_\beta^{\nu,\gamma+1} \circ \pi_\beta^\nu$$

as desired.

If $\mathrm{lh}(\mathcal{W}_\nu) = \beta+1$ or $\gamma+1 \in D^{\mathcal{U}}$, then $\mathrm{lh}(\mathcal{W}_{\gamma+1}) = \alpha+2$, so we are done. So suppose $\mathrm{lh}(\mathcal{W}_\nu) > \beta+1$, and $(\nu, \gamma+1]_U$ is not a drop of any kind in $\mathcal{U}$.

Inductive Case 1. $(\dagger)_{\xi+1}$ holds, and $\xi \geq \alpha+1$.

We must prove $(\dagger)$ at $\xi+2$. We are assuming $\xi+1 < \mathrm{lh}(\mathcal{W}_{\gamma+1})$. Let

$$E = E_\xi^{\mathcal{W}_{\gamma+1}}.$$

Let σ be the resurrection map for $\pi_\xi^{\gamma+1}(E)$ in $\mathbb{C}_\xi^{\gamma+1}$, that is,

$$\sigma = \sigma_{\mathrm{P}_\xi^{\gamma+1}}[P_\xi^{\gamma+1} | \mathrm{lh}(\pi_\xi^{\gamma+1}(E))].$$

Let

$$\begin{aligned} E^* &= B(\sigma \circ \pi_\xi^{\gamma+1}(E))^{\mathbb{C}_\xi^{\gamma+1}} \\ &= E_\xi^{\mathcal{W}^*_{\gamma+1}}. \end{aligned}$$

CLAIM 7.4.7. $E^* = E_\xi^{\mathcal{V}^*_{\gamma+1}}$.

PROOF. Since $\xi \geq \alpha+1$, we can write

$$\xi = u_{\nu,\gamma+1}(\rho),$$

where $\rho \geq \beta$. Let

$$D = E_\rho^{\mathcal{W}_\nu},$$

so that

$$E = t_\rho^{\nu,\gamma+1}(D).$$

Letting $H = \sigma \circ \pi_\xi^{\gamma+1}(E)$, we have

$$\begin{aligned} H &= \sigma \circ (\pi_\xi^{\gamma+1} \circ t_\rho^{\nu,\gamma+1}(D)) \\ &= \sigma \circ (\overset{*}{t}{}_\rho^{\nu,\gamma+1} \circ \pi_\rho^\nu(D)) \end{aligned}$$

by induction. Let τ be the resurrection map for $\pi_\rho^\nu(D)$ in $\mathbb{C}_\rho^\nu$, that is,

$$\tau = \sigma_{\mathrm{P}_\rho^\nu}[P_\rho^\nu | \mathrm{lh}(\pi_\rho^\nu(D))].$$

It is not hard to see that

$$\overset{*}{t}{}_D^{\nu,\gamma+1}(\tau) = \sigma.$$

This is because $\overset{*}{t}{}^{\nu,\gamma+1}_{\rho}(P^{\nu}_{\rho}) = P^{\gamma+1}_{\xi}$ by induction hypothesis (2)(a), and similarly $\overset{*}{t}{}^{\nu,\gamma+1}_{\rho}(\pi^{\nu}_{\rho}(D)) = \pi^{\gamma+1}_{\xi}(t^{\nu,\gamma+1}_{\rho}(D)) = \pi^{\gamma+1}_{\xi}(E)$. But then

$$\begin{aligned} E_{\xi}^{\mathcal{V}^*_{\gamma+1}} &= \overset{*}{t}{}^{\nu,\gamma+1}_{\rho}(E_{\rho}^{\mathcal{V}^*_{\nu}}) \\ &= \overset{*}{t}{}^{\nu,\gamma+1}_{\rho}(B(\tau(\psi^{\nu}_{\xi}(\bar{E})))^{\mathbb{C}^{\nu}_{\rho}}) \\ &= B(\overset{*}{t}{}^{\nu,\gamma+1}_{\rho}(\tau(\pi^{\nu}_{\rho}(D))))^{\mathbb{C}^{\gamma+1}_{\xi}} \\ &= B(\sigma(\overset{*}{t}{}^{\nu,\gamma+1}_{\rho}(\pi^{\nu}_{\rho}(D))))^{\mathbb{C}^{\gamma+1}_{\xi}} \\ &= B(H)^{\mathbb{C}^{\gamma+1}_{\xi}} \\ &= E^* \end{aligned}$$

as desired. ⊣

From Claim 7.4.7, we have that $\mathcal{W}^*_{\gamma+1} \restriction (\xi+2)$ is the unique quasi-normal continuation of $\mathcal{W}^*_{\gamma+1} \restriction (\xi+1) = \mathcal{V}^*_{\gamma+1} \restriction (\xi+1)$ via $E_{\xi}^{\mathcal{V}^*_{\gamma+1}}$. That is, $\mathcal{W}^*_{\gamma+1} \restriction (\xi+2) = \mathcal{V}^*_{\gamma+1} \restriction (\xi+2)$.

It remains to show, keeping our previous notation:

CLAIM 7.4.8. $\pi^{\gamma+1}_{\xi+1} \circ t^{\nu,\gamma+1}_{\rho+1} = \overset{*}{t}{}^{\nu,\gamma+1}_{\rho+1} \circ \pi^{\nu}_{\rho+1}$.

PROOF. Both maps act on $\mathcal{M}^{\mathcal{W}_{\nu}}_{\rho+1}$. The left side embeds it into $P^{\gamma+1}_{\xi+1}$ and the right side embeds it into $\overset{*}{t}{}^{\nu,\gamma+1}_{\rho+1}(P^{\nu}_{\rho+1})$. So first we show $(\dagger)_{\xi+1}(2)(a)$:

SUBCLAIM 7.4.8.1. $P^{\gamma+1}_{\xi+1} = \overset{*}{t}{}^{\nu,\gamma+1}_{\rho+1}(P^{\nu}_{\rho+1})$.

PROOF. Let

$$\begin{aligned} \theta &= \mathcal{W}_{\gamma+1}\text{-pred}(\xi+1) \\ &= \mathcal{V}^*_{\gamma+1}\text{-pred}(\xi+1) \\ &= \mathcal{W}^*_{\gamma+1}\text{-pred}(\xi+1). \end{aligned}$$

Recall that $E = E_{\xi}^{\mathcal{W}_{\gamma+1}} = t^{\nu,\gamma+1}_{\rho}(D)$.

Case 1. $\text{crit}(D) \geq \text{crit}(F)$, or $\theta < \beta$.

This is the case in which $u_{\nu,\gamma+1}$ preserves tree predecessor, that is, $\theta = u_{\nu,\gamma+1}(\rho) = u^*_{\nu,\gamma+1}(\rho)$ for $\bar{\theta} = \mathcal{W}_{\nu}\text{-pred}(\rho+1)$. We have

$$\mathcal{M}^{\mathcal{W}_{\nu}}_{\rho+1} = \text{Ult}(R, D),$$

where $R \trianglelefteq \mathcal{M}^{\mathcal{W}_{\nu}}_{\bar{\theta}}$. Let

$$S = t^{\nu,\gamma+1}_{\bar{\theta}}(R).$$

Quasi-normalization leads to

$$\mathcal{M}_{\xi+1}^{\mathcal{W}_{\gamma+1}} = \mathrm{Ult}(S, E).$$

Because $\mathcal{W}_{\gamma+1}^*$ is part of a conversion system,

$$\begin{aligned} P_{\xi+1}^{\gamma+1} &= i_{\theta,\gamma+1}^{\mathcal{W}_{\gamma+1}^*}(\mathrm{Res}_{\mathrm{P}_\theta^{\gamma+1}}[\pi_\theta^{\gamma+1}(S)]) \\ &= i_{\theta,\gamma+1}^{\mathcal{V}_{\gamma+1}^*}(\mathrm{Res}_{\mathrm{P}_\theta^{\gamma+1}}[\pi_\theta^{\gamma+1}(S)]). \end{aligned}$$

(The resurrection is in $\mathbb{C}_\theta^{\gamma+1}$.) Note that $\overset{*}{t}{}_{\bar\theta}^{\nu,\gamma+1}(P_{\bar\theta}^\nu) = P_\theta^{\gamma+1}$ by induction. Also, $\overset{*}{t}{}_{\bar\theta}^{\nu,\gamma+1}(\pi_{\bar\theta}^\nu(R)) = \pi_\theta^{\gamma+1}(t_{\bar\theta}^{\nu,\gamma+1}(R)) = \pi_\theta^{\gamma+1}(S)$. It follows that

$$\mathrm{Res}_{\mathrm{P}_\theta^{\gamma+1}}[\pi_\theta^{\gamma+1}(S)] = \overset{*}{t}{}_{\bar\theta}^{\nu,\gamma+1}(\mathrm{Res}_{\mathrm{P}_{\bar\theta}^\nu}[\pi_{\bar\theta}^\nu(R)]),$$

where the resurrections are in $\mathbb{C}_\theta^{\gamma+1}$ and $\mathbb{C}_{\bar\theta}^\nu$ respectively. Thus

$$\begin{aligned} P_{\xi+1}^{\gamma+1} &= i_{\theta,\xi+1}^{\mathcal{V}_{\gamma+1}^*}(\mathrm{Res}_{\mathrm{P}_\theta^{\gamma+1}}[\pi_\theta^{\gamma+1}(S)]) \\ &= i_{\theta,\xi+1}^{W_{\gamma+1}^*} \circ \overset{*}{\pi}{}_{\bar\theta}^{\nu,\gamma+1}(\mathrm{Res}_{\mathrm{P}_{\bar\theta}^\nu}[\pi_{\bar\theta}^\nu(R)]) \\ &= \overset{*}{\pi}{}_{\rho+1}^{\nu,\gamma+1} \circ i_{\bar\theta,\rho+1}^{\mathcal{V}_\nu^*}(\mathrm{Res}_{\mathrm{P}_{\bar\theta}^\nu}[\pi_{\bar\theta}^\nu(R)]) \\ &= \overset{*}{t}{}_{\rho+1}^{\nu,\gamma+1}(P_{\rho+1}^\nu), \end{aligned}$$

as desired.

Case 2. Otherwise.

In this case, we must have $\beta \leq \theta$ and $\mathrm{crit}(D) < \mathrm{crit}(F)$. It follows that $\theta = \beta$, and $\mathcal{W}_\nu$-pred$(\rho+1) = \mathcal{W}_{\gamma+1}$-pred$(\rho+1) = \beta$. The argument above works, with $\bar\theta = \theta = \beta$ and $R = S$, and $t_{\bar\theta}^{\nu,\gamma+1}$ and $\overset{*}{t}{}_{\bar\theta}^{\nu,\gamma+1}$ being replaced by the identity map. (If $\theta < \beta$ they are already the identity. This case is similar to the case $\theta < \beta$.) The relevant calculation is

$$\begin{aligned} P_{\xi+1}^{\gamma+1} &= i_{\beta,\xi+1}^{\mathcal{V}_{\gamma+1}^*}(\mathrm{Res}_{\mathrm{P}_\beta^{\gamma+1}}[\pi_\beta^{\gamma+1}(R)]^{\mathbb{C}_\beta^{\gamma+1}}) \\ &= i_{\beta,\xi+1}^{\mathcal{V}_{\gamma+1}^*}(\mathrm{Res}_{\mathrm{P}_\beta^\nu})[\pi_\beta^\nu(R)]^{\mathbb{C}_\beta^\nu}) \\ &= \overset{*}{t}{}_{\rho+1}^{\nu,\gamma+1} \circ i_{\beta,\rho+1}^{\mathcal{V}_\nu^*}(\mathrm{Res}_{\mathrm{P}_\beta^\nu}[\pi_\beta^\nu(R)]^{\mathbb{C}_\beta^\nu}) \\ &= \overset{*}{t}{}_{\rho+1}^{\nu,\gamma+1}(P_{\rho+1}^\nu). \end{aligned}$$

The first equation holds because $\mathcal{V}_{\gamma+1}^* \restriction (\xi+2) = \mathcal{W}_{\gamma+1}^* \restriction (\xi+2)$ is a conversion system. The second comes from the fact that $\mathcal{V}_{\gamma+1}^* \restriction (\beta+1) = \mathcal{V}_\nu^* \restriction (\beta+1)$. The third comes from properties of quasi-normalization. The last comes from $\mathcal{V}_\nu^*$ being a conversion system. ⊣

We now finish proving Claim 7.4.8. We keep the notation above. Let us assume that we are in Case 1. Let $x \in \mathcal{M}_{\rho+1}^{\mathcal{W}_\nu}$ be arbitrary, and let

$$x = [a, f]_D^R,$$

where $a \subseteq \mathrm{lh}(D)$ is finite and $f \in R$. (We assume $k(R) = 0$ for simplicity.) Then

$$\begin{aligned}\pi_{\xi+1}^{\gamma+1} \circ t_{\rho+1}^{\nu,\gamma+1}(x) &= \pi_{\xi+1}^{\gamma+1}(t_{\rho+1}^{\nu,\gamma+1}([a, f]_D^R)) \\ &= \pi_{\xi+1}^{\gamma+1}([t_\rho^{\nu,\gamma+1}(a), t_{\bar\theta}^{\nu,\gamma+1}(f)]_E^S),\end{aligned}$$

by the properties of embedding normalization, and the fact $t_{\bar\theta}^{\nu,\gamma+1}(R) = S$ and $\pi_\rho^{\nu,\gamma+1}(D) = E$. Thus

$$\pi_{\xi+1}^{\gamma+1} \circ t_{\rho+1}^{\nu,\gamma+1}(x) = [\sigma \circ \pi_\xi^{\gamma+1} \circ \pi_{\bar\xi}^{\nu,\gamma+1}(a), \phi \circ \psi_\theta^{\gamma+1} \circ \pi_{\bar\theta}^{\nu,\gamma+1}(f)]_{E^*}^{\mathcal{M}_\theta^{\mathcal{V}_{\gamma+1}^*}},$$

where σ resurrects $\pi_\xi^{\gamma+1}(E)$ in $\mathbb{C}_\xi^{\gamma+1}$ and ϕ resurrects $\pi_\theta^{\gamma+1}(S)$ in $\mathbb{C}_\theta^{\gamma+1}$.

On the other hand, letting

$$\sigma = \overset{*}{t}{}_\rho^{\nu,\gamma+1}(\bar\sigma), \quad \text{and} \quad \phi = \overset{*}{t}{}_{\bar\theta}^{\nu,\gamma+1}(\bar\phi),$$

we have

$$\begin{aligned}\overset{*}{t}{}_{\rho+1}^{\nu,\gamma+1} \circ \pi_{\rho+1}^\nu(x) &= \overset{*}{t}{}_{\rho+1}^{\nu,\gamma+1}(\pi_{\rho+1}^\nu([a, f]_D^R)) \\ &= \overset{*}{t}{}_{\rho+1}^{\nu,\gamma+1}([\bar\sigma \circ \pi_\rho^\nu(a), \bar\phi \circ \pi_{\bar\theta}^\nu(f)]_{E_\rho^{\mathcal{V}_\nu^*}}^{\mathcal{M}_{\bar\theta}^{\mathcal{V}_\nu^*}}) \\ &= [\overset{*}{t}{}_\rho^{\nu,\gamma+1} \circ \bar\sigma \circ \pi_\rho^\nu(a), \overset{*}{t}{}_{\bar\theta}^{\nu,\gamma+1} \circ \bar\phi \circ \pi_{\bar\theta}^\nu(f)]_{E_\xi^{\mathcal{V}_{\nu+1}^*}}^{\mathcal{M}_\theta^{\mathcal{V}_{\nu+1}^*}} \\ &= [\sigma \circ \overset{*}{t}{}_\rho^{\nu,\gamma+1} \circ \pi_\rho^\nu(a), \phi \circ \overset{*}{t}{}_{\bar\theta}^{\nu,\gamma+1} \circ \pi_{\bar\theta}^\nu(f)]_{E^*}^{\mathcal{M}_\theta^{\mathcal{V}_{\nu+1}^*}} \\ &= [\sigma \circ \pi_\xi^{\gamma+1} \circ t_\rho^{\nu,\gamma+1}(a), \rho \circ \pi_\theta^{\gamma+1} \circ t_{\bar\theta}^{\nu,\gamma+1}(f)]_{E^*}^{\mathcal{M}_\theta^{\mathcal{V}_{\nu+1}^*}}.\end{aligned}$$

The first 4 lines come from the way normalization and conversion work. The last line comes from our induction hypothesis.

We leave it to the reader to finish the proof in Case 2. This proves Claim 7.4.8. ⊣

Returning to the inductive proof of $(\dagger)_\xi$, we see that the limit case is trivial. We are left with

Inductive Case 2. ξ is a limit ordinal, and $(\dagger)_\xi$.

We must prove $(\dagger)_{\xi+1}$. We have $\mathcal{W}_{\gamma+1}^* \restriction \xi = \mathcal{V}_{\gamma+1}^* \restriction \xi$. Since Σ^* quasi-normalizes well, the branch $[0, \xi]_{\mathcal{V}_{\gamma+1}^*}$ of $\mathcal{V}_{\gamma+1}^*$ produced by normalization is equal to

$$\Sigma^*(\mathcal{W}_{\gamma+1}^* \restriction \xi).$$

Thus $\mathcal{W}^*_{\gamma+1} \restriction(\xi+1) = \mathcal{V}^*_{\gamma+1} \restriction(\xi+1)$. One can then prove $(\dagger)_{\xi+1}$ by looking at how the objects it deals with come from the $\mathcal{M}^{\mathcal{W}_\nu}_\tau$ and $\mathcal{M}^{\mathcal{V}^*_\nu}_\tau$ for $\tau <_{W_\gamma} u^{-1}_{\nu,\gamma+1}(\xi)$, and using our induction hypothesis $(\dagger)_\xi$. We omit further detail.

This completes our inductive proof of (1) and (2) of Lemma 7.4.3. We have already proved (3) of Lemma 7.4.3. We now prove (4). To simplify the notation a bit, let us assume that $[0,\gamma]_U$ does not drop, so that $\mathcal{W}^*_\nu = \mathcal{V}^*_\nu$ and $\mathcal{W}^*_{\gamma+1} = \mathcal{V}^*_{\gamma+1}$.

The following diagram summarizes the proof of (4).

$$
\begin{array}{ccccccc}
\mathcal{M}^{\mathcal{U}}_{\gamma+1} & \xrightarrow{\sigma_{\gamma+1}} & \mathcal{M}^{\mathcal{W}_{\gamma+1}}_{z(\gamma+1)} & \xrightarrow{\pi^{\gamma+1}_{z(\gamma+1)}} & P^{\gamma}_{z(\gamma+1)} & \in & \mathcal{M}^{\mathcal{V}^*_{\gamma+1}}_{z(\gamma+1)} \\
\uparrow{\scriptstyle i^{\mathcal{U}}_{\nu,\gamma+1}} & & \uparrow{\scriptstyle t^{\nu,\gamma+1}_{z(\nu)}} & & \uparrow{\scriptstyle {}^*t^{\nu,\gamma+1}_{z(\nu)}} & & \uparrow{\scriptstyle {}^*t^{\nu,\gamma+1}_{z(\nu)}} \\
\mathcal{M}^{\mathcal{U}}_{\nu} & \xrightarrow[\sigma_\nu]{} & \mathcal{M}^{\mathcal{W}_{\nu}}_{z(\nu)} & \xrightarrow[\pi^{\nu}_{z(\nu)}]{} & P^{\nu}_{z(\nu)} & \in & \mathcal{M}^{\mathcal{V}^*_{\nu}}_{z(\nu)}
\end{array}
$$

That the square on the right commutes is $(\dagger)_{z(\gamma+1)}$. It is a basic fact about normalization that the square on the left commutes. Also, ${}^*t^{\nu,\gamma+1}_{z(\nu)} = i^{\mathcal{U}^*}_{\nu,\gamma+1}$ holds, because the σ-maps are the identity in the case of coarse normalization.

We have that $\psi_\nu = \pi^\nu_{z(\nu)} \circ \sigma_\nu$ by induction. Further, the diagram

$$
\begin{array}{ccccc}
\mathcal{M}^{\mathcal{U}}_{\gamma+1} & \xrightarrow{\psi_{\gamma+1}} & P^{\gamma+1}_{z(\gamma)+1} \in \mathcal{M}^{\mathcal{V}^*_{\gamma+1}}_{z(\gamma+1)} & = & \mathcal{M}^{\mathcal{U}^*_{\gamma+1}} \\
\uparrow{\scriptstyle i^{\mathcal{U}}_{\nu,\gamma+1}} & & \uparrow{\scriptstyle i^{\mathcal{U}^*}_{\nu,\gamma+1}} & & \\
\mathcal{M}^{\mathcal{U}}_{\nu} & \xrightarrow[\psi_\nu]{} & P^{\nu}_{z(\nu)} \in \mathcal{M}^{\mathcal{W}^*_{\nu}}_{z(\nu)} & = & M^{\mathcal{U}^*_\nu}
\end{array}
$$

commutes, since it is part of the conversion of $\mathcal{U}$ to $\mathcal{U}^*$. So

$$
\begin{aligned}
\psi_{\gamma+1} \circ i^{\mathcal{U}}_{\nu,\gamma+1} &= i^{\mathcal{U}^*}_{\nu,\gamma+1} \circ \psi_\nu \\
&= {}^*t^{\nu,\gamma+1}_{z(\nu)} \circ \pi^{\nu}_{z(\nu)} \circ \sigma_\nu \\
&= \pi^{\gamma+1}_{z(\gamma+1)} \circ \sigma_{\gamma+1} \circ i^{\mathcal{U}}_{\nu,\gamma+1}.
\end{aligned}
$$

Thus $\psi_{\gamma+1}$ agrees with $\pi^{\gamma+1}_{z(\gamma+1)} \circ \sigma_{\gamma+1}$ on $\operatorname{ran} i^{\mathcal{U}}_{\nu,\gamma+1}$. But $\mathcal{M}^{\mathcal{U}}_{\gamma+1}$ is generated by $\operatorname{ran} i^{\mathcal{U}}_{0,\gamma+1}$ union $\varepsilon(E^{\mathcal{U}}_\gamma)$, where

$$
\varepsilon(G) = \begin{cases} \operatorname{lh}(G) & \text{if } G \text{ is of plus type,} \\ \lambda(G) & \text{otherwise.} \end{cases}
$$

So it is enough to show the two embeddings agree on $\varepsilon(E_\gamma^{\mathcal{U}})$. But

$$\begin{aligned}\psi_{\gamma+1}\restriction\varepsilon(E_\gamma^{\mathcal{U}}) &= \psi_\gamma\restriction\varepsilon(E_\gamma^{\mathcal{U}})\\ &= \pi^\gamma_{z(\gamma)}\circ\sigma_\gamma\restriction\varepsilon(E_\gamma^{\mathcal{U}}),\end{aligned}$$

by the agreement in conversion systems and our induction hypothesis. Since $\sigma_\gamma\restriction\varepsilon(E_\gamma^{\mathcal{U}}) = \sigma_{\gamma+1}\restriction\varepsilon(E_\gamma^{\mathcal{U}})$ (cf. Proposition 6.5.8), it is enough to show that

$$\pi^{\gamma+1}_{z(\gamma+1)}\restriction\varepsilon(F) = \pi^\gamma_{z(\gamma)}\restriction\varepsilon(F).$$

But

$$\begin{aligned}\pi^\gamma_{z(\gamma)}\restriction\varepsilon(F) &= \pi^\gamma_\alpha\restriction\varepsilon(F)\\ &= \pi^{\gamma+1}_\alpha\restriction\varepsilon(F)\\ &= \pi^{\gamma+1}_{z(\gamma+1)}\restriction\varepsilon(F).\end{aligned}$$

The first line holds because either $\mathrm{lh}(F) < \hat{\lambda}(E_\alpha^{\mathcal{W}_\gamma})$, or $E_\alpha^{\mathcal{W}_\gamma}$ is of plus type and $\mathrm{lh}(F) < \mathrm{lh}(E_\alpha^{\mathcal{W}_\gamma})$.[12] The second line holds because $\mathcal{W}_\gamma\restriction\alpha+1 = \mathcal{W}_{\gamma+1}\restriction\alpha+1$, and the third holds by the agreement of maps in $\mathrm{lift}(\mathcal{W}_{\gamma+1}, c)$.

This completes the proof of (4) in Lemma 7.4.3 in the case that $[0,\gamma+1]_U$ does not drop in model or degree. We leave the dropping case to the reader.

This completes the proof that if Lemma 7.4.3 holds at γ, then it holds at $\gamma+1$.

Now suppose γ is a limit ordinal. Let

$$\lambda = \sup\{\alpha_\xi \mid \xi < \gamma\}.$$

So $V(\mathcal{T},\mathcal{U}\restriction\gamma) = \mathcal{W}_\gamma\restriction\lambda$, and $\mathcal{V}_\gamma^*\restriction\gamma = \mathcal{W}_\gamma^*\restriction\lambda$. Since Σ^* quasi-normalizes well, $[0,\lambda)_{V_\gamma^*} = \Sigma^*(\mathcal{V}_\gamma^*\restriction\lambda)$. Thus

$$\mathcal{W}_\gamma^*\restriction(\lambda+1) = \mathcal{V}_\gamma^*\restriction(\lambda+1).$$

We now go on to prove $(\dagger)_\xi$, for $\xi\geq\lambda$, by induction. The proof is similar to the one above. Having $(\dagger)_\xi$ for $\xi = \mathrm{lh}(\mathcal{W}_\gamma)$, we go on to prove (4) as above. We omit further detail.

This proves Lemma 7.4.3. ⊣

Now let $\gamma+1 = \mathrm{lh}(\mathcal{U})$. By Lemma 7.4.3, $\mathcal{W}_\gamma^* = \mathcal{V}_\gamma^*\restriction z(\gamma)+1$. Since Σ^* quasi-normalizes well, $\mathcal{V}_\gamma^*$ is by Σ^*, so $\mathcal{W}_\gamma^*$ is by Σ^*, so $\mathcal{W}_\gamma$ is by $\Omega(c,\Sigma^*)$, as desired. We must also check the pullback clause, that is, that

$$\Sigma_{\langle\mathcal{T},\mathcal{U}\rangle} = (\Sigma_{\mathcal{W}_\gamma,R_\gamma})^{\sigma_\gamma},$$

where $R_\gamma = \mathcal{M}^{\mathcal{W}_\gamma}_{z(\gamma)}$. But

$$\begin{aligned}\Sigma_{\langle\mathcal{T},\mathcal{U}\rangle} &= \Omega(\mathcal{M}_\gamma^{\mathcal{U}},\psi_\gamma,P^\gamma_{z(\gamma)},\mathbb{C}^\gamma_{z(\gamma)},\Sigma^*_{\langle\mathcal{T}^*,\mathcal{U}^*\rangle})\\ &= \Omega(\mathcal{M}_\gamma^{\mathcal{U}},\pi^\gamma_{z(\gamma)}\circ\sigma_\gamma,P^\gamma_{z(\gamma)},\mathbb{C}^\gamma_{z(\gamma)},\Sigma^*_{\mathcal{W}_\gamma^*})\end{aligned}$$

[12]Here we use that we are only quasi-normalizing, that is, that α is $\alpha_0(\mathcal{W}_\gamma,F)$ rather than $\alpha(\mathcal{W}_\gamma,F)$.

$$= \Omega(R_\gamma, \pi^\gamma_{z(\gamma)}, P^\gamma_{z(\gamma)}, \mathbb{C}^\gamma_{z(\gamma)}, \Sigma^*_{\mathcal{W}^*_\gamma})^{\sigma_\gamma}$$
$$= (\Sigma_{\mathcal{W}_\gamma, R_\gamma})^{\sigma_\gamma},$$

as desired.

This finishes our proof of Theorem 7.4.1. ⊣

Strong unique iterability yields strategies for coarse premice that normalize well for infinite stacks. In particular, assuming AD^+, if (M,Σ^*) is a coarse Γ-Woodin pair, then Σ^* normalizes well for countable stacks. We believe that by extending the proof of Theorem 7.4.1 one can show that normalizing infinite stacks commutes with lifting to a background universe. Thus if we assume in the hypothesis of Theorem 7.4.1 that Σ^* normalizes well for infinite stacks, we can conclude that the induced strategies $\Omega(\mathbb{C}, M, \Sigma^*)$ normalize well for infinite stacks.

7.5. Fine strategies that condense well

We show that if Σ^* is an iteration strategy for V that has strong hull condensation, then the strategies for premice induced by Σ^* via a full background extender construction also have strong hull condensation. The proof is routine, but we include it for the sake of completeness. The corresponding result for ordinary hull condensation was proved by Sargsyan in [42].

THEOREM 7.5.1. *Let $c = \langle M, \varphi, Q, \mathbb{C}, S\rangle$ be a conversion stage, and suppose that Σ^* is a (λ,θ) iteration strategy for $(S, \in, w^{\mathbb{C}}, \mathcal{F}^{\mathbb{C}})$ that has strong hull condensation; then the induced strategy $\Omega(c,\Sigma^*)$ for M has strong hull condensation.*

PROOF. Let $\Sigma = \Omega(c,\Sigma^*)$. We show first that pseudo-hulls of plus trees by Σ are by Σ. The proof applies equally well to tails of Σ. We then deal with the pullback clause in the definition of strong hull condensaion.

Let $\mathcal{U}$ be a plus tree on M that is by Σ, and let $\Phi\colon \mathcal{T} \to \mathcal{U}$ be a tree embedding, with

$$\Phi = \langle u, v, \langle s_\beta \mid \beta < \mathrm{lh}(\mathcal{T})\rangle, \langle t_\beta \mid \beta+1 < \mathrm{lh}(\mathcal{T})\rangle\rangle.$$

We must see that $\mathcal{T}$ lifts to a tree by Σ^*. Let

$$\mathrm{lift}(\mathcal{T}, c) = \langle \mathcal{T}^*, \langle c_\alpha \mid \alpha < \mathrm{lh}(\mathcal{T})\rangle\rangle,$$

where

$$c_\alpha = \langle \mathcal{M}^{\mathcal{T}}_\alpha, \varphi_\alpha, Q_\alpha, \mathbb{C}_\alpha, \mathcal{M}^{\mathcal{T}^*}_\alpha\rangle$$

and

$$\mathrm{lift}(\mathcal{U}, c) = \langle \mathcal{U}^*, \langle d_\alpha \mid \alpha < \mathrm{lh}(\mathcal{U})\rangle\rangle,$$

where

$$d_\alpha = \langle \mathcal{M}^{\mathcal{U}}_\alpha, \psi_\alpha, X_\alpha, \mathbb{D}_\alpha, \mathcal{M}^{\mathcal{U}^*}_\alpha\rangle.$$

So $c_0 = d_0 = c$. Our plan, of course, is to construct a tree embedding $\Phi^* \colon \mathcal{T}^* \to \mathcal{U}^*$ by induction on its initial segments, so that the diagram

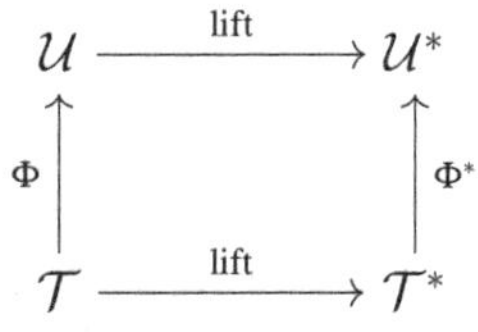

commutes, in the natural sense. The components of Φ^* will be given by

$$\Phi^* = \langle u, v, \langle r_\beta \mid \beta < \mathrm{lh}(\mathcal{T})\rangle, \langle w_\beta \mid \beta + 1 < \mathrm{lh}(\mathcal{T})\rangle\rangle.$$

Notice here that $u^{\Phi^*} = u = u^{\Phi}$, and similarly for v. Because Φ^* is to be a tree embedding, u completely determines the putative Φ^*, and what we have to show is just that the Φ^* it determines is a tree embedding of $\mathcal{T}^*$ into $\mathcal{U}^*$.

For $\gamma \le \mathrm{lh}(\mathcal{T})$, let

$$\Phi^*_\gamma = \Phi^* \upharpoonright \gamma = \langle u \upharpoonright \{\xi \mid \xi + 1 < \gamma\}, v \upharpoonright \gamma, \langle r_\beta \mid \beta < \gamma\rangle, \langle w_\beta \mid \beta + 1 < \gamma\rangle\rangle.$$

We show by induction on γ that

(1) $\Phi^* \upharpoonright \gamma$ is a tree embedding of $\mathcal{T}^* \upharpoonright \gamma$ into $\mathcal{U}^*$,
(2) for $\alpha < \gamma$, $\psi_{v(\alpha)} \circ s_\alpha = r_\alpha \circ \varphi_\alpha$, and
(3) for $\alpha < \gamma$, $r_\alpha(Q_\alpha) = X_{v(\alpha)}$.

Let $(*)_\gamma$ be the conjunction of (1)-(3). The following diagram illustrates the situation:

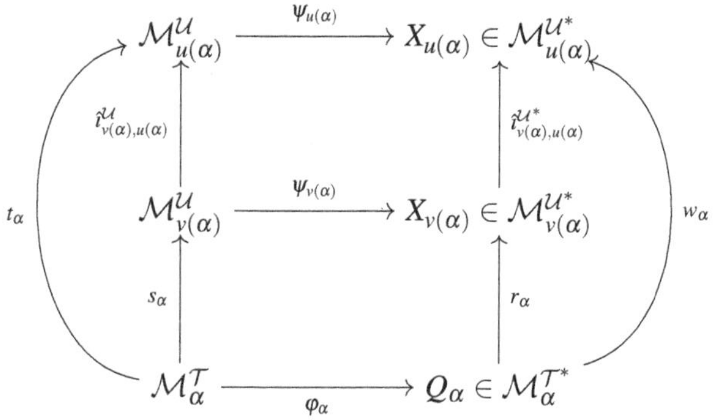

Some care is needed in reading this diagram. The bottom rectangle is just (2) and (3) of our induction hypotheses, and is always valid. The top rectangle involves only the conversion of $\mathcal{U}$ to $\mathcal{U}^*$, so our induction hypotheses are irrelevant. It is valid if and only if $(v(\alpha), u(\alpha)]_U$ does not drop (in model or degree), so that $i^{\mathcal{U}^*}_{v(\alpha),u(\alpha)}(X_{v(\alpha)}) = X_{u(\alpha)}$. In the case that $(v(\alpha), u(\alpha]_U$ drops, something like it is valid. We discuss that below.

To start with, Φ_1^* is given by setting $v(0) = 0$ and r_0 = identity map from $S = \mathcal{M}_0^{\mathcal{T}^*}$ to $S = \mathcal{M}_0^{\mathcal{U}^*}$.

If λ is a limit, and $(*)_\alpha$ for $\alpha < \lambda$, then

$$\Phi_\lambda^* = \bigcup_{\alpha<\lambda} \Phi_\alpha^*$$

in the obvious componentwise sense. It is clear that $(*)_\lambda$ holds.

If $\gamma = \lambda + 1$ for $\lambda < \mathrm{lh}(\mathcal{T})$ a limit such that $(*)_\lambda$, then $\Phi_{\lambda+1}^*$ is just Φ_λ^* together with the map r_λ, defined as follows. Recall that v preserves tree order, and

$$v(\lambda) = \sup_{\alpha<\lambda} v(\alpha).$$

For $\alpha <_T \lambda$ and $x \in \mathcal{M}_\alpha^{\mathcal{T}^*}$, we set

$$r_\lambda(i_{\alpha,\lambda}^{\mathcal{T}^*}(x)) = i_{v(\alpha),v(\lambda)}^{\mathcal{U}^*}(r_\alpha(x)).$$

Using (1) at $\gamma < \lambda$, we see that r_λ is well defined, elementary, and as required for $(*)_{\lambda+1}$.

Finally, suppose we have $\Phi_{\alpha+1}^*$ satisfying $(*)_{\alpha+1}$. The whole of $\Phi_{\alpha+2}^*$ is determined by $u(\alpha)$, which is already given to us, but we must see this choice works; that is, that $(*)_{\alpha+2}$ holds for the system it determines.

The extenders used in $\mathcal{T}^*$ and $\mathcal{U}^*$ are produced as follows. For any ξ, let

$$\begin{aligned}
\sigma_\xi &= \sigma_{Q_\xi}[Q_\xi | \mathrm{lh}(\varphi_\xi(E_\xi^{\mathcal{T}}))]^{\mathbb{C}_\xi},\\
R_\xi &= \mathrm{Res}_{Q_\xi}[Q_\xi | \mathrm{lh}(\varphi_\xi(E_\xi^{\mathcal{T}}))]^{\mathbb{C}_\xi},\\
\tau_\xi &= \sigma_{X_\xi}[X_\xi | \mathrm{lh}(\psi_\xi(E_\xi^{\mathcal{U}}))]^{\mathbb{D}_\xi},\\
Y_\xi &= \mathrm{Res}_{X_\xi}[X_\xi | \mathrm{lh}(\psi_\xi(E_\xi^{\mathcal{U}}))]^{\mathbb{D}_\xi}.
\end{aligned}$$

Then setting $G = E_\alpha^{\mathcal{T}}$, $G^* = E_\alpha^{\mathcal{T}^*}$, $H = E_{u(\alpha)}^{\mathcal{U}}$, and $H^* = E_{u(\alpha)}^{\mathcal{U}^*}$, we have

$$G^* = (B^{\mathbb{C}_\alpha} \circ \sigma_\alpha \circ \varphi_\alpha)(G),$$

and

$$H^* = (B^{\mathbb{D}_{u(\alpha)}} \circ \tau_\alpha \circ \psi_{u(\alpha)})(H).$$

Set now

$$w_\alpha = i_{v(\alpha),u(\alpha)}^{\mathcal{U}^*} \circ r_\alpha,$$

as we are forced to do. Note that $w_\alpha(\mathbb{C}_\alpha) = \mathbb{D}_{u(\alpha)}$. Lemma 8.2.3 below will tell us that the following claim is what we need.

CLAIM 7.5.2. *(a)* $\tau_{u(\alpha)} \circ \psi_{u(\alpha)} \circ i_{v(\alpha),u(\alpha)}^{\mathcal{U}} \circ s_\alpha \restriction (\mathrm{lh}(G)+1) = i_{v(\alpha),u(\alpha)}^{\mathcal{U}^*} \circ r_\alpha \circ \sigma_\alpha \circ \varphi_\alpha \restriction (\mathrm{lh}(G)+1)$.
(b) $w_\alpha(G^*) = H^*$.

PROOF. We prove (a). Suppose first that $(v(\alpha), u(\alpha]_U$ does not drop. In that case, $i^{\mathcal{U}*}_{v(\alpha),u(\alpha)}(X_{v(\alpha)}) = X_{u(\alpha)}$, so the top rectangle in the diagram above is valid. Expanding the diagram, we have

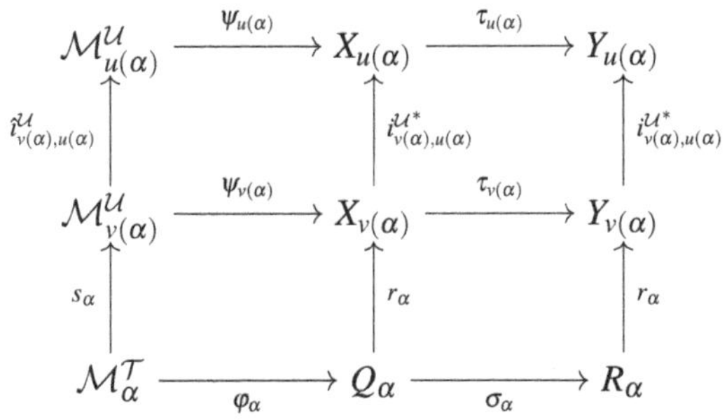

Notice that $r_\alpha(\sigma_\alpha) = \tau_{v(\alpha)}$. So the diagram commutes, and in particular the two routes from $\mathcal{M}^{\mathcal{T}}_\alpha$ to $Y_{u(\alpha)}$ around the outer edges are the same. This gives us (a).

Suppose now that $(v(\alpha), u(\alpha)]_U$ drops. Let $I = s_\alpha(G)$. Since $H = i^{\mathcal{U}}_{v(\alpha),u(\alpha)}(I)$ and $\mathcal{U}$ is λ-non-decreasing, all extenders used along $(v(\alpha), u(\alpha]_U$ have critical points less than or equal to the current image of λ_I.[13] For simplicity, let us assume there is just one such drop, at ξ, where $v(\alpha) <_U \xi \leq_U u(\alpha)$. Let $\theta = U\text{-pred}(\xi)$. We have the following diagram:

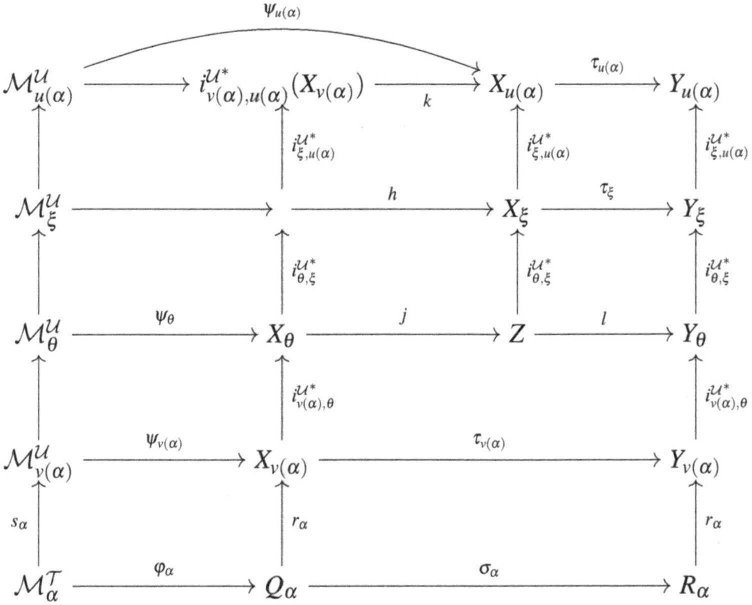

[13]It is possible that, for example, $v(\alpha) = U\text{-pred}(\xi+1)$, $\xi+1 = u(\alpha)$, and $\mathrm{lh}(I) < \mathrm{dom}(E^{\mathcal{U}}_\xi)$. In this case $\mathrm{lh}(E_{u(\alpha)}) < \mathrm{lh}(E^{\mathcal{U}}_\xi)$, so $\mathcal{U}$ is not normal.

In the diagram, $j = \sigma_{X_\theta}[\psi_\theta(\mathcal{M}_\xi^{*,\mathcal{U}}]^{\mathbb{D}_\theta}$ resurrects the drop in $\mathcal{U}$, and $\tau_\theta = l \circ j$. We have $X_\xi = i^{\mathcal{U}^*}_{\theta,\xi}(Z)$, and $\tau_\xi = i^{\mathcal{U}^*}_{\theta,\xi}(l)$. Also, $h = i^{\mathcal{U}^*}_{\theta,\xi}(j)$ and $k = i^{\mathcal{U}^*}_{\xi,u(\alpha)}(h)$. The unlabelled vertical arrows on the far left are the maps of $\mathcal{U}$. Finally, $r_\alpha(\sigma_\alpha) = \tau_{v(\alpha)}$.

The facts we have just enumerated imply that all parts of the diagram commute on the image of $\mathrm{lh}(G)+1$. (For the square at the bottom left, this is our induction hypothesis.) The reason for restricting to the image of $\mathrm{lh}(G)+1$ is that the resurrection maps j,h,k and the τ's and σ_α are partial, defined on initial segments of the models displayed above. But all are defined on the image of $\mathrm{lh}(G)+1$ in that model.

The fact that the two routes from $\mathcal{M}^{\mathcal{T}}_\alpha$ to $Y_{u(\alpha)}$ going along the outer edges are the same when restricted to $\mathrm{lh}(G)+1$ gives us part (a) of the claim.

Part (b) follows easily from the fact that the images of G in $Y_{u(\alpha)}$ along the two outer edges of the diagram are the same.

This proves Claim 7.5.2. ⊣

By Lemma 8.2.3, there is a unique tree embedding Ψ from $\mathcal{T}^* \upharpoonright (\alpha+2)$ to $\mathcal{U}^*$ that extends $\Phi^*_{\alpha+1}$ and satisfies $u^\Psi(\alpha) = u(\alpha)$. Let $\Phi^*_{\alpha+2}$ be this Ψ. We check now that $(*)_{\alpha+2}$ holds.

Let $\beta = T\text{-pred}(\alpha+1)$, and let $\tau = U\text{-pred}(u(\alpha)+1)$. Because Φ is a tree embedding, $\tau \in [v(\beta), u(\beta)]_U$. Let us assume for simplicity that there is no relevant dropping, that is,

(a) $(\alpha+1) \notin D^{\mathcal{T}}$, and

(b) $D^{\mathcal{U}} \cap [v(\beta), v(\alpha+1)] = \emptyset$.

So $\mathcal{M}^{\mathcal{T}}_{\alpha+1} = \mathrm{Ult}(\mathcal{M}^{\mathcal{T}}_\beta, G)$ and $\mathcal{M}^{\mathcal{U}}_{v(\alpha+1)} = \mathrm{Ult}(\mathcal{M}^{\mathcal{U}}_\tau, H)$. Let $\rho = i^{\mathcal{U}}_{v(\beta),\tau} \circ s_\beta$ and $\rho^* = i^{\mathcal{U}^*}_{v(\beta),\tau} \circ r_\beta$. The lifting construction yields $\mathcal{M}^{\mathcal{T}^*}_{\alpha+1} = \mathrm{Ult}(\mathcal{M}^{\mathcal{T}^*}_\beta, G^*)$ and $\mathcal{M}^{\mathcal{U}^*}_{v(\alpha+1)} = \mathrm{Ult}(\mathcal{M}^{\mathcal{U}^*}_\tau, H^*)$, moreover

$$X_{v(\alpha+1)} = i^{\mathcal{U}^*}_{v(\beta),v(\alpha+1)}(X_{v(\beta)}).$$

$r_{v(\alpha+1)}$ is given by the Shift Lemma:

$$r_{v(\alpha+1)}([a,f]^{\mathcal{M}^{\mathcal{T}^*}_\beta}_{G^*}) = [w_\alpha(a), \rho^*(f)]^{\mathcal{M}^{\mathcal{U}^*}_\tau}_{H^*}.$$

The diagram of the situation is shown in Figure 7.5.1. It resembles the diagram associated to our proof that copying commutes with embedding normalization. That is not an accident, of course. Embedding normalization yields tree embeddings, and lifting to a background universe is similar to copying.

We are asked to show that $\psi_{v(\alpha+1)} \circ s_{\alpha+1} = \varphi_{\alpha+1} \circ r_{\alpha+1}$, in other words, that the rectangle on the top face of the cube commutes. We argue just as we did in the proof of Lemma 6.8.2. The rectangle on the bottom commutes by our induction hypothesis. The rectangle in front commutes because $\mathcal{T}^*$ comes from lifting $\mathcal{T}$ to the background universe. The diagram on the back face commutes because $\mathcal{U}^*$ comes from lifting $\mathcal{U}$. The maps on the left face commute because Φ is a tree

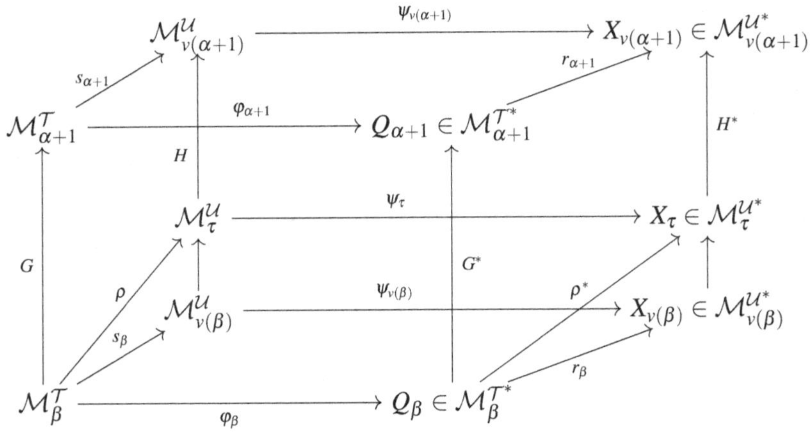

FIGURE 7.5.1

embedding of $\mathcal{T}$ into $\mathcal{U}$. The maps on the right face commute because we obtained $r_{\alpha+1}$ from the Shift Lemma. (This of course is where we used that $H^* = w_\alpha(G^*)$.)

It is clear from these facts that the top rectangle commutes on $\operatorname{ran}(i^{\mathcal{T}}_{\beta,\alpha+1})$. Since $\mathcal{M}^{\mathcal{T}}_{\alpha+1}$ is generated by $\operatorname{ran}(i^{\mathcal{T}}_{\beta,\alpha+1}) \cup \varepsilon(G)$, it is enough to see that the top square commutes on $\varepsilon(G)$. But

$$\begin{aligned}\psi_{v(\alpha+1)} \circ s_{\alpha+1} \restriction \varepsilon(G) &= \tau_{u(\alpha)} \circ \psi_{u(\alpha)} \circ i^{\mathcal{U}}_{v(\alpha),u(\alpha)} \circ s_\alpha \restriction \varepsilon(G)\\ &= i^{\mathcal{U}^*}_{v(\alpha),u(\alpha)} \circ r_\alpha \circ \sigma_\alpha \circ \varphi_\alpha \restriction \varepsilon(G)\\ &= r_{\alpha+1} \circ \varphi_{\alpha+1} \restriction \varepsilon(G).\end{aligned}$$

Line 1 comes from the facts that $s_{\alpha+1}$ agrees with $i^{\mathcal{U}}_{v(\alpha),u(\alpha)} \circ s_\alpha$ on $\varepsilon(G)$ by the way it is defined using the Shift Lemma (cf. Lemma 6.4.8(c)), and that $\psi_{v(\alpha+1)}$ agrees with $\tau_{u(\alpha)} \circ \psi_{u(\alpha)}$ on $\varepsilon(H)$ for a similar reason. Line 2 comes from Claim 7.5.2. Line 3 again comes from using the Shift Lemma, now at the level of $\mathcal{T}^*$ and $\mathcal{U}^*$.[14]

This completes the proof that Σ condenses well on plus trees. The proof that its tails do so as well is similar. Let us now consider the pullback condition, clause (b) of Definition 7.1.9. For this, let us keep our previous notation, but assume that $\operatorname{lh}(\mathcal{T}) = \alpha+1$, $\operatorname{lh}(\mathcal{U}) = \beta+1$, and that $v(\alpha) \leq_U \beta$ and Φ has been extended by adding the t-map

$$\pi = i^{\mathcal{U}}_{v(\alpha),\beta} \circ s_\alpha.$$

[14]In our definition of tree embeddings on plus trees, we allowed the possibility that G is not of plus type, and $H = t_\alpha(G)^+$. In this case, $\varepsilon(H) = t_\alpha(\varepsilon(G)) + 1$.

Let us assume $J \trianglelefteq \mathrm{dom}(\pi)$, and let $K = \pi(J)$. We need to see that $(\Sigma_{\mathcal{U},K})^\pi = \Sigma_{\mathcal{T},J}$. For that, consider the diagram

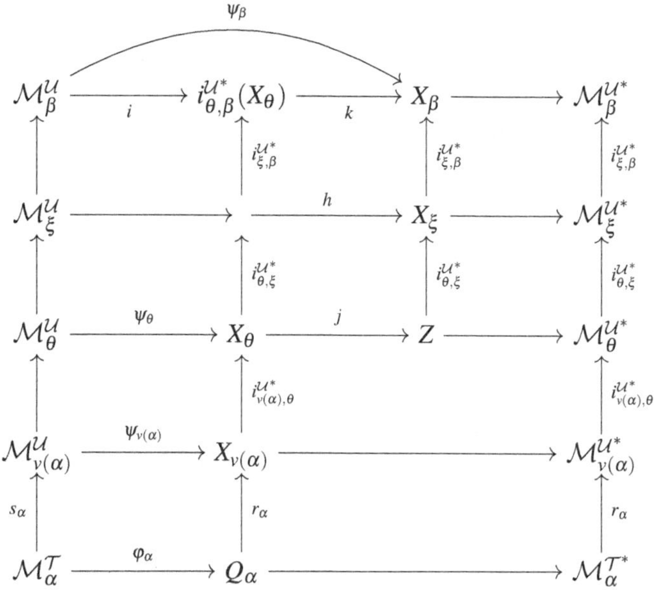

In the diagram, $j = \sigma_{\mu,n}[\psi_\theta(\mathcal{M}^{*,\mathcal{U}}_\xi)]^{\mathbb{D}_\theta}$, and h and k are its images under the $\mathcal{U}^*$ embeddings. We are assuming for definiteness that $\mathcal{U}$ dropped once on $(\nu(\alpha),\beta]_U$, at its step from θ to ξ. The maps j, h, and k are defined only on initial segments of the models displayed, but all are defined on the image of J in that model.

Let $L = \varphi_\alpha(J)$ and $P = \psi_\beta(K)$. Let also $N = i(K) = k^{-1}(P)$. By the commutativity of the left column in the diagram, it is enough to see that the $\mathbb{D}_\beta$-induced strategy of P pulls back under $k \circ i^{\mathcal{U}^*}_{\nu(\alpha),\beta} \circ r_\alpha$ to the $\mathbb{C}_\alpha$-induced strategy of L. The following claims show this. Put $Y = i^{\mathcal{U}^*}_{\nu(\alpha),\beta}(X_{\nu(\alpha)})$.

CLAIM 1. $\Omega(\mathbb{D}_\beta, Y, \Sigma^*_{\mathcal{U}^* \restriction \beta+1})_N = \Omega(\mathbb{D}_\beta, X_\beta, \Sigma^*_{\mathcal{U}^* \restriction \beta+1})^k_P$.

PROOF. This follows at once from Lemma 4.8.8. ⊣

CLAIM 2. $\Omega(\mathbb{D}_{\nu(\alpha)}, X_{\nu(\alpha)}, \Sigma^*_{\mathcal{U}^* \restriction \nu(\alpha)+1}) = \Omega(\mathbb{D}_\beta, Y, \Sigma^*_{\mathcal{U}^* \restriction \beta+1})^{i^{\mathcal{U}^*}_{\nu(\alpha),\beta}}$.

PROOF. Let $\pi = i^{\mathcal{U}^*}_{\nu(\alpha),\beta}$. Because Σ^* has strong hull condensation, it is pullback consistent, so $\Sigma^*_{\mathcal{U}^* \restriction \nu(\alpha+1)} = (\Sigma^*_{\mathcal{U}^* \restriction \beta+1})^\pi$. But

$$\Omega(\mathbb{D}_\beta, Y, \Sigma^*_{\mathcal{U}^* \restriction \beta+1})^\pi = \Omega(\mathbb{D}_{\nu(\alpha)}, X_{\nu(\alpha)}, (\Sigma^*_{\mathcal{U}^* \restriction \beta+1})^\pi)$$

by Corollary 5.1.3. ⊣

CLAIM 3. $\Omega(\mathbb{C}_\alpha, Q_\alpha, \Sigma^*_{\mathcal{T}^* \restriction \alpha+1}) = \Omega(\mathbb{D}_{\nu(\alpha)}, X_{\nu(\alpha)}, \Sigma^*_{\mathcal{U}^* \restriction \nu(\alpha)+1})^{r_\alpha}$.

PROOF. Since Σ^* has strong hull condensation, $\Sigma^*_{\mathcal{T}^*\restriction\alpha+1} = (\Sigma^*_{\mathcal{U}^*\restriction\nu(\alpha)+1})^{r_\alpha}$. We can therefore apply Corollary 5.1.3 again. ⊣

Let $\Lambda = \Omega(\mathbb{D}_\beta, X_\beta, \Sigma^*_{\mathcal{U}^*\restriction\beta+1})_P$. The claims imply that $\Omega(\mathbb{C}_\alpha, Q_\alpha, \Sigma^*_{\mathcal{T}^*\restriction\alpha+1})_L$ is the pullback of Λ under $k \circ i^{\mathcal{U}^*}_{\nu(\alpha),\beta} \circ r_\alpha$, and hence that $\Sigma_{\mathcal{T},J}$ is the pullback of Λ under $k \circ i^{\mathcal{U}^*}_{\nu(\alpha),\beta} \circ r_\alpha \circ \varphi_\alpha$. By commutativity, $\Sigma_{\mathcal{T},J}$ is the pullback of Λ under $\psi_\beta \circ i^{\mathcal{U}}_{\nu(\alpha),\beta} \circ s_\alpha$. But this means that it is the pullback of $\Sigma_{\mathcal{U}\restriction(\beta+1),K}$ under $i^{\mathcal{U}}_{\nu(\alpha),\beta} \circ s_\alpha$, as desired.

This completes the proof of Theorem 7.5.1. ⊣

7.6. Pure extender pairs

We have shown that if Σ^* is a strongly unique iteration strategy for some coarse premouse, then the iteration strategies Σ for premice that it induces via PFS constructions are internally lift consistent, quasi-normalize well, and have strong hull condensation. It seems that all of the nice behavior of iteration strategies one could wish for follows from these properties.[15] One explanation for that is that they imply that Σ can be compared with other such strategies. Because of this, the following is one of our central definitions.

DEFINITION 7.6.1. (P, Σ) is a *pure extender pair* with scope H_δ iff

(1) P is a pfs premouse of type 1, and $P \in H_\delta$,
(2) Σ is a complete (ω, δ)-iteration strategy for P, and
(3) Σ quasi-normalizes well, has strong hull condensation, and is internally lift consistent.

(P, Σ) is *strongly stable* iff P is strongly stable.

We have required that P be of type 1 because it simplifies some statements, and we do not need greater generality.

We are only interested in the case that Σ is absolutely definable. In the most important context, P is countable, Σ has scope H_{ω_1}, and its absolute definability is witnessed by membership in a model of AD^+. At other times we are working under hypotheses that allow us to reach something close to this AD^+ context in a generic extension.

It would be more natural to require that an iteration strategy with scope H_δ be a (δ, δ)-strategy, but then our existence proof for pure extender pairs would need a version of Theorem 7.4.1 that applies to normalizations of infinite stacks.[16] There

[15]But see Remark 7.6.11 below.

[16]The comparison proof only needs to deal with stacks of length 2, because by Lemma 7.6.5 the action of Σ on infinite stacks is determined by its action on single λ-separated trees. For the existence proof, one might try to quote the results of Schlutzenberg in [57] on strategy extension, but there seems

is such a theorem, but it is not needed for the analysis of HOD in models of AD^+, so we have elected not to go into it in this book.[17]

The following theorem summarizes much of our work so far.

THEOREM 7.6.2. *Let $c = \langle P, \varphi, Q, \mathbb{C}, S\rangle$ be a conversion stage, and suppose that Σ^* is a strongly unique (ω, δ)-iteration strategy for $(S, \in, w^{\mathbb{C}}, \mathcal{F}^{\mathbb{C}})$, and let $\Sigma = \Omega(c, \Sigma^*)$; then (P, Σ) is a pure extender pair with scope H_δ.*

PROOF. This follows at once from Theorems 5.4.5, 7.2.9, 7.4.1, and 7.5.1. ⊣

It follows immediately from the definitions that any iterate of a pure extender pair is also a pure extender pair. That is, if (P, Σ) is a pure extender pair with scope H_δ, and s is a P-stack by Σ with last model Q, then (Q, Σ_s) is a pure extender pair with scope H_δ. We have already in effect proved another useful basic fact, namely, that elementary submodels of pure extender pairs are pure extender pairs. More precisely,

LEMMA 7.6.3. *Let (M, Ω) be a pure extender pair with scope H_δ, and let $\pi \colon N \to M$ be nearly elementary, where N is a pfs premouse; then (N, Ω^π) is a pure extender pair with scope H_δ.*

PROOF. Clearly, Ω^π is a complete iteration strategy for N with scope H_δ. Ω^π normalizes well by Theorem 7.1.6, and has strong hull condensation by Proposition 7.1.11. Similar calculations show that internal lift consistency pulls back under π. ⊣

Another elementary fact is

LEMMA 7.6.4. *Let (M, Ω) be a pure extender pair; then*

(1) Ω is pullback consistent, and

(2) if s is a stack by Ω and $P \trianglelefteq N \trianglelefteq \mathcal{M}_\infty(s)$, then $(\Omega_{s,N})_P = \Omega_{s,P}$.

PROOF. We proved this in Lemma 7.1.10. ⊣

Concerning pairs with scope going beyond HC, the following lemmas will be useful. The first says that the strategy restricted to countable λ-separated trees determines the strategy on all trees.

LEMMA 7.6.5. *Let (P, Σ) and (P, Λ) be pure extender pairs with scope H_δ, and suppose that Σ and Λ agree on countable λ-separated plus trees; then $\Sigma = \Lambda$.*

PROOF. The two strategies are internally lift consistent, so they are determined by their action on finite, maximal stacks of plus trees. They quasi-normalize well, so in fact they are determined by their action on single plus trees. Since $\mathcal{T}$ is a pseudo-hull of its λ-separation $\mathcal{T}^{\text{sep}}$ and the strategies have strong hull condensation, they are determined by their action on λ-separated trees.

---

to be no way to show that the extended strategies are pushforward consistent if they are constructed by that method.

[17] See [57] and [64].

Suppose then we have a λ-separated tree $\mathcal{T}$ of limit length by both Σ and Λ, with $\Sigma(\mathcal{T}) = b$ and $\Lambda(\mathcal{T}) = c$, and $b \neq c$. Let H be countable and transitive, and

$$\pi\colon H \to V_\gamma$$

be elementary, with γ large and everything relevant in $\mathrm{ran}(\pi)$. Let $\bar{P}, \bar{\mathcal{T}}, \bar{b}, \bar{c}$ in H be the collapses of $P, \mathcal{T}, b, c$. So $\bar{b} \neq \bar{c}$. Letting

$$\mathcal{U} = \pi\bar{\mathcal{T}},$$

it is easy to see that $\mathcal{U}^\frown \bar{b}$ is a pseudo-hull of $\mathcal{T}^\frown b$. (For example, the relevant u-map is just $\pi \restriction \mathrm{lh}(\mathcal{U})$.) Similarly, $\mathcal{U}^\frown \bar{c}$ is a pseudo-hull of $\mathcal{T}^\frown c$. But by strong hull condensation, $\mathcal{U}^\frown \bar{b}$ is by Σ and $\mathcal{U}^\frown \bar{c}$ is by Λ, so $\bar{b} = \bar{c}$ because the strategies agree on countable λ-separated trees. This is a contradiction. ⊣

Remark 7.6.6. Assuming AD^+, if (P, Σ) is a pure extender pair with scope HC, then Σ is also determined by its action on countable λ-tight normal trees. The proof here involves strategy comparison.

The reader should compare the following lemma to Proposition 2.7.13.

LEMMA 7.6.7. *Let (P, Σ) be a pure extender pair with scope H_δ, and let $j\colon V \to M$ be elementary, where M is transitive and* $\mathrm{crit}(j) > |P|$; *then $j(\Sigma)$ and Σ agree on all trees in $j(H_\delta) \cap H_\delta$.*

PROOF. Otherwise we have a plus tree $\mathcal{T}$ with distinct cofinal branches b and c such that $\mathcal{T}^\frown b$ is by Σ and $\mathcal{T}^\frown c$ is by $j(\Sigma)$. As in the proof of the last lemma, this gives us a countable plus tree $\mathcal{U}$ on P with distinct cofinal branches $\bar{b}$ and $\bar{c}$ such that $\mathcal{U}^\frown \bar{b}$ is a pseudo-hull of $\mathcal{T}^\frown b$ and $\mathcal{U}^\frown \bar{c}$ is a pseudo-hull of $\mathcal{T}^\frown c$. Thus $\Sigma(\mathcal{U}) = \bar{b}$. But since $\mathcal{U}$ is countable, and M is wellfounded,

$$M \models \mathcal{U}^\frown c \text{ is a pseudo-hull of } \mathcal{T}^\frown c.$$

Thus $j(\Sigma)(\mathcal{U}) = \bar{c}$. But $\mathcal{U}$ is countable, hence fixed by j, so $\Sigma(\mathcal{U}) = \bar{c}$, a contradiction. ⊣

Returning to regularity properties, let us consider strategy coherence. Let (P, Σ) be a pure extender pair, $\mathcal{T}$ a plus tree by Σ, and $N \trianglelefteq \mathcal{M}^{\mathcal{T}}_\alpha$ and $N \trianglelefteq \mathcal{M}^{\mathcal{T}}_\beta$. Strategy coherence requires that $\Sigma_{\mathcal{T}\restriction\alpha+1,N} = \Sigma_{\mathcal{T}\restriction\beta+1,N}$. In Theorem 5.2.5 we proved an approximation to this directly in the case that Σ is induced by a strongly unique Σ^*. We can prove the same approximation abstractly for pure extender pairs, using the fact that they quasi-normalize well.

LEMMA 7.6.8. *Let (P, Σ) be a pure extender pair, let $\mathcal{T}$ be a plus tree on P by Σ and let N be an initial segment of its last model. Let $\nu + 1 < \mathrm{lh}(\mathcal{T})$, and suppose that either*

(a) $o(N) < \hat{\lambda}(E^{\mathcal{T}}_\nu)$, or

(b) $E^{\mathcal{T}}_\nu$ is of plus type, and $o(N) \leq \mathrm{lh}(E^{\mathcal{T}}_\nu)$;

then $\Sigma_{\mathcal{T}\restriction\nu+1,N} = \Sigma_{\mathcal{T},N}$.

PROOF. Let $R = \mathcal{M}_\nu^{\mathcal{T}}$ and $S = \mathcal{M}_\infty^{\mathcal{T}}$.

$$Q = \begin{cases} R || \mathrm{lh}(E_\nu^{\mathcal{T}}) & \text{if } E_\nu^{\mathcal{T}} \text{ is of plus type,} \\ R || \hat{\lambda}(E_\nu^{\mathcal{T}}) & \text{otherwise.} \end{cases}$$

Since $N \trianglelefteq Q$, it is enough to show that $\Sigma_{\mathcal{T} \restriction \nu+1, Q} = \Sigma_{\mathcal{T},Q}$, for then Lemma 7.6.4(b) implies that $\Sigma_{\mathcal{T} \restriction \nu+1,N} = \Sigma_{\mathcal{T},N}$. So let $\mathcal{U}$ be a plus tree of limit length on Q that is by both strategies.

Our plan is to show that $\langle \mathcal{T} \restriction \nu+1, \mathcal{U} \rangle$ and $\langle \mathcal{T}, \mathcal{U} \rangle$ have the same quasi-normalization. Unfortunately, this does not make literal sense, because $\mathcal{U}$ is on Q, not R or S, and we have not defined quasi-normalization for non-maximal stacks. So first we lift $\mathcal{U}$ to trees on R and S.

Let $\mathcal{U}_0$ and $\mathcal{U}_1$ be the lifts of $\mathcal{U}$ to R and S under the identity map, so that $\langle \mathcal{T} \restriction \nu+1, \mathcal{U}_0 \rangle$ and $\langle \mathcal{T}, \mathcal{U}_1 \rangle$ are by Σ by internal lift consistency. By internal lift consistency, it is enough to show that $\Sigma_{\mathcal{T} \restriction \nu+1,R}(\mathcal{U}_0) = \Sigma_{\mathcal{T},S}(\mathcal{U}_1)$. We shall show

(a) $V(\mathcal{T} \restriction \nu+1, \mathcal{U}_0) = V(\mathcal{T}, \mathcal{U}_1)$, and

(b) for any cofinal branch b of $\mathcal{U}$, $\mathrm{br}(b, \mathcal{T} \restriction \nu+1, \mathcal{U}_0) = \mathrm{br}(b, \mathcal{T}, \mathcal{U}_1)$.

This is enough: letting $a = \Sigma(V(\mathcal{T}, \mathcal{U}_1))$, there is a unique b such that $\mathrm{br}(b, \mathcal{T}, \mathcal{U}_1) = a$, moreover $\Sigma_{\mathcal{T},S}(\mathcal{U}_1) = b$ because Σ quasi-normalizes well. But $b = \Sigma_{\mathcal{T} \restriction \nu+1,R}(\mathcal{U}_0)$ by (a) and (b), as desired.

Since $o(Q)$ is a regular cardinal in M_ξ and $o(Q) \le \rho^-(S)$, $\mathcal{M}_\alpha^{\mathcal{U}} \trianglelefteq \mathcal{M}_\alpha^{\mathcal{U}_1}$ for all α and the lift map is the identity. Thus $\mathcal{U}_1$ uses the same extenders as $\mathcal{U}$. Using this it is easy to see that $V(\mathcal{T}, \mathcal{U}_1) = V(\mathcal{T} \restriction \nu+2, \mathcal{U}_1)$. So we may assume

$$\mathrm{lh}(\mathcal{T}) = \nu + 2.$$

For $\xi < \mathrm{lh}(\mathcal{U})$, let $\mathcal{W}_\xi$ be the ξ-th tree in the meta-tree associated to $V(\mathcal{T} \restriction \nu + 1, \mathcal{U}_0)$ and $\mathcal{V}_\xi$ be the ξ-th tree in the meta-tree associated to $V(\mathcal{T}, \mathcal{U}_1)$. So $\mathcal{W}_0 = \mathcal{T} \restriction \nu - 1$ and $\mathcal{V}_0 = \mathcal{T} = \mathcal{T} \restriction \nu + 2$. For $\xi <_U \eta$ let

$$\Phi_{\xi,\eta} : \mathcal{W}_\xi \to \mathcal{W}_\eta$$

and

$$\Psi_{\xi,\eta} : \mathcal{V}_\xi \to \mathcal{V}_\eta$$

be the (possibly partial) branch embeddings of the two meta-trees. Let $\mathrm{lh}(\mathcal{W}_\xi) = z^0(\xi) + 1$ and $\mathrm{lh}(\mathcal{V}_\xi) = z^1(\xi) + 1$, and let

$$R_\xi = \mathcal{M}_{z^0(\xi)}^{\mathcal{W}_\xi}$$

and

$$S_\xi = \mathcal{M}_{z^1(\xi)}^{\mathcal{V}_\xi}$$

be the two last models. We shall show by induction that $\mathcal{W}_\xi = \mathcal{V}_\xi \restriction z^0(\xi) + 1$, and either $z^1(\xi) = z(\xi) + 1$ or $z^1(\xi) = z^0(\xi)$. (The latter can only happen along branches $[0,\xi]_U$ of $\mathcal{U}_0$ that have dropped.) We shall also have that $\Phi_{\xi,\eta} = \Psi_{\xi,\eta} \restriction \mathrm{dom}(\Phi_{\xi,\eta})$ whenever $\xi <_U \eta$. Let

$$\sigma_\xi : \mathcal{M}_\xi^{\mathcal{U}_0} \to R_\xi$$

and

$$\tau_\xi : \mathcal{M}_\xi^{\mathcal{U}_1} \to S_\xi$$

be the final σ-maps of the two quasi-normalizations.

The lift maps from the models of $\mathcal{U}$ to those of $\mathcal{U}_1$ are the identity, but those to the models of $\mathcal{U}_0$ may not be. Let

$$\pi_\xi : \mathcal{M}_\xi^{\mathcal{U}} \to J_\xi \trianglelefteq \mathcal{M}_\xi^{\mathcal{U}_0}$$

be this map. Thus π_0 is the identity, and $\mathcal{M}_0^{\mathcal{U}} = J_0 = Q$. Let also

$$Q_\xi = \sigma_\xi(J_\xi).$$

As usual, if $J_\xi = \mathcal{M}_\xi^{\mathcal{U}_0}$, then $Q_\xi = R_\xi$. We shall also maintain by induction that $\tau_\xi = \sigma_\xi \circ \pi_\xi$. The following diagram summarizes the situation at stage ξ.

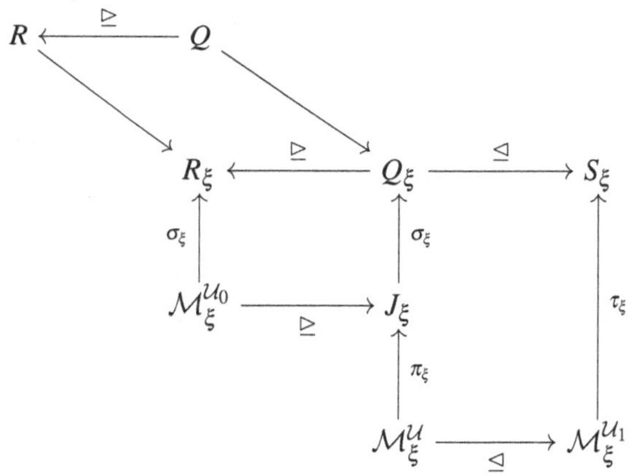

If $[0,\xi]_U \cap D^{\mathcal{U}} = \emptyset$, then the initial segments displayed are all proper. Otherwise parts of the diagram may collapse. Our inductive hypotheses are

$(\dagger)_\xi$

(i) $\mathcal{W}_\xi = \mathcal{V}_\xi \restriction z^0(\xi)+1$, and $z^1(\xi) \in \{z^0(\xi), z^0(\xi)+1\}$.

(ii) For $\xi <_U \eta$, $\Phi_{\xi,\eta} = \Psi_{\xi,\eta} \restriction \mathrm{dom}(\Phi_{\xi,\eta})$.

(iii) $\tau_\xi = \sigma_\xi \circ \pi_\xi$.

To prove $(\dagger)_{\gamma+1}$, where $\xi = U\text{-pred}(\gamma+1)$, we chase through the diagrams corresponding to $(\dagger)_\xi$ and $(\dagger)_\gamma$. The first thing to see is that the two meta-trees use the same extender as their F_γ. But the $\mathcal{W}$ system uses $\sigma_\gamma(E_\gamma^{\mathcal{U}_0}) = \sigma_\gamma \circ \pi_\gamma(E_\gamma^{\mathcal{U}})$, and the $\mathcal{V}$ system uses $\tau_\xi(E_\gamma^{\mathcal{U}_1}) = \tau_\xi(E_\gamma^{\mathcal{U}})$. These are the same by $(\dagger)_\gamma$.

Let $F = \sigma_\gamma(E_\gamma^{\mathcal{U}_0}) = \tau_\gamma(E_\gamma^{\mathcal{U}_1})$. The next thing to see is that

$$\alpha_0(\mathcal{V}_\gamma, F) = \alpha_0(\mathcal{W}_\gamma, F).$$

Here is precisely where we use the restrictions on $E_\nu^{\mathcal{T}}$ that we have imposed. (Without them, this could fail at $\gamma = 0$.) If $\mathcal{V}_\gamma = \mathcal{W}_\gamma$ we are done. Otherwise, $\mathcal{V}_\gamma = \mathcal{W}_\gamma^\frown\langle G\rangle$, where

$$G = \tau_\gamma \circ i_{0,\gamma}^{\mathcal{U}_1}(E_\nu^{\mathcal{T}})$$

and

$$Q_\gamma = \begin{cases} R_\gamma || \mathrm{lh}(G) & \text{if } G \text{ is of plus type} \\ R_\gamma || \hat{\lambda}(G) & \text{otherwise.} \end{cases}$$

F is on the sequence of Q_γ, so $\alpha_0(\mathcal{V}_\gamma, F) \le z^0(\gamma)$.[18] This implies $\alpha_0(\mathcal{V}_\gamma, F) = \alpha_0(\mathcal{W}_\gamma, F)$, as desired.

The remainder of the proof of $(\dagger)_{\gamma+1}$ consists of calculations like those we have done several times already, so we omit it. $(\dagger)_\lambda$ for λ a limit is a routine consequence of the commutativity clauses in $(\dagger)_\xi$ for $\xi < \lambda$. ⊣

As a corollary we get strategy coherence within λ-separated trees.

COROLLARY 7.6.9. *Let (P,Σ) be a pure extender pair. Suppose that $s^\frown\langle\mathcal{T}\rangle$ and $s^\frown\langle\mathcal{U}\rangle$ are stacks by Σ, and N is an initial segment of both last models. Suppose that $\mathcal{T}$ and $\mathcal{U}$ are λ-separated; then $\Sigma_{s^\frown\langle\mathcal{T}\rangle,N} = \Sigma_{s^\frown\langle\mathcal{U}\rangle,N}$.*

In light of Theorem 5.2.5, we could have made this approximation to strategy coherence part of the definition of pure extender pair without affecting the main results of the book.

We shall show in the next chapter that pure extender pairs can be compared. Here is some terminology related to the comparison process. It is appropriate to the comparison of strongly stable, type 1 pairs.

DEFINITION 7.6.10. Let (P,Σ) and (Q,Ψ) be pure extender pairs with common scope H_θ; then

(a) $(P,\Sigma) \trianglelefteq (Q,\Psi)$ iff $P \trianglelefteq Q$ and $\Sigma = \Psi_P$.
(b) $(P,\Sigma) \triangleleft (Q,\Psi)$ iff $P \triangleleft Q$ and $\Sigma = \Psi_P$.
(c) (P,Σ) *iterates past* (Q,Ψ) iff there is a λ-separated tree $\mathcal{T}$ on P by Σ with last model R such that $(Q,\Psi) \trianglelefteq (R, \Sigma_{\mathcal{T},R})$. If P-to-R drops, or if $Q \triangleleft R$, then we say that (P,Σ) *iterates strictly past* (Q,Ψ). If $Q = R$ and P-to-R does not drop, then we say (P,Σ) *iterates to* (Q,Ψ).

Note that if (P,Σ) iterates past (Q,Ψ), then the λ-separated tree $\mathcal{T}$ on P witnessing this is determined completely by Q and Σ: it comes from iterating away least

[18] Without our hypotheses, $\alpha_0(\mathcal{V}_\gamma, F) = z^0(\gamma) + 1$ would be possible.

extender disagreements, with the Q side never moving.[19] No strategy disagreements show up along the way, because there are no strategy disagreements at the end and (P,Σ) is strategy coherent.

We shall show that assuming AD^+, for any two strongly stable, type 1 pairs (P,Σ) and (Q,Ψ) with scope HC, there is a pair (R,Ω) such that either

(i) (P,Σ) iterates to (R,Ω), and (Q,Ψ) iterates past (R,Ω), or
(ii) (Q,Ψ) iterates to (R,Ω), and (P,Σ) iterates past (R,Ω).

We believe that it is possible to compare pairs are not strongly stable or not of type 1, but the possible termination patterns involve some complexity that we do not need to go into.[20]

Remark 7.6.11. There is one further property of the pure extender pairs that are produced in PFS constructions done in a strongly uniquely iterable background universe that, unlike the ones described above, does not seem to follow abstractly (even under AD^+) from the definition. Suppose that (P,Σ) is a pure extender pair and $i^{\mathcal{U}}: P \to Q$ is an iteration map by Σ. Suppose that $\mathcal{T}$ is a plus tree by Σ, and that $\mathcal{T} \in P$. Must $i^{\mathcal{U}}(\mathcal{T})$ be by $\Sigma_{\mathcal{U},Q}$? This is true when (P,Σ) is produced by a PFS construction as above, but it does not seem to follow abstractly.[21] We shall eventually summarize the proper general form of this property by saying that Σ is *pushforward consistent*.[22] It is crucial for a theory of strategy mice.

If we assume AD^+ and let (P,Σ) be a pure extender pair with scope HC, then by the comparison theorem of the next chapter, there is an iterate (Q,Λ) of (P,Σ) such that (Q,Λ) is pushforward consistent. But this does not seem to imply that (P,Σ) itself has the property.

---

[19] In the AD^+ context, $\mathcal{T}$ can also be taken to be λ-tight and normal.

[20] See Theorem 4.6.12.

[21] Pullback consistency implies that the pushforward $i^{\mathcal{U}}\mathcal{T}$ of $\mathcal{T}$ is by $\Sigma_{\mathcal{U},Q}$, but $i^{\mathcal{U}}\mathcal{T}$ is only a pseudo-hull of $i^{\mathcal{U}}(\mathcal{T})$, possibly a proper one.

[22] That more general form states that if $X,Y \in M$, and $X \subseteq \Sigma$ and $Y \cap \Sigma = \emptyset$, then $i^{\mathcal{U}}(X) \subseteq \Sigma_{\mathcal{U}}$ and $i^{\mathcal{U}}(Y) \cap \Sigma_{\mathcal{U}} = \emptyset$. In the case of strategy mice, we can take X and Y to be relative complements.

Chapter 8

COMPARING ITERATION STRATEGIES

The standard Comparison Theorem of inner model theory applies to mice. One statement of it is

THEOREM 8.0.12. *Let P and Q be premice of size $\leq \theta$, and suppose Σ and Ψ are $\theta^+ + 1$-iteration strategies for P and Q respectively; then there are normal trees by Σ and $\mathcal{U}$ by Ψ of size θ, with last models R and S, such that either*

(a) $R \trianglelefteq S$, and P-to-R does not drop, or

(b) $S \trianglelefteq R$, and Q-to-S does not drop.

This theorem, and the comparison process behind it, are the main engines driving inner model theory, but they have a clear defect. We have not really compared the data. We were given (P,Σ) and (Q,Ψ), and we only compared P with Q. Whether it is the P-side or the Q-side that comes out shorter could depend on which iteration strategies for P and Q we use. (See Proposition 9.3.11.)

The standard way to avoid this problem when it might arise is to make assumptions that imply P and Q can have at most one iteration strategy. This is good enough for practical purposes in many situations, but it is unnatural, and leads to somewhat awkward devices like the Weak Dodd-Jensen Lemma. The better response would be to strengthen the Comparison Theorem by finding a process which will compare all the data.

In this chapter, we shall do that. The resulting comparison process is the key to developing the theory of a class of strategy mice sufficiently rich to analyze HOD in models of $\mathsf{AD}_{\mathbb{R}} + \mathsf{NLE}$. This theory is the practical payoff for the work we do here, but one can see without knowing anything about HOD in models of determinacy that we are filling a gap in basic inner model theory.

We shall prove the main comparison theorem for strongly stable pure extender pairs (P,Σ). We believe that it is possible to compare pairs (P,Σ) that are not strongly stable or not of type 1, but this adds some complexity[1], and we do not need to do it in this book.[2] One can probably compare premouse pairs in ms-indexing in a very similar way. Once strategy mouse pairs have been properly

[1] See Theorem 4.6.12.

[2] Section 4.10 shows how to avoid such comparisons in the one place they seem relevant at first.

defined, the comparison argument of this chapter will apply to them with little change.

What really matters is that Σ quasi-normalizes well, is internally lift consistent, and has strong hull condensation. This good behavior of Σ is used heavily in the comparison argument, and it is unlikely that one could drop it as a hypothesis. It does not seem to be a restrictive hypothesis; for example, every iterable P has an iteration strategy with these properties. (See Proposition 9.3.10.)

By Lemma 7.6.5, iteration strategies with this good behavior are determined by their action on λ-separated trees. We shall make strong use of that in this chapter. In the λ-separated case, the agreement of maps in conversion systems and tree embeddings is better, in that it is tied to the lengths of the extenders that have been used up to some point, rather than to their λ's. A λ-separated tree is determined by its last model, together with the choice of branches at limit ordinals. If $\mathcal{T}$ is λ-separated, then

$$\begin{aligned} T\text{-pred}(\alpha+1) &= \text{least } \beta \text{ s.t. } \mathrm{dom}(E_\alpha^{\mathcal{T}}) \trianglelefteq \mathcal{M}_\beta^{\mathcal{T}} \\ &= \text{least } \beta \text{ s.t. } \mathrm{dom}(E_\alpha^{\mathcal{T}}) \trianglelefteq_0 \mathcal{M}_\beta^{\mathcal{T}}. \end{aligned}$$

$V(\mathcal{T},\mathcal{U}) = W(\mathcal{T},\mathcal{U})$ when $\mathcal{T}$ is λ-separated, and if $\mathcal{U}$ is also λ-separated, then $V(\mathcal{T},\mathcal{U})$ is λ-separated. None of this is true for plus trees in general. For that reason, when we compare (P,Σ) with (Q,Λ), what we shall compare directly are the actions of Σ and Λ on stacks of λ-separated trees.[3]

The first three sections contain some preliminary lemmas. The last contains the comparison argument.

8.1. Iterating into a backgrounded premouse

The idea that if one compares a countable mouse P with some level $M_{\nu,k}^{\mathbb{C}}$ of a background construction, then only the P side moves, goes back to Baldwin and Mitchell, and in some sense even to Kunen.[4] The proof is very much like the proof one learns now that least disagreement comparisons terminate. The Skolem-hull-of-V embedding is replaced by by some background extender embedding, and one gets thereby that no backgrounded extender ever particpates in a disagreement.

The argument has been used many times at the level of Woodin cardinals (cf. [48, Theorem 2.5] for example), but we know of no exposition in print of the very simple form we need in this book. So we give one here. We also check that the iteration tree on P can be taken to be λ-separated.[5]

We need to take some care when comparing pfs premice by this method, however. All $M_{\nu,k}^{\mathbb{C}}$ have type 1, but even if we start with a P of type 1, nondropping iterates

[3] If $\mathcal{T}$ is merely normal, it is possible that $\mathrm{crit}(E_\alpha^{\mathcal{T}}) = \hat{\lambda}(E_\beta^{\mathcal{T}})$, in which case the second equality dislpayed fails.

[4] See [27] and [2].

[5] See [60] for a variation on the method.

of it could have type 2. The net effect of this is that P could iterate to $\mathrm{Ult}(M^{\mathbb{C}}_{\nu,k}, D)$, for some order zero D on the sequence of $M^{\mathbb{C}}_{\nu,k}$, and not to any actual level of $\mathbb{C}$. Fortunately, we can arrange that this awkward case does not arise by restricting ourselves to strongly stable P.[6]

Recall here that M is strongly stable iff there is no M-total extender E on the M-sequence such that $\mathrm{crit}(E) = \eta^M_{k(M)}$. By Lemma 4.4.6, if M is a strongly stable type 1 pfs premouse, and $\mathcal{T}$ is a plus tree on M, then all $\mathcal{M}^{\mathcal{T}}_\alpha$ are type 1 pfs premice, and all branch embeddings $\hat{\imath}^{\mathcal{T}}_{\alpha,\beta}$ are elementary and exact.

DEFINITION 8.1.1. Let M and P be premice, and let Σ be an iteration strategy for P; then

(a) (P,Σ) *iterates past* M iff there is a λ-separated iteration tree $\mathcal{T}$ by Σ on P with last model Q such that $M \trianglelefteq Q$,
(b) (P,Σ) *iterates to* M iff there are $\mathcal{T}$ and Q as in (a), and moreover, $M = Q$, and the branch P-to-Q of $\mathcal{T}$ does not drop.
(c) (P,Σ) *iterates strictly past* M iff it iterates past M, but not to M.

LEMMA 8.1.2. *(Only the mouse moves.) Let $\mathbb{C}$ be a PFS construction such that $\mathcal{F}^{\mathbb{C}} \subseteq V_\delta$, where δ is inaccessible. Suppose that all extenders in $\mathcal{F}^{\mathbb{C}}$ have critical point $\geq \kappa$, and let P be a strongly stable pfs premouse such that $|P| < \kappa$. Let Σ be a $\mathfrak{E}$-iteration strategy for P, and suppose that whenever $E^* \in \mathcal{F}^{\mathbb{C}}$, then*

$$i_{E^*}(\Sigma) \subseteq \Sigma.$$

Let $M \in \mathrm{lev}(\mathbb{C})$, and suppose that (P,Σ) iterates strictly past N for all $N <_{\mathbb{C}} M$; then (P,Σ) iterates past M.

PROOF. We deal first with the case that M is a successor level of $\mathbb{C}$. This is the place where we use that M is strongly stable.

CLAIM 1. *For any $N \in \mathrm{lev}(\mathbb{C})$, if (P,Σ) iterates strictly past N, then (P,Σ) iterates past its core $\mathfrak{C}(N)$.*

PROOF. Let $\mathcal{T}$ with last model Q witness that (P,Σ) iterates strictly past N. If N and $\mathfrak{C}(N)$ are the same except for their distinguished soundness degrees, then $\mathcal{T}$ witnesses that (P,Σ) iterates past $\mathfrak{C}(N)$ (perhaps not strictly), as desired. Otherwise N is not sound. Because P is strongly stable, Q has type 1, and thus its proper initial segments are sound (not just almost sound). Thus $Q = N$. Lemma 4.4.6 then inplies that $\mathcal{T}$ dropped on the way to Q, and $\mathfrak{C}(N) = \mathcal{M}^{*,\mathcal{T}}_\xi$ for some ξ on the main branch of $\mathcal{T}$. This implies that $\mathcal{T} \restriction \xi + 1$ witnesses that (P,Σ) iterates past $\mathfrak{C}(N)$. ⊣

[6]In the solidity/universality proof for strategy mice, we may need to compare a type 1 M that is not strongly stable with the levels of some construction $\mathbb{C}$. But in that case, we can simply replace M with $N = \mathrm{Ult}_k(\bar{\mathfrak{C}}_k(M), D)$, where D is the order zero measure of M on η^M_k. It is easy to see that $\eta^N_k = \eta^M_k$, and N is a strongly stable premouse of type 1A. This is what we did in Section 4.10.

Next is the case that it is the ω-th level after some point.

CLAIM 2. *If* $\nu = \mu + 1$, *and* (P, Σ) *iterates strictly past* $M_{\mu,k}$ *for all* $k < \omega$; *then* (P, Σ) *iterates past* $M_{\nu,0}$.

PROOF. The bare premouse $\hat{M}_{\mu,k}$ is eventually constant as $k \to \omega$. Thus there is a fixed λ-separated tree $\mathcal{T}$ of minimal length witnessing that (P, Σ) iterates strictly past $M_{\mu,k}$ for all $k < \omega$. Letting Q be the last model of $\mathcal{T}$, we have $M_{\mu,k} \lhd Q$ for all sufficiently large k, and thus $M_{\mu+1,0} \unlhd Q$. ⊣

Next we have the case that $M = M^{\mathbb{C}}_{\nu,0}$ for some limit ordinal ν, and M is passive.

CLAIM 3. *If* ν *is a limit ordinal, and* (P, Σ) *iterates strictly past* $M_{\eta,j}$ *for all* $\eta < \nu$, *and* $M_{\nu,0}$ *is passive, then* (P, Σ) *iterates past* $M_{\nu,0}$.

PROOF. This is immediate. ⊣

By the claims, we may assume that $M = M_{\nu,0}$ is active, and (P, Σ) iterates past $M^{<\nu}$. Let E be the last extender of M, and let $E^* = F^{\mathbb{C}}_{\nu}$ be the background extender for E, and let $\mathcal{T}$ be the λ-separated tree by Σ on P of minimal length iterating it past $M||\text{lh}(E) = M^{<\nu}$. Since $\mathcal{T}$ is normal, it is completely determined by $M^{<\nu}$ and Σ: for each $\alpha + 1 < \text{lh}(\mathcal{T})$, $E^{\mathcal{T}}_{\alpha} = F^+$, where F is on the sequence of $\mathcal{M}^{\mathcal{T}}_{\alpha}$ and $\mathcal{M}^{\mathcal{T}}_{\alpha}||\text{lh}(F) = M^{<\nu}|\text{lh}(F)$. Since the lemma is failing, E gets used in the comparison of P with M. So setting $\alpha + 1 = \text{lh}(\mathcal{T})$, we have that

(i) $M||\text{lh}(E) = \mathcal{M}^{\mathcal{T}}_{\alpha}||\text{lh}(E)$,
(ii) $M|\text{lh}(E) \neq \mathcal{M}^{\mathcal{T}}_{\alpha}|\text{lh}(E)$, and
(iii) for all $\xi < \alpha$, $\text{lh}(E^{\mathcal{T}}_{\xi}) < \text{lh}(E)$.

Let $\kappa = \text{crit}(E)$, let $i_{E^*} : V \to N$ be the canonical embedding, and let $\mathcal{U} = i_{E^*}(\mathcal{T})$. Note that because $|P| < \kappa$ and κ is a (measurable) cardinal, $\kappa \leq \alpha$. Let $\lambda = i_{E^*}(\kappa)$.

CLAIM 4. $\kappa <_U \lambda$, $[\kappa, \lambda)_U$ *does not drop, and* $i_{E^*} \upharpoonright \mathcal{M}^{\mathcal{T}}_{\kappa} = i^{\mathcal{U}}_{\kappa,\lambda}$.

PROOF. If $\beta <_T \kappa$, then $\beta = i_{E^*}(\beta) <_U \lambda$. Since $[0, \lambda)_U$ is a closed set of ordinals, $\kappa \leq_U \lambda$. Since $[0, \kappa)_T$ has only finitely many drops, these are the same as the drops of $[0, \lambda)_U$, so $[\kappa, \lambda)_U$ does not drop. Finally, if $x \in \mathcal{M}^{\mathcal{T}}_{\kappa}$, then we have $\beta <_T \kappa$ and $\bar{x}$ such that $i^{\mathcal{T}}_{\beta,\kappa}(\bar{x}) = x$. But then

$$\begin{aligned} i_{E^*}(x) &= i_{E^*}(i^{\mathcal{T}}_{\beta,\kappa}(\bar{x})) \\ &= i_{E^*}(i^{\mathcal{T}}_{\beta,\kappa})(\bar{x}) \\ &= i^{\mathcal{U}}_{\beta,\lambda}(\bar{x}) \\ &= i^{\mathcal{U}}_{\kappa,\lambda}(i^{\mathcal{U}}_{\beta,\kappa}(\bar{x})) \\ &= i^{\mathcal{U}}_{\kappa,\lambda}(x), \end{aligned}$$

as desired. ⊣

CLAIM 5. $\mathcal{U}$ *is by* Σ, *and* $\mathcal{U} \upharpoonright \alpha + 1 = \mathcal{T}$.

PROOF. $\mathcal{U}$ is by $i_{E^*}(\Sigma)$. But $i_{E^*}(\Sigma) \subseteq \Sigma$, so $\mathcal{U}$ is by Σ. So in N, $\mathcal{U}$ is obtained by iterating P, using Σ, so as to remove least disagreements with $i_{E^*}(M)$. Since E^* certifies E, we have $i_{E^*}(M)|\operatorname{lh}(E) = \operatorname{Ult}(M|\operatorname{lh}(E), E)||\operatorname{lh}(E) = M||\operatorname{lh}(E)$. Thus the process that produces $\mathcal{U}$ is the same as the process that produced $\mathcal{T}$, until extenders with length $\geq \operatorname{lh}(E)$ are used, so $\mathcal{T} = \mathcal{U} \restriction \alpha + 1$. ⊣

Now let $G = E^{\mathcal{U}}_{\xi}$, where $\kappa = U\text{-pred}(\xi + 1)$ and $\xi + 1 <_U \lambda$. G is an initial segment of the extender of $i^{\mathcal{U}}_{\kappa,\lambda}$ because its generators (including $\hat{\lambda}(G)$) are not moved, so by Claim 4, G is compatible with E. Since G has plus type, G cannot be a proper initial segment of E, and since E is not of plus type, $G \neq E$. Thus E is an initial segment of G^-. But then E is on the sequence of $\mathcal{M}^{\mathcal{U}}_{\xi}$ and $\operatorname{lh}(E) \leq \operatorname{lh}(G^-)$. Since $\operatorname{lh}(E) \leq \operatorname{lh}(G^-)$, $\alpha \leq \xi$, and since $\operatorname{lh}(E^{\mathcal{U}}_{\alpha}) \geq \operatorname{lh}(E)$, E must be on the sequence of $\mathcal{M}^{\mathcal{U}}_{\alpha} = \mathcal{M}^{\mathcal{T}}_{\alpha}$. But this means that E was not part of the least disagreement between $\mathcal{M}^{\mathcal{T}}_{\alpha}$ and M, contradiction. ⊣

Remark 8.1.3. The proof also shows that there is a λ-tight normal tree $\mathcal{V}$ whereby (P, Σ) iterates past M.

We can use Lemma 8.1.2 to show that the output of a maximal construction done below a Woodin cardinal is universal for mice of size strictly less than its additivity. This argument has probably been known since the late 1980s, but we can find no appropriate reference. A stronger version involving partial background extenders and universality with respect to weasels traces back to the paper [34] by Mitchell and Schindler. The author adapted the stronger version to full background constructions, where the Woodin cardinal becomes necessary. See [68, Lemma 11.1] and [41].

THEOREM 8.1.4 (Universality at a Woodin cardinal). *Suppose that $\mathbb{C}$ is a PFS construction, and δ is Woodin, as witnessed by extenders in $\mathcal{F}^{\mathbb{C}}$. Let P be a strongly stable, type 1 pfs premouse such that $|P| < \operatorname{crit}(E)$ for all $E \in \mathcal{F}^{\mathbb{C}}$, and let Σ be a $\delta+1$-iteration strategy for P. Suppose that whenever $E^* \in \mathcal{F}^{\mathbb{C}}$, we have*

$$i_{E^*}(\Sigma) \subseteq \Sigma.$$

Then

(a) *If $\nu < \delta$ and $\mathbb{C}$ is not good at $\langle \nu, k \rangle$, then (P, Σ) iterates to $M^{\mathbb{C}}_{\eta,j}$ for some $\langle \eta, j \rangle <_{\text{lex}} \langle \nu, k \rangle$.*

(b) *If $\operatorname{lh}(\mathbb{C}) = \delta$, then (P, Σ) iterates to some $M^{\mathbb{C}}_{\eta,j}$ for some $\eta < \delta$.*

Remark 8.1.5. The theorem is stated in such a way that there is no iterability of the background universe assumed.

PROOF. We prove (a) first. Suppose $\mathbb{C}$ is not good at $\langle \nu, k \rangle$. If there is an $\langle \eta, j \rangle <_{\text{lex}} \langle \nu, k \rangle$ such that (P, Σ) does not iterate strictly past $M_{\eta,j}$, then for the lexicographically least such $\langle \eta, j \rangle$, (P, Σ) iterates to $M_{\eta,j}$, by Lemma 8.1.2, so we are done. Thus we may assume (P, Σ) iterates strictly past $M_{\eta,j}$ for all $\langle \eta, j \rangle <_{\text{lex}} \langle \nu, k \rangle$. By Lemma 8.1.2, we get that (P, Σ) iterates past $M_{\nu,k}$.

P is iterable, so its iterates are pfs premice. It follows that $M_{\nu,k}$ is parameter and projectum solid. Let us check the bicephalus clause in goodness. Let F and G be such that $(M^{<\nu}, F, G)$ is a nontrivial bicephalus, and F^* and G^* be background certificates for F and G. Let $\mathcal{T}$ be the shortest tree by which (P,Σ) iterates past $M_{\nu,0}||\text{lh}(F) = M_{\nu,0}||\text{lh}(G)$, and let $\alpha + 1 = \text{lh}(\mathcal{T})$. We now simply apply the proof of Lemma 8.1.2 to both F and G, and it shows that both of them are on the sequence of $\mathcal{M}^{\mathcal{T}}_{\alpha}$. Thus $F = G$, contradiction.

Finally, suppose $M^{\mathbb{C}}_{\nu,0} = (M^{<\nu}, F)$ is active, and $F^* = F^{\mathbb{C}}_{\nu}$. We must see that F^* backgrounds F^+. This too is implicit in the proof of Lemma 8.1.2. Let $\mathcal{T}$ be the λ-separated tree whereby (P,Σ) iterates past $M_{\nu,0}$. Let $\mathcal{U} = i_{F^*}(\mathcal{T})$, and let G be the first extender used in the branch $(\text{crit}(F^*), \lambda(F^*))_U$. We showed that G is compatible with F^*, so F^* backgrounds G. But G has plus type and $G^- = F$, so $G = F^+$. (G is the extender of G^- followed by $i_{G^-}(D)$ and F^+ is the extender of F followed by $i_F(D)$, where D is the order zero measure of $M_{\nu,0}$ on $\text{crit}(F)$.)

This proves (a). For (b), suppose toward contradiction that (P,Σ) iterates past $M_{\nu,k}$ for all $\nu < \delta$ and $k \leq \omega$. Let

$$M = (M^{<\delta})^{\mathbb{C}}$$

be the unique passive premouse such that $o(M) = \delta$ and for all $\xi < \delta$, $M|\xi \trianglelefteq M^{\mathbb{C}}_{\alpha,0}$ for all sufficiently large $\alpha < \delta$. Clearly, (P,Σ) iterates past M. Let $\mathcal{T}$ on P be the λ-separated tree by Σ that witnesses this. We have that $\text{lh}(\mathcal{T}) = \delta + 1$, $\delta(\mathcal{T}) = \delta$, and

$$M \triangleleft M^{\mathcal{T}}_{\delta},$$

because δ is inaccessible. Let $b = [0,\delta)_T$, and for $\beta < \delta$, let $f(\beta) = \min(b - (\beta + 1))$. Since δ is $\mathcal{F}^{\mathbb{C}}$-Woodin, we can find a nice extender $F^* \in \mathcal{F}^{\mathbb{C}}$ with critical point α and length η such that for $j = i_{F^*}$

(1) $f``\alpha \subseteq \alpha$, and $j(f)(\alpha) < \eta$,
(2) $M||\eta = j(M)||\eta$, and
(3) $j(b) \cap \eta = b \cap \eta$.

Let $\tau + 1 <_T \delta$ be such that $\alpha = T\text{-pred}(\tau + 1)$, and let $F = E^{\mathcal{T}}_{\tau}$. By (1) and (3), $\tau + 1 = j(f)(\alpha)$ is the first point in $j(b)$ above α. Letting $\mathcal{U} = j(\mathcal{T})$ and $\lambda = j(\alpha)$, we have as usual that $\mathcal{M}^{\mathcal{T}}_{\alpha} = \mathcal{M}^{\mathcal{U}}_{\alpha}$, and

$$j \upharpoonright \mathcal{M}^{\mathcal{T}}_{\alpha} = i^{\mathcal{U}}_{\alpha,\lambda}.$$

But in fact $\mathcal{T} \upharpoonright \eta = \mathcal{U} \upharpoonright \eta$ by (2) and the fact that $j(\Sigma) \subseteq \Sigma$. So $F = E^{\mathcal{U}}_{\tau}$, and $\alpha <_U \tau + 1 <_U \lambda \in j(b)$, which implies that F^* is a background certificate for F.

Let ν be the least stage of $\mathbb{C}$ such that $M||\text{lh}(F) \trianglelefteq M^{<\nu}$. Because $\text{lh}(F)$ is a cardinal of M, we must have $M^{<\nu} = M||\text{lh}(F)$. But then $M_{\nu,0} = (M^{<\nu}, F)$, because our construction is maximal. After $\langle \nu, 0\rangle$ the levels of $\mathbb{C}$ do not project strictly below λ_F, because $M \trianglelefteq \mathcal{M}^{\mathcal{T}}_{\delta}$. This implies that F is on the M-sequence, contrary to its being used in $\mathcal{T}$. ⊣

8.2. Extending tree embeddings

We shall prove an elementary lemma on the extendibility of tree embeddings. Its proof uses

PROPOSITION 8.2.1. *Let $\mathcal{S}$ be a λ-separated iteration tree, let $\delta \leq_S \eta$, and suppose that $P \trianglelefteq \mathcal{M}^{\mathcal{S}}_\eta$, but $P \ntrianglelefteq \mathcal{M}^{\mathcal{S}}_\sigma$ whenever $\sigma <_S \delta$. Suppose also that $P \in \operatorname{ran}(\hat{\imath}^{\mathcal{S}}_{\delta,\eta})$. Let*

$$\begin{aligned}\alpha &= \text{least } \gamma \text{ such that } P \trianglelefteq \mathcal{M}^{\mathcal{S}}_\gamma \\ &= \text{least } \gamma \text{ such that } o(P) < \operatorname{lh}(E^{\mathcal{S}}_\gamma) \text{ or } \gamma = \eta,\end{aligned}$$

then $\alpha \in [\delta, \eta]_S$, and

$$\alpha = \text{least } \gamma \in [0, \eta]_S \text{ such that } o(P) < \operatorname{crit}(\hat{\imath}^{\mathcal{S}}_{\gamma,\eta}) \text{ or } \gamma = \eta.$$

(We allow $\delta = \eta$, with the understanding $\hat{\imath}_{\delta,\delta}$ is the identity.)

PROOF. By normality, for any $\gamma < \eta$, $P \trianglelefteq \mathcal{M}^{\mathcal{S}}_\gamma$ iff $\operatorname{lh}(E^{\mathcal{S}}_\gamma) > o(P)$.[7] So the first two characterizations of α are equivalent. Let β be the least γ in $[0, \eta]_S$ such that $o(P) < \operatorname{crit}(\hat{\imath}^{\mathcal{S}}_{\gamma,\eta})$ or $\gamma = \eta$. Clearly, $P \trianglelefteq \mathcal{M}^{\mathcal{S}}_\beta$, so $\alpha \leq \beta$. We have that $o(P) \geq \operatorname{lh}(E^{\mathcal{S}}_\sigma)$ for all $\sigma <_S \delta$, and hence by normality, for all $\sigma <_S \delta$ whatsoever. So $\delta \leq \alpha$, and $\beta \in [\delta, \eta]_S$.

Suppose $\alpha < \beta$; then $o(P) < \operatorname{lh}(E^{\mathcal{S}}_\alpha)$, so $o(P) < \operatorname{lh}(E^{\mathcal{S}}_\sigma)$ where σ is least such that $\alpha \leq \sigma$ and $\sigma + 1 \leq_S \beta$. But $E^{\mathcal{S}}_\sigma$ has plus type and $P \in \operatorname{ran}(\hat{\imath}^{\mathcal{S}}_{\delta,\eta})$, so $o(P) < \operatorname{lh}(E^{\mathcal{S}}_\sigma)$ implies $o(P) < \operatorname{crit}(E^{\mathcal{T}}_\sigma)$. Thus $P \triangleleft \mathcal{M}^{\mathcal{S}}_\gamma$ where $\gamma = U\text{-pred}(\sigma + 1)$, contrary to our definition of β. Thus $\alpha = \beta$, as desired. ⊣

Remark 8.2.2. The lemma does need the hypothesis that $\mathcal{S}$ is λ-separated. Otherwise it is possible that the least $\gamma \in [0, \eta]_S$ such that $o(P) < \operatorname{crit}(\hat{\imath}^{\mathcal{S}}_{\gamma,\eta})$ or $\gamma = \eta$ is $\alpha + 1$, rather than α. In that case, $E^{\mathcal{S}}_\alpha$ is not of plus type, and $\lambda(E^{\mathcal{S}}_\alpha) \leq o(P) < \operatorname{lh}(E^{\mathcal{S}}_\alpha)$. This would cause trouble in various arguments to follow.

The simpler agreement pattern in λ-separated trees will be useful in this chapter. Recall that when E has plus type, $\varepsilon(E) = \operatorname{lh}(E) = \operatorname{lh}(E^-)$. If $\mathcal{T}$ is λ-separated, then

$$\begin{aligned}\varepsilon^{\mathcal{T}}_\alpha &= \sup\{\operatorname{lh}(E^{\mathcal{T}}_\xi) \mid \xi < \alpha\} \\ &= \sup\{\operatorname{lh}(E^{\mathcal{T}}_\xi) \mid \xi <_T \alpha\},\end{aligned}$$

moreover

$$\begin{aligned}T\text{-pred}(\alpha + 1) &= \text{least } \beta \text{ s.t. } \operatorname{crit}(E^{\mathcal{T}}_\alpha) < \hat{\lambda}(E^{\mathcal{T}}_\beta) \\ &= \text{least } \beta \text{ s.t. } \operatorname{crit}(E^{\mathcal{T}}_\alpha) < \operatorname{lh}(E^{\mathcal{T}}_\beta) \\ &= \text{least } \beta \text{ s.t. } \operatorname{dom}(E^{\mathcal{T}}_\alpha) \trianglelefteq \mathcal{M}^{\mathcal{T}}_\beta \\ &= \text{least } \beta \text{ s.t. } \operatorname{dom}(E^{\mathcal{T}}_\alpha) \trianglelefteq_0 \mathcal{M}^{\mathcal{T}}_\beta.\end{aligned}$$

[7] Recall here our conventions that $\operatorname{lh}(E) = \operatorname{lh}(E^+)$, and if $P \trianglelefteq N$, then $o(P)$ is not active in N.

Moreover, a λ-separated tree is determined by its last model, together with the choice of branches at limit ordinals.

On extending tree embeddings, we have

LEMMA 8.2.3. *Let* $\Phi = \langle u, v, \langle s_\beta \mid \beta \leq \alpha \rangle, \langle t_\beta \mid \beta < \alpha \rangle \rangle$ *be a tree embedding of* $\mathcal{T}$ *into* $\mathcal{U}$*, where* $\mathcal{T}$ *and* $\mathcal{U}$ *are* λ*-separated, and let* F *be a plus extender on the extended* $\mathcal{M}_\alpha^{\mathcal{T}}$*-sequence such that* $\mathrm{lh}(F) > \mathrm{lh}(E_\beta^{\mathcal{T}})$ *for all* $\beta < \alpha$*. Let* $\mathcal{T}^\frown\langle F \rangle$ *be the unique putative* λ*-separated tree* $\mathcal{S}$ *extending* $\mathcal{T}$ *such that* $F = E_\alpha^{\mathcal{S}}$*. Let* $\xi < \mathrm{lh}(\mathcal{U})$*; then the following are equivalent:*

(1) There is a tree embedding Ψ *of* $\mathcal{T}^\frown\langle F \rangle$ *into* $\mathcal{U}$ *such that* $\Phi \subseteq \Psi$ *and* $u^\Psi(\alpha) = \xi$*,*

(2) $v(\alpha) \leq_U \xi$*, and* $E_\xi^{\mathcal{U}} = \hat{\imath}_{v(\alpha),\xi}^{\mathcal{U}} \circ s_\alpha(F)$*.*

Moreover, there is at most one such Ψ*.*

PROOF. It is easy to see from Definition 6.4.1 that (1) implies (2). We show that (2) implies (1). Let us set $M_\nu = \mathcal{M}_\nu^{\mathcal{T}}$ and $N_\nu = \mathcal{M}_\nu^{\mathcal{U}}$.

Suppose that ξ witnesses that (2) holds. Set $u(\alpha) = \xi$ and $t_\alpha = \hat{\imath}_{v(\alpha),\xi}^{\mathcal{U}} \circ s_\alpha$. Clearly,

$$t_\alpha \restriction \varepsilon_\alpha^{\mathcal{T}} = s_\alpha \restriction \varepsilon_\alpha^{\mathcal{T}},$$

and

$$\mathrm{crit}(\hat{\imath}_{v(\alpha),\xi}^{\mathcal{U}}) \geq \varepsilon_{v(\alpha)}^{\mathcal{U}}.$$

Let $G = t_\alpha(F)$ and $v(\alpha+1) = \xi + 1$. We shall define $s_{\alpha+1}$ so that $\Psi = \langle u, v, \langle s_\beta \mid \beta \leq \alpha+1 \rangle, \langle t_\beta \mid \beta \leq \alpha \rangle \rangle$ is a tree embedding of $\mathcal{S} = \mathcal{T}^\frown\langle F \rangle$ into $\mathcal{U}$.

Let $\mu = \mathrm{crit}(F)$ and $\mu^* = \mathrm{crit}(G)$. Let

$$\beta = S\text{-pred}(\alpha+1) = \text{least } \eta \text{ s.t. } \mu < \varepsilon_{\eta+1}^{\mathcal{T}},$$

and

$$\beta^* = U\text{-pred}(\xi+1) = \text{least } \eta \text{ s.t. } \mu^* < \varepsilon_{\eta+1}^{\mathcal{U}}.$$

We have that $\mathrm{dom}(F) = M_\alpha|\gamma$, where $\gamma = (\mu^+)^{M_\alpha|\mathrm{lh}(F)}$, and $\mathrm{dom}(G) = N_\xi|\gamma^*$, where $\gamma^* = (\mu^{*,+})^{N_\xi|\mathrm{lh}(G)}$. Moreover, $t_\alpha(\mathrm{dom}(F)) = \mathrm{dom}(G)$. Because our trees are λ-separated,

$$\beta = \text{least } \eta \text{ s.t. } \mathrm{dom}(F) \trianglelefteq M_\eta,$$

and

$$\beta^* = \text{least } \eta \text{ s.t. } \mathrm{dom}(G) \trianglelefteq N_\eta.$$

$\mathrm{dom}(F)$ and $\mathrm{dom}(G)$ are passive levels in M_β and N_{β^*}. Suppose first that $\beta < \alpha$. We then have that $\mu < \lambda_\alpha^{\mathcal{T}}$, so

$$\begin{aligned}\mathrm{dom}(G) &= t_\alpha(\mathrm{dom}(F)) \\ &= s_\alpha(\mathrm{dom}(F))\end{aligned}$$

$$= t_\beta(\mathrm{dom}(F))$$
$$= \hat{\imath}^{\mathcal{U}}_{v(\beta),u(\beta)} \circ s_\beta(\mathrm{dom}(F)),$$

where the last equalities hold because $\mu < \hat{\lambda}_{E^{\mathcal{T}}_\beta}$. Thus $\mathrm{dom}(G)$ is in the range of $\hat{\imath}^{\mathcal{U}}_{v(\beta),u(\beta)}$. Proposition 8.2.1, with $\delta = v(\beta)$, $\eta = u(\beta)$, and $\mathrm{dom}(G)$ as its P then tells us that

$$\beta^* = \text{least } \eta \in [v(\beta), u(\beta)]_U \text{ such that } \mathrm{crit}\, \hat{\imath}^{\mathcal{U}}_{\eta,u(\beta)} > \hat{\imath}^{\mathcal{U}}_{v(\beta),\eta} \circ s_\beta(\mu).$$

Let Q be the first level of M_β beyond $\mathrm{dom}(F)$ that projects to or below μ, and let Q^* be the first level of N_{β^*} beyond $\mathrm{dom}(G)$ that projects to or below μ^*.[8] So $M_{\alpha-1} = \mathrm{Ult}(Q,F)$ and $N_{\xi+1} = \mathrm{Ult}(Q^*,G)$. Let

$$\pi = \hat{\imath}^{\mathcal{U}}_{v(\beta),\beta^*} \circ s_\beta.$$

We have that

$$\pi \upharpoonright \mathrm{dom}(F) = t_\beta \upharpoonright P = s_\alpha \upharpoonright P = t_\alpha \upharpoonright \mathrm{dom}(F).$$

Letting $k = k(Q)$, we let

$$s^0_{\alpha+1}([a,f]^{Q^k}_F) = [t_\alpha(a), \pi(f)]^{(Q^*)^k}_G,$$

and

$$s_{\alpha+1} = \text{completion of } s^0_{\alpha+1},$$

as we are required to do by Definition 6.4.1. Since $\langle \pi, t_\alpha \rangle \colon (Q,F) \to (Q^*,G)$, the Shift Lemma tells us that $s_{\alpha+1}$ as defined is well-defined, nearly elementary, and agrees with t_α as required in a tree embedding. That $s_{\alpha+1}$ is elementary follows from

CLAIM 8.2.4. $\langle \pi, t_\alpha \rangle \colon (Q,F) \xrightarrow{*} (Q^*,G)$.

PROOF. Suppose first that $\beta = \alpha$. We have the diagram

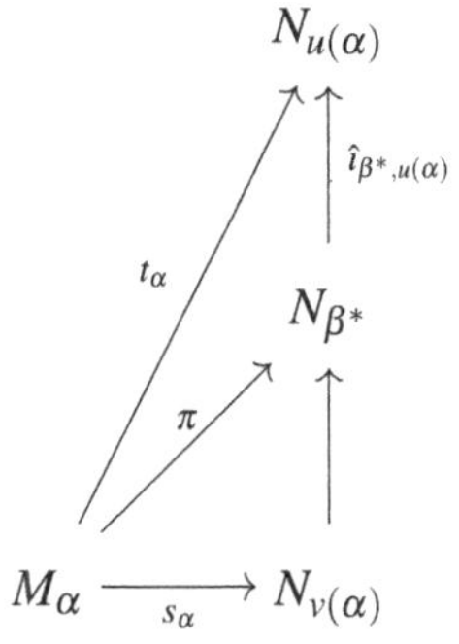

[8]The projecta are strictly below μ and μ^* by projectum solidity.

Here s_α is defined on all of M_α, but π may only be defined on Q, and t_α may only be defined on $M_\alpha|\text{lh}(F)$. We have $M_\alpha|\text{lh}(F) \trianglelefteq Q$ because $\mathcal{T}$ is maximal. Let $H = \pi(F)$ and $a \subseteq \varepsilon(F)$ be finite. F_a is Σ_1^Q in F and a, and π maps this definition to a definition of $H_{\pi(a)}$ over Q^*. Thus it is enough to see that $H_{\pi(a)} = G_{t_\alpha(a)}$. But $\hat{\imath}_{\beta^*,u(\alpha)}(H) = G$ and $\hat{\imath}_{\beta^*,u(\alpha)}(\pi(a)) = t_\alpha(a)$, so this is indeed the case.

Suppose next that $\beta < \alpha$ and F is very close to M_α. Let $\lambda = \hat{\lambda}(E_\beta^{\mathcal{T}})$, and let $a \subseteq \varepsilon(F)$ be finite. We have that $F_a \in M_\alpha|\lambda$, and $t_\alpha(F_a) = G_{t_\alpha(a)}$ because t_α is elementary. But $\pi(F_a) = t_\alpha(F_a)$ because $\pi \restriction \text{dom}(F) = t_\alpha \restriction \text{dom}(F)$, so we are done.

Finally, suppose $\beta < \alpha$ and F is not very close to M_α. By the Closeness Lemma, α is a special node in $\mathcal{T}^\frown\langle F\rangle$, so $\beta <_T \alpha$, and fixing η least in $(\beta,\alpha]_T$, $D^{\mathcal{T}} \cap (\eta,\alpha]_T = \emptyset$, $M_\eta^* \trianglelefteq Q$, M_η^* has a last extender H, and setting

$$i = i_{\eta,\alpha}^{\mathcal{T}} \circ i_\eta^{*,\mathcal{T}},$$

we have

$$i(H) = G$$

and

$$\text{dom}(H) < \text{crit}(i).$$

SUBCLAIM 8.2.4.1. $\beta^* <_U v(\eta)$.

PROOF. Since $\beta <_T \alpha$, $v(\beta) <_U v(\alpha) \leq_U u(\alpha)$. Let

$$K = s_\beta(H),$$

then

$$G = i_{v(\beta),u(\alpha)}^U(K).$$

Thus $\text{dom}(G) \in \text{ran}(\hat{\imath}_{v(\beta),u(\alpha)})$. By Proposition 8.2.1

$$\beta^* = \text{least } \xi \leq_U u(\alpha) \text{ s.t. } \text{crit}(\hat{\imath}^{\mathcal{U}}_{\xi,u(\alpha)}) > \hat{\imath}_{v(\beta),\xi}(\text{dom}(K)).$$

But let

$$P = \text{dom}(E_{\eta-1}^{\mathcal{T}}),$$
$$P^* = \text{dom}(E^{\mathcal{U}}_{u(\eta-1)}),$$

and

$$\gamma = U\text{-pred}(v(\eta)).$$

Then $\text{dom}(H) \triangleleft P \triangleleft M_\eta^*$, so $\text{dom}(K) \triangleleft s_\beta(P)$ and

$$P^* = i_{v(\beta),u(\alpha)}(s_\beta(P)) \in \text{ran}(i_{v(\beta),u(\alpha)}).$$

It follows from Proposition 8.2.1 that

$$\gamma = \text{least } \xi \leq_U u(\alpha) \text{ s.t. } \text{crit}(\hat{\imath}^{\mathcal{U}}_{\xi,u(\alpha)}) > \hat{\imath}_{v(\beta),\xi}(s_\beta(P)).$$

Thus $\beta^* \leq_U \gamma$. But $\gamma <_U v(\eta)$, so we have our subclaim. $\dashv$

Here is a diagram of the situation. Let $S = \pi(M^*_\eta)$.

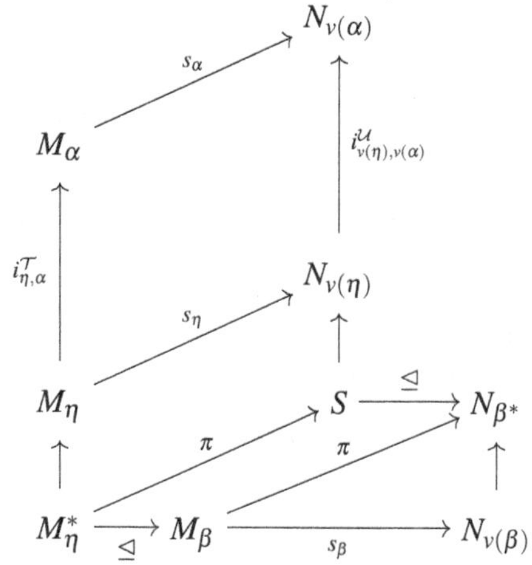

SUBCLAIM 8.2.4.2. *$u(\alpha)$ is special in $\mathcal{U}$.*

PROOF. Let γ be least such that $\beta^* <_U \gamma \leq_U \nu(\eta)$, and let

$$j = i^{\mathcal{U}}_{\gamma,u(\alpha)} \circ i^{*,\mathcal{U}}_\gamma.$$

We claim that $S = \mathcal{M}^{*,\mathcal{U}}_\gamma$ and $D^{\mathcal{U}} \cap (\gamma, u(\alpha)]_U = \emptyset$. This is true because $\pi(H)$ is the last extender of S, $j(\pi(H)) = G$, and $\rho(S) \leq \text{dom}(\pi(H)) < \text{crit}(j) < o(S)$. Thus the first extender used in j forces a drop at least as far as to S, but since j maps the last extender of S to G, the drop cannot be further than S, and there can be no further drops at all in $(\gamma, u(\alpha]_U$.

To finsih the proof of the subclaim, we must show that if $\gamma \leq_U \xi + 1 \leq_U u(\alpha)$, then $E^{\mathcal{U}}_\xi$ is very close to M_ξ and to $M^{*,\mathcal{U}}_{\xi+1}$. This follows from the proof of Lemma 4.5.8(a). If not, then ξ is special in $\mathcal{U}$, so $M^{*,\mathcal{U}}_{\xi+1}$ has a last extender with the same critical point as $E^{\mathcal{U}}_\xi$. But last extender of $M^{*,\mathcal{U}}_{\xi+1}$ has the same critical point as $\pi(H)$, and this is strictly less than $\text{crit}(E^{\mathcal{U}}_\xi)$. ⊣

Now let $I = I^{\mathcal{T}}_{\eta,\alpha}$ and $J = I^{\mathcal{U}}_{\gamma,u(\alpha)}$ be the branch extenders of i and j respectively. By Lemma 4.5.8(b), I is very close to Q and J is very close to Q^*. By Lemma 4.5.16

$$\langle \pi, t_\alpha \rangle \colon (Q, I) \overset{**}{\to} (Q^*, J).$$

But then for any finite $c \subseteq \varepsilon(F)$, I_c and H constitute a good code of F_c over $M^*_{\alpha+1}$. π moves this code to $\pi(I_c) = J_{t_\alpha(c)}$ and $\pi(H)$, which together code $G_{t_\alpha(c)}$ over Q^*. Thus

$$\langle \pi, t_\alpha \rangle \colon (Q, F) \overset{*}{\to} (Q^*, G),$$

as required by Claim 8.2.4. ⊣

Let us check that v preserves tree order. The new case involves F and G; we must see that $\gamma <_T \alpha+1$ iff $v(\gamma) <_U \xi+1$. But if $\gamma <_T \alpha+1$, then $\gamma \leq_T \beta$, so $v(\gamma) \leq_U v(\beta) \leq_U \beta^* <_U \xi+1$. Conversely, if $v(\gamma) <_U \xi+1$, then $v(\gamma) \leq_U \beta^*$. But $\operatorname{ran}(v) \cap (v(\beta), u(\beta)] = \emptyset$, so $v(\gamma) \leq_U v(\beta)$, so $\gamma \leq_T \beta$.

The case that $\alpha = \beta$ is similar. In this case, we apply the proposition to $\operatorname{dom}(G)$ with $\delta = v(\beta)$ and $\eta = \xi$. This gives us that

$$\beta^* = \text{least } \eta \in [v(\beta), \xi]_U \text{ such that } \operatorname{crit} \hat{\imath}^{\mathcal{U}}_{\eta,\xi} > \hat{\imath}^{\mathcal{U}}_{v(\beta),\eta} \circ s_\beta(\mu).$$

We leave the remaining details to the reader. ⊣

Remark 8.2.5. The proof gives a formula for the point of application of $E^{\mathcal{U}}_{u(\alpha)}$ under a tree embedding of $\mathcal{T}$ into $\mathcal{U}$, namely

$$U\text{-pred}(u(\alpha)+1) = \text{least } \eta \in [v(\beta), u(\beta)]_U \text{ such that}$$
$$\operatorname{crit} \hat{\imath}^{\mathcal{U}}_{\eta,u(\beta)} > \hat{\imath}^{\mathcal{U}}_{v(\beta),\eta} \circ s_\beta(\operatorname{crit}(E^{\mathcal{T}}_\alpha)),$$

where

$$\beta = T\text{-pred}(\alpha+1).$$

In the course of the proof we showed

COROLLARY 8.2.6. *Let $\Phi = \langle u, v, \langle s_\beta \mid \beta \leq \alpha \rangle, \langle t_\beta \mid \beta < \alpha \rangle \rangle$ be a tree embedding of $\mathcal{T}$ into $\mathcal{U}$, where $\mathcal{T}$ and $\mathcal{U}$ are λ-separated, and let F be a plus extender on the extended $\mathcal{M}^{\mathcal{T}}_\alpha$-sequence such that $\operatorname{lh}(F) > \operatorname{lh}(E^{\mathcal{T}}_\beta)$ for all $\beta < \alpha$. Let*

$$G = \hat{\imath}^{\mathcal{U}}_{v(\alpha),\xi} \circ s_\alpha(F),$$

and suppose $\operatorname{lh}(E^{\mathcal{U}}_\eta) < \operatorname{lh}(G)$ for all $\eta < \xi$. Let $\mathcal{S}$ and $\mathcal{W}$ be the unique putative λ-separated trees extending $\mathcal{T}$ and $\mathcal{U} \restriction \xi+1$ of lengths $\alpha+2$ and $\xi+2$ such that $F = E^{\mathcal{S}}_\alpha$ and $G = E^{\mathcal{W}}_\xi$; then there is a unique tree embedding from $\mathcal{S}$ to $\mathcal{W}$ that extends Φ and maps F to G.

COROLLARY 8.2.7. *Let $\Phi = \langle u, v, \langle s_\beta \mid \beta \leq \alpha \rangle, \langle t_\beta \mid \beta < \alpha \rangle \rangle$ be a tree embedding of $\mathcal{T}$ into $\mathcal{U}$, where $\mathcal{T}$ and $\mathcal{U}$ are λ-separated; then for any α*

$$\alpha \text{ is special in } \mathcal{T} \Longrightarrow u(\alpha) \text{ is special in } \mathcal{U}.$$

Moreover, if $\beta = T\text{-pred}(\alpha+1)$ and $\beta^* = U\text{-pred}(u(\alpha+1))$, and $Q = \mathcal{M}^{*,\mathcal{T}}_{\alpha+1}$ and $Q^* = \mathcal{M}^{*,\mathcal{U}}_{u(\alpha)+1}$, then setting $\pi = \hat{\imath}^{\mathcal{U}}_{v(\beta),\beta^*} \circ s_\beta$, we have

$$\langle \pi, t_\alpha \rangle \colon (\mathcal{M}^{*,\mathcal{T}}_{\alpha+1}, E^{\mathcal{T}}_\alpha) \xrightarrow{*} (\mathcal{M}^{*,\mathcal{U}}_{u(\alpha)+1}, E^{\mathcal{U}}_{u(\alpha)}).$$

8.3. Resurrection embeddings as branch embeddings

We prove a technical lemma on λ-separated iterations past levels of a background construction.

Let (P_0,Σ) be a pure extender pair with scope H_δ, where δ is inaccessible, and let $\mathbb{C}$ be a PFS construction such that $\mathrm{lh}(\mathbb{C}) \leq \delta$ and $\mathcal{F}^{\mathbb{C}} \subseteq V_\delta$. Suppose that $P_{\mathbb{C}}$ is strongly stable and type 1, and $|P_0| < \mathrm{crit}(E)$ for all $E \in \mathcal{F}^{\mathbb{C}}$. Let $M_{\nu,k} = M^{\mathbb{C}}_{\nu,k}$, fix $\langle \nu_0, k_0\rangle < \mathrm{length}(\mathbb{C})$, and suppose that whenever $\langle \nu,k\rangle <_{\mathrm{lex}} \langle \nu_0,k_0\rangle$, (P_0,Σ) does not iterate to $M_{\nu,k}$ via a λ-separated tree. By Lemma 8.1.2, whenever $\langle \nu,k\rangle \leq_{\mathrm{lex}} \langle \nu_0,k_0\rangle$, (P_0,Σ) iterates past $M_{\nu,k}$ via a λ-separated tree. Thus for $\langle \nu,k\rangle \leq_{\mathrm{lex}} \langle \nu_0,k_0\rangle$, we have

$$\mathcal{W}^*_{\nu,k} = \text{unique shortest } \lambda\text{-separated tree on } P_0 \text{ by } \Sigma$$
$$\text{with last model } Q \trianglerighteq M_{\nu,k}.$$

Our convention that $P \ntrianglelefteq Q$ when Q is active and $P = Q||o(Q)$ matters here: if $M_{\nu,k}$ is passive, then $o(M_{\nu,k})$ must be passive in the last model of $\mathcal{W}^*_{\nu,k}$.

Our technical lemma says that below $\langle \nu_0,k_0\rangle$, the resurrection embeddings of $\mathbb{C}$ are captured by branch embeddings of the $\mathcal{W}^*_{\nu,k}$. Let us write $\mathrm{Res}_{\eta,j}[P]$ for $\mathrm{Res}_{M_{\eta,j}}[P]$ and $\sigma_{\eta,j}[P]$ for $\sigma_{M_{\eta,j}}[P]$.

LEMMA 8.3.1. *Let $\langle \theta, j\rangle \leq \langle \nu_0,k_0\rangle$, $P \trianglelefteq M_{\theta,j}$,*

$$\alpha = \text{least } \xi \text{ such that } \mathcal{M}_\xi^{\mathcal{W}^*_{\theta,j}} \trianglerighteq P$$

and

$$M_{\theta_0,j_0} = \mathrm{Res}_{\theta,j}[P];$$

then

$$\mathcal{W}^*_{\theta,j} \upharpoonright (\alpha+1) = \mathcal{W}^*_{\theta_0,j_0} \upharpoonright (\alpha+1),$$

$\mathcal{W}^*_{\theta_0,j_0}$ has last model $\mathcal{M}_\xi^{\mathcal{W}^*_{\theta_0,j_0}} = M_{\theta_0,j_0}$, and $\alpha \leq_{\mathcal{W}^*_{\theta_0,j_0}} \xi$, and

$$\sigma_{\theta,j}[P] = \hat{\imath}_{\alpha,\xi}^{\mathcal{W}^*_{\theta_0,j_0}} \upharpoonright P.$$

Recall that if M is a premouse such that $k(M) > 0$, then M^- is the premouse that is equal to M, except that $k(M^-) = k(M) - 1$.

SUBLEMMA 8.3.1.1. *Suppose that $M_{\nu,k}$ is not $k+1$-sound. Let $\pi : M^-_{\nu,k+1} \to M_{\nu,k}$ be the anticore embedding. Let $\xi_0 + 1 = \mathrm{lh}(\mathcal{W}^*_{\nu,k+1})$ and $\xi_1 + 1 = \mathrm{lh}(\mathcal{W}^*_{\nu,k})$; then*

(a) *$\mathcal{W}^*_{\nu,k}$ has last model $M_{\nu,k}$,*

(b) *$\mathcal{W}^*_{\nu,k+1} = \mathcal{W}^*_{\nu,k} \upharpoonright (\xi_0+1)$,*

(c) *ξ_0 is the least γ such that $\mathrm{lh}(E_\gamma^{\mathcal{W}^*_{\nu,k}}) > \rho(M_{\nu,k})$, and*

(d) *$\xi_0 <_{\mathcal{W}^*_{\nu,k}} \xi_1$, and $\hat{\imath}_{\xi_0,\xi_1}^{\mathcal{W}^*_{\nu,k}} = \pi$.*

PROOF. By definition, $\mathcal{M}_{\xi_1}^{\mathcal{W}^*_{\nu,k}} \trianglerighteq M_{\nu,k}$. But $M_{\nu,k}$ is not $k+1$-sound, so $\mathcal{M}_{\xi_1}^{\mathcal{W}^*_{\nu,k}} = M_{\nu,k}$. This gives (a).

By Lemma 4.4.6, the iteration $\mathcal{W}^*_{\nu,k}$ from P_0 to $M_{\nu,k}$ must have dropped. The last drop had to be to $M_{\nu,k+1}$, and it lies on the branch to $M_{\nu,k}$. So we can fix η such that

$$M_{\nu,k+1} = \operatorname{dom} \hat{\imath}^{W^*_{\nu,k}}_{\eta,\xi_1}, \text{ and } \hat{\imath}^{\mathcal{W}^*_{\nu,k}}_{\eta,\xi_1} = \pi.$$

We have that $M_{\nu,k+1} \trianglelefteq M_\eta^{W^*_{\nu,k}}$.

Letting $\rho = \rho(M_{\nu,k})$, we have that $M_{\nu,k+1}$ agrees with $M_{\nu,k}$ to $\rho^{+M_{\nu,k}} = \rho^{+M_{\nu,k+1}}$. Thus $\mathcal{W}^*_{\nu,k+1}$ and $\mathcal{W}^*_{\nu,k}$ use the same extenders E such that $\mathrm{lh}(E) \le \rho$.

We claim that $\mathcal{W}^*_{\nu,k+1}$ uses no extenders E such that $\mathrm{lh}(E) > \rho$. For if $\mathcal{W}^*_{\nu,k+1}$ uses E such that $\mathrm{lh}(E) > \rho$, then the branch P_0-to-$\mathcal{M}_{\xi_0}^{\mathcal{W}^*_{\nu,k+1}}$ uses such an E, since $\xi_0 + 1 = \mathrm{lh}(W^*_{\nu,k+1})$. $\mathrm{lh}(E) \le o(M_{\nu,k+1})$ because $\mathcal{W}^*_{\nu,k+1}$ was of minimal length. But then $\rho \le \mathrm{crit}(E)$ is impossible, because $\mathrm{dom}(E) \subseteq M_{\nu,k+1}$, and $M_{\nu,k+1}$ is sound. However, $\mathrm{crit}(E) < \rho$ is also impossible, since no model on the branch $[0, \xi_0]$ after E can project into $(\mathrm{crit}(E), \mathrm{lh}(E))$.

So we have that $\mathcal{W}^*_{\nu,k+1} = \mathcal{W}^*_{\nu,k} \restriction \xi_0 + 1$. We have (a)-(c) of the sublemma already. For (d), we need to see $\xi_0 = \eta$. Since $M_{\nu,k+1} \trianglelefteq \mathcal{M}_\eta^{\mathcal{W}^*_{\nu,k}}$, $\xi_0 \le \eta$. Suppose toward contradiction that $\xi_0 < \eta$. We then have that

$$o(M_{\nu,k+1}) \le \mathrm{lh}(E_{\xi_0}^{\mathcal{W}^*_{\nu,k}})$$

because $M_{\nu,k+1}$ is an initial segment of both $\mathcal{M}_{\xi_0}^{\mathcal{W}^*_{\nu,k}}$ and $\mathcal{M}_\eta^{\mathcal{W}^*_{\nu,k}}$. But let $\theta + 1$ be the successor of η on the branch $[0, \xi_1]$ of $\mathcal{W}^*_{\nu,k}$, that is, $W^*_{\nu,k}$-pred$(\theta+1) = \eta$ and $\theta + 1 \le_{W^*_{\nu,k}} \xi_1$. Then $M_{\nu,k+1} = (\mathcal{M}^*_{\theta+1})^{\mathcal{W}^*_{\nu,k}}$, and so

$$\mathrm{lh}(E_\eta^{\mathcal{W}^*_{\nu,k}}) \le o(M_{\nu,k+1}) \le \mathrm{lh}(E_{\xi_0}^{\mathcal{W}^*_{\nu,k}}).$$

Thus $\eta \le \xi_0$, a contradiction. ⊣

PROOF OF LEMMA 8.3.1. We go by induction on $\langle \theta, j \rangle$. Suppose Lemma 8.3.1 holds for $\langle \theta', j' \rangle <_{\mathrm{lex}} \langle \theta, j \rangle$, as well as for all $Q \triangleleft P$, where $P \trianglelefteq M_{\theta,j}$. Let

$$S = A_n(M_{\theta,j}, P),$$

where $n = n(M_{\eta,j}, P)$, and

$$\rho = \rho^-(S).$$

S is the last element of the $(M_{\theta,j}, P)$ dropdown sequence. We can assume that $\rho < o(P)$, as otherwise $\tau =$ identity, and all is trivial. Thus $k(S) > 0$.

If $S \triangleleft M_{\theta,j}$, then by Lemma 4.7.13(a), $S = M_{\theta',j'}$ for some some $\langle \theta', j' \rangle <_{\mathrm{lex}} \langle \theta, j \rangle$. By Lemma 4.7.13(d), $\sigma_{\theta',j'}[S] \restriction P = \sigma_{\theta,j}[S] \restriction P = \sigma_{\theta,j}[P]$. So we can apply our induction hypothesis at θ', j'. Note that

$$\mathcal{W}^*_{\theta,j} \restriction (\alpha+1) = \mathcal{W}^*_{\theta',j'} \restriction (\alpha+1).$$

Thus we may assume $S = M_{\theta,j}$. So $j = k(S)$ and $j > 0$. If $\sigma_{\theta,j}[S] = \sigma_{\theta,j-1}[S]$, then as $\langle\theta, j-1\rangle <_{\text{lex}} \langle\theta, j\rangle$, our induction hypothesis carries the day. Otherwise, we have that $M_{\theta,j-1}$ is not sound. Moreover

$$\sigma_{\theta,j}[S] = \pi \circ \sigma_{\theta,j-1}[S],$$

where $\pi : M^-_{\theta,j} \to M_{\theta,j-1}$ is the anticore embedding.

Let $\alpha + 1 = \text{lh}(\mathcal{W}^*_{\theta,j})$ and $\beta + 1 = \text{lh}(\mathcal{W}^*_{\theta,j-1})$. By the sublemma, $S \trianglelefteq \mathcal{M}_\alpha^{\mathcal{W}^*_{\theta,j}}$ and $M_{\theta,j-1} = \mathcal{M}_\beta^{W^*_{\theta,j-1}}$, $\alpha \leq_{W^*_{\theta,j-1}} \beta$, and

$$\pi = \hat{\imath}_{\alpha,\beta}^{\mathcal{W}^*_{\theta,j}}.$$

Also, $\mathcal{W}^*_{\theta,j}$ uses only extenders of lh $\leq \rho$, so α is the least γ such that $P \trianglelefteq \mathcal{M}_\gamma^{W^*_{\theta,j}}$.

Let $P_1 = \pi(P)$. Let

$$\alpha_1 = \text{least } \gamma \text{ such that } P_1 \trianglelefteq \mathcal{M}_\gamma^{\mathcal{W}^*_{\theta,j-1}}.$$

We can assume $\text{crit}(\pi) \leq o(P)$, as otherwise $P \trianglelefteq M_{\theta,j-1}$ and $\sigma_{\theta,j}[P] = \sigma_{\theta,j-1}[P]$, so we are done by induction.

CLAIM. $\alpha <_{\mathcal{W}^*_{\theta,j-1}} \alpha_1 \leq_{W^*_{\theta,j-1}} \beta$.

PROOF. Let $\gamma \in (\alpha, \beta]_{\mathcal{W}^*_{\theta,j-1}}$ be least such that $o(P_1) < \text{crit}(\hat{\imath}_{\gamma,\beta}^{\mathcal{W}^*_{\theta,j-1}})$. We claim that $\alpha_1 = \gamma$. Certainly, $P_1 \trianglelefteq \mathcal{M}_\gamma^{\mathcal{W}^*_{\theta,j-1}}$. Also, $P_1 \ntrianglelefteq \mathcal{M}_\alpha^{\mathcal{W}^*_{\theta,j-1}}$. Since P_1 is in the range of $\hat{\imath}_{\alpha,\beta}^{\mathcal{W}^*_{\theta,j-1}}$, we get $\alpha_1 = \gamma$ from Proposition 8.2.1.[9] ⊣

The claim also showed that

$$\pi \restriction P = \hat{\imath}_{\theta,\alpha_1} \restriction P.$$

Now we apply our induction hypothesis to $P_1 \trianglelefteq M_{\theta,j-1}$. We get θ_0, j_0 such that

1. $\mathcal{W}^*_{\theta_0,j_0} \restriction (\alpha_1 + 1) = W^*_{\theta,j-1} \restriction (\alpha_1 + 1)$.
2. $\mathcal{W}^*_{\theta_0,j_0}$ has last model $M_{\theta_0,j_0} = \mathcal{M}_\xi^{\mathcal{W}^*_{\theta_0,j_0}}$, and
3. $\alpha_1 \leq_{W^*_{\theta_0,j_0}} \xi$, and $\sigma_{\theta,j-1}[P_1] = \hat{\imath}_{\alpha_1,\xi}^{\mathcal{W}^*_{\theta_0,j_0}}$.

But $\sigma_{\theta,j}[P] = \sigma_{\theta,j-1}[P_1] \circ \pi$. This yields

$$\sigma_{\theta,j}[P] = \hat{\imath}_{\alpha_1,\xi}^{\mathcal{W}^*_{\theta_0,j_0}} \circ \hat{\imath}_{\alpha,\alpha_1}^{\mathcal{W}^*_{\theta_0,j_0}} = \hat{\imath}_{\alpha,\xi}^{\mathcal{W}^*_{\theta_0,j_0}},$$

as desired.

This completes the proof of Lemma 8.3.1. ⊣

[9] If the tree $\mathcal{W}^*_{\theta,j-1}$ were not λ-separated, the proof of Lemma 8.3.1 would break down at this point.

8.4. Iterating into a backgrounded strategy

In this section we prove the basic comparison theorem for strongly stable pure extender pairs. In the next chapter we shall generalize it to least branch hod pairs, but all the main ideas occur in the pure extender proof.

The proof is based on proving $(*)(P,\Sigma)$, for such pairs. This involves iterating (P,Σ) to a level $(M_{\nu,k},\Omega_{\nu,k})$ of some background construction $\mathbb{C}$. In the statement of $(*)(P,\Sigma)$, $\mathbb{C}$ is the construction of some coarse Γ-Woodin background universe N^* that captures Σ, but here we shall assume somewhat less about $\mathbb{C}$.

DEFINITION 8.4.1. Let $\mathcal{F}$ be a set of nice extenders; then $\Omega^{\text{ubh}}_{\mathcal{F}}$ is the partial iteration strategy for V: if $\vec{\mathcal{T}}^\frown\langle\mathcal{U}\rangle$ is a finite stack of quasi-normal $\mathcal{F}$-trees by $\Omega^{\text{ubh}}_{\mathcal{F}}$ such that $\mathcal{U}$ has limit length, then

$$\Omega^{\text{ubh}}_{\mathcal{F}}(\vec{\mathcal{T}}^\frown\langle\mathcal{U}\rangle) = b \text{ iff } b \text{ is the unique cofinal, wellfounded branch of } \mathcal{U}.$$

So if V is strongly uniquely iterable for finite stacks of quasi-normal $\mathcal{F}$-trees, then $\Omega^{\text{ubh}}_{\mathcal{F}}$ is total, and it is the unique iteration strategy witnessing this. Moreover, $\Omega^{\text{ubh}}_{\mathcal{F}}$ quasi-normalizes well, and has strong hull condensation. The results of Chapter 7 show that this is the case if V is a coarse Γ-Woodin model, and $\mathcal{F} = \{E \mid E \text{ is nice}\}$, and under other hypotheses as well. But our notation allows the case that $\Omega^{\text{ubh}}_{\mathcal{F}}$ is partial. $\Omega^{\text{ubh}}_{\mathcal{F}}(\vec{\mathcal{T}}^\frown\langle\mathcal{U}\rangle)$ can fail to be defined because $\mathcal{U}$ has no cofinal wellfounded branch, or because it has more than one cofinal wellfounded branch.

DEFINITION 8.4.2. Let $\mathbb{C}$ be a PFS construction, and suppose $M_{\nu,k} = M^{\mathbb{C}}_{\nu,k}$ exists; then $\Omega^{\mathbb{C}}_{\nu,k}$ is the partial strategy for $M_{\nu,k}$ induced by $\Omega^{\text{ubh}}_{\mathcal{F}^{\mathbb{C}}}$, i.e.

$$\vec{\mathcal{T}} \text{ is by } \Omega^{\mathbb{C}}_{\nu,k} \text{ iff } \text{lift}(\vec{\mathcal{T}}, M_{\nu,k}, \mathbb{C})_0 \text{ is by } \Omega^{\text{ubh}}_{\mathcal{F}^{\mathbb{C}}},$$

whenever $\vec{\mathcal{T}}$ is a finite stack of plus trees on $M_{\nu,k}$.

So if V is strongly uniquely $(\omega,\theta,\mathcal{F}^{\mathbb{C}})$-iterable, then $\Omega^{\mathbb{C}}_{\nu,k}$ is a complete strategy with scope H_θ that quasi-normalizes well and has strong hull condensation.

The following is essentially Theorem 1.8.5, but in the pure extender model case.

THEOREM 8.4.3. *Let (P,Σ) be a strongly stable pure extender pair with scope H_δ, where δ is inaccessible. Let $\mathbb{C}$ be a PFS construction of length $\leq \delta$ such that $\mathcal{F}^{\mathbb{C}} \subseteq H_\delta$ and for all $E \in \mathcal{F}^{\mathbb{C}}$, $\text{crit}(E) > o(P)$. Let $\langle \nu,k\rangle < \text{lh}(\mathbb{C})$, and suppose that (P,Σ) iterates strictly past $(M^{\mathbb{C}}_{\eta,j},\Omega^{\mathbb{C}}_{\eta,j})$, for all $\langle\eta,j\rangle <_{\text{lex}} \langle\nu,k\rangle$; then (P,Σ) iterates past $(M^{\mathbb{C}}_{\nu,k},\Omega^{\mathbb{C}}_{\nu,k})$.*

Remark 8.4.4. Σ is total so if (P,Σ) iterates past $(M^{\mathbb{C}}_{\nu,k},\Omega^{\mathbb{C}}_{\nu,k})$, then $\Omega^{\mathbb{C}}_{\nu,k}$ is total. So although did not assume unique iterability in the hypothesis of Theorem 8.4.3, we got the $\Omega^{\mathbb{C}}_{\eta,l}$ are total, until we reach an $M_{\nu,k}$ that is beyond Σ. Before that point, $\mathbb{C}$-lifted trees have unique cofinal wellfounded branches.

Theorem 8.4.3 yields at once a comparison theorem for pure extender pairs. The following is the pure extender case of our main strategy comparison theorem.

THEOREM 8.4.5 (Pure extender mouse pair comparison). *Assume* AD^+*, and let* (P,Σ) *and* (Q,Ψ) *be strongly stable pure extender pairs, with scope* H_{ω_1}*; then there are countable* λ*-separated trees* $\mathcal{T}$ *on* P *and* $\mathcal{U}$ *on* Q *by* Ψ*, with last models* R *and* S *respectively, such that either*

1. *P-to-R does not drop,* $R \trianglelefteq S$*, and* $\Sigma_{\mathcal{T},R} = \Psi_{\mathcal{U},R}$*, or*
2. *Q-to-S does not drop,* $S \trianglelefteq R$*, and* $\Psi_{\mathcal{U},S} = \Sigma_{\mathcal{T},S}$*.*

PROOF. By the Basis Theorem of AD^+, we may assume that Code(Σ) and Code(Ψ) are Suslin and co-Suslin. (The paper [73] shows this directly, assuming only AD.) So we have a coarse Γ-Woodin tuple (N^*, w, S, T, Σ^*), where Γ is a pointclass big enough that Σ and Ψ are coded by sets of reals in Γ. We may assume P and Q are in N^*, and countable there. Working in N^*, let $E \in \mathcal{F}$ iff E is a nice extender and $i_E(w) \cap V_{\mathrm{lh}(E)+1} = w \cap V_{\mathrm{lh}(E)+1}$. It is easy to check that $(w, \mathcal{F})$ is a coherent pair. Let $\mathbb{C}$ be the unique (maximal) PFS construction of N^* of length δ such that $w = w^{\mathbb{C}}$ and $\mathcal{F} = \mathcal{F}^{\mathbb{C}}$. By Theorem 4.11.4, there is a unique such $\mathbb{C}$.

We now apply Theorem 8.1.4. This gives us a $\langle \nu, k \rangle$ such that $M^{\mathbb{C}}_{\nu,k}$ is a Σ-iterate of P, and P iterates by Σ past $M^{\mathbb{C}}_{\eta,j}$, for each $\langle \eta, j \rangle <_{\mathrm{lex}} \langle \nu, k \rangle$. Similarly we have $\langle \mu, l \rangle$ such that $M^{\mathbb{C}}_{\mu,l}$ is a Ψ-iterate of Q, and Q iterates by Ψ past $M^{\mathbb{C}}_{\eta,j}$, for each $\langle \eta, j \rangle <_{\mathrm{lex}} \langle \nu, k \rangle$. By Theorem 8.4.3, no strategy disagreements with the strategies in $\mathbb{C}$ show up in these iterations. So if $\langle \nu, k \rangle \le_{\mathrm{lex}} \langle \mu, l \rangle$, then by Theorem 8.4.3, we get conclusion (1), with $R = M^{\mathbb{C}}_{\nu,k}$ and $\Sigma_{\mathcal{T},R} = \Omega^{\mathbb{C}}_{\nu,k}$. If $\langle \mu, l \rangle \le_{\mathrm{lex}} \langle \nu, k \rangle$, then we get conclusion (2).

Let $\mathcal{T}$, $\mathcal{U}$, R, and S witness in N^* that either (1) or (2) holds. $\mathcal{T}$ and $\mathcal{U}$ are countable in V, and N^* is sufficiently correct that either (1) or (2) holds in V. ⊣

Remark 8.4.6. When we generalize the comparison theorem for pure extender pairs to strategy mouse pairs in Chapter 9, we shall have to re-organize the proof a bit. Lemma 8.1.2 and Theorem 8.1.4 do not help in the strategy mouse context, so in effect we must prove the analogs of both Theorem 8.4.3 and Lemma 8.1.2 as part of one induction.

Remark 8.4.7. Suppose that (P,Σ) is not strongly stable, and let D be its order zero measure on η^P_k, where $k = k(P)$. Let $Q = \mathrm{Ult}_k(\bar{\mathfrak{C}}_k(P), D)$ and Λ be the iteration strategy for Q we get from Σ. (Q,Λ) is a strongly stable pure extender pair of type 1A. We can compare arbitrary pure extender pairs by comparing the strongly stable pairs derived from them in this way, just as we did with premice in Section 4.10. See Section 9.6, where an indirect comparison of this sort is needed.

The rest of this chapter is devoted to the proof of Theorem 8.4.3.

PROOF OF THEOREM 8.4.3. Suppose that (P_0,Σ) iterates past $M^{\mathbb{C}}_{\nu,k}$ for all $\langle \nu,k\rangle <_{\text{lex}} \langle \nu_0,k_0\rangle$. For $\langle \nu,k\rangle \leq_{\text{lex}} \langle \nu_0,k_0\rangle$, let

$$\begin{aligned}\mathcal{W}^*_{\nu,k} &= \text{unique shortest } \lambda\text{-separated tree on } P_0 \text{ by } \Sigma\\ &\quad\text{with last model } Q \trianglerighteq M_{\nu,k}.\end{aligned}$$

Let $M = M_{\nu_0,k_0}$. We must show that $\Sigma_{\mathcal{W}^*_{\nu_0,k_0},M} = \Omega^{\mathbb{C}}_{\nu_0,k_0}$, and for this it will be enough to show that the two strategies agree on λ-separated trees.[10] So let $\mathcal{U}$ be a λ-separated tree on M that is of limit length, and is by both $\Sigma_{\mathcal{W}^*_{\nu_0,k_0},M}$ and $\Omega^{\mathbb{C}}_{\nu_0,k_0}$. Let

$$c = \langle M, \text{id}, M, \mathbb{C}, V\rangle,$$
$$\text{lift}(\mathcal{U},c) = \langle \mathcal{U}^*, \langle c_\alpha \mid \alpha < \text{lh}(\mathcal{U})\rangle\rangle,$$

and

$$c_\alpha = \langle \mathcal{M}^{\mathcal{U}}_\alpha, \psi_\alpha, Q_\alpha, \mathbb{C}_\alpha, S_\alpha\rangle.$$

LEMMA 8.4.8. *If b is a cofinal, wellfounded branch of $\mathcal{U}^*$, then*

$$\Sigma_{\mathcal{W}^*_{\nu_0,k_0},M}(\mathcal{U}) = b.$$

Lemma 8.4.8 implies that $\mathcal{U}^*$ has at most one cofinal wellfounded branch. Moreover, that branch is identified by Σ, if it exists, and Σ is universally Baire. So a simple reflection argument will then give that $\mathcal{U}^*$ has a cofinal, wellfounded branch. From this we get that $\Sigma_{\mathcal{W}^*_{\nu_0,k_0},M}(\mathcal{U}) = \Omega^{\mathbb{C}}_{\nu_0,k_0}(\mathcal{U})$.

PROOF OF LEMMA 8.4.8. We write $(\mathcal{W}^*_{\nu,k})^{S_\gamma}$ for $\langle \nu,k\rangle \leq_{\text{lex}} i^{\mathcal{U}^*}_{0,\gamma}(\langle \nu_0,k_0\rangle)$ to stand for $i^{\mathcal{U}^*}_{0,\gamma}(\langle \eta,l\rangle \mapsto \mathcal{W}^*_{\eta,l})_{\nu,k}$. Note that

$$i^{\mathcal{U}^*}_{0,\gamma}(\Sigma) \cap S_\gamma = \Sigma \cap S_\gamma,$$

by Lemma 7.6.7. Also $i^{\mathcal{U}^*}_{0,\gamma}(P_0) = P_0$. Thus $(\mathcal{W}^*_{\nu,k})^{S_\gamma}$ is by Σ.

$\mathcal{M}^{\mathcal{U}^*}_b$ is wellfounded, so we have a last conversion stage

$$c_b = \langle \mathcal{M}^{\mathcal{U}}_b, \psi_b, Q_b, \mathbb{C}_b, S_b\rangle$$

in $\text{lift}(\mathcal{U}^\frown b, c)$. For $\gamma < \text{lh}(\mathcal{U})$ or $\gamma = b$, let

$$\begin{aligned}\langle \eta_\gamma, l_\gamma\rangle &= \text{unique } \langle \eta,l\rangle \text{ such that } Q_\gamma = M^{\mathbb{C}_\gamma}_{\eta,l},\\ \mathcal{W}^*_\gamma &= (W^*_{\eta_\gamma,l_\gamma})^{S_\gamma}\\ z^*(\gamma) &= \text{lh}(\mathcal{W}^*_\gamma) - 1,\end{aligned}$$

and

$$N_\gamma = \mathcal{M}^{\mathcal{W}^*_\gamma}_{z^*(\gamma)}.$$

---

[10]If $\Omega_{\nu,k}$ is total, we can quote Theorem 7.6.5 here. In the general case, we can use the proof of Theorem 7.6.5.

Thus $Q_\gamma \trianglelefteq N_\gamma$. $\mathcal{W}_\gamma^*$ is the unique λ-separated tree by Σ that iterates P_0 past Q_γ. If $\nu <_U \gamma$ and $(\nu,\gamma]_U$ does not drop, then $i^{\mathcal{U}^*}_{\nu,\gamma}(\mathcal{W}_\nu^*) = \mathcal{W}_\gamma^*$. (This is not the case if we have a drop.)

Now let us look at the meta-tree associated to the embedding normalization $W(\langle \mathcal{W}_0^*, \mathcal{U}^+ \rangle)$. This is a maximal stack of λ-separated trees, so our theory of embedding normalization applies to it, and embedding normalization coincides with quasi-normalization. If $Q_0 = N_0$, then $\mathcal{U}^+ = \mathcal{U}$, but in any case, $\mathcal{U}$ and $\mathcal{U}^+$ have the same tree order. Set

$$\mathcal{W}_\gamma = W(\mathcal{W}_0^*, \mathcal{U}^+ \restriction (\gamma+1))$$

for $\gamma < \mathrm{lh}(\mathcal{U})$, and

$$\mathcal{W}_b = W(\mathcal{W}_0^*, (\mathcal{U}^+)^\frown b).$$

So $\mathcal{W}_0 = \mathcal{W}_0^*$. The $\mathcal{W}_\gamma$'s are all by Σ, because Σ normalizes well and $\mathcal{U}^+ \restriction (\gamma+1)$ is by Σ. Suppose that $\mathcal{W}_b$ is by Σ, and let $\Sigma(\langle \mathcal{W}_0, \mathcal{U}^+ \rangle) = c$; then $\mathcal{W}_c$ is by Σ because Σ normalizes well, so $\mathrm{br}(b, W_0, \mathcal{U}^+) = \mathrm{br}(c, W_0, \mathcal{U}^+)$, so $b = c$. Thus if $\mathcal{W}_b$ is by Σ, then $\Sigma(\langle \mathcal{W}_0, \mathcal{U}^+ \rangle) = b$, and hence $\Sigma(\langle \mathcal{W}_0, \mathcal{U} \rangle) = b$ by internal lift consistency. This is what we want, so it is enough to show that $\mathcal{W}_b$ is by Σ.

We shall show

SUBLEMMA 8.4.8.1. *$\mathcal{W}_b$ is pseudo-hull of $\mathcal{W}_b^*$.*

That is enough to yield Lemma 8.4.8, since $\mathcal{W}_b^*$ is by Σ, and Σ has strong hull condensation.

PROOF OF SUBLEMMA 8.4.8.1. We construct by induction on γ an extended tree embedding

$$\Phi_\gamma \colon \mathcal{W}_\gamma \to \mathcal{W}_\gamma^*.$$

We write $z(\gamma) + 1 = \mathrm{lh}(\mathcal{W}_\gamma)$, and

$$\Phi_\gamma = \langle u^\gamma, v^\gamma, \langle s_\beta^\gamma \mid \beta \le z(\gamma) \rangle, \langle t_\beta^\gamma \mid \beta \le z(\gamma) \rangle \rangle.$$

Let $p^\gamma \colon \mathrm{Ext}(\mathcal{W}_\gamma) \to \mathrm{Ext}(\mathcal{W}_\gamma^*)$ be the associated map on extenders, given by

$$p^\gamma(E_\alpha^{\mathcal{W}_\gamma}) = E_{u^\gamma(\alpha)}^{\mathcal{W}_\gamma^*}.$$

The domain of u^γ is $z(\gamma)$, and that of v^γ is $z(\gamma)+1$. Because Φ_γ is an extended tree embedding, we have $v^\gamma(z(\gamma)) \le_{W_\gamma^*} z^*(\gamma)$, and a last t-map

$$t^\gamma = t_{z(\gamma)}^\gamma = \hat{\imath}^{\mathcal{W}_\gamma^*}_{v^\gamma(z(\gamma)), z^*(\gamma)} \circ s_{z(\gamma)}^\gamma$$

from $\mathcal{M}_{z(\gamma)}^{\mathcal{W}_\gamma}$ to $\mathcal{M}_{z^*(\gamma)}^{\mathcal{W}_\gamma^*}$. We let

$$R_\gamma = \mathcal{M}_{z(\gamma)}^{\mathcal{W}_\gamma},$$

so that

$$t^\gamma\colon R_\gamma \to N_\gamma$$

is the last t-map of Φ_γ. As we noted above, the last t-map of an extended tree embedding determines the whole of the tree embedding.

The embedding normalization process gives us extended tree embeddings

$$\Psi_{\nu,\gamma}\colon \mathcal{W}_\nu \to \mathcal{W}_\gamma,$$

defined when $\nu <_U \gamma$. We use $\phi_{\nu,\gamma}$ for the u-map of $\Psi_{\nu,\gamma}$, so that $\phi_{\nu,\gamma} : \mathrm{lh}(\mathcal{W}_\nu) \to \mathrm{lh}(\mathcal{W}_\gamma)$, the map being total if $(\nu,\gamma]_U$ does not drop in model or degree. We write $\pi_\tau^{\nu,\gamma}$ for the t-map $t_\tau^{\Psi_{\nu,\gamma}}$, so that

$$\pi_\tau^{\nu,\gamma} : \mathcal{M}_\tau^{\mathcal{W}_\nu} \to \mathcal{M}_{\phi_{\nu,\gamma}(\tau)}^{\mathcal{W}_\gamma}$$

elementarily, for $\nu <_U \gamma$ and $\tau \in \operatorname{dom} \phi_{\nu,\gamma}$. Let also $e_{\nu,\gamma} = p^{\Psi_{\nu,\gamma}}$, so that

$$e_{\nu,\gamma}(E_\alpha^{\mathcal{W}_\nu}) = E_{\phi_{\nu,\gamma}(\alpha)}^{\mathcal{W}_\gamma},$$

is the natural partial map from $\mathrm{Ext}(\mathcal{W}_\nu)$ to $\mathrm{Ext}(\mathcal{W}_\gamma)$. Let also

$$\sigma_\eta^1 : \mathcal{M}_\eta^{\mathcal{U}^+} \to R_\eta$$

be the natural map from $\mathcal{M}_\eta^{\mathcal{U}^+}$ to the last model of $\mathcal{W}_\eta$, and

$$F_\eta = \sigma_\eta^1(E_\eta^{\mathcal{U}^+}),$$

so that

$$\mathcal{W}_{\eta+1} = W(\mathcal{W}_\xi, \mathcal{W}_\eta, F_\eta)$$

where $\xi = U\text{-pred}(\eta+1)$. Finally,

$$\alpha_\eta = \text{least } \alpha \text{ such that } F_\eta \text{ is on the } \mathcal{M}_\alpha^{\mathcal{W}_\eta} \text{ sequence.}$$

We also have an extended tree embedding $\Psi^*_{\nu,\gamma}\colon \mathcal{W}^*_\nu \to \mathcal{W}^*_\gamma$ defined when $\nu <_U \gamma$ and $(\nu,\gamma]_U$ does not drop. The maps of $\Psi^*_{\nu,\gamma}$ are all restrictions of $i^{\mathcal{U}^*}_{\nu,\gamma}$, so we do not need to give them special names. Part of what we want to maintain as we define the Φ_γ is that in this case, the diagram

$$\begin{array}{ccc}
\mathcal{W}_\gamma & \xrightarrow{\Phi_\eta} & \mathcal{W}^*_\gamma \\
\uparrow{\scriptstyle \Psi_{\nu,\gamma}} & & \uparrow{\scriptstyle \Psi^*_{\nu,\gamma}} \\
\mathcal{W}_\nu & \xrightarrow{\Phi_\nu} & \mathcal{W}^*_\nu
\end{array}$$

commutes, in the appropriate sense. The other inductive requirements have to do with the agreement between Φ_η and Φ_ξ for $\eta \leq \xi$, and the fact that σ_η factors into ψ_η. We spell the requirements out completely below.

Since $\mathcal{W}_0 = \mathcal{W}_0^*$, Φ_0 is trivial, consisting of identity embeddings.

Remark 8.4.9. Before going through the induction in technical detail, let us look at the definition of Φ_1 in a simple case. This case contains the main idea.

Let $F = E_0^{\mathcal{U}^+} = E_0^{\mathcal{U}} = \psi_0(E_0^{\mathcal{U}})$. Let G be the resurrection of F in $\mathbb{C}$, and suppose $G = F$ for simplicity. Let F^* be the background extender for F given by $\mathbb{C}$. Then $\mathcal{W}_1 = W(\mathcal{W}_0, F)$ and $\mathcal{W}_1^* = i_{F^*}(\mathcal{W}_0)$. Let $\alpha = \alpha(\mathcal{W}_0, F)$. The last model of $\mathcal{W}_1^*$ is $i_{F^*}(M)$, and $i_{F^*}(M)$ agrees with $\mathrm{Ult}(M,F)$ up to $\mathrm{lh}(F)+1$. (See Lemma 4.7.7. The "plus 1" part is important, and it is one reason we were careful about choosing our background extenders.) It follows that $\mathcal{W}_1^*$ uses F; in fact $\mathcal{W}_1 \upharpoonright (\alpha+2) = \mathcal{W}_1^* \upharpoonright (\alpha+2)$, with $F = E_{\alpha+1}^{\mathcal{W}_1} = E_{\alpha+1}^{\mathcal{W}_1^*}$. This gives us the desired tree embedding from $\mathcal{W}_1$ to $\mathcal{W}_1^*$. For example, the map $p^1\colon \mathrm{Ext}(\mathcal{W}_1) \to \mathrm{Ext}(\mathcal{W}_1^*)$ is given by:

$$p^1(E) = E, \text{ if } E = E_\xi^{\mathcal{W}_1} \text{ for some } \xi \leq \alpha+1,$$

and if there is no dropping at $\alpha+1$,

$$p^1(e_{0,1}(E)) = i_{F^*}(E).$$

This is typical of the general successor step. Various maps that are the identity in this special case are no longer so in the general case. In particular, the resurrection maps may not be the identity. But the key is still that if $\mathcal{W}_{\gamma+1} = W(\mathcal{W}_\nu, \mathcal{W}_\gamma, F)$, and $H = \psi_\gamma(E_\gamma^{\mathcal{U}})$ is the blowup of F in the last model of $\mathcal{W}_\gamma^*$, and G is the resurrection of H inside S_γ, then $\mathcal{W}_{\gamma+1}^* = i_{G^*}(\mathcal{W}_\nu^*)$, and G is used in $\mathcal{W}_{\gamma+1}^*$. [There is a small revision to the first part of the conclusion in the dropping case.] In showing this, we shall need to know that the map resurrecting H to G appears as a branch embedding inside a certain normal tree $\mathcal{W}_\gamma^{**}$ extending $\mathcal{W}_\gamma^*$.

Setting $p^{\gamma+1}(F) = G$ determines everything. For we certainly want $p^{\gamma+1}$ to agree with p^γ on the extenders used before F in $\mathcal{W}_{\gamma+1}$. Moreover, we need to take a limit of the Φ_η's along branches of $\mathcal{U}$ in order to get past limit ordinals, and this requires that $p^{\gamma+1} \circ e_{\nu,\gamma+1} = i_{\nu,\gamma+1}^{\mathcal{U}^*} \circ p^\nu$. But this accounts for all the extenders in $\mathrm{dom}(p^{\gamma+1})$, so we have completely determined $p^{\gamma+1}$, and hence $\Phi_{\gamma+1}$, from Φ_ν.

Remark 8.4.10. All the plus trees on premice that we are dealing with now are λ-separated, that is, use only extenders of plus type. We shall sometimes say that F is on the sequence of Q, or is its last extender, when we really mean that F^- is on the sequence of Q, or is its last extender. (I.e. F is on the extended Q sequence.) Similarly, if F is a plus extender on the (extended) Q-sequence and $\psi\colon Q \to N$, then $\psi(F) = \psi(F^-)^+$. Similarly, $B^{\mathbb{C}}(F) = B^{\mathbb{C}}(F^-)$ whenever $B^{\mathbb{C}}(F^-)$ is defined.

The following little lemma says something about how $i_{\nu,\gamma}^{\mathcal{U}^*}(\mathcal{W}_\nu^*)$ sits inside $\mathcal{W}_\gamma^*$. In the language of tree embeddings, the map l it describes is just $s_\beta^{\Psi_{\nu,\gamma}^*}$.

LEMMA 8.4.11. *Suppose $\nu <_U \gamma$, and $(\nu,\gamma]_U$ does not drop. Let $\beta \le z(\nu)$; then*

$$\sup i^{\mathcal{U}^*}_{\nu,\gamma}``\beta \le_{W^*_\gamma} i^{\mathcal{U}^*}_{\nu,\gamma}(\beta).$$

Moreover, setting $\theta = \sup i^{\mathcal{U}^*}_{\nu,\gamma}``\beta$, we have that $(\theta, i^{\mathcal{U}^*}_{\nu,\gamma}(\beta)]_{W^*_\gamma}$ does not drop, and there is a unique embedding $l\colon \mathcal{M}^{\mathcal{W}^*_\nu}_\beta \to \mathcal{M}^{\mathcal{W}^*_\gamma}_\theta$ such that

$$i^{\mathcal{W}^*_\gamma}_{\theta, i^{\mathcal{U}^*}_{\nu,\gamma}(\beta)} \circ l = i^{\mathcal{U}^*}_{\nu,\gamma} \restriction \mathcal{M}^{\mathcal{W}^*_\nu}_\beta.$$

PROOF. We have

$$i^{\mathcal{U}^*}_{\nu,\gamma}(\mathcal{W}^*_\nu) = \mathcal{W}^*_\gamma$$

because $(\nu,\gamma]_U$ did not drop. If β is a successor ordinal, or $i^{\mathcal{U}^*}_{\nu,\gamma}$ is continuous at β, then $\theta = i^{\mathcal{U}^*}_{\nu,\gamma}(\beta)$ and all is trivial. Otherwise, let $\tau <_{W^*_\nu} \beta$ be the site of the last drop; then $i^{\mathcal{U}^*}_{\nu,\gamma}(\tau)$ is the site of the last drop in $[0, i^{\mathcal{U}^*}_{\nu,\gamma}(\beta)]_{W^*_\nu}$, and $i^{\mathcal{U}^*}_{\nu,\gamma}(\tau) <_{\mathcal{W}^*_\gamma} \theta$. Finally, we can define l by: if $\eta \in (\tau,\beta)_{W^*_\nu}$ and

$$\mu = i^{\mathcal{U}^*}_{\nu,\gamma}(\eta),$$

then

$$l(i^{\mathcal{W}^*_\nu}_{\eta,\beta}(x)) = i^{\mathcal{W}^*_\gamma}_{\mu,\theta}(i^{\mathcal{U}^*}_{\nu,\gamma}(x)).$$

It is easy to see that this works. ⊣

The following diagram illustrates the lemma.

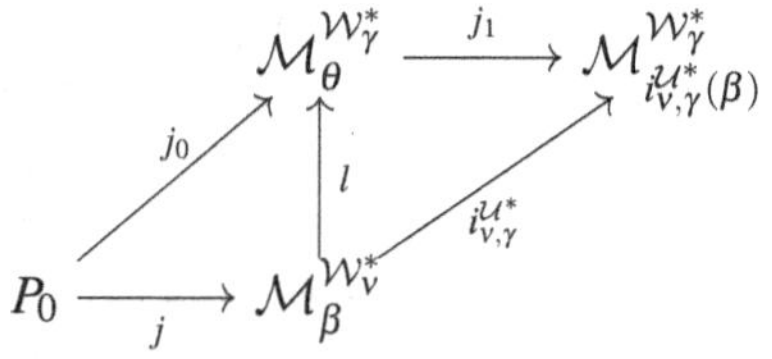

Here $j_1 \circ j_0 = i^{\mathcal{U}^*}_{\nu,\gamma}(j)$. (The diagram assumes j exists, which is of course not the general case.) j_0 is given by the downward closure of

$$\{i^{\mathcal{U}^*}_{\nu,\gamma}(E) \mid E \text{ is used in } [0,\beta)_{W^*_\nu}\}.$$

Again, l is just $s^{\Psi^*_{\nu,\gamma}}_\beta$.

We proceed to the general successor step. Suppose we are given Φ_η for $\eta \le \gamma$, and let us define $\Phi_{\gamma+1}$. For any $\gamma + 1 < \mathrm{lh}(\mathcal{U})$, let

- $H_\gamma = \psi_\gamma(E^{\mathcal{U}}_\gamma)$,
- $X_\gamma = Q_\gamma | \mathrm{lh}(H_\gamma) = N_\gamma | \mathrm{lh}(H_\gamma)$, and
- $\mathrm{res}_\gamma = \sigma_{Q_\gamma}[X_\gamma]^{\mathbb{C}_\gamma}$.

(Recall here the conventions of Remark 8.4.10.) So res_γ is the map resurrecting $\psi_\gamma(E^{\mathcal{U}}_\gamma)$ in $\mathbb{C}_\gamma$. Let also

- $Y_\gamma = \mathrm{Res}_{Q_\gamma}[X_\gamma]^{\mathbb{C}_\gamma}$,
- $G_\gamma = \mathrm{res}_\gamma(H_\gamma)$, and
- $G^*_\gamma = B^{\mathbb{C}_\gamma}(G_\gamma)$.

So $\mathrm{res}_\gamma \colon X_\gamma \to Y_\gamma$, G_γ is the last extender of Y_γ, and $G^*_\gamma = E^{\mathcal{U}^*}_\gamma$. Finally, let

$$\sigma^0_\gamma \colon \mathcal{M}^{\mathcal{U}}_\gamma \to K^0_\gamma \trianglelefteq \mathcal{M}^{\mathcal{U}^+}_\gamma$$

be the copy/lifting map, and set

$$\sigma_\gamma = \sigma^1_\gamma \circ \sigma^0_\gamma,$$

so that

$$\sigma_\gamma \colon \mathcal{M}^{\mathcal{U}}_\gamma \to K_\gamma \trianglelefteq R_\gamma.$$

To save notation below, we shall often just write $\sigma_\gamma \colon \mathcal{M}^{\mathcal{U}}_\gamma \to R_\gamma$. The reader will lose little by assuming that $\mathcal{U} = \mathcal{U}^+$ and $K_\gamma = R_\gamma$.

Our induction hypothesis is

Induction Hypothesis †.

$(\dagger)_\gamma$ (a) For $\xi < \eta \leq \gamma$, $\Phi_\xi \restriction (\alpha_\xi + 1) = \Phi_\eta \restriction (\alpha_\xi + 1)$.

(b) For all $\eta \leq \gamma$, t^η is well defined; that is, $v^\eta(z(\eta)) \leq_{W^*_\eta} z^*(\eta)$.

(c) For $\nu < \eta \leq \gamma$, $s^\eta_{z(\eta)} \restriction (\mathrm{lh}(F_\nu) + 1) = \mathrm{res}_\nu \circ t^\nu \restriction (\mathrm{lh}(F_\nu) + 1)$.

(d) Let $\nu < \eta \leq \gamma$, and $\nu <_U \eta$, and suppose that $(\nu, \eta]_U$ does not drop. Let $i^* = i^{\mathcal{U}^*}_{\nu,\eta}$, and let $\tau = \phi_{\nu,\eta}(\xi)$; then

(i) if $\xi < z(\nu)$, then $u^\eta(\tau) = i^*(u^\nu(\xi))$,

(ii) if $\xi < z(\nu)$, setting $j = i^{\mathcal{W}^*_\nu}_{v^\nu(\xi), u^\nu(\xi)}$ and $k = i^{\mathcal{W}^*_\eta}_{v^\eta(\tau), u^\eta(\tau)}$, there is an embedding $l \colon \mathcal{M}^{\mathcal{W}^*_\nu}_{v^\nu(\xi)} \to \mathcal{M}^{\mathcal{W}^*_\eta}_{v^\eta(\tau)}$ such that $k \circ l = i^* \circ j$, and $s^\eta_\tau \circ \pi^{\nu,\eta}_\xi = l \circ s^\nu_\xi$, and

(iii) if $\xi = z(\nu)$, then setting $j = i^{\mathcal{W}^*_\nu}_{v^\nu(\xi), z^*(\nu)}$ and $k = i^{\mathcal{W}^*_\eta}_{v^\eta(\tau), z^*(\eta)}$, there is an embedding $l \colon \mathcal{M}^{\mathcal{W}^*_\nu}_{v^\nu(\xi)} \to \mathcal{M}^{\mathcal{W}^*_\eta}_{v^\eta(\tau)}$ such that $k \circ l = i^* \circ j$, and $s^\eta_\tau \circ \pi^{\nu,\eta}_\xi = l \circ s^\nu_\xi$.

(e) For $\eta \leq \gamma$, $\psi_\eta = t^\eta \circ \sigma_\eta$.

(f) For all $\nu < \eta \leq \gamma$, Y_ν agrees with N_η strictly below $\mathrm{lh}\, G_\nu$. G_ν is on the extended Y_ν-sequence, but $\mathrm{lh}(G_\nu)$ is a cardinal of N_η.

There is one further induction hypothesis to come.
Items (a), (c), and (f) are our agreement hypotheses on the Φ_ν.
Clauses (c) and (f) should be read with clause (e) in mind. By (e), for all $\eta \leq \gamma$,

$$G_\eta = t^\eta(F_\eta).$$

For $\nu < \eta \le \gamma$, $\text{res}_\nu \circ t^\nu$ maps $R_\nu | \text{lh}(F_\nu)$ elementarily into Y_ν, and $s^\eta_{z(\eta)}$ maps $R_\eta \parallel \text{lh}(F_\nu)$ elementarily into $N_\eta \parallel \text{lh}(G_\nu)$. But dropping last extender predicates, the domain models are the same, and (f) says that the range models are the same. By (c), the maps agree on $\text{lh}(F_\nu)$. (This also uses (a), and the agreement between s and t maps in a tree embedding.) The upshot is that $(\dagger)_\gamma$ implies

$$\text{res}_\nu \circ t^\nu \restriction (R_\nu \parallel \text{lh}(F_\nu)) = s^\eta_{z(\eta)} \restriction (R_\gamma \parallel \text{lh}(F_\nu)),$$

for all $\nu < \eta \le \gamma$.

Remark 8.4.12. Literally speaking, $(\dagger)_\gamma$.(c) does not make sense, because

$$t^\nu(\text{lh}(F_\nu)) \notin \text{dom}(\text{res}_\nu).$$

As often, if $\sigma\colon P \to Q$, then we extend σ by setting $\sigma(o(P)) = o(Q)$.

Remark 8.4.13. $(\dagger)_\gamma$.(c) implies that if $\nu < \eta$, then

$$t^\eta \restriction \text{lh}(F_\nu) + 1 = \text{res}_\nu \circ t^\nu \restriction \text{lh}(F_\nu) + 1.$$

For letting $G_\nu = t^\eta_{\alpha_\nu}(F_\nu)$, we have that $\text{crit}(\hat{\imath}^{\mathcal{W}^*_\eta}_{\nu\eta(z(\eta)),z^*(\eta)}) \ge \lambda_{G_\nu}$, so $t^\eta = t^\eta_{z(\eta)}$ agrees with $s^\eta_{z(\eta)}$ on $\text{lh}(F_\nu) + 1$, and thus with $\text{res}_\nu \circ t^\nu$ on $\text{lh}(F_\nu) + 1$ by $(\dagger)_\gamma$.(c). We are using the λ-separation of $\mathcal{W}^*_\eta$ here; otherwise $\text{crit}(\hat{\imath}^{\mathcal{W}^*_\eta}_{\nu\eta(z(\eta)),z^*(\eta)}) = \lambda_{G_\nu}$ would be possible.

Remark 8.4.14. $(\dagger)_\gamma$ implies that for $\nu < \eta \le \gamma$,

$$t^\eta_{\alpha_\nu} \restriction \text{lh}(F_\nu) + 1 = \text{res}_\nu \circ t^\nu \restriction \text{lh}(F_\nu) + 1.$$

This is because $\alpha_\nu < z(\eta)$, and $F_\nu = E^{\mathcal{W}_\eta}_{\alpha_\nu}$. So on $\text{lh}(F_\nu) + 1$, $t^\eta_{\alpha_\nu}$ agrees with $s^\eta_{z(\eta)}$ by the agreement properties of tree embeddings (Lemma 6.4.8), and hence with $\text{res}_\nu \circ t^\nu$ by $(\dagger)_\gamma$.(c).

If $\alpha_\nu < z(\nu)$, then since Φ_ν is a tree embedding,

$$t^\nu \restriction \text{lh}(E^{\mathcal{W}_\nu}_{\alpha_\nu}) + 1 = t^\nu_{\alpha_\nu} \restriction \text{lh}(E^{\mathcal{W}_\nu}_{\alpha_\nu}) + 1.$$

But $\text{lh}(F_\nu) < \text{lh}(E^{\mathcal{W}_\nu}_{\alpha_\nu})$, so t^ν and $t^\nu_{\alpha_\nu}$ agree on $\text{lh}(F_\nu) + 1$.

Thus $t^\nu_{\alpha_\nu} \ne t^{\nu+1}_{\alpha_\nu}$ in general. (In fact, always.) The two maps agree up to $\text{lh}(F_\nu)$ if res_ν is the identity on $t^\nu_{\alpha_\nu}(\text{lh}(F_\nu))$, but they need not agree past that, and they do not agree below that if res_ν is not the identity that far. They may map into different models.

This is all consistent with $(\dagger)_\gamma$.(a), because $t^\nu_{\alpha_\nu}$ is not part of $\Phi_\nu \restriction (\alpha_\nu + 1)$. The map t^ξ_η is recording how the extender $E^{\mathcal{W}_\xi}_\eta$ is blown up into $\mathcal{W}^*_\xi$. As we go from ν to $\nu + 1$, $E^{\mathcal{W}_\nu}_{\alpha_\nu}$ is replaced by $F_\nu = E^{\mathcal{W}_{\nu+1}}_{\alpha_\nu}$. So the map blowing it up must be changed somewhat — even below $\text{lh}(F_\nu)$, if there is resurrection going on in S_ν. But $E^{\mathcal{W}_\nu}_{\alpha_\nu}$ is not part of $\mathcal{W}_\nu \restriction (\alpha_\nu + 1)$, so this does not affect (a).

Item (d) captures the commutativity hypothesis $\Phi_\eta \circ \Psi_{\nu,\eta} = \Psi^*_{\nu,\eta} \circ \Phi_\nu$. It is written out in terms of the component maps of these tree embeddings; the map l in part (d) is $(s_{v^\nu(\xi)})^{\Psi^*_{\nu,\eta}}$. $(\dagger)_\gamma$.(d)(i) says that $p^\eta(e_{\nu,\eta}(E)) = i^{\mathcal{U}^*}_{\nu,\eta}(p^\nu(E))$. Here is a diagram to go with the rest of this clause. In the diagram, $\tau = \phi_{\nu,\eta}(\xi)$. The far right assumes $u^\nu(\xi)$ exists, that is, $\xi < z(\nu)$.

$$\begin{CD}
\mathcal{M}^{\mathcal{W}_\eta}_\tau @>{s^\eta_\tau}>> \mathcal{M}^{\mathcal{W}^*_\eta}_{v^\eta(\tau)} @>{k}>> \mathcal{M}^{\mathcal{W}^*_\eta}_{u^\eta(\tau)} \\
@A{\pi^{\nu,\eta}_\xi}AA @A{l}AA @A{i^{\mathcal{U}^*}_{\nu,\eta}}AA \\
\mathcal{M}^{\mathcal{W}_\nu}_\xi @>{s^\nu_\xi}>> \mathcal{M}^{\mathcal{W}^*_\nu}_{v^\nu(\xi)} @>{j}>> \mathcal{M}^{\mathcal{W}^*_\nu}_{u^\nu(\xi)}
\end{CD}$$

Here j and k are the branch embeddings of $\mathcal{W}^*_\nu$ and $\mathcal{W}^*_\eta$. There is a similar diagram when $\xi = z(\nu)$, with $z^*(\nu)$ and $z^*(\eta)$ replacing $u^\nu(\xi)$ and $u^\eta(\tau)$.

Remark 8.4.15. The embedding along the bottom row of the diagram above is either t^ν_ξ or t^ν, depending on whether $\xi < z(\nu)$. The embedding along the top is either t^η_τ or t^η. So $(\dagger)_\gamma$.(d) implies that

$$t^\eta_{\phi_{\nu,\eta}(\xi)} \circ \pi^{\nu,\eta}_\xi = i^{\mathcal{U}^*}_{\nu,\eta} \circ t^\nu_\xi$$

if $\xi < z(\nu)$, and

$$t^\eta \circ \pi^{\nu,\eta}_{z(\nu)} = i^{\mathcal{U}^*}_{\nu,\eta} \circ t^\nu.$$

Here is a diagram to go with $(\dagger)_\gamma$(e).

$$\begin{array}{ccccccc}
 & & \mathcal{M}^{\mathcal{U}^+}_\gamma & \xrightarrow{\sigma^1_\gamma} & R_\gamma & \xrightarrow{t^\gamma} & N_\gamma \\
 & & \trianglelefteq & & \trianglelefteq & & \trianglelefteq \\
\mathcal{M}^{\mathcal{U}}_\gamma & \xrightarrow{\sigma^0_\gamma} & K^0_\gamma & \xrightarrow{\sigma^1_\gamma} & K_\gamma & \xrightarrow{t^\gamma} & Q_\gamma
\end{array}$$
$$\psi_\gamma : \mathcal{M}^{\mathcal{U}}_\gamma \to Q_\gamma$$

In diagrams below we shall condense this to

$$\mathcal{M}^{\mathcal{U}}_\gamma \xrightarrow{\sigma_\gamma} R_\gamma \xrightarrow{t^\gamma} N_\gamma, \qquad \psi_\gamma : \mathcal{M}^{\mathcal{U}}_\gamma \to N_\gamma$$

This is accurate if one remembers that the maps may only be nearly elementary as maps into proper initial segments of the target model.

In defining $\Phi_{\gamma+1}$, we shall make use of Lemma 8.3.1, which implies that res_γ is present in a branch embedding of some $(\mathcal{W}^*_{\nu,k})^{S_\gamma}$. Let

$$\tau_\gamma = \text{least } \xi \text{ such that } X_\gamma \trianglelefteq \mathcal{M}_\xi^{\mathcal{W}^*_\gamma}.$$

Let us also drop some subscripts for now, by setting

$$\langle F, H, G, G^*, X, \tau\rangle = \langle F_\gamma, H_\gamma, G_\gamma, G^*_\gamma, X_\gamma, \tau_\gamma\rangle.$$

CLAIM 8.4.16. *(a) If $\alpha_\gamma = z(\gamma)$, then $\tau \in [v^\gamma(\alpha_\gamma), z^*(\gamma)]_{W^*_\gamma}$,*
(b) If $\alpha_\gamma < z(\gamma)$, then $\tau \in [v^\gamma(\alpha_\gamma), u^\gamma(\alpha_\gamma)]_{W^*_\gamma}$.

PROOF. For (a): If $\alpha_\gamma = z(\gamma)$, then $v^\gamma(\alpha_\gamma) \leq_{W^*_\gamma} z^*(\gamma)$. $t^\gamma(F) = \hat{\imath}^{\mathcal{W}^*_\gamma}_{v(\alpha_\gamma),z^*(\gamma)} \circ s^\gamma_{z(\gamma)}(F)$ is on the sequence of $\mathcal{M}^{\mathcal{W}^*_\gamma}_{z^*(\gamma)}$. Since $\text{lh}(E^{\mathcal{W}_\gamma}_\xi) < \text{lh}(F)$ for all $\xi < \alpha_\gamma$, $\text{lh}(p^\gamma(E^{\mathcal{W}_\gamma}_\xi)) < \text{lh}(t^\gamma(F))$ for all $\xi < \alpha_\gamma$. Cofinally many extenders used in $[0, v(\alpha_\gamma))_{\mathcal{W}^*_\gamma}$ are in $\text{ran}\, p^\gamma$, which gives $\text{lh}(s^\gamma_{z(\gamma)}(F)) > \text{lh}(E^{\mathcal{W}^*_\gamma}_\xi)$ for all $\xi < v^\gamma(\alpha_\gamma)$. So $v^\gamma(\alpha_\gamma)$ is less than or equal to the least τ such that $t^\gamma(F)$ is on the $M^{\mathcal{W}^*_\gamma}_\tau$ sequence. That τ is the least η such that $t^\gamma(F) = \hat{\imath}^{\mathcal{W}^*_\gamma}_{v(\alpha_\gamma),\eta} \circ s^\gamma_{z(\gamma)}(F)$, so that $\tau \in [v^\gamma(\alpha_\gamma), z^*(\gamma)]_{W^*_\gamma}$. (See Proposition 8.2.1.)

For (b):If $\alpha_\gamma < z(\gamma)$, then $t^\gamma(F) = t^\gamma_{\alpha_\gamma}(F) = \hat{\imath}^{W^*_\gamma}_{v^\gamma(\alpha_\gamma),u^\gamma(\alpha_\gamma)} \circ s^\gamma_{\alpha_\gamma}(F)$. In this case

$$\tau = \text{least } \beta \in [v(\alpha_\gamma), u(\alpha_\gamma)]_{W^*_\gamma} \text{ such that } \text{crit}(\hat{\imath}_{\tau,u(\alpha_\gamma)}) > \hat{\imath}_{v(\alpha_\gamma),\tau}(\text{lh}(F)).$$

This can be shown as in the proof of (a). We omit the details. ⊣

By Lemma 8.3.1, there is a normal tree $\mathcal{W}^{**}_\gamma$ such that

(*i*) $\mathcal{W}^{**}_\gamma$ is by Σ, and extends $\mathcal{W}^*_\gamma \restriction (\tau+1)$,

(*ii*) letting $\xi_\gamma = \text{lh}\, \mathcal{W}^{**}_\gamma - 1$, G is on the extended $\mathcal{M}^{\mathcal{W}^{**}_\gamma}_{\xi_\gamma}$ sequence, and not on the $\mathcal{M}^{\mathcal{W}^{**}_\gamma}_\alpha$ sequence for any $\alpha < \xi_\gamma$,

(*iii*) $\tau \leq_{\mathcal{W}^{**}_\gamma} \xi_\gamma$, and $\hat{\imath}^{\mathcal{W}^{**}_\gamma}_{\tau,\xi_\gamma} \restriction (\text{lh}(H)+1) = \text{res}_\gamma \restriction (\text{lh}(H)+1)$.

Let

$$N^*_\gamma = \mathcal{M}^{\mathcal{W}^{**}_\gamma}_{\xi_\gamma}.$$

We shall show that $\mathcal{W}^{**}_\gamma$ is an initial segment of $\mathcal{W}^*_{\gamma+1}$, and that G is used in $\mathcal{W}^*_{\gamma+1}$. (So $G_\gamma = E^{\mathcal{W}^*_{\gamma+1}}_{\xi_\gamma}$.) By induction, the same has been true at all $\nu < \gamma$. That is, we have

Induction Hypothesis $(\dagger)_\gamma$.

$(\dagger)_\gamma$ (*g*). For all $\nu < \gamma$, $\mathcal{W}^{**}_\nu$ is an initial segment of $\mathcal{W}^*_\gamma \restriction (v^\gamma(\alpha_\gamma)+1)$. The last model of $\mathcal{W}^{**}_\nu$ is $N^*_\nu = \mathcal{M}^{\mathcal{W}^{**}_\nu}_{\xi_\nu}$, and $Y_\nu \trianglelefteq N^*_\nu$.

Here is a diagram showing where G came from, in the case that $\alpha_\gamma = z(\gamma)$.

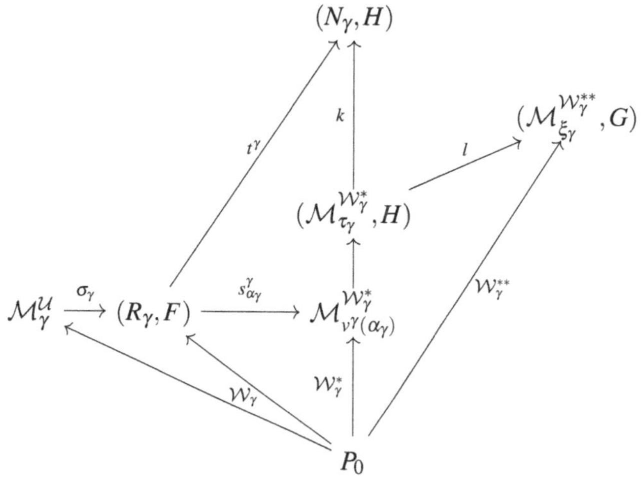

Here k is the branch embedding of $\mathcal{W}_\gamma^*$, and it is the identity on $\text{lh}(H)+1$. l is the branch embedding of $\mathcal{W}_\gamma^{**}$, and it agrees with res_γ on $\text{lh}(H)+1$.

If $\alpha_\gamma < z(\gamma)$, then the corresponding diagram is:

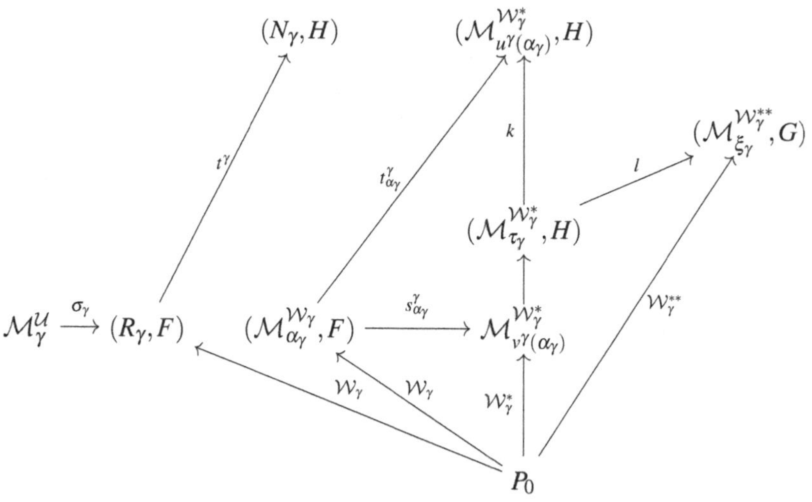

Here again, k is the branch embedding of $\mathcal{W}_\gamma^*$, and it is the identity on $\text{lh}(H)+1$. l is the branch embedding of $\mathcal{W}_\gamma^{**}$, and it agrees with res_γ on $\text{lh}(H)+1$. R_γ and $\mathcal{M}_{\alpha_\gamma}^{\mathcal{W}_\gamma}$ agree up to $\text{lh}(F)+1$, and t^γ agrees with $t_{\alpha_\gamma}^\gamma$ on $\text{lh}(F)+1$. (In fact, on $\text{lh}(E_{\alpha_\gamma}^{\mathcal{W}_\gamma})$.)

In either case, we get

CLAIM 8.4.17. $\text{res}_\gamma \circ t^\gamma$ *agrees with* $\hat{\imath}_{\nu^\gamma(\alpha_\gamma),\xi_\gamma}^{\mathcal{W}_\gamma^{**}} \circ s_{\alpha_\gamma}^\gamma$ *on* $\text{lh}(F)+1$.

PROOF. Suppose $\alpha_\gamma < z(\gamma)$. Let k and l be as in the diagram above. Then for $\eta \le \mathrm{lh}(F)$,

$$\begin{aligned}
\mathrm{res}_\gamma \circ t^\gamma(\eta) &= \mathrm{res}_\gamma \circ t^\gamma_{\alpha_\gamma}(\eta) \\
&= \mathrm{res}_\gamma \circ k \circ \hat{\imath}^{\mathcal{W}^*_\gamma}_{\nu^\gamma(\alpha_\gamma),\tau_\gamma} \circ s^\gamma_{\alpha_\gamma}(\eta) \\
&= \mathrm{res}_\gamma \circ \hat{\imath}^{\mathcal{W}^*_\gamma}_{\nu^\gamma(\alpha_\gamma),\tau_\gamma} \circ s^\gamma_{\alpha_\gamma}(\eta) \\
&= l \circ \hat{\imath}^{\mathcal{W}^*_\gamma}_{\nu^\gamma(\alpha_\gamma),\tau_\gamma} \circ s^\gamma_{\alpha_\gamma}(\eta) \\
&= \hat{\imath}^{\mathcal{W}^{**}_\gamma}_{\nu^\gamma(\alpha_\gamma),\xi_\gamma} \circ s^\gamma_{\alpha_\gamma}(\eta),
\end{aligned}$$

as desired. The calculation when $\alpha_\gamma = z(\gamma)$ is similar. ⊣

Now let $\nu = U\text{-pred}(\gamma+1)$. Thus we have $S_{\gamma+1} = \mathrm{Ult}(S_\nu, G^*)$, where G^* is the background extender for $G = G_\gamma$ provided by $\mathbb{C}_\gamma$. We write $i_{G^*} = i^{\mathcal{U}^*}_{\nu,\gamma+1}$ for the canonical embedding.

Case 1. $(\nu, \gamma+1]_U$ does not drop in model or degree.

In this case, we have

$$\begin{aligned}
Q_{\gamma+1} &= i_{G^*}(Q_\eta) \\
N_{\gamma+1} &= i_{G^*}(N_\nu)
\end{aligned}$$

and

$$\mathcal{W}^*_{\gamma+1} = i_{G^*}(\mathcal{W}^*_\nu).$$

Our goal is to define $\Phi_{\gamma+1}$, and with it $t^{\gamma+1}$, so that the following diagram is realized (among other things).

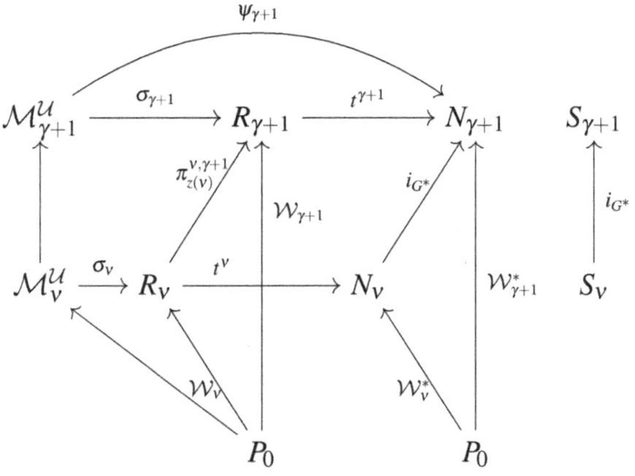

As we remarked in the case $\gamma+1 = 1$, it is important to see that the resurrection of the blowup of F, which is in our case G, is used in $\mathcal{W}^*_{\gamma+1}$.

CLAIM 8.4.18. *(a)* $\mathcal{W}^*_{\gamma+1} \restriction \xi_\gamma = \mathcal{W}^{**}_\gamma \restriction \xi_\gamma$.

(b) $G = E_{\xi_\gamma}^{\mathcal{W}^*_{\gamma+1}}$.

PROOF. Let $\mu = \text{crit}(F)$, where $F = F_\gamma$. Let $\sigma_\gamma(\bar{\mu}) = \mu$, where $\bar{\mu} = \text{crit}(E^{\mathcal{U}}_\gamma)$. Since $\mathcal{U}$ does not drop at $\gamma+1$, no level of $M^{\mathcal{U}}_\nu$ beyond $\text{lh}(E^{\mathcal{U}}_\nu)$ projects to or below $\bar{\mu}$. So no level of R_ν beyond $\text{lh}(F_\nu)$ projects to or below μ. So no level of N_ν beyond $\text{lh}(H_\nu)$ projects to or below $t^\nu(\mu)$. Thus res_ν is the identity on $t^\nu(\mu)^{+,N_\nu}$, and $N^*_\nu \restriction t^\nu(\mu)^{+,N^*_\nu} = N_\nu \restriction t^\nu(\mu)^{+,N_\nu}$. Also, $t^\nu(\mu)^{+,N^*_\nu} < \hat{\lambda}_{G_\nu}$. Thus

$$N_\nu | t^\nu(\mu)^{+,N_\nu} = N^*_\nu | t^\nu(\mu)^{+,N^*_\nu} = N_\gamma | t^\nu(\mu)^{+,N_\gamma}.$$

But also, if $\nu < \gamma$, then no proper initial segment of $\mathcal{M}^{\mathcal{U}}_\gamma$ projects to or below $\text{lh}(E^{\mathcal{U}}_\nu)$, so no proper initial segment of N_γ projects to or below $\text{lh}(G_\nu)$, so $\text{res}_\gamma = \text{id}$ on $\text{lh}(G_\nu)$, and $N_\gamma | t^\gamma(\mu)^{+,N_\gamma} = N^*_\gamma | t^\gamma(\mu)^{+,N^*_\gamma}$. Thus in both cases ($\nu < \gamma$ and $\nu = \gamma$),

$$N_\gamma | t^\gamma(\mu)^{+,N_\gamma} = N^*_\gamma | t^\gamma(\mu)^{+,N^*_\gamma}.$$

Letting $\lambda = t^\gamma(\mu)^{+,N^*_\gamma}$, we have then that $i_{G^*}(N_\gamma|\lambda) = i_{G^*}(N^*_\gamma|\lambda)$. But $\text{Ult}(N^*_\gamma, G)$ agrees with $i_{G^*}(N^*_\gamma|\lambda)$ up to $\text{lh}(G)+1$. (We chose G^* so that they would agree *at* $\text{lh}(G)$.) Thus

$$N_{\gamma+1} \parallel \text{lh}(G) = N^*_\gamma \parallel \text{lh}(G)$$

and $\text{lh}(G)$ is a cardinal in $N_{\gamma+1}$. Since $W^*_{\gamma+1}$ and W^{**}_γ are λ-separated trees by the same strategy Σ, we get Claim 8.4.18. ⊣

By Lemma 8.2.3, there is a unique tree embedding Ψ of $\mathcal{W}_{\gamma+1} \restriction (\alpha_\gamma + 2)$ into $\mathcal{W}^*_{\gamma+1}$ such that Ψ extends $\Phi_\gamma \restriction (\alpha_\gamma + 1)$, and $u^\Psi(\alpha_\gamma) = \xi_\gamma$, or equivalently, $p^\Psi(F) = G$. We let $\Phi_{\gamma+1} \restriction (\alpha_\gamma + 2)$ be the unique such Ψ.

In order to establish the proper notation related to $\Phi_{\gamma+1} \restriction (\alpha_\gamma + 2)$, as well as its relationship to Φ_ν, we shall now just run through the proof of Lemma 8.2.3 again.

Let us keep our notation $\mu = \text{crit}(F)$, and write

$$\mu^* = t^\nu(\mu) = t^\gamma(\mu) = \text{crit}(G).$$

Let

$$\beta = \beta^{\mathcal{W}_\nu, F},$$

so that F is applied to $\mathcal{M}^{\mathcal{W}_\nu}_\beta = \mathcal{M}^{\mathcal{W}_{\gamma+1}}_\beta$ in $\mathcal{W}_{\gamma+1}$. Let

$$\beta^* = W^*_{\gamma+1}\text{-pred}(\xi_\gamma + 1),$$

so that G is applied to $\mathcal{M}^{\mathcal{W}^*_{\gamma+1}}_{\beta^*} = \mathcal{M}^{\mathcal{W}^{**}_\gamma}_{\beta^*}$ in $\mathcal{W}^*_{\gamma+1}$.

CLAIM 8.4.19.

(a) $\beta^* \leq \tau_\nu$, *and* $\mathcal{M}^{\mathcal{W}^*_\nu}_{\beta^*} = \mathcal{M}^{\mathcal{W}^{**}_\nu}_{\beta^*} = \mathcal{M}^{\mathcal{W}^*_\gamma}_{\beta^*} = \mathcal{M}^{\mathcal{W}^{**}_\gamma}_{\beta^*} = \mathcal{M}^{\mathcal{W}^*_{\gamma+1}}_{\beta^*}$.

(b) $\beta^* = \mu^*$.
(c) *If* $\beta < z(\nu)$, *then* $\beta^* \in [v^\nu(\beta), u^\nu(\beta)]_{W^*_\nu}$.
(d) *If* $\beta = z(\nu)$, *then* $\beta^* \in [v^\nu(\beta), z^*(\nu)]_{W^*_\nu}$.

PROOF. Let P be the domain of F and P^* the domain of G; that is,

$$P = R_\gamma | \mu^{+,R_\gamma}$$

and

$$P^* = N_\gamma | t^\gamma(\mu)^{+,N_\gamma} = N^*_\gamma | t^\gamma(\mu)^{+,N^*_\gamma}.$$

(N_γ agrees with N^*_γ this far because we are not dropping when we apply F.) By the rules for λ-separated trees,

$$\beta^* = \text{least } \alpha \text{ such that } P^* = \mathcal{M}_\alpha^{\mathcal{W}^{**}_\gamma} | o(P^*).$$

Put another way, $\mathcal{W}^{**}_\gamma \restriction \beta^* + 1$ is unique shortest λ-separated tree on P_0 by Σ such that P^* is an initial segment of its last model, and $o(P^*)$ is passive in its last model. But we showed in the proof of Claim 8.4.18 that $P^* = N^*_\nu | o(P^*)$, and $o(P^*) < \hat{\lambda}(G_\nu)$. We also showed that $\text{res}_\nu \restriction P^* = \text{identity}$. Thus $P^* = N_\nu | o(P^*)$, and $o(P^*) < \hat{\lambda}(H_\nu)$. So P^* is a passive initial segment of the last models of $\mathcal{W}^*_\nu, \mathcal{W}^{**}_\nu, \mathcal{W}^*_\gamma, \mathcal{W}^{**}_\gamma$, and $\mathcal{W}^*_{\gamma+1}$. Thus all these trees agree up to $\beta^* + 1$. As $o(P^*) < \text{lh}(H_\nu)$, $\beta^* \le \tau_\nu$. This yields (*a*).

For (b), note that μ^* is a cardinal of S_γ, so $|\mathcal{M}_\alpha^{\mathcal{W}^*_\gamma}| < \mu^*$ in S_γ, for all $\alpha < \mu^*$. It follows that

$$\mu^* = \sup_{\alpha < \mu^*} \hat{\lambda}(E_\alpha^{\mathcal{W}^*_\gamma}) \le \beta^*.$$

Since $\mu^* = \text{crit}(G)$ and G is on the extended sequence of N_γ, $\mu^* \ne \hat{\lambda}(E_{\mu^*}^{\mathcal{W}^*_\gamma})$. Thus $\mu^* < \hat{\lambda}(E_{\mu^*}^{\mathcal{W}^*_\gamma})$, and hence $\beta^* \le \mu^*$. So $\beta^* = \mu^*$.

For (c): if $\beta < z(\nu)$, then $\mu < \hat{\lambda}(E_\beta^{\mathcal{W}_\nu})$, so

$$\begin{aligned} \mu^* &= t^\nu(\mu) = t^\nu_\beta(\mu) \\ &= i^{\mathcal{W}^*_\nu}_{v^\nu(\beta), u^\nu(\beta)} \circ s^\nu_\beta(\mu). \end{aligned}$$

Also, $\mu^* < \hat{\lambda}(E^{\mathcal{W}^*_\nu}_{u^\nu(\beta)})$, so $\beta^* \le u^\nu(\beta)$ and $P^* \lhd M^{\mathcal{W}^*_\nu}_{u^\nu(\beta)} \parallel \lambda(E^{\mathcal{W}^*_\nu}_{u^\nu(\beta)})$.

We claim that

$$\begin{aligned} \beta^* = \text{least } \alpha \in [v^\nu(\beta), u^\nu(\beta)]_{W^*_\nu} \text{ such that} \\ \text{crit}(i^{\mathcal{W}^*_\nu}_{\alpha, u^\nu(\beta)}) > i^{\mathcal{W}^*_\nu}_{v^\nu(\beta), \alpha}(s^\nu_\beta(\mu)) \text{ or } \alpha = u^\nu(\beta). \end{aligned}$$

This follows from Proposition 8.2.1, applied with $\mathcal{S} = \mathcal{W}^*_\nu$, $\delta = v^\nu(\beta)$, and $\eta = u^\nu(\beta)$. To see that the proposition applies, note that

$$P^*, \mu^* \in \text{ran}\, i^{\mathcal{W}^*_\nu}_{v^\nu(\beta), u^\nu(\beta)}.$$

Also $v^\nu(\beta) \le \beta^*$, since if $E = E_\eta^{\mathcal{W}_\nu}$ is used in $[0,\beta)_{W_\nu}$, then $\mathrm{lh}(E) \le \mu$, and thus $\mathrm{lh}(p^\nu(E)) = t_\eta^\nu(\mathrm{lh}(E)) \le t^\nu(\mathrm{lh}(E)) \le t^\nu(\mu) = \mu^*$. Finally, by the agreement of $\mathcal{W}_\gamma^{**}$ with $\mathcal{W}_\nu^*$ up to β^*+1, β^* is the least α such that $P^* = \mathcal{M}_\alpha^{\mathcal{W}_\nu^*}|o(P^*)$. Thus Proposition 8.2.1 applies, and we have proved our claim.

This proves (c). The proof of (d) is similar. ⊣

With regard to part (b) of the claim: it is perfectly possible that β is a successor ordinal. In this case $v^\nu(\beta) < \beta^* = \mu^*$, and $s_\beta^\nu(\mu) < \mu^*$ as well. So $\beta^* = \mu^*$ is strictly between $v^\nu(\beta)$ and either $u^\nu(\beta)$ or $z^*(\nu)$, as the case may be. This is a manifestation of the fact that the tree embeddings Φ_ν are very far from being onto when $\nu > 0$.

Our proof of Claim 8.4.19 actually showed

CLAIM 8.4.20. *(a) If $\beta < z(\nu)$, then $\beta^* =$ least $\alpha \in [v^\nu(\beta), u^\nu(\beta)]_{W_\nu^*}$ such that*

$$\mathrm{crit}(i_{\alpha,u^\nu(\beta)}^{\mathcal{W}_\nu^*}) > i_{v^\nu(\beta),\alpha}^{\mathcal{W}_\nu^*}(s_\beta^\nu(\mu)).$$

(b) If $\beta = z(\nu)$, then $\beta^* =$ least $\alpha \in [v^\nu(\beta), z^*(\nu)]_{W_\nu^*}$ such that

$$\mathrm{crit}(i_{\alpha,z^*(\nu)}^{\mathcal{W}_\nu^*}) > i_{v^\nu(\beta),\alpha}^{\mathcal{W}_\nu^*}(s_\beta^\nu(\mu)).$$

(c) In either case, the embeddings t^ν, $\mathrm{res}_\nu \circ t^\nu$, and $i_{v^\nu(\beta),\beta^*}^{\mathcal{W}_\nu^*} \circ s_\beta^\nu$ all agree on the domain of F.

PROOF. We have already proved (a) and (b). The following diagram illustrates the situation when $\beta < z(\nu)$.

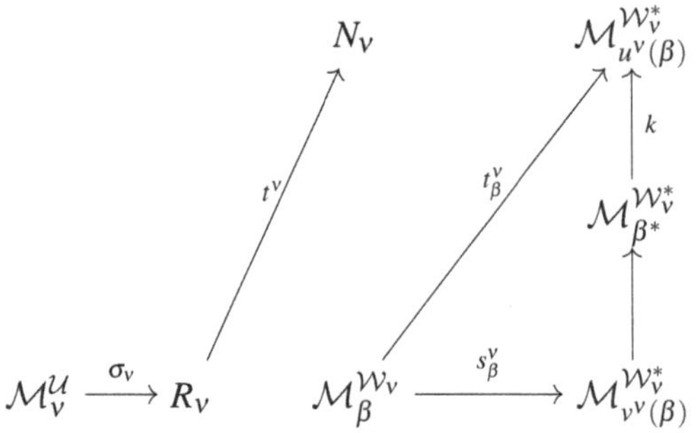

We have shown that both k and res_ν are the identity on the domain of G, that is, on $t^\nu(\mu)^+$ of $\mathcal{M}_{\beta^*}^{\mathcal{W}_\nu^*}$. The agreement of t^ν with t_β^ν on $\mathrm{lh}(E_\beta^{\mathcal{W}_\nu})$, which is strictly greater than $(\mu^+)^{R_\gamma}$, completes the proof of (c). The case that $\beta = z(\nu)$ is similar. ⊣

Now let

$$\rho = i_{v^\nu(\beta),\beta^*}^{\mathcal{W}_\nu^*} \circ s_\beta^\nu,$$

so that $\rho : \mathcal{M}_{\beta}^{\mathcal{W}_\nu} \to \mathcal{M}_{\beta^*}^{\mathcal{W}_\nu^*}$. On the domain of F, ρ agrees with t^ν and with $\mathrm{res}_\nu \circ t^\nu$. We can then define $\Phi_{\gamma+1}$ at $\alpha_\gamma + 1$. That is, we set

$$\begin{aligned} u^{\gamma+1} \restriction \alpha_\gamma &= u^\gamma \restriction \alpha_\gamma, \\ p^{\gamma+1} \restriction \mathrm{Ext}(W_\gamma \restriction \alpha_\gamma) &= p^\gamma \restriction \mathrm{Ext}(W_\gamma \restriction \alpha_\gamma), \\ v^{\gamma+1} \restriction \alpha_\gamma + 1 &= v^\gamma \restriction \alpha_\gamma + 1, \\ s_\eta^{\gamma+1} &= s_\eta^\gamma \quad \text{for } \eta \leq \alpha_\gamma, \end{aligned}$$

and

$$t_\eta^{\gamma+1} = t_\eta^\gamma \quad \text{for } \eta < \alpha_\gamma.$$

Then we set

$$\begin{aligned} u^{\gamma+1}(\alpha_\gamma) &= \xi_\gamma, \\ p^{\gamma+1}(F) &= G, \\ v^{\gamma+1}(\alpha_\gamma + 1) &= \xi_\gamma + 1, \end{aligned}$$

and let $s_{\alpha_\gamma+1}^{\gamma+1}$ be given by the Shift Lemma,

$$s_{\alpha_\gamma+1}^{\gamma+1}([a,f]_F^{\mathcal{M}_\beta^{\mathcal{W}_\nu}}) = [\mathrm{res}_\gamma \circ t^\gamma(a), \rho(f)]_G^{\mathcal{M}_{\beta^*}^{\mathcal{W}_\nu^*}}.$$

We have shown that ρ agrees with $\mathrm{res}_\nu \circ t^\nu$ on the domain of F. By $(\dagger)_\gamma$, ρ agrees with t^γ on the domain of F. Since res_γ is the identity on the domain of H (cf. Claim 8.4.18), ρ agrees with $\mathrm{res}_\gamma \circ t^\gamma$ on the domain of F, and we can apply the Shift Lemma here. Let us also set

$$t_{\alpha_\gamma}^{\gamma+1} = \hat{\imath}_{v^\gamma(\alpha_\gamma),\xi_\gamma}^{\mathcal{W}_\gamma^{**}} \circ s_{\alpha_\gamma}^\gamma.$$

Then $t^{\gamma+1} \colon \mathcal{M}_{\alpha_\gamma}^{\mathcal{W}_{\gamma+1}} \to \mathcal{M}_{u^{\gamma+1}(\alpha_\gamma)}^{\mathcal{W}_{\gamma+1}^*} = \mathcal{M}_{\xi_\gamma}^{\mathcal{W}_\gamma^{**}}$, and $t^{\gamma+1}$ agrees with $\mathrm{res}_\gamma \circ t^\gamma$ on $\mathrm{lh}(F) + 1$, by Claim 8.4.17.

This gives us $\Phi_{\gamma+1} \restriction (\alpha_\gamma + 2)$.

CLAIM 8.4.21. *$\Phi_{\gamma+1} \restriction (\alpha_\gamma + 2)$ is a tree embedding of $\mathcal{W}_{\gamma+1} \restriction (\alpha_\gamma + 2)$ into $\mathcal{W}_{\gamma+1}^* \restriction (\xi_\gamma + 2)$, and extends $\Phi_\gamma \restriction (\alpha_\gamma + 1)$.*

PROOF. We checked some of the tree embedding properties as we defined $\Phi_{\gamma+1}$. We must still check that $t_{\alpha_\gamma}^{\gamma+1}$ satisfies properties (e) and (f) of Definition 6.4.1. Noting that $E_{\alpha_\gamma}^{\mathcal{W}_\gamma} = F$ and that $t_{\alpha_\gamma}^{\gamma+1}$ agrees with $\mathrm{res}_\gamma \circ t^\gamma$ on $\mathrm{lh}(F) + 1$, this is easy to do. See the proof of Lemma 8.2.3. ⊣

We can define the remainder of the maps $u^{\gamma+1}$ and $p^{\gamma+1}$ of $\Phi_{\gamma+1}$ right now. If $\beta \leq \xi < z(\nu)$, then we set

$$u^{\gamma+1}(\phi_{\nu,\gamma+1}(\xi)) = i_{G^*}(u^\nu(\xi)), \quad \text{and} \quad p^{\gamma+1}(e_{\nu,\gamma+1}(E)) = i_{G^*}(p^\nu(E)),$$

for $E = E_\xi^{\mathcal{W}_\nu}$. Note that this then holds true for any E, since if $E = E_\xi^{\mathcal{W}_\nu}$ for some $\xi < \beta$, then $p^{\gamma+1}(e_{\nu,\gamma+1}(E)) = p^{\gamma+1}(E) = p^\nu(E) = i^{G^*}(p^\nu(E))$.

The definition of the s and t-maps of $\Phi_{\gamma+1}$, and the proof that everything fits together properly, must be done by induction.

As we define $\Phi_{\gamma+1}$, we shall check that it is a tree embedding, and we shall also check the applicable parts of $(\dagger)_{\gamma+1}$. We have $\Phi_{\gamma+1} \restriction (\alpha_\gamma + 1) = \Phi_\gamma \restriction (\alpha_\gamma + 1)$ by construction, which yields $(\dagger)_{\gamma+1}$(a). Claim 8.4.18 yields the agreement clauses (f) and (g) of $(\dagger)_{\gamma+1}$, so we are left with (b)-(e). The new cases in clauses (b), (c), and (e) have to do what happens at $z(\gamma+1)$, when the definition of $\Phi_{\gamma+1}$ is complete. The new case in (c) is $\eta = \gamma+1$, and it is enough to show that $s_{z(\gamma+1)}^{\gamma+1} \restriction \mathrm{lh}(F)+1 = \mathrm{res}_\gamma \circ t^\gamma \restriction \mathrm{lh}(F)+1$, since the rest of (c) follows by induction. But given that $\Phi_{\gamma+1}$ is a tree embedding, $s_{z(\gamma+1)}^{\gamma+1} \restriction \mathrm{lh}(F)+1 = s_{\alpha_\gamma+1}^{\gamma+1} \restriction \mathrm{lh}(F)+1$, and $s_{\alpha_\gamma+1}^{\gamma+1} \restriction \mathrm{lh}(F)+1 = \mathrm{res}_\gamma \circ t^\gamma \restriction \mathrm{lh}(F)+1$ by the Shift Lemma. So we can ignore (c).

The new case in (d) is $\eta = \gamma+1$. We can assume by induction that the $\nu <_U \gamma+1$ referred to in (d) is U-pred$(\gamma+1)$, that is, the ν we have already fixed. Clause (d)(i) then asserts that $u^{\gamma+1}(\phi_{\nu,\gamma+1}(\xi)) = i_{G^*}(u^\nu(\xi))$, which indeed is true by our definition above. So we can ignore (d)(i).

So as we define the s and t maps of $\Phi_{\gamma+1}$, we must check that $\Phi_{\gamma+1}$ is a tree embedding, that the commutativity clauses (d)(ii)(iii) hold for $\eta = \gamma+1$ and $\nu = U$-pred$(\gamma+1)$, and that (b) and (e) hold for $\eta = \gamma+1$ when we reach $z(\gamma+1)$.

We begin with

CLAIM 8.4.22. *$\Phi_{\gamma+1} \restriction (\alpha_\gamma+2)$ satisfies the applicable clauses of $(\dagger)_{\gamma+1}$.*

PROOF. Suppose that $(\dagger)_{\gamma+1}$(b) is applicable, that is, that $z(\gamma+1) = \alpha_\gamma+1$. So $z(\nu) = \beta$. We have $v^{\gamma+1}(\alpha_\gamma+1) = \xi_\gamma+1$. So what we must see is that $\xi_\gamma + 1 \leq_{W_{\gamma+1}^*} z^*(\gamma+1)$. That is, we must see that G is used on the branch to $z^*(\gamma+1)$. We are in the non-dropping case, so $z^*(\gamma+1) = i_{G^*}(z^*(\nu))$. The relevant diagram here is

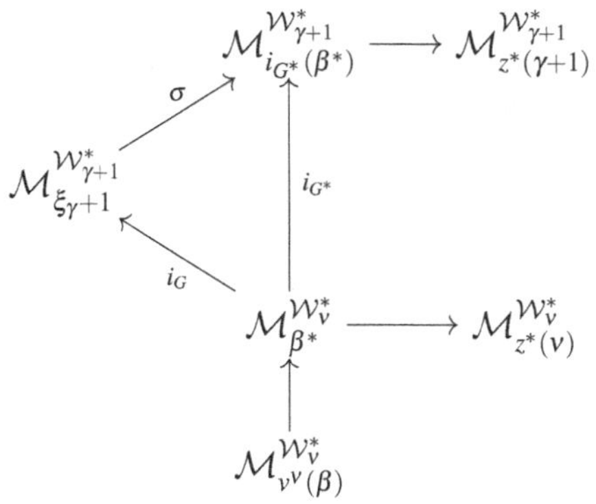

If s is the branch extender $s = e_{\beta^*}^{\mathcal{W}_\nu^*}$, then $i_{G^*}(s(i)) = s(i)$ for all $i \in \mathrm{dom}(s)$, and thus $s \subseteq e_{i_{G^*}(\beta^*)}^{\mathcal{W}_{\gamma+1}^*}$. It follows that

$$\mathcal{M}_{\beta^*}^{\mathcal{W}_\nu^*} = \mathcal{M}_{\beta^*}^{\mathcal{W}_{\gamma+1}^*},$$

and that

$$i_{G^*} \restriction \mathcal{M}_{\beta^*}^{\mathcal{W}_\nu^*} = i_{\beta^*, i_{G^*}(\beta^*)}^{\mathcal{W}_{\gamma+1}^*}.$$

The factor map σ in our diagram is the identity on the generators of G. It follows that G is compatible with the first extender used in $i_{\beta^*, i_{G^*}(\beta^*)}^{\mathcal{W}_{\gamma+1}^*}$, and thus G is that extender, as desired.

Turning to $(\dagger)_{\gamma+1}(\mathrm{d})$, the new applicable cases are (ii) and (iii), when $\xi = \beta$ and $\tau = \alpha_\gamma + 1$. Let us suppose that it is (ii) that applies, that is, that $\beta < z(\nu)$. The last paragraph showed that G is used on the branch to $i_{G^*}(\beta^*)$ in this case as well. We have the diagram

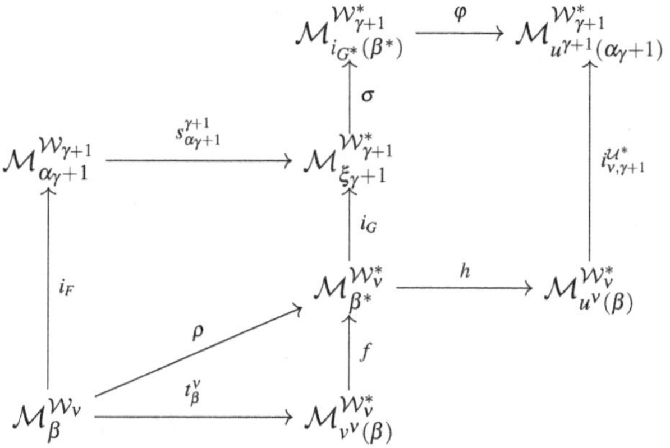

Here $\pi_\beta^{\nu,\gamma+1} = i_F^{\mathcal{M}_\beta^{\mathcal{W}_\nu}}$. The branch embeddings $\varphi \circ \sigma$ of $\mathcal{W}_{\gamma+1}^*$ and $h \circ f$ of $\mathcal{W}_\nu^*$ play the roles of k and j in $(\dagger)_\gamma.(\mathrm{d})$. The role of l is played by $i_G \circ f$. The diagram commutes, so we are done. The case $\beta = z(\nu)$ is similar.

We turn to $(\dagger)_{\gamma+1}(\mathrm{e})$, that $\psi_{\gamma+1} = t^{\gamma+1} \circ \sigma_{\gamma+1}$. This is applicable when

$$z(\gamma+1) = \alpha_\gamma + 1,$$

and hence since we did not drop, $z(\nu) = \beta$. So

$$\mathcal{M}_\beta^{\mathcal{W}_\nu} = R_\nu,$$

$$\mathcal{M}_{\alpha_\gamma+1}^{\mathcal{W}_{\gamma+1}} = R_{\gamma+1},$$

$$\mathcal{M}_{z^*(\nu)}^{\mathcal{W}_\nu^*} = N_\nu,$$

and

$$\mathcal{M}^{\mathcal{W}^*_{\gamma+1}}_{z^*(\gamma+1)} = N_{\gamma+1}.$$

Expanding the diagram immediately above a little, while making these substitutions, we get

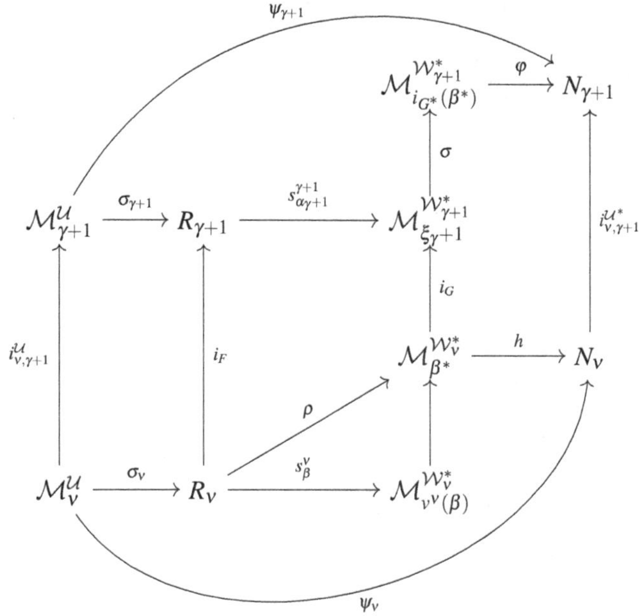

We have $t^{\gamma+1} = \varphi \circ \sigma \circ s^{\gamma+1}_{\alpha_\gamma+1}$ and $t^\nu = h \circ \rho$.

Note first that $\psi_{\gamma+1}$ agrees with $t^{\gamma+1} \circ \sigma_{\gamma+1}$ on $\mathrm{ran}(i^{\mathcal{U}}_{\nu,\gamma+1})$. This is because

$$\begin{aligned}
\psi_{\gamma+1} \circ i^{\mathcal{U}}_{\nu,\gamma+1} &= i^{\mathcal{U}^*}_{\nu,\gamma+1} \circ \psi_\nu \\
&= i^{\mathcal{U}^*}_{\nu,\gamma+1} \circ (h \circ \rho \circ \sigma_\nu) \qquad\qquad \text{(by } (\dagger)_\nu) \\
&= t^{\gamma+1} \circ \sigma_{\gamma+1} \circ i^{\mathcal{U}}_{\nu,\gamma+1}.
\end{aligned}$$

The last equality holds because of the commutativity of the non-ψ part of the diagram.

$\mathcal{M}^{\mathcal{U}}_{\gamma+1}$ is generated by $\mathrm{ran}(i^{\mathcal{U}}_{\nu,\gamma+1}) \cup \varepsilon$, where $\varepsilon = \mathrm{lh}(E^{\mathcal{U}}_\gamma)$. So it is now enough to show that $\psi_{\gamma+1}$ agrees with $t^{\gamma+1} \circ \sigma_{\gamma+1}$ on ε. But note

$$\begin{aligned}
\psi_{\gamma+1} \restriction \varepsilon &= \mathrm{res}_\gamma \circ \psi_\gamma \restriction \varepsilon \\
&= \mathrm{res}_\gamma \circ t^\gamma \circ \sigma_\gamma \restriction \varepsilon \\
&= t^{\gamma+1} \circ \sigma_\gamma \restriction \varepsilon \\
&= t^{\gamma+1} \circ \sigma_{\gamma+1} \restriction \varepsilon.
\end{aligned}$$

Line 2 follows from $(\dagger)_\gamma$, and line 3 holds because $t^{\gamma+1}$ agrees with $\text{res}_\gamma \circ t^\gamma$ on $\text{lh}(F)$. The last equality holds because σ_γ agrees with $\sigma_{\gamma+1}$ on $\text{lh}(F)+1$, by our earlier work on normalization. This proves $(\dagger)_{\gamma+1}$(e).

This proves Claim 8.4.22. ⊣

For the rest, we define $\Phi_{\gamma+1} \restriction \eta+1$, for $\alpha_\gamma + 1 < \eta \le z(\gamma+1)$, by induction on η, and verify that it is a tree embedding. At the same time, we prove those clauses in $(\dagger)_{\gamma+1}$ that make sense by stage η. We have already verified (a), (c), (f), and (g).

First, suppose we are given $\Phi_{\gamma+1} \restriction (\eta+1)$, where $\alpha_\gamma + 2 \le \eta + 1 < z(\gamma+1)$. We must define $\Phi_{\gamma+1} \restriction (\eta+2)$. Let

$$\phi_{\nu,\gamma+1}(\tau) = \eta,$$

$$E = E_\eta^{\mathcal{W}_{\gamma+1}},$$

and

$$K = E_\tau^{\mathcal{W}_\nu}.$$

Let

$$E^* = p^{\gamma+1}(E) \quad \text{and} \quad K^* = p^\nu(K).$$

We have already defined $p^{\gamma+1}$ so that $i_{G^*}(K^*) = E^*$, and $u^{\gamma+1}(\eta) = i_{G^*}(u^\nu(\tau))$. We can simply apply Lemma 8.2.3 to obtain $\Phi_{\gamma+1} \restriction (\eta+2)$ from $\Phi_{\gamma+1} \restriction (\eta+1)$. For we have the diagram from $(\dagger)_{\gamma+1}$(d).

$$\begin{array}{ccccc}
\mathcal{M}_\eta^{\mathcal{W}_{\gamma+1}} & \xrightarrow{s_\eta^{\gamma+1}} & \mathcal{M}_{v^{\gamma+1}(\eta)}^{\mathcal{W}^*_{\gamma+1}} & \longrightarrow & \mathcal{M}_{u^{\gamma+1}(\eta)}^{\mathcal{W}^*_{\gamma+1}} \\
\uparrow \scriptstyle{\pi_\tau^{\nu,\gamma+1}} & & \uparrow \scriptstyle{l} & & \uparrow \scriptstyle{i^{\mathcal{U}^*}_{\nu,\gamma+1}} \\
\mathcal{M}_\tau^{\mathcal{W}_\nu} & \xrightarrow{s_\tau^\nu} & \mathcal{M}_{v^\nu(\tau)}^{\mathcal{W}^*_\nu} & \longrightarrow & \mathcal{M}_{u^\nu(\tau)}^{\mathcal{W}^*_\nu}
\end{array}$$

Taking $\xi = u^{\gamma+1}(\eta)$, we see from the commutativity of this diagram that $E_\xi^{\mathcal{W}^*_{\gamma+1}} = i^{\mathcal{W}^*_{\gamma+1}}_{v^{\gamma+1}(\eta),\xi} \circ s_\eta^{\gamma+1}(E_\eta^{\mathcal{W}_{\gamma+1}})$. Thus the condition (2) in Lemma 8.2.3 is fulfilled, and we can let $\Phi_{\gamma+1} \restriction (\eta+2)$ be the unique tree embedding of $\mathcal{W}_{\gamma+1} \restriction (\eta+2)$ into $\mathcal{W}^*_{\gamma+1}$ that extends $\Phi_{\gamma+1} \restriction (\eta+1)$, and maps E to $i_{G^*}(p^\nu(K))$.

We now verify the applicable parts of $(\dagger)_{\gamma+1}$. The proofs are like the successor case $\eta = \alpha_\gamma$ that we have already done. We consider first clause (d). The new case to consider is $\xi = \tau + 1$. We have $\phi_{\nu,\gamma+1}(\tau+1) = \eta + 1$. Let $\sigma = W_\nu\text{-pred}(\tau+1)$ and $\theta = W_{\gamma+1}\text{-pred}(\eta+1)$ index the places K and E are applied. Let σ^* and θ^* index the models in $\mathcal{W}^*_\nu$ and $\mathcal{W}^*_{\gamma+1}$ to which K^* and E^* are applied. Let us write $i^* = i_{G^*}$. We have $i^*(K^*) = E^*$ and $i^*(\sigma^*) = \theta^*$.

For purposes of drawing the following diagram, we assume $\tau + 1 < z(\nu)$. The situation is

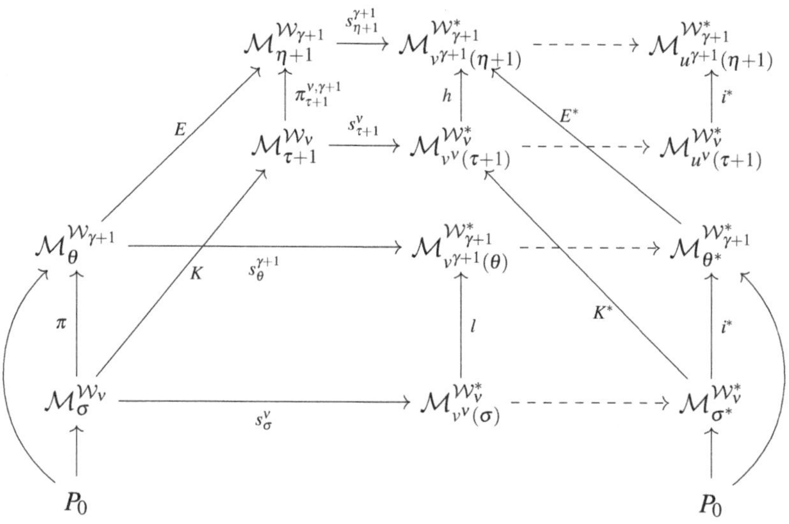

There are two cases being covered in this diagram:

Case A. $\operatorname{crit}(F) \leq \operatorname{crit}(K)$. In this case, $\theta = \phi_{\nu,\gamma+1}(\sigma)$, and $\pi = \pi_\sigma^{\nu,\gamma+1}$. The map l in our diagram is given by the part of $(\dagger)_{\gamma+1}$(d) we have already verified.

Case B. $\operatorname{crit}(K) < \operatorname{crit}(F)$. In this case, $\theta = \sigma \leq \beta$, where $\beta = \beta^{\mathcal{W}_\nu,F}$. Moreover, $\mathcal{W}_\nu \restriction (\sigma+1) = \mathcal{W}_{\gamma+1} \restriction (\theta+1)$, and π is the identity. Moreover, $\beta \leq \alpha_\nu$ by the way normalization works, so $(\dagger)_{\gamma+1}$(a) tells us that $s_\sigma^\nu = s_\theta^{\gamma+1}$, and $\mathcal{M}_{\nu^\nu(\sigma)}^{\mathcal{W}_\nu^*} = \mathcal{M}_{\nu^{\gamma+1}(\theta)}^{\mathcal{W}_{\gamma+1}^*}$. We take l to be the identity as well. In other words, the bottom left rectangle in the diagram above collapses to a line.

We also have $\operatorname{dom}(E) = \operatorname{dom}(K) < \operatorname{crit}(i^*)$ in this case (though $E \neq K$ is perfectly possible). So then $\operatorname{dom}(E^*) = \operatorname{dom}(K^*)$, which implies that $\mathcal{M}_{\sigma^*}^{\mathcal{W}_\nu^*} = \mathcal{M}_{\theta^*}^{\mathcal{W}_{\gamma+1}^*}$, and $i^* \restriction \mathcal{M}_{\sigma^*}^{\mathcal{W}_\nu^*}$ is the identity. Thus the bottom right rectangle collapses to a line. (It is however possible that $u^\nu(\sigma) \neq u^{\gamma+1}(\sigma)$ in this case.)

In both cases, our job is to define h so that it fits into the diagram as shown. Using the notation just established, we can handle the cases in parallel.

We define h using the Shift Lemma:

$$h([a,f]_{K^*}^{\mathcal{M}_{\sigma^*}^{\mathcal{W}_\nu^*}}) = [i^*(a), i^*(f)]_{E^*}^{\mathcal{M}_{\theta^*}^{\mathcal{W}_{\gamma+1}^*}}.$$

Note here that $i^*(u^\nu(\tau)) = u^{\gamma+1}(\eta)$ by our induction hypotheses, so i^* maps $\mathcal{M}_{u^\nu(\tau)}^{\mathcal{W}_\nu^*}$, the model where we found K^*, elementarily into $\mathcal{M}_{u^{\gamma+1}(\eta)}^{\mathcal{W}_{\gamma+1}^*}$, the model that had E^*. So the Shift Lemma gives us h, and that $h \circ i_{K^*} = i_{E^*} \circ i^*$.

We shall leave it to the reader to show that the rectangle on the upper right of our diagram commutes. If s is the branch extender of $[0, u^\nu(\tau+1)]_{\mathcal{W}_\nu^*}$ and t is the branch extender of $[0,u^{\gamma+1}(\eta+1)]_{W^*_{\gamma+1}}$, then $i^*(s)=t$. Moreover, if $s(a)=K^*$ and $t(b)=E^*$, then $i^*(s\restriction(a+1))=t\restriction(b+1)$. This implies that the upper right rectangle commutes.

So we are left to show that $h\circ s^{\nu}_{\tau+1}=s^{\gamma+1}_{\eta+1}\circ\pi^{\nu,\gamma+1}_{\tau+1}$. Let $x=[b,f]_K^{\mathcal{M}_\sigma^{\mathcal{W}_\nu}}$ be in $\mathcal{M}^{\mathcal{W}_\nu}_{\tau+1}$. Then

$$\begin{aligned} h\circ s^{\nu}_{\tau+1}(x) &= h(s^{\nu}_{\tau+1}([b,f]_K^{\mathcal{M}_\sigma^{\mathcal{W}_\nu}}))\\ &= h([t^{\nu}_{\tau}(b), i^{\mathcal{W}^*_\nu}_{v^\nu(\sigma),\sigma^*}\circ s^{\nu}_{\sigma}(f)]_{K^*}^{\mathcal{M}_{\sigma^*}^{\mathcal{W}^*_\nu}})\\ &= [i^*\circ t^{\nu}_{\tau}(b)), i^*\circ i^{\mathcal{W}^*_\nu}_{v^\nu(\sigma),\sigma^*}\circ s^{\nu}_{\sigma}(f)]_{E^*}^{\mathcal{M}_{\theta^*}^{\mathcal{W}^*_{\gamma+1}}}. \end{aligned}$$

The second step uses our definition of $s^{\nu}_{\tau+1}$. On the other hand,

$$\begin{aligned} s^{\gamma+1}_{\eta+1}\circ\pi^{\nu,\gamma+1}_{\tau+1}(x) &= s^{\gamma+1}_{\eta+1}(\pi^{\nu,\gamma+1}_{\tau+1}([b,f]_K^{\mathcal{M}_\sigma^{\mathcal{W}_\nu}}))\\ &= s^{\gamma+1}_{\eta+1}([\pi^{\nu,\gamma+1}_{\tau}(b),\pi(f)]_E^{\mathcal{M}_\theta^{\mathcal{W}_{\gamma+1}}})\\ &= [t^{\gamma+1}_{\eta}\circ\pi^{\nu,\gamma+1}_{\tau}(b), i^{\mathcal{W}^*_{\gamma+1}}_{v^{\gamma+1}(\theta),\theta^*}\circ s^{\gamma+1}_{\theta}\circ\pi(f)]_{E^*}^{\mathcal{M}_{\theta^*}^{\mathcal{W}^*_{\gamma+1}}}. \end{aligned}$$

Now let us compare the two expressions above. The function f is moved the same way in both cases because the bottom rectangles in the diagram above commute. That is,

$$i^*\circ i^{\mathcal{W}^*_\nu}_{v^\nu(\sigma),\sigma^*}\circ s^{\nu}_{\sigma}=i^{\mathcal{W}^*_{\gamma+1}}_{v^{\gamma+1}(\theta),\theta^*}\circ s^{\gamma+1}_{\theta}\circ\pi.$$

So we just need to see that

$$t^{\gamma+1}_{\eta}\circ\pi^{\nu,\gamma+1}_{\tau}=i^*\circ t^{\nu}_{\tau}.$$

But this follows from the part of $(\dagger)_{\gamma+1}$(d) that we have already verified. The relevant diagram is

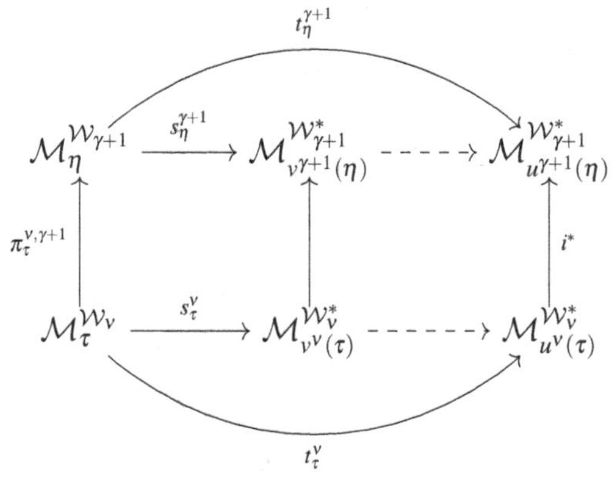

Thus we have verified the new case of $(\dagger)_{\gamma+1}$(d) that is applicable to $\Phi_{\gamma+1} \upharpoonright (\eta + 2)$.

We turn to $(\dagger)_{\gamma+1}$(e). If it is applicable, then $z(\gamma+1) = \eta+1$, and because we did not drop, $z(\nu) = \tau+1$. We must show that $\psi_{\gamma+1} = t^{\gamma+1} \circ \sigma_{\gamma+1}$. We have $R_{\gamma+1} = \mathcal{M}^{\mathcal{W}_{\gamma+1}}_{\eta+1}$, and $R_\nu = \mathcal{M}^{\mathcal{W}_\nu}_{\tau+1}$. Making these substitutions and expanding the upper part of the diagram above, we get

$$\begin{array}{ccccccc}
\mathcal{M}^{\mathcal{U}}_{\gamma+1} & \xrightarrow{\sigma_{\gamma+1}} & R_{\gamma+1} & \xrightarrow{s^{\gamma+1}_{z(\gamma+1)}} & \mathcal{M}^{\mathcal{W}^*_{\gamma+1}}_{\nu^{\gamma+1}(z(\gamma+1))} & \longrightarrow & \mathcal{M}^{\mathcal{W}^*_{\gamma+1}}_{z^*(\gamma+1)} = N_{\gamma+1} \\
\uparrow & & \uparrow {\scriptstyle \pi^{\nu,\gamma+1}_{z(\nu)}} & & \uparrow {\scriptstyle h} & & \uparrow {\scriptstyle i^*} \\
\mathcal{M}^{\mathcal{U}}_{\nu} & \xrightarrow{\sigma_{\nu}} & R_{\nu} & \xrightarrow{s^{\nu}_{z(\nu)}} & \mathcal{M}^{\mathcal{W}^*_{\nu}}_{\nu^{\nu}(z(\nu))} & \longrightarrow & \mathcal{M}^{\mathcal{W}^*_{\nu}}_{z^*(\nu)} = N_{\nu}
\end{array}$$

The embedding across the bottom row is $t^\nu \circ \sigma_\nu$, and hence by induction, it is ψ_ν. The embedding across the top row is $t^{\gamma+1} \circ \sigma_{\gamma+1}$. The diagram commutes, so

$$\begin{aligned}
\psi_{\gamma+1} \circ i^{\mathcal{U}}_{\nu,\gamma+1} &= i^{\mathcal{U}^*}_{\nu,\gamma} \circ \psi_\nu \\
&= i^* \circ t^\nu \circ \sigma_\nu. \\
&= t^{\gamma+1} \circ \sigma_{\gamma+1} \circ i^{\mathcal{U}}_{\nu,\gamma+1}.
\end{aligned}$$

Thus $t^{\gamma+1} \circ \sigma_{\gamma+1}$ agrees with $\psi_{\gamma+1}$ on $\mathrm{ran}(i^{\mathcal{U}}_{\nu,\gamma+1})$. So it will be enough to show the two embeddings agree on $\varepsilon = \mathrm{lh}(E^{\mathcal{U}}_\gamma)$. For that, we calculate exactly as we did in the case $\eta = \alpha_\gamma + 1$:

$$\begin{aligned}
\psi_{\gamma+1} \upharpoonright \varepsilon &= \mathrm{res}_\gamma \circ \psi_\gamma \upharpoonright \varepsilon \\
&= \mathrm{res}_\gamma \circ t^\gamma \circ \sigma_\gamma \upharpoonright \varepsilon \\
&= t^{\gamma+1} \circ \sigma_\gamma \upharpoonright \varepsilon \\
&= t^{\gamma+1} \circ \sigma_{\gamma+1} \upharpoonright \varepsilon.
\end{aligned}$$

The last equality holds because σ_γ agrees with $\sigma_{\gamma+1}$ on $\mathrm{lh}(F)+1$, by our earlier work on normalization. This proves $(\dagger)_{\gamma+1}$(e).

Finally, suppose that λ is a limit ordinal, and we have defined $\Phi_{\gamma+1} \upharpoonright \eta$ for all $\eta < \lambda$. Then we set

$$\Phi_{\gamma+1} \upharpoonright \lambda = \bigcup_{\eta<\lambda} \Phi_{\gamma+1} \upharpoonright \eta.$$

We are of course assuming $\Phi_{\gamma+1} \upharpoonright \eta$ is a subsystem of $\Phi_{\gamma+1} \upharpoonright \beta$ whenever $\eta < \beta$, and the tree embedding properties clearly pass through limits, so this gives us a tree embedding of $\mathcal{W}_{\gamma+1} \upharpoonright \lambda$ into $\mathcal{W}^*_{\gamma+1} \upharpoonright \lambda$.

In order to define $\Phi_{\gamma+1} \upharpoonright (\lambda+1)$, for $\lambda \le z(\gamma+1)$ a limit ordinal, let τ be such that

$$\lambda = \phi_{\nu,\gamma+1}(\tau).$$

Consider $r = \hat{p}^{\gamma+1}(e_\lambda^{\mathcal{W}_{\gamma+1}})$. Since $\Phi_{\gamma+1} \upharpoonright \lambda$ is a tree embedding, $\hat{p}^{\gamma+1}$ is $\subseteq$-preserving on $\mathcal{W}^{\mathrm{ext}}_{\gamma+1}$. Thus r is the extender of some branch b of $\mathcal{W}^*_{\gamma+1}$. In fact, b is the downward closure of $\{i_{G^*}(v^\nu(\xi)) \mid \xi <_{W_\nu} \tau\}$. Recall that the v-maps preserve tree order, so that $\{i_{G^*}(v^\nu(\xi)) \mid \xi <_{W_\nu} \tau\}$ is contained in the branch $[0, i_{G^*}(v^\nu(\tau))]_{W^*_{\gamma+1}}$ of $\mathcal{W}^*_{\gamma+1}$. So

$$v^{\gamma+1}(\lambda) = \sup\{i_{G^*}(v^\nu(\xi)) \mid \xi <_{W_\nu} \tau\}.$$

Moreover, we can define $s_\lambda^{\gamma+1} : \mathcal{M}_\lambda^{\mathcal{W}_{\gamma+1}} \to \mathcal{M}_{v^{\gamma+1}(\lambda)}^{\mathcal{W}^*_{\gamma+1}}$ using the commutativity given by (c) of Definition 6.4.1:

$$s_\lambda^{\gamma+1}(i_{\theta,\lambda}^{\mathcal{W}_{\gamma+1}}(x)) = i^{\mathcal{W}^*_{\gamma+1}}_{v^{\gamma+1}(\theta), v^{\gamma+1}(\lambda)}(s_\theta^{\gamma+1}(x)).$$

It is easy to verify the agreement of $s_\lambda^{\gamma+1}$ with earlier embeddings specified in clause (d) of Definition 6.4.1. Thus $\Phi_{\gamma+1} \upharpoonright (\lambda+1)$ is a tree embedding.

We must check that the applicable parts of $(\dagger)_{\gamma+1}$ hold. Let us keep the notation of the last paragraph. For part (b), we must consider the case $z(\gamma+1) = \lambda$. We have not dropped in $(\nu, \gamma+1]_U$, so $z(\nu) = \tau$, and $v^\nu(\tau) \leq_{W^*_\nu} z^*(\nu)$ by $(\dagger)_\nu$. We showed that $v^{\gamma+1}(\lambda) \leq_{W^*_{\gamma+1}} i_{G^*}(v^\nu(\tau))$ in the last paragraph. So $v^{\gamma+1}(\lambda) \leq_{W^*_{\gamma+1}} i_{G^*}(z^*(\nu)) = z^*(\gamma+1)$, as desired.

For $(\dagger)_{\gamma+1}$(d), the new case is $\xi = \tau$, and $\lambda = \phi_{\nu,\gamma+1}(\tau)$. Everything in sight commutes, so things work out. Let us work them out. Setting $i^* = i^{\mathcal{U}^*}_{\nu,\gamma+1}$, and letting k be the branch embedding from $\mathcal{M}^{\mathcal{W}^*_{\gamma+1}}_{v^{\gamma+1}(\lambda)}$ to $\mathcal{M}^{\mathcal{W}^*_{\gamma+1}}_{i^*(v^\nu(\tau))}$, the relevant diagram is

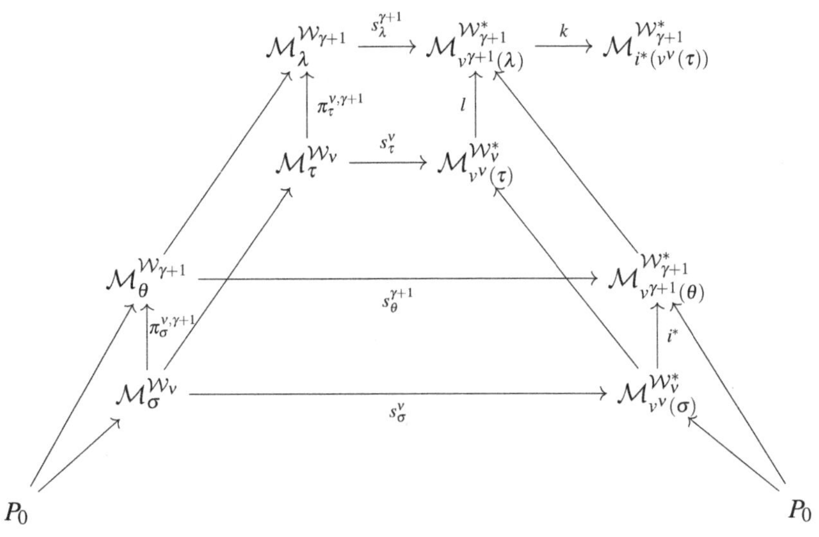

Here we are taking $\theta = \phi_{\nu,\gamma+1}(\sigma)$, where $\sigma <_{W_\nu} \tau$, and σ is sufficiently large that $\phi_{\nu,\gamma+1}$ preserves tree order above σ. We also take σ to be a successor ordinal, so that $i^*(v^\nu(\sigma)) = v^{\gamma+1}(\tau)$. The map l is defined by

$$l(i^{\mathcal{W}_\nu^*}_{v^\nu(\sigma),v^\nu(\tau)}(x)) = i^{\mathcal{W}_{\gamma+1}^*}_{v^{\gamma+1}(\theta),v^{\gamma+1}(\lambda)}(i^*(x)).$$

(Where of course we are taking the union over all such successor ordinals σ.) If we draw the same diagram with τ replaced by some sufficiently large $\tau_0 <_{W_\nu} \tau$ and λ replaced by $\lambda_0 = \phi_{\nu,\gamma+1}(\tau_0)$, then all parts of our diagram commute, because we have verified $(\dagger)_{\gamma+1}$ that far already. Since all these approximating diagrams commute, l is well-defined, and the diagram displayed commutes. Moreover, it is easy to check that $k \circ l = i^* \restriction \mathcal{M}^{\mathcal{W}_\nu^*}_{v^\nu(\tau)}$. Thus we have $(\dagger)_{\gamma+1}$(d).

The proof of $(\dagger)_{\gamma+1}$(e) is exactly the same as it was in the step from $\Phi_{\gamma+1} \restriction \eta+1$ to $\Phi_{\gamma+1} \restriction \eta+2$, so we omit it.

Remark 8.4.23. Actually, that proof seems to show that $(\dagger)_{\gamma}$.(e) is redundant, in that it follows from the other clauses.

This completes our work associated to the definition of $\Phi_{\gamma+1} \restriction \lambda+1$, for $\lambda > \alpha_\gamma$ a limit. Thus we have completed the definition of $\Phi_{\gamma+1}$, and the verification of $(\dagger)_{\gamma+1}$, in Case 1.

Case 2. $(\nu, \gamma+1]_U$ drops, in either model or degree.
Let

$$\begin{aligned}
\bar\mu &= \mathrm{crit}(E^{\mathcal{U}}_\gamma),\\
\bar P &= \mathrm{dom}(E^{\mathcal{U}}_\gamma),\\
\bar J &= \text{first level of } \mathcal{M}^{\mathcal{U}}_\nu \text{ beyond } \bar P \text{ that projects to or below } \bar\mu.
\end{aligned}$$

Since we are in the pfs hierarchy, $\rho(J) = \bar\mu$ is impossible. Also, $\mathcal{U}$ has dropped, so $\mathcal{M}^{\mathcal{U}}_{\gamma+1} = \mathrm{Ult}_k(\bar J, E^{\mathcal{U}}_\gamma)$, where k is $k(J)$ rather than $k(J)+1$. We have that

$$\bar P = \mathcal{M}^{\mathcal{U}}_\nu|(\bar\mu^+)^{\mathcal{M}^{\mathcal{U}}_\nu|\mathrm{lh}(E^{\mathcal{U}}_\nu)} = \mathcal{M}^{\mathcal{U}}_\gamma|(\bar\mu^+)^{\mathcal{M}^{\mathcal{U}}_\gamma|\mathrm{lh}(E^{\mathcal{U}}_\gamma)}.$$

Let

$$\begin{aligned}
\mu &= \sigma_\nu(\bar\mu) = \mathrm{crit}(F),\\
&= \sigma_\nu(\bar P) = \mathrm{dom}(F),\\
J &= \sigma_\nu(\bar J) = \text{first level of } R_\nu \text{ beyond } P \text{ that projects to or below } \mu.
\end{aligned}$$

Since σ_ν agrees with σ_γ on $\mathrm{lh}(F_\nu)$, we can replace σ_ν by σ_γ in the first two equations. (But if $\nu < \gamma$, then $\bar J \notin \mathrm{dom}(\sigma_\gamma)$.) We have that

$$P = R_\nu|(\mu^+)^{R_\nu|\mathrm{lh}(F_\nu)} = R_\gamma|(\mu^+)^{R_\gamma|\mathrm{lh}(F)}.$$

In this case, $z(\gamma+1) = \alpha_\gamma + 1$, and

$$\mathcal{W}_{\gamma+1} = \mathcal{W}_\gamma \restriction (\alpha_\gamma+1)^\frown\langle \mathrm{Ult}(J,F)\rangle.$$

CLAIM 8.4.24. $\mathrm{res}_\gamma \circ t^\gamma$ *agrees with* $\mathrm{res}_\nu \circ t^\nu$ *on* $\mathrm{lh}(F_\nu)$.

PROOF. This is clear if $\nu = \gamma$. But if $\nu < \gamma$, then t^γ agrees with $\mathrm{res}_\nu \circ t^\nu$ on $\mathrm{lh}(F_\nu)$ by $(\dagger)_\gamma$(b). (See the remarks after the statement of $(\dagger)_\gamma$.) But also, res_γ is the identity on $\mathrm{res}_\nu \circ t^\nu(\mathrm{lh}(F_\nu))$, because $\nu < \gamma$. This yields the claim. $\dashv$

We have $H = t^\gamma(F)$ and $G = \mathrm{res}_\gamma(G)$. We have that $\mathrm{res}_\gamma \colon N_\gamma|\mathrm{lh}(H) \to N^*_\gamma|\mathrm{lh}(G)$, and that res_γ agrees with $\hat{\imath}^{\mathcal{W}^{**}_\gamma}_{\tau_\gamma,\xi_\gamma}$ on $\mathrm{lh}(H)$. Let

$$\begin{aligned} J^* &= \mathrm{Res}_{\mathrm{Q}_\nu}[t^\nu(J)]^{\mathbb{C}_\nu}, \\ \sigma^* &= \sigma_{\mathrm{Q}_\nu}[t^\nu(J)]^{\mathbb{C}_\nu}, \\ \mu^* &= \sigma^*(t^\nu(\mu)), \text{ and} \\ P^* &= \sigma^*(t^\nu(P)). \end{aligned}$$

σ^* is a partial resurrection map at stage ν. We had $\mathrm{res}_\nu \colon N_\nu|\mathrm{lh}(H_\nu) \to N^*_\nu|\mathrm{lh}(G_\nu)$. σ^* resurrects more, namely $t^\nu(J)$, but does not trace it as far back in $\mathbb{C}_\nu$. Because no proper level of $t^\nu(J)$ projects to $t^\nu(\mu)$, σ^* agrees with res_ν on $t^\nu(P)$. So

$$\sigma^* \circ t^\nu \restriction P = \mathrm{res}_\nu \circ t^\nu \restriction P = \mathrm{res}_\gamma \circ t^\gamma \restriction P,$$

the last equality being Claim 8.4.24. The embeddings displayed also agree at P, where they have value P^*. Note that $P = \mathrm{dom}(F)$ and $P^* = \mathrm{dom}(G)$.

Letting $J^* = M^{\mathbb{C}_\nu}_{\theta,l}$, we have that J^* is the last model of $(W^*_{\theta,l})^{\mathbb{C}_\nu}$. Set

$$\mathcal{T}^* = (\mathcal{W}^*_{\theta,l})^{S_\nu}.$$

Lemma 8.3.1 tells us that $\mathcal{T}^*$ has the following form. Let ξ be least such that $t^\nu(J) \trianglelefteq \mathcal{M}^{\mathcal{W}^*_\nu}_\xi$. Then $\mathcal{T}^* \restriction \xi+1 = \mathcal{W}^*_\nu \restriction \xi+1$, and letting $\mathrm{lh}(\mathcal{T}^*) = \eta+1$, $\xi \leq_{\mathcal{T}^*} \eta$ and $\sigma^* = \hat{\imath}^{\mathcal{T}^*}_{\xi,\eta}$.

We have that

$$\mathcal{W}^*_{\gamma+1} = i_{G^*}(\mathcal{T}^*)$$

and

$$N_{\gamma+1} = Q_{\gamma+1} = i_{G^*}(J^*),$$

by the way that lifting to the background universe works in the dropping case. As in the non-dropping case, the key is

CLAIM 8.4.25. *(i)* $\mathcal{W}^*_{\gamma+1} \restriction \xi_\gamma + 1 = \mathcal{W}^{**}_\gamma \restriction \xi_\gamma + 1$, *and*

(ii) $G = E^{\mathcal{W}^*_{\gamma+1}}_{\xi_\gamma}$.

PROOF. We have that $\mathrm{dom}(G) = \mathrm{res}_\gamma \circ t^\gamma(P) = \mathrm{res}_\nu \circ t^\nu(P)$ by Claim 8.4.24, so $\mathrm{dom}(G) = \sigma^* \circ t^\nu(P) = P^* = J^*|(\mu^*)^{+,J^*}$. P is $\mathcal{M}^{\mathcal{W}_\gamma}_{\alpha_\gamma}|\mathrm{lh}(F)$ cut off at its μ^+. So P^* is $\mathrm{res}_\gamma \circ t^\gamma(\mathcal{M}^{\mathcal{W}_\gamma}_{\alpha_\gamma}|\mathrm{lh}(F))$, cut off at its $(\mu^*)^+$, that is, P^* is $\mathcal{M}^{\mathcal{W}^{**}_\gamma}_{\xi_\gamma}|\mathrm{lh}(G)$, cut off at $(\mu^*)^+$.

Thus J^* agrees with $\mathcal{M}_{\xi_\gamma}^{\mathcal{W}_\gamma^{**}} | \mathrm{lh}(G)$ up to their common value for $(\mu^*)^+$. It follows that $i_{G^*}(J^*)$ agrees with $\mathrm{Ult}(\mathcal{M}_{\xi_\gamma}^{\mathcal{W}_\gamma^{**}} | \mathrm{lh}(G), G)$ up to $\mathrm{lh}(G)+1$, with the agreement at $\mathrm{lh}(G)$ holding by our having chosen a minimal G^* for G. Claim 8.4.25 now follows from the fact that $\mathcal{W}_\gamma^{**}$ and $\mathcal{W}_{\gamma+1}^*$ are λ-separated trees by the same strategy. ⊣

We now get $\Phi_{\gamma+1}$ by setting $p^{\gamma+1}(F) = G$ and applying Lemma 8.2.3. We must see that $(\dagger)_{\gamma+1}$ holds. Part (a) is clear.

Let $\beta^* = W_{\gamma+1}\text{-pred}(\xi_\gamma)$.

CLAIM 8.4.26. *(i)* $\mathrm{lh}(\mathcal{T}^*) = \beta^* + 1$, *and* $J^* = \mathcal{M}_{\beta^*}^{\mathcal{W}_{\gamma+1}^*}$.
(ii) $\beta^* = \mu^*$, *and if* $s = s_{\mu^*}^{\mathcal{T}^*}$, *then* $s\colon \mu^* \to V_{\mu^*}$.

PROOF. By definition, β^* is the least α such that $\mathcal{M}_\alpha^{\mathcal{W}_{\gamma+1}^*} | o(P^*) = P^*$. But J^* is the last model of $\mathcal{T}^*$, and $P^* = J^* | o(P^*)$, so since $\mathcal{T}^*$ and $\mathcal{W}_{\gamma+1}^*$ are λ-separated trees by the same strategy, $\beta^* < \mathrm{lh}(\mathcal{T}^*)$ and $\mathcal{M}_{\beta^*}^{\mathcal{T}^*} = \mathcal{M}_{\beta^*}^{\mathcal{W}_{\gamma+1}^*}$. This gives (i).

Part (ii) is proved exactly as in Case 1. ⊣

Now consider $(\dagger)_{\gamma+1}$(b). We have $v^{\gamma+1}(\alpha_\gamma + 1) = \xi_\gamma + 1$, and $z^*(\gamma+1) = i_{G^*}(\mu^*)$. So we must see that $\xi_\gamma + 1 \le_{W_{\gamma+1}^*} i_{G^*}(\mu^*)$, that is, that G is used on the branch of $\mathcal{W}_{\gamma+1}^*$ to $i_{G^*}(\mu^*)$. But if $s = e_{\mu^*}^{\mathcal{T}^*}$, then $s = i_{G^*}(s) \upharpoonright \mu^*$, so μ^* is on the branch of $W_{\gamma+1}^*$ to $i_{G^*}(\mu^*)$. Moreover, $i_{G^*}(s)(\mu^*)$ is compatible with G, so it is equal to G, as desired.

$(\dagger)_{\gamma+1}$(d) is vacuous, because we have dropped. We shall leave the agreement conditions (c), (f), and (g) to the reader, and consider (e). That is, we show $\psi_{\gamma+1} = t^{\gamma+1} \circ \sigma_{\gamma+1}$. The relevant diagram is

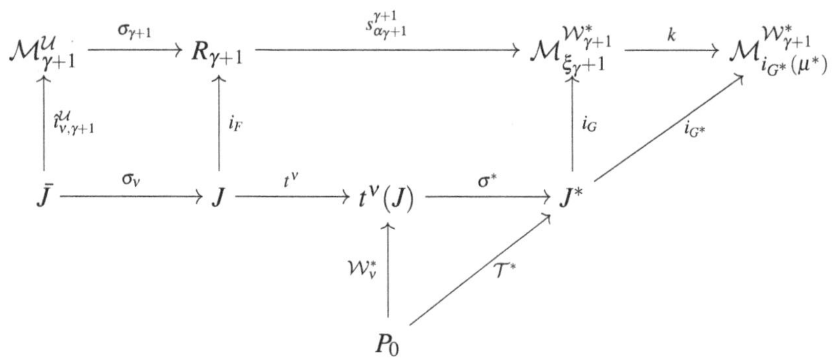

Here $k = i_{v^{\gamma+1}(\alpha_\gamma+1), z^*(\gamma+1)}^{\mathcal{W}_{\gamma+1}^*}$. Thus the embedding along the top row is $t^{\gamma+1} \circ \sigma_{\gamma+1}$. The lifting process defines $\psi_{\gamma+1}$ by

$$\psi_{\gamma+1}([a,f]_{E_\gamma^{\mathcal{U}}}^{\bar{J}}) = [\mathrm{res}_\gamma \circ \psi_\gamma(a), \sigma^* \circ \psi_\nu(f)]_{G^*}^{J^*},$$

where we have dropped a few superscripts for readability. Let us write $\hat{\imath}$ for $\hat{\imath}^{\mathcal{U}}_{\nu,\gamma+1}$. Then $\psi_{\gamma+1}$ agrees with $t^{\gamma+1}\circ\sigma_{\gamma+1}$ on $\mathrm{ran}(\hat{\imath})$, because

$$\begin{aligned} t^{\gamma+1}\circ\sigma_{\gamma+1}\circ\hat{\imath} &= i_{G^*}\circ\sigma^*\circ t^{\nu}\circ\sigma_{\nu} \\ &= i_{G^*}\circ\sigma^*\circ\psi_{\nu} \\ &= \psi_{\gamma+1}\circ\hat{\imath}. \end{aligned}$$

The first line comes from the commutativity of the diagram, the second from $(\dagger)_{\nu}$(e), and the last from the definition of $\psi_{\gamma+1}$.

So it is enough to see that $\psi_{\gamma+1}$ agrees with $t^{\gamma+1}\circ\sigma_{\gamma+1}$ on ε, where $\varepsilon=\mathrm{lh}(E^{\mathcal{U}}_{\gamma})$. But note that $t^{\gamma+1}=k\circ s^{\gamma+1}_{\alpha_{\gamma+1}}$, and $\mathrm{crit}(k)>\mathrm{lh}(G)$. So $t^{\gamma+1}$ agrees with the copy map $s^{\gamma+1}_{\alpha_{\gamma+1}}$ on $\mathrm{lh}(F)$. Thus $t^{\gamma+1}$ agrees with $\mathrm{res}_{\gamma}\circ t^{\gamma}$ on $\mathrm{lh}(F)$. So we can calculate

$$\begin{aligned} \psi_{\gamma+1}\restriction\varepsilon &= \mathrm{res}_{\gamma}\circ\psi_{\gamma}\restriction\varepsilon \\ &= \mathrm{res}_{\gamma}\circ t^{\gamma}\circ\sigma_{\gamma}\restriction\varepsilon \\ &= t^{\gamma+1}\circ\sigma_{\gamma+1}\restriction\varepsilon. \end{aligned}$$

The second line comes from $(\dagger)_{\gamma})$(e), and the third from our argument above, together with the fact $\sigma_{\gamma}\restriction\varepsilon=\sigma_{\gamma+1}\restriction\varepsilon$.

This finishes Case 2, and hence the definition of $\Phi_{\gamma+1}$ and verification of $(\dagger)_{\gamma+1}$.

We leave the detailed definition of Φ_{λ} and verification of $(\dagger)_{\lambda}$, for λ a limit ordinal or $\lambda=b$, to the reader. The normalization $\mathcal{W}_{\lambda}$ is a direct limit of the $\mathcal{W}_{\nu}$ for $\nu\in[0,\lambda)_U$. The tree $\mathcal{W}^*_{\lambda}$ is $i^{\mathcal{U}^*}_{\nu,\lambda}(\mathcal{W}^*_{\nu})$, for ν past the last drop. So it is a direct limit too. We define Φ_{λ} to be the direct limit of the Φ_{ν} for $\nu\in[0,\lambda)_U$ past the last drop. Part (d) of $(\dagger)$ tells us we can do that. We omit further detail.

This finishes our proof of Sublemma 8.4.8.1, that $\mathcal{W}_b$ is a pseudo-hull of $\mathcal{W}^*_b$. ⊣

That in turn proves Lemma 8.4.8 ⊣

LEMMA 8.4.27. *Let $M=M_{\nu_0,k_0}$, and let $\mathcal{U}$ be a λ-separated tree on M that is of limit length, and is by both $\Sigma_{\mathcal{W}^*_{\nu_0,k_0},M}$ and $\Omega^{\mathbb{C}}_{\nu_0,k_0}$. Let $\mathrm{lift}(\mathcal{U},M,\mathbb{C})_0=\mathcal{U}^*$; then $\mathcal{U}^*$ has a cofinal, wellfounded branch.*

PROOF. Let $\pi\colon H\to V_{\theta}$ be elementary, where H is countable and transitive, and θ is sufficiently large, and everything relevant is in $\mathrm{ran}(\pi)$. Let $\mathcal{S}=\pi^{-1}(\mathcal{U})$, $\mathcal{S}^*=\pi^{-1}(\mathcal{U}^*)$, and $\mathcal{T}=\pi^{-1}(\mathcal{W}^*_{\nu_0,k_0})$.

By the proof of Lemma 7.6.7, $\pi^{-1}(\Sigma)=\Sigma\cap H$, so $\langle\mathcal{T},\mathcal{S}\rangle$ is by Σ. Moreover, letting

$$b=\Sigma(\langle\mathcal{T},\mathcal{S}\rangle),$$

we have that $b\in H$. (Because $b\in H[g]$ for all g on $\mathrm{Col}(\omega,\tau)$, for $\tau\in H$ sufficiently large.) It will be enough to see that $\mathcal{M}^{\mathcal{S}^*}_b$ is wellfounded, as then the elementarity of π yields a cofinal wellfounded branch of $\mathcal{U}^*$.

By [30], $\mathcal{S}^*$ has a cofinal, wellfounded branch c. The proof of Sublemma 8.4.8.1 shows that $\mathcal{W}_c$ is a pseudo-hull of $\mathcal{W}_c^*$, where $\mathcal{W}_c = W(\mathcal{T}, \mathcal{S}^\frown c)$ and $\mathcal{W}_c^* = i_c^{\mathcal{S}^*}(\mathcal{T})$. That is because we can run the construction of Φ_c in H; we do not need $c \in H$ to do that. But then $\mathcal{W}_c^*$ is by Σ, so $\mathcal{W}_c$ is by Σ by strong hull condensation, and $c = \Sigma(\langle \mathcal{T}, \mathcal{S} \rangle)$ since Σ quasi-normalizes well. Thus $c = b$, and $\mathcal{M}_b^{\mathcal{S}^*}$ is wellfounded, as desired. ⊣

We can now finish the proof of Theorem 8.4.3. We have just shown that $\Sigma_{\mathcal{W}^*_{\nu_0,k_0},M}$ agrees with $\Omega^{\mathbb{C}}_{\nu_0,k_0}$ on λ-separated trees. By Lemma 7.6.5, they agree on finite stacks of plus trees, as desired.

This finishes the proof of Theorem 8.4.3. ⊣

Chapter 9

FINE STRUCTURE FOR THE LEAST BRANCH HIERARCHY

We now adapt the definitions and results of the previous sections to mice that are being told their own background-induced iteration strategy.

The particular kind of strategy mice dealt with in this book we call *least branch hod mice*. Paired with their iteration strategies, they become *least branch hod pairs*. Least branch hod pairs and pure extender pairs share many basic properties, and so we define a *mouse pair* to be a pair of one of the two varieties. Section 9.3 discusses some of the basic properties of mouse pairs.

The deeper results about least branch hod pairs require a comparison theorem. The proof of our comparison theorem for pure extender pairs generalizes in a straightforward way to least branch hod pairs, provided that we have background constructions for them that do not not break down that we can iterate our pairs into. The main problem is to show that.

One might worry that the proofs we gave in Chapter 4 that PFS constructions do not break down require a comparison, so we are being led into a vicious circle. But this is not a problem, because if (M,Σ) is a least branch hod pair, and $\mathbb{C}$ is the maximal hod pair construction of some coarse Γ-Woodin mouse that captures Σ, then $\mathbb{C}$ cannot break down until it has reached an iterate of (M,Σ). This means that, under the appropriate large cardinal or determinacy hypotheses, we have enough backgrounded hod pairs to prove termination for the comparisons needed to show that hod pair constructions are good at all $\langle \nu,k\rangle$.

But we do in fact confront a new problem in adapting the proof that PFS constructions are good everywhere (Theorem 4.11.4) to hod pair constructions. The main arguments all involved iterating away least disagreements in a phalanx comparison. For example, in the proof that $M_{\nu,k}$ is parameter solid, we compared a phalanx of the form (M,H,α) with M by iterating away least disagreements. Here we must compare strategies as well, and this forces us to compare (M,H,α) with M by iterating the two into levels of some common hod pair construction. The result is that disagreements will very often involve the two sides agreeing with each other, but not with the background. If we proceed naively, this renders invalid the usual argument that we cannot end up above M on the phalanx side. Our solution is to modify the way the phalanx is iterated, so that sometimes we move the whole phalanx up, including its exchange ordinal.[1]

Sections 9.1 and 9.2 lay out some elementary properties of least branch hod pairs. Section 9.3 contains some definitions which highlight the value of considering premouse and iteration strategy as a pair. Sections 9.4 through 9.6 are devoted to background constructions of least branch hod pairs, and the proof that all their levels are parameter solid. We shall finish that proof, and consider the other components of goodness for such constructions, in Chapter 10. This leads in Section 10.4 to our main existence theorems for least branch hod pairs.

9.1. Least branch premice

Let $\mathcal{L}_1$ be the language having the binary relation symbol $\in$, predicate symbols $\dot{E}, \dot{F}, \dot{\Sigma}, \dot{B}$, and constant symbol $\dot{\gamma}$. A *least branch premouse* (lpm) is a pair $(\hat{M}, k)$, where $k \leq \omega$, and

$$\hat{M} = (|\hat{M}|, \in, \dot{E}^M, \dot{F}^M, \dot{\Sigma}^M, \dot{B}^M, \dot{\gamma}^M)$$

is an $\mathcal{L}_1$ structure such that $|\hat{M}|$ is transitive and $\hat{M}$ has various first order properties described below. We call $\hat{M}$ the *bare premouse* associated to M and write $k = k(M)$. We often identify M with $\hat{M}$, and usually write $x \in M$ instead of $x \in |M|$.

If M is an lpm, then its predicates are amenable to M, and hence can be amalgamated in some fixed way into a single amenable $A = \dot{A}^M$. Considered this way, $M = (J^A_\alpha, \in, A)$ is an acceptable J-structure, so the basic fine structural notions described in [54] and Chapter 2 apply. However we shall use instead the projectum free spaces fine structure of Chapter 4, for the reasons described in Section 3.6. This amounts to adding certain parameters to our cores. In particular, we adopt wholesale the definitions and notation of Chapter 4 concerning projecta, cores, solidity, soundness, and elementarity and near elementarity of maps. The elementary results of Chapter 4 concerning these notions hold in the current context, with the same proofs. An lpm is just a pfs premouse expanded by one additional amenable predicate, used to describe an iteration strategy for it. Our focus in this chapter will be on the new elements this predicate introduces, and how to modify Chapter 4 so as to deal with them.

If M is k-sound lpm, then it is coded by its reduct M^k, and the $k+1$-st projectum, parameter, strong core $\bar{\mathfrak{C}}_{k+1}$, and core $\mathfrak{C}_{k+1}$ of M are given by

$$\rho_{k+1} = \rho_1(M^k),$$

$$p_{k+1} = p_1(M^k),$$

$$\bar{\mathfrak{C}}_{k+1} = \text{transitive collapse of } d^k \circ h^1_{M^k}``(\rho_{k+1} \cup p_{k+1}),$$

[1]W. H. Woodin and F. Schlutzenberg have, independently and earlier, developed and used this idea in other contexts. See [59].

$\bar{p}_{k+1} = \sigma^{-1}(p_{k+1})$, and

$\mathfrak{C}_{k+1} =$ transitive collapse of $d^k \circ h^1_{M^k}``(\rho_{k+1} \cup \{p_{k+1}, \rho_{k+1}\})$.

Here d^k decodes M^k, $\sigma\colon \bar{\mathfrak{C}}_{k+1} \to M$ is the anticollapse map, and $M^k = (M||\rho_k, A^k)$, where A^k codes $\mathrm{Th}_1^{M^{k-1}}(\rho_k \cup \{w_k\})$, for $w_k = \langle \eta_k, \rho_k, p_k \rangle$. Here η_k is the $r\Sigma_k$ cofinality of ρ_k. M is $k+1$-solid iff

(a) M^k is *parameter solid*; that is, p_{k+1} and $\bar{p}_{k+1}$ are solid and universal over M^k and $(\bar{\mathfrak{C}}_{k+1})^k$ respectively,
(b) M^k is *projectum solid*; that is, ρ_{k+1} is not measurable by the M-sequence, and either
  (i) $\bar{\mathfrak{C}}_{k+1} = \mathfrak{C}_{k+1}$, or
  (ii) $\mathfrak{C}_{k+1} = \mathrm{Ult}_k(\bar{\mathfrak{C}}_{k+1}, D)$, where D is the order zero measure of $\bar{\mathfrak{C}}_{k+1}$ on ρ_{k+1}, and $\sigma = \pi \circ i_D$,
(c) M^k is *stable*; that is, either $\eta_k^M < \rho_{k+1}$, or η_k^M is not measurable by the M-sequence, and
(d) M is *weakly ms-solid*; that is, if M is extender active, then the last extenders of $\bar{\mathfrak{C}}_1(M)$ and $\mathfrak{C}_1(M)$ satisfy the weak ms-ISC.

M is $k+1$-sound iff M is $k+1$-solid and $M = \mathfrak{C}_{k+1}(M)$.

If M is an lpm, then $o(M)$ is the ordinal height of M, and $\hat{o}(M)$ is the α such that $o(M) = \omega\alpha$. The index of M is

$$l(M) = \langle \hat{o}(M), k(M) \rangle.$$

If $\langle \nu, l \rangle \leq_{\mathrm{lex}} l(M)$, then $M|\langle \nu, l \rangle$ is the initial segment N of M with index $l(N) = \langle \nu, l \rangle$. (So $\dot{E}^N = \dot{E}^M \cap N$, $\dot{F}^N = \dot{E}^M_\nu$, $\dot{\Sigma}^N = \dot{\Sigma}^M \cap N$, and $\dot{B}^N$ is determined by $\dot{\Sigma}^M$ is a way that will become clear shortly.) In order that M be an lpm, all its initial segments N must be $k(N)$-sound in the projectum free spaces sense of Definition 4.1.10. If $\nu \leq \hat{o}(M)$, then we write $M|\nu$ for $M|\langle \nu, 0 \rangle$.

As with ordinary premice, if M is an lpm, then $\dot{E}^M$ is the sequence of extenders that go into constructing M, and $\dot{F}^M$ is either empty, or codes a new extender being added to our model by M. $\dot{F}^M$ must satisfy the Jensen conditions; that is, if $F = \dot{F}^M$ is nonempty (i.e., M is *extender active*), then $M \models \mathrm{crit}(F)^+$ exists, and for $\mu = \mathrm{crit}(F)^{+M}$, $o(M) = i^M_F(\mu)$. $\dot{F}^M$ is just the graph of $i^M_F \upharpoonright (M|\mu)$. M must satisfy the Jensen initial segment condition (ISC) in that the *whole* initial segments of $\dot{F}^M$ must appear in $\dot{E}^M$; moreover every extender in $\dot{E}^M$ must satisfy the weak ms-ISC. If there is a largest whole proper initial segment, then $\dot{\gamma}^M$ is its index in $\dot{E}^M$. Otherwise, $\dot{\gamma}^M = 0$. Finally, an lpm M must be *coherent*, in that $i^M_F(\dot{E}^M) \upharpoonright o(M) + 1 = \dot{E}^{M\frown} \langle \emptyset \rangle$.

In other words, the conditions for adding extenders to M are just as in Jensen's work. The structure $(|M|, \in, \dot{E}^M, \dot{F}^M, \dot{\gamma}^M)$ would be a pure extender pfs premouse in the sense of Definition 4.1.11, except that $\dot{\Sigma}^M$ has been used in generating $|M|$.

The definitions related to plus trees and iteration strategies defined on stacks of them given in Sections 4.4 and 4.6 extend to the case that the base model M is an lpm without change. The predicates $\dot{\Sigma}^M$ and $\dot{B}^M$ are used to record information about an iteration strategy Ω for M. The strategy Ω will be determined by its action on λ-separated trees, in an absolute way, so that we need only tell the model we are building how Ω acts on λ-separated trees, and then the model itself can recover the action of Ω on the finite stacks of plus trees that it sees. This is what we shall do.[2]

Let us write $M|\langle \nu, -1\rangle$ for $(M|\langle \nu, 0\rangle)^-$; that is, for $M|\langle \nu, 0\rangle$ with its last extender predicate set to $\emptyset$.

DEFINITION 9.1.1. An *M-tree* is a triple $s = \langle \nu, k, \mathcal{T}\rangle$ such that

(1) $\langle \nu, k\rangle \leq_{\text{lex}} l(M)$, and
(2) $\mathcal{T}$ is a λ-separated iteration tree on $M|\langle \nu, k\rangle$.

We allow here $\mathcal{T}$ to be empty. The case $k = -1$ allows us to drop by throwing away a last extender predicate. Given an M-tree s we write $s = \langle \nu(s), k(s), \mathcal{T}(s)\rangle$. We write $M_\infty(s)$ for the last model of $\mathcal{T}(s)$, if it has one. We say $\text{lh}(\mathcal{T}(s))$ is the length of s.

What we shall feed into an lpm M is information about how its iteration strategy acts on M-trees.

$\dot{\Sigma}^M$ is a predicate that codes the strategy information added at earlier stages, with $\dot{\Sigma}^M(s,b)$ meaning that $\mathcal{T}(s)$ is a λ-separated tree on $M|\langle \nu(s), k(s)\rangle$ of limit length, and $\mathcal{T}(s)^\frown b$ is according to the strategy. We write $\Sigma^M_{\nu,k}$ for the partial iteration strategy for $M|\langle \nu, k\rangle$ determined by $\dot{\Sigma}^M$. We write

$$\begin{aligned}\Sigma^M(s) = b &\text{ iff } \dot{\Sigma}^M(s,b)\\ &\text{ iff } \Sigma^M_{\nu(s),k(s)}(\mathcal{T}(s)) = b.\end{aligned}$$

We say that s is according to Σ^M iff $\mathcal{T}(s)$ is according to $\Sigma^M_{\nu(s),k(s)}$.

We now describe how strategy information is coded into the $\dot{B}^M$ predicate. Here we use the $\mathfrak{B}$-operator discovered by Schlutzenberg and Trang in [61]. In the original version of this paper, we made use of a different coding, one that has fine-structural problems. The authors of [81] discovered those problems. The discussion to follow is taken from [81].

DEFINITION 9.1.2. M is *branch active* (or just B-active) iff

(a) there is a largest $\eta < o(M)$ such that $M|\eta \models \mathsf{KP}$, and letting $N = M|\eta$,
(b) there is a $<_N$-least N-tree $s \in N$ such that s is by Σ^N, $\mathcal{T}(s)$ has limit length, and $\Sigma^N(s)$ is undefined.
(c) for N and s as above, $o(M) \leq o(N) + \text{lh}(\mathcal{T}(s))$.

Note that being branch active can be expressed by a Σ_2 sentence in $\mathcal{L}_0 - \{\dot{B}\}$. This contrasts with being extender active, which is not a property of the premouse

[2]Our comparison method only applies directly to strategies acting on stacks of λ-separated trees, so we must restrict the part of Ω being inserted into M at least that much.

with its top extender removed. In contrast with extenders, we know when branches must be added before we do so.

DEFINITION 9.1.3. Suppose that M is branch active. We set

$$\eta^M = \text{the largest } \eta \text{ such that } M|\eta \models \mathsf{KP},$$
$$b^M = \{\alpha \mid \eta^M + \alpha \in \dot{B}^M\},$$
$$s^M = \text{least } M|\eta^M\text{-tree } s \text{ such that } \dot{\Sigma}^{M|\eta^M}(s) \text{ is undefined, and}$$
$$\nu^M = \text{unique } \nu \text{ such that } \eta^M + \nu = o(M).$$

Moreover, for $s = s^M$,

(1) M is a *potential lpm* iff b^M is a cofinal branch of $\mathcal{T}(s) \restriction \nu^M$.
(2) M is *honest* iff $\nu^M = \text{lh}(\mathcal{T}(s))$, or $\nu^M < \text{lh}(\mathcal{T}(s))$ and $b^M = [0, \nu^M)_{T(s)}$.
(3) M is an *lpm* iff M is an honest potential lpm.
(4) M is *strategy active* iff $\nu^M = \text{lh}(\mathcal{T}(s))$.

We demand of an lpm M that if M is not $\dot{B}$-active, then $\dot{B}^M = \emptyset$.

The $\dot{\Sigma}$ predicate of an lpm grows at strategy active stages. More precisely, let

$$\hat{\Sigma}^M = \begin{cases} \dot{\Sigma}^M \cup \{\langle s^M, b^M \rangle\} & \text{if } M \text{ is strategy active, and} \\ \dot{\Sigma}^M & \text{otherwise.} \end{cases}$$

Suppose that $\hat{o}(Q)$ is a successor ordinal, and $M = Q|(\hat{o}(Q) - 1)$; then in order for Q to be an lpm, we must have

$$\dot{\Sigma}^Q = \hat{\Sigma}^M.$$

That is, $\dot{\Sigma}^Q = \hat{\Sigma}^M \cup \{\langle s, b^M \rangle\}$ if M is strategy active, and $\dot{\Sigma}^Q = \dot{\Sigma}^M$ otherwise. If $\hat{o}(Q)$ is a limit ordinal, then we require that $\dot{\Sigma}^Q = \bigcup_{\eta < \hat{o}(Q)} \dot{\Sigma}^{Q|\eta}$. We see then that if M is an lpm and $\nu < \hat{o}(M)$, then $\dot{\Sigma}^{M|\nu} \subseteq \dot{\Sigma}^M$, and $M|\nu$ is strategy active iff $\dot{\Sigma}^{M|\nu} \neq \dot{\Sigma}^M$.

This completes our definition of what it is for M to be a least branch premouse of type 1, the definition being by induction on the hierarchy of M.

DEFINITION 9.1.4. M is a least branch premouse (lpm) of type 1 iff M is an acceptable J structure meeting the requirements stated above.

Type 2 lpms arise in the same way type 2 pfs premice did, via $r\Sigma_k$ ultrapowers that are discontinuous at ρ_k. See Section 4.2. In the end, we shall avoid them in the same way that we did in Section 4.10. Type has to do with soundness properties, not the bare premice. If $k(M) = 0$, then M has type 1.

Notice that if M is an lpm, then no level of M is both $\dot{B}$-active and extender active, because $\dot{B}$-active stages are additively decomposable.

Returning to the case that M is branch active, note that η^M is a Σ_0^M singleton, because it is the least ordinal in $\dot{B}^M$ (because 0 is in every branch of every iteration tree), and thus s^M is also a Σ_0^M singleton. We have separated honesty from the other conditions because it is not expressible by a Q-sentence, whereas the rest is. Honesty is expressible by a Boolean combination of Σ_2 sentences. See Lemma 9.1.9 below.

The original version of this book required that when $o(M) < \eta^M + \mathrm{lh}(\mathcal{T}(s))$, $\dot{B}^M$ is empty, whereas here we require that it code $[0, o(M))_{T(s)}$, in the same way that $\dot{B}^M$ will have to code a new branch when $o(M) = \eta^M + \mathrm{lh}(\mathcal{T}(s))$. Of course, $[0, \nu^M)_{T(s)} \in M$ when $o(M) < \eta^M + \mathrm{lh}(\mathcal{T}(s))$ and M is honest, so the current $\dot{B}^M$ seems equivalent to the original $\dot{B}^M = \emptyset$. However, $\dot{B}^M = \emptyset$ leads to Σ_1^M being too weak, with the consequence that a Σ_1 hull of M might collapse to something that is not an lpm. (The hull could satisfy $o(H) = \eta^H + \mathrm{lh}(\mathcal{T}(s^H))$, even though $o(M) < \eta^M + \mathrm{lh}(\mathcal{T}(s^M))$. But then being an lpm requires $\dot{B}^H \neq \emptyset$.) Our current choice for $\dot{B}^M$ solves that problem.

Remark 9.1.5. Suppose N is an lpm, and $N \models \mathsf{KP}$. It is very easy to see that $\dot{\Sigma}^N$ is defined on all N-trees s that are by $\dot{\Sigma}^N$ iff there are arbitrarily large $\xi < o(N)$ such that $N|\xi \models \mathsf{KP}$. So if M is branch active, then η^M is a successor admissible; moreover, we do add branch information, related to exactly one tree, at each successor admissible. Waiting until the next admissible to add branch information is just a convenient way to make sure we are done coding in the branch information for a given tree before we move on to the next one. One could go faster.

We say that an lpm M is (fully) passive if $\dot{F}^M = \emptyset$ and $\dot{B}^M = \emptyset$.

We would like to see that being a bare lpm is preserved by the appropriate embeddings. Q-formulae are useful for that.

DEFINITION 9.1.6. A *rQ-formula* of $\mathcal{L}_1$ is a conjunction of formulae of the form

(a) $\forall u \exists v (u \subseteq v \wedge \varphi)$, where φ is a Σ_1 formula of $\mathcal{L}_0$ such that u does not occur free in φ,

or of the form

(b) "$\dot{F} \neq \emptyset$, and for $\mu = \mathrm{crit}(\dot{F})^+$, there are cofinally many $\xi < \mu$ such that ψ", where ψ is Σ_1.

Formulae of type (a) are usually called Q-formulae. Being a passive bare lpm can be expressed by a Q-sentence, but in order to express being an extender-active bare lpm, we need type (b) clauses, in order to say that the last extender is total. *rQ* formulae are Π_2, and hence preserved downward under Σ_1-elementary maps. They are preserved upward under Σ_0 maps that are *strongly cofinal.*

DEFINITION 9.1.7. Let M and N be $\mathcal{L}_1$-structures and $\pi\colon M \to N$ be Σ_0 and cofinal. We say that π is *strongly cofinal* iff M and N are not extender active, or M and N are extender active, and $\pi``\mathrm{crit}(\dot{F})^{+,M}$ is cofinal in $\mathrm{crit}(\dot{F})^{+,N}$.

It is easy to see that

LEMMA 9.1.8. *rQ formulae are preserved downward under Σ_1-elementary maps, and upward under strongly cofinal Σ_0-elementary maps.*

LEMMA 9.1.9. *(a) There is a Q-sentence φ of $\mathcal{L}_1$ such that for all transitive $\mathcal{L}_1$ structures M, $M \models \varphi$ iff M is a passive bare lpm.*

(b) There is a rQ-sentence φ of $\mathcal{L}_1$ such that for all transitive $\mathcal{L}_1$ structures M, $M \models \varphi$ iff M is an extender active bare lpm.

(c) There is a Q-sentence φ of $\mathcal{L}_1$ such that for all transitive $\mathcal{L}_0$ structures M, $M \models \varphi$ iff M is a potential branch active bare lpm.

PROOF SKETCH. We omit the proofs of (a) and (b). For (c), note that "$\dot{B} \neq \emptyset$" is Σ_1. One can go on then to say with a Σ_1 sentence that if η is least in $\dot{B}$, then $M|\eta$ is admissible, and s^M exists. One can say with a Π_1 sentence that $\{\alpha \mid \dot{B}(\eta+\alpha)\}$ is a branch of $\mathcal{T}(s)$, perhaps of successor order type. One can say that $\dot{B}$ is cofinal in the ordinals with a Q-sentence. Collectively, these sentences express the conditions on potential lpm-hood related to $\dot{B}$. That the rest of M constitutes an extender passive lpm can be expressed by a Π_1 sentence. ⊣

COROLLARY 9.1.10. *(a) If M is a passive (resp. extender active, potential branch active) bare lpm, and* $\mathrm{Ult}_0(M,E)$ *is wellfounded, then* $\mathrm{Ult}_0(M,E)$ *is a passive (resp. extender active, potential branch active) bare lpm.*

(b) Suppose that M is a passive (resp. extender-active, potential branch active) bare lpm, and $\pi\colon H \to M$ is Σ_1-elementary; then H is a passive (resp. potential branch active) bare lpm.

(c) Let $k(M) = k(H) = 0$, and $\pi\colon H \to M$ be Σ_2 elementary; then H is a branch active bare lpm iff M is a branch active bare lpm.

PROOF. rQ-sentences are preserved upward by strongly cofinal Σ_0 embeddings, so we have (a). They are Π_2, hence preserved downward by Σ_1- elementary embeddings, so we have (b).

It is easy to see that honesty is expressible by a Boolean combination of Σ_2 sentences, so we get (c). ⊣

Part (c) of Corollary 9.1.10 is not particularly useful. In general, our embeddings will preserve honesty of a potential branch active lpm M because $\dot{\Sigma}^M$ and $\dot{B}^M$ are determined by a complete iteration strategy for M that has strong hull condensation. So the more useful preservation theorem in the branch active case applies to *hod pairs*, rather than to hod premice. See Lemma 9.2.3 below.

Remark 9.1.11. The following examples show that the preservation reults of Corollary 9.1.10 are optimal in certain respects.

(1) Let M be an extender active lpm, and $N = \mathrm{Ult}_0(M,E)$, where E is a long extender over M whose space is $\mathrm{crit}(\dot{F})^{+,M}$, so that the canonical embedding $\pi\colon M \to N$ is discontinuous at $\mathrm{crit}(\dot{F})^{+,M}$. Then π is cofinal and Σ_0, so that M and N satisfy the same Q-sentences, but N is not an lpm, because its last extender is not total. π is not strongly cofinal, of course.

(2) The interpolation arguments in [49] yield examples of $\pi: M \to N$ being Σ_0 elementary, and N being an extender active bare lpm, but M not being a bare lpm. Again, M falls short in that its last extender is not total.

The copying construction, and the lifting argument in the iterability proof, do give rise to maps that are only nearly elementary. However, in those cases we know the structures on both sides are lpms for other reasons. On the other hand, core maps and ultrapower maps are fully elementary, so we can apply (a) and (b) of Corollary 9.1.10 to them. We do need to do this.

9.2. Least branch hod pairs

We are interested in least branch premice M that have well-behaved iteration strategies, that is, strategies Ω that are internally lift consistent, quasi-normalize well and have strong hull condensation. Another aspect of the good behavior of Ω is that all Ω-iterates of M are least branch premice whose strategy predicate is consistent with the appropriate tail of Ω. It is really the pair (M, Ω) to which our results apply.

DEFINITION 9.2.1. M is a least branch premouse, and let Ω be a complete iteration strategy for M with scope H_δ; then (M, Ω) is *pushforward consistent* iff whenever s is a stack by Ω and has last model N, then N is an lpm, and $\hat{\Sigma}^N \subseteq \Omega_s$.

Recall here that if N is strategy active, then $\hat{\Sigma}^N$ includes the new branch information present in $\dot{B}^N$.

DEFINITION 9.2.2. (M, Ω) is a *least branch hod pair* (lbr hod pair) with scope H_δ iff

(1) M is a least branch premouse of type 1, and Ω is a complete iteration strategy for M with scope H_δ,[3]
(2) Ω quasi-normalizes well,
(3) Ω is internally lift consistent and has strong hull condensation, and
(4) (M, Ω) is pushforward consistent.

We have made it part of the definition that M has type 1 because it is convenient, and we do not need more generality.

Definition 9.2.2 assumes we have made sense of quasi-normalization and tree embeddings as they apply to iteration trees on least branch premice. The definitions and basic results that apply to pure extender premice go over word-for-word, so we shall simply assume this has been done.

There is one small difference in the two situations, in that the class of bare lpms is not closed under Σ_0 ultrapowers or Σ_1 elementary embeddings, because of the branch-honesty requirement. But we will always be dealing with hulls or iterates

[3]That is, Ω is a winning strategy for II in $G^+(M, \omega, \delta)$. See Section 4.6.

of *pairs*, and lpm-hood is preserved in that context. For iterates, that is just part of clause (4) of Definition 9.2.2. In the case of hulls, it is part of the following lemma.

LEMMA 9.2.3 (Downward Extension for pairs). *Let (M,Ω) be a least branch hod pair with scope H_δ, and $k = k(M)$. Suppose that $\pi\colon N \to M$ is the completion of $\pi_0\colon (P,B) \to M^k$, where either*

(a) π_0 is Σ_2 elementary, or

(b) π_0 is cofinal and Σ_1 elementary, or

(c) π_0 is Σ_0 elementary and N is an lpm;

then setting $k(N) = k$, (N,Ω^π) is a lbr hod pair with scope H_δ and soundness degree k.

PROOF. Let us prove (a) and (b). Assume first that $k = 0$. N is then an lpm by Corollary 9.1.10, except perhaps when M is branch active and π_0 is only cofinal and Σ_1 elementary. In this case, N is a potential branch active lpm, and we must see that N is honest.

So let $\nu = \nu^N$, $b = b^N$, and $\mathcal{T} = \mathcal{T}(s^N)$. If $\nu = \mathrm{lh}(\mathcal{T})$, there is nothing to show, so assume $\nu < \mathrm{lh}(\mathcal{T})$. We must show that $b = [0,\nu)_{\mathcal{T}}$. We have by induction that for $Q = N|\eta^N$, (Q,Ω^π_Q) is an lbr hod pair, and in particular, that $\dot{\Sigma}^Q \subseteq \Omega^\pi$. Thus $\mathcal{T}$ is by Ω^π, and so we just need to see that for $\mathcal{U} = \mathcal{T}\restriction\nu$, $\mathcal{U}^\frown b$ is by Ω^π, or equivalently, that $\pi\mathcal{U}^\frown b$ is by Ω. But it is easy to see that $\pi\mathcal{U}^\frown b$ is a pseudo-hull of $\pi(\mathcal{U})^\frown b^M$, and Ω has strong hull condensation, so we are done.

Thus N is an lpm. Ω^π is a complete iteration strategy defined on all N-stacks in H_δ, where H_δ is the scope of (M,Ω). Ω^π quasi-normalizes well by the proof of Theorem 7.1.6, and has strong hull condensation by the proof of Proposition 7.1.11. It is easy to see that Ω^π is also internally lift consistent.

Finally, we must show that (N,Ω^π) is pushforward consistent. Let P be a Ω^π iterate of N, via the stack s. Let Q be the corresponding iterate of M via πs, and let $\tau\colon P \to Q$ be the copy map. Then

$$\begin{aligned}\mathcal{U} \text{ is by } \hat{\Sigma}^P &\Longrightarrow \tau(\mathcal{U}) \text{ is by } \hat{\Sigma}^Q\\ &\Longrightarrow \tau(\mathcal{U}) \text{ is by } \Omega_{\pi s,Q}\\ &\Longrightarrow \tau\mathcal{U} \text{ is by } \Omega_{\pi s,Q}\\ &\Longrightarrow \mathcal{U} \text{ is by } (\Omega^\pi)_{s,P},\end{aligned}$$

as desired.[4]

This proves (a) and (b) when $k = 0$, and the proof also gave (c).

If $k > 0$, then we must also show that N is k-sound. The proof of the Downward Extension Lemma 4.3.5 does this. ⊣

Definition 9.2.2 records the properties of the hod pairs we construct needed to prove the comparison theorem and the existence of cores. The other properties one

[4]If P is strategy active, then $\mathcal{U}$ may be of the form $\mathcal{T}^\frown b^P$. In this case, by $\pi(\mathcal{U})$ we mean $\pi(\mathcal{T})^\frown b^Q$. Copy maps are nearly elementary, so $\tau\mathcal{U}$ is a pseudo-hull of $\tau(\mathcal{U})$ in this case too.

might hope for seem to follow from these, as they did in the case of pure extender pairs, and by the same proofs. For example, from the proofs of Lemma 7.1.10, Corollary 5.2.6, and Lemma 7.6.5, we get

LEMMA 9.2.4. *Let (M,Ω) be an lbr hod pair with scope H_θ; then*

(a) (M,Ω) is pullback consistent and strategy coherent on stacks of λ-separated trees[5]*, and*

(b) if (M,Ψ) is an lbr hod pair with scope H_θ such that Ψ and Ω agree on λ-separated trees, then $\Psi = \Omega$.

Inspired by these and many other similarities, we define

DEFINITION 9.2.5. (M,Ω) is a *mouse pair* iff (M,Ω) either a pure extender pair, or an lbr hod pair.

The reader will naturally ask whether there are other classes of strategy pairs (M,Σ) which behave like the two classes we have isolated here. The answer is positive. The remarks to follow were stimulated by a suggestion by Hugh Woodin.

One can vary how much of Σ gets encoded into $\dot{\Sigma}^M$, and when that is done. One can think each of these variations as associated to some Σ_1 formula $\varphi(v)$. Roughly, a φ-premouse M starts to encode a branch for $\mathcal{T}$ when it reaches some α such that $M|\alpha \models \varphi[\mathcal{T}]$. Pure extender premice are φ-premice, for $\varphi =$ "$v \neq v$". Least branch premice are φ-premice, for φ a Σ_1 formula that can be abstracted from Section 5.1. Other Σ_1 formulae would lead to classes that might be called "φ-mouse pairs". The requirements of normalizing well, strong hull condensation, and pushforward consistency are the same for all classes of φ-mouse pairs. What varies is how much of the strategy Σ is encoded into M, and when that is done.

We should note that the rigidly layered hod pairs of [42] are not φ-mouse pairs, because the condition governing branch insertion is not first order. φ-mouse pairs have the condensation properties of pure extender pairs, while rigidly layered hod pairs do not.

The analysis of HOD in models of AD^+ that do not satisfy $\mathsf{AD}_{\mathbb{R}}$ may need φ-mouse pairs, for φ not one of the two formulae we have given privileged status in Definition 9.2.5. But this is speculation right now, and we have no real applications for classes of mouse pairs beyond those identified in Definition 9.2.5, so we have avoided the extra generality.[6]

---

[5]Namely, if $s^\frown\langle\mathcal{T}\rangle$ and $s^\frown\langle\mathcal{U}\rangle$ are λ-separated stacks by Ω and N is an initial segment of both last models, then $\Omega_{s^\frown\langle\mathcal{T}\rangle,N} = \Sigma_{s^\frown\langle\mathcal{U}\rangle,N}$. This is what we proved for pure extender pairs in Corollary 5.2.6.

[6]In order to analyze HOD in such determinacy models, it seems one must use pairs (M,Σ) such that only the *short tree component* of Σ is inserted into $\dot{\Sigma}^M$. See also [43], [44], and [80]. To our knowledge, there is as of now no general fine structure theory for such pairs.

9.3. Mouse pairs and the Dodd-Jensen Lemma

Mouse is generally taken to mean *iterable premouse*, and the Comparison Lemma is taken to say that any two mice M and N can be compared as to how much information they contain. But in fact, how M and N are compared depends on which iteration strategies witnessing their iterability are chosen. There is no mouse order on iterable premice, even of the pure extender variety, unless we make restrictive assumptions which imply that the iteration strategy is unique. The canonical information levels of the mouse order are occupied not by mice, but by mouse pairs. These pairs are the objects to which the Comparison Lemma, the Dodd-Jensen Lemma, and the other basic results of inner model theory apply. In the special case that M can have at most one strategy, we do not need to make the pair explicit, but in general, we do.

Let us introduce some terminology that reflects this point of view. We have already used some of it as it applies to pure extender pairs. (See Definition 7.6.10.)

DEFINITION 9.3.1. Let (P,Σ) and (Q,Ω) be mouse pairs.

(a) $(P,\Sigma) \trianglelefteq (Q,\Omega)$ iff $P \trianglelefteq Q$ and $\Sigma = \Omega_P$.

(b) $\pi\colon (P,\Sigma) \to (Q,\Omega)$ is *elementary* (resp. *nearly elementary*) iff π is elementary (resp. nearly elementary) as a map from P to Q, and $\Sigma = \Omega^\pi$,

(c) An *iteration tree on* (P,Σ) is a an iteration tree $\mathcal{T}$ on P such that $\mathcal{T}$ is by Σ. The αth *pair of* $\mathcal{T}$ is $(\mathcal{M}_\alpha^\mathcal{T}, \Sigma_{\mathcal{T}\upharpoonright\alpha+1})$.

(d) A (P,Σ)-*stack* is a P-stack by Σ. If s is a (P,Σ)-stack with last model Q, then the *last pair* of s is $(Q, \Sigma_{s,Q})$.

(e) (Q,Ψ) is an *iterate of* (P,Σ) iff there is a (P,Σ)-stack with last pair (Q,Ψ). If s can be taken to be a single normal (resp. λ-separated, λ-tight) tree, then (Q,Ψ) is a *normal (resp. λ-separated, λ-tight) iterate* of (P,Σ). If s can be taken so that P-to-Q in s does not drop, then (Q,Ψ) is a *non-dropping iterate* of (P,Σ).

(f) $(P,\Sigma) \le^* (Q,\Omega)$ iff there is an iterate (R,Ψ) of $Q,\Omega)$ and an elementary $\pi\colon (P,\Sigma) \to (R,\Psi)$. We call $\le^*$ the *mouse pair order*.

Notice that the natural agreement of pairs in a stack of λ-separated trees on (P,Σ) follows at once from strategy coherence. Here are some further elementary facts stated in this language.

LEMMA 9.3.2. *Let (P,Σ) be a mouse pair with scope H_δ, and let (Q,Ω) be an iterate of (P,Σ) such that Q has type 1; then (Q,Ω) is a mouse pair with scope H_δ.*

PROOF. Quasi-normalizing well, internal lift consistency, strong hull condensation, and pushforward consistency are defined so that they pass to tail strategies. If P is an lpm, then Q is an lpm by clause (4) of Definition 9.2.2. ⊣

In the mouse pair language, the elementarity of iteration maps amounts to pullback consistency. So we have

LEMMA 9.3.3. *Let (P,Σ) be a mouse pair, and let s be a (P,Σ)-stack; then the iteration maps of s are elementary in the category of mouse pairs. That is, if $Q = \mathcal{M}_\alpha^{\mathcal{T}_m(s)}|\langle \nu, k\rangle$ and $\pi\colon Q \to \mathcal{M}_\infty(s)$ is the iteration map of s, then for $t = s\restriction(m-1)^\frown\langle(\nu_m(s), k_m(s), \mathcal{T}_m(s)\restriction(\alpha+1))\rangle$, π is elementary as a map from $(Q,\Sigma_{t,Q})$ to $(\mathcal{M}_\infty(s),\Sigma_s)$.*

The appropriate statement of the Dodd-Jensen Lemma on the minimality of iteration maps is:

THEOREM 9.3.4 (Dodd-Jensen Lemma). *Let (P,Σ) be an mouse pair, let (Q,Ω) be an iterate of (P,Σ) via the stack s, and let $\pi\colon (P,\Sigma) \to (Q,\Omega)$ be nearly elementary; then*

(a) the branch P-to-Q of s does not drop, and
(b) letting $i_s\colon P \to Q$ be the iteration map, for all $\eta < o(P)$, $i_s(\eta) \le \pi(\eta)$.

We omit the well known proof. Notice that it requires the assumption that $\Sigma^\pi_{s,Q} = \Sigma$. This was at one time a nontrivial restriction on the applicability of the Dodd-Jensen Lemma, and led to the Weak Dodd-Jensen Lemma of [39]. Now that we can compare iteration strategies, the restriction is less important.

We get the Dodd-Jensen corollary on the uniqueness of iteration maps.

COROLLARY 9.3.5. *Let (P,Σ) be a mouse pair, (Q,Ω) a non-dropping iterate of (P,Σ) via the stack s, and suppose $(Q,\Omega) \trianglelefteq (R,\Psi)$, where (R,Ψ) is an iterate of (P,Σ) via the stack t; then*

(a) $(Q,\Omega) = (R,\Psi)$, and the branch P-to-R of tdoes not drop, and
(b) letting i_s and i_t be the two iteration maps, $i_s = i_t$.

In the language of mouse pairs, the Comparison Lemma reads

THEOREM 9.3.6 (Comparison Lemma). *Assume AD^+, and let (P,Σ) and (Q,Ψ) be strongly stable mouse pairs with scope* HC *of the same kind; then there are iterates (R,Λ) of (P,Σ) and (S,Ω) of (Q,Ψ), obtained via λ-separated trees $\mathcal{T}$ on P and $\mathcal{U}$ on Q, such that either*

(1) $(R,\Lambda) \trianglelefteq (S,\Omega)$ and P-to-R does not drop, or
(2) $(S,\Omega) \trianglelefteq (R,\Lambda)$ and Q-to-S does not drop.

We proved this for pure extender pairs in Theorem 8.4.5, and we shall give the proof for least branch hod pairs in Theorem 9.5.10. For now let us assume it. We get

COROLLARY 9.3.7. *Assume AD^+; then*

(a) For (P,Σ) and (Q,Ψ) mouse pairs with scope HC *of the same kind,*

$$(P,\Sigma) <^* (Q,\Psi) \iff \exists (R,\Omega)\exists\pi[(R,\Omega) \text{ is a dropping iterate of } (Q,\Psi) \text{ and } \pi\colon (P,\Sigma) \to (R,\Omega) \text{ is nearly elementary}].$$

(b) *When restricted to a fixed kind,* $\le^*$ *is a prewellorder of the strongly stable mouse pairs with scope* HC.

PROOF. The left-to-right direction of (a) follows from the Comparison Lemma. The right-to-left direction follows from Dodd-Jensen. For (b), the Comparison Lemma implies that $\le^*$ is linear. That it is wellfounded follows from (a), using the proof of the Dodd-Jensen Lemma. ⊣

For the record

DEFINITION 9.3.8. Let (P,Σ) be a mouse pair; then Σ is *positional* iff whenever (Q,Ψ) and (R,Ω) are iterates of (P,Σ), and $Q = R$, then $\Psi = \Omega$.

The property is clearly related to what is called being positional in [42]. In the present context it implies strategy coherence.

[64] proves

LEMMA 9.3.9. *Assume* AD^+, *and let* (P,Σ) *be a strongly stable mouse pair with scope* HC; *then* Σ *is positional.*

Fortunately, this lemma is not needed in the proof of the Comparison Lemma 9.3.6. Its proof instead relies on a comparison argument.

Here are two propositions that explain the relationship between pure extender mice and pure extender pairs.

PROPOSITION 9.3.10. *Assume* AD^+, *and let* P *be a countable, strongly stable,* ω_1*-iterable pure extender premouse; then there is a* Σ *such that* (P,Σ) *is a pure extender pair.*

PROOF. Let Ψ be an arbitrary ω_1 iteration strategy for P. We may assume Ψ is Suslin and co-Suslin by Woodin's Basis Theorem. (See [69], Theorem 7.1.) Thus there is a coarse strategy pair[7] $((N^*,\in,w,\mathcal{F},\Sigma),\Sigma^*)$ that captures Ψ. Working in N^*, we get that P iterates by Ψ to a level (Q,Ω) of the pure extender pair construction of N^*. Since Σ^* exists, Ω can be extended to to have scope HC. Let $\pi\colon P \to Q$ be the iteration map; then (P,Ω^π) is a pure extender pair. ⊣

PROPOSITION 9.3.11. *Assume* AD^+, LEC, *and* $\theta_0 < \theta$; *then there are strongly stable pure extender pairs* (P,Σ) *and* (P,Ω) *such that* $(P,\Sigma) <^* (P,\Omega)$.

PROOF SKETCH. By LEC, there is a pure extender pair (P,Ω) such that Ω is not ordinal definable from a real. Fix such a pair. By the Basis Theorem, there is a Σ such that (P,Σ) is a pure extender pair, and Σ is ordinal definable from a real. Suppose toward contradiction that $(P,\Omega) \le^* (P,\Sigma)$; then

$$\Omega = (\Sigma_s)^\pi$$

for some stack s and iteration map π. Thus Ω is ordinal definable from a real, contradiction. ⊣

[7] See Definition 9.4.14.

It follows that under the hypotheses of Proposition 9.3.11, there are pure extender pairs (P,Σ) and (P,Ω) such that for some R, P iterates normally by Σ to a proper initial segment of R, and normally by Ω to a proper extension of R.

The proofs of Lemmas 7.6.5 and 7.6.7 go over from pure extender pairs to least branch hod pairs with no change. We get

LEMMA 9.3.12. *Let (P,Σ) and (P,Λ) be mouse pairs with scope H_δ, and suppose that Σ and Λ agree on countable λ-separated trees; then $\Sigma = \Lambda$.*

LEMMA 9.3.13. *Let (P,Σ) be a mouse pair pair with scope H_δ, and let $j\colon V \to M$ be elementary, where M is transitive and* $\text{crit}(j) > |P|$; *then $j(\Sigma)$ and Σ agree on all trees in $j(H_\delta) \cap H_\delta$.*

Weak Dodd-Jensen. Because we can compare iteration strategies, we shall be able to use the full Dodd-Jensen Lemma instead of the weak one in the proof of solidity and universality of standard parameters. Nevertheless, let us state the weak one, for the sake of completeness.[8]

Note that the proofs we have given that background induced strategies are internally lift consistent, quasi-normalize well and have strong hull condensation actually yield (ω_1,ω_1) strategies Ω such that each Ω_s^*, for $\text{lh}(s) < \omega_1$, is internally lift consistent, quasi-normalizes well and has strong hull condensation. Here Ω_s^* is the complete strategy, defined on finite stacks t, given by $\Omega_s^*(t) = \Omega(s^\frown t)$. The same will be true for pushforward consistency. This is used in the Weak Dodd-Jensen argument.

Just as in the pure extender case:

DEFINITION 9.3.14. An iteration strategy Ω for an lpm M has the *Weak Dodd-Jensen property* relative to an enumeration $\vec{e}$ of its universe in order type ω iff whenever $N = M_\infty(s)$ for some stack s by Ω, then

(1) if there is a nearly elementary embedding from M to an initial segment of N, then the branch M-to-N of s does not drop, and the iteration map i^s is $\vec{e}$-minimal, and
(2) if M has type 1A, $k = k(M)$, and there is a Σ_0 elementary map from M_0^k to N_0^k, then the branch M-to-N of s does not drop in model.

LEMMA 9.3.15 (Weak Dodd-Jensen). *Let (M,Ω) be a mouse pair with scope H_δ, and let $\vec{e}$ be an enumeration of the universe of M in order type ω. Suppose that Ω is defined on all countable M-stacks s from H_δ, and that for any such s having a last model, $(M_\infty(s),\Omega_s)$ is a mouse pair; then there is a countable M-stack s by Ω having last model $N = M_\infty(s)$, and a nearly elementary $\pi\colon M \to N$, such that*

(1) $(N,(\Omega_s)^\pi)$ is a mouse pair, and
(2) $(\Omega_s)^\pi$ has the Weak Dodd-Jensen property relative to $\vec{e}$.

[8]We shall need it in the proof of Theorem 10.2.3.

PROOF. The proof from [39] goes over verbatim. Notice here that any such $(N, (\Omega_s)^\pi)$ is an lbr hod pair, by Lemmas 9.3.2 and 9.2.3. ⊣

We have stated the elementary results about lbr hod pairs in this section as results about mouse pairs, because that is their natural context. We are mainly interested in lbr hod pairs for the rest of this book, so we shall return to that level of generality.

9.4. Background constructions

It is easy to modify the background constructions of pure extender premice described in Section 4.7 so that they produce least branch hod pairs. The background conditions for adding an extender are unchanged. If we have reached the stage at which $M_{\nu,0}$ is to be defined, then our construction, together with an iteration strategy for the background universe, will have provided us with complete iteration strategies $\Omega_{\eta,l}$ for $M_{\eta,l}$, for all $\eta < \nu$. We must assume here that the background universe knows how to iterate itself for trees that are of the form $\text{lift}(\mathcal{T}, M_{\eta,l}, \mathbb{C})_0$. Each $(M_{\eta,l}, \Omega_{\eta,l})$ will be a least branch hod pair. If $M_{\nu,0}$ is to be branch active according to the lpm requirements, then we use the appropriate $\Omega_{\eta,l}$ to determine $\dot{B}^{M_{\nu,0}}$.

The additional strategy predicates in our structures affect what we mean by cores and resurrection, but otherwise nothing much changes. We shall therefore go quickly.

The simplest sort of iterability hypothesis under which we can carry out a least branch construction is the following.

DEFINITION 9.4.1. $\mathsf{IH}_{\kappa,\delta}$ is the assertion: if $(w, \mathcal{F})$ is a coherent pair such that $\mathcal{F} \subseteq V_\delta$ and $\forall E \in \mathcal{F}(\text{crit}(E) > \kappa)$, then for all θ, $(V, \in, w, \mathcal{F})$ is strongly uniquely (θ, θ)- iterable.

Assuming AD^+, we have by Corollary 7.2.8 that whenever $(N^*, \delta, S, T, w, \Sigma^*)$ is a coarse Γ-Woodin tuple, then $L(N^*, w, S, T) \models \mathsf{IH}_{\omega,\delta}$, where δ is the Γ-Woodin of N^*. So we could be doing our background construction inside this model.

What we actually need from unique iterability is an iteration strategy for the background universe that quasi-normalizes well, has strong hull condensation, and is pushforward consistent.

DEFINITION 9.4.2. A *coarse strategy premouse* is a structure $(M, \in, w, \mathcal{F}, \Sigma)$ such that $(M, \in, w, \mathcal{F})$ is a coarse extender premouse, and $\Sigma \in M$, and letting $\delta = \delta(w)$, the following hold in M:

(a) δ is inaccessible and
(b) Σ is a $(\delta, \delta, \mathcal{F})$-iteration strategy for V that quasi-normalizes well and has strong hull condensation, and

(c) Σ is pushforward consistent, in that if $i\colon V \to N$ is an iteration map associated to the stack s by Σ, then $i(\Sigma) \subseteq \Sigma_s$.

We showed in Chapter 7:

LEMMA 9.4.3. *Assume* $\mathsf{IH}_{\kappa,\delta}$ *and that* δ *is inaccessible. Let* $(w,\mathcal{F})$ *be a coherent pair such that* $\delta = \delta(w)$ *and* $\forall E \in \mathcal{F}(\text{crit}(E) > \kappa)$. *Let* Σ *be the unique* (δ,δ) *iteration strategy for* $(V,\in,w,\mathcal{F})$*; then for any transitive* M *such that* $V_\delta \cup \{w,\mathcal{F}\} \subseteq M$ *and* $M \models \mathsf{ZFC}$, $(M,\in,w,\mathcal{F},\Sigma)$ *is a coarse strategy premouse.*

PROOF. Notice that since δ is inaccessible, $\Sigma \in M$, and Σ witnesses strong unique iterability for V_δ in M. The rest follows from Theorem 7.2.9 and Lemma 7.3.9. ⊣

Now suppose that $(V,\in,w,\mathcal{F},\Sigma)$ is a coarse strategy premouse. We shall define what it is to be a $(w,\mathcal{F},\Sigma)$ *construction*. This is a tuple

$$\mathbb{C} = (w,\mathcal{F},\langle(M_{\nu,k},\Omega_{\nu,k}),F_\nu \mid \langle\nu,k\rangle <_{\text{lex}} \text{lh}(\mathbb{C})\rangle)$$

whose levels $(M_{\nu,k},\Omega_{\nu,k})$ are lbr hod pairs. We require that $\mathbb{C}$ add an extender to the current $(M^{<\nu},\emptyset)$ whenever there is one with a suitable background in $\mathcal{F}$. F_ν is then the minimal such background extender. The $\Omega_{\nu,k}$ are the strategies induced by Σ and $\mathbb{C}$. There is at most one $(w,\mathcal{F},\Sigma)$-construction of length $\langle\eta,j\rangle$, but there may be none, if the unique attempt at such a construction breaks down before it reaches this stage. We say that $\mathbb{C}$ is *good at* $\langle\nu,k\rangle$ iff it does not break down at $\langle\nu,k\rangle$.

DEFINITION 9.4.4. Let $\mathbb{C}$ be a least branch construction, $\langle\nu,k\rangle < \text{lh}(\mathbb{C})$, and $k \geq 0$; then $\mathbb{C}$ is *good at* $\langle\nu,k\rangle$ iff $M^{\mathbb{C}}_{\nu,k}$ is solid (that is, $k+1$-solid).

The other clauses in goodness apply at stages of the form $\langle\nu,-1\rangle$ such that $M_{\nu,0}$ is extender active. We shall state them below. In general, if $\mathbb{C}$ is not good at $\langle\nu,k\rangle$, then $M_{\nu,k+1}$ is undefined, and $\text{lh}(\mathbb{C}) = \langle\nu,k+1\rangle$.

The first level of $\mathbb{C}$ must be (M,Ω), where M is the passive lpm with universe V_ω, and Ω is its unique iteration strategy. Given $\langle\nu,k\rangle < \text{lh}(\mathbb{C})$ such that $k \geq 0$ and $\mathbb{C}$ is good at at $\langle\nu,k\rangle$, we set

$$\begin{aligned} M_{\nu,k+1} &= \mathfrak{C}(M_{\nu,k}) \\ &= \text{cHull}^{M_{\nu,k}}_{k+1}(\rho_{k+1} \cup \{p_{k+1},\rho_{k+1},w_k\}). \end{aligned}$$

We are using here the notation of Section 4.1: $\rho_{k+1} = \rho(M_{\nu,k})$, $p_{k+1} = p(M_{\nu,k}) = p_1(M^k_{\nu,k})$, and $w_k = \langle\eta_k,\rho_k,p_k\rangle$, where $\rho_k = \rho^-(M_{\nu,k})$ and $\eta_k = \eta_k^{M_{\nu,k}}$ is the $r\Sigma_k$ cofinality of ρ_k.

$M_{\nu,k+1}$ is an lpm of soundness degree $k+1$. We require that

$$\Omega_{\nu,k+1} = \Omega(\mathbb{C},M_{\nu,k+1},\Sigma)$$

be the induced strategy. Let us review briefly what that means.

$\mathbb{C}$ determines resurrection maps $\mathrm{Res}_{\nu,k}$ and $\sigma_{\nu,k}$ for $\langle \nu, k\rangle <_{\mathrm{lex}} \mathrm{lh}(\mathbb{C})$ as in Section 4.7:

1. If $N = M_{\nu,k+1}$, then $\mathrm{Res}_{\nu,k+1}[N] = N$ and $\sigma_{\nu,k+1}[N] = $ identity.
2. If $N \lhd M_{\nu,k+1}$, then letting $\pi : M_{\nu,k+1} \to M_{\nu,k}$ be the anticore map,
$$\mathrm{Res}_{\nu,k+1}[N] = \mathrm{Res}_{\nu,k}[\pi(N)]$$
and
$$\sigma_{\nu,k+1} = \sigma_{\nu,k}[\pi(N)] \circ \pi.$$
3. For $N \lhd M_{\nu,0}$, $\mathrm{Res}_{\nu,0}[N]$ is the common value of $\mathrm{Res}_{\eta,j}[N]$ for all sufficiently large $\langle \eta, j\rangle <_{\mathrm{lex}} \langle \nu, 0\rangle$, and similarly for $\sigma_{\nu,0}[N]$.

Because we are using the pfs fine structure, clauses (2) and (3) are appropriate, and our resurrection maps are consistent with one another. If $Q = M_{\nu,k}$, we often write $\mathrm{Res}_Q[N]$ and $\sigma_Q[N]$ for $\mathrm{Res}_{\nu,k}[N]$ and $\sigma_{\nu,k}[N]$. This is justified because there is at most one $\langle \nu, k\rangle$ such that $Q = M_{\nu,k}$.

A *conversion stage* is a tuple
$$c = \langle M, \psi, Q, \mathbb{D}, S\rangle$$
such that S is a coarse strategy premouse, $\mathbb{D}$ is a least branch construction in the sense of S, $Q = M^{\mathbb{D}}_{\nu,k}$ for some ν, k, and $\psi \colon M \to Q$ is nearly elementary. If $\mathcal{T}$ is a plus tree on M, then $\mathrm{lift}(\mathcal{T}, c)$ is the conversion system
$$\mathrm{lift}(\mathcal{T}, c) = \langle \mathcal{T}^*, \langle c_\alpha \mid \alpha < \gamma\rangle\rangle,$$
where
$$c_\alpha = \langle M^{\mathcal{T}}_\alpha, \psi_\alpha, Q_\alpha, \mathbb{D}_\alpha, \mathcal{M}^{\mathcal{T}^*}_\alpha\rangle$$
is the α-th stage in the conversion, defined exactly as in Section 4.8. We resurrect $\psi_\alpha(E^{\mathcal{T}}_\alpha)$ inside $\mathbb{D}_\alpha$ in order to obtain $E^{\mathcal{T}^*}_\alpha$. Either $\gamma = \mathrm{lh}(\mathcal{T})$ or $\mathcal{M}^{\mathcal{T}^*}_{\gamma-1}$ is illfounded, in which case conversion stops. It is important for conversion that $F^{\mathbb{D}}_\nu$ backgrounds F^+, for F the last extender of $M^{\mathbb{D}}_{\nu,0}$. This is part of $\mathbb{D}$ being good at $\langle \nu, -1\rangle$. If Σ^* is an iteration strategy for S and $\mathcal{T}$ is a plus tree on M, then
$$\mathcal{T} \text{ is by } \Omega(c, \Sigma^*) \text{ iff } \mathrm{lift}(\mathcal{T}, c) \text{ is by } \Sigma^*.$$
Of course, $S = (|S|, \in, w, \mathcal{F}, \Sigma)$ is a coarse strategy premouse, with its internal strategy Σ. We shall only be interested in the case that $\Sigma \subseteq \Sigma^*$. So if $|S| = V$, then $\Sigma = \Sigma^*$ is the interesting case. Stacks of plus trees are converted in succession, as before. In the special case $M = Q$ and $\psi = \mathrm{id}$, we write
$$\Omega(\mathbb{D}, M, \Sigma^*) = \Omega(\langle M, \mathrm{id}, M, \mathbb{D}, S\rangle, \Sigma^*),$$
in the case that S can be understood from context.

In our case, $S = (V, \in, w, \mathcal{F}, \Sigma)$, so $\Sigma = \Sigma^*$, and $\mathbb{D}$ is what we are calling $\mathbb{C}$. Thus $\Omega(\mathbb{C}, M_{\nu,k+1}, \Sigma)$ does indeed exist, and it is a (θ, θ) iteration strategy. We can then let $\Omega_{\nu,k+1}$ be its restriction to finite, maximal stacks of plus trees, as we did above.

LEMMA 9.4.5. *Let $(V, \in, w, \mathcal{F}, \Sigma)$ be a coarse strategy premouse with scope V_θ, and $\mathbb{C}$ a least branch $(w, \mathcal{F}, \Sigma)$ construction that is good at $\langle \nu, k \rangle$, where $k \geq 0$; then*

(1) $(M_{\nu,k+1}, \Omega_{\nu,k+1})$ is a least branch hod pair with scope V_θ, and
(2) setting $\rho = \rho(M_{\nu,k})$ and $\gamma = (\rho^+)^{M_{\nu,k}}$, $(\Omega_{\nu,k})_{\langle \gamma,0 \rangle} = (\Omega_{\nu,k+1})_{\langle \gamma,0 \rangle}$.

PROOF. Part (2) is an immediate consequence of the fact that for $\xi < (\rho^+)^{M_{\nu,k}}$ and $Q = M_{\nu,k}|\langle \xi, l \rangle$, $\mathrm{Res}_{\nu,k}[Q] = \mathrm{Res}_{\nu,k+1}[Q]$ and $\sigma_{\nu,k}[Q] = \sigma_{\nu,k+1}[Q]$.

For part (1), we repeat the proofs that background induced strategies are internally lift consistent, quasi-normalize well and have strong hull condensation (Theorems 7.4.1 and 7.5.1) that we gave in the pure extender model case. (See Theorems 5.4.5, 7.4.1, and 7.5.1.) What is left is to show that $(M_{\nu,k+1}, \Omega_{\nu,k+1})$ is pushforward consistent.

Let $(M, \Omega) = (M_{\nu,k+1}, \Omega_{\nu,k+1})$ and let s be a stack on M by Ω, with last model N. Let $\langle \eta, l, \mathcal{T} \rangle \in \hat{\Sigma}^N$ and $P = N|\langle \eta, l \rangle$. We must see that $\mathcal{T}$ is by $\Omega_{s,P}$. Let

$$s^* = \mathrm{lift}(s, M, \mathbb{C})_0,$$

and let R be the last model of s^*. (R is a coarse strategy premouse.) Let

$$\langle N, \psi, Q, \mathbb{D}, R \rangle = \text{last conversion stage in } \mathrm{lift}(s, M, \mathbb{C}),$$

then

$$\Omega_{s,P} = (\Omega(Q, \mathbb{D}, \Sigma_{s^*,R})_{\psi(P)})^{\psi}.$$

Thus it will suffice to show that $\psi\mathcal{T}$ is by $\Omega(Q, \mathbb{D}, \Sigma_{s^*,R})_{\psi(P)}$. Since $\Sigma_{s^*,R}$ has strong hull condensation, so does $\Omega(Q, \mathbb{D}, \Sigma_{s^*,R})$, by the proof of Theorem 7.5.1. Strong hull condensation passes to the tail strategy $\Omega(Q, \mathbb{D}, \Sigma_{s^*,R})_{\psi(P)}$, and hence it will suffice to show that $\psi(\mathcal{T})$ is by $\Omega(Q, \mathbb{D}, \Sigma_{s^*,R})_{\psi(P)}$.

By the near elementarity of ψ, $\psi(\mathcal{T}) \in \hat{\Sigma}^Q_{\psi(P)}$.[9] Letting

$$R = (|R|, \in, u, \mathcal{G}, \Phi),$$

we have that $\psi(\mathcal{T})$ is by $\Omega(\mathbb{D}, Q, \Phi)_{\psi(P)}$ as computed inside R. But Σ is pushforward consistent, so

$$\Phi \subset \Sigma_{s^*,R},$$

and thus $\pi(\mathcal{T})$ is by $\Omega(\mathbb{D}, Q, \Sigma_{s^*,R})_{\psi(P)}$, as desired. ⊣

Suppose now that $\mathbb{C}$ is good at $\langle \nu, k \rangle$ for all $k < \omega$. For $k < \omega$ sufficiently large, $\rho(M_{\nu,k}) = \rho^-(M_{\nu,k})$, so the bare premice associated to $M_{\nu,k}$ and $M_{\nu,k+1}$ are the same, and $\Omega_{\nu,k} = (\Omega_{\nu,k+1})_{\langle \nu,k \rangle}$.[10] We set

$$\hat{M}_{\nu,\omega} = \text{eventual value of } \hat{M}_{\nu,k} \text{ as } k \to \omega,$$

[9] If N is strategy active, then $\mathcal{T}$ may be the tree to which N is adding a branch via $\dot{B}^N$. In this case, $\pi(\mathcal{T})$ is the corresponding tree defined over Q.

[10] For sufficiently large k, $\Omega_{\nu,k} = \Omega_{\nu,k+1}$, up to minor notational differences in the way the iteration trees on which they act are presented.

and

$$M^{<\nu+1} = \text{rud closure of } \hat{M}_{\nu,\omega} \cup \{\hat{M}_{\nu,\omega}\},$$
$$\text{arranged as an } \mathcal{L}_0 - \{\dot{F}, \dot{B}\} \text{ structure,}$$

and for $\xi \leq \nu$ and $l < \omega$,

$$(\Omega^{<\nu+1})_{\langle \xi, l \rangle} = \text{eventual value of } (\Omega_{\nu,k})_{\langle \xi, l \rangle} \text{ as } k \to \omega.$$

Case 1. $M^{<\nu+1}$ is branch active.
Let $s^{M^{<\nu+1}} = \langle \xi, l, \mathcal{T} \rangle$ be the critical tree, and $b = \Omega^{<\nu+1}_{\langle \xi, l \rangle}(\mathcal{T})$; then

$$M_{\nu,0} = (M^{<\nu+1}, \emptyset, B),$$

where $B = \{\eta^M + \gamma \mid \gamma \in b \wedge \eta^M + \gamma < o(M^{<\nu+1})\}$.

Case 2. $M^{\nu+1}$ is branch passive.
In this case,

$$M_{\nu+1,0} = (M^{<\nu+1}, \emptyset, \emptyset).$$

In both cases, the induced strategy

$$\Omega_{\nu+1,0} = \Omega(\mathbb{C}, M_{\nu+1,0}, \Sigma)$$

is essentially the same as $\Omega^{<\nu+1}$. Any iteration tree on $M_{\nu+1,0}$ is essentially equivalent to an iteration tree on $M_{\nu,\omega}$.

The proof of Lemma 9.4.5 gives

LEMMA 9.4.6. *Let $(V, \in, w, \mathcal{F}, \Sigma)$ be a coarse strategy premouse with scope H_θ, and $\mathbb{C}$ a least branch $(w, \mathcal{F}, \Sigma)$ construction that is good at $\langle \nu, k \rangle$ for all $k < \omega$; then then $(M_{\nu+1,0}, \Omega_{\nu+1,0})$ is an lbr hod pair with scope H_θ.*

Finally, if ν is a limit, put

$$M^{<\nu} = \text{unique } \mathcal{L}_0 - \{\dot{F}, \dot{B}\} \text{ structure } P \text{ such that for all lpms } N,$$
$$N \lhd P \text{ iff } N \lhd M_{\alpha,j} \text{ for all sufficiently large } \langle \alpha, j \rangle < \langle \nu, 0 \rangle.$$
$$\Omega^{<\nu}_{\langle \xi, l \rangle} = \text{eventual value of } (\Omega_{\alpha,j})_{\langle \xi, l \rangle} \text{ as } \alpha \to \nu.$$

Case 1. $M^{<\nu}$ is branch active.
Let $M = M^{<\nu}$, $s^M = \langle \xi, l, \mathcal{T} \rangle$, and $b = \Omega^{<\nu}_{\langle \xi, l \rangle}(\mathcal{T})$; then

$$M_{\nu,0} = (M^{<\nu}, \emptyset, B),$$

where $B = \{\eta^M + \gamma \mid \gamma \in b \wedge \eta^M + \gamma < o(M)\}$.

Case 2. There is an F such that $(M^{<\nu}, F, \emptyset)$ is an lpm, $\text{crit}(F) \geq \kappa$, and there is a $\mathbb{C}$-certificate F^* for F, in the following sense.

DEFINITION 9.4.7. Let $\mathbb{C}$ be a $(w, \mathcal{F}, \Sigma)$-construction, $M^{<\nu} = (M^{<\nu})^{\mathbb{C}}$, and $(M^{<\nu}, F, \emptyset)$ be an lpm; then F^* is a *$\mathbb{C}$-certificate for F* iff

(i) $F^* \in \mathcal{F}$, and

(ii) $F^* \restriction \lambda_F \cap M^{<\nu} = F \restriction \lambda_F$.
(iii) $\lambda_F < \mathrm{lh}(F^*)$ and $\forall \tau < \nu\ (\mathrm{lh}(F_\tau^{\mathbb{C}}) < \mathrm{lh}(F^*))$.

The $\mathbb{C}$-*minimal certificate for* F is the unique certificate for F that is minimal in the Mitchell order among all certificates for F, and w-least among all Mitchell order minimal certificates for F.

DEFINITION 9.4.8. $\mathbb{C}$ is *good at* $\langle \nu, -1\rangle$ iff ν is a successor ordinal, or ν is a limit ordinal and

(a) whenever $(M^{<\nu}, F, \emptyset)$ is an lpm and F^* is a $\mathbb{C}$-minimal certificate for F, then $F^+ \subseteq F^*$ and $\mathrm{lh}(F)$ is a cardinal in $i_{F^*}(M^{<\nu})$, and
(b) whenever $(M^{<\nu}, F, \emptyset)$ and $(M^{<\nu}, G, \emptyset)$ are lpms such that F and G have $\mathbb{C}$-certificates, then $F = G$.

We say that $\mathbb{C}$ is *plus consistent at* ν iff (a) holds, and that $\mathbb{C}$ is *extender unique at* ν iff (b) holds. We say that $\mathbb{C}$ *breaks down at* $\langle \nu, k\rangle$ iff $\mathbb{C}$ is good at all $\langle \eta, j\rangle <_{\mathrm{lex}} \langle \nu, k\rangle$, but is not good at $\langle \nu, k\rangle$.

If $\mathbb{C}$ breaks down at $\langle \nu, -1\rangle$ then we stop the construction, leaving $M_{\nu,0}$ undefined. Otherwise let F be as in the case hypothesis, and set

$$M_{\nu,0} = (M^{<\nu}, F, \emptyset)$$
$$F_\nu^{\mathbb{C}} = \text{unique minimal } \mathbb{C}\text{-certificate } F^* \text{ for } F.$$

Case 3. Otherwise.
Then we set

$$M_{\nu,0} = (M^{<\nu}, \emptyset, \emptyset),$$
$$\Omega_{\nu,0} = \Omega(\mathbb{C}, M_{\nu,0}, \Sigma).$$

In Cases 1 and 3, $\Omega_{\nu,0}$ is essentially the same as $\Omega^{<\nu}$. In Case 2 it is not, since iterations that use the new extender are now allowed.

The proof of Lemma 9.4.5 yields

LEMMA 9.4.9. *Let* $(V, \in, w, \mathcal{F}, \Sigma)$ *be a coarse strategy premouse with scope* H_θ, *and* $\mathbb{C}$ *a be least branch* $(w, \mathcal{F}, \Sigma)$ *construction. Let* ν *be a limit ordinal such that* $\mathbb{C}$ *is is good at* $\langle \nu, -1\rangle$; *then* $(M_{\nu,0}, \Omega_{\nu,0})$ *is an lbr hod pair with scope* H_θ.

DEFINITION 9.4.10. $\mathbb{C}$ is a *least branch* $(w, \mathcal{F}, \Sigma)$*-construction* iff $(V, \in, w, \mathcal{F}, \Sigma)$ is a coarse strategy premouse, and

$$\mathbb{C} = (w, \mathcal{F}, \langle M_{\nu,k}, \Omega_{\nu,k}, F_\nu \mid \langle \nu, k\rangle < \mathrm{lh}(\mathbb{C})\rangle),$$

where $\langle M_{\nu,k}, \Omega_{\nu,k}, F_\nu \mid \langle \nu, k\rangle < \mathrm{lh}(\mathbb{C})\rangle$ satisfies the conditions above. $\mathbb{C}$ is a *least branch construction* iff $\mathbb{C}$ is a least branch $(w, \mathcal{F}, \Sigma)$ construction, for some $w, \mathcal{F}$, and Σ.

The elementary lemmas of Section 4.7 on the agreement between levels of a construction and on the coherence properties of a construction go through. The agreement and coherence apply to successive pairs, not just successive premice.

LEMMA 9.4.11. *Let $\mathbb{C}$ be a least branch construction, with levels $(M_{\nu,k},\Omega_{\nu,k})$.*

(a) Let $\langle\mu,l\rangle <_{\text{lex}} \langle\nu,k\rangle < \text{lh}(\mathbb{C})$, and suppose that whenever

$$\langle\mu,l\rangle \leq_{\text{lex}} \langle\eta,j\rangle \leq_{\text{lex}} \langle\nu,k\rangle,$$

then $\rho^-(M_{\mu,l}) \leq \rho^-(M_{\eta,j})$; then $(M_{\mu,l},\Omega_{\mu,l}) \lhd (M_{\nu,k},\Omega_{\nu,k})$.

(b) Let $\gamma < o(M_{\nu,k})$ be a cardinal of $M_{\nu,k}$ such that $\gamma \leq \rho^-(M_{\nu,k})$, and suppose $P \trianglelefteq M_{\nu,k}$ is such that $\rho^-(P) = \gamma$; then

(i) there is a unique $\langle\mu,l\rangle \leq_{\text{lex}} \langle\nu,k\rangle$ such that $P = M_{\mu,l}$, moreover

(ii) if $P = M_{\mu,l}$, then $\gamma \leq \rho^-(M_{\eta,j})$ whenever $\langle\mu,l\rangle \leq_{\text{lex}} \langle\eta,j\rangle \leq_{\text{lex}} \langle\nu,k\rangle$.

COROLLARY 9.4.12. *Let $\mathbb{C}$ be a least branch construction; then for any lpm N, there is at most one $\langle\nu,k\rangle$ such that $N = M^{\mathbb{C}}_{\nu,k}$.*

LEMMA 9.4.13. *Let $\mathbb{C}$ be a least branch construction and $M^{<\nu} = (M^{<\nu})^{\mathbb{C}}$. Suppose that $(M^{<\nu},F,\emptyset)$ is an lpm, and F^* is a $\mathbb{C}$-minimal certificate for F. Let $\mathbb{D} = i_{F^*}(\mathbb{C})$; then*

(1) $\text{lh}(F^*)$ is the least inaccessible η such that $\lambda_F < \eta$ and for all $\tau < \nu$, $\text{lh}(F^{\mathbb{C}}_\tau) < \eta$,

(2) λ_F is not measurable in $i_{F^*}(M^{<\nu})$,

(3) $\mathbb{D}\restriction\nu = \mathbb{C}\restriction\nu$,

(4) $M^{\mathbb{D}}_{\nu,0} \neq (M^{<\nu},F,\emptyset)$, and if $\mathbb{C}$ is extender unique at $\langle\nu,-1\rangle$, then $M^{\mathbb{D}}_0 = (M^{<\nu},\emptyset,\emptyset)$,

(5) $(M^{<\nu},\emptyset) \lhd i_{F^*}(M^{<\nu})$, and

(6) if $\xi < \nu$, and $\mathbb{C}\restriction\xi$ has last model N such that $o(N) < \text{crit}(F^*)$, then $\mathbb{C}\restriction\xi \in V_{\text{crit}(F^*)}$.

PROOF. See the proofs of Lemmas 3.1.9 and 3.1.11. ⊣

The proof of the lemma uses that $(w^{\mathbb{C}},\mathcal{F}^{\mathbb{C}})$ is a coherent pair, that $\mathbb{C}$ is maximal relative to $(w,\mathcal{F}^{\mathbb{C}})$, and that Σ is pushforward consistent.

What we have said so far concerns the construction determined by a coarse strategy premouse M as viewed from within M. We shall also need to look at it from the outside, in a model of AD^+ where M is countable and its internal strategy can be extended to all countable iteration trees. The next definition is meant to be considered in this AD^+ context.

DEFINITION 9.4.14. A *coarse strategy pair* is a pair $\langle(M,\in,w,\mathcal{F},\Sigma),\Sigma^*\rangle$ such that

(a) $(M,\in,w,\mathcal{F},\Sigma)$ is a countable coarse strategy premouse,

(b) Σ^* is a complete (ω_1,ω_1) iteration strategy for $(M,\in,w,\mathcal{F},\Sigma)$ that normalizes well and has strong hull condensation, and

(c) if $i\colon M\to N$ is the iteration map associated to a stack s by Σ^*, then $i(\Sigma) \subseteq \Sigma^*_{s,N}$.

Remark 9.4.15. Suppose that $\langle (M, \in, w, \mathcal{F}, \Sigma), \Sigma^* \rangle$ is a coarse strategy pair, and

$$(M, \in, w, \mathcal{F}, \Sigma) \models \mathbb{C} \text{ is a least branch } (w, \mathcal{F}, \Sigma)\text{-construction.}$$

Let $\langle \nu, k \rangle < \mathrm{lh}(\mathbb{C})$ and $\Psi = \Omega^{\mathbb{C}}_{\nu,k}$. By definition, the scope of Ψ is H^M_δ, where $\delta = \delta(w)$, but we can extend Ψ to a strategy with scope HC^V in a natural way. This is because $\mathrm{lift}(\vec{\mathcal{T}}, M_{\nu,k}, \mathbb{C}, M)$ is defined even when $\vec{\mathcal{T}} \notin M$, $\Sigma \subseteq \Sigma^*$, and Σ^* has scope HC^V. In this context we shall usually write $\Omega^{\mathbb{C}}_{\nu,k}$ for the extended strategy. That is

$$\vec{\mathcal{T}} \text{is by } \Omega_{\nu,k} \text{ iff } \mathrm{lift}(\vec{\mathcal{T}}, M_{\nu,k}, \mathbb{C})_0 \text{ is by } \Sigma^*,$$

for all $\vec{\mathcal{T}} \in \mathrm{HC}^V$.

The existence of coarse strategy pairs under AD^+ is implicit in the results of Section 7.2 on the existence of iterable Γ-Woodin models.

THEOREM 9.4.16. *Assume* AD^+*, and let* (M, Σ^*) *be a coarse* Γ*-Woodin pair, and* U *the tree of a universal* Γ *set. Let* $\delta = \delta^M$*,* w *be a wellorder of* V^M_δ *such that* $w \in C_\Gamma(V^M_\delta)$*, and*

$$N = L_\alpha[V^M_\delta, U],$$

where $L_\alpha[V^M_\delta, U] \models \mathsf{ZFC}$*. Let* $\mathcal{F}$ *be such that* $(w, \mathcal{F})$ *is a maximal coherent pair in* N*, and let* $\Sigma = \Sigma^* \cap V^M_\delta$*. Then*

(a) $\langle (N, \in, w, \mathcal{F}, \Sigma), \Sigma^* \rangle$ *is a coarse strategy pair,*
(b) $N \models$ *"*δ *is Woodin",*
(c) for any set $A \in \Gamma \cap \check{\Gamma}$*, there is a term* $\tau \in N$ *such that* $(N, \delta, \tau, \Sigma^*)$ *captures* A*.*

PROOF. Since Σ^* is guided by C_{Γ_1} Q-structures, for some Γ_1 properly beyond Γ, Σ^* witnesses strong unique iterability on M in V. (See Corollary 7.2.8.) By Theorem 7.2.9, Σ^* normalizes well and has strong hull condensation as an (ω_1, ω_1) strategy for N. $\Sigma^* \cap M$ witnesses strong unique iterability for M inside M, and hence $\Sigma^* \cap V^M_\delta$ is definable over V^M_δ, and witnesses strong unique (δ, δ) iterability in N. So by Theorem 7.2.9 in N, $(N, \in, w, \mathcal{F}, \Sigma)$ is a coarse strategy premouse. Thus we have (a).

δ is Γ-Woodin in M, so $N \models$ "δ is Woodin". Finally, let A be in $\Gamma \cap \check{\Gamma}$, and let U_0 and U_1 be trees in $L[V^M_\delta, U]$ such that $p[U_0] = A$ and $p[U_1] = \mathbb{R} - A$ hold in V. Letting τ be the natural term in $N^{\mathrm{Col}(\omega,\delta)}$ for $p[U_0]$, and using that $p[W] \subseteq p[i(W)]$ for any tree W and elementary i in the standard way, we get that $(N, \delta, \tau, \Sigma^*)$ captures A. ⊣

If we are starting with ZFC and very large cardinals, together with $\mathsf{IH}_{\kappa,\delta}$, we can use

THEOREM 9.4.17. *Assume* ZFC *plus* $\mathsf{IH}_{\kappa,\delta}$*, where* $\kappa < \delta < \theta < \alpha$ *for some inaccessible* θ *and* α*. Suppose also that there are* $\lambda < \mu < \kappa$ *such that* λ *is a limit of Woodin cardinals, and* μ *is measurable. Let* $(w, \mathcal{F})$ *be a coherent pair such* *that $\mathcal{F} \subseteq V_\delta$ and $\forall E \in \mathcal{F}(\text{crit}(E) > \kappa)$, and let Ω be the unique $(\theta, \theta, \mathcal{F})$-iteration strategy for V; then $(V_\alpha, \in, w, \mathcal{F}, \Omega)$ is a coarse strategy premouse. Moreover, whenever*

$$\pi\colon (M, \in, u, \mathcal{G}, \Sigma) \to (V_\alpha, \in, w, \mathcal{F}, \Omega)$$

is elementary and M is countable and transitive, then letting

$$\Sigma^* = \Omega^\pi \restriction \text{HC},$$

(a) *$L(\mathbb{R}, \Sigma^*) \models \mathsf{AD}^+$, and*
(b) *$L(\mathbb{R}, \Sigma^*) \models$ "$\langle (M, \in, u, \mathcal{G}, \Sigma), \Sigma^* \rangle$ is a coarse strategy pair."*

PROOF. We showed in Section 7.3 that $(V_\alpha, \in, w, \mathcal{F}, \Omega)$ is a coarse strategy premouse. Since π is elementary, $(M, \in, u, \mathcal{G}, \Sigma)$ is a coarse strategy premouse. $\langle (M, \in, u, \mathcal{G}, \Sigma), \Sigma^* \rangle$ is a coarse strategy pair because strong hull condensation, normalizing well, and pushforward consistency pull back under π. Finally, Σ^* is κ-uB by $\mathsf{IH}_{\kappa,\delta}$.[11] Since we have λ and μ, we get that $L(\mathbb{R}, \Sigma^*) \models \mathsf{AD}^+$. ⊣

Theorem 9.4.17 makes theorems about the constructions of coarse strategy pairs proved assuming AD^+ applicable in the ZFC context. Whatever was true of $\mathbb{C}$ in V_α is true of $\pi^{-1}(\mathbb{C})$ in M.

The proofs of Lemmas 9.4.5, 9.4.6, and 9.4.9 show

THEOREM 9.4.18. *Let $\langle (M, \in, w, \mathcal{F}, \Sigma), \Sigma^* \rangle$ be a coarse strategy pair, and let $\mathbb{C}$ be a $(w, \mathcal{F}, \Sigma)$-construction done in M; then*

(1) *if $\mathbb{C}$ is good at $\langle \nu, k \rangle$ in M and $k \geq 0$, then $(M_{\nu,k+1}, \Omega_{\nu,k+1})$ is a least branch hod pair with scope* HC *in V, and setting $\rho = \rho(M_{\nu,k})$ and $\gamma = (\rho^+)^{M_{\nu,k}}$, $(\Omega_{\nu,k})_{\langle \gamma, 0 \rangle} = (\Omega_{\nu,k+1})_{\langle \gamma, 0 \rangle}$,*
(2) *if $\mathbb{C}$ is good in M at $\langle \nu, k \rangle$ for all $k < \omega$, then $(M_{\nu+1,0}, \Omega_{\nu+1,0})$ is a least branch hod pair with scope* HC *in V, and*
(3) *if ν is a limit ordinal and $\mathbb{C}$ is good at $\langle \nu, -1 \rangle$ in M, then $(M_{\nu,0}, \Omega_{\nu,0})$ is a least branch hod pair with scope* HC.

The pairs referred to in the conclusions of (1)-(3) have scope all of HC, and are hod pairs with that scope in V, even though they come from a construction done in the countable model M. The goodness at $\langle \nu, k \rangle$ in the hypotheses is understood in the sense of M, but since Σ^* normalizes well and has strong hull condensation in V, we can make this step.

We shall show that assuming AD^+, if $\langle (M, \in, w, \mathcal{F}, \Sigma), \Sigma^* \rangle$ is a coarse strategy pair, and $\mathbb{C}$ is a least branch $(w, \mathcal{F}, \Sigma)$-construction in M, then $\mathbb{C}$ is good at all $\langle \nu, k \rangle < \text{lh}(\mathbb{C})$. This is done in Theorem 9.6.11 on the existence of cores, in the Bicephalus Lemma 10.1.3, and in Theorem 10.2.3 on the backgrounding of F^+ by F^* Theorem 9.4.18 then shows that any such construction produces least branch hod pairs with scope HC in V.

[11] The key to this folklore result is that a countable nice tree $\mathcal{T}$ on M is by Σ^* iff $\pi\mathcal{T}$ is continuously illfounded off the branches that it chooses. In fact, one can show directly that Σ^* is κ-homogeneously Suslin.

9.5. Comparison and the hod pair order

We adapt Theorem 8.4.3 to hod pairs.

DEFINITION 9.5.1. Let (M,Σ) and (N,Ω) be mouse pairs with scope H_θ; then

(a) (M,Σ) *iterates past* (N,Ω) iff there is a λ-separated iteration tree $\mathcal{T}$ by Σ on M with last model Q such that $N \trianglelefteq Q$, and $\Sigma_{\mathcal{T},N} = \Omega$.

(b) (M,Σ) *iterates to* (N,Ω) iff there are $\mathcal{T}$ and Q as in (a), and moreover, $N = Q$, and the branch M-to-Q of $\mathcal{T}$ does not drop.

(c) (M,Σ) *iterates strictly past* (N,Ω) iff it iterates past (N,Ω), but not to (N,Ω).

We shall not need to compare pairs with different scopes. The λ-separated tree $\mathcal{T}$ above is completely determined by N and Σ; we must have $E^{\mathcal{T}}_\alpha = F^+$, where F is on the sequence of $\mathcal{M}^{\mathcal{T}}_\alpha$, $\mathcal{M}^{\mathcal{T}}_\alpha || \mathrm{lh}(F) = N | \mathrm{lh}(F)$ (and hence $\mathrm{lh}(F)$ is not an extender active level of N). Since (M,Σ) and (N,Ω) are strategy coherent and pushforward consistent, (M,Σ) iterates past (N,Ω) if and only if

(i) no strategy disagreements show up as we iterate,

(ii) no non-empty extenders from N participate in least disagreements, so that N does not move, and

(iii) N is an initial segment of the final model on the M-side.

The following notation is convenient: let $\mathbb{C}$ be a construction such that $M^{\mathbb{C}}_{\nu,0}$ is extender active; then

$$(M^{\mathbb{C}}_{\nu,-1}, \Omega^{\mathbb{C}}_{\nu,-1}) = (M^{<\nu}, \Omega^{<\nu}).$$

Setting $\gamma = \hat{o}(M^{\mathbb{C}}_{\nu,0})$, we can write this

$$(M^{\mathbb{C}}_{\nu,-1}, \Omega^{\mathbb{C}}_{\nu,-1}) = (M^{\mathbb{C}}_{\nu,0} | \langle \gamma, -1 \rangle, (\Omega^{\mathbb{C}}_{\nu,0})_{\langle \gamma, -1 \rangle}).$$

Adapting the proof of Theorem 8.4.3, we get

THEOREM 9.5.2. *Let (P,Σ) be a strongly stable least branch hod pair with scope H_δ, where δ is inaccessible. Let $(V, \in, w, \mathcal{F}, \Psi)$ be a coarse strategy premouse such that $|P| < \mathrm{crit}(E)$ for all $E \in \mathcal{F}$, and let $\mathbb{C}$ be the maximal $(w, \mathcal{F}, \Psi)$-construction. Let $\langle \nu, k \rangle < \mathrm{lh}(\mathbb{C}) \leq \langle \delta, 0 \rangle$, and suppose that (P,Σ) iterates strictly past $(M^{\mathbb{C}}_{\eta,j}, \Omega^{\mathbb{C}}_{\eta,j})$, for all $\langle \eta, j \rangle <_{\mathrm{lex}} \langle \nu, k \rangle$; then (P,Σ) iterates past $(M^{\mathbb{C}}_{\nu,k}, \Omega^{\mathbb{C}}_{\nu,k})$.*

PROOF SKETCH. Chapter 8 was devoted the proof of this theorem in the case that (P,Σ) is a pure extender pair. The proof for least branch hod pairs is essentially the same, the only difference being that we cannot separate the proof that no extenders on the $M_{\nu,k}$-sequence ever participate in a least disagreement from the proof that no strategy disagreements show up.[12] Instead, both facts are proved in an induction on the construction of the λ-separated $\mathcal{T}$ whereby (P,Σ) iterates

[12] Notice that $i_E(\Sigma) \subseteq \Sigma$ for all $E \in \mathcal{F}$ by Lemma 9.3.13.

past $(M_{\nu,k}, \Omega_{\nu,k})$. For the strategy agreement part, suppose that $M \trianglelefteq M_{\nu,k}$, and α is least such that $M \trianglelefteq \mathcal{M}^{\mathcal{T}}_{\alpha}$. The proof of Theorem 8.4.3 shows that

$$(\Omega_{\nu,k})_M = \Sigma_{\mathcal{T}\restriction\alpha+1,M}.$$

In that proof we were able to assume that $M = M_{\nu,k}$ because we could consider extender disagreements separately, but this is not necessary. For if $M \lhd M_{\nu,k}$, then letting $\pi\colon M \to M_{\eta,j}$ be the resurrection map of $\mathbb{C}$, $\pi = \hat{\imath}^{\mathcal{W}}_{\alpha,\beta}$ for the λ-separated $\mathcal{W}$ by Σ extending $\mathcal{T}\restriction\alpha+1$ whose last model is $M_{\eta,j}$. That such a $\mathcal{W}$ exists is the content of Lemma 8.3.1, whose proof adapts easily to least branch constructions. Since Σ is pullback consistent,

$$\Sigma_{\mathcal{T}\restriction\alpha+1,M} = \Sigma^{\pi}_{\mathcal{W}} = \Omega^{\pi}_{\eta,j} = (\Omega_{\nu,k})_M,$$

as desired.

The proof that no extender from the $M_{\nu,k}$-sequence is part of a least disagreement does not change. ⊣

Remark 9.5.3. Suppose that $F^{\mathbb{C}}_{\nu} \neq \emptyset$. It is not possible that (P,Σ) iterates strictly past $(M^{\mathbb{C}}_{\eta,j}, \Omega^{\mathbb{C}}_{\eta,j})$ for all $\eta < \nu$ and (P,Σ) iterates to $(M^{\mathbb{C}}_{\nu,-1}, \Omega^{\mathbb{C}}_{\nu,-1})$. For if so, then letting $i = i_{F^{\mathbb{C}}_{\nu}}$, in $i(V)$ $(i(P), i(\Sigma))$ would iterate strictly past $(M^{i(\mathbb{C})}_{\nu,0}, \Omega^{i(\mathbb{C})}_{\nu,0})$. But $i(P) = P$, $i(\Sigma) \subseteq \Sigma$, and $(M^{i(\mathbb{C})}_{\nu,0}, \Omega^{i(\mathbb{C})}_{\nu,0}) = (M^{\mathbb{C}}_{\nu,-1}, \Omega^{\mathbb{C}}_{\nu,-1})$ by Lemma 9.4.13. So (P,Σ) iterates strictly past $(M^{\mathbb{C}}_{\nu,-1}, \Omega^{\mathbb{C}}_{\nu,-1})$ in V.

Remark 9.5.4. The work in Section 8.3 on realizing resurrection embeddings as branch embeddings shows that if (P,Σ) iterates to $(M_{\nu,l+1}, \Omega_{\nu,l+1})$, then it iterates strictly past $(M_{\nu,l}, \Omega_{\nu,l})$. This terminology might be a bit confusing at first, because the iteration tree $\mathcal{T}$ from P to $M_{\nu,l+1}$ is an initial segment of the tree $\mathcal{U}$ from P to $M_{\nu,l}$. If $\hat{M}_{\nu,l+1} = \hat{M}_{\nu,l}$ then $\mathcal{T} = \mathcal{U}$, and $(M_{\nu,l}, \Omega_{\nu,l}) \lhd (M_{\nu,l+1}, \Omega_{\nu,l+1}) =$ last model of $\mathcal{U}$ in virtue of the degree change. If $\hat{M}_{\nu,l} \neq \hat{M}_{\nu,l+1}$, then along the branch of $\mathcal{U}$ from P to $M_{\nu,l}$ we dropped at $M_{\nu,l+1}$, from degree $l+1$ to degree l. That drop meant that M iterates past, but not to, $M_{\nu,l}$.

Remark 9.5.5. We do not know whether there can be more than one $\langle \nu, k\rangle$ such that (P,Σ) iterates to $(M^{\mathbb{C}}_{\nu,k}, \Omega^{\mathbb{C}}_{\nu,k})$.

The proof of Theorem 9.5.2 yields a small variant that applies to coarse strategy pairs.

THEOREM 9.5.6. *Let $\langle (M, \in, w, \mathcal{F}, \Sigma), \Sigma^*\rangle$ be a coarse strategy pair, and $\delta = \delta(w)$ be inaccessible. Let (P,Λ) be a strongly stable least branch hod pair with scope* HC, *and suppose that P is countable in M, and that for some term τ, $\langle (M, \in, \delta, \tau), \Sigma^*\rangle$ captures* Code(Λ). *Let $\mathbb{C}$ be a least branch construction $(w, \mathcal{F}, \Sigma)$ construction in M such that $\mathcal{F}^{\mathbb{C}} \subseteq V_{\delta}$ such that $|P|^M < \mathrm{crit}(E)$ for all $E \in \mathcal{F}^{\mathbb{C}}$. Let $\langle \nu, k\rangle < \mathrm{lh}(\mathbb{C}) \leq \langle \delta, 0\rangle$, and suppose that in M, (P,Σ) iterates strictly past $(M^{\mathbb{C}}_{\eta,j}, \Omega^{\mathbb{C}}_{\eta,j}) \cap M$, for all $\langle \eta, j\rangle <_{\mathrm{lex}} \langle \nu, k\rangle$; then in V, (P,Σ) iterates past $(M^{\mathbb{C}}_{\nu,k}, \Omega^{\mathbb{C}}_{\nu,k})$.*

PROOF. M satisfies "$(P,\Lambda\cap M)$ is a least branch hod pair with scope H_δ^M". The comparison trees involved in iterating (P,Λ) to or past $(M_{\nu,k}^{\mathbb{C}},\Omega_{\nu,k}^{\mathbb{C}})$ are all in M. Let $\mathcal{T}$ be this tree, that is, $\mathcal{T}=\mathcal{W}_{\nu,k}^*$ in the notation of the proof of Theorem 8.4.3, and let N be a common initial segment of $M_{\nu,k}^{\mathbb{C}}$ and the last model of $\mathcal{T}$. We must show that $\Lambda_{\mathcal{T},N}=(\Omega_{\nu,k}^{\mathbb{C}})_N$, that is, that the two strategies agree on all stacks in HC^V, not just all stacks in H_δ^M. But Σ^* normalizes well and has strong hull condensation in V, moreover if i is an iteration map by Σ^* in V, then $i(\mathcal{T})$ is by Λ. This is what is required by the proof of Theorem 8.4.3, so $\Lambda_{\mathcal{T},N}=(\Omega_{\nu,k}^{\mathbb{C}})_N$ holds in V. ⊣

In order to apply Theorem 9.5.2 or Theorem 9.5.6, we need to know that, given (P,Σ), there is a coarse strategy premouse whose construction does not break down before it reaches an iterate of (P,Σ). The following lemma will help with that.

LEMMA 9.5.7. *Let (P,Σ) be a strongly stable least branch hod pair with scope H_δ, where δ is inaccessible. Let $\mathbb{C}$ be a least branch construction such that $\mathcal{F}^{\mathbb{C}}\subseteq V_\delta$ and $|P|<\mathrm{crit}(E)$ for all $E\in\mathcal{F}^{\mathbb{C}}$ then for any ν:*

(a) if (P,Σ) iterates strictly past all $(M_{\mu,l}^{\mathbb{C}},\Omega_{\mu,l}^{\mathbb{C}})$ such that $\mu<\nu$, then $\mathbb{C}$ is good at $\langle\nu,-1\rangle$, and

(b) if $k\geq 0$ and (P,Σ) iterates strictly past $(M_{\nu,k}^{\mathbb{C}},\Omega_{\nu,k}^{\mathbb{C}})$, then $\mathbb{C}$ is good at $\langle\nu,k\rangle$.

PROOF. For (a), suppose first that $\mathbb{C}$ reaches $M^{<\nu}$, and that $(M^{<\nu},F)$ is an extender active lpm, and that F^* is the unique $\mathbb{C}$-minimal certificate for F. We must see that $F^+\subseteq F^*$ and $\mathrm{lh}(F)$ is a cardinal in $i_{F^*}(M^{<\nu})$. Let $\mathcal{T}$ be the λ-separated tree whereby (P,Σ) iterates past $(M^{<\nu},F)$, and let $\mathcal{U}=i_{F^*}(\mathcal{T})$. In $\mathrm{Ult}(V,F^*)$, $\mathcal{U}$ is the unique λ-separated tree whereby $(P,i_{F^*}(\Sigma))$ iterates past $i_{F^*}(M^{<\nu})$. Letting $\kappa=\mathrm{crit}(F)$, the proof of Lemma 8.1.2 show that the first extender G used in $\mathcal{U}$ along the branch $(\kappa,i_{F^*}(\kappa))_U$ is compatible with F^*, and moreover, $G=F^+$. So $F^+\subseteq F^*$. Also, $\mathrm{lh}(F)$ is a cardinal in $\mathcal{M}_{i_{F^*}(\kappa)}^{\mathcal{U}}$ because G was used in $\mathcal{U}$, so it is a cardinal in $i_{F^*}(M^{<\nu})$ by the agreement between the two.

Next we consider extender uniqueness. Suppose toward contradiction that $F_0\neq F_1$, and for $i\in\{0,1\}$, $(M^{<\nu},F_i,\emptyset)$ is an lpm, $\mathrm{crit}(F_i)\geq\kappa$, and F_i has a certificate in the sense of Definition 3.1.2. It follows that for $i\in\{0,1\}$ there is a construction $\mathbb{C}_i$ such that $M_{\nu,0}^{\mathbb{C}_i}=(M^{<\nu},F_i,\emptyset)$, and for all $\mu<\nu$ and k, $(M_{\mu,k}^{\mathbb{C}_i},\Omega_{\mu,k}^{\mathbb{C}_i})=(M_{\mu,k}^{\mathbb{C}},\Omega_{\mu,k}^{\mathbb{C}})$. It follows from Theorem 9.5.2 that (P,Σ) iterates past both $(M_{\nu,0}^{\mathbb{C}_0},\Omega_{\nu,0}^{\mathbb{C}_0})$ and $(M_{\nu,0}^{\mathbb{C}_1},\Omega_{\nu,0}^{\mathbb{C}_1})$.[13] This is impossible, for it has to be the same iteration, but $F_0\neq F_1$.

For (b), we have a λ-separated tree $\mathcal{T}$ on P by Σ, with last model $N=\mathcal{M}_\gamma^{\mathcal{T}}$, such that either

(i) $M_{\nu,k}^{\mathbb{C}}$ is a proper initial segment of N, or

(ii) $M_{\nu,k}^{\mathbb{C}}=N$, and $[0,\gamma]_T$ drops (in model or degree).

[13]We do not need that $\mathbb{C}$ is good at $\langle\nu,-1\rangle$ here.

In both cases there has been a drop (in the first, it is a gratuitous one at the end). Since solidity is preserved by iteration maps, $M^{\mathbb{C}}_{\nu,k}$ is solid in either case. ⊣

From this we get

THEOREM 9.5.8. *Assume* AD^+, *and let* (P,Σ) *be a strongly stable lbr hod pair with scope* HC. *Let* $\langle(M,\in,w,\mathcal{F},\Lambda),\Lambda^*\rangle$ *be a coarse strategy pair that captures* Code(Σ) *at* $\delta=\delta(w)$, *and let* $\mathbb{C}$ *be the maximal least branch construction of* M; *then there is an* $\langle\nu,k\rangle$ *such that*

(i) $\nu<\delta$,

(ii) $(M^{\mathbb{C}}_{\nu,k},\Omega^{\mathbb{C}}_{\nu,k})$ *exists (that is, the construction has not broken down yet), and*

(iii) (P,Σ) *iterates to* $(M^{\mathbb{C}}_{\nu,k},\Omega^{\mathbb{C}}_{\nu,k})$, *and strictly past* $(M^{\mathbb{C}}_{\eta,j},\Omega^{\mathbb{C}}_{\eta,j})$ *for all* $\langle\eta,j\rangle<_{\text{lex}}\langle\nu,k\rangle$.

PROOF. If not, then by applying Theorem 9.5.2 and Lemma 9.5.7 in M, we have that $\mathbb{C}$ does not break down at all, and P iterates past $M^{\mathbb{C}}_{\delta,0}$ in M. The proof of universality at a Woodin cardinal in the pure extender premouse case (Lemma 3.1.4) then leads to a contradiction. ⊣

We can now show that under AD^+, any two least branch hod pairs are comparable. First, some notation for cutpoint initial segments:

DEFINITION 9.5.9. For M and N lpms, we write $M\trianglelefteq^{\text{ct}}N$ iff $M\trianglelefteq N$, and whenever E is on the N-sequence and $\text{lh}(E)\geq o(M)$, then $\text{crit}(E)>o(M)$.

THEOREM 9.5.10. *Assume* AD^+, *and let* (P,Σ) *and* (Q,Ψ) *be strongly stable lbr hod pairs with scope* HC; *then there are countable* λ*-separated trees* $\mathcal{T}$ *and* $\mathcal{U}$ *by* Σ *and* Ψ *respectively, with last models* R *and* S *respectively, such that either*

(a) $R\trianglelefteq^{\text{ct}}S$, $\Sigma_{\mathcal{T},R}=\Psi_{\mathcal{U},R}$, *and the branch* P*-to-*R *of* $\mathcal{T}$ *does not drop, or*

(b) $S\trianglelefteq^{\text{ct}}R$, $\Psi_{\mathcal{U},S}=\Sigma_{\mathcal{T},S}$, *and the branch* Q*-to-*S *of* $\mathcal{U}$ *does not drop.*

PROOF. Assume AD^+. The failure of the theorem is a Σ^2_1 statement, so if it fails, there is a counterexample such that Code(Σ) and Code(Ψ) are Suslin and co-Suslin, and thus in $\Gamma\cap\check{\Gamma}$ for some lightface Γ with the scale property. From Theorem 7.2.4 and Theorem 9.4.16 we get a coarse strategy pair $\langle(M,\in,w,\mathcal{F},\Lambda),\Lambda^*\rangle$ such that P and Q are countable in M, and (M,Λ^*) captures both Code(Σ) and Code(Ψ) at $\delta=\delta(w)$. Letting $\mathbb{C}$ be the maximal least branch $(w,\mathcal{F},\Lambda)$- construction of M, we have by Theorem 9.5.8 that there are $\langle\nu,k\rangle$ and $\langle\mu,l\rangle$ such that (P,Σ) iterates to $(M^{\mathbb{C}}_{\nu,k},\Omega^{\mathbb{C}}_{\nu,k})$ and strictly past all earlier pairs, while (Q,Ψ) iterates to $(M^{\mathbb{C}}_{\mu,l},\Omega^{\mathbb{C}}_{\mu,l})$ and strictly past all earlier pairs. If say $\langle\nu,k\rangle\leq_{\text{lex}}\langle\mu,l\rangle$, then $(Q,\Psi))$ iterates past $(M^{\mathbb{C}}_{\nu,k},\Omega^{\mathbb{C}}_{\nu,k})$, and the latter is a λ-separated, nondropping iterate of (P,Σ). By perhaps using one more extender on the Q-side, we can arrange that $M^{\mathbb{C}}_{\nu,k}$ is a cutpoint of the last model. This yields a successful comparison of type (a). If $\langle\mu,l\rangle\leq_{\text{lex}}\langle\nu,k\rangle$, then we have a successful comparison of type (b). ⊣

Theorem 9.5.10 was phrased in the language of mouse pairs in Theorem 9.3.6. We get at once

COROLLARY 9.5.11. *Assume* AD^+, *and let* (M,Ω) *be a strongly stable lbr hod pair with scope* HC; *then every real in* M *is ordinal definable.*

It is natural to ask whether M satisfies "every real is ordinal definable". Borrowing Lemma 11.1.1 from the future, we have

THEOREM 9.5.12. *Assume* AD^+, *and let* (M,Ω) *be an lbr hod pair with scope* HC. *Suppose* $M \models \mathsf{ZFC} +$ "δ *is Woodin*". *Working in* M, *let* uB *be the collection of* δ*-universally Baire sets; then*

$$M \models \textit{there is a } (\Sigma^2_1)^{\mathrm{uB}} \textit{ wellorder of } \mathbb{R}.$$

PROOF. Let $<^*$ be the order of construction in M. Working in M, we claim that for $x, y \in \mathbb{R}$,

$$\begin{aligned} x <^* y \text{ iff } \exists N \exists \Omega [&(N,\Omega) \text{ is a strongly stable lbr hod pair} \\ &\text{with scope HC} \wedge \mathrm{Code}(\Omega) \in \mathrm{uB} \wedge N \models x <^* y]. \end{aligned}$$

For suppose $x <^* y$. Let N be a strongly stable countable initial segment of M such that $N \models x <^* y$. Clearly (N, Ω_N) is an lbr hod pair in M, moreover, by Lemma 11.1.1, Ω_N is δ-uB in M.

Conversely, let (N,Ψ) be a strongly stable lbr hod pair in M such that $N \models x <^* y$ and Ψ is δ-uB in M. Suppose toward contradiction that $y \leq^* x$, and let $S \trianglelefteq M$ be a strongly stable countable initial segment of M such that $S \models y \leq^* x$. Thus (S, Ω_S) is an lbr hod pair in M, and by Lemma 11.1.1, Ω_S is δ-uB in M. We apply Theorem 9.5.2 in M. Letting $\mathbb{C}$ be the maximal construction below δ in M, neither side can iterate past $M_{\langle \delta,0 \rangle}$ because δ is Woodin, so in lev($\mathbb{C}$) there are iterates (P,Σ) and (Q,Λ) of (N,Ψ) and (S,Ω_S). $P \models x <^* y$ and $Q \models y \leq^* x$ because the iteration maps are the identity on $\mathbb{R}$. On the other hand, anticore maps are also the identity on $\mathbb{R}$, so no two levels of $\mathbb{C}$ can have incompatible versions of $<^*$ on $\mathbb{R}$. This is a contradiction. ⊣

Theorem 9.5.12 stands in contrast to the situation with pure extender mice, which can satisfy "not all reals are ordinal definable". (See for example [51].) We shall show in Chapter 11 that $V = \mathrm{HOD}$ holds in any hod mouse with arbitrarily large Woodin cardinals, and in fact, a version of $V = K$ holds true. In this respect, hod mice are more natural than pure extender mice; they are self-contained in a way that pure extender mice with Woodin cardinals are not.

One feature of our comparison process is that we may often use the same extender on both sides. That does not happen in an ordinary comparison of premice by iterating least disagreements. This feature can be awkward. What we gain is that we never encounter strategy disagreements in our comparison process. A comparison process that involves iterating away strategy disagreements as we encounter them (such as the process of [42]) will also often use the same extender on both sides. But such a process (if we knew one in general) might have some advantages. For example, it might be possible to get by without assuming the existence of a Γ-Woodin background universe, where Σ_0 and Σ_1 are in Γ. It might also give better bounds on the lengths of comparisons between uncountable pairs.

For example, Grigor Sargsyan has pointed out that our results leave the following question open. Suppose that (P,Σ) and (Q,Ψ) are pure extender pairs with scope H_δ, where δ is Woodin, and that $o(P) = o(Q) = \omega_1$. Our results show that (P,Σ) and (Q,Ψ) have a common iterate (R,Λ) such that one of P-to-R and Q-to-R does not drop. Can we find such an (R,Λ) with $o(R) = \omega_1$? The standard "weasel comparison" proof shows that one can find iterates (R,Λ_0) and (R,Λ_1) such that $o(R) = \omega_1$, but if one demands that $\Lambda_0 = \Lambda_1$, the question is open, and our strategy-comparison theorem does not answer it.

9.6. The existence of cores

As in the case of ordinary premice, we can formulate our solidity and universality results abstractly, in a theorem about least branch premice having sufficiently good iteration strategies. The problems with ultrapowers that are discontinuous at $\rho^-(M)$ that we faced in Section 4.10 occur here too, and the same indirect argument that we used to deal with them in Section 4.10 will work here too. In this section we shall focus on the case that M is strongly stable, so that such ultrapowers do not occur.[14]

There are new difficulties even in that case, caused by the fact that we cannot compare the relevant phalanx $((M,\Psi),(H,\Psi^\pi),\alpha)$ with (M,Ψ) directly, by iterating away least disagreements, but only indirectly, by iterating both into the levels of some construction. In this section we shall focus on the solution to these new difficulties. It is present in the proof of Lemma 9.6.2 below.

Our proof is simpler than that in Section 4.10 and [35] in one respect, namely, we do not need the Weak Dodd-Jensen property. This is because we shall compare iteration strategies, and that makes the Dodd-Jensen Lemma 9.3.4 applicable. It also leads to a stronger conclusion, in that we get condensation for the full, external strategy of M, not just the part that is coded in the strategy predicate of M.

We begin with a useful lemma on pulling back solidity under ultrapower maps. It is due to F. Schlutzenberg.[15] Recall that an extender E over a premouse M is *weakly amenable to M* iff $P(\kappa_E) \cap M = P(\kappa_E) \cap \mathrm{Ult}(M,E)$, or equivalently, $E_a \cap M|\alpha \in M$ whenever $\alpha < \kappa_E^{+,M}$ and $a \subseteq \mathrm{lh}(E)$ is finite. . If E is close to M, then it is weakly amenable to M.

LEMMA 9.6.1 (Schlutzenberg). *Let M be an lpm and $k = k(M)$. Let $N = \mathrm{Ult}_k(M,E)$, where E is weakly amenable to M, and let $i\colon M \to N$ be the canonical embedding.*

[14] With one minor exception, described in Remark 4.10.3.

[15] See [58, Lemma 2.21] and [59, Lemma 3.7]. See also the proof of [35, Theorem 6.2].

(a) *If* $A \subseteq \alpha < \rho_k(M)$, *and* $A \cap \beta \in M$ *for all* $\beta < \alpha$, *and* $\bigcup_{\beta<\alpha} i(A \cap \beta) \in N$, *then* $A \in M$.
(b) *If* $\alpha < \rho_k(M)$ *and* $Th_1^{N^k}(\sup i``\alpha \cup \{i(q)\}) \in N$, *then* $Th_1^{M^k}(\alpha \cup \{q\}) \in M$.
(c) *If* $\mathrm{crit}(E) < \rho(M)$, *then* $\rho(N) = \sup i``\rho(M)$.

PROOF. Let $\kappa = \mathrm{crit}(E)$. Part (a) is trivial if $\alpha < \kappa$, so suppose $\kappa \leq \alpha$. Let $\theta = \sup i``\alpha$ and

$$\begin{aligned} B &= \bigcup_{\alpha<\rho} i(A \cap \alpha) \\ &= [a, f]_E^{M^k}, \end{aligned}$$

where $k = k(M)$. For $\gamma < \theta$ let

$$X_\gamma = \{u \in [\kappa]^{|a|} \mid B \cap \gamma = i(f)(u) \cap \gamma\}.$$

Clearly $\gamma < \xi$ implies $X_\xi \subseteq X_\gamma$. For $\beta < \alpha$, $X_{i(\beta)} \in E_a$, because $A \cap \beta = f(u) \cap \beta$ for E_a a.e. u, and if $A \cap \beta = f(u) \cap \beta$, then $B \cap i(\beta) = i(f)(u) \cap i(\beta)$. Since the X_γ are decreasing, $X_\gamma \in E_a$ for all $\gamma < \theta$. The function $\gamma \mapsto X_\gamma$ is in $\mathrm{Ult}(M, E)$.

Let $\mu = \mathrm{cof}_0^M(\alpha)$, and suppose first that $\mu \leq \kappa$. Let $g\colon \mu \to \alpha$ witness this. Put

$$\langle \eta, u \rangle \in C \text{ iff } \eta < \mu \wedge u \in X_{i(g)(\eta)}.$$

$C \in \mathrm{Ult}(M, E)$, and $P(\kappa) \cap M = P(\kappa) \cap \mathrm{Ult}(M, E)$ because E is weakly amenable to M, so $C \in M$. But then M can use C, g, and f to compute A, since for $\eta < \mu$

$$Z = A \cap g(\eta) \text{ iff } \exists u(\langle \eta, u \rangle \in C \wedge Z = f(u) \cap g(\eta)).$$

Suppose next that $\kappa < \mu$. It follows that i is continuous at μ and α. But then X_γ is eventually constant as $\gamma \to \theta$. Letting $X \in E_a$ be such that for all sufficiently large γ, $X = X_\gamma$, we have that $B = i(f)(u) \cap \theta$ for all $u \in X$.

But then pick any $u \in X$. For all $\xi < \alpha$,

$$\begin{aligned} \xi \in A &\text{ iff } i(\xi) \in B \\ &\text{ iff } i(\xi) \in i(f)(u) \\ &\text{ iff } \xi \in f(u). \end{aligned}$$

So $f(u) \cap \alpha = A$, so $A \in M$. This proves (a).

Let us prove (b). Let $\alpha < \rho_k(M)$ and for $\beta \leq \alpha$,

$$\begin{aligned} A_\beta &= \mathrm{Th}_1^{M^k}(\beta \cup \{q\}), \\ A &= A_\alpha, \end{aligned}$$

and for $\beta \leq \sup i``\alpha$,

$$\begin{aligned} B_\beta &= \mathrm{Th}_1^{N^k}(\beta \cup \{i(q)\}), \\ B &= B_{\sup i``\alpha}. \end{aligned}$$

Suppose that $B \in N$; we wish to show that $A \in M$. By induction, we may assume that for all $\beta < \alpha$, $A_\beta \in M$.

By part (a), if $i(A_\beta) = B_{i(\beta)}$ for all $\beta < \alpha$, then $A \in M$. So we may assume that $\gamma < \alpha$ is such that $i(A_\gamma) \neq B_{i(\gamma)}$. For any $\beta \leq \alpha$ let $\leq_\beta$ be the natural prewellorder of A_β based on where in M^k the Σ_1 truth is verified. $\leq_\beta$ is coded into A_β, so for $\beta < \alpha$, $\leq_\beta \in M^k$. For $x \in A_\beta$, let

$$A_\beta^x = \{y \mid y <_\beta x\},$$
$$\leq_\beta^x = \leq_\beta \restriction A_\beta^x;$$

then

$$B_{i(\beta)} = \bigcup_{x \in A_\beta} i(A_\beta^x).$$

Since $i(A_\gamma) \neq B_{i(\gamma)}$, $i(\leq_\gamma)$ is a proper end extension of $\bigcup_{x \in A_\gamma} i(A_\gamma^x)$, so $\mathrm{cof}_0^M(\leq_\gamma) = \kappa$. Let $h\colon \kappa \to A_\gamma$ be total, order-preserving, continuous, and cofinal, with $h \in M$. Note that $A_\gamma \subseteq A_\beta$ for all $\beta \geq \gamma$, and $\mathrm{ran}(h)$ is also cofinal in A_β, for $\beta \geq \gamma$.

Now let

$$B = [b, f]_E^{M^k}.$$

We may assume that $\kappa = \min(b)$.We may also assume that for all $u \in [\kappa]^{<\omega}$, $f(u)$ is a Σ_1 theory of parameters in $\alpha \cup \{q\}$. For $\beta < \alpha$, let

$$f_\beta(u) = f(u) \text{ restricted to parameters from } \beta \cup \{q\}.$$

CLAIM 1. *Let $\gamma \leq \beta < \alpha$; then for E_b a.e. u, $f_\beta(u) = A_\beta^{h(\min(u))}$.*

PROOF. For such β, we have

$$\begin{aligned} [b, f_\beta] &= B_{i(\beta)} \\ &= \bigcup_{\xi < \kappa} i(A_\beta^{h(\xi)}) \\ &= i(A_\beta)^{i(h)(\kappa)} \\ &= [b, \lambda u. A_\beta^{h(u_0)}]. \end{aligned}$$

So by Łoś's Theorem, $f_\beta(u) = A_\beta^{h(\min(u))}$ for E_b a.e. u. ⊣

CLAIM 2. *Let $\gamma \leq \beta < \alpha$; then for all $u \in [\kappa]^{|b|}$, $f_\beta(u) \subseteq A_\beta$ iff for E_b a.e. v, $f_\beta(u) \subseteq f_\beta(v)$.*

PROOF. This follows at once from Claim 1. ⊣

Now let

$$R(\beta, u, v) \text{ iff } \gamma \leq \beta < \alpha \wedge f_\beta(u) \subseteq f_\beta(v).$$

Note $R \in M$. Let $R_\beta = \{\langle u, v\rangle \mid R(\beta, u, v)\}$ and $R_\beta^u = \{v \mid R_\beta(u, v)\}$. Clearly

$$\beta < \xi \Longrightarrow R_\beta \supseteq R_\xi.$$

We now break into cases based on whether $\mathrm{cof}_0^M(\alpha) = \kappa$, just as in the proof of (a).

Case 1. $\mathrm{cof}_0^M(\alpha) > \kappa$.

PROOF. Since $R_\beta \subseteq ([\kappa]^{|b|} \times [\kappa]^{|b|})$, R_β is eventually constant as $\beta \to \alpha$. Let S be the eventually constant value. Let

$$X = \{u \mid S^u \in E_b\}.$$

By weak amenability, $X \in M$. But then whenever $R_\beta = S$,

$$A_\beta = \bigcup_{u \in X} f_\beta(u),$$

by Claim 2. Thus we can compute the map $\beta \mapsto A_\beta$ on all sufficiently large $\beta < \alpha$ in M. So $A \in M$. ⊣

Case 2. $\mathrm{cof}_0^M(\alpha) = \kappa$.

PROOF. Let $g \colon \kappa \to \alpha$ witness this. Let

$$\langle \xi, u \rangle \in X \text{ iff } \xi < \kappa \wedge R^u_{g(\xi)} \in E_b.$$

Again, $X \in M$ by weak amenability. But then

$$A = \bigcup_{\langle \xi, u \rangle \in X} f_{g(\xi)}(u).$$

So once again, $A \in M$. ⊣

This finishes the proof of part (b) of the lemma. For (c), letting $\rho = \rho(M)$, we show first that $\sup i``\rho \leq \rho(N)$. For let $\beta < \rho$ and $q = i(f)(a) \in N^k$, where $a \in [\varepsilon(E)]^{<\omega}$. Since $\mathrm{Th}_1^{M^k}(\beta \cup \{f\}) \in M^k$, we have $\mathrm{Th}_1^N(i(\beta) \cup \{i(f)\}) \in N^k$ by the usual proof for solidity witnesses. Choosing $\beta \geq \kappa$, we get that $a \subset i(\beta)$ so using $\mathrm{Th}_1^N(i(\beta) \cup \{i(f)\})$ we can compute $\mathrm{Th}_1^N(i(\beta) \cup \{q\})$ in N^k, as desired.

Let q be such that $\mathrm{Th}_1^{M^k}(\rho \cup \{q\}) \notin M$. By (b), $\mathrm{Th}_1^N(\sup i``\rho \cup \{i(q)\}) \notin N$. Thus $\rho(N) \leq \sup i``\rho$, and we have proved (c). ⊣

We are ready to move to the main argument. The following is a counterpart to Lemma 4.10.2.

LEMMA 9.6.2. *Assume* AD^+, *and let* (M, Ψ) *be a strongly stable lbr hod pair with scope* HC; *then*

(a) M is parameter solid, and

(b) if $\rho(M)$ *is not measurable by the M-sequence, then M is projectum solid.*

PROOF. Let $k = k(M)$, and

$$\rho = \rho_1(M^k) = \rho_{k+1}(M),$$
$$r = p_1(M^k) = p_{k+1}(M).$$

Let q be the longest solid initial segment of r in its decreasing enumeration, and let

$$r = s \cup q,$$

where either $s = \emptyset$ or $\max(s) < \min(q)$. Let

$$\alpha_0 = \text{least } \beta \text{ such that } \mathrm{Th}^M_{k+1}(\beta \cup q) \notin M.$$
$$= \text{least } \beta \text{ such that } \mathrm{Th}^{M^k}_1(\beta \cup q) \notin M.$$

We may assume that $\alpha_0 \in M^k$, as otherwise $r = \emptyset$ and $\alpha_0 = \rho_{k+1}(M) = \rho_k(M)$, in which case the theorem is trivially true.[16] Let

$$H = \mathrm{cHull}^M_{k+1}(\alpha_0 \cup q),$$

and let

$$\pi\colon H \to M$$

be the anticollapse map. Equivalently, $H = \mathrm{Dec}((P,B))$, where

$$(P,B) = \mathrm{cHull}^{M^k}_1(\alpha_0 \cup q).$$

Note that $k(H) = k(M) = k$, and π is cofinal and elementary by the Downward Extension Lemma.[17] Also, $\eta^M_k \in \mathrm{ran}(\pi)$, so $\pi^{-1}(\eta^M_k) = \eta^H_k$, so H is stable.

CLAIM 0. *(a) If $q = r$, then $\rho = \alpha_0$.*
(b) If $q \neq r$, then $\pi \neq$ id, and $\rho < \alpha_0 \leq \mathrm{crit}(\pi) \leq \max(s)$.
(c) $H \models \alpha_0$ is a cardinal.

PROOF. The proof is the same as the proof of Claim 0 in the proof of Lemma 4.10.2. ⊣

In view of Claim 0, we may assume that $\pi \neq$ id, and

$$\mathrm{crit}(\pi) < \rho_k(H).$$

Again, the proof is the same as that in Lemma 4.10.2.

As in the proof of Lemma 4.10.2, we show now that if $q \neq r$, then $\mathrm{Th}^M_{k+1}(\alpha_0 \cup q) \in M$. This implies $q = r$, so r is solid over M^k, and (H, Ψ^π) is the strong core of (M, Ψ). We then show that r is universal over M^k and that M is projectum solid. The argument is based on comparing the phalanx $((M,\Psi),(H,\Psi^\pi),\alpha_0)$ with M.

In the comparison argument, we iterate both M and (M,H,α_0) into the models of a common background construction. Additional phalanxes (N,L,β) may appear above (M,H,α_0) in its tree. The relevant background construction is the following. By Theorems 7.2.4 and 9.4.16 we have a coarse strategy pair $\langle (N^*, \in, w, \mathcal{F}, \Phi), \Phi^* \rangle$ such that M is countable in N^* and and (N^*, Φ^*) captures $\mathrm{Code}(\Psi)$ at $\delta = \delta(w)$. We shall work entirely within N^* for most of the proof. Let $\mathbb{C}$ be the maximal least branch $(w, \mathcal{F}, \Phi)$- construction of N^*, and let

$$M_{\eta,l} = M^{\mathbb{C}}_{\eta,l} \text{ and } \Omega_{\eta,l} = \Omega^{\mathbb{C}}_{\eta,l},$$

16 For $X \subseteq M^k$, $\mathrm{Th}^{M^k}_1(X)$ and $\mathrm{Th}^M_{k+1}(X)$ are very simply interdefinable.

17 Let $\sigma\colon (P,B) \to M^k$ be the anticollapse map. σ is cofinal because $\mathrm{Th}_1(\alpha_0 \cup q) \notin M$.

By Theorem 9.5.8 we can fix ν_0, $\mathcal{U}$, and $\mathcal{U}_{\langle \nu,l\rangle}$ such that[18]

- (M,Ψ) iterates to $(M_{\nu_0,k},\Omega_{\nu_0,k})$ via the λ-separated tree $\mathcal{U}_{\nu_0,k}$, and
- for all $\langle \nu,l\rangle <_{\text{lex}} \langle \nu_0,k\rangle$, (M,Ψ) iterates strictly past $(M_{\nu,l},\Omega_{\nu,l})$ via the λ-separated tree $\mathcal{U}_{\nu,l}$.

We now want to compare $((M,\Psi),(H,\Psi^{\pi}),\alpha_0)$ with each $(M_{\nu,l},\Omega_{\nu,l})$ for

$$\langle \nu,l\rangle \leq_{\text{lex}} \langle \eta_0,k\rangle.$$

For each such $\langle \nu,l\rangle$ we shall define a λ-separated "pseudo iteration tree" $\mathcal{S}_{\nu,l}$ on (M,H,α_0). We shall have complete strategies attached to the models of $\mathcal{S}_{\nu,l}$, and as before, the key will be that no strategy disagreements with $\Omega_{\nu,l}$ show up, and that $M_{\nu,l}$ does not move.

The rules for forming $\mathcal{S}_{\nu,l}$ will be the usual ones for iterating a phalanx, with the exception that at certain steps we are allowed to move the whole phalanx up. (We do not throw away the phalanxes we had before, we just create a new one.) Whenever we introduce a new phalanx, we continue the construction of $\mathcal{S}$ by looking at the least disagreement between its second model and $M_{\nu,l}$.

Fix ν and l. Let us write $\mathcal{U} = \mathcal{U}_{\nu,l}$. At the same time that we define $\mathcal{S} = \mathcal{S}_{\nu,l}$, we shall copy it to a λ-separated tree $\mathcal{T} = \mathcal{T}_{\nu,l}$ on M that is by Ψ. Let

$$P_\theta = \mathcal{M}^{\mathcal{S}}_\theta,$$
$$P^*_\theta = \mathcal{M}^{\mathcal{T}}_\theta,$$
$$Q_\theta = \mathcal{M}^{\mathcal{U}}_\theta,$$

and

$$\pi_\theta : P_\theta \to P^*_\theta$$

be the copy map. π_θ will be nearly elementary as a map into some $N_\theta \trianglelefteq P^*_\theta$, but we shall generally suppress mention of N_θ, as above. We allow a bit of padding in $\mathcal{T}$; that is, occasionally $P^*_\theta = P^*_{\theta+1}$. The (possibly partial) branch embeddings of $\mathcal{S},\mathcal{T}$, and $\mathcal{U}$ are

$$i_{\alpha,\beta} = \hat{\imath}^{\mathcal{S}}_{\alpha,\beta},$$
$$i^*_{\alpha,\beta} = \hat{\imath}^{\mathcal{T}}_{\alpha,\beta},$$

and

$$j_{\alpha,\beta} = \hat{\imath}^{\mathcal{U}}_{\alpha,\beta}.$$

The copy maps π_θ will have the usual commutativity and agreement properties. We should write $\pi^{\nu,l}_\theta$, etc., but will omit the superscripts when we can. The strategies attached to P_θ, P^*_θ, and Q_θ are

$$\Sigma^*_\theta = \Psi_{\mathcal{T}\restriction\theta+1},$$

[18]As in Section 9.5, the iteration strategies involved here extend to V, and the equalities between them hold in V, not just N^*.

$$\Sigma_\theta = (\Sigma_\theta^*)^{\pi_\theta},$$

and

$$\Lambda_\theta = \Psi_{\mathcal{U}\restriction\theta+1}.$$

$(P_\theta,\Sigma_\theta),(P_\theta^*,\Sigma_\theta^*)$, and $(Q_\theta,\Lambda_\theta)$ will be lbr hod pairs. Finally, we shall have ordinals $\varepsilon_\theta = \varepsilon_\theta^{\mathcal{S}}$ for each $\theta < \mathrm{lh}(\mathcal{S})$ that measure agreement between the models of $\mathcal{S}$, and tell us which one we should apply the next extender to.[19]

We start with

$$P_0 = M, P_1 = H, \text{ and } \varepsilon_0 = \alpha_0,$$

and

$$P_0^* = P_1^* = M.$$

We let π_0 be the identity and let $\pi_1 = \pi$. Since $\mathrm{crit}(\pi_1) \geq \alpha_0 = \varepsilon_0$, π_0 and π_1 agree up to the relevant exchange ordinal. We think of 0 and 1 as distinct roots of $\mathcal{S}$. One additional root will be created each time we move a phalanx up, and only then.

As we proceed, we define what it is for a node θ of $\mathcal{S}$ to be *unstable*.[20] We shall have that if θ is unstable, then $0 \leq_S \theta$ and $[0,\theta]_S$ does not drop. We then set

$$\alpha_\theta = \sup i_{0,\theta}\text{``}\alpha_0.$$

The idea is that θ is unstable iff $(P_\theta, P_{\theta+1}, \alpha_\theta)$ is a phalanx that we are allowed to move up. If θ is unstable, then $\theta+1$ is stable, and a new root in $\mathcal{S}$, that is, there are no $\xi <_S \theta+1$. These are the only roots, except for 0. Our first unstable node is 0, and 1 is stable.

The padding in $\mathcal{T}$ corresponds exactly to the unstable nodes of $\mathcal{S}$, in that θ is unstable iff $P_\theta^* = P_{\theta+1}^*$.

We maintain by induction on the construction of $\mathcal{S}$ that the current last model is stable, and conversely, every stable model is the last model at some stage. So really, we are defining $\mathcal{S}^\eta$, which has a stable last model, by induction on η, sometimes adding two models at once, and taking $\mathcal{S} = \bigcup_\eta \mathcal{S}^\eta$. We shall suppress the superscript η, however. All extenders used in $\mathcal{S}$ will be taken from stable nodes. We maintain by induction

Induction hypotheses. If θ is unstable, then

(i) $0 \leq_S \theta$, the branch $[0,\theta]_S$ does not drop in model or degree,

(ii) $\varepsilon_\theta \leq \alpha_\theta \leq \rho_k(P_\theta)$,

(iii) every $\xi \leq_S \theta$ is unstable,

---

[19]Earlier we defined $\varepsilon_\alpha^{\mathcal{T}}$, for $\mathcal{T}$ a λ-separated, to be the sup of the lengths of extenders used on the branch $[0,\alpha)_T$. Our use of the notation now is a different one. Psuedo-trees are not normal trees, so there is not a literal conflict. But if $\mathcal{S}$ is a pseudo-tree, then $\varepsilon_\alpha^{\mathcal{S}}$ corresponds to $\varepsilon_{\alpha+1}^{\mathcal{T}}$ in the λ-separated case, and not to $\varepsilon_\alpha^{\mathcal{T}}$.

[20]The terminology has nothing to do with stability of premice.

(iv) there is a τ such that $(P_\theta, \Sigma_\theta) = (Q_\tau, \Lambda_\tau)$, $[0,\tau]_U$ does not drop in model or degree, and $i_{0,\theta} = j_{0,\tau}$,

(v) $\rho(P_\theta) = \sup i_{0,\theta}``\rho$ and $i_{0,\theta}(q)$ is an initial segment of $p(P_\theta)$, and

(vi) $\alpha_\theta =$ least β such that $\mathrm{Th}_1^{P_\theta^{pk}}(\beta \cup i_{0,\theta}(q)) \notin P_\theta$.

Item (ii) explains why $[0,\theta]_S$ does not drop in model or degree, for an extender applied to P_θ must have critical point $< \varepsilon_\theta$, and $k = k(M)$.

The following notation will be useful. For any node γ of $\mathcal{S}$, let

$$\mathrm{st}(\gamma) = \text{least stable } \theta \text{ such that } \theta \leq_S \gamma,$$

and

$$\mathrm{rt}(\gamma) = \begin{cases} \mathcal{S}\text{-pred}(\mathrm{st}(\gamma)) & \text{if } \mathcal{S}\text{-pred}(\mathrm{st}(\gamma)) \text{ exists} \\ \mathrm{st}(\gamma) & \text{otherwise.} \end{cases}$$

Note that if θ is unstable and $\theta + 1 \leq_S \gamma$, then $\mathrm{rt}(\gamma) = \theta + 1$. If θ is the largest unstable ordinal $\leq_S \gamma$, then $\mathrm{rt}(\gamma) = \theta$. Finally, if there are unstable ordinals $\leq_S \gamma$, but no largest one, then $\mathrm{rt}(\gamma) = \sup\{\theta \mid \theta \leq_S \gamma \text{ and } \theta \text{ is unstable }\}$.

The construction of $\mathcal{S}$ ends iff we reach a stable θ such that

(1) $M_{\nu,l} \lhd P_\theta$, or
(2) $P_\theta \unlhd M_{\nu,l}$, and $[\mathrm{rt}(\theta), \theta]_S$ does not drop in model or degree.

In both cases, the full external strategies will be lined up, by Lemma 9.6.5 below. Case 2 constitutes a successful comparison of $((M,\Psi),(H,\Psi^\pi),\alpha_0)$ with (M,Ψ), which iterated past $(M_{\nu,l},\Omega_{\nu,l})$ via $\mathcal{U}$. So in case 2, we leave $\mathcal{S}_{\eta,m}$ undefined for all $\langle \eta, m\rangle >_{\text{lex}} \langle \nu, l\rangle$. In case 1 our phalanx has iterated strictly past $(M_{\nu,l},\Omega_{\nu,l})$, and so we go on to define $\mathcal{S}_{\nu,l+1}$.

In both case 1 and case 2, the last model of $\mathcal{S}$ is P_θ.

CLAIM 1. *Induction hypotheses (i)-(vi) hold for $\theta = 0$ and $\theta = 1$.*

PROOF. (i)-(vi) are trivial for $\theta = 0$, and vacuous for $\theta = 1$. ⊣

The rules for extending $\mathcal{S}$ at successor steps are the following. Suppose P_γ is the current last model of $\mathcal{S}$, so that γ is stable, and suppose the construction is not required to stop by (1) or (2) above. So we have a least disagreement between $(P_\gamma, \Sigma_\gamma)$ and $(M_{\nu,l},\Omega_{\nu,l})$. Suppose the least disagreement involves only an extender E from the P_γ sequence. By this we mean: letting $\tau = \mathrm{lh}(E)$,

- $M_{\nu,l}|\langle \tau, 0\rangle = P_\gamma|\langle \tau, -1\rangle$, and
- $(\Omega_{\nu,l})_{\langle \tau,0\rangle} = (\Sigma_\gamma)_{\langle \tau,-1\rangle}$.

Lemma 9.6.5 below proves that this is the case. Set

$$E_\gamma^{\mathcal{S}} = E^+,$$
$$E_\gamma^{\mathcal{T}} = \pi_\gamma(E^+),$$

and

$$\varepsilon_\gamma = \mathrm{lh}(E).$$

Letting ξ be least such that $\text{crit}(E) < \varepsilon_\xi$, we declare that S-pred$(\gamma+1) = \xi$.

We shall avoid the anomalous cases described in Remarks 4.10.3 and 4.10.4 by assuming

Simplifying assumption: If S-pred$(\gamma+1)$ is unstable, then $\gamma+1 \notin D^S$.

Making this assumption, we set

$$\begin{aligned} N = \mathcal{M}^{*,\mathcal{S}}_{\gamma+1} &= \text{least } S \trianglelefteq P_\xi \text{ such that } \rho(S) \le \text{crit}(E), \\ P_{\gamma+1} &= \text{Ult}(N, E^+), \\ P^*_{\gamma+1} &= \text{Ult}(\pi_\xi(N), \pi_\gamma(E^+)), \end{aligned}$$

and

$$\pi_{\gamma+1} = \text{copy map associated to } (\pi_\xi, \pi_\gamma, E^+).$$

The simplifying assumption, coupled with our assumption that M is strongly stable, implies that the branch embeddings of $\mathcal{S}$, $\mathcal{T}$, and $\mathcal{U}$ are all elementary and exact, and the copy maps π_η are all elementary from P_η to P^*_η. The anomalies that we are avoiding by making this assumption can be dealt with just as we did in the proof of Lemma 4.10.2. See Remarks 4.10.3 and 4.10.4.[21]

If ξ is stable or $N \triangleleft P_\xi$, then we declare $\gamma+1$ to be stable, and we just go on now to look at least disagreement between $P_{\gamma+1}$ and $M_{\nu,l}$. Nothing unusual has happened.[22] Induction hypotheses (i)-(vi) concern only unstable nodes, so they are vacuously true at $\theta = \gamma+1$.

Now suppose ξ is unstable and $N = P_\xi$. We look to see whether there is a β such that $(P_{\gamma+1}, \Sigma_{\gamma+1}) = (Q_\beta, \Lambda_\beta)$, $[0,\beta]_U$ does not drop in model or degree, and $i_{0,\gamma+1} = j_{0,\beta}$. If not, then again we declare $\gamma+1$ to be stable, and go on. Our new last node $\gamma+1$ is stable, so (i)-(vi) are vacuous for $\theta = \gamma+1$.

Finally, suppose that there is a β such that $(P_{\gamma+1}, \Sigma_{\gamma+1}) = (Q_\beta, \Lambda_\beta)$, $[0,\beta]_U$ does not drop in model or degree, and $i_{0,\gamma+1} = j_{0,\beta}$. We then declare $\gamma+1$ to be unstable, and $\gamma+2$ to be stable, and set

$$P_{\gamma+2} = \text{cHull}^{P_{\gamma+1}}(\alpha_{\gamma+1} \cup i_{0,\gamma+1}(q)).$$

Let also $\sigma_{\gamma+1} : P_{\gamma+2} \to P_{\gamma+1}$ be the anticollapse map, and

$$P^*_{\gamma+2} = P^*_{\gamma+1},$$

and

$$\pi_{\gamma+2} = \pi_{\gamma+1} \circ \sigma_{\gamma+1}.$$

[21] If ξ is unstable but $\gamma+1 \in D^S$, then the definition of $P^*_{\gamma+1}$ does not change. However, as in Remark 4.10.3, if N has type 1B and $\eta_{k(N)}(N) = \text{crit}(E)$, then in order to avoid type 2 premice, we set $P_{\gamma+1} = \text{Ult}_{k(N)}(\mathfrak{C}_n(N), \pi_{\gamma+1}(E^+))$. Also, $\pi_{\gamma+1}$ now maps into $i_{\xi,\gamma+1}(N)$ or $i_{\xi,\gamma+1}(\mathfrak{C}_{k(N)})$, as appropriate. Finally, it is possible that $P_{\gamma+1}$ is not even a potential lpm, because it is extender active but does not satisfy the Jensen initial segment condition. That happens iff ξ is unstable, and for some F from the P_ξ-sequence, $\alpha_\xi = \text{lh}(F)$ and $\text{crit}(E) = \lambda(F)$. In this case $k(N) = 0$, so it is distinct from the case in which we replace N by its strong core.

[22] In general, if ξ is unstable and $\gamma+1 \in D^S$, then $\gamma+1$ is stable.

Our new last node is stable. Our induction hypothesis (i) holds for $\theta = \gamma + 1$ because it held for $\theta = \xi$, and because $\varepsilon_\xi \leq \alpha_\xi$. (iii) is clear. For (ii), we must define $\varepsilon_{\gamma+1}$. Suppose that there is a least disagreement between $P_{\gamma+2}$ and $M_{\nu,l}$, and Lemma 9.6.5 applies to it, so it involves only some F from the sequence of $P_{\gamma+2}$. If there is no such F, $P_{\gamma+2}$ is the last model of $\mathcal{S}$, and we just leave $\varepsilon_{\gamma+1}$ undefined. If F exists, we set

$$\varepsilon_{\gamma+2} = \mathrm{lh}(F) = \varepsilon(F^+),$$

and

$$\varepsilon_{\gamma+1} = \inf(\varepsilon_{\gamma+2}, \alpha_{\gamma+1}).$$

This insures that (ii) holds at $\theta = \gamma + 1$. It also insures that $\varepsilon_\gamma < \varepsilon_{\gamma+1} \leq \varepsilon_{\gamma+2}$, so that the ε's remain nondecreasing, which is something we want. $\pi_{\gamma+2}$ agrees with $\pi_{\gamma+1}$ on $\varepsilon_{\gamma+1}$, as required. $(P_{\gamma+1}, P_{\gamma+2}, \alpha_{\gamma+1})$ is the result of moving up the phalanx $(P_\xi, P_{\xi+1}, \alpha_\xi)$.

Remark 9.6.3. It is possible that $\varepsilon_{\gamma+1} = \varepsilon_{\gamma+2}$ and $\mathrm{lh}(F) < \alpha_{\gamma+1}$. Indeed, this will happen a lot. In this case, if $\varepsilon_\gamma \leq \mathrm{crit}(F)$, then F will immediately move the phalanx $(P_{\gamma+1}, P_{\gamma+2}, \alpha_{\gamma+1})$ up again. Since $\varepsilon_{\gamma+1} = \varepsilon_{\gamma+2}$, no extender ever gets applied to $P_{\gamma+2}$. It is a "dead node". The phalanx $(P_{\gamma+1}, P_{\gamma+2}, \alpha_{\gamma+1})$ may get moved up repeatedly, along various branches, but that does not really involve $P_{\gamma+2}$. After contributing F, it became irrelevant.

Induction hypothesis (iv) is clear. Let us verify (v) and (vi).

CLAIM 2. *Let $\theta = \gamma + 1$ be unstable; then*

(a) $\rho(P_\theta) = \sup i_{0,\theta}``\rho$ and $i_{0,\theta}(q)$ is an initial segment of $p(P_\theta)$, and

(b) $\alpha_\theta =$ least β such that $\mathrm{Th}_1^{P_\theta^{pk}}(\beta \cup i_{0,\theta}(q)) \notin P_\theta$.

PROOF. Let $i = i_{\xi,\theta} = i_{E^+}^{P_\xi}$. One can show that E^+ is close to P_ξ. Thus if $\rho(P_\xi) \leq \mathrm{crit}(E^+)$, then by the proof of Lemma 4.3.11, $\rho(P_\theta) = \rho(P_\xi) = \sup i_{0,\xi}``\rho = \sup i_{0,\theta}``\rho$. Moreover, $i_{0,\xi}(q)$ is a solid initial segment of $p(P_\xi)$, so $i(i_{0,\xi}(q))$ is a solid initial segment of $p(P_\theta)$.

On the other hand, if $\mathrm{crit}(E^+) < \rho(P_\xi)$, then $\rho(P_\theta) = \sup i``\rho(P_\xi)$ by Lemma 9.6.1(c). $i_{0,\theta}(q)$ is solid over P_θ by the usual argument on preservation of witnesses. But by Lemma 9.6.1(b), $\mathrm{Th}_1^{P_\theta^{pk}}(\rho(P_\theta) \cup i(p(P_\xi))) \notin P_\theta$, and $i_{0,\theta}(q)$ is an initial segment of $i(p(P_\xi))$ by induction. It follows that $i_{0,\theta}(q)$ is an initial segment of $p(P_\theta)$.[23]

This proves (a). Part (b) of the claim is an immediate consequence of Lemma 9.6.1(b) and our induction hypothesis (vi) at ξ. ⊣

[23] It does not follow that $i(p(P_\xi)) = p(P_\theta)$. The standard parameter could move down in its non-solid part.

Now let θ be a limit ordinal, and let $b = \Psi(\mathcal{T} \restriction \theta)$ be the branch of $\mathcal{T}$ chosen by Ψ. b may have pairs of the form $\gamma, \gamma+1$ in it where $P^*_\gamma = P^*_{\gamma+1}$; this occurs precisely when $\gamma \in b$ is unstable. By construction, the set of such pairs is an initial segment of b that is closed as a set of ordinals.

Suppose first

Case 1. There is a largest $\eta \in b$ such that η is unstable.

Fix this η. There are two subcases.

Subcase 1(a). For all $\gamma \in b - (\eta+1)$, $\mathrm{rt}(\gamma) = \eta + 1$. In this case, $b - (\eta+1)$ is a branch of $\mathcal{S}$. We let $\mathcal{S}$ choose this branch, that is,

$$[\eta+1, \theta)_S = b - (\eta+1),$$

and let P_θ be the direct limit of the P_γ for $\gamma \in b - (\eta+1)$ sufficiently large. The branch embeddings $\hat{\imath}^S_{\gamma,\theta}$, for $\gamma \geq \eta$ in b, are as usual. $\pi_\theta : P_\theta \to P^*_\theta$ is given by the fact that the copy maps commute with the branch embeddings. We declare θ to be stable.

Subcase 1(b). For all $\gamma \in b - (\eta+1)$, $\mathrm{rt}(\gamma) = \eta$. We let $\mathcal{S}$ choose

$$[0, \theta)_S = (b - \eta) \cup [0, \eta]_S,$$

and let P_θ be the direct limit of the P_γ for $\gamma \in b$ sufficiently large. The branch embeddings $\hat{\imath}^S_{\gamma,\theta}$, for $\gamma \geq \eta$ in b, are as usual. $\pi_\theta : P_\theta \to P^*_\theta$ is given by the fact that the copy maps commute with the branch embeddings. Again, we declare θ to be stable.

In this case, θ is stable, so (i)-(vi) still hold.

Case 2. There are boundedly many unstable ordinals in b, but no largest one.

Let η be the sup of the unstable ordinals in b. We let $\mathcal{S}$ choose

$$[0, \theta_S] = (b - \eta) \cup [0, \eta]_S,$$

etc. Again, we declare θ to be stable, and (i)-(vi) still hold.

Case 3. There are arbitrarily large unstable ordinals in b.

In this case b is a disjoint union of pairs $\{\gamma, \gamma+1\}$ such that γ is unstable and $\gamma+1$ is stable. That is, in $\mathcal{S}$ we have been moving our phalanx up all along b. We set

$$[0, \theta)_S = \{\xi \in b \mid \xi \text{ is unstable }\},$$

and let P_θ be the direct limit of the P_ξ for $\xi \in b$ unstable. There is no dropping of any kind in $[0, \theta)_S$. The branch embeddings $i^S_{\gamma,\theta}$ and the copy map π_θ are as usual. If P_θ is not a model of $\mathcal{U}$, then we declare θ to be stable. Otherwise, we declare θ to be unstable, and set

$$P_{\theta+1} = \mathrm{cHull}^{P_\theta}(\alpha_\theta \cup i_{0,\theta}(q)).$$

ε_θ is defined as it was in the unstable successor case: first we set $\varepsilon_{\theta+1} = \text{lh}(E^{\mathcal{S}}_{\theta+1})$, then set

$$\varepsilon_\theta = \inf(\varepsilon_{\theta+1}, \alpha_\theta).$$

Let also $\sigma_\theta : P_{\theta+1} \to P_\theta$ be the anticollapse map, and

$$P^*_{\theta+1} = P^*_\theta,$$

and

$$\pi_{\theta+1} = \pi_\theta \circ \sigma_\theta.$$

$\pi_{\theta+1}$ agrees with π_θ on ε_θ, as desired.

(i)-(iv) are clear. Items (v) and (vi) are routine.

We shall use the following proposition in the next section.

PROPOSITION 9.6.4. *Let θ be a limit ordinal such that θ is stable in $\mathcal{S}_{\nu,l}$, but every $\xi <_{\mathcal{S}_{\nu,l}} \theta$ is unstable in $\mathcal{S}_{\nu,l}$; then* $\text{cof}(\theta) = \omega$.

PROOF. Let $t = e^{\mathcal{S}_{\nu,l}}_\theta$ be the sequence of extenders used in $[0,\theta)_S$, and $\lambda = \text{dom}(t)$. By hypothesis, $t \restriction \eta \in \mathcal{U}^{\text{ext}}_{\nu,l}$ for all $\eta < \lambda$, but $t \notin \mathcal{U}^{\text{ext}}_{\nu,l}$. For $\eta < \lambda$, let ξ_η be such that

$$t \restriction \eta = e^{\mathcal{U}_{\nu,l}}_{\xi_\eta}.$$

Then $\eta < \gamma$ implies $e^{\mathcal{U}}_{\xi_\eta} \subseteq e^{\mathcal{U}}_{\xi_\gamma}$, and hence $\xi_\eta <_U \xi_\gamma$. Letting $\mu = \sup(\{\xi_\eta \mid \eta < \lambda\})$, and b be the branch of $\mathcal{U} \restriction \mu$ determined by the ξ_η's, we have that t is the branch extender of b in $\mathcal{U}$, so $b \neq e^{\mathcal{U}}_\mu$, so $b \neq [0,\mu)_U$. This implies $\text{cof}(\mu) = \omega$, so $\text{cof}(\lambda) = \omega$, so $\text{cof}(\theta) = \omega$, as desired. ⊣

This finishes our construction of the pseudo-tree $\mathcal{S}_{\nu,l}$, and its lift $\mathcal{T}_{\nu,l}$. Notice that every extender used in $\mathcal{S}$ was taken from the sequence of a stable node. Every stable node, except the last model of $\mathcal{S}$, contributes exactly one extender to be used. The last model of $\mathcal{S}$ is stable.

Recall that we assumed that the construction never reached a strategy disagreement between the current model of $\mathcal{S}_{\nu,l}$ and $(M_{\nu,l}, \Omega_{\nu,l})$, and that the extender disagreements involved only empty extenders on the $M_{\nu,l}$ side. Let us record this in a lemma.

LEMMA 9.6.5. *Let $\gamma < \text{lh}(\mathcal{S})$, where $\mathcal{S} = \mathcal{S}_{\nu,l}$ is defined as above; then either*

(1) $(\mathcal{M}^{\mathcal{S}}_\gamma, \Sigma_\gamma) \trianglelefteq (M_{\nu,l}, \Omega_{\nu,l})$, or

(2) $(M_{\nu,l}, \Omega_{\nu,l}) \triangleleft (\mathcal{M}^{\mathcal{S}}_\gamma, \Sigma_\gamma)$, or

(3) there is a nonempty extender E on the $\mathcal{M}^{\mathcal{S}}_\gamma$ sequence such that, setting $\tau = \text{lh}(E)$,

(i) $\dot{E}^{M_{\nu,l}}_\tau = \emptyset$, and

(ii) $(\Sigma_\gamma)_{\langle \tau, -1 \rangle} = (\Omega_{\nu,l})_{\langle \tau, 0 \rangle}$.

So far as we can see, the lemma can only be proved by going back through the proof of Theorem 8.4.3, and extending the arguments so that they apply to $\mathcal{S}_{\nu,l}$. That involves generalizing strong hull condensation to pseudo-trees like $\mathcal{S}$, and normalizing well to stacks $\langle \mathcal{S},\mathcal{U}\rangle$, where $\mathcal{U}$ is a normal tree on the last model of $\mathcal{S}$. Then we need to run the construction of Theorem 8.4.3, showing that $W(\mathcal{S},\mathcal{U}^\frown b)$ is a pseudo-hull of $i^*_b(\mathcal{S})$, where b is the branch of $\mathcal{U}$ chosen by $\Omega_{\nu,l}$. There is nothing fundamentally new in these arguments, but it does not seem possible to get by with quoting our earlier results. We therefore defer the proof of Lemma 9.6.5 to the next section.

CLAIM 3. *For some $\langle \nu,l\rangle \leq_{\text{lex}} \langle \nu_0,k\rangle$, the construction of $\mathcal{S}_{\nu,l}$ stops for reason 2; that is, $\text{lh}(\mathcal{S}_{\nu,l}) = \theta+1$, where $\mathcal{M}_\theta^{\mathcal{S}_{\nu,l}} \trianglelefteq M_{\nu,l}$, and the branch of $\mathcal{S}_{\nu,l}$ ending at $\mathcal{M}_\theta^{\mathcal{S}_{\nu,l}}$ does not drop in model or degree.*

PROOF. If not, then the construction of $\mathcal{S} = \mathcal{S}_{\nu_0,k}$ must reach some P_θ such that $M_{\nu_0,k}$ is a proper initial segment of P_θ. By Lemma 9.6.5, the strategies agree, that is,

$$(M_{\nu_0,k},\Omega_{\nu_0,k}) \lhd (P_\theta,\Sigma_\theta).$$

But $M_{\nu_0,k},\Omega_{\nu_0,k})$ is an iterate of (M,Ψ) via a branch of $\mathcal{U}_{\nu_0,k}$ that does not drop. Letting j be the iteration map,

$$j\colon (M,\Psi) \to (M_{\nu_0,k},\Omega_{\nu_0,k})$$

is elementary in the category of mouse pairs, because Ψ is pullback consistent. Moreover

$$\pi_\theta\colon (P_\theta,\Sigma_\theta) \to (P^*_\theta,\Sigma^*_\theta)$$

is nearly elementary in the category of mouse pairs by construction. Letting $Q = \pi_\theta(M_{\nu_0,k})$, we get that

$$\pi_\theta \circ j\colon (M,\Psi) \to (Q,(\Sigma^*_\theta)_Q)$$

is nearly elementary in the category of mouse pairs, and $(Q,(\Sigma^*_\theta)_Q)$ is an initial segment of an iterate of (M,Ψ) along a branch that has dropped (perhaps only at its last model). This contradicts the Dodd-Jensen Lemma 9.3.4, as applied to (M,Ψ). ⊣

Let us now fix ν,l as in Claim 3, and let $\mathcal{S} = \mathcal{S}_{\nu,l}$, $\mathcal{U} = \mathcal{U}_{\nu,l}$, and $\mathcal{T} = \mathcal{T}_{\nu,l}$. Let $\text{lh}(\mathcal{S}) = \theta+1$. We have that $[\text{rt}(\theta),\theta]_S$ does not drop in model or degree. If $0 \leq_S \theta$, this implies that $[0,\theta]_S$ does not drop in model or degree. Let (P_η,Σ_η), (P^*_η,Σ^*_η), and (Q_η,Λ_η) be the η-th lbr hod pairs of $\mathcal{S},\mathcal{T}$, and $\mathcal{U}$. Let $i_{\alpha,\beta}, i^*_{\alpha,\beta}$, and $j_{\alpha,\beta}$ be their branch embeddings. Let $\pi_\eta : P_\eta \to P^*_\eta$ be the lifting map, and when η is unstable, let $\sigma_\eta : P_{\eta+1} \to P_\eta$ be the anticollapse. There is a diagram of the relationships we are aiming to establish now after Claim 8.

The following version of induction hypotheses (v) and (vi) holds in $\mathcal{U}$.

CLAIM 4. *Suppose* $[0,\eta]_U$ *does not drop in model or degree, and let* $j = j_{0,\eta}$*; then*

(a) *for any* $\beta < \alpha_0$, $Th_1^{Q_\eta^k}(j(\beta)\cup j(q)) \in Q_\eta$,
(b) $\sup j``\rho = \rho(Q_\eta)$, *and*
(c) *if* $q \neq r$, *then* $Th_1^{Q_\eta^k}(\rho(Q_\eta)\cup j(q)) \in Q_\eta$.

PROOF. Part (a) holds because $j(\mathrm{Th}_1^{M^k}(\beta\cup q))$ can be used to compute

$$\mathrm{Th}_1^{Q_\eta^k}(j(\beta)\cup j(q)).$$

Part (b) follows by repeated application of Lemma 9.6.1(c) and Lemma 4.3.11 along the branch $[0,\eta]_U$. If $q \neq r$, then $\rho < \alpha_0$, and $\rho(Q_\eta) \leq j_{0,\eta}(\rho)$, so we get (c) by using (a) with $\beta = \rho$. ⊣

The next claim is the analog of Claim 1 in the proof of Lemma 4.10.2, that the comparison ends above H rather than M on the phalanx side. The Dodd-Jensen Lemma is the crucial ingredient.

CLAIM 5. *For some unstable* ξ, $\mathrm{rt}(\theta) = \xi + 1$.

PROOF. If not, then $0 \leq_S \theta$, and $[0,\theta]_S$ does not drop. We are lining up iteration strategies, so

$$(P_\theta, \Sigma_\theta) \trianglelefteq (M_{\nu,l}, \Omega_{\nu,l}) \trianglelefteq (Q_\delta, \Lambda_\delta)$$

for some δ. Notice that $i_{0,\theta}$ is elementary as a map from (M,Ψ) to (P_θ,Σ_θ), because

$$\begin{aligned}\Psi &= (\Sigma_\theta^*)^{i_{0,\theta}^*} \\ &= (\Sigma_\theta^*)^{\pi_\theta \circ i_{0,\theta}} \\ &= ((\Sigma_\theta^*)^{\pi_\theta})^{i_{0,\theta}} \\ &= (\Sigma_\theta)^{i_{0,\theta}}.\end{aligned}$$

The first line holds because Ψ is pullback consistent. Since $i_{0,\theta}$ is elementary in the category of mouse pairs, the Dodd-Jensen Lemma applies, and

$$(P_\theta, \Sigma_\theta) = (Q_\delta, \Lambda_\delta),$$

$[0,\delta]_U$ does not drop, and for all $\eta \in M$

$$j_{0,\delta}(\eta) \leq i_{0,\theta}(\eta).$$

On the other hand, $\pi_\theta \circ j_{0,\delta} \colon (M,\Psi) \to (P_\theta^*, \Sigma_\theta^*)$ is nearly elementary in the category of mouse pairs[24], so again by Dodd-Jensen, for all $\eta \in M$

$$\pi_\theta \circ i_{0,\theta}(\eta) = i_{0,\theta}^*(\eta) \leq \pi_\theta \circ j_{0,\delta}(\eta).$$

[24]Our simplifying assumption actually implies that it is elementary.

Applying π_θ^{-1} to both sides, we get that $i_{0,\theta}(\eta) \le j_{0,\delta}(\eta)$, and hence

$$i_{0,\theta} = j_{0,\delta}.$$

But the sequence of extenders used in each of these branches can be recovered from the embeddings[25], so $e_\theta^{\mathcal{S}} = e_\delta^{\mathcal{U}}$.

Now let η be least such that η is stable and $\eta \le_S \theta$. Then $e_\eta^{\mathcal{S}} = e_\theta^{\mathcal{S}} \upharpoonright \gamma = e_\delta^{\mathcal{U}} \upharpoonright \gamma$, for some γ. But there is τ such that $e_\tau^{\mathcal{U}} = e_\delta^{\mathcal{U}} \upharpoonright \gamma$. Thus

$$P_\eta = Q_\tau,$$

and since $i_{\eta,\theta} = j_{\tau,\delta}$,

$$\begin{aligned}\Sigma_\eta &= (\Sigma_\theta)^{i_{\eta,\theta}} \\ &= (\Lambda_\delta)^{j_{\tau,\delta}} \\ &= \Lambda_\tau.\end{aligned}$$

If η is a limit ordinal, then by the rules in limit case 3, η was declared unstable, contradiction. If S-pred$(\eta) = \mu$, then μ is unstable, and our rules in the successor case declare η to be unstable. So in any case, we have a contradiction. ⊣

Fix ξ as in Claim 5. Since ξ is unstable, we can fix τ such that $(Q_\tau, \Lambda_\tau) = (P_\xi, \Sigma_\xi)$. Fix also $\delta \ge \tau$ such that $M_{\nu,l} \trianglelefteq Q_\delta$, and hence $(P_\theta, \Sigma_\theta) \trianglelefteq (Q_\delta, \Lambda_\delta)$. Set

$$\mu = \rho(P_{\xi+1}),$$

and

$$t = \sigma_\xi^{-1}(i_{0,\xi}(q)).$$

CLAIM 6. *Either*

(i) $\mu = \alpha_\xi$, *or*

(ii) $\mu < \alpha_\xi \le \mathrm{crit}(\sigma_\xi)$, *and* $\mathrm{crit}(\sigma_\xi) = (\mu^+)^{P_{\xi+1}}$.

PROOF. By induction hypothesis (vi), $\mathrm{Th}_1^{P^k_{\xi+1}}(\alpha_\xi \cup t) \notin P_{\xi+1}$, and therefore $\mu \le \alpha_\xi$.

Suppose $\mu < \alpha_\xi$. We can then find some finite $p \subset \alpha_\xi$ such that $\mathrm{Th}_1^{P^k_{\xi+1}}(\mu \cup p \cup t) \notin P_{\xi+1}$. Since $\max(p) < \alpha_\xi$, we get from (vi) that $R = \mathrm{Th}_1^{P^k_\xi}(\mu \cup p \cup i_{0,\xi}(q)) \in P_\xi$. Clearly $R \notin P_{\xi+1}$. Since R is essentially a subset of μ, we get (ii) of Claim 6. ⊣

CLAIM 7. $\mu = \rho(P_\theta)$.

PROOF. This follows easily from the fact that all extenders used in $[\xi+1, \theta]_S$ are close to the model to which they are applied, and $\mathrm{crit}(i_{\xi+1,\theta}) \ge \alpha_\xi$. ⊣

CLAIM 8. *(i) If* $\tau \le \eta < \delta$, *then* $\mathrm{lh}(E_\eta^{\mathcal{U}}) \ge \alpha_\xi$, *and* $P(\alpha_\xi) \cap Q_\eta \subseteq Q_\tau$.

[25] See Remark 2.6.8.

(ii) $P_\theta = Q_\delta$.
(iii) $\tau \leq_U \delta$, $[0,\delta]_U \cap D^{\mathcal{U}} = \emptyset$, *and* $\alpha_\xi \leq \text{crit}(j_{\tau,\delta})$.
(iv) $i_{\xi+1,\theta}$ *is exact.*

PROOF. Note that if $E^{\mathcal{S}}_{\xi+1}$ exists (i.e., $\theta \neq \xi+1$), then $\text{lh}(E^{\mathcal{S}}_{\xi+1}) \geq \alpha_\xi$. This is because otherwise $\varepsilon_\xi = \varepsilon_{\xi+1}$, so $\xi+1$ is a dead node of $\mathcal{S}$, and $\xi+1 <_S \theta$ is impossible. So in any case, P_θ agrees with P_ξ below α_ξ. It follows that Q_δ agrees with P_ξ below α_ξ, and hence with Q_τ below α_ξ. Thus all $E^{\mathcal{U}}_\eta$ for $\tau \leq \eta < \delta$ have length $\geq \alpha_\xi$. This implies that $P(\alpha_\xi) \cap Q_\eta \subseteq Q_\tau$ for all such η. This proves (i).

Set

$$\begin{aligned} A &= \text{Th}_1^{P^k_\xi}(\alpha_\xi \cup i_{0,\xi}(q)) \\ &= \text{Th}_1^{P^k_{\xi+1}}(\alpha_\xi \cup t). \end{aligned}$$

A is essentially a subset of α_ξ, and $A \notin P_\xi$ by induction hypothesis (vi). Since $[\xi+1,\theta]_S$ does drop in model or degree and $\text{crit}(i_{\xi+1,\theta}) \geq \alpha_\xi$, A is definable over P_θ. Thus if $P_\theta \in Q_\delta$, then $A \in Q_\delta$, so $A \in Q_\tau$ by part (i). But $Q_\delta = P_\xi$ and $A \notin P_\xi$, contradiction. It follows that P_θ and Q_δ are the same as bare premice. Moreover, P_θ is k-sound, so

$$P_\theta = Q_\delta \downarrow k.$$

If $[0,\delta]_U \cap D^{\mathcal{U}} \neq \emptyset$, then $k = k(Q_\delta)$, so $P_\theta = Q_\delta$.

So let us prove (iii). Clearly, we may assume $\tau < \delta$. Let η be least such that $\tau \leq \eta$ and $\eta+1 \leq_U \delta$, and let $G = E^{\mathcal{U}}_\eta$. Suppose toward contradiction that $\text{crit}(G) < \alpha_\xi$. Since G has plus type and $\text{lh}(G) \geq \alpha_\xi$,[26]

$$\mu = \rho_{k+1}(Q_\delta) \notin (\text{crit}(G), \alpha_\xi].$$

It follows from Claim 6 that $\mu = \text{crit}(G)$ and $\alpha_\xi = \mu^{+,P_{\xi+1}} < \mu^{+,P_\xi} = \mu^{+,Q_\tau}$. If $U\text{-pred}(\eta+1) = \beta < \tau$, then $P(\mu) \cap Q_\tau = P(\mu) \cap Q_\beta || \text{lh}(E^{\mathcal{U}}_\beta) = P(\mu) \cap Q_\tau || \alpha_\xi$, contradiction. Thus

$$\alpha_\xi \leq \text{crit}(E^{\mathcal{U}}_\eta)$$

and

$$\tau = U\text{-pred}(\eta+1), \text{ where } \eta+1 \leq_U \delta.$$

Suppose now toward contradiction that $D^{\mathcal{U}} \cap [0,\delta]_U \neq \emptyset$. Since Q_δ is k-sound and $\rho_{k+1}(Q_\delta) = \alpha_\xi$, the drop must occur at $\eta+1$, and there can be no further dropping in $(\eta+1,\delta]_U$. Let $J = \mathcal{M}^{*,\mathcal{U}}_{\eta+1}$; then $k(J) = k(Q_\delta) = k = k(Q_\tau)$, so $J \in Q_\tau$. But $A \subseteq \text{crit}(E^{\mathcal{U}}_\eta$ and A is boldface $\Sigma_1^{Q^k_\delta}$, so A is boldface $\Sigma_1^{J^k}$, so $A \in Q_\tau$, contradiction.

[26]Here we use that G has plus type to see that $\rho_{k+1}(Q_\delta) = \alpha_\xi$ is impossible. In the proof of Lemma 4.10.2 we needed a more roundabout argument at this point.

This proves (iii) and the rest of (ii). For (iv) we must see that $i_{\xi+1,\theta}(w_k(P_{\xi+1})) = w_k(P_\theta)$. But note that

$$w_k(P_{\xi+1}) = \sigma_\xi^{-1}(w_k(P_\xi)) = \sigma_\xi^{-1}(i_{0,\xi}(w_k(M))),$$

so $\eta_k^{P_{\xi+1}}$ is not measurable by the $P_{\xi+1}$-sequence. This implies that

$$i_{\xi+1,\theta}(w_k(P_{\xi+1})) = w_k(P_\theta). \qquad \dashv$$

Here is a diagram of the situation.

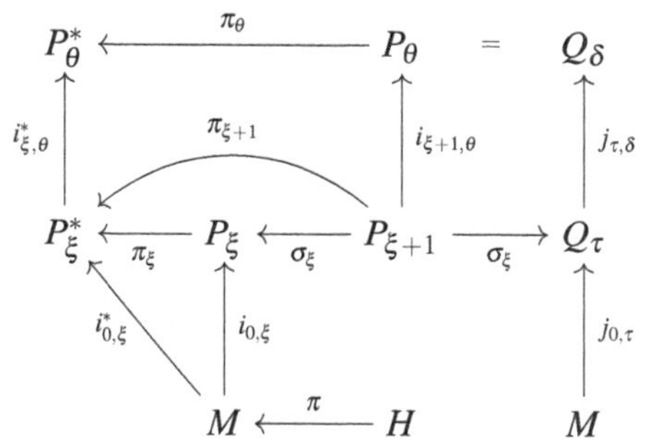

CLAIM 9. $\mu = \alpha_\xi$.

PROOF. By Claim 8, either $\tau = \delta$ or $\text{crit}(j_{\tau,\delta}) \geq \alpha_\xi$. In either case

$$(\mu^+)^{P_\xi} = (\mu^+)^{Q_\tau} = (\mu^+)^{Q_\delta} = (\mu^+)^{P_\theta} = (\mu^+)^{P_{\xi+1}},$$

and all models displayed agree to their common value for μ^+. In particular,

$$P_\xi|(\mu^+)^{P_\xi} = P_{\xi+1}|(\mu^+)^{P_{\xi+1}}.$$

It follows then from Claim 6 that $\mu = \alpha_\xi$. $\dashv$

CLAIM 10. $i_{\xi+1,\theta}(t) = j_{0,\delta}(q)$.

PROOF. Let β be the first (i.e., largest) element of q such that $j_{0,\delta}(\beta) \neq i_{\xi+1,\theta} \circ \sigma_\xi^{-1} \circ i_{0,\xi}(\beta)$. If

$$j_{0,\delta}(\beta) < i_{\xi+1,\theta} \circ \sigma_\xi^{-1} \circ i_{0,\xi}(\beta),$$

then

$$\pi_\theta \circ j_{0,\delta}(\beta) < \pi_\theta \circ i_{\xi+1,\theta} \circ \sigma_\xi^{-1} \circ i_{0,\xi}(\beta) = i^*_{0,\theta}(\beta).$$

But $\pi_\theta \circ j_{0,\delta}$ is nearly elementary, in the category of mouse pairs, from (M,Ψ) to $(P_\theta^*, \Sigma_\theta^*)$, and $i^*_{0,\theta}$ is an iteration map of (M,Ψ), so this contradicts the Dodd-Jensen Lemma. On the other hand, suppose

$$j_{0,\delta}(\beta) > i_{\xi+1,\theta} \circ \sigma_\xi^{-1} \circ i_{0,\xi}(\beta).$$

Let $\gamma = \sigma_\xi^{-1} \circ i_{0,\xi}(\beta)$, and $u = t - (\gamma + 1)$. Since q is solid at β, and $i_{\xi+1,\theta}(u) = j_{0,\delta}(q - (\beta + 1))$ and $i_{\xi+1,\theta}$ is exact, we get that

$$\mathrm{Th}_1^{P_\theta^k}(i_{\xi+1,\theta}((\gamma+1) \cup i_{0,\xi}(u)) \in P_\theta.$$

It follows that $\mathrm{Th}_1^{P_\theta^k}(\alpha_\xi \cup i_{\xi+1,\theta}(t)) \in P_\theta$. But the theory is a subset of $\alpha_x i$, and it is equal to $\mathrm{Th}_1^{P_{\xi+1}^k}(\alpha_\xi \cup t)$. So $\mathrm{Th}_1^{P_\xi}(\alpha_\xi \cup i_{0,\xi}(q)) \in P_\xi$, contradiction. ⊣

CLAIM 11. *r is solid; that is, $q = r$.*

PROOF. If not, then $\rho(M) < \alpha_0$. It follows that

$$\rho(Q_\tau) < \sup j_{0,\tau}``\alpha_0 = \sup i_{0,\xi}``\alpha_0 = \alpha_\xi = \mu = \rho(P_\theta) = \rho(Q_\delta).$$

However, $\mathrm{crit}(j_{\tau,\delta}) \geq \alpha_\xi$ or $\delta = \tau$, so $\rho(Q_\tau) = \rho(Q_\delta)$. This is a contradiction. ⊣

By Claim 11, $\alpha_0 = \rho$. It follows from (v) and (vi) that for all unstable η, $\alpha_\eta = \rho(P_\eta)$. Moreover, by the usual preservation of solid parameters, $i_{0,\eta}(r)$ is the standard parameter of P_η. In particular, this is true when $\eta = \xi$. That tells us that the parameter of P_ξ is universal:

CLAIM 12. *$i_{0,\xi}(r)$ is universal over P_ξ^k.*

PROOF. Let $\eta = \alpha_\xi^{+,P_\xi}$. Since $\mathrm{crit}(i_{\xi+1,\theta}) \geq \alpha_\xi$ and $\mathrm{crit}(j_{0,\delta}) \geq \alpha_\xi$, and neither branch drops, we get

$$P_\xi|\eta = Q_\tau|\eta = Q_\delta|\eta = P_\theta|\eta = P_{\xi+1}|\eta.$$

A parallel chain of equalities shows that whenever $A \subseteq \alpha_\xi$ and A is boldface $\Sigma_1^{P_\xi}$, then A is boldface $\Sigma_1^{P_{\xi+1}}$. ⊣

CLAIM 13. *$M|\rho^{+,M} = H|\rho^{+,M}$.*

PROOF. By Claim 12, we may assume $\xi > 0$. We have already assumed $\rho < \rho_k(M)$, as we may.

Suppose first that ρ is regular in M.[27] Let $N \lhd M|(\rho^+)^M$, $\rho(N) = \rho$, and $B \subseteq \rho$ code $\mathrm{Th}_n^N(\rho(N) \cup p(N))$ for $n = k(N)$. We must show $N \lhd H$, and that is equivalent to

(∗) For some Σ_1 formula φ, some $b < \rho$, and some $\sigma < \rho_k(M)$, there is a unique $\langle P, C\rangle$ such that:
(a) $P \lhd M^k \restriction \sigma$ and $C \subseteq \rho(P)$ codes $\mathrm{Th}_n^P(\rho(P) \cup p(P))$ for $n = k(P)$, and
(b) $M^k \restriction \sigma \models \varphi[P, C, b, r]$.
Moreover, for the unique such $\langle P, C\rangle$, we have $C \cap \rho = B$.

[27] $\mathrm{cof}_k^M(\rho) = \mathrm{cof}_0^M(\rho)$ because $\rho < \rho_k(M)$.

Here $M^k \restriction \sigma = M^k | \sigma$ if $k = 0$ and M is passive. Otherwise $M^k \restriction \sigma = (M||\sigma, A \cap M||\sigma)$, where $A = A^k_M$ if $k > 0$, and A is $\dot{F}^M$ or $\dot{B}^M$ if $k(M) = 0$ and M is active.

We can express $(*)$ as

$$M \models \psi[B, \rho, r],$$

where ψ is Σ_1. Let $i = i_{0,\xi}$, and note that $i\colon M^k \to P^k_\xi$ is cofinal and Σ_1-elementary. Moreover, $i(\rho) = \sup i``\rho = \alpha_\xi$, because ρ is regular in M^k. By Claim 12

$$P_\xi \models \psi[i(B), i(\rho), i(r)].$$

Thus $M \models \psi[B, \rho, r]$, as desired.

Now assume that ρ is singular in M. It will then be enough to show that $P(\rho)^M \subseteq H$. This is because if $\pi\colon H \to M$ is the collapse map, then $\mathrm{crit}(\pi) > \rho$, as otherwise $\mathrm{crit}(\pi) = \rho$ is regular in H, and hence regular in M because $P(\rho)^M \subseteq H$. It follows that $\mathrm{crit}(\pi) \geq \rho^{+,H} = \rho^{+,M}$, which yields Claim 13.

Suppose toward contradiction that $B \subseteq \rho$, $B \in M$, and $B \notin H$. We show by induction on $\eta \leq_S \xi$ that $i_{0,\eta}(B) \cap \alpha_\eta \notin P_{\eta+1}$. The case η is a limit ordinal is easy, so assume S-pred$(\eta) = \beta$, let $E = E^S_{\eta-1}$, and let $A = i_{0,\beta}(B) \cap \alpha_\beta$. So $A \notin P_{\beta+1}$. Let us write i_E for $i_{\beta,\eta}$, and let $s = i_{0,\beta}(r)$. Suppose toward contradiction that $i_E(A) \cap \alpha_\eta \in P_{\eta+1}$; then we have some $b < \alpha_\eta$, some C, and some Σ_1 formula φ such that

$$P^k_\eta \models C \text{ is the unique } D \text{ such that } \varphi(D, b, i_E(s)),$$

and $C \cap \alpha_\eta = i_E(A) \cap \alpha_\eta$. Fix b, C, and φ. There are cofinally many ordinals in P^k_β that are Σ_1 definable from parameters in $\alpha_\beta \cup s$, so we can find such an ordinal σ such that

$$P^k_\eta \restriction i_E(\sigma) \models C \text{ is the unique } D \text{ such that } \varphi(D, b, i_E(s)),$$

But now let

$$b = [a, f]^{P_\beta}_E.$$

For E_a almost every u,

$$P^k_\beta \restriction \sigma \models \text{there is a unique } D \text{ such that } \varphi(D, f(u), s).$$

Let C_u be the unique such D, when it exists. The function $u \mapsto C_u$ is definable over $P^k_\beta \restriction \sigma$ from f and s. Since $\alpha_\eta = \sup i_E``\alpha_\beta$, we may assume that $f \in P_\beta | \alpha_\beta$. ($\alpha_\beta = \rho(P_\beta)$ is a singular cardinal of P^k_β in the present case.) Since $\mathrm{crit}(E) < \alpha_\beta$, $\mathrm{dom}(E) < \alpha_\beta$, so $E_a \in P_\beta | \alpha_\beta$ because E is close to P_β. But then for $\gamma < \alpha_\beta$,

$$\gamma \in A \iff \text{for } E_a \text{ a.e. } u,\ \gamma \in C_u.$$

This defines A over $P_\beta \restriction \sigma$ from f, s, and E_a. That implies $A \in P_{\beta+1}$, a contradiction.

Thus $i_{0,\xi}(B) \cap \alpha_\xi \notin P_{\xi+1}$, contrary to Claim 12. This proves Claim 13. ⊣

CLAIM 14. *r is universal over M.*

PROOF. Let $A \subseteq \rho$ be boldface $\Sigma_1^{M^k}$. We must show that A is boldface $\Sigma_1^{H^k}$. Say that for $\gamma < \rho$

$$\gamma \in A \text{ iff } M^k \models \exists v \varphi[v, \gamma, z],$$

where φ is Σ_0. For $0 \leq_S \eta \leq_S \xi$, and $\gamma < \alpha_\eta$, let

$$\gamma \in A_\eta \text{ iff } P_\eta^k \models \exists v \varphi[v, \gamma, i_{0,\eta}(z)].$$

By Łoś's Theorem, whenever $\beta \leq_S \eta \leq_S \xi$, $\gamma \in A_\beta$ iff $i_{\beta,\eta}(\gamma) \in A_\eta$, and let $(*)_\eta$ be the assertion: there is a Σ_0 formula ψ and a parameter $x \in P_\eta | \alpha_\eta$ such that for all $\gamma < \alpha_\eta$,

$$\gamma \in A_\eta \text{ iff } P_\eta \models \exists v \psi[v, \gamma, x, i_{0,\eta}(r)].$$

By Claim 12, $(*)_\xi$ holds, and our goal is to prove that $(*)_0$ holds.

Let η be least such that $(*)_\eta$ holds, and assume toward contradiction that $\eta > 0$. It is easy to see that η is not a limit ordinal, since if $\psi, i_{\beta,\eta}(x)$ witness $(*)_\eta$, then ψ, x witness $(*)_\beta$. So let $\beta = S\text{-pred}(\eta)$ and $E = E_{\eta-1}^{S}$, and let

$$\psi, [a, f]_E^{P_\beta^k}$$

witness $(*)_\eta$. As in the proof of Claim 13, we may assume that f and E_a belong to $P_\beta | \alpha_\beta$. But then by Łoś's Theorem,

$$\gamma \in A_\beta \text{ iff } P_\beta^k \models \exists X \in E_a \exists g \forall u \in X \psi[g(u), \gamma, f(u), i_{0,\beta}(r)].$$

Thus $(*)_\beta$ holds, contradiction. ⊣

CLAIM 15. *If ρ is not measurable by the M-sequence, then M is projectum solid.*

PROOF. We may assume that $\text{crit}(\pi) = \rho$, as otherwise projectum solidity is vacuous. This implies that ρ is regular in H, hence regular in M by Chaim 13. Thus $i_{0,\xi}(\rho) = \alpha_\xi = \rho(P_\xi)$. It is easy then to see that $\text{crit}(\sigma_\xi) = \alpha_\xi$.

P_ξ is projectum solid by the proof of Claim 11 in Lemma 4.10.2, which in turn traces back to the proof of Theorem 3.7.1. We can then use $i_{0,\xi}$ to pull this back to M, just as in the proof of Claim 5 in the proof of Lemma 4.10.8. The reader should see the proofs of Lemmas 4.10.2 and 4.10.8 for the relevant details. ⊣

This completes the proof of Lemma 9.6.2, modulo Lemma 9.6.5. ⊣

The proof of Lemma 9.6.2 shows that the pullback strategy for $\bar{\mathfrak{C}}(M)$ agrees with the iteration strategy for M on all trees based on $M|\rho(M)^{+,M}$, not just those trees belonging to M.

COROLLARY 9.6.6. *Assume* AD^+, *and let (M, Ψ) be a strongly stable lbr hod pair with scope* HC. *Let $H = \bar{\mathfrak{C}}(M)$ be the strong core of M, and $\pi\colon H \to M$ be the anticore map. Let $\rho = \rho(M)$ and $N = M|\rho^{+,M} = H|\rho^{+,H}$; then $\Psi_N = (\Psi^\pi)_N$.*

PROOF. Let us adopt all the notation in the proof of Lemma 9.6.2. So $\mathcal{S} = \mathcal{S}_{\nu_0,k}$ has last model $(P_\theta, \Sigma_\theta)$, and $\mathcal{U}$ has last model $(Q_\delta, \Lambda_\delta) = (P_\theta, \Sigma_\theta)$. Let $\mathrm{rt}(\theta) = \xi + 1$, so that ξ is unstable, and we have τ such that

$$(P_\xi, \Sigma_\xi) = (Q_\tau, \Lambda_\tau)$$

and

$$i_{0,\xi} = j_{0,\tau}.$$

It is easy to see that $\mathrm{ran}(i_{0,\xi} \circ \pi) \subseteq \mathrm{ran}(\sigma_\xi)$, so we may set $\psi = \sigma_\xi^{-1} \circ i_{0,\xi} \circ \pi$, and we have the diagram

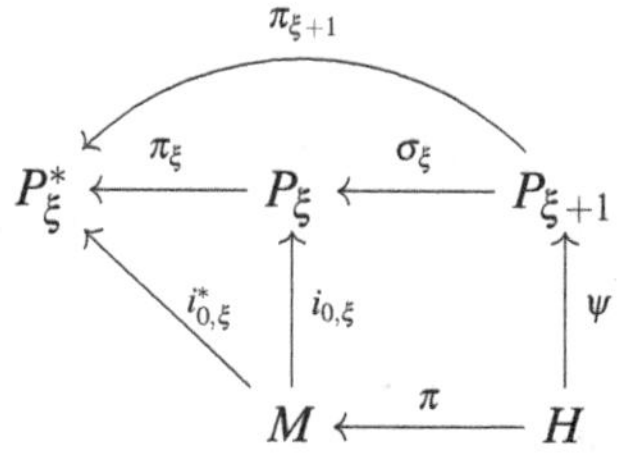

From the diagram, we see that

$$\begin{aligned}\Psi &= (\Sigma^*_\xi)^{i^*_{0,\xi}} \\ &= (\Sigma^*_\xi)^{\pi_\xi \circ i_{0,\xi}} \\ &= \Sigma_\xi^{i_{0,\xi}}.\end{aligned}$$

Line 1 uses the pullback consistency of Ψ. Using it again, we get

$$\begin{aligned}\Psi^\pi &= (\Sigma^*_\xi)^{i^*_{0,\xi} \circ \pi} \\ &= (\Sigma^*_\xi)^{\pi_{\xi+1} \circ \psi} \\ &= \Sigma_{\xi+1}^{\psi}.\end{aligned}$$

Since r is solid, $\rho = \rho(M) = \alpha_0$. We may assume that $\mathrm{crit}(\pi) = \rho$, since otherwise $\pi \restriction N = \mathrm{id}$ and the corollary is trivial. This implies that ρ is regular in M, so

$$i_{0,\xi}(\rho) = \rho(P_\xi) = \alpha_\xi$$

and

$$i_{0,\xi}(N) = P_\xi | \alpha_\xi^{+,P_\xi} = P_{\xi+1} | \alpha_\xi^{+,P_{\xi+1}}.$$

Let $R = i_{0,\xi}(N)$. We then have

$$\begin{aligned}(\Sigma_{\xi+1})_R &= (\Omega_{\nu_0,k})_R \\ &= (\Lambda_\tau)_R \\ &= (\Sigma_\xi)_R.\end{aligned}$$

Line 1 holds by Lemma 9.6.5, and line 2 holds by Theorem 9.4.18.

Notice that $\psi \restriction N = i_{0,\xi} \restriction N$, because for $A \subseteq \rho$ with $A \in N$, $A = \pi(A) \cap \rho$, so $i_{0,\xi}(A) = i_{0,\xi} \circ \pi(A) \cap \alpha_\xi = \sigma_\xi^{-1} \circ i_{0,\xi} \circ \pi(A) \cap \alpha_\xi$. But then

$$\begin{aligned}\Psi_N^\pi &= (\Sigma_{\xi+1}^\psi)_N \\ &= ((\Sigma_{\xi+1})_R)^\psi \\ &= ((\Sigma_\xi)_R)^{i_{0,\xi}} \\ &= (\Sigma_\xi^{i_{0,\xi}})_N \\ &= \Psi_N,\end{aligned}$$

as desired. ⊣

Remark 9.6.7. The proof of Corollary 9.6.6 shows how comparison of strategies can lead to condensation properties of mouse pairs that are stronger than those we get directly from a background construction.

We can adapt the rest of the solidity/universality proof in Section 4.10 in a similar way. Let us just record the main steps.

Let $k = k(M)$, and M_0^k be M^k restricted to the language without a symbol for $w_k(M)$, and let

$$\mathfrak{D}_{k+1}(M) = \text{decoding of } \mathrm{cHull}_1^{M_0^k}(\rho_1(M_0^k) \cup p_1(M_0^k)).$$

$\rho_1(M_0^k) = \rho_1(M^k)$, but the standard parameters of the two structures may differ. We show that if M is strongly stable and has type 1A, then $p_1(M_0^k)$ behaves well:

LEMMA 9.6.8. *Assume* AD^+*, and let* (M, Ψ) *be a strongly stable lbr hod pair such that* M *has type 1A. Let* $k = k(M)$*, and suppose that* $\eta_k^M < \rho_{k+1}(M)$ *and* $\rho_{k+1}(M)$ *is not measurable by the* M*-sequence. Let* $\pi\colon \mathfrak{D}_{k+1}(M) \to M$ *be the anticore map; then*

(a) $p_1(M_0^k)$ *is solid and universal over* M_0^k*,*
(b) if $\mathrm{crit}(\pi) = \rho_{k+1}(M) = \rho$*, then letting* $D = (E_\pi)_\rho$*,* D *is the order zero measure of* $\mathfrak{D}_{k+1}(M)$ *on* ρ*,*
(c) $\rho_k(M) = \pi(\rho_k(\mathfrak{D}))$*, and*
(d) $\eta_k^M = \eta_k^{\mathfrak{D}}$*, where* $\mathfrak{D} = \mathfrak{D}_{k+1}(M)$*.*

Parts (a) and (b) are proved in the same way that we just proved Lemma 9.6.2. See the proof of Lemma 4.10.7 for the proofs of (c) and (d).

Next we use Lemma 9.6.8 by pulling back its conclusions under an ultrapower map.

LEMMA 9.6.9. *Assume* AD^+*, and let* (M, Ψ) *be an lbr hod pair such that* M *is stable and of type 1. Let* $k = k(M)$*, and suppose that* $\rho_{k+1}(M)$ *is not measurable by the* M*-sequence. Suppose that* M *is not strongly stable. Let* $\pi\colon \mathfrak{D} \to \bar{\mathfrak{C}}_k(M)$ *be the anticore map, where* $\mathfrak{D} = \mathfrak{D}_{k+1}(M)$*; then*

(a) $p_1(M_0^k)$ *is solid and universal over* M_0^k*,*

(b) *if* $\text{crit}(\pi) = \rho_{k+1}(M) = \rho$, *then letting* $D = (E_\pi)_\rho$, D *is the order zero measure of* $\mathfrak{D}$ *on* ρ,
(c) $\pi(\rho_k(\mathfrak{D})) = \rho_k(M)$, *and*
(d) $\eta_k^{\mathfrak{D}} = \eta_k^M$.

PROOF SKETCH. We prove the lemma by setting

$$N = \text{Ult}_k(\bar{\mathfrak{C}}_k(M), D),$$

where D is the order zero measure on η_k^M. i_D is discontinuous at $\rho_k(M)$, so N is an lpm of type 1A. Let $\tau\colon N \to \text{Ult}_k(M,D)$ be the copy map associated to (π, π, D), where $\pi\colon \bar{\mathfrak{C}}_k(M) \to M$ is the anticore map. Let

$$\Phi = (\Psi_{\langle D \rangle})^\tau.$$

Since τ is elementary and $(\text{Ult}_k(M,D), \Psi_{\langle D \rangle})$ is an lbr hod pair, (N, Φ) is an lbr hod pair. Thus Lemma 9.6.8 applies to (N, Φ). We obtain the conclusions of Lemma 9.6.9 by pulling pack those of Lemma 9.6.8 under i_D. ⊣

The proof of weak ms-solidity elaborates on the one we gave for pure extender mice in Section 4.10 in a parallel fashion.

LEMMA 9.6.10. *Assume* AD^+, *and let* (M, Ψ) *be an lbr hod pair; then* M *is weakly ms-solid.*

PROOF SKETCH. This is similar to the proof of Lemma 4.10.11, with an overlay of phalanx-lifting made necessary by iteration into a background construction. See the proof of Theorem 9.6.15 below. ⊣

We can put these lemmas together just as we did in Theorem 4.10.9.

THEOREM 9.6.11. *Assume* AD^+, *and let* (M, Ψ) *be an lbr hod pair such that* M *is stable and* $\rho(M)$ *is not measurable by the* M*-sequence; then* M *is solid.*

PROOF SKETCH. M is weakly ms-solid by Lemma 9.6.10. We must see that M is parameter solid and projectum solid. If M is strongly stable, this follows from Lemma 9.6.2, so assume not. Let $k = k(M)$. By Lemma 9.6.9, M is projectum solid and $p_1(M_0^k)$ behaves well. We deduce that $p_1(M^k)$ behaves well by translating between it and $p_1(M_0^k)$, just as we did in the proof of Theorem 4.10.9. The translation is possible because $\pi(\rho_k(\mathfrak{D})) = \rho_k(M)$ and $\eta_k^{\mathfrak{D}} = \eta_k^M$, where $\mathfrak{D} = \mathfrak{D}_{k+1}(M)$ and $\pi\colon \mathfrak{D} \to M$ is the anticore map. See the proof of Theorem 4.10.9 for more detail. ⊣

For a level (M, Ω) of a background construction, amenable closure implies that M is stable and $\rho(M)$ is not measurable by the M-sequence. The proof is identical those of Lemma 4.10.1, Lemma 3.7.1, and Theorem 3.8.2, so we omit it.

LEMMA 9.6.12. *Assume* AD^+, *and let* $\langle (N, \in, w, \mathcal{F}, \Sigma), \Sigma^* \rangle$ *be a coarse strategy pair. Let* $\mathbb{C}$ *be a* $(w, \mathcal{F}, \Sigma)$*-construction done in* N *and let* $(M, \Omega) \in \text{lev}(\mathbb{C})$; *then* M *is stable, and* $\rho(M)$ *is not measurable by the* M*-sequence.*

From Theorem 9.6.11 and Lemma 9.6.12 we get at once, *modulo Lemma 9.6.5*,

THEOREM 9.6.13. *Assume* AD^+, *and let* $\langle (M, \in, w, \mathcal{F}, \Sigma), \Sigma^* \rangle$ *be a coarse strategy pair. Let* $\mathbb{C}$ *be an* $(w, \mathcal{F}, \Sigma)$*-construction done in* M $\langle \nu, k \rangle < \mathrm{lh}(\mathbb{C})$ *be such that* $k \geq 0$*; then* $\mathbb{C}$ *is good at* $\langle \nu, k \rangle$*, that is,*$M^{\mathbb{C}}_{\nu,k}$ *is solid.*

Thus in order to see that least branch constructions do not break down, we must prove Lemma 9.6.5, and we must to show that they are good at all pairs of the form $\langle \nu, -1 \rangle$. We shall do this in the next chapter.

Combining Theorem 9.6.13 with Theorem 9.4.17, we get a parallel to Theorem 9.6.13 whose hypotheses are (we believe) consistent with the Axiom of Choice.

COROLLARY 9.6.14. *Assume* ZFC *plus* $\mathsf{IH}_{\kappa,\delta}$, *where* $\kappa < \delta < \theta < \alpha$ *for some inaccessible* θ *and* α. *Suppose also that there are* $\lambda < \mu < \kappa$ *such that* λ *is a limit of Woodin cardinals, and* μ *is measurable. Let* $(w, \mathcal{F})$ *be a coherent pair such that* $\mathcal{F} \subseteq V_\delta$ *and* $\forall E \in \mathcal{F}(\mathrm{crit}(E) > \kappa)$*, let* Σ *be the unique* $(\theta, \theta, \mathcal{F})$*-iteration strategy for* V*, and let* $\mathbb{C}$ *be a* $(w, \mathcal{F}, \Sigma)$ *construction; then for any* $\langle \nu, k \rangle < \mathrm{lh}(\mathbb{C})$ *such that* $k \geq 0$, $\mathbb{C}$ *is good at* $\langle \nu, k \rangle$.

PROOF. This follows at once from Theorem 9.4.17, Theorem 9.6.13, and the fact goodness at $\langle \nu, k \rangle$ is first order. ⊣

We now prove a condensation lemma for lbr hod pairs by the same method. Rather than attempt a general statement, we shall content ourselves with the following simple one, since it is what we need in this book.[28]

THEOREM 9.6.15 (Condensation lemma). *Assume* AD^+, *and let* (M, Ψ) *be an lbr hod pair such that* $k(M) = 0$. *Let*

$$\pi \colon H \to M$$

be elementary, with $\mathrm{crit}(\pi) = \rho(H) < \rho(M)$*, and* H *being sound. Suppose also that* $\rho(H)$ *is a limit cardinal of* H*; then* $(H, \Psi^\pi) \trianglelefteq (M, \Psi)$.

PROOF SKETCH. Since $k(M) = k(H) = 0$, both H and M are strongly stable. We proceed as in the proof of Theorem 9.6.2. Let $\mathcal{N} = (N^*, \in, w, \mathcal{F}, \Phi), \Phi^*)$ be a coarse strategy pair such that M is countable in N^* and $\mathcal{N}$ captures $\mathrm{Code}(\Psi)$ at $\delta(w)$. Let $\mathbb{C}$ be the maximal $(w, \mathcal{F}, \Phi)$-construction of $\mathcal{N}$, with levels $(M_{\nu,l}, \Omega_{\nu,l})$. By Theorem 9.5.8 we can fix ν_0 so that (M, Ψ) iterates to $(M_{\nu_0,0}, \Omega_{\nu_0,0})$, and strictly past $(M_{\nu,l}, \Omega_{\nu,l})$ whenever $\langle \nu, l \rangle <_{\mathrm{lex}} \langle \nu_0, 0 \rangle$. Let $\mathcal{U}_{\nu,l}$ be the λ-separated trees on (M, Ψ) witnessing this.

For $\langle \nu, l \rangle \leq_{\mathrm{lex}} \langle \nu_0, 0 \rangle$ we define a λ-separated pseudo iteration tree $\mathcal{S}_{\nu,l}$ on $(M, H, \rho(H))$, by iterating away least extender disagreements with $M_{\nu,l}$. $\mathcal{S}_{\nu,l}$ is defined exactly as it was in the proof of Theorem 9.6.2, with one exception with regard to how we move phalanxes up. Again, we show that only the phalanx

[28]The author and Nam Trang have proved a stronger condensation theorem in [81], and used it to generalize the Schimmerling-Zeman characterization of $\{\kappa \mid M \models \square_\kappa\}$ to the case that M is a least branch hod mouse. The proof in [81] is given in much greater detail than we give here.

moves, and no strategy disagreements appear. Let $\mathcal{T}_{v,l}$ be the lift of $\mathcal{S}_{v,l}$ to a tree on (M,Ψ), defined as before, and let (P_η,Σ_η), (P^*_η,Σ^*_η), and (Q_η,Λ_η) be the mouse pairs of $\mathcal{S}_{v,l}, \mathcal{T}_{v,l}$ and $\mathcal{U}_{v,l}$, where we are suppressing the dependence on $\langle v,l\rangle$ whenever we can. Let us adopt the rest of the notation of Theorem 9.6.2 for the various branch embeddings and lifting maps.

Note that because $\rho(H) < \rho(M)$, we have $H \in M$.(The theory coding H is a bounded Σ_1^M subset of $\rho(M)$, hence in M. Since $M|\rho(M) \models \mathsf{KP}$, $H \in M|\rho(M)$.) Now let $\mathcal{S} = \mathcal{S}_{v,l}$, and suppose $\gamma+1$ is an unstable node of $\mathcal{S}$, and $\xi = \mathcal{S}\text{-pred}(\gamma+1)$. We have $P_{\gamma+1} = \mathrm{Ult}_0(P_\xi, E^{\mathcal{S}}_\gamma)$ as before. We then set

$$P_{\gamma+2} = i_{0,\gamma+1}(H),$$

and

$$\alpha_{\gamma+1} = i_{0,\gamma+1}(\rho(H)).$$

So it can happen that $\sup i_{0,\gamma+1}``\rho(H) < \alpha_{\gamma+1}$. Otherwise we proceed as before:

$$\sigma_{\gamma+1} : P_{\gamma+2} \to P_{\gamma+1}$$

is determined by: $\sigma_{\gamma+1} \restriction \alpha_{\gamma+1}$ is the identity, and $\sigma_{\gamma+1} \circ i_{0,\gamma+1}(p(H)) = i_{0,\gamma+1} \circ \pi(p(H))$. There is a similar change at unstable limit ordinals θ. We set $P_{\theta+1} = i_{0,\theta}(H)$ and $\alpha_\theta = i_{0,\theta}(\rho(H))$, etc.

CLAIM 0. *If ξ is unstable and $(P_{\xi+1},\Sigma_{\xi+1}) \trianglelefteq (P_\xi,\Sigma_\xi)$, then $(H,\Psi^\pi) \trianglelefteq (M,\Psi)$.*

PROOF. $H \lhd M$ by the elementarity of $i_{0,\xi}$. That $\Psi_H = \Psi^\pi_H$ is proved just as in the proof of Corollary 9.6.6. Using the pullback consistency of Ψ, we get that $\Psi = \Sigma_\xi^{i_{0,\xi}}$ and $\Psi^\pi == \Sigma^\psi_{\xi+1}$, where $\psi = \sigma_\xi^{-1} \circ i_{0,\xi} \circ \pi$. But $\psi = i_{0,\xi} \restriction H$, so setting $R = P_{\xi+1}$,

$$\begin{aligned}\Psi^\pi_H &= (\Sigma^\psi_{\xi+1})_H \\ &= ((\Sigma_{\xi+1})_R)^\psi \\ &= ((\Sigma_\xi)_R)^{i_{0,\xi}} \\ &= (\Sigma_\xi^{i_{0,\xi}})_H \\ &= \Psi_H. \end{aligned}$$

⊣

So if our initial phalanx is bad, in that (H,Ψ^π) is not an initial segment of (M,Ψ), then its images at higher unstable levels of $\mathcal{S}$ remain bad.

The rest of the construction of $\mathcal{S}_{v,l}$, and its conditions for termination, are the same as in the proof of Theorem 9.6.2. Again, the key lemma is the counterpart of Lemma 9.6.5, according to which no strategy disagreements show up, and least extender disagreements involve only empty extenders on the $M^{\mathbb{C}}_{v,l}$ side. We shall prove this lemma in the next section.

We argue as before that for some v,l, the construction of $\mathcal{S}_{v,l}$ terminates at a stable θ such that $P_\theta \trianglelefteq Q_\delta = \mathcal{M}_\delta^{\mathcal{U}_{v,l}}$. (We no longer have $Q_\delta \trianglelefteq P_\theta$, as the proof of that in Theorem 9.6.2 used that $H \notin M$, whereas $H \in M$.) Using the Dodd-Jensen Lemma, we get that for some unstable ξ, $\mathrm{rt}(\theta) = \xi + 1$.

Let $Q_\tau = P_\xi$. We have that $\mathrm{lh}(E_\tau^{\mathcal{U}}) \geq \alpha_\xi$, as otherwise $\xi + 1$ would have been dead. But in the present case, α_ξ is a limit cardinal of $P_\xi = Q_\tau$, so $\mathrm{lh}(E_\tau^{\mathcal{U}}) > \alpha_\xi$.

Now we simply follow the proofs of Claims 1-4 in the proof of Theorem 8.2 of [35]. We get that $\xi + 1 = \theta$ and

$$(P_{\xi+1}, \Sigma_{\xi+1}) \lhd (Q_\delta, \Lambda_\delta).$$

This implies there are no cardinals of Q_δ strictly between α_ξ and $o(P_{\xi+1})$ Thus $\mathrm{lh}(E_\tau^{\mathcal{U}}) \geq o(P_{\xi+1})$, so

$$(P_{\xi+1}, \Sigma_{\xi+1}) \lhd (Q_\tau, \Lambda_\tau) = (P_\xi, \Sigma_\xi).$$

But then by Claim 0, $(H, \Psi^\pi) \lhd (M, \Psi)$. ⊣

We get at once analogs of Theorem 9.6.13 and Corollary 9.6.14. The levels of a least branch construction subject to the hypotheses of Theorem 9.6.13 or Corollary 9.6.14 have the condensation property described in Theorem 9.6.15.

Chapter 10

PHALANX ITERATION INTO A CONSTRUCTION

In this chapter we complete the proof that if $\mathbb{C}$ is a least branch hod pair construction done in an appropriate environment, then $\mathbb{C}$ is good at all $\langle \nu, k \rangle$. In Section 10.1 we prove the uniqueness of the extender F added by $\mathbb{C}$ at stage ν, and in Section 10.2 we show that its background F^* actually backgrounds F^+. This shows that $\mathbb{C}$ is good at $\langle \nu, -1 \rangle$. In Section 10.3 we prove Lemma 9.6.5, thereby completing the proof in Section 9.6 that $\mathbb{C}$ is good at $\langle \nu, k \rangle$ whenever $k \geq 0$.

The proofs of these results involve showing that certain bicephali and phalanxes iterate into background constructions in the same way that ordinary lbr hod pairs do. That is our main new burden.

Section 10.4 puts the results of this and the previous chapters together in two theorems: Theorem 10.4.1, which states that under AD^+, the least branch construction of a coarse strategy pair does not break down, and Theorem 10.4.6, which states that assuming ZFC, sufficiently large cardinals, and UBH there are least branch constructions that produce strategy mice with subcompact cardinals.

In Section 10.5 we use a more elaborate phalanx comparison argument to show that if (M, Ω) is a least branch hod pair such that $M \models \mathsf{ZFC} +$ "there are arbitrarily large Woodin cardinals", then whenever g is $\mathbb{P}$-generic over M, $M[g] \models$ "UBH holds for all nice, normal iteration trees that use extenders from $\dot{E}^M$ with critical points strictly above $|\mathbb{P}|^M$". This implies that M has a term for the action of Ω in generic extensions of M. We shall use this in Chapter 11 to show that if λ is a limit of cutpoint Woodin cardinals in M, and N is a derived model of M below λ, then HOD^N is an Ω-iterate of M.

10.1. The Bicephalus Lemma

DEFINITION 10.1.1. An *lpm-bicephalus* is a structure

$$\mathcal{B} = (B, \in, \dot{E}^{\mathcal{B}}, \dot{\Sigma}^{\mathcal{B}}, F, G)$$

such that both $(B, \in, \dot{E}^{\mathcal{B}}, \dot{\Sigma}^{\mathcal{B}}, F, \emptyset)$ and $(B, \in, \dot{E}^{\mathcal{B}}, \dot{\Sigma}^{\mathcal{B}}, G, \emptyset)$ are extender-active least branch premice. We say that $\mathcal{B}$ is *nontrivial* iff $F \neq G$.

We shall usually drop "lpm" from "lpm-bicephalus".

We think of $\mathcal{B}$ as a structure in the language with $\in$ and predicate symbols $\dot{\Sigma}, \dot{E}, \dot{F}$, and $\dot{G}$. We let

$$\mathcal{B}^- = (B, \in, \dot{E}^{\mathcal{B}}, \dot{\Sigma}^{\mathcal{B}}, \emptyset)$$

be the lpm obtained by removing both top extenders. $\mathcal{B}^-$ is a passive lpm. The degree of $\mathcal{B}$ is zero, i.e., $k(\mathcal{B}) = 0$. For $\nu < o(\mathcal{B}) = \hat{o}(\mathcal{B})$, we set $\mathcal{B}|\langle \nu, l\rangle = \mathcal{B}^-|\langle \nu, l\rangle$. The extender sequence of $\mathcal{B}$ is $\dot{E}^{\mathcal{B}}$ together with $\dot{F}^{\mathcal{B}}$ and $\dot{G}^{\mathcal{B}}$; it is not actually a sequence.

A *$\mathcal{B}$-tree* is a tuple $\langle \nu, k, \mathcal{T}\rangle$ such that $\langle \nu, k\rangle \leq_{\text{lex}} \langle \hat{o}(\mathcal{B}), 0\rangle$, and $\mathcal{T}$ is a λ-separated plus tree on $\mathcal{B}|\langle \nu, k\rangle$. That is, $\mathcal{M}_0^{\mathcal{T}} = \mathcal{B}|\langle \nu, k\rangle$, the extenders used in $\mathcal{T}$ are length-increasing and nonoverlapping along branches, and $E_\alpha^{\mathcal{T}} = F^+$ for some F on the sequence of $\mathcal{M}_\alpha^{\mathcal{T}}$. A *$\mathcal{B}$-stack* is a sequence $\langle (\nu_i, k_i, \mathcal{T}_i) \mid i \leq n\rangle$ such that $\langle \nu_0, k_0, \mathcal{T}_0\rangle$ is a $\mathcal{B}$-tree, and $\langle \nu_{i+1}, k_{i+1}, \mathcal{T}_{i+1}\rangle$ is a $\mathcal{M}_\infty(\mathcal{T}_i)$-tree. A *complete strategy* for $\mathcal{B}$ is a strategy Ω defined on all $\mathcal{B}$-stacks s by Ω such that $s \in N$, for some set N. N is called the scope of Ω.[1]

DEFINITION 10.1.2. A *bicephalus pair* is a pair $(\mathcal{B}, \Omega)$ such that $\mathcal{B}$ is an lpm-bicephalus, and Ω is a complete strategy for $\mathcal{B}$.

Tail strategies are given by $\Omega_s(t) = \Omega(s^\frown t)$. We use $\Omega_{s,N}$ and Ω_N as before. We write Ω^- for $\Omega_{\mathcal{B}^-}$, the complete strategy for $\mathcal{B}^-$ induced by Ω.

We define normalizing well, strong hull condensation, internal lift consistency, and pushforward consistency for for bicephalus pairs just as we did for lbr hod pairs. If $(\mathcal{B}, \Omega)$ has these properties, then its natural projections $((\mathcal{B}^-, \dot{F}^{\mathcal{B}}), \Omega_0)$ and $((\mathcal{B}^-, \dot{G}^{\mathcal{B}}), \Omega_1)$ are lbr hod pairs, but the converse is of course not true, because Ω must behave well on trees that use images of both top extenders in $\mathcal{B}$.

If $\mathcal{B}$ is a bicephalus and M is an lpm, then we define

$$M \trianglelefteq \mathcal{B} \text{ iff } M \trianglelefteq \mathcal{B}^-.$$

In other words, we do not regard the projections $(\mathcal{B}^-, \dot{F}^{\mathcal{B}})$ and $(\mathcal{B}^-, \dot{G}^{\mathcal{B}})$ as initial segments of $\mathcal{B}$. If (M, Ω) is an lbr hod pair and $(\mathcal{B}, \Psi)$ is a bicephalus pair, then

$$(M, \Omega) \trianglelefteq (\mathcal{B}, \Psi) \text{ iff } M \trianglelefteq \mathcal{B} \wedge \Omega = \Psi_M.$$

Since $k(\mathcal{B}) = 0$, $\mathcal{B}$ is strongly stable, and hence the branch embeddings in any $\mathcal{B}$-tree are elementary and exact, and all its models have type 1. Along a branch that has dropped, the branch embedding from the last drop onward is the corresponding anticore map.

The main theorem about bicephalus pairs is that there are not any interesting ones.

THEOREM 10.1.3. *Let $(\mathcal{B}, \Psi)$ be a bicephalus pair such that Ψ has scope* HC *and $L(\Psi, \mathbb{R}) \models \mathsf{AD}^+$. Suppose also that Ψ normalizes well, has strong hull condensation, and is internally lift consistent and pushforward consistent; then $\dot{F}^{\mathcal{B}} = \dot{G}^{\mathcal{B}}$.*

[1]The requirement that $\mathcal{B}$-trees use only extenders of the form F^+ applies also in the case that F is one of the top extenders of a bicephalus.

PROOF. Let us assume toward contradiction that $\dot{F}^{\mathcal{B}} \neq \dot{G}^{\mathcal{B}}$.

We work in $L(\Psi, \mathbb{R})$. By the Basis Theorem, we may assume Code(Ψ) is Suslin and co-Suslin, and hence we can fix a coarse strategy pair $((N^*, \in, w, \mathcal{F}, \Phi), \Phi^*)$ that captures Code(Ψ). Let

$$\mathbb{C} = \text{maximal } (w, \mathcal{F}, \Phi)\text{-construction of } (N^*, \in, w, \mathcal{F}, \Phi).$$

Let $\langle \eta_0, l_0 \rangle$ be the lex-least pair $\langle \nu, k \rangle < \langle \delta(w), 0 \rangle$ such that $\mathbb{C}$ is not good at $\langle \nu, k \rangle$ if there is one, and $\langle \eta_0, l_0 \rangle = \langle \delta(w), 0 \rangle$ otherwise. We write

$$M_{\nu,l} = M^{\mathbb{C}}_{\nu,l} \text{ and } \Omega_{\nu,l} = \Omega^{\mathbb{C}}_{\nu,l},$$

for $\langle \nu, l \rangle \leq \langle \eta_0, l_0 \rangle$.

We now compare $(\mathcal{B}, \Psi)$ with itself, by comparing two copies of it with

$$(M_{\nu,l}, \Omega_{\nu,l}).$$

The result will be two trees $\mathcal{S}_{\nu,l}$ and $\mathcal{T}_{\nu,l}$, each on $\mathcal{B}$ and by Ψ. We show that only the two $\mathcal{B}$ sides move in our coiteration, and that no strategy disagreement with $\Omega_{\nu,l}$ shows up. This is done by induction on $\langle \nu, l \rangle$. It is not possible for our coiterations to terminate because $\mathcal{B}$ is nontrivial, so we end up with $\mathcal{B}$ iterating past $M^{\mathbb{C}}_{\eta_0, l_0}$. This leads to a contradiction.

Let $\mathcal{C}$ be a premouse. For $\eta < \hat{o}(\mathcal{C})$, we let $E^{\mathcal{C}}_\eta = \dot{E}^{\mathcal{C}}_\eta$, and for $\eta = \hat{o}(\mathcal{C})$, we let $E^{\mathcal{C}}_\eta = \dot{F}^{\mathcal{C}}$. If $\mathcal{C}$ is a bicephalus, and $\eta < \hat{o}(\mathcal{C})$, then we set $E^{\mathcal{C}}_\eta = \dot{E}^{\mathcal{C}}_\eta$. If $\eta = \hat{o}(\mathcal{C})$, we leave $E^{\mathcal{C}}_\eta$ undefined.

Fix $\langle \nu, l \rangle$, and suppose we have defined $\mathcal{S}_{\mu,j}$ and $\mathcal{T}_{\mu,j}$ for all $\langle \mu, j \rangle <_{\text{lex}} \langle \nu, l \rangle$. (The trees are empty until $\mathbb{C}$ has gone well past $0^\sharp$.) We define normal trees $\mathcal{S} = \mathcal{S}_{\nu,l}$ and $\mathcal{T} = \mathcal{U}_{\nu,l}$ on $\mathcal{B}$ by induction. At stage α, we have $\mathcal{S}^\alpha$ and $\mathcal{T}^\alpha$ with last models

$$\mathcal{C} = \mathcal{M}^{\mathcal{S}^\alpha}_\infty \text{ and } \mathcal{D} = \mathcal{M}^{\mathcal{T}^\alpha}_\infty.$$

We are not padding the trees, so $\text{lh}(\mathcal{S}^\alpha) \neq \text{lh}(\mathcal{T}^\alpha)$ is possible. We shall show that if there is a disagreement between $M_{\nu,l}$, $\mathcal{C}$, and $\mathcal{D}$, then the least one is an extender disagreement at a stage that is passive in $M_{\nu,l}$.

CLAIM 10.1.4. *(a) Suppose that $M \trianglelefteq M_{\nu,l}$, $M \trianglelefteq \mathcal{C}$, and $M \trianglelefteq \mathcal{D}$; then $(\Omega_{\nu,l})_M = \Psi_{\mathcal{S}^\alpha, M} = \Psi_{\mathcal{T}^\alpha, M}$.*

(b) Suppose that $M_{\nu,l}|\gamma$ is extender active, and that $M_{\nu,l}||\gamma \trianglelefteq \mathcal{C}$ and $M_{\nu,l}||\gamma \trianglelefteq \mathcal{D}$; then $M_{\nu,l}|\gamma \trianglelefteq \mathcal{C}$ and $M_{\nu,l}|\gamma \trianglelefteq \mathcal{D}$.

Claim 10.1.4 is of course the main point. Before proving it, let us assume it and complete the proof of the theorem.

Case 1. $M_{\nu,l} \trianglelefteq \mathcal{C}$ and $M_{\nu,l} \trianglelefteq \mathcal{D}$.

By the claim, we must have that $(M_{\nu,l}, \Omega_{\nu,l}) \triangleleft (\mathcal{C}, \Psi_{\mathcal{S}^\alpha, \mathcal{C}})$ and $(M_{\nu,l}, \Omega_{\nu,l}) \triangleleft (\mathcal{D}, \Psi_{\mathcal{T}^\alpha, \mathcal{D}})$. (Claim 10.1.4 implies we never get "half" of a bicephalus lining up with an $M_{\nu,l}$.) We stop the construction of $\mathcal{S}_{\nu,l}$ and $\mathcal{T}_{\nu,l}$, and go on to $\mathcal{S}_{\nu,l+1}$ and $\mathcal{T}_{\nu,l+1}$.

Case 2. Otherwise.

By Claim 10.1.4, the least disagreement between $M_{\nu,l}$ and one of $\mathcal{C}$ and $\mathcal{D}$ occurs at γ, where

(a) $M_{\nu,l}|\langle\gamma,0\rangle$ is extender-passive,
(b) $M_{\nu,l}|\langle\gamma,0\rangle = \mathcal{C}|\langle\gamma,-1\rangle = \mathcal{D}|\langle\gamma,-1\rangle$, and

$$(\Omega_{\nu,l})_{\langle\gamma,0\rangle} = \Psi_{\mathcal{S}^\alpha,\langle\gamma,-1\rangle} = \Psi_{\mathcal{T}^\alpha,\langle\gamma,-1\rangle},$$

and
(c) at least one of $\mathcal{C}|\langle\gamma,0\rangle$ and $\mathcal{D}|\langle\gamma,0\rangle$ is extender-active.

We get $\mathcal{S}^{\alpha+1}$ and $\mathcal{T}^{\alpha+1}$ as follows. Let $\eta = o(M_{\nu,l}|\langle\gamma,0\rangle)$. Let

$$\mathcal{C} = \mathcal{M}_\xi^{\mathcal{S}^\alpha} \text{ and } \mathcal{D} = \mathcal{M}_\tau^{\mathcal{T}^\alpha}.$$

Suppose $\eta < o(\mathcal{C})$, or $\eta = o(\mathcal{C})$ but $\mathcal{C}$ is not a bicephalus, because $[0,\xi]_S$ dropped. If $E_\eta^{\mathcal{C}} \neq \emptyset$, then we set

$$E_\xi^{\mathcal{S}^{\alpha+1}} = (E_\eta^{\mathcal{C}})^+,$$

and $\mathcal{S}^{\alpha+1}$ is then determined by normality. If $E_\eta^{\mathcal{C}} = \emptyset$, then $\mathcal{S}^{\alpha+1} = \mathcal{S}^\alpha$. Similarly, if $\eta < o(\mathcal{D})$ or $\mathcal{D}$ is not a bicephalus, and $E_\eta^{\mathcal{D}} \neq \emptyset$, then we set

$$E_\tau^{\mathcal{T}^{\alpha+1}} = (E_\eta^{\mathcal{D}})^+,$$

and $\mathcal{T}^{\alpha+1}$ is determined by normality. If $E_\eta^{\mathcal{D}} = \emptyset$, then $\mathcal{T}^{\alpha+1} = \mathcal{T}^\alpha$.

If $\eta = o(\mathcal{C})$ and $\mathcal{C}$ is a bicephalus, then if $E_\tau^{\mathcal{T}^{\alpha+1}}$ has already been determined, we let

$$E_\eta^{\mathcal{S}^{\alpha+1}} = G^+,$$

where

$$G = \begin{cases} \dot{F}^{\mathcal{D}} & \text{if } \dot{F}^{\mathcal{D}} \neq E_\tau^{\mathcal{T}^{\alpha+1}}, \\ \dot{G}^{\mathcal{D}} & \text{otherwise.} \end{cases}$$

If $E_\tau^{\mathcal{T}^{\alpha+1}}$ is not yet determined, then $o(\mathcal{D}) = \eta$ and $\mathcal{D}$ is also a bicephalus, and we set

$$E_\xi^{\mathcal{S}^{\alpha+1}} = (\dot{F}^{\mathcal{C}})^+,$$

and

$$E_\tau^{\mathcal{T}^{\alpha+1}} = \begin{cases} (\dot{F}^{\mathcal{D}})^+ & \text{if } \dot{F}^{\mathcal{D}} \neq \dot{F}^{\mathcal{C}}, \\ (\dot{G}^{\mathcal{D}})^+ & \text{otherwise.} \end{cases}$$

Our definitions guarantee that if one of $(E_\xi^{\mathcal{S}})^-$ and $(E_\tau^{\mathcal{T}})^-$ is a top extender of a bicephalus, then $E_\xi^{\mathcal{S}} \neq E_\tau^{\mathcal{T}}$.

This finishes the definition of $\mathcal{S}^{\alpha+1}$ and $\mathcal{T}^{\alpha+1}$. The limit steps in the construction of $\mathcal{S}_{\nu,l}$ and $\mathcal{T}_{\nu,l}$ are determined by Ψ. Note that $\alpha < \beta \Longrightarrow \gamma(\alpha) < \gamma(\beta)$; that is, the common lined up part keeps lengthening.

Eventually, we reach Case 1 above, and the construction of $\mathcal{S}_{\nu,l}$ and $\mathcal{T}_{\nu,l}$ stops. $(\mathcal{B},\Psi)$ has iterated strictly past $(M_{\nu,l},\Omega_{\nu,l})$, in two ways. As in the proof of Lemma 9.5.7, this implies that $\mathbb{C}$ is good at $\langle \nu, l\rangle$. (When $l = -1$ as well.) It follows then that

$$\eta_0 = \delta(w) \quad \text{and} \quad l_0 = 0.$$

However, $(\mathcal{B},\Psi)$ cannot iterate past $M_{\delta(w),0}$, by the proof of Lemma 8.1.4. Note here that $\mathbb{C}$ is good at $\langle \nu, -1\rangle$ for all $\nu < \delta(w)$, so the extenders added to the $M_{\nu,-1}$ are unique, and the universality argument applies. This contradiction completes the proof, modulo Claim 10.1.4.

PROOF OF CLAIM 10.1.4. (Sketch.) We repeat the proof of Theorem 8.4.3. Very little changes, so we shall just discuss the main points.

The main change is the following. We used many times in the proof of Theorem 8.4.3 that for premice Q and R, and Σ an iteration strategy for Q, there is at most one λ-separated plus tree $\mathcal{T}$ by Σ such that $R \trianglelefteq M_\alpha(\mathcal{T})$ for $\alpha+1 = \mathrm{lh}(\mathcal{T})$, and $R \ntrianglelefteq \mathcal{M}_\alpha^{\mathcal{T}}$ whenever $\alpha+1 < \mathrm{lh}(\mathcal{T})$. This uniqueness of iterations past a given R clearly fails for bicephali; let $Q = \mathcal{B}$ and $R = \mathrm{Ult}(\mathcal{B}, \dot{F}^{\mathcal{B}})$. What saves us is that in our situation, with $Q = \mathcal{B}$ and R some initial segment of $M_{\nu,l}$, the trees $\mathcal{S}_{\nu,l}$ and $\mathcal{T}_{\nu,l}$ are being defined together in a way that completely specifies which extender to use at each step on both sides, whether that extender is from the top pair of a bicephalus or not. Moreover, this specification is absolute.

DEFINITION 10.1.5. Let R be a premouse, and suppose $\mathcal{S}$ and $\mathcal{T}$ are λ-separated plus trees on $\mathcal{B}$ of lengths $\alpha+1$ and $\beta+1$ respectively such that

(a) α is the least ξ such that $R \trianglelefteq \mathcal{M}_\xi^{\mathcal{S}}$,

(b) β is the least ξ such that $R \trianglelefteq \mathcal{M}_\xi^{\mathcal{T}}$,

(c) $\mathcal{S}$ and $\mathcal{T}$ are by Ψ, and

(d) the extenders used in $\mathcal{S}$ and $\mathcal{T}$ are chosen according to the rules above, with R playing the role of $M_{\nu,l}$.

Then we call $(\mathcal{S},\mathcal{T})$ the (R,Ψ)- *coiteration.*

SUBCLAIM A.

(1) If $R_0 \trianglelefteq R_1$, and $(\mathcal{S}_i,\mathcal{T}_i)$ is the (R_i,Ψ)-coiteration, then $\mathcal{S}_0$ is an initial segment of $\mathcal{S}_1$ and $\mathcal{T}_0$ is an initial segment of $\mathcal{T}_1$.

(2) If S_0 and S_1 are transitive models of ZFC such that $\mathcal{B}, R \in S_i$ and $\Psi \cap S_i \in S_i$ for $i = 0,1$, and $S_0 \models (\mathcal{S},\mathcal{T})$ is the $(R,\Psi\cap S_0)$-coiteration, then $S_1 \models (\mathcal{S},\mathcal{T})$ is the $(R,\Psi\cap S_1)$-coiteration.

PROOF. This is obvious. ⊣

Let us assume that Claim 10.1.4 is true at $\langle\mu,k\rangle$, for all $\langle\mu,k\rangle <_{\text{lex}} \langle\nu,l\rangle$. We deal first with the extender agreement claim at $\langle\nu,l\rangle$.

SUBCLAIM B. If Claim 10.1.4(a) holds at $\langle\nu,l\rangle$, then Claim 10.1.4(b) holds at $\langle\nu,l\rangle$.

PROOF. Suppose that $M_{\nu,l}|\gamma$ is extender active, $M_{\nu,l}||\gamma \trianglelefteq \mathcal{C}$ and $M_{\nu,l}||\gamma \trianglelefteq \mathcal{D}$. Let F be the last extender of $M_{\nu,l}|\gamma$. We must show that F is on the sequences of $\mathcal{C}$ and $\mathcal{D}$, and not as one of the two top extenders in a bicephalus. Assume otherwise.

We claim first that $l = 0$. For suppose $l = k+1$. F cannot be on the sequence of $M_{\nu,k}$, since otherwise $\mathcal{S}_{\nu,k}$ would agree with $\mathcal{S}_{\nu,l}$ on all extenders used with length $< \text{lh}(F)$, and similarly for $\mathcal{T}_{\nu,k}$ and $\mathcal{T}_{\nu,l}$. But this would mean Claim 10.1.4(b) failed at $\langle\nu,k\rangle$, contrary to our induction hypothesis. It follows that $M_{\nu,k}$ is not sound. That implies that $M_{\nu,k}$ is the last model of $\mathcal{S}_{\nu,k}$, along a branch that dropped to $M_{\nu,l}$. Similarly, $M_{\nu,k}$ is the last model of $\mathcal{T}_{\nu,k}$, along a branch that dropped to $M_{\nu,l}$. Let α be least such that $M_{\nu,l} \trianglelefteq \mathcal{M}_\alpha^{\mathcal{S}_{\nu,k}}$ and β be least such that $M_{\nu,l} \trianglelefteq \mathcal{M}_\beta^{\mathcal{T}_{\nu,k}}$. From Subclaim A(1), we see that $\mathcal{S}_{\nu,l} = \mathcal{S}_{\nu,k} \upharpoonright (\alpha+1)$ and $\mathcal{T}_{\nu,l} = \mathcal{T}_{\nu,k} \upharpoonright (\beta+1)$. Thus $M_{\nu,l}$ is the last model of $\mathcal{S}_{\nu,l}$ and $\mathcal{T}_{\nu,l}$, contradiction.

So $l = 0$. But then F must be the last extender of $M_{\nu,0}$, for otherwise F is on the sequence of some $M_{\mu,k}$ with $\mu < \nu$, and Claim 10.1.4(b) fails at $\langle\mu,k\rangle$, contrary to induction hypothesis.

Thus $M_{\nu,0}$ is extender active with last extender F, $\mathcal{S} = \mathcal{S}^\alpha_{\nu,0}$ and $\mathcal{T} = \mathcal{T}^\alpha_{\nu,0}$ have last models $\mathcal{C}$ and $\mathcal{D}$ respectively, and

$$(M_{\nu,-1}, \Omega_{\nu,-1}) = (\mathcal{C}|\langle\nu,-1\rangle, \Psi_{\mathcal{S},\langle\nu,-1\rangle}) = (\mathcal{D}|\langle\nu,-1\rangle, \Psi_{\mathcal{T},\langle\nu,-1\rangle}).$$

$(\mathcal{S},\mathcal{T})$ is the $(M_{\nu,-1},\Psi)$-coiteration. We want to show that F is on the sequences of $\mathcal{C}$ and $\mathcal{D}$, and not as a top extender of a bicephalus in either case. For this, let

$$j\colon V \to \text{Ult}(V, F_\nu^{\mathbb{C}})$$

be the canonical embedding, and $\kappa = \text{crit}(j)$. ($V = N^*$ at this moment.) We have that $M_{\nu,-1} \trianglelefteq j(M_{\nu,-1})$ by coherence. Note that $\mathbb{C}$ is good at $\langle\nu,-1\rangle$ by Lemma 9.5.7, so $j(M_{\nu,-1})|\nu$ is branch and extender passive by Lemma 9.4.13. $j(\mathcal{S},\mathcal{T})$ is the $(j(M_{\nu,-1}),\Psi)$ coiteration, because $j(\Psi) \subseteq \Psi$. So by Subclaim A, $\mathcal{S}$ is an initial segment of $j(\mathcal{S})$ and $\mathcal{T}$ is an initial segment of $j(\mathcal{T})$.

We have that $\mathcal{M}_\kappa^{\mathcal{S}} = \mathcal{M}_\kappa^{j(\mathcal{S})}$ and $j \upharpoonright \mathcal{M}_\kappa^{\mathcal{S}} = i^{j(\mathcal{S})}_{\kappa,j(\kappa)}$, so F is compatible with the first extender G used in $[\kappa, j(\kappa)]_{j(\mathcal{S})}$. $M_{\nu,-1} \triangleleft \mathcal{M}^{j(\mathcal{S})}_{j(\kappa)}$, so G^- cannot be a proper initial segment of F. But F is not on the sequence of $\mathcal{M}^{j(\mathcal{S})}_{j(\kappa)}$, so F cannot be a proper initial segment of G^-. Hence $F = G^-$. Since $\mathcal{S} = j(\mathcal{S}) \upharpoonright (\xi+1)$, where $\mathcal{C} = \mathcal{M}_\xi^{\mathcal{S}}$, we have that F is on the sequence of $\mathcal{C}$.

Similarly, F is on the sequence of $\mathcal{D}$, and F^+ is used in $j(\mathcal{T})$. But then F cannot be one of the top extenders in a bicephalus in either $j(\mathcal{S})$ or $j(\mathcal{T})$, because our process makes sure that such extenders are part of an extender disagreement.

Thus F is on the sequences of $\mathcal{C}$ and $\mathcal{D}$, but not as one of the top extenders in a bicephalus in either case. ⊣

Finally, we must see that no strategy disagreement shows up in the $(M_{\nu,l},\Psi)$ coiteration of $(\mathcal{B},\Psi)$.

SUBCLAIM C. Claim 10.1.4(a) holds at $\langle \nu, l\rangle$.

PROOF. Suppose that $M \trianglelefteq M_{\nu,l}$, $M \trianglelefteq \mathcal{C}$, and $M \trianglelefteq \mathcal{D}$, where $\mathcal{C}$ and $\mathcal{D}$ are the last models of $\mathcal{S} = \mathcal{S}^{\alpha}_{\nu,l}$ and $\mathcal{T} = \mathcal{T}^{\alpha}_{\nu,l}$ respectively. Let $\Omega = \Omega_{\nu,l}$. We must show that $\Omega_M = \Psi_{\mathcal{S},M} = \Psi_{\mathcal{T},M}$. The situation is symmetric, so it is enough to show

$$\Omega_M = \Psi_{\mathcal{T},M}.$$

We consider first the case that

$$M = M_{\nu,l},$$

and then reduce to this case using the pullback consistency of Ψ. The proof is very close to the proof of Theorem 8.4.3, and we shall keep our notation close to the notation there.

Ω_M and $\Psi_{\mathcal{T},M}$ are determined by their action on λ-separated trees, by the proof of Lemma 7.6.5. So suppose that $\mathcal{U}$ is a λ-separated tree on M of limit length that is by both Ω and $\Psi_{\mathcal{T},M}$, and

$$\Omega(\mathcal{U}) = b.$$

We must show that $\Psi(\langle \mathcal{T}, \mathcal{U}^+\rangle) = b$, for then $\Psi(\langle \mathcal{T}, \mathcal{U}\rangle) = b$ by internal lift consistency. We do that by looking at the embedding normalizations

$$\mathcal{W}_\gamma = W(\mathcal{T}, \mathcal{U}^+ \restriction (\gamma+1))$$

and

$$\mathcal{W}_b = W(\mathcal{T}, (\mathcal{U}^+)^\frown b).$$

These are defined just as they were for trees on premice of the ordinary or least branch variety. $\mathcal{T}$ and $\mathcal{U}$ are λ-separated, so embedding normalization coincides with quasi-normalization. The fact that some models in the trees are bicephali affects nothing. Let us adopt all the previous notation associated to the embedding normalization meta-tree. For example, R_γ is the last model of $\mathcal{W}_\gamma$, and $\sigma_\gamma \colon \mathcal{M}^{\mathcal{U}}_\gamma \to R_\gamma$ is the natural map.

Ω is defined by lifting to V. Let

$$c = c_0 = \langle M, \mathrm{id}, M, \mathbb{C}, V\rangle$$

and

$$\mathrm{lift}(\mathcal{U}^\frown b, c) = \langle \mathcal{U}^*, \langle c_\alpha \mid \alpha < \mathrm{lh}(\mathcal{U})+1\rangle\rangle,$$

where

$$c_\alpha = \langle \mathcal{M}^{\mathcal{U}}_\alpha, \psi_\alpha, Q_\alpha, \mathbb{C}_\alpha, S_\alpha\rangle.$$

So $S_\alpha = \mathcal{M}_\alpha^{\mathcal{U}^*}$, $\mathbb{C}_\alpha = i_{0,\alpha}^{\mathcal{U}^*}(\mathbb{C})$, $Q_\alpha \in \mathrm{lev}(\mathbb{C}_\alpha)$, and $\psi_\alpha : \mathcal{M}_\alpha^{\mathcal{U}} \to Q_\alpha$ is nearly elementary. We write $\psi_b : \mathcal{M}_b^{\mathcal{U}} \to Q_b$ for the last ψ-map, etc.

For $Q <_{\mathbb{C}} M$, let

$$(\mathcal{V}_Q^*, \mathcal{W}_Q^*) = \text{the } (Q,\Psi)\text{-coiteration of } \mathcal{B},$$

For $\gamma < \mathrm{lh}(\mathcal{U})$ or $\gamma = b$, let

$$(\mathcal{V}_\gamma^*, \mathcal{W}_\gamma^*) = (\mathcal{V}_{Q_\gamma}^*, \mathcal{W}_{Q_\gamma}^*)^{S_\gamma}.$$

So if $[0,\gamma]_U$ does not drop in model or degree, $(\mathcal{V}_\gamma^*, \mathcal{W}_\gamma^*) = i_{0,\gamma}^{\mathcal{U}^*}((\mathcal{S},\mathcal{T}))$.

We define by induction tree embeddings Φ_γ from $\mathcal{W}_\gamma$ into $\mathcal{W}_\gamma^*$, for $\gamma < \mathrm{lh}(\mathcal{U})$ or $\gamma = b$, just as before. It is enough to do this, because then

(i) $i_{0,b}^{\mathcal{U}^*}(\Psi) \subseteq \Psi$, so $\mathcal{W}_b^*$ is by Ψ, so

(ii) W_b is by Ψ, so

(iii) $\langle \mathcal{T}, (\mathcal{U}^+)^\frown b\rangle$ is by Ψ.

(i) comes from Code(Ψ) being Universally Baire[2], (ii) from strong hull condensation for Ψ, and (iii) from the fact that Ψ normalizes well.[3]

Let

$$\Phi_\gamma = \langle u^\gamma, v^\gamma, \langle s_\beta^\gamma \mid \beta \le z(\gamma)\rangle, \langle t_\beta^\gamma \mid \beta < z(\gamma)\rangle\rangle.$$

Let us just say a few words about how to obtain $\Phi_{\gamma+1}$, since this is where the only new point lies. We assume we are given the Φ_β for $\beta \le \gamma$, and that the analog of the induction hypotheses $(\dagger)_\gamma$ in the proof of Theorem 8.4.3 holds.

Let $t^\gamma = t_{z(\gamma)}^\gamma$. We have $t^\gamma : R_\gamma \to N_\gamma$, where N_γ is the last model of $\mathcal{W}_\gamma^*$. Let $F = \sigma_\gamma(E_\gamma^{\mathcal{U}})$, and let $\mu = U\text{-pred}(\gamma+1)$. (Sadly, we cannot use "ν" for this ordinal.) So $\mathcal{W}_{\gamma+1} = W(\mathcal{W}_\mu, F)$. Let us assume for simplicity that $(\mu, \gamma+1]_U$ is not a drop in model or degree. Let

$$\mathrm{res}_\gamma = (\sigma_{Q_\gamma}[Q_\gamma | \mathrm{lh}(\psi_\gamma(E_\gamma^{\mathcal{U}}))])^{S_\gamma},$$

and let

$$G = \mathrm{res}_\gamma(t^\gamma(F)) = \mathrm{res}_\gamma(\psi_\gamma(E_\gamma^{\mathcal{U}})).$$

We use here that $t^\gamma \circ \sigma_\gamma = \psi_\gamma^{\mathcal{U}}$ by $(\dagger)_\gamma$. Let G^* be the background extender for G provided by $\mathbb{C}_\gamma$, so that

$$S_{\gamma+1} = \mathrm{Ult}(S_\mu, G^*).$$

Since we are not dropping,

$$\mathcal{W}_{\gamma+1}^* = i_{G^*}(\mathcal{W}_\mu^*),$$

where $i_{G^*} = i_{\mu,\gamma+1}^{\mathcal{U}^*}$. The main thing we need to see is that G is used in $\mathcal{W}_{\gamma+1}^*$.

---

[2]Or one can use Lemma 7.6.7.

[3]Pushforward consistency is used everywhere.

Let $P = Q_\gamma | \mathrm{lh}(t^\gamma(F))$, θ be least such that $P \trianglelefteq \mathcal{M}_\theta^{\mathcal{V}_\gamma^*}$, and τ least such that $P \trianglelefteq \mathcal{M}_\tau^{\mathcal{W}_\gamma^*}$. Let $(\mathcal{V}_\gamma^{**}, \mathcal{W}_\gamma^{**})$ be the $(\mathrm{res}_\gamma(P), \Psi)$-coiteration of $\mathcal{B}$. By the counterpart of Lemma 8.3.1,

(i) $\mathcal{W}_\gamma^{**}$ extends $\mathcal{W}_\gamma^* \restriction (\tau + 1)$,

(ii) letting $\xi = \mathrm{lh}(\mathcal{W}_\gamma^{**}) - 1$, G is on the $\mathcal{M}_\xi^{\mathcal{W}_\gamma^{**}}$ sequence, and not on the $\mathcal{M}_\alpha^{\mathcal{W}_\gamma^{**}}$ sequence for any $\alpha < \xi$,

(iii) $\tau \leq_{W_\gamma^{**}} \xi$, and $\hat{\imath}_{\tau,\xi}^{\mathcal{W}_\gamma^{**}} \restriction \mathrm{lh}(t^\gamma(F)) + 1 = \mathrm{res}_\gamma \restriction \mathrm{lh}(t^\gamma(F)) + 1$, and

(iv) similarly for $\mathcal{V}_\gamma^{**}$ vis-a-vis $\mathcal{V}_\gamma^*$.

P, $\mathrm{res}_\gamma(P)$, and Q_μ all agree up to $\mathrm{dom}(G)$, so

$$\mathrm{res}_\gamma(P) || \mathrm{lh}(G) \trianglelefteq i_{G^*}(Q_\mu) = Q_{\gamma+1},$$

and $i_{G^*}(Q_\mu) | \mathrm{lh}(G)$ is extender-passive, by coherence. Thus $o(\mathrm{res}_\gamma(P)) = \mathrm{lh}(G)$, and

$$\mathrm{res}_\gamma(P) || \mathrm{lh}(G) = Q_{\gamma+1} | \mathrm{lh}(G).$$

$(\mathcal{V}_\gamma^{**}, \mathcal{W}_\gamma^{**})$ is the $(\mathrm{res}_\gamma(P), \Psi)$-coiteration of $\mathcal{B}$ in S_γ, so by the absoluteness in Subclaim A(2), $(\mathcal{V}_\gamma^{**}, \mathcal{W}_\gamma^{**})$ is the $(\mathrm{res}_\gamma(P), \Psi)$-coiteration of $\mathcal{B}$ in $S_{\gamma+1}$. Since $(\mathcal{V}_{\gamma+1}^*, \mathcal{W}_{\gamma+1}^*)$ is the $(Q_{\gamma+1}, \Psi)$-coiteration of $\mathcal{B}$ in $S_{\gamma+1}$, we get that $\mathcal{V}_\gamma^{**}$ is an initial segment of $\mathcal{V}_{\gamma+1}^*$, $\mathcal{W}_\gamma^{**}$ is an initial segment of $\mathcal{W}_{\gamma+1}^*$ and G is used in both $\mathcal{V}_{\gamma+1}^*$ and $\mathcal{W}_{\gamma+1}^*$. It matters here that $\mathrm{res}_\gamma(P)$ is a premouse, not a bicephalus, so both trees are forced to use G by our rules.(Recall that $F = t^\gamma(E_\gamma^{\mathcal{U}})$, and $\mathcal{U}$ is a tree on the premouse M.)

This completes our sketch of the case $M = M_{\nu,l}$. The only part of the proof going beyond the proof of Theorem 8.4.3 is the use of the absoluteness of our coiteration process in the preceding paragraph.

Now let $M \triangleleft M_{\nu,l}$, and

$$\langle \nu_0, l_0 \rangle = \mathrm{Res}_{M_{\nu,l}}[M],$$
$$\pi = \sigma_{M_{\nu,l}}[M].$$

$(\Omega_{\nu,l})_M$ is defined by $(\Omega_{\nu,l})_M = \Omega_{\nu_0,l_0}^\pi$. By induction, the (M_{ν_0,l_0}, Ψ)-coiteration is a pair $(\mathcal{V}^*, \mathcal{W}^*)$ such that M_{ν_0,l_0} is the last model of $\mathcal{W}^*$, and $\Omega_{\nu_0,l_0} = \Psi_{\mathcal{W}^*, M_{\nu_0,l_0}}$. By the counterpart of Lemma 8.3.1, the last drop along the main branch of $\mathcal{W}^*$ was to M, and the branch embedding is the resurrection map π, that is,

$$\pi = \hat{\imath}_{\xi,\theta}^{\mathcal{W}^*} : M \to M_{\nu_0,l_0}.$$

Here ξ is least such that $M \trianglelefteq \mathcal{M}_\xi^{\mathcal{W}^*}$, so the (M, Ψ) coiteration $(\mathcal{S}, \mathcal{T})$ of $\mathcal{B}$ is such that

$$\mathcal{W}^* \restriction (\xi + 1) = \mathcal{T}.$$

But then

$$\begin{aligned}\Psi_{\mathcal{T},M} &= (\Psi_{\mathcal{W}^*,M_{\nu_0,l_0}})^{\hat{\imath}^{\mathcal{W}^*}_{\xi,\theta}} \\ &= (\Omega_{\nu_0,l_0})^{\pi} \\ &= (\Omega_{\nu,l})_M.\end{aligned}$$

The first equality holds because Ψ normalizes well and has strong hull condensation, and is therefore pullback consistent. ⊣

The proof of Subclaim C completes our proof of Claim 10.1.4, and that in turn completes the proof of Theorem 10.1.3. ⊣

⊣

The theorem implies that background constructions done in an appropriate environment do not admit nontrivial bicephali. See Theorem 10.4.1.

10.2. The Pseudo-premouse Lemma

The definition of pseudo-premice in Section 4.9 goes over without change to least branch premice. A *pseudo-lpm* is a structure (N,A) such that

(1) N is a passive lpm having a largest cardinal λ, and A is an N amenable predicate coding an extender $G = G(N,A)$ over N by means of its fragments,
(2) λ is a limit cardinal of N, $\lambda = \lambda(G \restriction \lambda)$, and λ is a generator of G,
(3) $N = \mathrm{Ult}(N,G)|\lambda^{+,\mathrm{Ult}(N,G)}$, and λ is not measurable by the $\mathrm{Ult}(N,G)$-sequence,
(4) letting $\gamma = \gamma(G,\lambda) = \lambda^{+,\mathrm{Ult}(N,G\restriction\lambda)}$, it is not the case that $\gamma < o(N)$ and E^N_γ is the trivial completion of $G \restriction \lambda$.

We shall usually identify A with the extender $G = G(N,A)$ of length $o(N)$ that is coded by A. We also write $G = \dot{G}^{(N,A)}$ when $G = G(N,A)$. Notice that both A and N can be recovered from G. Thus if (N,G) is a pseudo-lpm, we may set

$$\begin{aligned}\mathrm{lh}(G) &= o(N), \\ \lambda(G) &= \text{largest cardinal of } N, \\ \gamma(G) &= \lambda(G)^{+,\mathrm{Ult}(N,G\restriction\lambda(G))}, \\ F(G) &= \text{Jensen completion of } G\restriction\lambda(G).\end{aligned}$$

If (N,G) is a pseudo-lpm, then $\lambda(G)$ is a generator of G.[4] If $\gamma(G) = \mathrm{lh}(G)$, then $\lambda(G)$ is the largest generator of G. If not, then $\gamma(G)$ is the largest generator of G.

If (M,H) is another pseudo-lpm, then we say that (N,G) is an *initial segment of* (M,H) iff $N = M||o(M)$ and $G = H\restriction o(N)$.

[4]Using $\lambda(G)$ for the largest cardinal of N conflicts with our general convention for extenders that $\lambda(G) = i_G(\mathrm{crit}(G))$. Let us just drop this general convention in the special case that G is the last extender of a pseudo-premouse.

Let us say

$$(N,G) \text{ is } \textit{bad} \text{ iff } G \neq F(G)^+.$$

The definitions and results associated to iteration trees and iteration strategies all go over to trees on pseudo-lpms. The fact that the top extender is coded somewhat differently, and more importantly, does not satisfy the initial segment condition, does not affect anything important. We restrict our iteration strategies to stacks of λ-separated trees. One thing to note is that if $\mathcal{T}$ is a plus tree on (N,G) and $[0,\alpha)_T = \emptyset$ and $H = G(\mathcal{M}_\alpha^\mathcal{T})$ is the top extender of $\mathcal{M}_\alpha^\mathcal{T}$, then we do not allow $E_\alpha^\mathcal{T} = H^+$. $\mathcal{T}$ is λ-separated iff it always uses plus extenders, subject to this restriction. That is, we treat the top extender of a pseudo-lpm as if it were already a plus extender.

A *pseudo-lpm pair* is a pair $((N,G),\Omega)$ such that (N,G) is a pseudo-lpm and Ω is a complete strategy defined on finite stacks of λ-separated trees. The notions of normalizing well, having strong hull condensation, being internally lift consistent, and being pushforward consistent extend routinely to such pairs.

We get a bad pseudo-lpm pair from the failure of plus consistency in some least branch construction. Namely, suppose that $\mathbb{C}$ is a least branch construction, ξ is a limit ordinal, and $\mathbb{C}$ is good at all $\eta < \xi$. Suppose that for $M = (M^{<\xi})^\mathbb{C}$, (M,F) is an lpm, and F^* is the $\mathbb{C}$-minimal certificate for F. Let $\kappa = \mathrm{crit}(F)$, and set

$$\phi = i_{F^*}^V \restriction M$$

and

$$G = \begin{cases} E_\phi \restriction \mathrm{lh}(F) & \text{if } \mathrm{lh}(F) = \lambda_F^{+,\phi(M|\kappa)}, \\ E_\phi \restriction \mathrm{lh}(F)+1 & \text{if } \mathrm{lh}(F) < \lambda_F^{+,\phi(M|\kappa)}, \end{cases}$$

$$N = \mathrm{Ult}(M,G)||(\lambda_F^+)^{\mathrm{Ult}(M,G)}.$$

G has an N-amenable code obtained by coding its fragments, and (N,G) is then a pseudo-lpm, with $F = F(G)$. Moreover, (N,G) has a complete iteration strategy Ψ induced by the "pseudo-construction" $\mathbb{C}^\frown\langle (N,G),F^*\rangle$.

DEFINITION 10.2.1. In the situation just described, we write $\Psi = \Omega(\mathbb{C},F)$, $G = G(\mathbb{C},F)$, and $N = N(\mathbb{C},F)$. We say that $((N,G),\Psi)$ is a *pseudo-lpm pair associated to* $\mathbb{C}$ *at* ξ. We say that $((N,G),\Psi)$ is bad iff (N,G) is bad. We also say that $F(G)$ is $\mathbb{C}$-*bad at* ξ in this case.

It follows easily from the definitions that $\mathbb{C}$ is plus consistent at ξ iff there is no bad pseudo-lpm pair associated to $\mathbb{C}$ at ξ.

LEMMA 10.2.2. *Assume* AD^+, *and let* $\mathbb{C}$ *be the maximal construction of a coarse strategy pair, and* $((N,G),\Psi)$ *be a pseudo-lpm pair associated to* $\mathbb{C}$; *then* Ψ *normalizes well for stacks of* λ*-separated trees and has strong hull condensation, and* $((N,G),\Psi)$ *is internally lift consistent and pushforward consistent.*

PROOF SKETCH. The proof we have given for the lbr hod pairs associated to $\mathbb{C}$ work here too. Notice here that the $\mathbb{C}$-minimal certificate F^* for $F(G)$ backgrounds G in such a way that λ-separated trees on (N,G) can use images of G without giving rise to the background coherence problem in the proof that Ψ normalizes well. ⊣

The proof that there are no iterable bad pseudo-premice given in Section 4.9 generalizes as follows.

THEOREM 10.2.3. *Assume* AD^+, *and let* $((N,G),\Psi)$ *be a pseudo-lpm pair such that N is countable,* Ψ *has scope* HC, *and* $((N,G),\Psi)$ *normalizes well, has strong hull condensation, and is internally lift consistent and pushforward consistent; then* (N,G) *is not bad.*

PROOF. The Weak Dodd-Jensen Lemma adapts to pseudo-pairs, so we may assume that Ψ has the Weak Dodd-Jensen property relative to the enumeration $\vec{e}$ of N.[5] Similarly, we may assume that $((N,G),\Psi)$ is *minimally bad*, in that whenever $((M,H),\Lambda)$ is an iterate of $((N,G),\Psi)$, and $(M||\gamma,H\restriction\gamma)$ is a proper initial segment of (M,H), then $(M||\gamma,H\restriction\gamma)$ is not a bad pseudo-lpm.

Following the pattern in Section 4.9, we compare G with $F(G)^+$ by comparing the phalanx $(((N,G),\Psi),\mathrm{Ult}_0(((N,G),\Psi),F(G)),\lambda(G))$ with $((N,G),\Psi)$. Because we are comparing iteration strategies, this must be done by comparing both with the levels $M_{\nu,l}$ of a construction done in a universe that captures Code(Ψ), and that involves re-starting the comparison at lifted versions of this phalanx many times. The model $\mathrm{Ult}_0((N,G),F(G))$ ends up being irrelevant; only various lifts of it matter. Hence we shall drop it from the notation.

By the Basis Theorem, we may assume Code(Ψ) is Suslin and co-Suslin, and hence we can fix a coarse strategy pair $((N^*,\in,w,\mathcal{F},\Phi),\Phi^*)$ that captures Code(Ψ). Let $\mathbb{C}$ be the maximal $(w,\mathcal{F},\Phi)$-construction of $(N^*,\in,w,\mathcal{F},\Phi)$ of length $\leq\langle\delta(w),1\rangle$. We write

$$M_{\nu,l}=M^{\mathbb{C}}_{\nu,l} \text{ and } \Omega_{\nu,l}=\Omega^{\mathbb{C}}_{\nu,l},$$

for $\langle\nu,l\rangle<\mathrm{lh}(\mathbb{C})$, and let

$$Q_0=(N,G).$$

Let us assume toward contradiction that Q_0 is bad.

We shall be considering the comparison of (Q_0,Ψ) with $(M_{\nu,l},\Omega_{\nu,l})$ in which we iterate away least extender disagreements, forming a λ-separated tree $\mathcal{U}_{\nu,l}$ on (Q_0,Ψ). We shall show that no strategy disagreements show up, and the $M_{\nu,l}$ side does not move. In this connection, if the current last model $((P,H),\Lambda)$ of $\mathcal{U}_{\nu,l}$ is a non-dropping iterate of (Q_0,Ψ) and $P=M_{\nu,l}||o(P)$, then we require that the next extender used in $\mathcal{U}_{\nu,l}$ be H. Thus (Q_0,Ψ) cannot iterate to any $(M_{\nu,l},\Omega_{\nu,l})$, simply

[5]Here we are assuming that Ψ is defined on all countable stacks of λ-separated trees, and that each tail Ψ_s normalizes well for finite stacks. We cannot use the full Dodd-Jensen Lemma unless we compare the iteration strategy components of pseudo-lpm pairs. We are not going to do that.

by the rules for $\mathcal{U}_{\nu,l}$. We shall show that what happens in this case is that for some $\xi < \delta(w)$, $\mathbb{C}$ breaks down at $\langle \xi, -1 \rangle$, and (Q_0, Ψ) iterates into the unique proof that $\mathbb{C}$ is not plus consistent at ξ.

CLAIM 10.2.4. *Suppose that $(M^{<\nu}, F)$ is an lpm and F^* is a $\mathbb{C}$-minimal certificate for F. Let $\mathcal{U}$ be a λ-separated tree whereby (Q_0, Ψ) iterates strictly past $((M^{<\nu}, \emptyset), \Omega^{<\nu})$, and α be least such that $M^{<\nu} \trianglelefteq_0 \mathcal{M}^{\mathcal{U}}_\alpha$; then either*

(i) F is not $\mathbb{C}$-bad at ν and F is on the $\mathcal{M}^{\mathcal{U}}_\alpha$-sequence, or

(ii) F is $\mathbb{C}$-bad at ν, $[0, \alpha)_U \cap D^{\mathcal{U}} = \emptyset$ and $\mathcal{M}^{\mathcal{U}}_\alpha = (N(\mathbb{C}, F), G(\mathbb{C}, F))$.

PROOF. Let $\kappa = \text{crit}(F)$, $i^* = i_{F^*}$, and $\mathcal{V} = i^*(\mathcal{U})$. We have that $\mathcal{U} \upharpoonright \alpha + 1 = \mathcal{V} \upharpoonright \alpha + 1$ and

$$\hat{\imath}^{\mathcal{V}}_{\kappa, i^*(\kappa)} = i^* \upharpoonright \mathcal{M}^{\mathcal{U}}_\kappa,$$

as usual. Let $\eta + 1 <_V i^*(\kappa)$ and $V\text{-pred}(\eta + 1) = \kappa$. Let

$$H = E^{\mathcal{V}}_\eta.$$

H is compatible with the extender of i^*, and hence $H \upharpoonright \lambda_F = F \upharpoonright \lambda_F$.

Let us call τ a *top node* of $\mathcal{V}$ iff $[0, \tau]_V \cap D^{\mathcal{V}} = \emptyset$ and $E^{\mathcal{V}}_\tau$ is the top extender of $\mathcal{M}^{\mathcal{V}}_\tau$. Similarly for $\mathcal{U}$.

Case 1. η is a not a top node of $\mathcal{V}$.

In this case $\mathcal{M}^{\mathcal{V}}_\eta | \text{lh}(H)$ is an ordinary lpm with last extender H^-. If $\lambda(H) < \lambda_F$, then H^- is on the sequence of $M^{<\nu}$, and hence on the sequence of $\mathcal{M}^{\mathcal{V}}_{i^*(\kappa)}$, contradiction. If $\lambda_F < \lambda(H)$, then F is on the sequence of $\mathcal{M}^{\mathcal{V}}_{i^*(\kappa)}$, and hence on the sequence of $i^*(M^{<\nu})$, contradiction. So $\lambda(H) = \lambda_F$, and hence $F = H^-$, and F is on the sequence of $\mathcal{M}^{\mathcal{V}}_\eta$. By coherence, $\eta = \alpha$, and F is on the sequence of $\mathcal{M}^{\mathcal{U}}_\alpha$.

$E^{\mathcal{V}}_\eta = H = F^+$, so F^+ is an initial segment of the extender of i^*. Moreover, $\text{lh}(F)$ is a cardinal of $\mathcal{M}^{\mathcal{V}}_{i^*(\kappa)}$, and hence a cardinal of $i^*(M^{<\nu})$. So in this case we have alternative (i) of Claim 10.2.4.

Case 2. η is a top node of $\mathcal{V}$.

Let $\beta = \lambda(H)$ be the largest cardinal of $\mathcal{M}^{\mathcal{V}}_\eta$. We claim that $\beta = \lambda_F$. If $\beta < \lambda_F$, then the initial segment condition for $(M^{<\nu}, F)$ implies that $F(H)$ is on the sequence of $M^{<\nu}$. But $\mathcal{M}^{\mathcal{V}}_\eta$ is a pseudo-lpm, so $F(H)$ is not on the sequence of $\text{Ult}(\mathcal{M}^{\mathcal{V}}_\eta, H)$, and hence by coherence it is not on the sequence of $\mathcal{M}^{\mathcal{V}}_{i^*(\kappa)}$, and hence not that of $M^{<\nu}$, contradiction.

Suppose toward contradiction that $\lambda_F < \beta$. Let $\beta_0 = \lambda(G)$ be the largest cardinal of Q_0. The whole initial segments of G below β_0 are on the Q_0-sequence, so the whole initial segments of H below $\sup i^{\mathcal{V}}_{0,\eta}$"$\beta_0$ are on the $\mathcal{M}^{\mathcal{V}}_\eta$-sequence. Since F is not on that sequence, $\sup i^{\mathcal{V}}_{0,\eta}$"$\beta_0 \le \lambda_F$. If $\text{lh}(F) \le \beta$ then $i^{\mathcal{V}}_{0,\eta}$ is discontinuous at β_0, and we get that $\text{lh}(F)$ is not a cardinal in $\mathcal{M}^{\mathcal{V}}_\eta$ by the same argument as in the proof of Theorem 4.9.1.[6] Thus $\mathrm{lh}(F)$ is not a cardinal in $i_{F^*}(M^{<\nu})$, so that $(N(\mathbb{C},F),G(\mathbb{C},F))$ exists and is bad. But $H\restriction\gamma=(F^*\cap[\gamma]^{<\omega}\times M^{<\nu})$, so $(\mathcal{M}^{\mathcal{V}}_\eta||\gamma,H\restriction\gamma)=(N(\mathbb{C},F),G(\mathbb{C},F))$. Thus $\mathcal{M}^{\mathcal{V}}_\eta$ has a bad proper initial segment, contrary to the minimality of $(N,G),\Psi)$.

Thus $\beta\leq\lambda_F$. It follows that $F=F(H)$, and $\mathrm{lh}(H)=\beta^{+,\mathcal{M}^{\mathcal{V}}_{i^*(\kappa)}}=\lambda_F^{+,i^*(M^{<\nu})}$. If $\alpha<\eta$ then there is a cardinal of $\mathcal{M}^{\mathcal{V}}_\eta$ between λ_F and $\mathrm{lh}(H)$, but this is not the case, so $\alpha=\eta$. $\mathcal{M}^{\mathcal{V}}_\alpha=\mathcal{M}^{\mathcal{U}}_\eta$ is a bad pseudo-lpm because badness is preserved by non-dropping iterations[7], moreover $G(\mathbb{C},F)=H$ because $H\subseteq F^*$. It follows that F is $\mathbb{C}$-bad and $(N(\mathbb{C},F),G(\mathbb{C},F))=\mathcal{M}^{\mathcal{U}}_\alpha$. So in this case we have alternative (ii) of the claim. ⊣

CLAIM 10.2.5. $\mathrm{lh}(\mathbb{C})=\langle\nu,0\rangle$ *for some* $\nu<\delta(w)$ *such that* $\mathbb{C}$ *is not plus consistent at* ν*; moreover,*

(a) there is a unique F *such that* F *is* $\mathbb{C}$*-bad at* ν*,*
(b) (Q_0,Ψ) *iterates to* $((N(\mathbb{C},F),G(\mathbb{C},F),\Omega(\mathbb{C},F))$*, and*
(c) (Q_0,Ψ) *iterates strictly past* $(M_{\eta,j},\Omega_{\eta,j})$*, for all* $\eta<\nu$ *and* $j<\omega$*.*

The iterations in (b) and (c) are via λ*-separated trees.*

PROOF. We show by induction on ν that if $\mathbb{C}$ is plus consistent at ν, then $\mathbb{C}$ does not break down at $\langle\nu,l\rangle$ and (Q_0,Ψ) iterates via a λ-separated $\mathcal{U}_{\nu,l}$ strictly past $(M_{\nu,l},\Omega_{\nu,l})$, for all $l<\omega$. The proof that no strategy disagreements show up is essentially the same as that of Theorem 9.5.2, the bulk of which is in Section 8.4, so we omit the numerous further details. Claim 10.2.4 implies that $\mathbb{C}$ is extender unique at ν, and that for any $l<\omega$, no extenders from $M_{\nu,l}$ participate in a least disagreement. The iteration $\mathcal{U}_{\nu,l}$ must go strictly past $(M_{\nu,l},\Omega_{\nu,l})$ because Q_0 is bad, and since it does go strictly past, $\mathbb{C}$ does not break down at $\langle\nu,l\rangle$ for any l such that $0\leq l<\omega$.

$\mathbb{C}$ must break down at some $\langle\nu,l\rangle<\langle\delta(w),0\rangle$ by the universality argument in the proof of Lemma 8.1.4. By the last paragraph, the breakdown is a failure of plus consistency at ν. By Claim 10.2.4, this failure is unique, and has the form described in (a) and (b). ⊣

Fix ν_0 having the properties of ν in Claim 10.2.5. For $\langle\nu,l\rangle\leq_{\mathrm{lex}}\langle\nu_0,0\rangle$, let $\mathcal{U}_{\nu,l}$ be the λ-separated tree on (Q_0,Ψ) given by (b) or (c) of the claim.

Following the proof of Theorem 4.9.1, we should now define a tree $\mathcal{S}_{\nu,l}$ that compares the phalanx $(((N,G),\Psi),\mathrm{Ult}_0(((N,G),\Psi),F(G)),\lambda(G))$ with $(M_{\nu,l},\Omega_{\nu,l})$. It is notationally more convenient, however, to think of $\mathcal{S}_{\nu,l}$ as an iteration tree on $((N,G),\Psi)$ that is allowed to use extenders of the form $F(\dot{G}^{P_\alpha})$ at certain special stages α. This will not actually happen when $\alpha=0$, because the least disagreement

[6]See the proof Theorem 4.9.1, Claim 2, Subcase 1B.
[7]See Claim 0 in the proof of Theorem 4.9.1.

between (N,G) and $M_{\nu,l}$ will be at the least measurable of N, below $\lambda(G)$.[8] The construction of $\mathcal{S}_{\nu,l}$ guarantees that at most one such special extender is used along any branch.

More precisely, fix ν and l, and let $\mathcal{U} = \mathcal{U}_{\nu,l}$. At the same time that we define $\mathcal{S} = \mathcal{S}_{\nu,l}$, we copy it to a λ-separated tree $\mathcal{T} = \mathcal{T}_{\nu,l}$ on N that is by Ψ. Let

$$P_\theta = \mathcal{M}_\theta^{\mathcal{S}},$$
$$P_\theta^* = \mathcal{M}_\theta^{\mathcal{T}},$$
$$Q_\theta = \mathcal{M}_\theta^{\mathcal{U}},$$

and

$$\pi_\theta : P_\theta \to P_\theta^*$$

be the copy map. So $P_0 = P_0^* = Q_0 = (N,G)$, and π_0 is the identity. π_θ will be elementary by the results of Section 4.5, which generalize routinely to our current context. The (possibly partial) branch embeddings of $\mathcal{S}, \mathcal{T}$, and $\mathcal{U}$ are

$$i_{\alpha,\beta} = \hat{\imath}^{\mathcal{S}}_{\alpha,\beta},$$
$$i^*_{\alpha,\beta} = \hat{\imath}^{\mathcal{T}}_{\alpha,\beta},$$

and

$$j_{\alpha,\beta} = \hat{\imath}^{\mathcal{U}}_{\alpha,\beta}.$$

Since $k(P_0) = 0$, all branch embeddings are elementary and exact. The copy maps π_θ have the usual commutativity and agreement properties. The strategies attached to P_θ, P_θ^*, and Q_θ are

$$\Sigma_\theta^* = \Psi_{\mathcal{T}\restriction\theta+1},$$
$$\Sigma_\theta = (\Sigma_\theta^*)^{\pi_\theta},$$

and

$$\Lambda_\theta = \Psi_{\mathcal{U}\restriction\theta+1}.$$

$(P_\theta, \Sigma_\theta), (P_\theta^*, \Sigma_\theta^*)$, and $(Q_\theta, \Lambda_\theta)$ will be pseudo-lpm pairs along non-dropping branches, and lbr hod pairs otherwise. Finally, we have ordinals

$$\varepsilon_\theta = \lambda(E_\theta^{\mathcal{S}})$$

that tell us which model we should apply the next extender to. We shall have $\mathrm{lh}(E_\xi^{\mathcal{S}}) < \mathrm{lh}(E_\gamma^{\mathcal{S}})$ and $P_\xi || \mathrm{lh}(E_\xi^{\mathcal{S}}) = P_\gamma | \mathrm{lh}(E_\xi^{\mathcal{S}})$ whenever $\xi < \gamma$. There are three possibilities for $E_\theta^{\mathcal{S}}$: either

(i) $[0,\theta]_S \cap D^{\mathcal{S}} = \emptyset$ and $E_\theta^{\mathcal{S}} = \dot{G}^{P_\theta} = G(P_\theta)$, or

---

[8] So if we thought of $\mathcal{S}_{\nu,l}$ as a tree on $(((N,G),\Psi), \mathrm{Ult}_0(((N,G),\Psi), F(G)), \lambda(G))$, the first thing it would do is move this phalanx up via an extender from N, and no later extender would be applied to $\mathrm{Ult}_0(N, F(G))$. Thus $\mathrm{Ult}_0(N,G)$ would play no role in $\mathcal{S}_{\nu,l}$.

(ii) $[0,\theta]_S \cap D^{\mathcal{S}} = \emptyset$ and $E_\theta^{\mathcal{S}} = F(\dot{G}^{P_\theta})$, or
(iii) $E_\theta = F^+$, where F is on the sequence of P_θ, and if $[0,\theta]_S \cap D^{\mathcal{S}} = \emptyset$, then $F \neq \dot{G}^{P_\theta}$.

We say that θ is *special* iff (ii) holds.

As we go, we define what it is for a node θ of $\mathcal{S}$ to be stable.[9] 0 is unstable, and 1 is stable. We maintain by induction:

Induction hypotheses.

(a) If θ is unstable, then
  (i) $0 \leq_S \theta$, the branch $[0,\theta]_S$ does not drop in model or degree,
  (ii) every $\xi \leq_S \theta$ is unstable,
  (iii) there is a τ such that $(P_\theta, \Sigma_\theta) = (Q_\tau, \Lambda_\tau)$, $[0,\tau]_U$ does not drop in model or degree, and $i_{0,\theta} = j_{0,\tau}$,
(b) if θ is special, then θ is unstable and $\theta+1$ is stable.

0 is unstable. Note that if θ is special, then $\mathrm{dom}(E_\theta^{\mathcal{S}}) \in \mathrm{ran}(i_{0,\theta})$, so S-pred$(\theta+1) \leq_S \theta$, so S-pred$(\theta+1)$ is unstable by (b) and (a)(ii).

Suppose first that we have defined $\mathcal{S} \restriction \theta$, where θ is a limit ordinal. We then set

$$[0,\theta]_S = \Psi(\mathcal{T} \restriction \theta),$$

as we must, and let $(P_\theta, \Sigma_\theta)$ be the resulting pair. We say that θ is unstable iff there is a τ such that $(P_\theta, \Sigma_\theta) = (Q_\tau, \Lambda_\tau)$, $[0,\tau]_U$ does not drop in model or degree, and $i_{0,\theta} = j_{0,\tau}$. Otherwise θ is stable. This defines $\mathcal{S} \restriction \theta+1$ and its notion of stability. It is easy to see that our induction hypotheses still hold.

Now suppose we have defined $\mathcal{S} \restriction \theta+1$, and let us define $\mathcal{S} \restriction \theta+2$.

Case 1. θ is stable and $M_{\nu,l} \lhd P_\theta$.

In this case the construction of $\mathcal{S}_{\nu,l}$ is finished, and we move on to $\mathcal{S}_{\nu,l+1}$. No strategy disagreements show up, so $\Omega_{\nu,l} = (\Sigma_\theta)_{M_{\nu,l}}$.

Case 2. θ is unstable, and for $\lambda = \lambda(\dot{G}^{P_\theta})$, $M_{\nu,l}|\lambda = P_\theta|\lambda$.

In this case θ is special. We let

$$E_\theta^{\mathcal{S}} = F = F(\dot{G}^{P_\theta}).$$

Since θ is unstable, $P_\theta = Q_\tau$ for some τ, and since Q_0 iterates past $M_{\nu,l}$ via $\mathcal{U}$ (in the sense of Claim 10.2.5), either $M_{\nu,l} \lhd Q_\tau || \mathrm{lh}(F)$ or $M_{\nu,l} || \mathrm{lh}(F) = Q_\tau || \mathrm{lh}(F)$.

Let β be least such that $\mathrm{crit}(E_\theta^{\mathcal{S}}) < \varepsilon_\beta$; then since $\mathrm{dom}(E_\theta^{\mathcal{S}}) \in \mathrm{ran}(i_{0,\theta})$, $\beta \leq_S \theta$, so β is unstable. Moreover, $\mathrm{dom}(E_\theta^{\mathcal{S}}) = \mathrm{dom}(\dot{G}^{P_\beta})$, so $E_\theta^{\mathcal{S}}$ is total over P_β. Thus

$$P_{\theta+1} = \mathrm{Ult}(P_\beta, E_\theta^{\mathcal{S}}).$$

We declare that $\theta+1$ is stable. It is easy to verify the induction hypotheses.

[9]This has nothing to do with stability for premice.

Remark 10.2.6. In Case 2, if $M_{\nu,l} \trianglelefteq P_\theta || \mathrm{lh}(F)$, then $M_{\nu,l} \triangleleft P_{\theta+1}$ and $\theta+1$ is stable, so Case 1 applies and $\mathcal{S}_{\nu,l}$ has length $\theta+2$. It could be that $M_{\nu,l} \triangleleft P_\theta$, but we are insisting that $\mathcal{S}_{\nu,l}$ end at a stable node.

If $P_\theta \trianglelefteq M_{\nu,l}$, then again the construction of $\mathcal{S}_{\nu,l}$ is finished. We shall derive a contradiction in this case. Let us assume now that $P_\theta \not\trianglelefteq M_{\nu,l}$ and neither Case 1 nor Case 2 applies. Thus there is a least disagreement between $(P_\theta, \Sigma_\theta)$ and $(M_{\nu,l}, \Omega_{\nu,l})$. We need to show that the least disagreement is an extender disagreement, and that it is active on the P_θ side and passive on the $M_{\nu,l}$ side.

CLAIM 10.2.7. *There is an extender F on the P_θ-sequence such that*

(i) $M_{\nu,l}|\mathrm{lh}(F) = P_\theta||\mathrm{lh}(F)$, and

(ii) $(\Omega_{\nu,l})_R = (\Sigma_\theta)_R$, where $R = M_{\nu,l}||\mathrm{lh}(F)$.

The proof is similar to that of Theorem 8.4.3 and Claim 10.1.4. We shall describe the the adaptations it requires insofar as extender disagreement goes in Claims 10.2.8 and 10.2.10. We describe the adaptations relevant to strategy disagreements at the end of this section. For now, let us fix F witnessing the claim and go on. Since Case 2 does not apply, if θ is unstable then $\mathrm{lh}(F) < \gamma(\dot{G}^{P_\theta})$. So we must be in Case 3:

Case 3. θ is stable, or θ is unstable and $\mathrm{lh}(F) < \gamma(\dot{G}^{P_\theta})$.

We let

$$E_\theta^{\mathcal{S}} = \begin{cases} F & \text{if } F = \dot{G}^{P_\theta} \text{ and } [0,\theta]_S \cap D^{\mathcal{S}} = \emptyset, \\ F^+ & \text{otherwise.} \end{cases}$$

The rules for plus trees determine $\mathcal{S} \restriction \theta+2$: $\beta = \mathcal{S}\text{-pred}(\theta+1)$ is least such that $\mathrm{crit}(E_\theta^{\mathcal{S}}) < \varepsilon_\beta$, and $P_{\theta+1} = \mathrm{Ult}(R, E_\theta^{\mathcal{S}})$ where R is the longest possible initial segment of P_β. If β is unstable, $R = P_\beta$, and there is a τ such that $(P_{\theta+1}, \Sigma_{\theta+1}) = (Q_\tau, \Lambda_\tau)$, $[0,\tau]_U$ does not drop in model or degree, and $i_{0,\theta} = j_{0,\tau}$, then we declare $\theta+1$ to be unstable. Otherwise $\theta+1$ is stable. It is easy to verify the induction hypotheses.

This finishes the definition of $\mathcal{S}_{\nu,l}$. Our claims regarding the $\mathcal{U}_{\nu,l}$ hold for $\mathcal{S}_{\nu,l}$.

CLAIM 10.2.8. *Let $\nu \leq \nu_0$, and suppose that $(M^{<\nu}, F)$ is an lpm and F^* is a $\mathbb{C}$-minimal certificate for F. Let $\mathcal{S} = \mathcal{S}_{\nu,0}$, and α be least such that $M^{<\nu} \trianglelefteq_0 \mathcal{M}_\alpha^{\mathcal{U}}$; then either*

(i) F is not $\mathbb{C}$-bad at ν and F is on the $\mathcal{M}_\alpha^{\mathcal{S}}$-sequence, or

(ii) F is $\mathbb{C}$-bad at ν, $[0,\alpha)_S \cap D^{\mathcal{S}} = \emptyset$ and $\mathcal{M}_\alpha^{\mathcal{S}} = (N(\mathbb{C},F), G(\mathbb{C},F))$.

PROOF. Let $\kappa = \mathrm{crit}(F)$, $\mathcal{V} = i_{F^*}(\mathcal{S})$, and $\eta+1 <_V i_{F^*}(\kappa)$ be such that

$$V\text{-pred}(\eta+1) = \kappa.$$

The main new thing we must see is that η is not special in $\mathcal{V}$. But if η is special in $\mathcal{V}$, then there are arbitrarily large $\gamma < \kappa$ such that $\gamma+1 <_S \kappa$ and γ is special in $\mathcal{U}$.

This is impossible because if γ is special, then $\gamma+1$ is stable and S-pred$(\gamma+1)$ is unstable. Thus if ξ and δ are special, then $\xi+1 \not<_S \delta+1$

The rest of the proof of Claim 10.2.4 goes through without change. Alternative (i) corresponds to the case that η is not a top node of $\mathcal{V}$, and alternative (ii) corresponds to the case that it is. ⊣

Remark 10.2.9. If we had allowed $\mathcal{S}$ to use extenders of the form $F(\dot{G}^{P_\eta})$ at arbitrary η then the proof of Claim 10.2.8 would break down, as would the rest of our argument.

CLAIM 10.2.10. *(i) If $\nu < \nu_0$, then (P_0, Ψ) iterates strictly past $(M_{\nu,l}, \Omega_{\nu,l})$ via $\mathcal{S}_{\nu,l}$.*

(ii) If $\nu = \nu_0$, then (P_0, Ψ) iterates to $((N(\mathbb{C},F), G(\mathbb{C},F)), \Omega(\mathbb{C},F))$ via $\mathcal{S}_{\nu,0}$, where F is the unique extender that is $\mathbb{C}$-bad at ν.

PROOF. Let $\nu < \nu_0$. We are assuming that no strategy disagreements show up in the construction of $\mathcal{S}_{\nu,l}$. Claim 10.2.8(i) implies that if $l = 0$, then extender disagreements involve only extenders from the $\mathcal{S}_{\nu,l}$ side. One can reduce the case $l > 0$ to the case $l = 0$ as before. The following analog of Sublemma 8.3.1.1 is what we need for that.

SUBCLAIM 10.2.10.1. *Suppose that $M_{\nu,l}$ is not $l+1$-sound, and let*

$$\pi : M^-_{\nu,l+1} \to M_{\nu,l}$$

be the anticore embedding. Let $\xi_0 + 1 = \mathrm{lh}(\mathcal{S}_{\nu,l+1})$ and $\xi_1 + 1 = \mathrm{lh}(\mathcal{S}_{\nu,l})$; then

(i) $\mathcal{S}_{\nu,l}$ has last model $M_{\nu,l}$,

(ii) $\mathcal{S}_{\nu,l+1} = \mathcal{S}_{\nu,l} \restriction (\xi_0 + 1)$,

(iii) ξ_0 is the least stable γ such that $\mathrm{lh}(E_\gamma^{\mathcal{S}_{\nu,l}}) > \rho(M_{\nu,l})$, and

(iv) $\xi_0 <_{\mathcal{S}_{\nu,l}} \xi_1$, and $\hat{i}^{\mathcal{S}_{\nu,l}}_{\xi_0,\xi_1} = \pi$.

PROOF SKETCH. This is identical to Sublemma 8.3.1.1 except for the restriction in (iii) to *stable* $\gamma < \mathrm{lh}(\mathcal{S}_{\nu,l})$. Without that restriction the subclaim is not true, because the background coherence problem, which we have avoided in Sublemma 8.3.1.1 by using only plus extenders, could show up at special nodes of $\mathcal{S}_{\nu,l}$. By restricting (iii) to stable γ and demanding that the last node in $\mathcal{S}_{\nu,l+1}$ be stable we have avoided those problems, and the proof of Sublemma 8.3.1.1 goes through. ⊣

So if $\nu < \nu_0$, only the P_0 side moves in its comparison with $M_{\nu,l}$. Letting $\gamma+1 = \mathrm{lh}(\mathcal{S}_{\nu,l})$, it cannot be that $P_\gamma \lhd M_{\nu,l}$, for then $[0,\gamma)_{S_{\nu,l}}$ does not drop, so P_γ is a pseudo-lpm rather than an lpm. Similarly, if $P_\gamma = M_{\nu,l}$ then $[0,\gamma)_{S_{\nu,l}}$ drops. Thus (P_0, Ψ) iterates strictly past $(M_{\nu,l}, \Omega_{\nu,l})$ via $\mathcal{S}_{\nu,l}$.

If $\nu = \nu_0$ then (P_0, Ψ) iterates to $(N(\mathbb{C},F), \Omega(\mathbb{C},F))$ by Claim 10.2.8(ii). ⊣

Now let $\mathcal{S} = \mathcal{S}_{\nu_0,0}$ and $\mathcal{U} = \mathcal{U}_{\nu_0,0}$. Let $\gamma+1 = \mathrm{lh}(\mathcal{S})$ and $\theta+1 = \mathrm{lh}(\mathcal{U})$, so that

$$P_\gamma = N(\mathbb{C},F) = Q_\theta,$$

and neither branch drops. The usual Weak Dodd-Jensen argument shows that $i_{0,\gamma} = j_{0,\theta}$.[10]

Let $s = e^{\mathcal{S}}_{\gamma}$ and $u = e^{\mathcal{U}}_{\theta}$ be the sequences of extenders used on the two branches.

CLAIM 10.2.11. $s = u$.

PROOF. Assume not, and let η be least such that $s(\eta) \neq u(\eta)$. Let

$$e^{\mathcal{S}}_{\xi} = s \upharpoonright \eta = u \upharpoonright \eta = e^{\mathcal{U}}_{\tau},$$

so that $P_\xi = Q_\tau$, and ξ is unstable in $\mathcal{S}$. Let $D = s(\eta)$ and $E = u(\eta)$. Since $i_{\xi,\gamma} = j_{\tau,\theta}$, D and E are compatible. Let $D = E^{\mathcal{S}}_{\alpha}$ and $E = E^{\mathcal{U}}_{\beta}$.

Suppose first that E is a proper initial segment of D. E is not on the Q_θ-sequence, so not on the P_γ-sequence, so the initial segment condition fails for D. This means that $D = \dot{G}^{P_\alpha}$ and $[0,\alpha)_S$ does not drop. If also $E = \dot{G}^{Q_\beta}$ and $[0,\beta)_U$ does not drop, then Q_β is a bad proper initial segment of P_α, so $\pi_\alpha(Q_\beta)$ is a bad proper initial segment of P^*_α, contrary to our minimality assumption on (P_0, Ψ). Thus $E = F^+$ for some ordinary extender F on the Q_β-sequence.

$\mathrm{lh}(F)$ is a cardinal of Q_θ, and hence a cardinal of $P_{\alpha+1}$. It follows that $\lambda(G) < \mathrm{lh}(F)$[11], and thus $F = F(D)$ and $\mathrm{lh}(F) = o(P_\alpha)$. But then $E = F(D)^+$ is an initial segment of D and $\mathrm{lh}(F(D)) = o(P_\alpha)$, so P_α is not bad. Badness is preserved by non-dropping iterations, so this is a contradiction.

Thus D is a proper initial segment of E, and the initial segment condition fails for E. So $E = \dot{G}^{Q_\beta}$ where $[0,\beta)_U$ does not drop. The minimality of $((N,G),\Psi)$ implies that $[0,\alpha)_S$ drops or $D \neq \dot{G}^{P_\alpha}$. If $D = F^+$ for some ordinary extender from the P_α-sequence, then we get the same contradiction as in the last paragraph. Since we are on the $\mathcal{S}$ side now, there is one further possibility, namely, that α is special and $D = F(\dot{G}^{P_\alpha})$.

Arguing as above, we get that $D = F(E)$ in this case, moreover $\mathrm{lh}(D)$ is a cardinal in P_γ, and hence in $Q_{\beta+1}$, so $\mathrm{lh}(D) = \mathrm{lh}(E)$. Moreover, $\lambda(E)$ is not measurable by the $Q_{\beta+1}$-sequence, hence not measurable by the P_γ-sequence. $\lambda(E) = \lambda(D)$ is measurable by the $P_{\alpha+1}$-sequence, so $\mathrm{crit}(i_{\alpha+1,\gamma}) = \lambda(E)$ and the first extender used in $i_{\alpha+1,\gamma}$ is the order zero measure of $P_{\alpha+1}$ on $\lambda(E)$. Thus $D^+ = E$, so Q_β is not bad, contradiction. ⊣

Claim 10.2.11 contradicts the stability of γ. For let $\delta \leq_S \gamma$ be least such that δ is stable. $e^{\mathcal{S}}_{\delta} = e^{\mathcal{U}}_{\tau}$ for some τ because $s = u$. If δ is a limit ordinal, then δ would be declared unstable, so $\delta = \alpha + 1$ for some α. Let $\beta + 1 = \tau$, so that $E^{\mathcal{S}}_{\alpha} = E^{\mathcal{U}}_{\beta}$. If α is not special, then since $e^{\mathcal{S}}_{\delta} = e^{\mathcal{U}}_{\tau}$, $\alpha + 1$ was declared unstable, contradiction. But if α is special, then $E^{\mathcal{S}}_{\alpha}$ is not the sort of extender that is used in $\mathcal{U}$; that is, it is not a plus extender or the top extender of a pseudo-lpm.

---

[10] If n is least such that $i_{0,\gamma}(e_n) \neq j_{0,\theta}(e_n)$, then $j_{0,\theta}(e_n) < i_{0,\gamma}(e_n)$ because $j_{0,\theta}$ is an iteration map of Ψ, and Ψ has the weak Dodd-Jensen property. But $i^*_{0,\gamma}(e_n) = \pi_\gamma(i_{0,\gamma}(e_n)) < \pi_\gamma(j_{0,\theta}(e_n))$ for the same reason, so $i_{0,\gamma}(e_n) < j_{0,\theta}(e_n)$, contradiction.

[11] See the proof Theorem 4.9.1, Claim 2, Subcase 1B.

This contradiction completes the proof of Theorem 10.2.3, modulo a proof that no strategy disagreements show up in the construction of $\mathcal{S}_{\nu,l}$. ⊣

10.2.1. No strategy disagreements. We sketch a proof of Claim 10.2.7. The proof is parallel to the proof of Theorem 8.4.3, and we are going to give more detail as to how the slightly new part goes in the more complicated setting of Lemma 9.6.5 in Section 10.3.

We can think of $\mathcal{S}_{\nu,l}$ as a play of an iteration game $\mathcal{G}_1$ according to a winning strategy for II that is induced by Ψ. In $\mathcal{G}_1$, player I not only plays the extenders $E^{\mathcal{S}}_\theta$, but also decides whether nodes are unstable. We demand that if I declares θ unstable, then he must have declared all $\tau <_S \theta$ unstable, and $0 \leq_S \theta$, and $[0,\theta]_S$ does not drop in model or degree. He must follow the rules (i)-(iii) for choosing $E^{\mathcal{S}}_\theta$, and if his choice makes θ special, then he must declare $\theta+1$ to be stable.

$\mathcal{S}_{\nu,l}$ is a play of $\mathcal{G}_1$ in which I declares nodes θ to be unstable according to the rule in induction hypothesis (iv), that is, iff $\mathcal{M}^{\mathcal{S}_{\nu,l}} = \mathcal{M}^{\mathcal{U}_{\nu,l}}_\tau$ for some τ, and so on. But for now, we do not assume I is playing in any such special way. We call a play of $\mathcal{G}_1$ in which II has not lost yet a *$\mathcal{G}_1$-tree*.

Ψ induces an winning strategy $\hat{\Psi}$ for II in $\mathcal{G}_1$ in the way we have described above. If $\mathcal{S}$ is by $\hat{\Psi}$ with models P_θ, then we have a tree $\mathcal{T}$ by Ψ with the same tree order and models P^*_θ, together with elementary maps $\pi_\theta : P_\theta \to P^*_\theta$ obtained by copying/lifting. $\hat{\Psi}$ has strong hull condensation in the natural sense.

We turn now to normalization. If $\mathcal{S}$ is a $\mathcal{G}_1$ tree with last model P, we set

$$N^{\mathcal{S}}_\infty = \begin{cases} P & \text{if } P_0\text{-to-}P \text{ drops,} \\ P||o(P) & \text{otherwise.} \end{cases}$$

In other words, $N^{\mathcal{S}}_\infty$ is the last model of $\mathcal{S}$ unless the last model is a pseudo-lpm, in which case it is the last model with its top extender predicate removed. Let $\mathcal{G}_2$ be the game in which I and II play $\mathcal{G}_1$ until someone loses, or I decides that they should play the usual game for producing finite stacks of λ-separated trees on $N^{\mathcal{S}}_\infty$, where $\mathcal{S}$ is the current position. Clearly, we can pull back Ψ to a winning strategy for II in this game. We again call this strategy $\hat{\Psi}$.

Let $\mathcal{V}$ be a $\mathcal{G}_1$-tree on P_0, and $s = \langle(\nu_i, k_i, \mathcal{U}_i) | i \leq n\rangle$ a finite stack of λ-separated trees on $N^{\mathcal{V}}_\infty$. We can define the embedding normalization $\mathcal{W} = W(\mathcal{V}, s)$ in essentially the same way that we did when $\mathcal{V}$ was an ordinary plus tree. We only need to do this when the last node of $\mathcal{V}$ is stable, and that simplifies a few things, so let us assume that.

For example, suppose that s consists of just one normal tree $\mathcal{U}$ on $R = N^{\mathcal{V}}_\infty$. Being the last model, R has been declared stable in $\mathcal{V}$. We define

$$\mathcal{W}_\gamma = W(\mathcal{V}, \mathcal{U}\restriction(\gamma+1))$$

by induction on γ. Each $\mathcal{W}_\gamma$ is a $\mathcal{G}_1$-tree with last model R_γ, and we have $\sigma_\gamma : \mathcal{M}^{\mathcal{U}}_\gamma \to R_\gamma$. We shall have that $z(\gamma)$ is stable in $\mathcal{W}_\gamma$, where $z(\gamma) = \mathrm{lh}(\mathcal{W}_\gamma) - 1$.

We also have tree embeddings

$$\Psi_{\nu,\gamma}\colon \mathcal{W}_\nu \to \mathcal{W}_\gamma,$$

defined when $\nu <_U \gamma$. $\Psi_{\nu,\gamma}$ is partial iff $(\nu,\gamma]_U$ drops somewhere. We call its u-map $\phi_{\nu,\gamma}$, and its t-maps are $\pi_\xi^{\nu,\gamma}$.

We set $\mathcal{W}_0 = \mathcal{V}$, and $R = R_0$ or $R = R_0||o(R_0)$. The successor step is given by

$$\mathcal{W}_{\gamma+1} = \mathcal{W}_\gamma \restriction (\theta+1)^\frown \langle F\rangle^\frown i_F\text{``}(\mathcal{W}_\nu^{\geq\beta}),$$

where $F = \sigma_\gamma(E_\gamma^{\mathcal{U}})$. By the way we defined R, $E_\gamma^{\mathcal{U}}$ is not the last extender of a pseudo-lpm. We let

$$\alpha_F = \text{least } \textit{stable } \theta < \text{lh}(\mathcal{W}_\gamma) \text{ such that } F \text{ is on the } \mathcal{M}_\theta^{\mathcal{W}_\gamma}\text{-sequence},$$

and $\nu = U\text{-pred}(\gamma+1)$. Let

$$\beta(\mathcal{W}_\nu, \mathcal{W}_\gamma, F) = \begin{cases} \text{least } \eta \text{ such that } \text{crit}(F) < \varepsilon_\eta^{\mathcal{W}_\nu} & \text{if there is such an } \eta, \\ \text{lh}(\mathcal{W}_\nu) - 1 & \text{otherwise.} \end{cases}$$

Set $\beta = \beta(\mathcal{W}_\nu, \mathcal{W}_\gamma, F)$. It is easy to see that $\beta \leq \alpha_F$, and

$$\mathcal{W}_\nu \restriction \beta+1 = \mathcal{W}_\gamma \restriction \beta+1 = \mathcal{W}_{\gamma+1} \restriction \beta+1.$$

This is because between ν and γ, all the $\mathcal{W}_\eta$ used the same extenders E such that $\lambda(E) < \text{lh}(F_\nu)$.

Let us assume for simplicity that $(\nu,\gamma+1]_U$ does not drop. We have

$$\phi\colon \text{lh}(\mathcal{W}_\nu) \to \text{lh}(\mathcal{W}_{\gamma+1})$$

given by

$$\phi(\xi) = \begin{cases} \xi & \text{if } \xi < \beta, \\ (\alpha_F+1)+(\xi-\beta) & \text{otherwise.} \end{cases}$$

For $\eta \leq_U \nu$, we let $\phi_{\eta,\gamma+1} = \phi \circ \phi_{\eta,\nu}$. A node η of $\mathcal{W}_{\gamma+1}$ is stable in $\mathcal{W}_{\gamma+1}$ just in case $\eta \leq \alpha_F$ and η is stable as a node of $\mathcal{W}_\gamma$, or $\eta = \phi(\xi)$, where ξ is stable as a node of $\mathcal{W}_\nu$. For $\xi < \beta$, $\pi_\xi^{\nu,\gamma+1}$ is the identity. We define by induction on $\xi \geq \beta$ the models $\mathcal{M}_{\varphi(\xi)}^{\mathcal{W}_{\gamma+1}}$ and maps $\pi_\xi : \mathcal{M}_\xi^{\mathcal{W}_\nu} \to \mathcal{M}_{\varphi(\xi)}^{\mathcal{W}_{\gamma+1}}$ as before.

It is not hard to see that $\mathcal{W}_{\gamma+1}$ is a $\mathcal{G}_1$-tree with stable last model. Note that for $\eta \leq \alpha_F$, η is special in $\mathcal{W}_\gamma$ iff η is special in $\mathcal{W}_{\gamma+1}$. If $\eta < \alpha_F$ this is trivial, and if $\eta = \alpha_F$, then since α_F is stable in $\mathcal{W}_\gamma$, it is not special in $\mathcal{W}_\gamma$, and since F^- is on the $\mathcal{M}_\eta^{\mathcal{W}_{\gamma+1}}$-sequence, η is not special in $\mathcal{W}_{\gamma+1}$. For $\alpha_F < \eta$, η is special in $\mathcal{W}_{\gamma+1}$ iff $\phi_{\nu,\gamma+1}^{-1}(\eta)$ is special in $\mathcal{W}_\nu$.

This gives us $\mathcal{W}_{\gamma+1}$ and $\Psi_{\nu,\gamma+1} : \mathcal{W}_\nu \to \mathcal{W}_{\gamma+1}$. At limit ordinals λ, we let $\mathcal{W}_\lambda$ be the direct limit of the $\mathcal{W}_\nu$ for $\nu <_U \lambda$, under the $\Psi_{\nu,\mu}$. Finally, $W(\mathcal{V},\mathcal{U}) = \mathcal{W}_\gamma$, where $\gamma+1 = \text{lh}(\mathcal{U})$.

If $\mathcal{V}$ is a $\mathcal{G}_1$-tree and with stable last node, and $s^\frown\langle\mathcal{U}\rangle$ is a maximal stack of λ-separated trees on $N_\infty^{\mathcal{V}}$, then

$$W(\mathcal{V}, s^\frown\langle\mathcal{U}\rangle) = W(W(\mathcal{V}, s), \sigma\mathcal{U}),$$

where σ is the natural embedding from the last model of $\mathcal{V}$ to the last model of $W(\mathcal{V}, s)$ that we get from the normalization process. That is, we normalize stacks "bottom up".

This finishes our discussion of the normalization $W(\mathcal{V}, s)$. We say that strategy Σ for the game $\mathcal{G}_1$ *normalizes well* iff whenever $\langle\mathcal{V}, s\rangle$ is according to Σ, then $W(\mathcal{V}, s)$ is according to Σ.

LEMMA 10.2.12. *$\hat{\Psi}$ normalizes well.*

PROOF SKETCH. If $\mathcal{V}$ is a $\mathcal{G}_1$-tree on P_0 by $\hat{\Psi}$, then let us write $\mathcal{V}^*$ for the tree on P_0 by Ψ that we get by lifting $\mathcal{V}$. If $\mathcal{U}$ is a λ-separated tree on $N_\infty^{\mathcal{V}}$, let us write

$$\langle\mathcal{V}, \mathcal{U}\rangle^* = \langle\mathcal{V}^*, \sigma\mathcal{U}\rangle,$$

where σ is the copy map acting on the last model of $\mathcal{V}$. Just for the space of this proof, to keep things straight, let us write $\hat{W}$ for the embedding normalization operation defined above.

Ψ itself normalizes well. But normalizing commutes with copying in this context, as it did in the case of ordinary iteration trees. That is

$$\hat{W}(\mathcal{V}, \mathcal{U})^* = W(\langle\mathcal{V}, \mathcal{U}\rangle^*).$$

So

$$\begin{aligned} \hat{W}(\mathcal{V}, \mathcal{U}) \text{ is by } \hat{\Psi} &\iff \hat{W}(\mathcal{V}, \mathcal{U})^* \text{ is by } \Psi \\ &\iff W(\langle\mathcal{V}, \mathcal{U}\rangle^*) \text{ is by } \Psi \\ &\iff \langle\mathcal{V}^*, \sigma\mathcal{U}\rangle \text{ is by } \Psi \\ &\iff \langle\mathcal{V}, \mathcal{U}\rangle \text{ is by } \hat{\Psi}, \end{aligned}$$

as desired. See the proof of Theorem 7.1.6. ⊣

We can now complete the proof of Claim 10.2.7. Let $\mathcal{S} = \mathcal{S}_{\nu,l} \restriction \theta + 1$, and suppose that $R \trianglelefteq P_\theta$ and $R \trianglelefteq M_{\nu,l}$. We must show that $(\Sigma_\theta)_R = \Omega_R$, where $\Omega = \Omega_{\nu,l}$. We follow closely the proof of Theorem 8.4.3, making use of the absoluteness of our coiteration rules in the same way that the proof of Claim 10.1.4 did.

Note that $(R, (\Sigma_\theta)_R)$ and (R, Ω_R) are ordinary lbr hod pairs. Using internal lift consistency, we can reduce to the case that either $R = P_\theta$, or θ is a top node and $R = P_\theta || o(P_\theta)$. Let us assume this. Let $\mathcal{U}$ be a λ-separated tree on R that is by both $(\Sigma_\theta)_R$ and Ω_R, and let

$$\begin{aligned} b &= \Omega_R(\mathcal{U}), \\ \mathcal{U}^* &= \text{lift}(\mathcal{U}, R, \sigma_{\nu,l}[R], \text{Res}_{\nu,l}[R], \mathbb{C}, N^*)_0, \\ i_b^* &= i_{0,\text{lh}(\mathcal{U})+1}^{\mathcal{U}^*}, \end{aligned}$$

and

$$\mathcal{W}_b = W(\mathcal{S}, \mathcal{U}^\frown b).$$

Let us assume for simplicity that b does not drop; we show then that $\mathcal{W}_b$ is a pseudo-hull of $i_b^*(\mathcal{S})$. Since $\hat{\Psi}$ normalizes well and has strong hull condensation, this implies that $(\Sigma_\theta)_R(\mathcal{U}) = b$, as desired.

To do this, we define by induction tree embeddings

$$\Phi_\gamma \colon \mathcal{W}_\gamma \to \mathcal{W}_\gamma^*,$$

where $\mathcal{W}_\gamma = W(\mathcal{S}, \mathcal{U} \restriction \gamma + 1)$, and $\mathcal{W}_\gamma^* = i_{0,\gamma}^{\mathcal{U}^*}(\mathcal{S})$ in the case that $[0,\gamma]_U$ does not drop.[12] Φ_γ is defined for $\gamma < \mathrm{lh}(\mathcal{U})$ or $\gamma = b$.

The step from Φ_γ to $\Phi_{\gamma+1}$ goes as in Theorem 8.4.3 and Claim 10.1.4. We let G_γ be the resurrection of $\psi_\gamma^{\mathcal{U}}(E_\gamma^{\mathcal{U}})$, where $\psi_\gamma^{\mathcal{U}}$ is the lift map at γ. Let $Y = M_{\eta,j}^{\mathbb{C}_\gamma}$ have last extender G_γ, and let $\mathcal{W}_\gamma^{**}$ be $\mathcal{S}_{\eta,j}$ as computed in $\mathcal{M}_\gamma^{\mathcal{U}^*}$. The main thing one must show is that $(\mathcal{W}_\gamma^{**})^\frown\langle G_\gamma \rangle$ is an initial segment of $\mathcal{W}_{\gamma+1}^*$. This is true because N^* captures Code(Ψ), P_0 is countable in N^*, and the coiteration rules that give rise to $\mathcal{S}_{\eta,j}$ in $\mathcal{M}_\gamma^{\mathcal{U}^*}$ depend on Y^- and Ψ in an absolute way.

This completes our sketch of the proof of Claim 10.2.7.

Combining the results of this and the last section, we get

THEOREM 10.2.13. *Assume* AD^+, *and let* $\mathbb{C}$ *be the maximal construction of a coarse strategy pair. Let* ν *be a limit ordinal; then* $\mathbb{C}$ *is good at* $\langle \nu, -1 \rangle$.

PROOF. . $\mathbb{C}$ is plus consistent at $\langle \nu, -1 \rangle$ by Lemma 10.2.2 and Theorem 10.2.3.

Suppose that $\mathbb{C}$ is not extender unique at $\langle \nu, -1 \rangle$, and let $\mathcal{B} = (M^{<\nu}, F, G)$ be a nontrivial bicephalus such that F and G have $\mathbb{C}$-minimal certificates F^* and G^*. Let Ψ be the strategy for $\mathcal{B}$ induced by $\mathbb{C}$, F^*, and G^*. Since $\mathbb{C}$ is plus consistent at $\langle \nu, -1 \rangle$, F^* and G^* background F^+ and G^+, so Ψ is indeed defined on all $\mathcal{B}$ trees. Ψ normalizes well for stacks of $\mathcal{B}$ trees; the background coherence problem does not arise because $\mathcal{B}$-trees are λ-separated. One can show by proofs parallel to the ones for the $(M_{\xi,k}^{\mathbb{C}}, \Omega_{\xi,k}^{\mathbb{C}})$ that Ψ has strong hull condensation, and that $(\mathcal{B}, \Psi)$ is internally lift consistent and pushforward consistent. The Bicephalus Lemma (10.1.3) now yields a contradiction. ⊣

10.3. Proof of Lemma 9.6.5

Let us assume AD^+ throughout this section. Our proof of Lemma 9.6.5 follows closely the proof of Theorem 8.4.3. We begin by discussing tree embeddings and normalization for pseudo-trees.

Let (M, Λ) be an lbr hod pair, and let

$$H = \mathrm{cHull}_{k+1}^M(\alpha_0 \cup q),$$

[12] We of course have to define $\mathcal{W}_\gamma^*$ and $\mathcal{W}_b^*$ in the dropping case too. See the proof of Theorem 8.4.3.

where $k = k(M)$, $\rho_{k+1}(M) \leq \alpha_0$, and q is a finite set of ordinals. Let

$$\pi\colon H \to M$$

be the anticollapse map. So π is elementary, and $\alpha_0 \leq \mathrm{crit}(\pi)$. The assumptions of Lemma 9.6.5 imply that α_0 is a cardinal of H, so we assume this. We have a pullback iteration strategy

$$\Sigma = \Lambda^{(\mathrm{id},\pi)}$$

for (M,H,α_0), obtained by using $\mathrm{id}\colon M \to M$ and $\pi\colon H \to M$ to lift $\mathcal{S}$ on (M,H,α_0) to a tree $\mathcal{T} = (\mathrm{id},\pi)\mathcal{S}$ on M, then choosing the branch chosen by Λ. That is

$$\Sigma(\mathcal{S}) = \Lambda((\mathrm{id},\pi)\mathcal{S}).$$

Σ is actually a strategy for a stronger iteration game than the usual game producing a plus tree on a phalanx. Namely, Σ wins $\mathcal{G}_0$, where in $\mathcal{G}_0$ the opponent, player I, plays not just the extenders $E_\gamma^{\mathcal{S}}$, but also decides whether nodes are unstable. We demand that if I declares θ unstable, then he must have declared all $\tau <_S \theta$ unstable, and $0 \leq_S \theta$, and $[0,\theta]_S$ does not drop in model or degree. We then set $\alpha_\theta = \sup i_{0,\theta}^{\mathcal{S}}``\alpha_0$ and let

$$\mathcal{M}_{\theta+1}^{\mathcal{S}} = \mathrm{cHull}_{k+1}^{\mathcal{M}_\theta^{\mathcal{S}}}(\alpha_\theta \cup i_{0,\theta}^{\mathcal{S}}(q)).$$

I must then declare $\theta+1$ to be stable, and take his next extender from $\mathcal{M}_{\theta+1}^{\mathcal{S}}$. The rest of $\mathcal{G}_0$ is as in the iteration game $G^+(M,\omega_1)$. We need only consider normal, that is, length-increasing plays of the game. Let us call a play $\mathcal{V}$ of $\mathcal{G}_0$ in which no one has yet lost and the extenders played have strictly increasing lengths a *pseudo iteration tree* on (M,H,α_0).

The pseudo-tree occurring in the proof of Theorem 9.6.2 was a play of $\mathcal{G}_0$ in which I followed certain rules for picking his extenders and declaring nodes unstable. But for now, we do not assume I is playing in any such special way.

Remark 10.3.1. One can generalize $\mathcal{G}_0$ further, to a game in which I is allowed to gratuitously drop to Skolem hulls whenever he pleases. With some minimal conditions, Ψ will pull back to a strategy for this game. We do not need that generality, so we shall not go into it.

Let us define strong hull condensation. The changes we need to make in order to accomodate pseudo-trees are straightforward, but we may as well spell them out.

If $\mathcal{T}$ is a pseudo-tree on (M,H,α_0), then we set

$$\mathrm{stab}(\mathcal{T}) = \{\beta < \mathrm{lh}(\mathcal{T}) \mid \beta \text{ is } \mathcal{T}\text{-stable}\}.$$

We let $\mathrm{Ext}(\mathcal{T})$ be the set of extenders used, and $\mathcal{T}^{\mathrm{ext}}$ the extender tree of $\mathcal{T}$. $\mathcal{T}$ is determined by $\mathrm{stab}(\mathcal{T})$ and $\mathrm{Ext}(\mathcal{T})$. (Pseudo-trees are normal, and their last nodes are stable, by definition.). If β is an unstable node of $\mathcal{T}$, we write $\alpha_\beta^{\mathcal{T}} = \sup(i_{0,\beta}^{\mathcal{T}}``\alpha_0)$.

DEFINITION 10.3.2. For $\mathcal{T}$ a pseudo-tree, we put $\xi \leq_T^* \eta$ iff

(a) $\xi \leq_T \eta$, or
(b) there is a $\gamma \leq_T \eta$ such that ξ and γ are stable roots of $\mathcal{T}$, and $\xi - 1 \leq_T \gamma - 1$.

In case (b), we let $i^{\mathcal{T}}_{\xi,\eta}: \mathcal{M}^{\mathcal{T}}_{\xi} \to \mathcal{M}^{\mathcal{T}}_{\eta}$ be given by

$$\hat{\imath}^{\mathcal{T}}_{\xi,\eta} = \hat{\imath}^{\mathcal{T}}_{\gamma,\eta} \circ (\tau^{-1} \circ i^{\mathcal{T}}_{\xi-1,\gamma-1} \circ \sigma),$$

where $\sigma: \mathcal{M}_{\xi} \to \mathcal{M}_{\xi-1}$ and $\tau: \mathcal{M}_{\gamma} \to \mathcal{M}_{\gamma-1}$ are the maps from the Skolem hulls.

Notice that $i^{\mathcal{T}}_{\xi,\gamma} = (\tau^{-1} \circ i^{\mathcal{T}}_{\xi-1,\gamma-1} \circ \sigma)$ is total in case (b), because $i^{\mathcal{T}}_{0,\gamma-1}(q)$ is in $\operatorname{ran}(\tau)$. Recall here that for θ unstable,

$$\alpha^{\mathcal{T}}_{\theta} = \alpha_{\theta} = \sup i^{\mathcal{T}}_{0,\theta}\text{``}\alpha,$$

and

$$\mathcal{M}_{\theta+1} = \text{collapse of } h_{\mathcal{M}_{\theta}}\text{``}(\alpha_{\theta} \cup i_{0,\theta}(q)).$$

So in case (b), we also get that $i_{\xi-1,\gamma-1} \restriction \alpha_{\xi-1} = i_{\xi,\gamma} \restriction \alpha_{\xi-1}$. Here is a diagram:

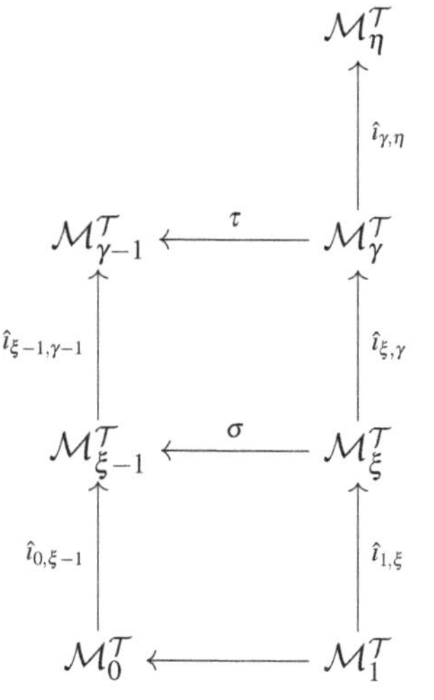

Thus the stable roots of $\mathcal{T}$ have a branch structure themselves, with 1 at its root.

As before, a tree embedding will have u, v, t, and s maps. The u maps connect exit extenders, but we shall also define them at unstable $\alpha < \operatorname{lh}(\mathcal{T})$ as well, and at $\alpha = \operatorname{lh}(\mathcal{T}) - 1$) if it exists.[13] $v(\alpha)$ is the least ξ on the branch in $\leq^*_T$ to $u(\alpha)$ such that $\mathcal{M}^{\mathcal{T}}_{\alpha}$ is naturally embedded into $\mathcal{M}^{\mathcal{U}}_{\xi}$. The t and s maps are the corresponding maps on models.

[13] If $\operatorname{lh}(\mathcal{T}) - 1$ exists then it must be stable, but it has no exit extender. So we are generalizing what was called an extended tree embedding in Section 6.4.

DEFINITION 10.3.3. Let $\mathcal{T}$ and $\mathcal{U}$ be pseudo-iteration trees on (M, K, α_0). An *(extended) tree embedding* of $\mathcal{T}$ into $\mathcal{U}$ is a system

$$\langle u, v, \langle s_\beta \mid \beta < \mathrm{lh}\,\mathcal{T}\rangle, \langle t_\beta \mid \beta + 1 < \mathrm{lh}\,\mathcal{T} \wedge \beta \in \mathrm{stab}(\mathcal{T})\rangle\rangle$$

such that

1. u: $\mathrm{lh}(\mathcal{T}) \to \mathrm{lh}(\mathcal{U})$, for all α, β, $\alpha < \beta \implies u(\alpha) < u(\beta)$, and for all $\alpha < \mathrm{lh}(\mathcal{T})$,
   (a) $\alpha \in \mathrm{stab}(\mathcal{T}) \iff u(\alpha) \in \mathrm{stab}(\mathcal{U})$,
   (b) $\alpha + 1 = \mathrm{lh}(\mathcal{T}) \implies u(\alpha) + 1 = \mathrm{lh}(\mathcal{U})$, and
   (c) if $\alpha \notin \mathrm{stab}(\mathcal{T})$, then $u(\alpha) = \mathrm{rt}(u(\alpha + 1)) - 1$.
2. $v : \mathrm{lh}(\mathcal{T}) \to \mathrm{lh}(\mathcal{U})$ is given by $v(0) = 0$, $v(\lambda) = \sup_{\alpha<\lambda} v(\alpha)$ for λ a limit, and

$$v(\alpha+1) = \begin{cases} u(\alpha)+1 & \text{if } \alpha \in \mathrm{stab}(\mathcal{T}), \\ v(\alpha)+1 & \text{otherwise.} \end{cases}$$

Moreover, v preserves $\leq^*_T$, and for all α
   (i) $\alpha \in \mathrm{stab}(\mathcal{T}) \iff v(\alpha) \in \mathrm{stab}(\mathcal{U})$, and
   (ii) $v(\alpha) \leq^* u(\alpha)$.

3. For any β,

$$s_\beta : M^{\mathcal{T}}_\beta \to M^{\mathcal{U}}_{v(\beta)}$$

is total and elementary. Moreover, for $\alpha <^*_T \beta$,

$$s_\beta \circ \hat{\imath}^{\mathcal{T}}_{\alpha,\beta} = \hat{\imath}^{\mathcal{U}}_{v(\alpha),v(\beta)} \circ s_\alpha.$$

In particular, the two sides have the same domain. Further, if β is unstable, then

$$s_\beta(\alpha^{\mathcal{T}}_\beta) = \alpha^{\mathcal{U}}_{v(\beta)}.$$

4. For $\alpha < \mathrm{lh}(\mathcal{T})$,

$$t_\alpha = \hat{\imath}^{\mathcal{U}}_{v(\alpha),u(\alpha)} \circ s_\alpha,$$

and if $\alpha \in \mathrm{stab}(\mathcal{T})$, then

$$E^{\mathcal{U}}_{u(\alpha)} = t_\alpha(E^{\mathcal{T}}_\alpha).$$

5. If $\beta \notin \mathrm{stab}(\mathcal{T})$,

$$s_{\beta+1} = \sigma^{-1} \circ s_\beta \circ \tau,$$

where τ: $\mathcal{M}^{\mathcal{T}}_{\beta+1} \to \mathcal{M}^{\mathcal{T}}_\beta$ and σ: $\mathcal{M}^{\mathcal{U}}_{v(\beta)+1} \to \mathcal{M}^{\mathcal{U}}_{v(\beta)}$ are the Skolem hull maps. (Note $v(\beta+1) = v(\beta)+1$ when β is unstable, by (2) above.) In other words, $s_{\beta+1}$ agrees with s_β on $\alpha^{\mathcal{T}}_\beta$, and maps the collapse of $i^{\mathcal{T}}_{0,\beta}(q)$ to the collapse of $i^{\mathcal{U}}_{0,v(\beta)}(q)$.

6. If $\beta = T$-pred$(\alpha+1)$ (and hence $\alpha \in \mathrm{stab}(\mathcal{T}) \cap \mathrm{dom}(u)$), then letting $\beta^* = U$-pred$(u(\alpha)+1)$,

$$v(\beta) \leq_U^* \beta^* \leq_U^* u(\beta),$$

and

$$s_{\alpha+1}([a,f]^P_{E^{\mathcal{T}}_\alpha}) = [t_\alpha(a), i^{\mathcal{U}}_{v(\beta),\beta^*} \circ s_\beta(f)]^{P^*}_{E^{\mathcal{U}}_{u(\alpha)}},$$

where $P \trianglelefteq M^{\mathcal{T}}_\beta$ is what $E^{\mathcal{T}}_\alpha$ is applied to, and $P^* \trianglelefteq M^{\mathcal{U}}_{\beta^*}$ is what $E^{\mathcal{U}}_{u(\alpha)}$ is applied to.

Figure 10.3.1 goes with the first clause of the definition, and Figure 10.3.2 goes with the last one.

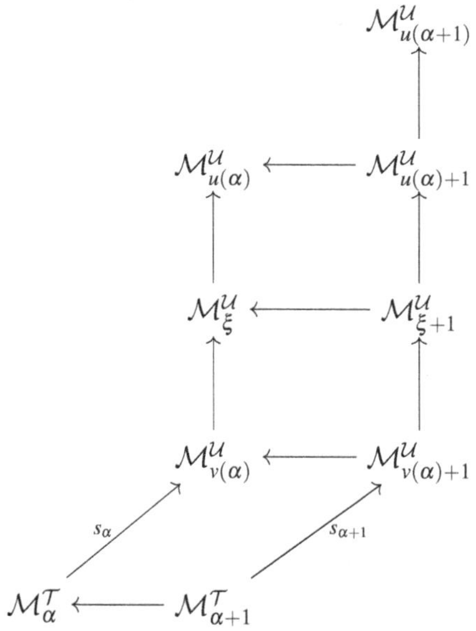

FIGURE 10.3.1. Clause 1 in the definition of tree embedding. α is unstable in $\mathcal{T}$, $u(\alpha)$ is unstable in $\mathcal{U}$, and $u(\alpha)+1 = \mathrm{rt}(u(\alpha+1))$.

Remark 10.3.4. The slightly new feature is the following. If $\Psi\colon \mathcal{T} \to \mathcal{U}$ is a tree embedding of pseudo-trees, and N is a stable root of $\mathcal{T}$, and P is a stable root immediately above it in $\mathcal{T}$, then we want Ψ to lift the process whereby we got the embedding from N to P of $\mathcal{T}$. This embedding came from an ultrapower of the backup model M for N. To make it possible to copy such ultrapowers, Ψ must have associated (M,N) to (M^*,N^*), where N^* is a stable root of $\mathcal{U}$, and M^* is its backup model. This leads to clause 6 in the definition.

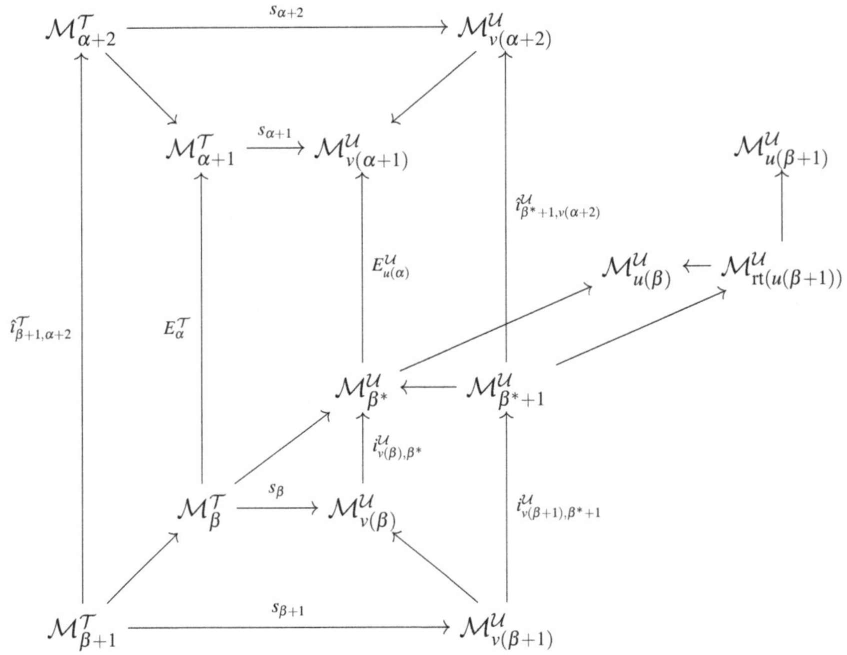

FIGURE 10.3.2. Extending a tree embedding of pseudo-trees at a successor step, in the case that $\alpha+1$ and β are both $\mathcal{T}$-unstable.

The agreement of maps in a tree embedding is given by

LEMMA 10.3.5. *Let* $\langle u, v, \langle s_\beta \mid \beta < \mathrm{lh}\,\mathcal{T}\rangle, \langle t_\beta \mid \beta+1 < \mathrm{lh}\,\mathcal{T} \wedge \beta \in \mathrm{stab}(\mathcal{T})\rangle\rangle$ *be a tree embedding of* $\mathcal{T}$ *into* $\mathcal{U}$*; then for* $\xi < \beta < \mathrm{lh}(\mathcal{T})$

(a) if $\xi \in \mathrm{stab}(\mathcal{T})$*, then* $s_\beta \restriction \mathrm{lh}(E^{\mathcal{T}}_\xi)+1 = t_\xi \restriction \mathrm{lh}(E^{\mathcal{T}}_\xi)+1$.

(b) if $\xi \notin \mathrm{stab}(\mathcal{T})$ *and* $\xi+1 < \beta$*, then*

$$s_\beta \restriction \inf(\alpha^{\mathcal{T}}_\xi, \mathrm{lh}(E^{\mathcal{T}}_{\xi+1})+1) = t_\xi \restriction \inf(\alpha^{\mathcal{T}}_\xi, \mathrm{lh}(E^{\mathcal{T}}_{\xi+1})+1).$$

The proof is an easy induction that we omit here. Part (a) comes from the fact that s_β agrees with the map from $\mathrm{lh}(E^{\mathcal{T}}_\xi)+1$ to $\mathrm{lh}(E^{\mathcal{U}}_{u(\xi)})+1$ that is an input for the Shift Lemma. In case (a), that map is t_ξ. In case (b) the same proof shows that s_β agrees with $t_{\xi+1}$ on $\mathrm{lh}(E^{\mathcal{T}}_{\xi+1})+1$. But it is easy to see that t_ξ agrees with $t_{\xi+1}$ on $\alpha^{\mathcal{T}}_\xi$ when ξ is unstable. (They may disagree at $\alpha^{\mathcal{T}}_\xi$.) This gives us (b).

DEFINITION 10.3.6. Let Σ be a winning strategy for II in $\mathcal{G}_0$; then Σ has *strong hull condensation* iff whenever $\mathcal{U}$ is a pseudo-tree according to Σ, and there is a tree embedding from $\mathcal{T}$ into $\mathcal{U}$, then $\mathcal{T}$ is according to Σ.

LEMMA 10.3.7. *Let (M,Λ) be an lbr hod pair, let $H = \mathrm{cHull}^M_{k(M)+1}(\alpha \cup q)$ where q is a finite set of ordinals and $\rho(M) \leq \alpha$, and let $\pi\colon H \to M$ be the anticollapse map. Let $\Sigma = \Lambda^{(id,\pi)}$ be the pullback strategy for* II *in the game $\mathcal{G}_0$ on (M,H,α); then Σ has strong hull condensation.*

PROOF SKETCH. This is like the proof of Proposition 7.1.11. If $\mathcal{U}$ is a play by Σ, and $\mathcal{T}$ is a pseudo-hull of $\mathcal{U}$, then $(\mathrm{id},\pi)\mathcal{T}$ is a pseudo-hull of $(\mathrm{id},\pi)\mathcal{U}$. ⊣

Definition 10.3.6 does not have the clause on pullback strategies that is part of the definition of strong hull condensation for ordinary strategies. This is just because we do not have a use for it. We believe that Lemma 10.3.7 holds for the stronger property.

We turn now to normalization.

Let $\mathcal{G}$ be the game in which I and II play $\mathcal{G}_0$ until someone loses, or I decides that they should play the game $G^+(N,\omega,\omega_1)$ for producing finite stacks of plus trees on the last model N of their play of $\mathcal{G}_0$. Clearly, we can pull back Λ via (id,π) to a winning strategy for II in this game. We again call this strategy Σ, and write $\Sigma = \Lambda^{(\mathrm{id},\pi)}$ for it.

Let M,H,α_0,q, and π be as above. Let $\mathcal{V}$ be a pseudo-tree on (M,H,α_0) with last model N, and $s = \langle(\nu_i,k_i,\mathcal{U}_i) | i \leq n\rangle$ an N-stack. We can define the embedding normalization $\mathcal{W} = W(\mathcal{V},s)$ in essentially the same way that we did when no pseudo-trees were involved. By Lemma 7.6.5 we shall only need to consider the case that $\mathcal{V}$ is λ-separated and s consists of a single λ-separated tree on the last model of $\mathcal{V}$, so let us assume that. This guarantees that embedding normalization coincides with quasi-normalization.

Let N be the last model of $\mathcal{V}$. Being the last model, N has been declared stable in $\mathcal{V}$. We define

$$\mathcal{W}_\gamma = W(\mathcal{V},\mathcal{U}\restriction(\gamma+1))$$

by induction on γ. Each $\mathcal{W}_\gamma$ is a pseudo-tree with last model R_γ, and we have $\sigma_\gamma\colon \mathcal{M}^{\mathcal{U}}_\gamma \to R_\gamma$. We also have extended tree embeddings

$$\Psi_{\nu,\gamma}\colon \mathcal{W}_\nu \to \mathcal{W}_\gamma,$$

defined when $\nu <_U \gamma$. $\Psi_{\nu,\gamma}$ is partial iff $(\nu,\gamma]_U$ drops somewhere. We call its maps $u_{\nu,\gamma}$, $v_{\nu,\gamma}$, $s^{\nu,\gamma}_\xi$, and $t^{\nu,\gamma}_\xi$.

We set $\mathcal{W}_0 = \mathcal{W}$. The successor step is given by

$$\mathcal{W}_{\gamma+1} = \mathcal{W}_\gamma\restriction(\theta+1)^\frown\langle F\rangle^\frown i_F``(\mathcal{W}_\nu^{\geq\beta}),$$

where $F = \sigma_\gamma(E^{\mathcal{U}}_\gamma)$, $\theta = \alpha_F$ is the least *stable* node of $\mathcal{W}_\gamma$ such that F is on the extended $\mathcal{M}^{\mathcal{W}_\gamma}_\theta$-sequence, and $\nu = U\text{-pred}(\gamma+1)$. Let

$$\beta(\mathcal{W}_\nu,\mathcal{W}_\gamma,F) = \begin{cases} \text{least } \eta \text{ such that } \mathrm{crit}(F) < \varepsilon^{\mathcal{W}_\nu}_\eta & \text{if there is such an } \eta, \\ \mathrm{lh}(\mathcal{W}_\nu)-1 & \text{otherwise.} \end{cases}$$

Set $\beta = \beta(\mathcal{W}_\nu, \mathcal{W}_\gamma, F)$. It is easy to see that $\beta \le \theta$, and

$$\mathcal{W}_\nu \restriction \beta+1 = \mathcal{W}_\gamma \restriction \beta+1 = \mathcal{W}_{\gamma+1} \restriction \beta+1.$$

This is because between ν and γ, all the $\mathcal{W}_\eta$ used the same extenders E such that $\mathrm{lh}(E) < \mathrm{lh}(F_\nu)$.

Let us assume for simplicity that $(\nu, \gamma+1]_U$ does not drop. We have

$$u, v\colon \ \mathrm{lh}(\mathcal{W}_\nu) \to \mathrm{lh}(\mathcal{W}_{\gamma+1})$$

given by, for $\xi < \mathrm{lh}(\mathcal{W}_\nu)$,

$$u(\xi) = \begin{cases} \xi & \text{if } \xi < \beta, \\ (\theta+1)+(\xi-\beta) & \text{if } \beta \le \xi, \end{cases}$$

and

$$v(\xi) = \begin{cases} \xi & \text{if } \xi \le \beta, \\ (\theta+1)+(\xi-\beta) & \text{if } \beta+1 < \xi, \end{cases}$$

and

$$v(\beta+1) = \begin{cases} \beta+1 & \text{if } \beta \notin \mathrm{stab}(\mathcal{W}_\nu), \\ \theta+2 & \text{otherwise.} \end{cases}$$

In the case $v(\beta+1) = \beta+1$, the phalanx at $(\beta, \beta+1)$ in $\mathcal{W}_\nu$ is the same as the one at $(\beta, \beta+1)$ in $\mathcal{W}_{\gamma+1}$. In $\mathcal{W}_{\gamma+1}$ it will be moved up to a phalanx at $(\theta+1, \theta+2)$.

For $\eta \le_U \nu$, we let $u_{\eta,\gamma+1} = u \circ \phi_{\eta,\nu}$, and similarly for v. A node η of $\mathcal{W}_{\gamma+1}$ is stable just in case $\eta \le \theta$ and η is stable as a node of $\mathcal{W}_\gamma$, or $\eta = u(\xi)$, where ξ is stable as a node of $\mathcal{W}_\nu$. The stable nodes are just those having exit extenders together with the last node, so there is no other reasonable choice here. u may not preserve the property of being a stable root, the exception being when β is a stable root in $\mathcal{W}_\nu$, so that $u(\beta)$ is stable but $\mathrm{rt}(u(\beta)) = \beta$ in $\mathcal{W}_{\gamma+1}$. Nevertheless, clause (1)(c) of Definition 10.3.3 still holds in this case: if $\alpha+1 = \beta$, then $u(\alpha) = \alpha = \mathrm{rt}(u(\alpha+1)) - 1$. The rest of the requirements on u and v are easy to check.

For $\xi < \beta$, $t_\xi^{\nu,\gamma+1}$ is the identity. We define by induction on $\xi \ge \beta$ the models $\mathcal{M}_{u(\xi)}^{\mathcal{W}_{\gamma+1}}$ and maps $t_\xi : \mathcal{M}_\xi^{\mathcal{W}_\nu} \to \mathcal{M}_{u(\xi)}^{\mathcal{W}_{\gamma+1}}$ as before. s_ξ is the identity for $\xi \le \beta$, and $s_\xi = t_\xi$ for $\xi > \beta+1$.

For example, suppose $\xi = \beta$. We let

$$\mathcal{M}_{\theta+1}^{\mathcal{W}_{\gamma+1}} = \mathrm{Ult}(\mathcal{M}_\beta^{\mathcal{W}_\nu}, F),$$

and let t_β be the canonical embedding, so that

$$t_\beta = i_{\beta,\theta+1}^{\mathcal{W}_{\gamma+1}}.$$

If β is stable in $\mathcal{W}_\nu$, then $E_{\theta+1}^{\mathcal{W}_{\gamma+1}} = t_\beta(E_\beta^{\mathcal{W}_\nu})$, and

$$\mathcal{M}_{\theta+2}^{\mathcal{W}_{\gamma+1}} = \text{Ult}(P, E_{\theta+1}^{\mathcal{W}_{\gamma+1}}),$$

where P is the appropriate initial segment of some $\mathcal{M}_\tau^{\mathcal{W}_\nu}$. We determine $t_{\beta+1}$ using the Shift Lemma as before. (I.e., $t_{\beta+1}([a,f]) = [t_{\theta+1}(a), t_\tau(f)]$ if $\tau \neq \beta$, or if $\tau = \beta$ and $\text{crit}(F) \leq \text{crit}(E_{\theta+1}^{\mathcal{W}_{\gamma+1}})$. Otherwise, $t_{\beta+1}([a,f]) = [t_{\theta+1}(a), f]$.) So nothing changes.

On the other hand, if β is unstable in $\mathcal{W}_\nu$, then $\theta+1$ is unstable in $\mathcal{W}_{\gamma+1}$. We set

$$(\alpha_{\theta+1})^{\mathcal{W}_{\gamma+1}} = \sup i_{0,\theta+1}^{\mathcal{W}_{\gamma+1}}\text{``}(\alpha_0),$$

and as we must,

$$\mathcal{M}_{\theta+2}^{\mathcal{W}_{\gamma+1}} = \text{cHull}_{k+1}^{\mathcal{M}_{\theta+1}^{\mathcal{W}_{\gamma+1}}}(\alpha_{\theta+1} \cup i_{0,\theta+1}^{\mathcal{W}_{\gamma+1}}(q)).$$

Let σ be the anticollapse map. Let $\tau \colon \mathcal{M}_{\beta+1}^{\mathcal{W}_\nu} \to \mathcal{M}_\beta^{\mathcal{W}_\nu}$ be the anticollapse map. Note that $\mathcal{W}_\nu \restriction (\beta+2) = \mathcal{W}_\gamma \restriction (\beta+2) = \mathcal{W}_{\gamma+1} \restriction (\beta+2)$ in the present case. We set

$$s_{\beta+1} = \text{id}$$

and

$$t_{\beta+1} = \sigma^{-1} \circ t_\beta \circ \tau.$$

If $\beta+2 = \text{lh}(\mathcal{W}_\nu)$ then we are done defining $\mathcal{W}_{\gamma+1}$ and $\Psi_{\nu,\gamma+1}$. If not, we set $E_{\theta-2}^{\mathcal{W}_{\gamma+1}} = t_{\beta+1}(E_{\beta+1}^{\mathcal{W}_\nu})$. We have $\varepsilon_\beta^{\mathcal{W}_\nu} = \inf(\alpha_\beta^{\mathcal{W}_\nu}, \varepsilon(E_{\beta+1}^{\mathcal{W}_\nu}))$, and we set

$$\varepsilon_{\theta+1}^{\mathcal{W}_{\gamma+1}} = \inf(\alpha_{\theta+1}^{\mathcal{W}_{\gamma+1}}, \varepsilon(E_{\theta+2}^{\mathcal{W}_{\gamma+1}})).$$

It is easy to see that $\mathcal{M}_{\theta+2}^{\mathcal{W}_{\gamma+1}} | \varepsilon_{\theta+1} = \mathcal{M}_{\theta+1}^{\mathcal{W}_{\gamma+1}} | \varepsilon_{\theta+1}$. (We are ignoring the anomalous case here.) We also have

$$t \restriction \varepsilon_\beta^{\mathcal{W}_\nu} = t_{\beta+1} \restriction \varepsilon_\beta^{\mathcal{W}_\nu},$$

which is the agreement we need to continue defining $\mathcal{W}_{\gamma+1}$ and $\Psi_{\nu,\gamma+1}$.

Let us check that $\Psi_{\nu,\gamma+1}$ satisfies the clauses in Definition 10.3.3 that are relevant so far. These involve the behavior of its maps at $\alpha \leq \beta$ if β is stable in $\mathcal{W}_\nu$, and at $\alpha \leq \beta+1$ otherwise.

For (1), clearly u is order-preserving and preserves stability. If $\xi \neq \beta$, then ξ is a root in $\mathcal{W}_\nu$ iff $u(\xi)$ is a root in $\mathcal{W}_{\gamma+1}$, so (1)(c) is clear except when $\alpha+1 = \beta$ and α is unstable in $\mathcal{W}_\nu$. As we remarked above, this case is ok too.

For (2), the case to check is at β and possibly $\beta+1$. But $v(\beta) = \beta$, and $\beta \in \text{stab}(\mathcal{W}_\nu)$ iff $\beta \in \text{stab}(\mathcal{W}_{\gamma+1})$, so (i) holds. (ii) holds at β because $v(\beta) = \beta <_{\mathcal{W}_{\gamma+1}} \theta+1 = u(\beta)$. If $\beta \notin \text{stab}(\mathcal{W}_\nu)$, then $v(\beta+1) = \beta+1$, and $\beta+1 \leq^*_{W_{\gamma+1}} \theta+2 = u(\beta+1)$, so (ii) holds at $\beta+1$.

Clause (3) is trivial at this stage, because $s_\alpha = \text{id}$ for $\alpha \le \beta$, and for $\alpha = \beta+1$ if β is unstable. Clause (4) is also trivial at $\alpha < \beta$, because all maps are then the identity. At β, it only applies if β is stable, and then it amounts to $t_\beta = i^{\mathcal{W}_{\gamma+1}}_{\beta,\theta+1}$, which is indeed how we defined t_β. If β is unstable in $\mathcal{W}_\nu$, then clause (4) requires that $t_{\beta+1} = \hat{\imath}^{\mathcal{W}_{\gamma+1}}_{\beta+1,\theta+2} \circ s_{\beta+1}$. But $s_{\beta+1} = \text{id}$, and $t_{\beta+1} = \sigma^{-1} \circ \hat{\imath}^{\mathcal{W}_{\gamma+1}}_{\beta,\theta+1} \circ \tau = \hat{\imath}^{\mathcal{W}_{\gamma+1}}_{\beta+1,\theta+2}$, so (4) is satisfied.

The rest of the definition proceeds as above, defining $t_\xi : \mathcal{M}^{\mathcal{W}_\nu}_\xi \to \mathcal{M}^{\mathcal{W}_{\gamma+1}}_{u(\xi)}$ using the Shift Lemma and the appropriate earlier t_τ. If ξ is unstable in $\mathcal{W}_\nu$, we then go on to define $t_{\xi+1} : \mathcal{M}^{\mathcal{W}_\nu}_{\xi+1} \to \mathcal{M}^{\mathcal{W}_{\gamma+1}}_{u(\xi+1)}$ as we did above. At limit steps, we take direct limits.

This gives us $\mathcal{W}_{\gamma+1}$ and $\Psi_{\nu,\gamma+1} : \mathcal{W}_\nu \to \mathcal{W}_{\gamma+1}$. At limit ordinals λ, we let $\mathcal{W}_\lambda$ be the direct limit of the $\mathcal{W}_\nu$ for $\nu <_U \lambda$, under the $\Psi_{\nu,\mu}$. Finally, $W(\mathcal{V},\mathcal{U}) = \mathcal{W}_\gamma$, where $\gamma+1 = \text{lh}(\mathcal{U})$.

This finishes our discussion of the normalization $W(\mathcal{V},\mathcal{U})$, for $\mathcal{V}$ a λ-separated pseudo-tree on (M,H,α), and $\mathcal{U}$ a a λ-separated tree on the last model of $\mathcal{V}$. We say that strategy Σ for the game $\mathcal{G}$ *normalizes well* iff whenever $\langle \mathcal{V},\mathcal{U}\rangle$ is according to Σ, then $W(\mathcal{V},\mathcal{U})$ is according to Σ.

LEMMA 10.3.8. *Let (M,Λ) be an lbr hod pair, and H,π,q,α_0 be as above. Let $\Sigma = \Lambda^{(id,\pi)}$; then Σ normalizes well.*

PROOF SKETCH. If $\mathcal{V}$ is a pseudo-tree, and $\mathcal{U}$ is a λ-separated tree on the last model of $\mathcal{V}$, let us write

$$(\text{id},\pi)\langle \mathcal{V},\mathcal{U}\rangle = \langle (\text{id},\pi)\mathcal{V}, \sigma\mathcal{U}\rangle,$$

where σ is the copy map acting on the last model of $\mathcal{V}$. Just for the space of this proof, to keep things straight, let us write $\hat{W}$ for the embedding normalization operation on pseudo-trees defined above.

Λ itself normalizes well. But normalizing commutes with copying in this context, as it did in the case of ordinary iteration trees. That is

$$(\text{id},\pi)\hat{W}(\mathcal{V},\mathcal{U}) = W((\text{id},\pi)\langle \mathcal{V},\mathcal{U}\rangle).$$

So

$$\begin{aligned} \hat{W}(\mathcal{V},\mathcal{U}) \text{ is by } \Sigma &\iff (\text{id},\pi)\hat{W}(\mathcal{V},\mathcal{U}) \text{ is by } \Lambda \\ &\iff W((\text{id},\pi)\mathcal{V}, \sigma\mathcal{U}) \text{ is by } \Lambda \\ &\iff \langle (\text{id},\pi)\mathcal{V}, \sigma\mathcal{U}\rangle \text{ is by } \Lambda \\ &\iff \langle \mathcal{V},\mathcal{U}\rangle \text{ is by } \Sigma\,, \end{aligned}$$

as desired. See the proof of Theorem 7.1.6. ⊣

Let us turn now to the proof of Lemma 9.6.5. We were given an lbr hod pair (M,Σ), but it works better with the current notation to call that pair (M,Λ), so let us make that switch. $k = k(M)$. We are also given K, α_0, π, and q with

$K = \mathrm{cHull}^M_{k+1}(\alpha_0 \cup q)$, $\rho(M) \leq \alpha_0$, q a finite set of ordinals, and $\pi\colon K \to M$ the anticollapse.[14] We have the pullback strategy

$$\Sigma = \Lambda^{(\mathrm{id},\pi)}$$

for $\mathcal{G}$ on (M, K, α_0), and Σ normalizes well and has strong hull condensation. $\langle (N^*, \in, w, \mathcal{F}, \Phi), \Phi^* \rangle$ is a coarse strategy pair that captures $\mathrm{Code}(\Lambda)$ and such that M is countable in N^*, and $\mathbb{C}$ is the maximal $(w, \mathcal{F})$ construction of $(N^*, \in, w, \mathcal{F}, \Phi)$. (M, Λ) iterates to $(M^{\mathbb{C}}_{\eta_0, j_0}, \Omega^{\mathbb{C}}_{\eta_0, j_0})$. For $\langle \eta, j \rangle \leq \langle \eta_0, j_0 \rangle$, we set[15]

$$\begin{aligned} \mathcal{V}_{\eta,j} = {} & \lambda\text{-separated tree of minimal length} \\ & \text{whereby } (M, \Lambda) \text{ iterates past } (M_{\eta,j}, \Omega_{\eta,j}). \end{aligned}$$

We also had λ-separated pseudo-trees $\mathcal{S}_{\eta,j}$ on (M, K, α_0) formed by certain rules.

DEFINITION 10.3.9. For an lpm R, we say that $(\mathcal{T}, \mathcal{V})$ is the (Σ, Λ, R)-*coiteration* (of (M, K, α_0) with M) iff

(a) $\mathcal{T}$ is a λ-separated pseudo-tree by Σ on (M, K, α_0) with last model P,
(b) $\mathcal{V}$ is a λ-separated tree by Λ on M with last model Q,
(c) $R \trianglelefteq P$ and $R \trianglelefteq Q$, and $\mathcal{T}$ and $\mathcal{V}$ are of minimal length such that this is true, and
(d) stability (and hence the next model) in $\mathcal{T}$ is determined by the rules we have given: θ is unstable iff $[0, \theta]_T$ does not drop, and $e^{\mathcal{T}}_\theta = e^{\mathcal{V}}_\tau$ for some τ.

We remark that the internal strategy $\dot{\Sigma}^R$ is relevant in (c), but no external strategy agreement is relevant. (c) tells us that $\mathcal{V}$ and $\mathcal{W}$ proceed by hitting the least extender disagreement with R, and that the corresponding R-extenders are all empty.

We had fixed $\langle \nu_0, k_0 \rangle \leq_{\mathrm{lex}} \langle \eta_0, j_0 \rangle$ such that for each $\langle \eta, j \rangle <_{\mathrm{lex}} \langle \nu_0, k_0 \rangle$, the $(\Sigma, \Lambda, M_{\eta,j})$-coiteration (M, K, α_0) with M exists, and moreover, the last model on both sides is strictly longer than $M_{\eta,j}$, and no external strategy disagreements show up on either side.[16] We are trying to show that the $(\Sigma, \Lambda, M_{\nu_0,k_0})$-coiteration exists, and that no external strategy disagreements show up on the (M, K, α_0) side. That is,

LEMMA 10.3.10. *Let $\mathcal{T}$ be an initial segment of $\mathcal{S}_{\nu_0,k_0}$ with stable last node, and let R_0 be the last model of $\mathcal{T}$; then either*

(1) $(R_0, \Sigma_{\mathcal{T}}) \trianglelefteq (M_{\nu_0,k_0}, \Omega_{\nu_0,k_0})$, *or*
(2) $(M_{\nu_0,k_0}, \Omega_{\nu_0,k_0}) \triangleleft (R_0, \Sigma_{\mathcal{T}})$, *or*
(3) there is a nonempty extender E on the R_0 sequence such that, setting $\tau = \mathrm{lh}(E)$,
   (i) $\dot{E}^{M_{\nu_0,k_0}}_\tau = \emptyset$, *and*
   (ii) $(\Sigma_{\mathcal{T}})_{\langle \tau, -1 \rangle} = (\Omega_{\nu_0,k_0})_{\langle \tau, 0 \rangle}$.

[14] K was called H before, but we want to reserve "H" for something else now.
[15] We called it $\mathcal{U}_{\eta,j}$ before, but $\mathcal{V}_{\eta,j}$ works better now. j_0 was formerly k_0.
[16] We called this pair $\langle \nu, l \rangle$ before, but we want to free up those letters for other use below.

PROOF. Suppose $\mathcal{T}$ and R_0 are a counterexample. Since (1) and (2) fail, there is a least disagreement between $(R_0, \Sigma_{\mathcal{T}})$ and $(M_{\nu_0,k_0}, \Omega_{\nu_0,k_0})$, and since (3) fails, the least disagreement either involves a nonempty extender from M_{ν_0,k_0}, or is a strategy disagreement.

Suppose first that (3) fails because there is a nonempty extender on the M_{ν_0,k_0} side at the least disagreement between $(R_0, \Sigma_{\mathcal{T}})$ with $(M_{\nu_0,k_0}, \Omega_{\nu_0,k_0})$. As in the proof of the Bicephalus Lemma, we can reduce to the case that $k_0 = 0$, and the least disagreement involves $F = \dot{F}^{M_{\nu_0,0}}$, with $F \neq \emptyset$. Letting $\mathcal{V} = \mathcal{V}_{\nu_0,0}$, we then have that $(\mathcal{T}, \mathcal{V})$ is the $(\Sigma, \Lambda, M_{\nu_0,-1})$- coiteration. Let P and Q be the last models of $\mathcal{T}$ and $\mathcal{V}$. So

$$(M_{\nu_0,-1}, \Omega_{\nu_0,-1}) = (P|\langle \nu_0, -1\rangle, \Sigma_{\mathcal{T},\langle \nu_0,-1\rangle}) = (Q, \Lambda_{\mathcal{V},\langle \nu_0,-1\rangle}).$$

Let

$$j \colon N^* \to \mathrm{Ult}(N^*, F^{\mathbb{C}}_{\nu_0})$$

be the canonical embedding, and $\kappa = \mathrm{crit}(j)$. We have that

$$M_{\nu_0,-1} = j(M_{\nu_0,-1})|\,\mathrm{lh}(F)$$

by coherence. (Note $j(M_{\nu_0,-1})|\,\mathrm{lh}(F)$ is extender passive and $\mathrm{lh}(F)$ is a cardinal in $j(M_{\nu_0,-1})$ because $\mathbb{C}$ is good at $\langle \nu_0, -1\rangle$.) $j(\mathcal{T}, \mathcal{V})$ is the $(\Sigma, \Lambda, j(M_{\nu_0,-1}))$ coiteration, because $j(\Lambda) \subseteq \Lambda$, and hence $j(\Sigma) \subseteq \Sigma$. So $\mathcal{V}$ is an initial segment of $j(\mathcal{V})$. But then $\mathcal{T}$ is an initial segment of $j(\mathcal{T})$, because the relevant conditions for declaring stability are the same in N^* and $j(N^*)$.

We have that $\mathcal{M}^{\mathcal{T}}_\kappa = \mathcal{M}^{j(\mathcal{T})}_\kappa$ and $j \restriction \mathcal{M}^{\mathcal{T}}_\kappa = i^{j(\mathcal{T})}_{\kappa, j(\kappa)}$, so F is compatible with the first extender G used in $[\kappa, j(\kappa)]_{j(\mathcal{T})}$. $M_{\nu_0,-1} \lhd \mathcal{M}^{j(\mathcal{T})}_{j(\kappa)}$, so G cannot be a proper initial segment of F. But F is not on the sequence of $\mathcal{M}^{j(\mathcal{T})}_{j(\kappa)}$, so F cannot be a proper initial segment of G, unless $F = G^-$. Hence $F = G^-$, so F is on the sequence of $\mathcal{M}^{j(\mathcal{T})}_\xi$, where $G = E^{j(\mathcal{T})}_\xi$. But $\mathcal{T} = j(\mathcal{T}) \restriction (\xi + 1)$ by coherence, so F is on the sequence of $\mathcal{M}^{\mathcal{T}}_\xi$, contradiction.

So we may assume that we have $J \trianglelefteq M_{\nu_0,k_0}$ such that

$$J \trianglelefteq R_0,$$

but there is a strategy disagreement, that is

$$(\Omega_{\nu_0,k_0})_J \neq \Sigma_{\mathcal{T},J}.$$

Note that $(J, (\Omega_{\nu_0,k_0})_J)$ and $(J, \Sigma_{\mathcal{T},J})$ are lbr hod pairs. (In the case of $(J, \Sigma_{\mathcal{T},J})$, this is because the pair is elementarily embedded, as a mouse pair, into some iterate of (M, Λ).) Thus the two strategies are determined by their actions on λ-separated trees by Lemma 7.6.5, and we can fix a single λ-separated tree $\mathcal{U}$ on J of limit length such that $(\Omega_{\nu_0,k_0})_J(\mathcal{U}) \neq \Sigma_{\mathcal{T},J}(\mathcal{U})$.

We can reduce to the case that $J = M_{\nu_0,k_0}$ by using the pullback consistency of $\Sigma_{\mathcal{T},J}$, just as we did in the proof of Theorem 9.5.2, so let us assume that. Let

$$\Omega = \Omega_{\nu_0,k_0},$$

and

$$b = \Omega_J(\mathcal{U}).$$

Let $\mathcal{U}^+$ be the lift of $\mathcal{U}$ to a tree on R_0, and

$$\sigma^0_\gamma \colon \mathcal{M}^{\mathcal{U}}_\gamma \to J^0_\gamma \trianglelefteq \mathcal{M}^{\mathcal{U}^+}_\gamma$$

be the lifting map. So $J = J^0_0$ and $\sigma^0_0 = \text{id}$. Since $\mathcal{U}$ is by $\Sigma_{\mathcal{T},J}$ and Σ is internally lift consistent, $\mathcal{U}^+$ is by $\Sigma_{\mathcal{T},R_0}$. It is enough to show that $\Sigma_{\mathcal{T},R_0}(\mathcal{U}^+) = b$.

We do this by repeating the proof of Theorem 8.4.3. Large stretches of that proof can be simply copied, and that is basically what we are going to do. We shall try to condense things enough that the new points stand out. We have set up the notation to mimic that in the proof of Theorem 8.4.3. To make the correspondence better, let us now set

$$\mathcal{W}^*_{\eta,j} = \mathcal{S}_{\eta,j},$$

so as to free up $\mathcal{S}$ for other use.

Let us look at how $\Omega(\mathcal{U})$ is defined. Let c_0 be the conversion stage

$$c_0 = \langle M_{\nu_0,k_0}, \text{id}, M_{\nu_0,k_0}, \mathbb{C}, \mathcal{N}^* \rangle$$

where $\mathcal{N}^* = (N^*, \in, w, \mathcal{F}, \Phi)$, and

$$\text{lift}(\mathcal{U}^\frown b, c_0) = \langle \mathcal{U}^*, \langle c_\alpha \mid \alpha < \text{lh}(\mathcal{U}) \vee \alpha = b \rangle \rangle,$$

where

$$c_\alpha = \langle \mathcal{M}^{\mathcal{U}}_\alpha, \psi_\alpha, Q_\alpha, \mathbb{C}_\alpha, S_\alpha \rangle.$$

For $\gamma + 1 < \text{lh}(\mathcal{U})$, let res_γ be the map resurrecting $\psi_\gamma(E^{\mathcal{U}}_\gamma)^-$ inside $\mathbb{C}_\gamma$, namely

$$\text{res}_\gamma = \sigma_{Q_\gamma}[Q_\gamma | \langle \text{lh}(\psi_\gamma(E^{\mathcal{U}}_\gamma)), 0 \rangle].$$

We have $M_{\nu_0,k_0} = \mathcal{M}^{\mathcal{U}}_0 = J^0_0$, and $\psi_0 = \text{id}$. For $\langle \eta, j \rangle \leq_{\text{lex}} i^{\mathcal{U}^*}_{0,\gamma}(\langle \nu_0, k_0 \rangle)$, we let $(\mathcal{W}^*_{\eta,j})^{S_\gamma} = i^{\mathcal{U}^*}_{0,\gamma}(\langle \mu, l \rangle \mapsto \mathcal{W}^*_{\mu,l})_{\eta,j}$ and $(\mathcal{V}_{\eta,j})^{S_\gamma} = i^{\mathcal{U}^*}_{0,\gamma}(\langle \mu, l \rangle \mapsto \mathcal{V}_{\mu,l})_{\eta,j}$. Note that $i^{\mathcal{U}^*}_{0,\gamma}(\Lambda) \cap S_\gamma = \Lambda \cap S_\gamma$, so that the $\mathcal{V}^{S_\gamma}_{\eta,j}$ and $(\mathcal{W}^*_{\eta,j})^{S_\gamma}$ are by Λ and Σ, respectively. Let

$$Q_\gamma = M^{\mathbb{C}_\gamma}_{\eta_\gamma, l_\gamma}$$

and

$$(\mathcal{W}^*_\gamma, \mathcal{V}_\gamma) = (\mathcal{W}^*_{\eta_\gamma, l_\gamma}, \mathcal{V}_{\eta_\gamma, l_\gamma})^{S_\gamma},$$

for $\gamma < \text{lh}(\mathcal{U})$ or $\gamma = b$. Let $z^*(\gamma) + 1 = \text{lh}(\mathcal{W}^*_\gamma)$, and put

$$N_\gamma = \mathcal{M}^{\mathcal{W}^*_\gamma}_{z^*(\gamma)}.$$

So $\mathcal{W}_0^* = \mathcal{T}$ is our pseudo-tree on (M, K, α_0) by Σ. Its last model is $R_0 = N_0$, and $M_{\nu_0,k_0} = J = J_0^0 \trianglelefteq R_0$. $\mathcal{U}$ is a tree on J and $\mathcal{U}^+$ is on R_0. Let us look at the meta-tree associated to $W(\mathcal{T}, \mathcal{U}^+)$. Set

$$\mathcal{W}_\gamma = W(\mathcal{T}, \mathcal{U}^+ \restriction (\gamma+1))$$

for $\gamma < \mathrm{lh}(\mathcal{U})$, and

$$\mathcal{W}_b = W(\mathcal{T}, (\mathcal{U}^+)^\frown b).$$

So $\mathcal{W}_0 = \mathcal{W}_0^* = \mathcal{T}$. The $\mathcal{W}_\gamma$'s are all by Σ, because Σ normalizes well and $\mathcal{U}^+ \restriction (\gamma+1)$ is by $\Sigma_{\mathcal{T},R_0}$. Since Σ normalizes well, it is enough to show that $\mathcal{W}_b$ is by Σ, for then $\Sigma_{\mathcal{T},R_0}(\mathcal{U}^+) = b$, so $\Sigma_{\mathcal{T},J}(\mathcal{U}) = b$, as desired. Since Σ has strong hull condensation, it is enough to show

SUBLEMMA 10.3.10.1. *$\mathcal{W}_b$ is pseudo-hull of $\mathcal{W}_b^*$.*

PROOF. As before, we define by induction on γ, for $\gamma < \mathrm{lh}(\mathcal{U})$ or $\gamma = b$, tree embeddings

$$\Phi_\gamma \colon \mathcal{W}_\gamma \to \mathcal{W}_\gamma^*.$$

Let

$$\Phi_\gamma = \langle u^\gamma, v^\gamma, \langle s_\beta^\gamma \mid \beta \leq z(\gamma)\rangle, \langle t_\beta^\gamma \mid \beta < z(\gamma)\rangle\rangle.$$

Φ_γ can be extended, in that $v^\gamma(z(\gamma)) \leq^*_{\mathcal{W}_\gamma^*} z^*(\gamma)$, and we let

$$t^\gamma = i^{\mathcal{W}_\gamma^*}_{v^\gamma(z(\gamma)), z^*(\gamma)} \circ s^\gamma_{z(\gamma)}$$

be the final t-map of the extended tree embedding. Letting $R_\gamma = \mathcal{M}^{\mathcal{W}_\gamma}_{z(\gamma)}$ we have that

$$t^\gamma \colon R_\gamma \to N_\gamma.$$

Again, the rest of Φ_γ is actually determined by t^γ. It is also determined by u^γ.

The embedding normalization process gives us extended tree embeddings

$$\Psi_{\nu,\gamma} \colon \mathcal{W}_\nu \to \mathcal{W}_\gamma,$$

defined when $\nu <_U \gamma$. We use $\phi_{\nu,\gamma}$ for the u-map of $\Psi_{\nu,\gamma}$, so that $\phi_{\nu,\gamma} \colon \mathrm{lh}(\mathcal{W}_\nu) \to \mathrm{lh}(\mathcal{W}_\gamma)$, the map being total if $(\nu, \gamma]_U$ does not drop in model or degree. We let $\pi_\tau^{\nu,\gamma}$ be the t-map $t_\tau^{\Psi_{\nu,\gamma}}$, so that

$$\pi_\tau^{\nu,\gamma} : \mathcal{M}_\tau^{\mathcal{W}_\nu} \to \mathcal{M}^{\mathcal{W}_\gamma}_{\phi_{\nu,\gamma}(\tau)}$$

elementarily, for $\nu <_U \gamma$ and $\tau \in \mathrm{dom}\, \phi_{\nu,\gamma}$. Let also:

- $\sigma_\eta^1 : \mathcal{M}_\eta^{\mathcal{U}^+} \to R_\eta$ be the embedding normalization map,
- $\sigma_\eta = \sigma_\eta^1 \circ \sigma_\eta^0 \colon \mathcal{M}_\eta^{\mathcal{U}} \to J_\eta^1 \trianglelefteq R_\eta$, where $J_\eta^1 = \sigma_\eta^1(J_\eta^0)$,
- $F_\eta = \sigma_\eta(E_\eta^{\mathcal{U}})$, and

- $\bar{\xi}_\eta$ = least α such that F_η^- is on the $\mathcal{M}_\alpha^{\mathcal{W}_\eta}$ sequence.[17]

Thus $\mathcal{W}_{\gamma+1} = W(\mathcal{W}_\nu, \mathcal{W}_\gamma, F_\gamma)$, where $\nu = U\text{-pred}(\gamma+1)$.

We also have an extended tree embedding $\Psi^*_{\nu,\gamma}\colon \mathcal{W}^*_\nu \to \mathcal{W}^*_\gamma$ defined when $\nu <_U \gamma$ and $(\nu,\gamma]_U$ does not drop. The maps of $\Psi^*_{\nu,\gamma}$ are all restrictions of $i^{\mathcal{U}^*}_{\nu,\gamma}$, so we do not give them special names. As before, we maintain by induction that the diagram

$$\begin{array}{ccc} \mathcal{W}_\gamma & \xrightarrow{\quad\Phi_\eta\quad} & \mathcal{W}^*_\gamma \\ \uparrow \scriptstyle{\Psi_{\nu,\gamma}} & & \uparrow \scriptstyle{\Psi^*_{\nu,\gamma}} \\ \mathcal{W}_\nu & \xrightarrow{\quad\Phi_\nu\quad} & \mathcal{W}^*_\nu \end{array}$$

commutes, in the appropriate sense.

Our induction hypothesis is

Induction Hypothesis $(\dagger)_\gamma$.

(1) (a) For $\nu < \eta \leq \gamma$, $\Phi_\nu \restriction (\bar{\xi}_\nu + 1) = \Phi_\eta \restriction (\bar{\xi}_\nu + 1)$.
(b) For all $\eta \leq \gamma$, t^η is well defined; that is, $v^\eta(z(\eta)) \leq^*_{\mathcal{W}^*_\eta} z^*(\eta)$.
(c) For $\nu < \eta \leq \gamma$, $s^\eta_{z(\eta)} \restriction \mathrm{lh}(F_\nu) + 1 = \mathrm{res}_\nu \circ t^\nu \restriction \mathrm{lh}(F_\nu) + 1$.

(2) Let $\nu < \eta \leq \gamma$, and $\nu <_U \eta$, and suppose that $(\nu,\eta]_U$ does not drop; then $\Phi_\eta \circ \Psi_{\nu,\eta} = \Psi^*_{\nu,\eta} \circ \Phi_\nu$.

(3) For $\xi \leq \gamma$, $\psi_\xi = t^\xi \circ \sigma_\xi$.

(4) For all $\nu < \gamma$, N^*_ν agrees with N_γ strictly below $\mathrm{lh}(G_\nu)$. G_ν is on the N^*_ν-sequence, but $\mathrm{lh}(G_\nu)$ is a cardinal of N_γ. W^{**}_ν is an initial segment of $W^*_\gamma \restriction (v^\gamma(\bar{\xi}_\gamma) + 1)$.

We shall explain the terms in clause (4) shortly. The precise meaning of clause (2) can be given by writing it out in terms of the component maps, as we did in Theorem (d) in the proof of Theorem 8.4.3. We leave it to the reader to do that.

Here is a diagram illustrating clause (3).

$$\begin{array}{ccccccc} & & \mathcal{M}^{\mathcal{U}^+}_\xi & \xrightarrow{\quad\sigma^1_\xi\quad} & R_\xi & \xrightarrow{\quad t^\xi\quad} & N_\xi \\ & & \triangledown\!\mid & & \triangledown\!\mid & & \triangledown\!\mid \\ \mathcal{M}^{\mathcal{U}}_\xi & \xrightarrow{\quad\sigma^0_\xi\quad} & J^0_\xi & \xrightarrow{\quad\sigma^1_\xi\quad} & J^1_\xi & \xrightarrow{\quad t^\xi\quad} & Q_\xi \end{array}$$

$$\psi_\xi \colon \mathcal{M}^{\mathcal{U}}_\xi \to Q_\xi$$

[17] We called this ordinal α_η before, but that would clash with our notation for exchange ordinals in pseudo-trees.

We now describe how to obtain $\Phi_{\gamma+1}$ from the Φ_α for $\alpha \le \gamma$.

We have $t^\gamma\colon R_\gamma \to N_\gamma$, where N_γ is the last model of W^*_γ. Let $F = F_\gamma$, and let $\nu = U\text{-pred}(\gamma+1)$. So $\mathcal{W}_{\gamma+1} = W(\mathcal{W}_\nu, W_\gamma, F)$. Let us assume for simplicity that $(\nu, \gamma+1]_U$ is not a drop in model or degree. Let[18]

- $H = H_\gamma = t^\gamma(F)$,
- $X = X_\gamma = Q_\gamma | \langle \mathrm{lh}(H), 0\rangle$,
- $G = G_\gamma = \mathrm{res}_\gamma(H)$,
- $Y = Y_\gamma = \mathrm{Res}_{Q_\gamma}[X]$, and
- $G^* = B(G^-)^{\mathbb{C}_\gamma}$.[19]

We have $t^\gamma \circ \sigma_\gamma = \psi_\gamma$, so $G = \mathrm{res}_\gamma(\psi_\gamma(E^{\mathcal{U}}_\gamma))$, so

$$S_{\gamma+1} = \mathrm{Ult}(S_\nu, G^*).$$

Since we are not dropping, $W^*_{\gamma+1} = i_{G^*}(W^*_\nu)$, where $i_{G^*} = i^{\mathcal{U}^*}_{\nu,\gamma+1}$. The first thing we need to see is that G is used in $W^*_{\gamma+1}$.

Lemma 8.3.1 on capturing resurrection embeddings works also for our system of pseudo-trees:

CLAIM 10.3.11. *Let τ be least in* $\mathrm{stab}(\mathcal{W}^*_\gamma)$ *such that $X \trianglelefteq \mathcal{M}^{\mathcal{W}^*_\gamma}_\tau$, and θ least such that $X \trianglelefteq \mathcal{M}^{\mathcal{V}_\gamma}_\theta$. Let $(\mathcal{W}^{**}_\gamma, \mathcal{V}^{**}_\gamma)$ be the (Σ, Λ, Y)-coiteration of (M, K, α_0) with M; then*

(i) $\mathcal{W}^{**}_\gamma$ extends $\mathcal{W}^*_\gamma \restriction (\tau+1)$,

(ii) letting $\xi = \mathrm{lh}(\mathcal{W}^{**}_\gamma) - 1$, G^- is on the $\mathcal{M}^{\mathcal{W}^{**}_\gamma}_\xi$ sequence, and not on the $\mathcal{M}^{\mathcal{W}^{**}_\gamma}_\alpha$ sequence for any $\alpha < \xi$,

(iii) $\tau \le_{W^{**}_\gamma} \xi$, and $\hat{\imath}^{W^{**}_\gamma}_{\tau,\xi} \restriction (\mathrm{lh}(t^\gamma(F))+1) = \mathrm{res}_\gamma \restriction (\mathrm{lh}(t^\gamma(F))+1)$, and

(iv) similarly for $\mathcal{V}^{**}_\gamma$ vis-a-vis $\mathcal{V}_\gamma$.

PROOF SKETCH. Part (iv) literally follows from Lemma 8.3.1[20], because the $\mathcal{V}_{\eta,j}$ do not depend on the $\mathcal{W}^*_{\eta,j}$. For parts (i)-(iii), one simply repeats the proof of Lemma 8.3.1.

Item (i) includes the agreement on stability declarations and next models. The point is that the (Σ, Λ, Y)-coiteration reaches models extending X on both sides by the proof of Lemma 8.3.1. Let η be least such that $\eta \le_{W^{**}_\gamma} \xi$ and $X \trianglelefteq \mathcal{M}^{\mathcal{W}^{**}_\gamma}_\eta$. We have that from the proof of Lemma 8.3.1 that

$$\hat{\imath}^{W^{**}_\gamma}_{\eta,\xi} \restriction (\mathrm{lh}(t^\gamma(F))+1) = \mathrm{res}_\gamma \restriction (\mathrm{lh}(t^\gamma(F))+1).$$

The proof also shows that either $\eta = \xi$, or the first ultrapower taken in $(\eta, \xi]_{W^{**}_\gamma}$ involves a drop in model or degree. In either case, η is stable in $\mathcal{W}^{**}_\gamma$. Let also δ

[18]The conventions of Remark 8.4.10 apply here and in what follows.

[19]G is a plus extender and G^* is the background for G^- in $\mathbb{C}_\gamma$. G^* also backgrounds G.

[20]Apart from the fact that we are now dealing with a least branch construction.

be least such that $X \trianglelefteq \mathcal{M}_\delta^{\mathcal{V}_\gamma^{**}}$. We then have that $(\mathcal{W}_\gamma^{**} \upharpoonright (\eta+1), \mathcal{V}_\gamma^{**} \upharpoonright (\delta+1))$ is the (Σ, Λ, X) coiteration. But $X \trianglelefteq N_\gamma$, so this is an initial segment of the $(\Sigma, \Lambda, N_\gamma)$ coiteration, that is, of $(\mathcal{W}_\gamma^*, \mathcal{V}_\gamma)$. This implies $\eta = \tau$ and $\delta = \theta$. ⊣

Let

- $\xi_\gamma = \text{lh}(\mathcal{W}_\gamma^{**}) - 1$
- $\tau_\gamma =$ least τ such that $X \trianglelefteq \mathcal{M}_\tau^{\mathcal{W}_\gamma^*}$ and τ is stable in $\mathcal{W}_\gamma^*$, and
- $N_\gamma^* = \mathcal{M}_{\xi_\gamma}^{\mathcal{W}_\gamma^{**}}$.

With these definitions, clause (4) of $(\dagger)_\gamma$ now makes sense. Note $\mathcal{W}_\gamma^* \upharpoonright \tau_\gamma + 1 = \mathcal{W}_\gamma^{**} \upharpoonright \tau_\gamma + 1$, and ξ_γ is the least stable α of $\mathcal{W}_\gamma^{**}$ such that G^- is on the sequence of $\mathcal{M}_\alpha^{\mathcal{W}_\gamma^{**}}$.

X, Y, and N_ν all agree up to $\text{dom}(G)$, so

$$\begin{aligned} Y||o(Y) &= i_G(Y)|o(Y) \\ &= i_{G^*}(Y)|o(Y) \\ &= i_{G^*}(N_\nu)|o(Y) \\ &= N_{\gamma+1}|o(Y). \end{aligned}$$

Note here that $i_{G^*}(Y)$ is passive at $o(Y)$ because G^* backgrounds G, and not just G^-. $(\mathcal{W}_\gamma^{**}, \mathcal{V}_\gamma^{**})$ is the Y coiteration, so G^- is on the sequence of the last model on both sides. We then get that $\mathcal{V}_\gamma^{**}$ is an initial segment of $\mathcal{V}_{\gamma+1}$, $\mathcal{W}_\gamma^{**}$ is an initial segment of $\mathcal{W}_{\gamma+1}^*$ and G is used in both $\mathcal{V}_{\gamma+1}$ and $\mathcal{W}_{\gamma+1}^*$.

We define

$$\Phi_{\gamma+1} \upharpoonright \bar{\xi}_\gamma + 1 = \Phi_\gamma \upharpoonright \bar{\xi}_\gamma + 1,$$

and this is ok because $\mathcal{W}_\gamma \upharpoonright (\bar{\xi}_\gamma + 1) = \mathcal{W}_{\gamma+1} \upharpoonright (\bar{\xi}_\gamma + 1)$ and $\mathcal{W}_\gamma^* \upharpoonright v^\gamma(\bar{\xi}_\gamma + 1) = \mathcal{W}_{\gamma+1}^* \upharpoonright v^{\gamma+1}(\bar{\xi}_\gamma + 1)$. $\Phi_{\gamma+1} \upharpoonright \bar{\xi}_\gamma + 1$ is a tree embedding of $\mathcal{W}_{\gamma+1} \upharpoonright \bar{\xi}_\gamma + 1$ into $\mathcal{W}_{\gamma+1}^* \upharpoonright \xi_\gamma + 1$. Let

$$\begin{aligned} \bar{\xi} &= \bar{\xi}_\gamma, \\ \xi &= \xi_\gamma, \end{aligned}$$

and define

$$u^{\gamma+1}(\bar{\xi}) = \xi.$$

Thus $F = E_{\bar{\xi}}^{\mathcal{W}_{\gamma+1}}$ and $G = E_\xi^{\mathcal{W}_{\gamma+1}^*}$. Let

$$\beta = \beta(\mathcal{W}_\nu, \mathcal{W}_\gamma, F) = W_{\gamma+1}\text{-pred}(\bar{\xi}+1),$$

and

$$\beta^* = W_{\gamma+1}^*\text{-pred}(\xi+1).$$

Let us verify that β^* is located where it should be in $\mathcal{W}^*_{\gamma+1}$ according to Definition 10.3.3. Basically, we just run through the proof of Sublemma 8.4.8.1, taking into account the stability structure now present. So let

- $\bar{\kappa} = \text{crit}(E^{\mathcal{U}}_\gamma)$, and $\bar{P} = \mathcal{M}^{\mathcal{U}}_\gamma|\bar{\kappa}^+ = \mathcal{M}^{\mathcal{U}}_\nu|\bar{\kappa}^+ = \text{dom}(E^{\mathcal{U}}_\gamma)$,
- $\kappa = \text{crit}(F)$, and $P = \mathcal{M}^{\mathcal{W}_\nu}_\beta|\kappa^+ = \mathcal{M}^{\mathcal{W}_\gamma}_\beta|\kappa^+ = \mathcal{M}^{\mathcal{W}_{\gamma+1}}_\beta|\kappa^+ = \text{dom}(F)$, and
- $\kappa^* = \text{crit}(G)$, and $P^* = \text{dom}(G)$.

In these formulae, the successor cardinals are evaluated in the corresponding models, of course. Recall here that $\mathcal{W}_\nu \upharpoonright \beta+1 = \mathcal{W}_\gamma \upharpoonright \beta+1 = \mathcal{W}_{\gamma+1} \upharpoonright \beta+1$.

CLAIM 10.3.12. *σ_ν agrees with σ_γ on* $\text{lh}(E^{\mathcal{U}}_\nu)$, *and* $\sigma_\nu(\bar{P}) = \sigma_\gamma(\bar{P}) = P$.

PROOF. We have that σ^0_ν agrees with σ^0_γ on $\text{lh}(E^{\mathcal{U}}_\nu)$ by the agreement of copy maps, and σ^1_ν agrees with σ^1_γ on $\text{lh}(\sigma^0(E^{\mathcal{U}}_\nu))$ by the agreement of the embedding normalization maps in $W(\mathcal{T},\mathcal{U}^+)$. (Cf. Proposition 6.5.8.) This proves the first part. But $\bar{P} \lhd \mathcal{M}^{\mathcal{U}}_\nu|\hat{\lambda}(E^{\mathcal{U}}_\nu)$, so $\sigma_\nu(\bar{P}) = \sigma_\gamma(\bar{P})$, and $\sigma_\gamma(E^{\mathcal{U}}_\gamma) = F$, so $\sigma_\gamma(\bar{P}) = P$. ⊣

CLAIM 10.3.13. $t^\nu(P) = t^\gamma(P) = P^*$, *and* $t^\nu \upharpoonright P = t^\gamma \upharpoonright P$.

PROOF. Because $[\nu,\gamma+1)_U$ does not drop, whenever $\mathcal{M}^{\mathcal{U}}_\nu|\text{lh}(E^{\mathcal{U}}_\nu) \trianglelefteq Z \lhd \mathcal{M}^{\mathcal{U}}_\nu$, then $\rho(Z) > \bar{\kappa}$. This implies that whenever $R_\nu|\text{lh}(F_\nu) \trianglelefteq Z \lhd R_\nu$, then $\rho(Z) > \kappa$. It follows that

$$\text{res}_\nu \upharpoonright (t^\nu(P) \cup \{t^\nu(P)\}) = \text{id}.$$

If $\nu < \gamma$, we also get for the same reason

$$\text{res}_\gamma \upharpoonright (t^\gamma(P) \cup \{t^\gamma(P)\}) = \text{id}.$$

This implies

$$t^\gamma(P) = \text{res}_\gamma \circ t^\gamma(P) = \text{dom}(\text{res}_\gamma \circ t^\gamma(F)) = \text{dom}(G) = P^*.$$

But also $\psi_\gamma \upharpoonright \varepsilon(E^{\mathcal{U}}_\nu) = \text{res}_\nu \circ \psi_\nu \upharpoonright \varepsilon(E^{\mathcal{U}}_\nu)$ by the properties of conversion systems. So we get

$$\begin{aligned} t^\gamma(P) &= t^\gamma \circ \sigma_\gamma(\bar{P}) \\ &= \psi_\gamma(\bar{P}) \\ &= \text{res}_\nu \circ \psi_\nu(\bar{P}) \\ &= \text{res}_\nu \circ t^\nu \circ \sigma_\nu(\bar{P}) \\ &= t^\nu(P). \end{aligned}$$

The same calculation shows that $t^\gamma \upharpoonright P = t^\nu \upharpoonright P$. ⊣

CLAIM 10.3.14. *If β is stable in $\mathcal{W}_\nu$, and $\beta < z(\nu)$, then* $v^\nu(\beta) \leq^*_{W^*_\nu} \beta^* \leq^*_{W^*_\nu} u^\nu(\beta)$.

PROOF. Consider the diagram

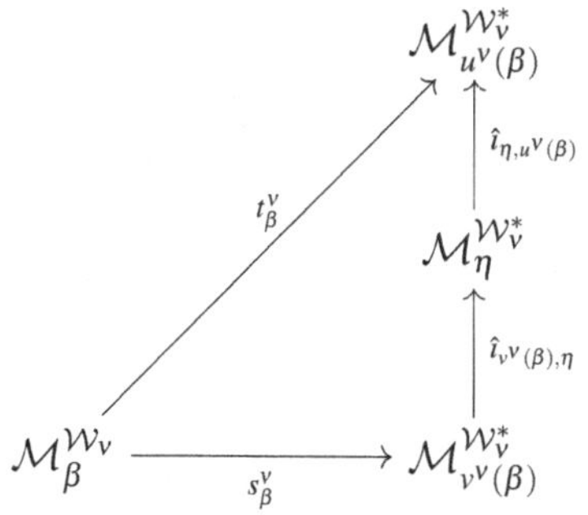

By Lemma 10.3.5, t^ν agrees with t^ν_β on $\mathcal{M}^{\mathcal{W}_\nu}_\beta|\varepsilon(E^{\mathcal{W}_\nu}_\beta)$, and $P \lhd \mathcal{M}^{\mathcal{W}_\nu}_\beta|\varepsilon(E^{\mathcal{W}_\nu}_\beta)$, so

$$t^\nu_\beta(P) = P^*.$$

Let $\eta \in [v^\nu(\beta), u^\nu(\beta)]_{W_\nu}$ be least such that either $\eta = u^\nu(\beta)$ or $\mathrm{crit}(\hat{\imath}_{\eta,u^\nu(\beta)}) > \kappa^*$. Thus

$$\hat{\imath}_{v^\nu(\beta),\eta} \circ s^\nu_\beta(P) = P^*,$$

and all extenders used in $\mathcal{W}_\nu^* \restriction \eta + 1$ have length $< \kappa^*$.

We claim that $\hat{\lambda}(E^{\mathcal{W}_\nu^*}_\eta) > \kappa^*$. If $\eta = u^\nu(\beta)$, this holds because $\kappa < \hat{\lambda}(E^{\mathcal{W}_\nu}_\beta)$, and t^ν_β preserves that fact. If $\eta < u^\nu(\beta)$, then $\kappa^* < \mathrm{crit}(\hat{\imath}_{\eta,u^\nu(\beta)}) < \hat{\lambda}(E^{\mathcal{W}_\nu^*}_\eta)$, so again our claim is correct. The claim tells us that $\beta^* \leq \eta$.

On the other hand, if $\alpha < \eta$ and α is stable in $\mathcal{W}_\nu^*$, then $\mathrm{lh}(E^{\mathcal{W}_\nu^*}_\alpha) < \kappa^*$. This is true by definition for those α such that $\alpha + 1 \leq^*_{W_\nu^*} \eta$, but the lengths of these special E_α are cofinal in $\{\mathrm{lh}(E^{\mathcal{W}_\nu^*}_\alpha) \mid \alpha < \eta \wedge \alpha \in \mathrm{stab}(\mathcal{W}_\nu^*)\}$. This tells us that if $\alpha < \eta$ and α is stable, then $\alpha < \beta^*$.

We claim $\eta = \beta^*$. What is left to rule out is that β^* is unstable, and $\beta^* + 1 = \eta$. Supposing this holds, we get that $\beta = \theta + 1$, where θ is unstable in $\mathcal{W}_\nu$. We have $\alpha^{\mathcal{W}_\nu}_\theta \leq \kappa$ because F is applied to $\mathcal{M}^{\mathcal{W}_\nu}_\beta$. Thus $s^\nu_\beta(\kappa) \leq s^\nu_\beta(\alpha^{\mathcal{W}_\nu}_\theta) = \alpha^{\mathcal{W}_\nu^*}_{v^\nu(\theta)}$. But $\alpha^{\mathcal{W}_\nu^*}_{\beta^*} = \sup(i^{\mathcal{W}_\nu^*}_{v^\nu(\theta),\beta^*}``\alpha^{\mathcal{W}_\nu^*}_{v^\nu(\theta)})$. It follows that

$$\alpha^{\mathcal{W}_\nu^*}_{\beta^*} \leq i^{\mathcal{W}_\nu^*}_{v^\nu(\theta),\beta^*} \circ s^\nu_\theta(\kappa) = \kappa^*.$$

But then G is not applied to $\mathcal{M}^{\mathcal{W}_\nu^*}_{\beta^*}$ in $\mathcal{W}^*_{\gamma+1}$, contradiction. ⊣

CLAIM 10.3.15. *If $\beta = z(\nu)$, then $v^\nu(\beta) \leq^*_{W_\nu^*} \beta^* \leq^*_{W_\nu^*} z^*(\nu)$.*

PROOF. If $\beta = z(\nu)$, then β must be stable. The proof of Claim 10.3.14 then works with small changes. ⊣

Note that Claims 10.3.14 and 10.3.15 imply that if β is stable in $\mathcal{W}_\nu$, then β^* is stable in $\mathcal{W}_\nu^*$.

CLAIM 10.3.16. *If β is unstable in $\mathcal{W}_\nu$ and $\beta+1<z(\nu)$, then β^* is unstable in $\mathcal{W}^*_\nu$, and $v^\nu(\beta)\leq^*\beta^*\leq^* u^\nu(\beta)$ in $\mathcal{W}^*_\nu$.*

PROOF. Let $\varepsilon=\varepsilon_\beta^{\mathcal{W}_\nu}=\inf(\alpha_\beta^{\mathcal{W}_\nu},\varepsilon_{\beta+1}^{\mathcal{W}_\nu})$. By Lemma 10.3.5, t^ν agrees with t^ν_β on $\mathcal{M}_\beta^{\mathcal{W}_\nu}|\varepsilon$. Since $P\lhd\mathcal{M}_\beta^{\mathcal{W}_\nu}|\varepsilon$, we have again $t^\nu_\beta(P)=P^*$. Let $\eta\in[v^\nu(\beta+1),u^\nu(\beta+1)]_{W_\nu}$ be least such that either $\eta=u^\nu(\beta+1)$ or $\mathrm{crit}(\hat{\imath}_{\eta,u^\nu(\beta+1)})>\kappa^*$. Thus

$$\hat{\imath}_{v^\nu(\beta+1),\eta}\circ s^\nu_{\beta+1}(P)=P^*,$$

and all extenders used in $\mathcal{W}^*_\nu\restriction\eta+1$ have length $<\kappa^*$.

Note that $s^\nu_\beta\restriction\alpha_\beta^{\mathcal{W}_\nu}=s_{\beta+1}^{\mathcal{W}_\nu}$, and $\kappa<\alpha_\beta^{\mathcal{W}_\nu}$. All extenders used in $[v^\nu(\beta+1),\eta]_{W^*_\nu}$ have critical point below the current image of $s^\nu_{\beta+1}(\kappa)$, hence below the current image of $s^\nu_\beta(\alpha_\beta^{\mathcal{W}_\nu})$. Thus all these extenders are moving up the current image of the phalanx indexed at $(v^\nu(\beta),v^\nu(\beta+1))$. It follows that $\eta=\gamma+1$, where γ is unstable in $\mathcal{W}^*_\nu$, and $v^\nu(\beta)\leq^*\gamma\leq^* u^\nu(\beta)$.

It is now easy to see that $\gamma=\beta^*$, so that Claim 10.3.16 holds. ⊣

CLAIM 10.3.17. *If β is unstable in $\mathcal{W}_\nu$ and $\beta+1=z(\nu)$, then β^* is unstable in $\mathcal{W}^*_\nu$, and $v^\nu(\beta)\leq^*_{W^*_\nu}\beta^*\leq^*_{W^*_\nu}z^*(\nu)-1$.*

PROOF. The proof of Claim 10.3.16 works here. ⊣

We let $v^{\gamma+1}(\bar{\xi}+1)=\xi+1$. We need to see

CLAIM 10.3.18. $\bar{\xi}+1\in\mathrm{stab}(\mathcal{W}_{\gamma+1})$ *if and only if* $\xi+1\in\mathrm{stab}(\mathcal{W}^*_{\gamma+1})$.

PROOF. We have that

$$\begin{aligned}\bar{\xi}+1\in\mathrm{stab}(\mathcal{W}_{\gamma+1}) &\iff \beta\in\mathrm{stab}(\mathcal{W}_\nu)\\ &\iff \beta^*\in\mathrm{stab}(\mathcal{W}^*_\nu)\\ &\iff \xi+1\in\mathrm{stab}(\mathcal{W}^*_{\gamma+1}).\end{aligned}$$

The first line holds because $\Phi_{\nu,\gamma+1}$ is a tree embedding. The second line was proved in Claims 10.3.14 to 10.3.17. Toward the last line, suppose first that $\beta^*\in\mathrm{stab}(\mathcal{W}^*_\nu)$. Since $\mathcal{W}^*_\nu\restriction\beta^*+1=\mathcal{W}^*_{\gamma+1}\restriction\beta^*+1$, and $\mathcal{V}_\nu$ uses the same extenders of length $<o(P^*)$ as $\mathcal{V}_{\gamma+1}$ does, we get that $\beta^*\in\mathrm{stab}(\mathcal{W}^*_{\gamma+1})$. But $\beta^*\leq^*\xi+1$ in $\mathcal{W}^*_{\gamma+1}$, so $\xi+1\in\mathrm{stab}(\mathcal{W}^*_{\gamma+1})$.

Conversely, suppose β^* is unstable in $\mathcal{W}^*_\nu$. The agreement noted in the last paragraph shows that β^* is unstable in $\mathcal{W}^*_{\gamma+1}$. Now recall that $(\mathcal{W}^{**}_\gamma,\mathcal{V}^{**}_\gamma)$ is the (Σ,Λ,Y) coiteration. Letting $\rho+1=\mathrm{lh}(\mathcal{V}^{**}_\gamma)$, we have that G is on the sequence of $\mathcal{M}_\rho^{\mathcal{V}^{**}_\gamma}$, but not on the sequence of any earlier model. It follows that

$$\mathcal{V}_{\gamma+1}\restriction(\rho+1)=\mathcal{V}^{**}_\gamma,$$

and

$$E_\rho^{\mathcal{V}_{\gamma+1}}=G.$$

Since β^* is unstable in $\mathcal{W}^*_{\gamma+1}$, we have τ such that

$$\mathcal{M}_\tau^{\mathcal{V}_{\gamma+1}} = \mathcal{M}_{\beta^*}^{\mathcal{W}^*_{\gamma+1}}.$$

But then G must be applied to $\mathcal{M}_\tau^{\mathcal{V}_{\gamma+1}}$ in $\mathcal{V}_{\gamma+1}$, leading to

$$\mathcal{M}_{\tau+1}^{\mathcal{V}_{\gamma+1}} = \mathcal{M}_{\xi+1}^{\mathcal{W}^*_{\gamma+1}},$$

so that $\xi+1$ is unstable in $\mathcal{W}^*_{\gamma+1}$, as desired. ⊣

The map $s^{\gamma+1}_{\bar{\xi}+1}$ of $\Phi_{\gamma+1}$ is given by the Shift Lemma as the definition of tree embeddings requires. If $\bar{\xi}+1$ is unstable in $\mathcal{W}_{\gamma+1}$, this also determines $s^{\gamma+1}_{\bar{\xi}+2}$. So we have now defined $\Phi_{\gamma+1} \upharpoonright \bar{\xi}+2$, and verified that it is a tree embedding in the pseudo-tree sense.

The rest of $\Phi_{\gamma+1}$ is determined by

$$u^{\gamma+1}(\phi_{\nu,\gamma+1}(\eta)) = i_{G^*}(u^\nu(\eta)).$$

$u^{\gamma+1}$ preserves stability, because u^ν and $\phi_{\nu,\gamma+1}$ do, and i_{G^*} is elementary.

One must check that the associated $v^{\gamma+1}$ also preserves stability at $\tau > \bar{\xi}+1$. Such τ are of the form $\phi(\eta)$, where $\phi = \phi_{\nu,\gamma+1}$ and $\eta > \beta$.

If η is unstable in $\mathcal{W}_\nu$, then $u^{\gamma+1}(\phi(\eta))$ is unstable in $\mathcal{W}^*_{\gamma+1}$, and

$$v^{\gamma+1}(\phi(\eta)) \leq_{\mathcal{W}^*_{\gamma+1}} u^{\gamma+1}(\phi(\eta)),$$

so $v^{\gamma+1}(\phi(\eta))$ is unstable in $\mathcal{W}^*_{\gamma+1}$, as desired.

Suppose that η is stable in $\mathcal{W}_\nu$; we want to see that $v^{\gamma+1}(\phi(\eta))$ is stable in $\mathcal{W}^*_{\gamma+1}$. By induction, we may assume that η is a stable root in $\mathcal{W}_\nu$. Suppose first that $\eta = \tau+1$ where τ is unstable, then $v^{\gamma+1}(\phi(\eta)) = v^{\gamma+1}(\phi(\tau)+1) = u^{\gamma+1}(\phi(\tau))+1$ and since $u^{\gamma+1}(\phi(\tau))$ is unstable in $\mathcal{W}^*_{\gamma+1}$, $u^{\gamma+1}(\phi(\tau))+1$ is stable there, as desired. So the case to worry about is that η is a stable root and also a limit ordinal. Here we use Proposition 9.6.4. In general,

$$v^{\gamma+1}(\phi(\eta)) = \sup i_{G^*}\text{``}v^\nu(\eta).$$

However, if η is stable in $\mathcal{W}_\nu$ and a limit of unstable $\theta <_{W_\nu} \eta$, then by Proposition 9.6.4, $\mathrm{cof}(\eta) = \omega$, so $\mathrm{cof}(\phi(\eta)) = \omega$. But then $\mathrm{cof}(v^\nu(\eta)) = \omega$, so i_{G^*} is continuous at $v^\nu(\eta)$. Thus

$$v^{\gamma+1}(\phi(\eta)) = i_{G^*}(v^\nu(\eta)),$$

hence $v^{\gamma+1}(\phi(\eta))$ is stable in $\mathcal{W}^*_{\gamma+1}$ by the elementarity of i_{G^*} and the fact that v^ν preserves stability.

This proves Sublemma 10.3.10.1. ⊣

That in turn proves Lemma 10.3.10, or what is the same, Lemma 9.6.5. ⊣

10.4. Some successful background constructions

Let us collect our results to the effect that least branch constructions do not break down.

THEOREM 10.4.1. *Assume* AD^+, *and let* $\mathbb{C}$ *be the maximal least branch construction of some coarse strategy pair; then* $\mathbb{C}$ *is good at all* $\langle \nu, k\rangle$.

PROOF. This was proved in Theorem 9.6.13 for $k \geq 0$, modulo Lemma 9.6.5, which we proved in the last section. The case that $k = -1$ is covered by Theorem 10.2.13. ⊣

COROLLARY 10.4.2. *Assume* ZFC *plus* $\mathsf{IH}_{\kappa,\delta}$, *where* $\kappa < \delta < \theta < \alpha$ *for some inaccessible* θ *and* α. *Suppose also that there are* $\lambda < \mu < \kappa$ *such that* λ *is a limit of Woodin cardinals, and* μ *is measurable. Let* $(w, \mathcal{F})$ *be a coherent pair such that* $\mathcal{F} \subseteq V_\delta$ *and* $\forall E \in \mathcal{F}(\text{crit}(E) > \kappa)$, *let* Σ *be the unique* $(\theta, \theta, \mathcal{F})$*-iteration strategy for* V, *and let* $\mathbb{C}$ *be a* $(w, \mathcal{F}, \Sigma)$ *construction; then* $\mathbb{C}$ *is good at all* $\langle \nu, k\rangle$.

PROOF. This follows at once from Theorem 9.4.17, Theorem 10.4.1, and the fact goodness at $\langle \nu, k\rangle$ is first order. ⊣

We have shown that least branch constructions done in a coarse Γ Woodin model do not break down, but we are missing a proof that such constructions go far enough; that is, a proof of HPC. We do get

THEOREM 10.4.3. *Assume* AD^+; *then* LEC *implies* HPC.

PROOF. It is enough to show that whenever (P, Σ) is a pure extender mouse pair with scope HC, then there is an lbr hod pair (Q, Ψ) with scope HC such that Σ is definable from parameters over $(\text{HC}, \in, \Psi)$.

So fix (P, Σ), and let $((N^*, \in, w, \mathcal{F}, \Phi), \Phi^*)$ be a coarse strategy pair that captures $\text{Code}(\Sigma)$, with P countable in N^*. Let $\mathbb{C}$ be the maximal $(w, \mathcal{F})$- construction of $L[N^*, \in, w, \mathcal{F}, \Phi]$, with last pair

$$(Q, \Psi) = (M^{\mathbb{C}}_{\delta,0}, \Omega^{\mathbb{C}}_{\delta,0}).$$

By Theorem 10.4.1, $\mathbb{C}$ does not break down. Since Φ^* has scope all of HC, it induces an extension of Ψ with scope HC. We call this extension Ψ^*.

Let w_1 be the order of construction in Q, and

$$\mathcal{F}_1 = \{E \restriction \eta \mid E \text{ is on the } Q\text{-sequence and } Q \models E \restriction \eta \text{ is nice}\}.$$

It is not hard to see that $((Q, \in, w_1, \mathcal{F}_1, \Psi), \Psi^*)$ is a coarse strategy pair. Let $\mathbb{D}$ be the maximal pure extender pair construction of $(Q, \in, w_1, \mathcal{F}_1, \Psi)$. Each $(M^{\mathbb{D}}_{\nu,k}, \Omega^{\mathbb{D}}_{\nu,k})$ is a pure extender pair in Q, and hence can be canonically extended to such a pair in N^*. Working in N^*, we can compare (P, Σ) with each $(M^{\mathbb{D}}_{\nu,k}, \Omega^{\mathbb{D}}_{\nu,k})$. Because the background extenders of $\mathbb{D}$ are assigned background extenders over N^* by $\mathbb{C}$, we can repeat the proof of Theorem 8.4.3, so (P, Σ) iterates past $(M^{\mathbb{D}}_{\nu,k}, \Omega^{\mathbb{D}}_{\nu,k})$, provided it iterates strictly past all earlier levels of $\mathbb{D}$.

By the Q-filtered backgrounding again, (P,Σ) cannot iterate past $(M^{\mathbb{D}}_{\delta,0},\Omega^{\mathbb{D}}_{\delta,0})$. It follows that (P,Σ) iterates to some $(M^{\mathbb{D}}_{\nu,k},\Omega^{\mathbb{D}}_{\nu,k})$. This is true in N^*, but it is also true in V of (P,Σ) and the canonical extension (M,Ω) of $(M^{\mathbb{D}}_{\nu,k},\Omega^{\mathbb{D}}_{\nu,k})$, because N^* is sufficiently correct. But then Σ is projective in Ω, and Ω is projective in Ψ, so we are done. $\dashv$

Remark 10.4.4. We do not see how to show that under AD^+, HPC implies LEC.

We now look at constructions done under strong large cardinal hypotheses. Here we must assume unique iterability. We shall show that under such assumptions, least branch constructions can produce hod pairs (M,Ω) such that $M \models$ "there is a subcompact cardinal".

DEFINITION 10.4.5. A cardinal κ is *subcompact* iff for all $A \subseteq H_{\kappa^+}$, there are μ, B, and j such that

(a) $\mu < \kappa$ and $B \subseteq H_{\mu^+}$,
(b) $j\colon (H_{\mu^+},\in,B) \to (H_{\kappa^+},\in,A)$ is elementary, and
(c) $\mu = \mathrm{crit}(j)$.

Subcompactness was introduced by Jensen. It is interesting in part because it can be represented by short extenders[21], but it is strong enough that if κ is subcompact, then $\neg\Box_\kappa$. The main theorem of [49] is that in iterable pure extender models, $\neg\Box_\kappa$ if and only if κ is subcompact. If κ is subcompact, then the set

$$S = \{i_E(\mu^+) \mid E \text{ is a superstrong } (\mu,\kappa)\text{-extender}\}$$

is stationary in κ^+.[22] Jensen showed that in iterable pure extender models, the stationarity of S is equivalent to subcompactness. (See [49].)

Subcompactness is close to the limit of the large cardinal properties that can be represented by short extenders, and it is thus close to the limit of the large cardinal properties exhibited in the strategy mice whose theory is developed in this book.

The large cardinal hypothesis of the following theorem is just beyond those that can be captured by short extenders.

THEOREM 10.4.6. *Suppose*

(i) $j\colon V \to N$ *is elementary,* $\kappa = \mathrm{crit}(j)$*, and* $\delta = j(\kappa)$*,*
(ii) $V_\delta \cup \{E_j \restriction \delta\} \subseteq N$*,*
(iii) $\mathsf{IH}_{\mu,\delta}$ *holds, where* $\mu < \kappa$*, and*
(iv) w_0 *is a wellorder of* V_κ*,* $\mathcal{F}_0$ *is the set of all nice* $E \in V_\kappa$ *such that* $\mu < \mathrm{crit}(E)$ *and* $i_E(w_0) \cap V_{\mathrm{lh}(E)} = w_0 \cap V_{\mathrm{lh}(E)}$*,* $(w,\mathcal{F}) = j((w_0,\mathcal{F}_0))$*, and* $\mathbb{C}$ *is the maximal least branch* $(w,\mathcal{F})$*-construction.*

Then $\mathbb{C}$ *is good at all* $\langle \nu, k\rangle$*, and*

[21] Let E be the (μ,κ)-extender of j; then i_E also satisfies (b) and (c) of Definition 10.4.5.
[22] To see this, let A be a given club, apply the definition to get j and μ, and then let $E = E_j \restriction \kappa$.

(a) $M^{\mathbb{C}}_{\delta,0} \models \kappa$ *is subcompact, and*

(b) $M^{\mathbb{C}}_{\delta,0} \models$ *there are arbitrarily large superstrong cardinals.*

PROOF. $\mathbb{C}$ does not break down by Corollary 10.4.2. We show first that κ is subcompact in M.

Let $A \subseteq (\kappa^+)^M$ and $A \in M$. It will be enough to show that δ is $j(A)$-subcompact in $j(M)$.

Our choice of w guarantees that $j(w) \cap V_\delta = w$. It follows then that $j(\mathbb{C}) \upharpoonright \langle \delta, 0 \rangle = \mathbb{C}$. Thus

$$M = M^{j(\mathbb{C})}_{\delta,0}.$$

But this implies that

$$M = j(M) | \langle \delta, 0 \rangle.$$

To see that, let us call η a β-closure point of $\mathbb{C}$ iff $\eta = o(M^{\mathbb{C}}_{\eta,0})$, $\eta < \beta$, and η is a cardinal of $M^{\mathbb{C}}_{\beta,0}$. Note that this implies $M^{\mathbb{C}}_{\eta,0} \trianglelefteq M^{\mathbb{C}}_{\beta,0}$. The set B_β of β-closure points of $\mathbb{C}$ is closed in β. If β is a cardinal of V, it is club in β. But then

$$\begin{aligned} B^{\mathbb{C}}_\kappa &= j(B^{\mathbb{C}}_\kappa) \cap \kappa \\ &= B^{j(\mathbb{C})}_\delta \cap \kappa \\ &= B^{\mathbb{C}}_\delta \cap \kappa, \end{aligned}$$

so $\kappa \in B^{\mathbb{C}}_\delta$, so $\delta \in B^{j(\mathbb{C})}_{j(\delta)}$, or in other words, δ is a closure point of $j(\mathbb{C})$. That implies $M = j(M) | \langle \delta, 0 \rangle$.

Let

$$E = \{(a, X) \mid a \in [\delta]^{<\omega} \wedge X \in P([\kappa]^{|a|})^M \wedge a \in j(X)\}$$

be the length δ extender of j, restricted to M.

Claim. If $\eta \le \delta$ and $E \upharpoonright \eta$ is whole, then the trivial completion of $E \upharpoonright \eta$ is on the $j(M)$-sequence.

PROOF. We prove this by induction on η. Suppose we know it for $\beta < \eta$, and let F be the trivial completion of $E \upharpoonright \eta$, and $\gamma = i^M_F(\kappa^{+,M})$. Assume first that $\eta < \delta$. We have that $\mathrm{Ult}(M, F) = \mathrm{Ult}(M, E \upharpoonright \eta)$, and there is a natural factor embedding

$$\sigma \colon \mathrm{Ult}(M, F) \to \mathrm{Ult}(M, E)$$

such that $\sigma \upharpoonright \eta = \mathrm{id}$, and $\sigma(\eta) = \delta$. Since η is a limit cardinal of $\mathrm{Ult}(M, F)$, we have that η is a limit cardinal of M. Using the Condensation Lemma 9.6.15 applied to σ, we get that

$$\mathrm{Ult}(M, F) | \langle \gamma, -1 \rangle = \mathrm{Ult}(M, E) \upharpoonright \langle \gamma, -1 \rangle = M \upharpoonright \langle \gamma, -1 \rangle.$$

Since η is a cardinal of M, there must be a stage of $\mathbb{C}$ at which we have $M | \langle \eta, 0 \rangle = M^{\mathbb{C}}_{\nu,0}$. After this stage, no projectum drops strictly below η, and stages which project to η are initial segments of M. Thus there is a ν such that

$$(M^{<\nu})^{\mathbb{C}} = M|\langle\gamma, -1\rangle.$$

But then $(M^{<\nu}, F, \emptyset)$ is an lpm. (Coherence we verified above, and the Jensen initial segment condition holds by our induction hypothesis.) Moreover, F has a background certificate that shifts w to itself, namely $E_j \restriction \mu$, for μ the least inaccessible cardinal strictly greater than η. By the Bicephalus Lemma,

$$M^{\mathbb{C}}_{\nu,0} = (M^{<\nu}, F, \emptyset).$$

Since η is a cardinal of M and $M^{\mathbb{C}}_{\nu,0}$ projects to η, $M^{\mathbb{C}}_{\nu,0} \lhd M$. Thus F is on the M-sequence. Since $\eta < \delta$, it is on the $j(M)$-sequence.

Now we take the case $\eta = \delta$, that is, $F = E$. Again, let $\gamma = i^M_E(\kappa^{+,M}) = i^{M|\delta}_E(\kappa^{+,M})$ be the length of the Jensen completion of E. The factor embedding from $\mathrm{Ult}(M,E)$ to $j(M)$ has critical point $\geq \gamma$, and thus $\mathrm{Ult}(j(M)|\gamma, E)$ agrees with $j(M)$ strictly below γ. E satisfies the Jensen initial segment condition by the claim applied to $\eta < \delta$. To get a background certificate E^* for E in N, simply take

$$E^* = j_1(E_j \restriction \delta) \restriction \lambda,$$

where $j_1 = j(j)$ and λ is the least inaccessible of N above δ. This clearly works, so by the Bicephalus Lemma, E is on the sequence of $j(M)$. ⊣

Let $i_E\colon (M|\kappa^{+,M}, A) \to \mathrm{Ult}((M|\kappa^{+,M}, A), E) = (j(M)||\mathrm{lh}(E), B)$ be the canonical fully elementary embedding. Let $\sigma\colon \mathrm{Ult}((M,A),E) \to (j(M), j(A))$ be the factor embedding. Since $\mathrm{crit}(\sigma) = \mathrm{lh}(E)$ and σ is elementary, we see that

$$(j(M)||\mathrm{lh}(E), B) \prec (j(M)||\delta^{+,j(M)}, j(A)).$$

Thus E witnesses that δ is $j(A)$-subcompact in N.

To see that δ is a limit of superstrong cardinals in M, it is enough to see that $M|\kappa \models$ "there are arbitrarily large superstrong cardinals", for then we can apply j to this fact. But κ is subcompact in M, and it is quite easy to see that if κ is subcompact, then $V_\kappa \models$ "there are arbitrarily large superstrong cardinals". ⊣

10.5. UBH holds in hod mice

We shall outline a proof that whenever (M, Ω) is an lbr hod pair with scope HC, and Ω is Suslin-co-Suslin in some model of AD^+, then UBH for nice, normal iteration trees holds in M. This was proved in [67] for pure extender mice, and our proof here involves a similar comparison of phalanxes of the form $\Phi(\mathcal{T}^\frown b)$ and $\Phi(\mathcal{T}^\frown c)$. The new difficulties involve moving phalanxes up in a pseudo iteration tree generated by this comparison.

The new difficulties show up in the proof that the iteration strategies for $\Phi(\mathcal{T}^\frown b)$ and $\Phi(\mathcal{T}^\frown c)$ used to produce these pseudo-trees normalize well and have strong hull condensation in the appropriate sense, and that therefore no strategy disagreements show up when they are compared with a common background construction. We shall go into very little detail in this part of the proof. There is a full account of it in [64]. Here we shall simply describe the comparison via pseudo iteration trees iterating into a common background construction, and show that granted these facts about the associated pseudo iteration strategies, it does its job.[23]

We shall use this theorem to show that if (M,Ω) is as above, and λ is a limit of Woodin cardinals in M, then for each $\xi < \lambda$ there is a term $\tau \in M$ such that for all g generic over M for a poset belonging to $M|\lambda$,

$$\tau^g = \Omega_{M|\xi} \cap (M|\lambda)[g].$$

This generic interpretability result is important in showing that the HOD of the derived model of M below λ is an iterate of $M|\lambda$. It has other uses as well.

DEFINITION 10.5.1. Let M be a premouse such that $M \models \mathsf{ZFC}^-$; then an *M-nice tree* is a normal iteration tree $\mathcal{T}$ on M such that for all $\alpha < \mathrm{lh}(\mathcal{T})$,

(1) $\mathcal{M}_\alpha^\mathcal{T} \models$ "$E_\alpha^\mathcal{T}$ is a nice extender", and
(2) $E_\alpha^\mathcal{T} = F \restriction \mathrm{lh}(E_\alpha^\mathcal{T})$, for some F on the sequence of $\mathcal{M}_\alpha^\mathcal{T}$.

Notice here that if $\mathcal{T}$ is M-nice, then $E_\alpha^\mathcal{T}$ cannot be on the sequence of $\mathcal{M}_\alpha^\mathcal{T}$, because in Jensen-indexed premice, the extenders of the sequence are never nice. Nevertheless, if (M,Σ) is a mouse pair and $\mathcal{T}$ is an M-nice tree on (M,Σ), then the pairs in $\mathcal{T}$ have the extender and strategy agreement properties that a tree using extenders from the sequences would have. That is, is $\alpha < \beta$, then $(\mathcal{M}_\alpha, \Sigma_{\mathcal{T}\restriction\alpha+1})|\mathrm{lh}(E_\alpha^\mathcal{T}) = (\mathcal{M}_\beta, \Sigma_{\mathcal{T}\restriction\beta+1})|\mathrm{lh}(E_\alpha^\mathcal{T})$.[24]

THEOREM 10.5.2. *Assume* AD^+, *and let* (M,Ω) *be a least branch hod pair with scope* HC. *Suppose* $M \models \mathsf{ZFC}^-$, *and* Ω *is coded by a Suslin-co-Suslin set of reals. Let* δ *be a cutpoint of* M, $\mu > \delta$ *a regular cardinal of* M, *and let* $\mathcal{T}$ *be an* M*-nice tree such that*

(a) $\mathcal{T}$ *has all critical points* $> \delta$, *and*
(b) $\mathcal{T} \in (M|\mu)[g]$, *for some* g *that is* M*-generic over* $\mathrm{Col}(\omega,\delta)$*;*

then

$$M[g] \models \mathcal{T} \textit{ has at most one cofinal, wellfounded branch.}$$

Remark 10.5.3. Our proof of this theorem extends without much more work to cover plus two trees $\mathcal{T}$, as does the theorem of [67] it generalizes. We do not see how to make it work for arbitrary non-dropping trees.

PROOF. Suppose not. Let $\dot{\mathcal{T}} \in M|\mu$ be the M-least name such that 1 forces $\dot{\mathcal{T}}$ to be a counterexample. Let g be M-generic over $\mathrm{Col}(\omega,\delta)$, and $\mathcal{T} = \dot{\mathcal{T}}^g$. $\mathcal{T}$ is

[23]In [64] the pseudo-trees are replaced by closely related meta-trees. See Remark 10.5.6.
[24]$\mathrm{lh}(E_\alpha^\mathcal{T})$ is inaccessible in $\mathcal{M}_\alpha^\mathcal{T}$, so it is not an index.

countable in $M|\mu[g]$. Let

$$\pi\colon N \to M|\mu$$

be elementary, and such that $\text{crit}(\pi) > \delta$, and N is pointwise definable from ordinals $\leq \delta$. Thus $\dot{\mathcal{T}} \in \text{ran}(\pi)$. Let

$$\hat{\pi}\colon N[g] \to (M|\mu)[g]$$

be the canonical extension of π, and let

$$\hat{\pi}(\mathcal{S}) = \mathcal{T}.$$

By assumption, $\mathcal{T}$ has distinct, cofinal, wellfounded branches in $(M|\mu)[g]$, so we have b, c such that

$$N[g] \models b \text{ and } c \text{ are distinct cofinal, wellfounded branches of } \mathcal{S}.$$

Let $\Phi(\mathcal{S}^\frown b)$ be the phalanx $((\langle \mathcal{M}^{\mathcal{S}}_\alpha \mid \alpha < \text{lh}(\mathcal{S})\rangle^\frown \langle \mathcal{M}^{\mathcal{S}}_b\rangle, \langle \text{lh}(E^{\mathcal{S}}_\alpha) \mid \alpha < \text{lh}(\mathcal{S})\rangle)$. We get an iteration strategy for $\Phi(\mathcal{S}^\frown b)$ by finding maps with sufficient agreement that embed its models into M.

CLAIM 10.5.4. *There are $\pi_\alpha, \gamma_\alpha$ for $\alpha < \text{lh}(\mathcal{S})$, and π_b, such that π_b and the π_α are the identity on $\delta + 1$, and for all α,*

(1) $\pi_b\colon \mathcal{M}^{\mathcal{S}}_b \to M|\mu$,
(2) $\pi_\alpha\colon \mathcal{M}^{\mathcal{S}}_\alpha \to M|\gamma_\alpha$, *and*
(3) $\pi_\alpha \restriction \text{lh}(E^{\mathcal{S}}_\alpha) = \pi_b \restriction \text{lh}(E^{\mathcal{S}}_\alpha)$.

PROOF. The proof is given, under slightly different strength hypotheses on the $E^{\mathcal{T}}_\chi$, in [67, §3]. See especially the proof of Theorem 3.3.[25] ⊣

By Claim 10.5.4 we can lift iteration trees on the phalanx $\Phi(\mathcal{S}^\frown b)$ to M using the π_α, for $\alpha < \text{lh}(\mathcal{S})$ or $\alpha = b$, so we can pull back Ω to a strategy Ψ_0 for trees based on $\Phi(\mathcal{S}^\frown b)$. We shall need an extension of Ψ_0 defined not just on ordinary λ-separated trees, but on pseudo-trees formed by rules we shall describe below. We shall call this strategy Ψ.

Similarly, we have

CLAIM 10.5.5. *There are $\sigma_\alpha, \xi_\alpha$ for $\alpha < \text{lh}(\mathcal{S})$, and σ_c, such that σ_c and the σ_α are the identity on $\delta + 1$, and for all α,*

(1) $\sigma_c\colon \mathcal{M}^{\mathcal{S}}_c \to M|\mu$,
(2) $\sigma_\alpha\colon \mathcal{M}^{\mathcal{S}}_\alpha \to M|\xi_\alpha$, *and*
(3) $\sigma_\alpha \restriction \text{lh}(E^{\mathcal{S}}_\alpha) = \sigma_c \restriction \text{lh}(E^{\mathcal{S}}_\alpha)$.

[25] Here is a sketch. The copy maps $\psi_\alpha\colon \mathcal{M}^{\mathcal{S}}_\alpha \to \mathcal{M}^{\mathcal{T}}_\alpha|\mu$ are all restrictions of π, as is the copy map $\psi_b\colon \mathcal{M}^{\mathcal{S}}_b \to \mathcal{M}^{\mathcal{T}}_b|\mu$. ($\mu$ is fixed by the maps of $\mathcal{T}$.) Letting $\nu_\alpha = \sup \psi_\alpha``\text{lh}(E^{\mathcal{S}}_\alpha)$, we have $\nu_\alpha < \text{lh}(E_\alpha)^{\mathcal{T}}$. Using Condensation inside $\mathcal{M}^{\mathcal{T}}_\alpha$, we then get $\xi_\alpha < \text{lh}(E^{\mathcal{T}}_\alpha)$ and $\phi_\alpha\colon \mathcal{M}^{\mathcal{S}}_\alpha \to \mathcal{M}^{\mathcal{T}}_b|\xi_\alpha$ such that ϕ_α agrees with ψ_α, and hence π, on $\text{lh}(E^{\mathcal{S}}_\alpha)$. Each of the ϕ_α is in $\mathcal{M}^{\mathcal{T}}_b$. An absoluteness argument done in the wellfounded model $\mathcal{M}^{\mathcal{T}}_b$ then gives us Claim 10.5.4, but with $\mathcal{M}^{\mathcal{T}}_b$ replacing M. Pulling back under $i^{\mathcal{T}}_b$, we get the Claim 10.5.4 itself.

Claim 10.5.5 will yield an iteration strategy for the relevant pseudo-trees on the phalanx $\Phi(\mathcal{S}^\frown c)$. We shall call this strategy Σ.

Let $((N^*, \in, w, \mathcal{F}, \Phi), \Phi^*)$ be a coarse strategy pair that captures Code(Λ). We assume that the various countable objects we have encountered so far are countable in N^*. In particular, $M[g]$, $\Phi(\mathcal{S}^\frown b)$, $\Phi(\mathcal{S}^\frown c)$, and the maps from Claims 10.5.4 and 10.5.5 are countable in N^*. Let $\mathbb{C}$ be the maximal w-construction of N^*. We compare $\Phi(\mathcal{S}^\frown b)$ with $\Phi(\mathcal{S}^\frown c)$ by defining, for each ν, l,

$$(\mathcal{U}_{\nu,l}{}^\frown \mathcal{W}_{\nu,l}, \mathcal{U}_{\nu,l}{}^\frown \mathcal{V}_{\nu,l}) = \text{the } (\Psi, \Sigma, M^{\mathbb{C}}_{\nu,l})\text{-coiteration of } \Phi(\mathcal{S}^\frown b) \text{ with } \Phi(\mathcal{S}^\frown c).$$

This is a pair of pseudo-trees according to Ψ and Σ respectively, obtained by iterating away least disagreements with $M^{\mathbb{C}}_{\nu,l}$, as in the proof of Theorem 9.6.2. The process of moving a phalanx up is somewhat different, however, so let us look at it.

The first phase in the coiteration consists in moving $\Phi(\mathcal{S}^\frown b)$ and $\Phi(\mathcal{S}^\frown c)$ up by an ordinary λ-separated plus tree on $M|\delta = N|\delta$. Note δ is a cutpoint of $N = \mathcal{M}^{\mathcal{S}}_0$, and Ψ and Σ both agree with $\Omega_{N|\delta}$ for trees on $N|\delta$ because all the π_α and σ_α are the identity on $N|\delta$. We let $\mathcal{U} = \mathcal{U}_{\nu,l}$ be the unique λ-separated tree on $N|\delta$ that is by $\Omega_{N|\delta}$ and has last model $P = M_{\nu,l}|\langle \delta_0, 0\rangle$, with the strategy agreement $\Omega_{\mathcal{U},P} = (\Omega^{\mathbb{C}}_{\nu,l})_{\langle \delta_0, 0\rangle}$. There is such a $\mathcal{U}$ by Theorem 9.5.2. We assume here that $\langle \nu, l\rangle$ is large enough that $(N|\delta, \Omega_{N|\delta})$ does not iterate past $(M_{\nu,l}, \Omega_{\nu,l})$. We wish now to define $\mathcal{W} = \mathcal{W}_{\nu,l}$ and $\mathcal{V} = \mathcal{V}_{\nu,l}$.

Thinking of $\mathcal{U}$ as a tree on N, its last model is

$$Q = \mathcal{M}^{\mathcal{U}}_{\tau_0} = \mathcal{M}^{\mathcal{W}}_0 = \mathcal{M}^{\mathcal{V}}_0.$$

$P = Q|\delta_0$ is a cutpoint initial segment of Q, and Q is pointwise definable from the ordinals $< \delta_0$. (In most cases, $\tau_0 = \delta_0$.) Letting E be the branch extender of $i^{\mathcal{U}}_{0,\tau_0}$, we move up our two phalanxes by setting, for $\alpha < \theta$,

$$\mathcal{M}^{\mathcal{W}}_\alpha = \mathcal{M}^{\mathcal{V}}_\alpha = \text{Ult}(\mathcal{M}^{\mathcal{S}}_\alpha, E),$$

$$\rho_\alpha = i_E^{\mathcal{M}^{\mathcal{S}}_\alpha}(\text{lh}(E^{\mathcal{S}}_\alpha)),$$

$$\mathcal{M}^{\mathcal{W}}_\theta = \text{Ult}(\mathcal{M}^{\mathcal{S}}_b, E),$$

$$\mathcal{M}^{\mathcal{V}}_\theta = \text{Ult}(\mathcal{M}^{\mathcal{S}}_c, E).$$

We can also think of $\mathcal{U}$ as a tree $\mathcal{U}^+$ on M, with last model

$$R = \mathcal{M}^{\mathcal{U}^+}_{\tau_0},$$

and lift the π_α to

$$\psi_\alpha \colon \mathcal{M}^{\mathcal{W}}_\alpha \to J_\alpha \lhd R.$$

We get that $\psi_\alpha \restriction \rho_\alpha = \psi_\theta \restriction \rho_\alpha$ for $\alpha < \theta$. So we are in our initial position, but we have replaced N and M by Q and R.

The rest of $\mathcal{W}$ and $\mathcal{V}$ will be pseudo-trees on the phalanxes $(\langle \mathcal{M}_\xi^{\mathcal{W}} \mid \xi \leq \theta \rangle, \langle \rho_\xi \mid \xi < \theta \rangle)$ and $(\langle \mathcal{M}_\xi^{\mathcal{V}} \mid \xi \leq \theta \rangle, \langle \rho_\xi \mid \xi < \theta \rangle)$. A *root* of $\mathcal{W}$ or $\mathcal{V}$ is an ordinal $\xi \leq \theta$. If $\xi = \theta$, the root is stable, and if $\xi < \theta$ the root is unstable. A *phalanx interval* in $\mathcal{W}$ or $\mathcal{V}$ is an interval of the form $[\alpha, \beta]$ such that β is stable, every γ in $[\alpha, \beta)$ is unstable, and there are arbitrarily large stable $\gamma < \alpha$. (So if $\alpha = \xi + 1$, then ξ is stable.) The first phalanx interval in both trees is $[0, \theta]$. $\mathrm{lh}(\mathcal{W})$ and $\mathrm{lh}(\mathcal{V})$ will be partitioned into phalanx intervals, often of length 1. Let us say that $[\alpha, \beta]$ is nontrivial iff $\alpha < \beta$. The main new thing is that we may create a new nontrivial phalanx interval by lifting a tail of some previous one, rather than the whole of it.

For every node $\alpha < \mathrm{lh}(\mathcal{W})$ there is a unique root $r^W(\alpha) \leq \theta$ such that $r^W(\alpha) \leq_W \alpha$. If α is unstable, then $r^W(\alpha) < \theta$, and the branch $[r^W(\alpha), \alpha]_W$ does not drop in model or degree. Letting β be the least stable node $> \alpha$, we shall have

$$\beta = \alpha + (\theta - r^W(\alpha))$$

and

$$r^W(\alpha + \xi) = r^W(\alpha) + \xi$$

for all $\xi \leq \theta - r^W(\alpha)$. Similarly on the $\mathcal{V}$ side.

We define $\mathcal{W}$ and $\mathcal{V}$ by induction, one phalanx interval at a time; that is, at any stage the current last models of $\mathcal{W}$ and $\mathcal{V}$ are stable. If $\mathcal{M}_\gamma^{\mathcal{W}}$ is the current last model of $\mathcal{W}$ at some stage, then we let $E_\gamma^{\mathcal{W}} = E^+$, where E is the first extender on its sequence that is part of a disagreement with $M_{\nu,l}$. Similarly on the $\mathcal{V}$ side. We show that the corresponding extender on $M_{\nu,l}$ is empty, and no strategy disagreements ever show up. If there is no disagreement, the construction of $\mathcal{W}_{\nu,l}$ is complete, and similarly on the $\mathcal{V}$ side.

We shall also have ordinals $\varepsilon_\alpha^{\mathcal{W}}$ and $\varepsilon_\alpha^{\mathcal{V}}$ that tell us what model in $\mathcal{W}$ or $\mathcal{V}$ we should apply a given extender to. If α is stable in $\mathcal{W}$ and $E_\alpha^{\mathcal{W}}$ exists (that is, the construction of $\mathcal{W}$ is not finished), then

$$\varepsilon_\alpha^{\mathcal{W}} = \varepsilon(E_\alpha^{\mathcal{W}}) = \mathrm{lh}(E_\alpha^{\mathcal{W}}).$$

(Recall that all extenders in $\mathcal{W}$ or $\mathcal{V}$ have plus type.) If α is unstable in $\mathcal{W}$, then there is a least stable $\gamma \geq \alpha$. Suppose again $E_\gamma^{\mathcal{W}}$ exists, as otherwise the construction of $\mathcal{W}$ is done. Since α is unstable, we will have a unique unstable root $\eta < \theta$ such that $\eta \leq_W \alpha$, and $[\eta, \alpha]_W$ will not drop.[26] We then set

$$\varepsilon_\alpha^{\mathcal{W}} = \inf(i_{\eta,\alpha}^{\mathcal{W}}(\rho_\eta), \varepsilon(E_\gamma^{\mathcal{W}})).$$

Similarly for $\varepsilon_\alpha^{\mathcal{V}}$. The extenders used in $\mathcal{W}$ have increasing lengths, so if $\alpha < \beta$, then $\mathcal{M}_\alpha^{\mathcal{W}} | \varepsilon_\alpha^{\mathcal{W}} = \mathcal{M}_\beta^{\mathcal{W}} | \varepsilon_\alpha^{\mathcal{W}}$.[27]

---

[26] Where before we had $\gamma = \alpha + 1$, now we have $\gamma \leq \alpha + (\theta - \eta)$.

[27] For example, suppose there is a ξ such that $\varepsilon(E_\theta^{\mathcal{W}}) \leq \rho_\xi$, and let ξ be the least such. Then $\varepsilon(E_\theta^{\mathcal{W}}) < \rho_\xi$, and $E_\theta^{\mathcal{W}}$ is actually on the extended $\mathcal{M}_\xi^{\mathcal{W}}$ sequence. If $\xi < \alpha \leq \theta$, then $\varepsilon_\alpha^{\mathcal{W}} = \varepsilon_\xi^{\mathcal{W}}$, and the net effect of our definition of the ε's is that no extender will ever be applied later to $\mathcal{M}_\alpha^{\mathcal{W}}$.

Now let us look at the general successor step. Suppose $\mathcal{M}_\gamma^\mathcal{W}$ is the current last model of $\mathcal{W}$, and hence is stable. Let

$$E = E_\gamma^\mathcal{W}$$

be such that E^- is the least disagreement between $\mathcal{M}_\gamma^\mathcal{W}$ and $M_{\nu,l}$. Again, we are assuming that such a disagreement exists, it is not a strategy disagreement, and it does not involve an extender on the $M_{\nu,l}$-sequence. Set

$$\varepsilon_\gamma^\mathcal{W} = \varepsilon(E),$$

and for unstable α such that γ is the least stable above α, let $\varepsilon_\alpha^\mathcal{W}$ be defined as above. Let $\kappa = \text{crit}(E)$, and α be least such that $\kappa < \varepsilon_\alpha^\mathcal{W}$.[28] We set $\alpha = W\text{-pred}(\gamma+1)$ and

$$\mathcal{M}_{\gamma+1}^\mathcal{W} = \text{Ult}(\mathcal{M}_{\gamma+1}^{*,\mathcal{T}}, E)$$

as usual. If α is stable, so is $\gamma+1$. If α is unstable, then we have

(†) for some ξ and η,

  (i) $r^W(\alpha) = r^V(\xi) = \eta$, and

  (ii) $\mathcal{M}_\alpha^\mathcal{W} = \mathcal{M}_\xi^\mathcal{V}$ and $i_{\eta,\alpha}^\mathcal{W} = i_{\eta,\xi}^\mathcal{V}$.

If E is not also used in $\mathcal{V}$, then we declare $\gamma+1$ stable in $\mathcal{W}$ and go on.[29]

Now suppose that α is unstable and that E is also used in $\mathcal{V}$. We declare that $\alpha+1$ is unstable in $\mathcal{W}$. Let ξ and η be as in (†) and $E = E_\delta^\mathcal{V}$. One can check that $V\text{-pred}(\delta+1) = \xi$ and $\delta+1$ is unstable in $\mathcal{V}$. $r^W(\gamma+1) = r^V(\delta+1) = \eta$. For $\beta \leq \theta - \eta$, we set

$$\mathcal{M}_{\gamma+1+\beta}^\mathcal{W} = \text{Ult}(\mathcal{M}_{\alpha+\beta}^\mathcal{W}, E)$$

and

$$\mathcal{M}_{\delta+1+\xi}^\mathcal{V} = \text{Ult}(\mathcal{M}_{\xi+\beta}^\mathcal{V}, E).$$

$\alpha+\beta \leq_W \gamma+1+\beta$, and $i_{\alpha+\beta,\gamma+1+\beta}^\mathcal{W} = i_E$. Similarly for $\mathcal{V}$. For $\beta < \theta - \eta$, $\gamma+1+\beta$ is unstable, in $\mathcal{W}$, and (†) holds, with its mate in $\mathcal{V}$ being $\delta+1+\beta$. For $\beta = \theta - \eta$, $\gamma+1+\beta$ and $\delta+1+\beta$ are stable in $\mathcal{W}$ and $\mathcal{V}$, and index the new last models.

Here is a diagram of the phalanx lifting. In it we omit the superscript $\mathcal{W}$ everywhere.

---

[28] Since the extenders used have plus type and our mice are projectum solid, this is equivalent to $\kappa < \hat{\lambda}(E_\alpha^\mathcal{W})$.

[29] At this point, we already know what extenders with length $\leq \text{lh}(E)$ are used in $\mathcal{V}$.

$$\begin{array}{ccccc}
 & & & M_{\rho+1} & \\
 & & & \uparrow{\scriptstyle E_\rho} & \\
 & & M_{\gamma+1} & M_\sigma & M_\tau \\
 & & \uparrow{\scriptstyle E_\gamma} & \uparrow{\scriptstyle E_\gamma} & \uparrow{\scriptstyle E_\gamma} \\
 & M_{\alpha+1} & M_\varepsilon & M_\delta & M_\beta \\
 & \uparrow{\scriptstyle E_\alpha} & \uparrow{\scriptstyle E_\alpha} & \uparrow{\scriptstyle E_\alpha} & \uparrow{\scriptstyle E_\alpha} \\
M_0 & M_{r(\alpha+1)} & M_{r(\gamma+1)} & M_{r(\rho+1)} & M_\theta
\end{array}$$

The rows in the diagram are phalanx intervals.

Remark 10.5.6. Our process of moving phalanxes in $\mathcal{W}$ and $\mathcal{V}$ up amounts to a step of full normalization. We could have used a step of embedding normalization instead, and thereby arranged that our $\mathcal{W}$ and $\mathcal{V}$ are actually normal iteration trees on N. $\mathcal{W}$ and $\mathcal{V}$ would then be *meta-iterates* of $\mathcal{S}^\frown b$ and $\mathcal{S}^\frown c$, in the sense of [64]. That paper contains a proof of Theorem 10.5.2 that rests on the theory of meta-iteration trees.

As we define $\mathcal{W}$, we lift it to a λ-separated tree $\mathcal{X}$ on R such that $\mathcal{X}$ is by $\Omega_{\mathcal{U}^+,R}$. $\mathcal{X}$ is padded at the beginning, because $\mathcal{M}_\alpha^\mathcal{X} = R$ for all $\alpha \leq \theta$. The further padding corresponds precisely to the phalanx intervals in $\mathcal{W}$:

$$[\alpha,\beta)_W \cap \mathrm{stab}(\mathcal{W}) = \emptyset \text{ iff } M_\alpha^\mathcal{X} = \mathcal{M}_\beta^\mathcal{X}.$$

We shall have

$$\psi_\alpha \colon \mathcal{M}_\alpha^\mathcal{W} \to J_\alpha \trianglelefteq \mathcal{M}_\alpha^\mathcal{X}$$

for all α, with the agreement and commutativity needed to keep copying. For example, in the successor step just described, letting

$$F = \psi_\gamma(E),$$

and supposing that $\alpha = W\text{-pred}(\gamma+1)$, and $\gamma+1$ is unstable,

$$\mathcal{M}_{\gamma+1+\beta}^\mathcal{X} = \mathrm{Ult}(\mathcal{M}_\alpha^\mathcal{X}, F),$$
$$J_{\gamma+1+\beta} = i_F(J_{\alpha+\beta}),$$

and

$$\psi_{\gamma+1+\beta}([a,f]_E^P) = [\psi_\gamma(a), \psi_{\alpha+\beta}(f)]_F^J,$$

where $P = \mathcal{M}_{\alpha+\beta}^\mathcal{W}$ and $J = J_{\alpha+\beta}$.

At limit stages μ in the construction of $\mathcal{W}$, we use $\Omega_{\mathcal{U}_{\nu,l},R}$ to pick a cofinal branch a of $\mathcal{X}$. The branch is a disjoint union of phalanx intervals,

$$a = \bigcup_{\xi<\mu} [\alpha_\xi, \beta_\xi],$$

where $\xi < \eta \Longrightarrow \beta_\xi < \alpha_\eta$. If $\alpha_\xi = \beta_\xi$, then $\alpha_\eta = \beta_\eta$ for all $\eta > \xi$. In this case, letting $\lambda = \sup(a)$,

$$\begin{aligned} d &= [r^W(\beta_\xi), \beta_\xi]_W \cup \{\beta_\eta \mid \xi < \eta < \mu\} \\ &= [r^W(\lambda), \lambda)_W \end{aligned}$$

is the branch of $\mathcal{W}$ chosen by Ψ. All sufficiently large $\beta \in d$ are stable, so λ is stable.

If $\alpha_\xi < \beta_\xi$ for all $\xi < \mu$, then $\{\beta_\xi \mid \xi < \mu\}$ is a branch of $\mathcal{W}$, with least element $\beta_0 = \theta$. Let $\eta \leq \theta$ be least such that for unboundedly many (equivalently, all) $\xi < \mu$, $\exists \gamma \in [\alpha_\xi, \beta_\xi](r^W(\gamma) = \eta)$. In this case, Ψ chooses

$$\begin{aligned} d &= \{\gamma \mid \exists \xi(\gamma \in [\alpha_\xi, \beta_\xi]) \wedge r^W(\gamma) = \eta\} \\ &= [\eta, \lambda)_W. \end{aligned}$$

If some $\gamma \in d$ is stable, then we declare λ to be stable. If all $\gamma \in d$ are unstable (so $\eta < \theta$) then we declare λ to be unstable iff (†) holds, that is, there is a ξ such that $r^V(\xi) = \eta$, $\mathcal{M}_\lambda^{\mathcal{W}} = \mathcal{M}_\xi^{\mathcal{V}}$ and $i_{\eta,\lambda}^{\mathcal{W}} = i_{\eta,\xi}^{\mathcal{V}}$. The information about $\mathcal{V}$ relevant to this question is available at the current stage. If λ is stable, we are finished with the limit stage.

If λ is unstable in $\mathcal{W}$, we continue the limit stage by lifting the tail of $\mathcal{W}$ from η to θ along d. For $\beta \leq \theta - \eta$, let

$$\begin{aligned} d_\beta &= \{\gamma \mid \exists \xi(\gamma \in [\alpha_\xi, \beta_\xi]) \wedge r^W(\gamma) = \beta\} \\ &= [\beta, \lambda)_W, \end{aligned}$$

and

$$\mathcal{M}_{\lambda+\beta}^{\mathcal{W}} = \lim_{\gamma \in d_\beta} \mathcal{M}_\gamma^{\mathcal{W}},$$

where the limit is under the branch embeddings of $\mathcal{W}$. One can easily check that all d_β have the same branch extender as d. If $\beta < \theta - \eta$ then $\lambda + \beta$ is unstable, and if $\beta = \theta - \eta$, then $\lambda + \beta$ is stable, and the new last node of $\mathcal{W}$. We define the ε_β as before. That is, if $E_{\lambda+(\theta-\eta)}^{\mathcal{W}}$ exists (i.e., $\mathcal{W}_{\nu,l}$ is not completely defined), then we set

$$\begin{aligned} \varepsilon &= \mathrm{lh}(E_{\lambda+(\theta-\eta)}) \\ &= \varepsilon_{\lambda+(\theta-\eta)}, \end{aligned}$$

and for $\beta < \theta - \eta$

$$\varepsilon_{\lambda+\beta} = \inf(\varepsilon, i_{\beta,\lambda+\beta}^{\mathcal{W}}(\rho_\beta)).$$

If $E^{\mathcal{W}}_{\lambda+(\theta-\eta)}$ does not exist, then $\mathcal{W}_{\nu,l}$ is completely defined, and we go on to $\mathcal{W}_{\nu,l+1}$, unless the last model of $\mathcal{W}_{\nu,l}$ is $M_{\nu,l}$, and the branch of $\mathcal{W}_{\nu,l}$ to it has not dropped. In this latter case, we shall see that last model of $\mathcal{V}_{\nu,l}$ is also $M_{\nu,l}$, and the branch of $\mathcal{V}_{\nu,l}$ to it has not dropped. In this case, the construction of the $\mathcal{W}_{\nu,l}$ and $\mathcal{V}_{\nu,l}$ is over.

The construction of $\mathcal{V}_{\nu,l}$ proceeds in completely parallel fashion; indeed, nothing in our situation has distinguished b from c. Although the constructions of $\mathcal{W}_{\nu,l}$ and $\mathcal{V}_{\nu,l}$ determine stability by looking at each other, the reader can check that there is no circularity: when it comes time to determine whether γ is stable in $\mathcal{W}$, the relevant part of $\mathcal{V}$ is already determined.

As in Section 9.6, the maps ψ_α, for $\alpha \leq \theta$ yield a pullback strategy

$$\Psi = \Omega^{\vec{\psi}}_{\mathcal{U},R}$$

for a more general iteration game on $\mathcal{W} \restriction \theta + 1$, via the lifting process we have defined above. In the more general game, I makes stability declarations and creates new models according to the rules above. Of course, there are no $M_{\nu,l}$ and $\mathcal{V}$ in the setting of the general game. I picks the next extender E freely (subject to normality), and if E is to be applied to an unstable $\mathcal{M}_\alpha$, I may decide whether $\mathrm{Ult}(\mathcal{M}_\alpha, E)$ is stable as he pleases. If he decides against stability, he must create new models as above. At limit γ such that the branch to γ chosen by II consists of unstable nodes, I is again free to decide whether γ is stable. If he decides against stability, he must create new models in the way we have described.

Similarly, the σ_α for $\alpha < \mathrm{lh}(\mathcal{S})$ or $\alpha = c$ get lifted to

$$\varphi_\alpha \colon \mathcal{M}^{\mathcal{V}}_\alpha \to K_\alpha \trianglelefteq R$$

for $\alpha \leq \theta$, and this yields the pullback strategy

$$\Sigma = \Omega^{\vec{\varphi}}_{\mathcal{U},R}$$

for the more general game on $\mathcal{V} \restriction \theta + 1$.

Let us consider how the coiteration can terminate. Let

$$Z = \mathrm{Th}^N(\delta),$$

and

$$Z_0 = \mathrm{Th}^{Q}(\delta_0) = i^{\mathcal{U}}_{0,\tau_0}(Z).$$

Q is pointwise definable from ordinals $< \delta_0$, so it is completely determined by Z_0. All critical points in $\mathcal{S}$ are above δ, so $Z = \mathrm{Th}^{\mathcal{M}^{\mathcal{S}}_\alpha}(\delta)$ for all $\alpha < \mathrm{lh}(\mathcal{S})$, and also for $\alpha = b$ or $\alpha = c$. Thus for all $\xi \leq \theta$,

$$Z_0 = \mathrm{Th}^{\mathcal{M}^{\mathcal{W}}_\xi}(\delta_0) = \mathrm{Th}^{\mathcal{M}^{\mathcal{V}}_\xi}(\delta_0).$$

Moreover, for all η, the critical points of $E^{\mathcal{W}}_\eta$ or $E^{\mathcal{V}}_\eta$ (if they exist) are $> \delta_0$.

Motivated by this, let us call $\langle \nu, l \rangle$ *relevant* iff

(a) $(Q|\delta_0, \Omega_{\mathcal{U}_{\nu,l},\langle\delta_0,0\rangle}) = (M^{\mathbb{C}}_{\nu,l}|\langle\delta_0,0\rangle, (\Omega^{\mathbb{C}}_{\nu,l})_{\langle\delta_0,0\rangle})$,

(b) δ_0 is a cardinal cutpoint of $M^{\mathbb{C}}_{\nu,l}$, and

(c) for no proper initial segment S of $M^{\mathbb{C}}_{\nu,l}$ do we have $Z_0 = \mathrm{Th}^S(\delta_0)$.

Let us call $\langle \nu, l\rangle$ *terminal* iff it is relevant, and $Z_0 = \mathrm{Th}^{M^{\mathbb{C}}_{\nu,l}}(\delta_0)$.

If $\langle \nu, l\rangle$ is relevant, then neither $\mathcal{W}_{\nu,l}$ nor $\mathcal{V}_{\nu,l}$ can reach a last model that is a proper initial segment of $M_{\nu,l}$. Let us state explicitly the lemma on stationarity of background constructions we have been using

LEMMA 10.5.7. *If $\langle \nu, l\rangle$ is relevant, then in the $(\Psi, \Sigma, M^{\mathbb{C}}_{\nu,l})$ coiteration, no strategy disagreements show up, and no nonempty extender on the $M^{\mathbb{C}}_{\nu,l}$ side is part of a least disagreement.*

PROOF SKETCH. This proof is like the proofs of Theorem 8.4.3 and Lemma 10.3.10 we gave earlier. We show that the strategies Ψ and Σ normalize well and have strong hull condensation, in the appropriate senses. We then show there are no strategy disagreements. For example, if $\mathcal{Z}$ is a candidate disagreement on a stable $\mathcal{M}^{\mathcal{W}}_\gamma$ with $\mathcal{M}^{\mathcal{W}_{\nu,l}}_\gamma = M_{\nu,l}$, and $d = (\Omega_{\nu,l})(\mathcal{Z})$, we show that the normalization of $\langle \mathcal{W}\restriction\gamma+1, \mathcal{Z}^\frown d\rangle$ tree-embeds into a pseudo-tree by Ψ. This involves an inductive construction like that in the proofs of Sublemmas 8.4.8.1 and 10.3.10.1.[30] ⊣

CLAIM 10.5.8. *There is a terminal $\langle \nu, l\rangle <_{\mathrm{lex}} \langle \delta(w), 0\rangle$.*

PROOF. Otherwise $\langle \delta(w), 0\rangle$ is relevant, so the $(\Psi, \Sigma, \mathcal{M}^{\mathbb{C}}_{\delta(w),0})$ coiteration produces $(\mathcal{W}, \mathcal{V})$ with last models extending $\mathcal{M}^{\mathbb{C}}_{\delta(w),0}$. This contradicts the universality of $\mathcal{M}^{\mathbb{C}}_{\delta(w),0}$. ⊣

Now let $\langle \nu, l\rangle$ be the unique terminal pair. Z_0 contains statements which collectively assert that $\rho_\omega = \mathrm{OR}$, and $\mathrm{Th}^{M_{\nu,l}}(\delta_0) = Z_0$, so $l = 0$. We have also that $M_{\nu,0} \models \mathsf{ZFC}^-$. Z_0 is Σ_1 over $M_{\nu+1,0}$, so $\rho(M_{\nu+1,0}) = \delta_0$.

Let $\mathcal{W} = \mathcal{W}_{\nu,0}$ and $\mathcal{V} = \mathcal{V}_{\nu,0}$ have lengths γ_0 and γ_1.

CLAIM 10.5.9. *$\mathcal{M}^{\mathcal{W}}_{\gamma_0} = \mathcal{M}^{\mathcal{V}}_{\gamma_1} = M_{\nu,0}$; moreover, the branches of $\mathcal{W}$ and $\mathcal{V}$ to γ_0 and γ_1 do not drop.*

PROOF. Neither side can iterate to a proper initial segment of $M_{\nu,0}$ because $\langle \nu, 0\rangle$ is relevant. Neither side can iterate strictly past $M_{\nu,0}$ because $\langle \nu, 0\rangle$ is terminal. ⊣

Let $\eta_0 \leq_W \gamma_0$ and $\eta_1 \leq_V \gamma_1$ be the roots of the two trees below γ_0 and γ_1. Let

$$i_0 \colon Q \to \mathcal{M}^{\mathcal{W}}_{\eta_0} \text{ and } i_1 \colon Q \to \mathcal{M}^{\mathcal{V}}_{\eta_1}$$

be the embeddings given by the fact that $Z_0 = \mathrm{Th}^{\mathcal{M}^{\mathcal{W}}_{\eta_0}}(\delta_0) = \mathrm{Th}^{\mathcal{M}^{\mathcal{V}}_{\eta_1}}(\delta_0)$. These are just the lifts under $i^{\mathcal{U}}_{0,\tau_0}$ of the branch embeddings $i^{\mathcal{S}}_{0,\eta_0}$ and $i^{\mathcal{S}}_{0,\eta_1}$. We have that

$$i^{\mathcal{W}}_{\eta_0,\gamma_0} \circ i_0 = i^{\mathcal{V}}_{\eta_1,\gamma_1} \circ i_1,$$

[30]The many details in the argument are covered in [64].

since both embeddings are the embedding given by Q being the transitive collapse of $\mathrm{Hull}^{M_{\nu,0}}(\delta_0)$.

We now get a contradiction using the hull and definability properties in $M_{\nu,0}$ as usual.

DEFINITION 10.5.10. For M an lpm, we say that M has the *definability property at α* iff α is first order definable over M from some ordinals $d \in [\alpha]^{<\omega}$, and write $\mathrm{Def}(M, \alpha)$ in this case. We say that M has the *hull property at α* iff whenever $A \subset \alpha$ and $A \in M$, there is a $B \in M$ such that B is definable over M from some $d \in [\alpha]^{<\omega}$, and $B \cap \alpha = A$. We write $\mathrm{Hp}(M, \alpha)$ in this case.

CLAIM 10.5.11. $\eta_0 = \eta_1$.

PROOF. Suppose otherwise. Let

$$j_0 \colon \mathcal{M}^{\mathcal{S}}_{\eta_0} \to \mathrm{Ult}(\mathcal{M}^{\mathcal{S}}_{\eta_0}, E) = \mathcal{M}^{\mathcal{W}}_{\eta_0},$$

and

$$j_1 \colon \mathcal{M}^{\mathcal{S}}_{\eta_1} \to \mathrm{Ult}(\mathcal{M}^{\mathcal{S}}_{\eta_1}, E) = \mathcal{M}^{\mathcal{V}}_{\eta_1}$$

be the canonical embeddings. Suppose first that η_0 and η_1 are incomparable in $\mathcal{S}$, and let $F = E^{\mathcal{S}}_{\alpha}$ and $G = E^{\mathcal{S}}_{\beta}$, where $\alpha + 1 \leq_S \eta_0$, $\beta + 1 \leq_S \eta_1$, $\alpha \neq \beta$, and $S\text{-pred}(\alpha + 1) = S\text{-pred}(\beta + 1) = \xi$. We may assume $\mathrm{lh}(F) < \mathrm{lh}(G)$, or equivalently, $\alpha < \beta$. Let $\lambda = \sup\{\mathrm{lh}(E^{\mathcal{S}}_{\nu}) \mid \nu + 1 \leq_S \xi\}$. Letting $\kappa_0 = \mathrm{crit}(F)$, we have

$$\kappa_0 = \text{least } \mu \geq \lambda \text{ such that } \neg\mathrm{Def}(\mathcal{M}^{\mathcal{S}}_{\eta_0}, \mu).$$

Because the generators of j_0 (i.e., the generators of E) are contained in δ_0, we get

$$\begin{aligned} j_0(\kappa_0) &= \text{least } \mu \geq j_0(\lambda) \text{ such that } \neg\mathrm{Def}(\mathcal{M}^{\mathcal{W}}_{\eta_0}, \mu) \\ &= \text{least } \mu \geq j_0(\lambda) \text{ such that } \neg\mathrm{Def}(M_{\nu,0}, \mu). \end{aligned}$$

To see the first line, note that $\neg\mathrm{Def}(\mathcal{M}^{\mathcal{S}}_{\eta_0}, \kappa_0)$ because F was used on the branch to η_0, and j_0 is fully elementary so it preserves this. On the other hand, any $\mu < j_0(\kappa_0)$ is of the form $j_0(f)(a)$, where f is definable over $\mathcal{M}^{\mathcal{S}}_{\xi}$ from ordinals $< \lambda$, and $a \in [\delta_0]^{<\omega}$. The second line comes from using $i^{\mathcal{W}}_{\eta_0,\gamma_0}$ to move up to $\mathcal{M}^{\mathcal{W}}_{\gamma_0} = M_{\nu,0}$. Note for this that $j_0(\kappa_0) < j_0(\mathrm{lh}(F)) = \rho_\alpha$, and $\rho_\alpha \leq \mathrm{crit}(i^{\mathcal{W}}_{\eta_0,\gamma_0})$ because $\alpha < \eta_0$. Similarly, letting $\kappa_1 = \mathrm{crit}(G)$, we get

$$\begin{aligned} j_1(\kappa_1) &= \text{least } \mu \geq j_1(\lambda) \text{ such that } \neg\mathrm{Def}(\mathcal{M}^{\mathcal{V}}_{\eta_1}, \mu) \\ &= \text{least } \mu \geq j_1(\lambda) \text{ such that } \neg\mathrm{Def}(M_{\nu,0}, \mu). \end{aligned}$$

So $j_0(\kappa_0) = j_1(\kappa_1)$. But $\kappa_0, \kappa_1 < \mathrm{lh}(F)$, and $j_0 \restriction (\mathrm{lh}(F) + 1) = j_1 \restriction (\mathrm{lh}(F) + 1)$, so $\kappa_0 = \kappa_1$.

It is not hard to see that

$$\mathrm{lh}(F) = \text{least cardinal } \mu > \kappa_0 \text{ such that } \mathrm{Hp}(\mathcal{M}^{\mathcal{S}}_{\eta_0}, \mu),$$

and

$$\text{lh}(G) = \text{least cardinal } \mu > \kappa_0 \text{ such that } \text{Hp}(\mathcal{M}^{\mathcal{S}}_{\eta_1}, \mu).$$

Here cardinals are in the sense of $\mathcal{M}^{\mathcal{S}}_{\eta_0}$ and $\mathcal{M}^{\mathcal{S}}_{\eta_1}$, of course.[31] Using $i^{\mathcal{W}}_{\eta_0,\gamma_0} \circ j_0$ and $i^{\mathcal{V}}_{\eta_1,\gamma_1} \circ j_1$ to move up to $M_{\nu,0}$, and considering the hull property there, we get as above that $j_0(\text{lh}(F)) = j_1(\text{lh}(G))$. But $j_0(\text{lh}(F)) = j_1(\text{lh}(F))$, so $\text{lh}(F) = \text{lh}(G)$. However, G was used strictly after F in $\mathcal{S}$, so $\text{lh}(F) < \text{lh}(G)$, contradiction.[32]

We are left to consider the case $\eta_0 <_S \eta_1$. Let G be the extender used in $[0, \eta_1)_S$ and applied to $\mathcal{M}^{\mathcal{S}}_{\eta_0}$. Let $\kappa_1 = \text{crit}(G)$, and let $\lambda = \sup\{\text{lh}(E^{\mathcal{S}}_\alpha) \mid \alpha + 1 \leq_S \eta_0\}$ be the sup of the generators of $\mathcal{M}^{\mathcal{S}}_{\eta_0}$. Then again,

$$\begin{aligned} j_1(\kappa_1) &= \text{least } \mu \geq j_1(\lambda) \text{ such that } \neg\text{Def}(\mathcal{M}^{\mathcal{V}}_{\eta_1}, \mu) \\ &= \text{least } \mu \geq j_1(\lambda) \text{ such that } \neg\text{Def}(\mathcal{M}^{\mathcal{V}}_{\gamma_1}, \mu). \end{aligned}$$

Note that γ_0 is stable and η_0 is unstable in $\mathcal{W}$, so $\eta_0 <_W \gamma_0$. Let H be the extender used in $[\eta_0, \gamma_0)_W$ and applied to $\mathcal{M}^{\mathcal{W}}_{\eta_0}$. Let

$$\mu = \text{crit}(H).$$

If $\mu < j_1(\lambda)$, then $\mu < \rho_\alpha$ for some $\alpha < \eta_0$, so H should have been applied to an earlier model of $\mathcal{W}$. Thus $j_1(\lambda) \leq \mu$, and since $\mathcal{M}^{\mathcal{W}}_{\eta_0}$ has the definability property everywhere above $j_1(\lambda)$, using $i^{\mathcal{W}}_{\eta_0,\gamma_0}$ we see that μ is the least $\gamma \geq j_1(\lambda)$ such that $\neg\text{Def}(M_{\nu,0}, \gamma)$. Thus

$$\mu = j_1(\kappa_1).$$

But $H = E^{\mathcal{W}}_\xi$ for some $\xi \geq \theta$, so H is a plus extender and $\hat{\lambda}(H) > \rho_\alpha$ for all $\alpha < \theta$. Thus

$$j_1(\text{lh}(G)) < \sup_{\alpha<\theta} \rho_\alpha < \hat{\lambda}(H).$$

An easy induction shows that $\mathcal{M}^{\mathcal{W}}_\xi$ does not project strictly below $\sup_{\alpha<\theta} \rho_\alpha$, so we get that $H \restriction j_1(\text{lh}(G)) \in \text{Ult}(\mathcal{M}^{\mathcal{W}}_{\eta_0}, H)$, so the hull property fails in

$$\text{Ult}(\mathcal{M}^{\mathcal{W}}_{\xi+1}, H)$$

at $j_1(\text{lh}(G))$. Moving up by $i^{\mathcal{W}}_{\xi+1,\gamma_0}$, the hull property fails in $M_{\nu,0}$ at $j_1(\text{lh}(G))$.

---

[31] Proof: for μ a cardinal such that $\kappa_0 < \mu < \text{lh}(F)$, $F \restriction \mu$ is coded by a a subset of μ that is not definable in $\mathcal{M}^{\mathcal{S}}_{\alpha+1} = \text{Ult}(\mathcal{M}^{\mathcal{S}}_\xi, F)$ from ordinals $< \mu$, as otherwise the factor embedding would show $F \restriction \mu$ is in its own ultrapower. On the other hand, every point in $\mathcal{M}^{\mathcal{S}}_{\alpha+1}$ is definable from ordinals $< \text{lh}(F)$. Since $\text{crit}(i^{\mathcal{S}}_{\alpha+1,\eta_0}) > \text{lh}(F)$, we get the first line displayed. The second is proved in parallel fashion.

[32] We could also identify $\text{lh}(F)$ as the least ordinal $> \kappa_0$ definable in $\mathcal{M}^{\mathcal{S}}_{\eta_0}$ from ordinals $< \kappa_0$. This uses that $\text{lh}(F)$ is not a critical point in $\mathcal{T}$, which follows from niceness. That would let us avoid the hull property in proving Theorem 10.5.2. The hull property seems to be needed in proving the plus-two version of Theorem 10.5.2.

However, $\mathcal{M}^{\mathcal{S}}_{\eta_1}$ does have the hull property at $\mathrm{lh}(G)$. This gives

$$\mathrm{Hp}(M^{\mathcal{V}}_{\eta_1}, j_1(\mathrm{lh}(G))),$$

and thus $\mathrm{Hp}(M_{\mathcal{V},0}, j_1(\mathrm{lh}(G)))$, noting here that $\mathrm{crit}(i^{\mathcal{V}}_{\eta_1,\gamma_1}) \geq j_1(\mathrm{lh}(G))$. This is a contradiction. ⊣

CLAIM 10.5.12. $\eta_0 < \theta$.

PROOF. Otherwise $\eta_0 = \eta_1 = \theta$. Let F be the first extender used in $b - c$ and G the first extender used in $c - b$. We get a contradiction just as we did in the proof of Claim 10.5.11, in the case η_0 and η_1 were S-incomparable. ⊣

Now let s be the increasing enumeration of the extenders used in $(\eta_0, \gamma_0)_{\mathcal{W}}$ and t the increasing enumeration of the extenders used in $(\eta_0, \gamma_1)_{\mathcal{V}}$. We show by induction on ξ that $s(\xi) = t(\xi)$. For given that $s \restriction \xi = t \restriction \xi$, we have that $\mathrm{Ult}(\mathcal{M}^{\mathcal{W}}_{\eta_0}, s \restriction \xi)$ is pointwise definable from $\sup_{\alpha<\xi} \varepsilon(s(\alpha))$, so

$$s(\xi)^- = H,$$

where H is the least whole initial segment of the extender derived from the factor embedding from $\mathrm{Ult}(\mathcal{M}^{\mathcal{W}}_{\eta_0}, s \restriction \xi)$ to $M_{\mathcal{V},0}$ such that $H \notin M_{\mathcal{V},0}$. Similarly, $t(\xi)^- = H$, so $s(\xi) = t(\xi)$.

Thus $s = t$. But this implies that γ_0 and γ_1 are unstable, a contradiction. That completes the proof of Theorem 10.5.2. ⊣

Chapter 11

HOD IN THE DERIVED MODEL OF A HOD MOUSE

In this chapter, we show that if D is the derived determinacy model associated to a hod pair (M,Σ), then HOD^D is a least branch premouse. This is Theorem 11.3.2 below. The proof also shows that HOD^D is an initial segment of an iterate of M. This implies that, under an iterability hypothesis, there are determinacy models whose HOD has a fine structure, and yet is rich enough to satisfy "there is a subcompact cardinal". This is Theorem 11.3.13 below.

We must assume here some of the basic facts about homogeneously Suslin sets and derived determinacy models. The material covered in [69] is more than sufficient. See also [28].

We show in Section 11.4 that reasonably closed hod mice satisfy $\mathsf{V}=\mathsf{K}$, in a certain natural sense. We then close the chapter with a short survey of further results on the structure of HOD in determinacy models that have been proved by the methods of this book.

11.1. Generic interpretability

We shall need the following *generic interpretability* theorem. Its proof follows the same basic outline as Sargsyan's proof of the corresponding fact for rigidly layered hod pairs below LSA. (See [42] and [44].)[1]

THEOREM 11.1.1 (Generic interpretability). *Assume* AD^+, *and let* (P,Σ) *be an lbr hod pair with scope* HC, *and such that* Σ *is coded by a Suslin-co-Suslin set of reals. Let*

$$P \models \mathsf{ZFC}^- + \text{“}\delta \text{ is Woodin”};$$

then there is a term $\tau \in P$ *such that whenever* $i\colon P \to Q$ *is the iteration map associated to a non-dropping* P*-stack* s *by* Σ, *and* g *is* $\mathrm{Col}(\omega, < i(\delta))$*-generic over*

[1]The main difference is that our mice may have extenders overlapping Woodin cardinals, which means we cannot use Q-structures to determine Σ on small generic extensions of (M,Σ) in the way Sargsyan did. It is at this point that we use Theorem 10.5.2 on UBH in $M[g]$. The proof of that theorem used a phalanx comparison, as any proof of generic interpretability at the level of extenders overlapping Woodin cardinals would probably need to do.

Q, then

$$i(\tau)^g = \Sigma_{s,Q|i(\delta)} \restriction \mathrm{HC}^{Q[g]}.$$

PROOF. For $\xi < \eta < \delta$ we shall define a term $\tau_{\xi,\eta}$ such that whenever g is P-generic over $\mathrm{Col}(\omega,\eta)$, then $\tau^g_{\xi,\eta} = \Sigma_{P|\xi} \restriction \mathrm{HC}^{P[g]}$. We then take τ to be the join of the $\tau_{\xi,\eta}$. Clearly then $\tau^g = \Sigma_{P|\delta} \restriction \mathrm{HC}^{P[g]}$ whenever g is $\mathrm{Col}(\omega,<\delta)$ generic over P. It will be clear that this property of τ is preserved by Σ-iteration.

So fix $\xi < \eta < \delta$, and let g be P-generic over $\mathrm{Col}(\omega,\eta)$. Let

$$N = P|\xi.$$

We shall define $\Sigma_N \restriction \mathrm{HC}^{P[g]}$ from $\xi, P|\delta$, and g. The definition will be uniform in g, giving us the desired term.

Let $\mu = (\eta^+)^P$. We may assume that μ is a cutpoint of P. For if not, let E be the first extender on the P-sequence such that $\mathrm{crit}(E) < \mu < \mathrm{lh}(E)$, and set $Q = \mathrm{Ult}(P,E)$. Then μ is a cutpoint of Q, $\mathrm{HC}^{P[g]} = \mathrm{HC}^{Q[g]}$, and by strategy coherence, $\Sigma_{\langle E\rangle,N} = \Sigma_N$. A definition of $\Sigma_{\langle E\rangle,N} \restriction \mathrm{HC}^{Q[g]}$ from $Q|i_E(\delta),\xi$, and g will then give the desired definition of $\Sigma_N \restriction \mathrm{HC}^{P[g]}$. So we assume μ is a cutpoint of P.

Let w be the canonical wellorder of $P|\delta$, and working in P, let $\mathcal{F}$ be the set of all nice extenders of the form $E \restriction \gamma$ for some E on the P sequence. Let $\mathbb{C}$ be the maximal $(w,\mathcal{F})$ construction of P. Our background condition has the consequence that for any $\mathcal{T}$ on $M^{\mathbb{C}}_{\nu,k}$, the iteration tree $\mathcal{T}^*$ on P that is part of $\mathrm{lift}(\mathcal{T},M_{\nu,k},\mathbb{C})$ is a P-nice tree. So by Theorem 10.5.2, if $\mathcal{T} \in P[g]$, then UBH holds for $\mathcal{T}^*$.

We also have CBH for P-nice trees $\mathcal{S}$ on P such that $\mathcal{S} \in P$. This is because $\mathcal{S}$ induces naturally a tree $\mathcal{S}^+$ with the same tree order that uses extenders from the P-sequence. We have that $b = \dot{\Sigma}^P(\mathcal{S}^+)$ is defined, in P, and wellfounded as a branch of $\mathcal{S}^+$. But then b is wellfounded as a branch of $\mathcal{S}$.[2] Thus in P, the $\Omega^{\mathbb{C}}_{\nu,l}$ are total. In P, they are induced by $\dot{\Sigma}^P$, but $\dot{\Sigma}^P \subseteq \Sigma$, and Σ is total on V. So Σ induces a total-on-V strategy $\Omega^*_{\nu,l}$ for $M_{\nu,l}$ such that $\Omega^{\mathbb{C}}_{\nu,l} \subseteq \Omega^*_{\nu,l}$. The $\Omega^*_{\nu,l}$ are Suslin-co-Suslin in V because Σ is. Since they are induced by Σ, they have strong hull condensation and normalize well. In fact, each $(M^{\mathbb{C}}_{\nu,l},\Omega^*_{\nu,l})$ is an lbr hod pair in V. Moreover, $V \models \mathsf{AD}^+$, so in V we can carry out the comparisons needed to see each $(M_{\nu,l},\Omega^*_{\nu,l})$ has a core. Thus $(M_{\nu,l},\Omega_{\nu,l})$ has a core in P, and $\mathbb{C}$ does not break down in P.

CLAIM 11.1.2. *In P, there is a $\nu < \delta$ such that $(N,\dot{\Sigma}^P_N)$ iterates to $(M^{\mathbb{C}}_{\nu,k},\Omega^{\mathbb{C}}_{\nu,k})$.*

PROOF. Suppose not. Working in P, we apply Theorem 9.5.2, and we get that for all $\langle \nu,l\rangle$ such that $\nu < \delta$, $(N,\dot{\Sigma}^P_N)$ iterates strictly past $(M_{\nu,l},\Omega_{\nu,l})$. But then $(N,\dot{\Sigma}^P_N)$ iterates past $M_{\delta,0}$ in P. This contradicts universality at Woodin cardinals, Theorem 8.1.4. ⊣

---

[2]One can show that the trivial completion of $E^{\mathcal{S}}_\alpha$ is on the sequence of $\mathcal{M}^{\mathcal{S}}_\alpha$, so that we can take $\mathcal{S}^+ = \mathcal{S}$.

Let $\mathcal{T}$ be the normal tree by $\dot{\Sigma}_N^P$ whose last model is $M_{\nu,k}^{\mathbb{C}}$ given by Claim 11.1.2, and let $i\colon N \to M_{\nu,k}$ be its canonical embedding.

CLAIM 11.1.3. $\Sigma_{\mathcal{T},M_{\nu,k}} = \Omega_{\nu,k}^*$.

PROOF. The proof that the two strategies agree on all trees in P actually shows that they agree on all trees in V. (By Lemma 7.6.5, it is enough that they agree on λ-separated trees. Let $\mathcal{U}$ be λ-separated and by both strategies, and $b = \Omega_{\nu,k}^*(\mathcal{U})$. Let $\mathcal{U}^*$ be the tree according to Σ that is part of lift$(\mathcal{U}, M_{\nu,k}, \mathbb{C})$; again, we do not need $\mathcal{U} \in P$ to make sense of lifting. Then $W(\mathcal{T}, \mathcal{U}^\frown b)$ is a pseudo-hull of $i_b^{\mathcal{U}^*}(\mathcal{T})$ by our previous calculations. However, $i_b^{\mathcal{U}^*}(\mathcal{T})$ is by $\Sigma_{(\mathcal{U}^*)^\frown b,N}$ by the elementarity of $i_b^{\mathcal{U}^*}$, and $\Sigma_{(\mathcal{U}^*)^\frown b,N} = \Sigma_N$ by strategy coherence for (P,Σ), applied in V. So $i_b^{\mathcal{U}^*}(\mathcal{T})$ is by Σ_N, so $W(\mathcal{T}, \mathcal{U}^\frown b)$ is by Σ_N by strong hull condensation, so so $b = \Sigma_{\mathcal{T},M_{\nu,k}}(\mathcal{U})$ because Σ_N quasi-normalizes well.) ⊣

Now let $\mathcal{U}$ be a λ-separated tree on N of limit length that is according to Σ_N, and such that $\mathcal{U}$ is countable in $P[g]$. We wish to find $\Sigma_N(\mathcal{U})$ in $P[g]$, and define it from the relevant parameters.[3] But Σ_N is pullback consistent, so

$$\begin{aligned}\Sigma_N(\mathcal{U}) = b &\text{ iff } \Sigma_{\mathcal{T},M_{\nu,k}}(i\mathcal{U}) = b\\ &\text{ iff } \Omega_{\nu,k}^*(i\mathcal{U}) = b.\end{aligned}$$

So it will be enough to show

CLAIM 11.1.4. *If $\mathcal{S}$ is λ-separated and countable in $P[g]$, of limit length, and by $\Omega_{\nu,k}^*$, and $b = \Omega_{\nu,k}^*(\mathcal{S})$, then $b \in P[g]$. Moreover, b is uniformly definable over $P[g]$ from $\mathcal{S}$ and $\mathbb{C}$.*

PROOF. Let $\mathcal{S}^*$ be the P-nice tree on P that it part of lift$(\mathcal{S}, M_{\nu,k}, \mathbb{C})$. We know from Theorem 10.5.2 that in $P[g]$, $\mathcal{S}^*$ has at most one cofinal, wellfounded branch. Since all critical points in $\mathcal{S}^*$ are strictly above μ, we can think of $\mathcal{S}^*$ as a P-nice tree on $P[g]$. Then by [30], since $\mathcal{S}^*$ is countable in $P[g]$, it has exactly one cofinal wellfounded branch b in $P[g]$. Moreover, again by [30], $\mathcal{S}^*$ is continuously illfounded off b. It follows that $b = \Sigma(\mathcal{S}^*)$, and therefore $b = \Omega_{\nu,k}^*(\mathcal{S})$, as desired. ⊣

This completes the proof of Lemma 11.1.1. ⊣

11.2. Mouse limits

Assume AD^+, and let (M,Ω) be a strongly stable mouse pair with scope HC. Suppose s and t are stacks by Ω on M with last models P and Q such that M-to-P and M-to-Q do not drop. By Theorem 9.5.10 and Dodd-Jensen, we can then find stacks u and v by Ω_s and Ω_t with a common last model such that neither stack

[3] Again, by Lemma 7.6.5 this is enough.

drops getting to N, and such that $\Omega_{s^\frown u} = \Omega_{t^\frown v}$. (In fact, we can take u and v to consist of single λ-separated trees.) By Dodd-Jensen, for any such s, t, u, and v, $i_u \circ i_s = i_v \circ i_t$, where these are the iteration maps in question. Thus we have a well-defined direct limit system.

DEFINITION 11.2.1. Let (P, Σ) be a mouse pair; then

(1) $\mathcal{F}(P, \Sigma)$ is the collection of all (Q, Ψ) such that there is an P-stack s by Σ with last model Q, such that P-to-Q does not drop, and $\Psi = \Sigma_s$.
(2) For $(Q, \Psi) \in \mathcal{F}(P, \Sigma)$, $\pi_{(P,\Sigma),(Q,\Psi)} : P \to Q$ is the unique iteration map given by any and all stacks by Σ.
(3) $M_\infty(P, \Sigma)$ is the direct limit of $\mathcal{F}(P, \Sigma)$ under the $\pi_{(Q,\Psi),(R,\Phi)}$.
(4) $\pi_{(P,\Sigma),\infty} : P \to M_\infty(P, \Sigma)$ is the direct limit map.

Of course, $M_\infty(P, \Sigma) = M_\infty(Q, \Psi)$ for all $(Q, \Psi) \in \mathcal{F}(P, \Sigma)$. The iterates of (P, Σ) by single λ-separated trees are cofinal in $\mathcal{F}(P, \Sigma)$. Clearly, if $(P, \Sigma) \equiv^* (Q, \Psi)$, then $M_\infty(P, \Sigma) = M_\infty(Q, \Psi)$.[4] Thus $M_\infty(P, \Sigma) \in \text{HOD}$, being definable from the rank of (P, Σ) in the mouse order. In fact, this is true uniformly, in the sense that letting

(1) $m_e(\alpha) = X$ iff there is a pure extender pair (P, Σ) of mouse rank α such that $X = M_\infty(P, \Sigma)$, and
(2) $m_h(\alpha) = X$ iff here is a least branch hod pair (P, Σ) of mouse rank α such that $X = M_\infty(P, \Sigma)$,

we have

$$m_e, m_h \in \text{HOD}.$$

Assuming $\mathsf{AD}_{\mathbb{R}} + \mathsf{HPC}$, one can show that $\text{HOD} = L[m_h]$. This is not a very useful representation however, as it does not seem to lead to a fine structure for HOD. We do not know whether $L[m_e]$ has any natural identity, assuming say $\mathsf{AD}_{\mathbb{R}} + \mathsf{LEC}$.

Another simple fact worth noting is

PROPOSITION 11.2.2. *(AD^+) Let (P, Σ) and (P, Ψ) be strongly stable mouse pairs with scope* HC *such that (P, Σ) is mouse-equivalent to (P, Ψ) and $\pi_{(P,\Sigma),\infty} = \pi_{(P,\Psi),\infty}$; then $\Sigma = \Psi$.*

PROOF. By our comparison theorems, the two pairs have a common iterate (Q, Ω). Let $i : (P, \Sigma) \to (Q, \Omega)$ and $j : (P, \Psi) \to (Q, \Omega)$ be the two iteration maps. Then

$$\begin{aligned} \pi_{(Q,\Omega),\infty} \circ i &= \pi_{(P,\Sigma),\infty} \\ &= \pi_{(P,\Psi),\infty} \\ &= \pi_{(Q,\Omega),\infty} \circ j. \end{aligned}$$

This implies that $i = j$. But then by pullback consistency, $\Sigma = \Omega^i = \Omega^j = \Psi$, as desired. ⊣

[4] The converse is also true; see [73, Proposition 2.2].

Thus assuming AD^+, every mouse pair with scope HC is ordinal definable from a countable sequence of ordinals. On the other hand, a mouse pair (P,Σ) such that $\theta_0 \le o(M_\infty(P,\Sigma))$ cannot be ordinal definable from a real.

In order to compute HOD, we must relate different mouse limits. The concept of fullness helps do that.

DEFINITION 11.2.3. Assume AD^+, and let (P,Σ) be a mouse pair with scope HC. We say that (P,Σ) is *full* iff Σ is Suslin-co-Suslin, and

(a) $P \models \mathsf{ZFC}^-$, P has a largest cardinal δ, and $k(P) = 0$, and

(b) whenever s is a P-stack by Σ with last model Q, and the branch P-to-Q of s does not drop, and $i_s \colon P \to Q$ is the iteration map, then there is no mouse pair (R,Φ) such that R is solid, Φ is Suslin-co-Suslin, $Q \trianglelefteq^{\mathrm{ct}} R$, $\rho(R) \le i_s(\delta)$, and $\Phi_Q = \Sigma_s$.

This notion is sometimes called *mouse-fullness*.[5,6] The following lemma explains its importance in relating mouse limits to one another.

Notice that the solidity requirement on R in Definition 11.2.3(b) could be replaced by projectum solidity and stability, since these imply parameter solidity in the presence of iterability.

Remark 11.2.4. If there is a (s,Q,R,Φ) that witnesses that (P,Σ) is not full, then there is such a witness in which R is strongly stable. For let $k = k(R)$, and suppose that R itself is not strongly stable. Let $\eta = \eta_k^R$ and let D be the order zero measure of R on η. Since R is stable, $\eta < \rho_{k+1}(R)$, so D is on the sequence of both R and Q with index $< i_s(\delta)$. We can now just replace R with $R_1 = \mathrm{Ult}_k(\bar{\mathfrak{C}}_k(R),D)$, and Φ with the strategy Φ_1 for R_1 that it induces. Setting $s_1 = s^\frown\langle D\rangle$ and $Q_1 = \mathrm{Ult}(Q,D)$, one can check that (s_1,Q_1,R_1,Φ_1) is a counterexample to the fullness of (P,Σ), and that R_1 is strongly stable.

LEMMA 11.2.5. *Let (P,Σ) and (N,Ψ) be mouse pairs of the same type such that $(P,\Sigma) \le^* (N,\Psi)$, and suppose that (P,Σ) is full; then letting $\gamma = o(M_\infty(P,\Sigma))$*

$$M_\infty(P,\Sigma) = M_\infty(N,\Psi)|\gamma,$$

and γ is a successor cardinal cutpoint of $M_\infty(N,\Psi)$.

PROOF. Let (P,Σ) be full, and suppose that $(P,\Sigma) \le^* (N,\Psi)$. Comparing the two leads to (Q,Λ) a nondropping, normal iterate of (P,Σ) and (R,Φ) a normal iterate of (N,Ψ) such that $(Q,\Lambda) \trianglelefteq (R,\Phi)$. By perhaps taking one additional ultrapower on the N side, we can arrange that Q is a cutpoint of R. But then $o(Q) \le \rho(R)$, and if N-to-R drops in $\mathcal{U}$, then R is solid and $\rho(R) < o(Q)$, contrary

[5]It is customary to define fullness for P itself, and then say that Σ is *fullness-preserving* iff (P,Σ) is full in the sense of our definition.

[6]OD-*fullness* is the intensionally stronger requirement that whenever $(Q,\Psi_{s,Q})$ is as in (b) of Definition 11.2.3, and A is a bounded subset of $o(Q)$ that is ordinal definable from $(Q,\Psi_{s,Q})$, then $A \in Q$. Under an appropriate mouse capturing hypothesis, the two are equivalent.

to fullness. So neither iteration drops, and we have $M_\infty(P,\Sigma) = M_\infty(Q,\Lambda)$ and $M_\infty(N,\Psi) = M_\infty(R,\Phi)$.

But $o(Q)$ is a successor cardinal cutpoint of R, and $o(Q) \leq \rho(R)$. Also, $\Lambda = \Phi_Q$. It follows then that $M_\infty(Q,\Lambda)$ is a successor cardinal cutpoint of $M_\infty(R,\Phi)$, and that $o(M_\infty(Q,\Lambda) \leq \rho(M_\infty(R,\Phi))$. ⊣

COROLLARY 11.2.6. *(AD^+) Let (P,Σ) and (N,Ψ) be full mouse pairs; then*

$$(P,\Sigma) \leq^* (N,\Psi) \text{ iff } o(M_\infty(P,\Sigma)) \leq o(M_\infty(N,\Psi))$$
$$\text{iff } M_\infty(P,\Sigma) \trianglelefteq^{ct} M_\infty(N,\Psi).$$

11.3. HOD as a mouse limit

We shall show that in the derived model of a hod mouse, HOD can be represented as a mouse limit.

We shall need the following notions associated to derived models. Working in ZFC, suppose that λ is a limit of Woodin cardinals. Let g be $\mathrm{Col}(\omega, <\lambda)$-generic over V. We set

$$\mathbb{R}^*_g = \bigcup\{\mathbb{R} \cap V[g(\restriction \omega \times \alpha)] \mid \alpha < \lambda\},$$

and

$$\mathrm{Hom}^*_g = \{p[T] \cap \mathbb{R}^*_g \mid \exists \alpha < \lambda (V[g \restriction (\omega \times \alpha)] \models T \text{ is } < \lambda\text{-absolutely complemented})\}.$$

The symmetry of the forcing tells us that $\mathbb{R}^*_g = \mathbb{R} \cap L(\mathbb{R}^*_g, \mathrm{Hom}^*_g)$. The sets in Hom^*_g are those that have $<\lambda$-homogeneously Suslin representations in some intermediate collapse, which is is equivalent to having a $<\lambda$-uB representation in some intermediate collapse because λ is a limit of Woodin cardinals. Homogeneous Suslinity implies determinacy for sets in Hom^*_g, and with more work, that every set in Hom^*_g has a scale in Hom^*_g. Stationary tower forcing helps us pass from absolute definitions to absolutely complementing trees. In the end, we get

THEOREM 11.3.1 (Woodin; ZFC). *Suppose λ is a limit of Woodin cardinals, and let g be $\mathrm{Col}(\omega, <\lambda)$-generic over V; then*

$$L(\mathbb{R}^*_g, \mathrm{Hom}^*_g) \models \mathsf{AD}^+,$$

and

$$A \in \mathrm{Hom}^*_g \iff (L(\mathbb{R}^*_g, \mathrm{Hom}^*_g) \models \text{“}A \text{ is Suslin and co-Suslin”}),$$

for all $A \subseteq \mathbb{R}^*_g$.

The theorem was proved by Woodin in the late 1980s, as part of a more general theorem known as the Derived Model Theorem.[7]

[7] See for example [69]. That every Hom^*_g set is Suslin in $L(\mathbb{R}^*_g, \mathrm{Hom}^*_g)$ is due to the author.

We want to look at the derived model construction in the case that our ground model is a least branch hod mouse. What we get is

THEOREM 11.3.2. *Assume* AD^+, *and let* (M,Ψ) *be an lbr hod pair with scope* HC, *and such that* Ψ *is coded by a Suslin-co-Suslin set of reals. Suppose*

$$M \models \mathsf{ZFC} + \text{“}\lambda \text{ is a limit of Woodin cardinals”}.$$

Let g *be* $\mathrm{Col}(\omega, <\lambda)$*-generic over* M*; then*

$$L(\mathbb{R}^*_g, \mathrm{Hom}^*_g) \models \mathsf{AD}_{\mathbb{R}}.$$

and

(a) if λ *is a limit of cutpoints in* M*, then there is an iteration map* $i\colon M \to M_\infty(s)$ *coming from a stack* s *on* $M|\lambda$ *by* Ψ *such that*

$$\mathrm{HOD}^{L(\mathbb{R}^*_g, \mathrm{Hom}^*_g)} = L[M_\infty(s)|i(\lambda)],$$

and

(b) if $\kappa < \lambda$ *is least so that* $o(\kappa) \geq \lambda$ *in* M*, then there is an iteration map* $i\colon M \to M_\infty(s)$ *coming from a stack* s *on* $M|\lambda$ *by* Ψ *such that*

$$\mathrm{HOD}^{L(\mathbb{R}^*_g, \mathrm{Hom}^*_g)} = L[M_\infty(s)|i(\kappa)].$$

PROOF. The techniques here are pretty well known. Let (M,Ψ) and g be as in the hypotheses. For $\nu < \lambda$, let

$$\Psi^g_\nu = \Psi_{M|\nu} \restriction \mathrm{HC}^{M(\mathbb{R}^*_g)}.$$

Fixing a coding of elements of HC by reals, we can identify Ψ^g_ν with a subset of $\mathbb{R}^*_g$. Our first two claims say that the Ψ^g_ν witness that HPC holds in $L(\mathbb{R}^*_g, \mathrm{Hom}^*_g)$.

CLAIM 11.3.3. *If* $\nu < \lambda$*, then* $\Psi^g_\nu \in \mathrm{Hom}^*_g$.

PROOF. Let $h = g \cap \mathrm{Col}(\omega, <\nu^+)$. In $M[h]$ we have, for each $\mu < \lambda$, a term τ such that for all l that are $\mathrm{Col}(\omega,\mu)$-generic over $M[h]$,

$$\tau^l = \Psi_\nu \restriction \mathrm{HC}^{M[h][l]}.$$

For the specific such term τ given to us by Theorem 11.1.1, it is not hard to see that for all sufficiently large γ,

$$M[h] \models \text{there are club many generically } \tau\text{-correct hulls of } V_\gamma.$$

That is, in $M[h]$, whenever N is countable and transitive, and

$$\pi\colon N[h] \to (M|\gamma)[h]$$

is elementary, and everything relevant is in $\mathrm{ran}(\pi)$, and

$$\pi(\langle \bar{\tau}, \bar{\mu} \rangle) = \langle \tau, \mu \rangle,$$

then for any l that is $\mathrm{Col}(\omega, \bar{\mu})$-generic over N,

$$\bar{\tau}^l = \Psi_\nu \cap \mathrm{HC}^{N[l]}.$$

The proof of this is similar to the proof of Theorem 5.1 of [68]. Working in M, let $\mathbb{C}$ be the background construction and

$$i\colon M|\nu \to M^{\mathbb{C}}_{\eta,0}$$

be the iteration map by Ψ_ν that is described in the term τ. Let $\bar{\mathbb{C}} = \pi^{-1}(\mathbb{C})$ and $\bar{i} = \pi^{-1}(i)$, etc. So these are described in $\bar{\tau}$. Suppose $\mathcal{U}$ is according to $\bar{\tau}^l$. Let

$$\mathcal{W} = \mathrm{lift}^N(\bar{i}\mathcal{U})_0$$

be the nice tree on N that is given to us by $\bar{\tau}^l$. $\mathcal{W}$ is countable and nice in $N[h,l]$, so by Theorem 10.5.2, it picks unique cofinal wellfounded branches there. This implies that $\mathcal{W}$ is continuously illfounded off the branches it chooses. But then $\pi\mathcal{W}$ is continuously illfounded off the branches it chooses, so $\pi\mathcal{W}$ is by Ψ. But lifting commutes with copying, so

$$\begin{aligned} \pi\mathcal{W} &= \pi\,\mathrm{lift}^N(\bar{i}\mathcal{U})_0 \\ &= \mathrm{lift}^M((\pi\circ\bar{i})\mathcal{U})_0 \\ &= \mathrm{lift}^M(i\mathcal{U})_0. \end{aligned}$$

Note here that π is the identity on the base model of $\mathcal{U}$, so $\pi\circ\bar{i}$ agrees with $\pi(\bar{i}) = i$ on the base model of $\mathcal{U}$. This gives the last equality.

So $\mathrm{lift}^M(i\mathcal{U})_0$ is by Ψ, and hence $i\mathcal{U}$ is by $(\Omega^{\mathbb{C}}_{\eta,0})^{h,l}$. But we saw in the proof of Theorem 11.1.1 that this means $i\mathcal{U}$ is by the tail strategy$(\Psi_\nu)_{\mathcal{T},M^{\mathbb{C}}_{\eta,0}}$, where $\mathcal{T}$ is the tree giving rise to i. Since Ψ_ν is pullback consistent, $\mathcal{U}$ is by Ψ_ν, as desired.

It is easy to go from a club of $<\lambda$-generically τ-correct hulls to a $<\lambda$-absolutely complemented tree projecting to τ^h whenever h is $<\lambda$-generic. (See [69].) This proves the claim. ⊣

CLAIM 11.3.4. *The Ψ^g_ν, for $\nu < \lambda$ are Wadge-cofinal in* Hom^*_g.

PROOF. Let $\eta < \lambda$ and

$$M[g\restriction(\omega\times\eta)] \models \text{“}T \text{ and } T^* \text{ are } <\lambda\text{-absolute complements”}.$$

Let $\eta < \delta < \lambda$, and $M \models \delta$ is Woodin. We may assume that it is forced in $\mathrm{Col}(\omega,\delta)$ that $p[T] = p[T\cap(\omega\times\delta^+)]$ and $p[T^*] = p[T^*\cap(\omega\times\delta^+)]$.

Let $\mu = (\delta^{++})^M$, and put $\pi \in \mathcal{I}$ iff there is a non-dropping, normal iteration tree $\mathcal{U}$ on $M|\mu$ such that

(i) $\mathcal{U}$ is by Ψ^g_μ, with last model N,
(ii) all critical points in $\mathcal{U}$ are strictly above η, and
(iii) $\pi\colon M[g\restriction(\omega\times\eta)] \to N[g\restriction(\omega\times\eta)]$ is the lift of the iteration map.

Standard arguments[8] show that for $x \in \mathbb{R}^*_g$,

$$x \in p[T] \iff \exists \pi \in \mathcal{I}(x \in p[\pi(T \cap (\omega \times \delta^{+,M}))]).$$

This shows that $p[T]$ is projective in Ψ^g_μ. This easily implies the claim. $\dashv$

CLAIM 11.3.5. *Let η be a successor cardinal of M, and $\eta < \lambda$; then $(M|\eta, \Psi^g_\eta)$ is a full lbr hod pair in $L(\mathbb{R}^*_g, \mathrm{Hom}^*_g)$.*

PROOF. $(M|\eta, \Psi_\eta)$ is an lbr hod pair in V, so $(M|\eta, \Psi^g_\eta)$ is an lbr hod pair in $L(\mathrm{Hom}^*_g, \mathbb{R}^*_g)$. We must see that $(M|\eta, \Psi_\eta)$ is full. In short, this is true because non-dropping iterations of $M|\eta$ carry the rest of M along on top, and the resulting iterates of M can compute truth in the derived model of M by consulting their own derived models.[9]

Let us fill in our sketch. Suppose toward contradiction that in $L(\mathbb{R}^*_g, \mathrm{Hom}^*_g)$ we have

(i) an $M|\eta$-stack s by Ψ_η with last model Q, such that the branch $M|\eta$-to-Q of s does not drop, and
(ii) an lbr hod pair (R, Φ) such that R is solid and strongly stable, Φ is Suslin-co-Suslin, $Q \trianglelefteq^{\mathrm{ct}} R$, $\rho(R) < o(Q)$, and $\Phi_Q = \Psi_{s,Q}$.

We can assume strong stability by Remark 11.2.4. Let (R, Φ) be the minimal such pair in the mouse order, and let

$$T_R = \mathrm{Th}^R_{k+1}(\rho(R) \cup p(R)),$$

where $k = k(R)$, be the theory coding the core of R.

Since η is a cardinal of M, s is in fact an M-stack, and regarding it this way, it has a last model S such that $Q \trianglelefteq S$, and the branch M-to-S of s does not drop. Since $o(Q)$ is a cardinal of S and $\rho(R) < o(Q)$, if $T_R \in S$ then $T_R \in Q$. But then $T_R \in R$, contradiction. We conclude that $T_R \notin S$.

However, working in V now, we can find an $\mathbb{R}^*_g$-genericity iteration of $S|\lambda$ by Ψ_s so that all its critical points are strictly above $o(Q)$. Let W be the final model of this genericity iteration; then we have h being $\mathrm{Col}(\omega, <\lambda)$ generic over W so that

$$\mathbb{R}^*_h = \mathbb{R}^*_g.$$

Moreover, as in Claim 11.3.4, the strategies $(\Psi_s)^h_\nu$ for $\nu < \lambda$ are Wadge cofinal in Hom^*_h, and clearly $(\Psi_s)^h_\nu = (\Psi_s)^g_\nu$. It follows that

$$\mathrm{Hom}^*_h = \mathrm{Hom}^*_g.$$

Thus we realized our derived model of M as a derived model of its iterate W.

---

[8]If $x \in p[T]$, then let π come from an iteration making x $\mathrm{Col}(\omega, \pi(\delta))$ generic over N. Then $x \in p[\pi(T)]$, so $x \in p[\pi(T \cap (\omega \times \delta^+))]$. Conversely, if x and π are as on the right, $x \notin p[\pi(T^*)]$, so $x \notin p[T^*]$, so $x \in p[T]$.

[9]We are showing that $(M|\eta, \Psi_\eta)$ is not just mouse-full, but OD-full. But we are in the derived model of a mouse, where the two are equivalent, so that is not surprising.

We shall show that T_R is ordinal definable in $L(\mathbb{R}^*_g, \mathrm{Hom}^*_g)$ from Q and $(\Psi_Q)^h$. But by generic interpretability, $(\Psi_Q)^h$ is definable in $W(\mathbb{R}^*_h)$ from parameters in W. By the homogeneity of the forcing, we then get that $T_R \in W$, and hence $T_R \in S$, contradiction.

So working in $L(\mathbb{R}^*_g, \mathrm{Hom}^*_g)$, let (P,Σ) be an lbr hod pair of minimal mouse rank such that P is solid and strongly stable, $Q \trianglelefteq^{\mathrm{ct}} P$, $\Sigma_Q = \Psi_{s,Q}$, and $\rho(P) < o(Q)$. Let $T_P = \mathrm{Th}^P_{k+1}(\rho(P) \cup p(P))$. The following claim finishes our proof.

SUBCLAIM 11.3.5.1. $T_P = T_R$.

PROOF. We work in $L(\mathbb{R}^*_g, \mathrm{Hom}^*_g)$. Since (R,Φ) and (P,Σ) are mouse minimal with respect to the same property, they have a common iterate (N,Λ), via normal trees $\mathcal{T}$ and $\mathcal{U}$ that do not drop along their main branches. Because neither side drops, we have

$$k(R) = k(N) = k(P).$$

Let k be the common value. Let $i = i^{\mathcal{T}}$ and $j = i^{\mathcal{U}}$ be the two main branch embeddings. Because Q is a cutpoint on both sides, and $o(Q)$ is $r\Sigma_k$-regular on both sides[10], we get that

$$i \restriction o(Q) = j \restriction o(Q).$$

But then $\rho(R) =$ least α such that $i(\alpha) \geq \rho(N) =$ least α such that $j(\alpha) \geq \rho(N)$. So $\rho(R) = \rho(P)$. Also

$$i(p(R)) = p(N) = j(p(P)).$$

Since i and j are elementary (hence $r\Sigma_{k+1}$ elementary), we get that $i``T_R = T_N = j``T_P$, so $T_R = T_P$. ⊣

This proves Claim 11.3.5. ⊣

We define in $L(\mathbb{R}^*_g, \mathrm{Hom}^*_g)$:

$$\mathcal{F} = \{(P,\Sigma) \mid (P,\Sigma) \text{ is a full lbr hod pair.}\}$$

For $(P,\Sigma), (Q,\Psi) \in \mathcal{F}$,

$$(P,\Sigma) \prec^* (Q,\Psi) \text{ iff } \exists (R,\Phi)[(R,\Phi) \trianglelefteq^{\mathrm{ct}} (Q,\Psi) \wedge (P,\Sigma) \text{ iterates to } (R,\Phi)].$$

If $(P,\Sigma) \prec^* (Q,\Psi)$, then

$$\pi_{(P,\Sigma),(Q,\Psi)} : P \to R \trianglelefteq^{\mathrm{ct}} Q$$

is the iteration map. By Dodd-Jensen, it is well-defined, that is, independent of the choice of stack witnessing that (P,Σ) iterates to some $(R,\Phi) \trianglelefteq^{\mathrm{ct}} (Q,\Psi)$. The π's commute, and $\prec^*$ is directed by Lemma 11.2.5, so we have a direct limit system. Set

$$M_\infty = \text{direct limit of } (\mathcal{F}, \prec^*) \text{ under the } \pi_{(P,\Sigma),(Q,\Psi)},$$

[10] Otherwise $\rho_k(R) < o(Q)$ or $\rho_k(P) < o(Q)$, contrary to minimality.

and let

$$\pi_{(P,\Sigma),\infty}\colon P \to M_\infty$$

be the direct limit map. Another way to characterize M_∞ is that it is the lpm N of minimal height such that for all $(P,\Sigma) \in \mathcal{F}$, $M_\infty(P,\Sigma) \trianglelefteq^{\text{ct}} M_\infty$. Our two definitions of $\pi_{(P,\Sigma),\infty}$ are consistent with one another.

Let us write

$$\begin{aligned}\Theta &= o(\text{Hom}^*_g)\\ &= \sup\{|W| \mid W \text{ is a prewellorder of } \mathbb{R}^*_g \text{ in } \text{Hom}^*_g\}.\end{aligned}$$

Θ is also the Wadge ordinal of Hom^*_g.

CLAIM 11.3.6. $o(M_\infty) \leq \Theta$.

PROOF. This follows immediately from Claim 11.3.3. ⊣

Clearly $\Theta \leq \theta^{L(\mathbb{R}^*_g,\text{Hom}^*_g)}$. In fact

CLAIM 11.3.7. $o(M_\infty) = \Theta = \theta^{L(\mathbb{R}^*_g,\text{Hom}^*_g)}$.

PROOF. We need only show that $\theta^{L(\mathbb{R}^*_g,\text{Hom}^*_g)} \leq o(M_\infty)$. The proof is essentially due to G. Hjorth. (See [16].)

Let $\tau < \theta^{L(\mathbb{R}^*_g,\text{Hom}^*_g)}$, and let $f\colon \mathbb{R}^*_g \to \tau$ be a surjection. f is ordinal definable in $L(\mathbb{R}^*_g,\text{Hom}^*_g)$ from some set of reals $A \in \text{Hom}^*_g$, and by Claim 11.3.4, we can take our A to be Wadge reducible via z to $\text{Code}(\Psi^g_\eta)$, for some cardinal η of M and some real $z \in \mathbb{R}^*_g$. By amalagamating the f_z associated to all possible z, we can eliminate z from the definition.

So we can fix

$$B = \Psi^g_\eta,$$

where η is a cardinal of M, and

$$f\colon \mathbb{R}^*_g \to \tau$$

a surjection, and a formula $\varphi(u,v,w)$ and ordinal α such that

$$f(x) = \xi \text{ iff } L_\alpha(\mathbb{R}^*_g,\text{Hom}^*_g) \models \varphi[x,\xi,B].$$

Let $M_0 = \text{Ult}(M,E)$, for E the first extender on M overlapping η, if there is one. Let $M_0 = M$ otherwise. Let

$$\delta_0 = \text{least } \delta > \eta \text{ such that } M_0 \models \delta \text{ is Woodin}.$$

So η and δ_0 are cutpoints of M_0. Letting $N = M_0|(\delta_0^+)^{M_0}$ and $\Phi = \Psi_{\langle E\rangle,N}$ or $\Phi = \Psi_N$ as appropriate, we have that $(N,\Phi) \in \mathcal{F}$. We shall show that

$$\pi_{(N,\Phi),\infty}(\delta_0) \geq \tau.$$

Remark 11.3.8. Let $\theta(B)$ be the sup of the lengths of OD(B) prewellorders of $\mathbb{R}$, in $L(\mathbb{R}^*_g, \mathrm{Hom}^*_g)$ of course. Since α and φ are arbitrary so far, we are showing that $\pi_{(N,\Phi),\infty}(\delta_0) \geq \theta(B)$. We believe that a little more work shows that $\pi_{(N,\Phi),\infty}(\delta_0) = \theta(B)$. See [73] for more along these lines.

To see this, it is more convenient to consider the relativised direct limit system $\mathcal{F}^\eta(N,\Phi)$, in which all iterations must be strictly above η. It is not hard to see that $\mathcal{F}^\eta(N,\Phi)$ is directed. Let $M^\eta_\infty(N,\Phi)$ be its direct limit, and $\pi^\eta_{(N,\Phi),\infty}$ be the direct limit map. We shall show

$$\tau \leq \pi^\eta_{(N,\Phi),\infty}(\delta_0).$$

Since $\mathcal{F}^\eta(N,\Phi)$ is a subsystem of the full $\mathcal{F}(N,\Phi)$, this is enough.

Working in V, let

$$\mathbb{R}^*_g = \{x_i \mid i < \omega\},$$

and let s be a run of $G^+(N,\omega,\omega_1)$ by Φ that is cofinal in $\mathcal{F}^\eta(N,\Phi)$, so that

$$N_\omega = M^\eta_\infty(N,\Phi),$$

where N_ω is the direct limit along s, and $i^s_{0,\omega} = \pi^\eta_{(N,\Phi),\infty}$. Let $N_0 = N$, and N_k be the last model of $s \restriction k$, for $k > 0$. Let $\delta_k = i^s_{0,k}(\delta_0)$. We can arrange that whenever $i < k$, then $x_i \in N_k[H]$, for some H that is generic over N_k for the extender algebra at δ_k.

We have $N_0 \trianglelefteq^{\mathrm{ct}} M_0$. The stack s is according to Ψ_{M_0}, so thinking of s as a stack on M_0, and letting M_k be the last model of $s \restriction k$ in this context, we have

$$N_k \trianglelefteq^{\mathrm{ct}} M_k,$$

and

$$i_{k,l} \colon M_k \to M_l$$

the iteration map given by s, for $k,l \leq \omega$.

Now we do the usual dovetailed $\mathbb{R}^*_g$- genericity iterations, iterating each

$$(M_k, \Psi_{s \restriction k, M_k}),$$

strictly above δ_k to (Q_k, Ω_k), and arranging that $L(\mathrm{Hom}^*_g, \mathbb{R}^*_g)$ is also a derived model of Q_k. Let

$$j_k \colon M_k \to Q_k$$

be the map of the $\mathbb{R}^*_g$ genericity iteration, and let

$$\sigma_{k,l} \colon Q_k \to Q_l$$

be the copy map, which exists because we dovetailed the genericity iterations together. (See for example the proof of Theorem 6.29 of [82] for the details of this well-known construction.) Here is a diagram.

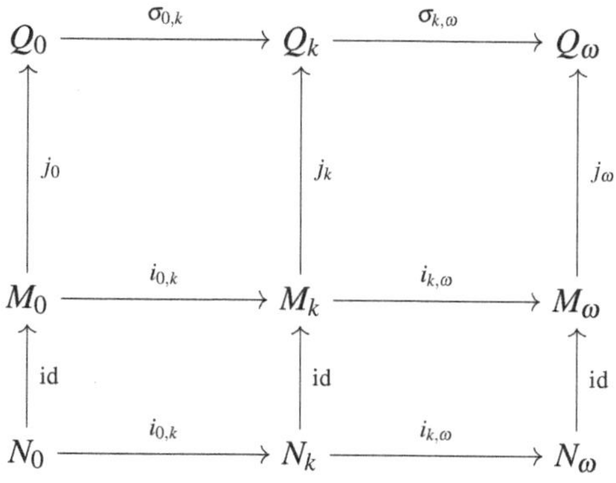

We have for each $k<\omega$ a Q_k-generic h_k such that $\mathbb{R}^*_{h_k}=\mathbb{R}^*_g$ and $\mathrm{Hom}^*_{h_k}=\mathrm{Hom}^*_g$. The latter holds because for each $\xi<\lambda$, the critical points in j_k are eventually above $j_k(\xi)$, and the initial segment of the iteration that gets us to this point acts only on some $M|\gamma$ for $\gamma<\lambda$. This tells us that $(\Omega_k)^{h_k}_{\langle j_k(\xi),0\rangle}$ is projective in Ψ^g_γ. That implies $\mathrm{Hom}^*_{h_k}\subseteq\mathrm{Hom}^*_g$. The reverse inclusion comes from the fact that each Ψ_γ is a pullback of some $\Omega_{\langle\xi,0\rangle}$.

Note that we have for each $k<\omega$ a term $\dot{B}_k\in Q_k$ such that

$$\dot{B}_k^{Q_k[l]}=B$$

for all l that are $\mathrm{Col}(\omega,<\lambda)$ generic over Q_k and such that $\mathbb{R}^*_l=\mathbb{R}^*_g$. Moreover,

$$\sigma_{k,n}(\dot{B}_k)=\dot{B}_n$$

for $k<n<\omega$. This follows from Lemma 11.1.1, the fact that all embeddings in the diagram above have critical point $>\eta$, and strategy coherence. Let $\mathbb{W}_k$ be the extender algebra of Q_k at δ_k, and put

$$\begin{aligned}\xi\in Y_k \text{ iff } Q_k\models \exists b\in\mathbb{W}_k[b\Vdash (&\mathrm{Col}(\omega,<\lambda)\Vdash\check{\xi}\text{ is the least}\\ &\gamma\text{ such that } L_{\check{\alpha}}(\mathrm{Hom}^*_{\dot{G}},\mathbb{R}^*_{\dot{G}})\models\varphi[\dot{x},\gamma,\dot{B}_k])]\end{aligned}$$

Because $\mathbb{W}_k$ has the δ_k-chain condition in Q_k,

$$Q_k\models |Y_k|<\delta_k.$$

Now we define an order preserving map

$$p\colon \tau\to\pi^\eta_{(N,\Phi),\infty}(\delta_0)=i_{0,\omega}(\delta_0).$$

Let $\xi<\tau$, and pick any x such that $f(x)=\xi$. Let $k<\omega$ be sufficiently large that

(i) $x=x_i$ for some $i<k$, and

(ii) for $k\le m\le n<\omega$, $\sigma_{m,n}(\alpha)=\alpha$ and $\sigma_{m,n}(\xi)=\xi$.

Since Q_ω is wellfounded, we can find such a k. By (i), x is $\mathbb{W}_k$-generic over Q_k. It follows that $\xi\in Y_k$; say that

$$\xi=\text{the }\gamma\text{-th element of }Y_k$$

in its increasing enumeration. We then set

$$p(\xi) = i_{k,\omega}(\gamma) = \sigma_{k,\omega}(\gamma).$$

We must check that $p(\xi)$ is independent of the choice of x, and that p is order preserving. For this, let $f(y) = \tau$. Let $k_{x,\xi}$ and $k_{y,\tau}$ be as in (i) and (ii) above, for (x,ξ) and (y,τ) respectively. Let $\gamma_{x,\xi}$ and $\gamma_{y,\tau}$ be the corresponding γ's. Taking $n \geq \max(k_{x,\xi}, k_{y,\tau})$, we have $\xi, \tau \in Y_n$, and

$$\xi = \text{the } \sigma_{k_{x,\xi},n}(\gamma_{x,\xi})\text{-th element of } Y_n.$$

This is because $\sigma_{k_{x,\xi},n}(\xi) = \xi$. Similarly,

$$\tau = \text{the } \sigma_{k_{y,\tau},n}(\gamma_{y,\tau})\text{-th element of } Y_n.$$

So

$$\begin{aligned} \xi \leq \tau \text{ iff } & i_{k_{x,\xi},n}(\gamma_{x,\xi}) \leq i_{k_{y,\tau},n}(\gamma_{y,\tau}) \\ \text{iff } & i_{k_{x,\xi},\omega}(\gamma_{x,\xi}) \leq i_{k_{y,\tau},\omega}(\gamma_{y,\tau}), \end{aligned}$$

as desired. This proves Claim 11.3.7. ⊣

From the fact that $\Theta = \theta^{L(\mathbb{R}^*_g, \mathrm{Hom}^*_g)}$ we get at once that $L(\mathbb{R}^*_g, \mathrm{Hom}^*_g) \cap P(\mathbb{R}^*_g) = \mathrm{Hom}^*_g$. Thus in $L(\mathbb{R}^*_g, \mathrm{Hom}^*_g)$, all sets are Suslin, and therefore we get

CLAIM 11.3.9. $L(\mathbb{R}^*_g, \mathrm{Hom}^*_g) \models \mathsf{AD}_{\mathbb{R}}$.

Suppose that $(P,\Sigma) \in \mathcal{F}$, and let $\tau = o(M_\infty(P,\Sigma))$. The proof of Claim 11.3.7 showed that for some $\gamma < \lambda$, $(M|\gamma, \Psi^g_\gamma) \in \mathcal{F}$ and $\tau < o(M_\infty(M|\gamma, \Psi^g_\gamma))$. But this implies that $(P,\Sigma) \leq^* (M|\gamma, \Psi^g_\gamma)$. It follows then that the iterates of proper initial segments of $(M|\lambda, \Psi_{M|\lambda})$ are $\prec^*$-cofinal in $\mathcal{F}$.

This gives

CLAIM 11.3.10. *There is a stack s on $M|\lambda$ of length ω that does not drop along its main branch, with canonical embedding $i_s \colon M \to M_\infty(s)$, such that*

(a) for $n < \omega$, $s \upharpoonright n \in (\mathrm{HC})^{L(\mathbb{R}^*_g, \mathrm{Hom}^*_g)}$,
(b) $M_\infty \trianglelefteq M_\infty(s)$, and
(c) if λ is a limit of cutpoints in M, then $i_s(\lambda) = o(M_\infty)$, and
(d) if κ is the least $< \lambda$-strong of M, then $i_s(\kappa) = o(M_\infty)$.

PROOF. Let $\langle (P_i, \Lambda_i) \mid i < \omega \rangle$ be $\prec^*$-increasing and cofinal in $\mathcal{F}$. Let $(M, \Psi) = (Q_0, \Phi_0)$. Given $s \upharpoonright i$ with last model (Q_i, Φ_i), let $s(i)$ be a normal tree on Q_i that comes from comparing (P_i, Λ_i) with some cardinal initial segment below λ of (Q_i, Φ_i) that is strictly greater that than (P_i, Λ_i) in the mouse order. There is such an initial segment by the remarks above. Let (Q_{i+1}, Φ_{i+1}) be the last pair of $s(i)$.

We do the comparison in such a way that (P_i, Λ_i) iterates to a cutpoint (N, Ω) of (Q_{i+1}, Φ_{i+1}). It follows that $i_{s \upharpoonright i+1, \infty}$ agrees with the iteration map $\pi_{(N,\Omega),\infty}$ on N. This tells us that

$$\pi_{(P_i,\Lambda_i),\infty}``o(P_i) \subseteq i_{s \upharpoonright i+1, \infty}(o(N)).$$

This implies that $M_\infty \trianglelefteq M_\infty(s)$. Also, N is a cutpoint, so $o(N)$ is below the least $<\lambda$-strong of Q_{i+1}, if there is one. Thus $o(M_\infty) \leq i_{s\restriction 0,\infty}(\kappa)$, where κ is the least $<\lambda$-strong, if there is one.

The cutpoint successor cardinal initial segments (N,Ω) of (Q_i,Φ_i) below λ are all in $\mathcal{F}$, and so $o(M_\infty)(N,\Omega) = i_{s\restriction i,\infty}(o(N)) < o(M_\infty)$ for such (N,Ω). It follows that

$$o(M_\infty) = \sup\{i_{s\restriction i,\infty}(o(N)) \mid i < \omega \wedge N \trianglelefteq^{\mathrm{ct}} Q_i|\lambda\}.$$

So if λ is a limit of cutpoints in M, and hence in each Q_i, then we get $i_s(\lambda) = o(M_\infty)$. If κ is the least strong to λ in M, we get $i_s(\kappa) = o(M_\infty)$. ⊣

CLAIM 11.3.11. $\mathrm{HOD}^{L(\mathbb{R}^*_g,\mathrm{Hom}^*_g)} = L[M_\infty]$.

PROOF. Let us write HOD for $\mathrm{HOD}^{L(\mathbb{R}^*_g,\mathrm{Hom}^*_g)}$, and θ for $\theta^{L(\mathbb{R}^*_g,\mathrm{Hom}^*_g)}$. It is clear that $M_\infty \in \mathrm{HOD}$, so what we must show is that $\mathrm{HOD} \subseteq L[M_\infty]$.

We use here

LEMMA 11.3.12 (Woodin). *Assume* $\mathrm{AD}_\mathbb{R} + V = L(P(\mathbb{R}))$; *then there is a definable (from no parameters) set* $A \subseteq \theta$ *such that* $\mathrm{HOD} = L[A]$.

Fix A as in the lemma, and let $\varphi(v)$ be such that

$$\xi \in A \text{ iff } L(\mathbb{R}^*_g,\mathrm{Hom}^*_g) \models \varphi[\xi].$$

It is enough to show that $A \in L[M_\infty]$. For that, let s be a stack as in Claim 11.3.10, and let (Q_i,Φ_i) be the last model of $s\restriction i$. Let κ_i be the least $<\lambda$-strong of Q_i if there is one, and otherwise let $\kappa_i = \lambda$. We define $A_i \subseteq \kappa_i$ by

$$\xi \in A_i \text{ iff } i_{s\restriction i,\infty}(\xi) \in A.$$

We claim that A_i is definable over $L[Q_i|\kappa_i]$, uniformly in i. The definition is displayed in the following equivalence: for any ξ,

$$\xi \in A_i \text{ iff } L[Q_i|\kappa_i] \models \varphi_0[\xi],$$

where $\varphi_0(v)$ is the formula

$$\forall\alpha, h[(h \text{ is } \mathrm{Col}(\omega,<\kappa_i)\text{-generic and } \alpha \text{ is a cardinal cutpoint of } Q_i|\kappa_i)$$
$$\Longrightarrow (L(\mathbb{R}^*_h,\mathrm{Hom}^*_h) \models \varphi[\pi(v)], \text{where } \pi = \pi^{L(\mathbb{R}^*_h,\mathrm{Hom}^*_h)}_{(Q_i|\alpha,\Lambda),\infty} \text{ for } \Lambda = (\dot{\Sigma}^{Q_i})^h_{Q_i|\alpha})].$$

We give the well-known proof of the equivalence. Let $\alpha > \xi$ be a cutpoint of Q_i. Via an $\mathbb{R}^*_g$-genericity iteration of Q_i above α, we can find

$$\sigma\colon Q_i \to S$$

and h generic over S for $\mathrm{Col}(\omega,<\sigma(\kappa_i))$ such that

$$L(\mathbb{R}^*_g,\mathrm{Hom}^*_g) = L(\mathbb{R}^*_h,\mathrm{Hom}^*_h).$$

The only slight wrinkle here is that if $\kappa_i < \lambda$, our genericity iteration must weave in infinitely many steps at which we move the image of κ_i up by an extender with that critical point. Note $S|\alpha = Q_i|\alpha$, and the two are assigned the same strategy in their respective pairs. Call that strategy Λ.

We then get that

$$\begin{aligned} L[Q_i|\kappa_i] \models \varphi_0[\xi] &\text{ iff } L[S|\sigma(\kappa_i)] \models \varphi_0[\xi] \\ &\text{ iff } \pi_{(Q_i|\alpha,\Lambda),\infty}(\xi) \in A \\ &\text{ iff } \xi \in A_i, \end{aligned}$$

as desired.

Since $i_{s\restriction k,s\restriction l}$ is elementary, we get that $i_{s\restriction k,s\restriction l}(A_k) = A_l$ whenever $k < l$. This implies that $i_{s\restriction k,\infty}(A_k) = A$ for all k. But then A is definable over $L[M_\infty(s)|i_{s\restriction 0,\infty}(\kappa_0)]$ by the same formula that defined A_0 over $L[Q_0|\kappa_0]$. So $A \in L[M_\infty]$, as desired. ⊣

This completes the proof of Theorem 11.3.2. ⊣

By combining Theorem 11.3.2 with our earlier results on the existence of hod pairs with large cardinals, we get

THEOREM 11.3.13. *Suppose there is* $j\colon V \to M$ *with* $\text{crit}(j) = \kappa$ *and* $V_{j(\kappa)+1} \subseteq M$. *Suppose* $\mathsf{IH}_{\mu,j(\kappa)}$ *hold for some* $\mu < \kappa$, *and that there are* $\lambda < \nu < \kappa$ *such that* λ *is a limit of Woodin cardinals, and* ν *is measurable. Then there is a Wadge cut* Γ *in* $\text{Hom}_{<\lambda}$ *such that* $L(\Gamma,\mathbb{R}) \models \mathsf{AD}_{\mathbb{R}}$, *and*

$$\text{HOD}^{L(\Gamma,\mathbb{R})} \models \mathsf{GCH} + \textit{there is a subcompact cardinal.}$$

PROOF. Under the hypotheses of Theorem 11.3.13, we have shown in Theorem 10.4.6 that there is an lbr hod pair (M,Ψ) with scope HC such that for some λ, $M \models$ "λ is a limit of cutpoint Woodins, and there is a subcompact $< \lambda$." Moreover, we have that Code(Ψ) is $\text{Hom}_{<\lambda}$. So we can apply Theorem 11.3.2, and we get that the HOD of the derived model $D(M, <\lambda)$ is an iterate of M, and satisfies "there is a subcompact cardinal". But then via an $\mathbb{R}$-genericity iteration M-to-M^*, we can realize $D(M^*, <\lambda)$ as $L(\Gamma,\mathbb{R})$, for some $\Gamma \subsetneq \text{Hom}_{<\lambda}$. This proves the theorem. ⊣

11.4. HOD mice satisfy $V = K$

We shall show that if (H,Ω) is an lbr hod pair such that $H \models \mathsf{ZFC}+$ "there are arbitrarily large Woodin cardinals", then in a certain natural sense, $H \models V = K$. This sense derives from the definition of K below one Woodin cardinal that uses thick sets at a regular cardinal, as in [22]. The definition has a generically absolute version, so that in a certain sense, $H = K^{H[g]}$, whenever g is set-generic over H.

Pure extender mice do not in general satisfy even $V = \text{HOD}$, much less $V = K$. The basic problem is that they may not know how to iterate themselves.[11] In this respect, strategy mice are more natural; they know who they are, so to speak.

[11]See [50], [51], and [58] for results on the extent to which $V = K$ and $V = \text{HOD}$ hold in pure extender mice.

DEFINITION 11.4.1. Let α be a regular cardinal, and P be a premouse; then we say P is α^+*-universal* iff

(1) $P \models$ "α is the largest cardinal",
(2) $o(P) = \alpha^+$, and
(3) $\{\eta \mid E^P_\eta \neq \emptyset\}$ is not stationary in α^+.

Of course, P determines α, so we write $\alpha = \alpha^P$. P must be passive, since otherwise $\alpha^+ = o(P) = \mathrm{lh}(\dot{F}^P)$ would be singular. We say that P is universal iff P is α^+-universal, where $\alpha = \alpha^P$. One could make these definitions in the case α is singular, or α is subcompact, but then some complicating cases arise.

DEFINITION 11.4.2. Let P be universal, and $\alpha = \alpha^P$. Then

(1) Γ is *thick* iff there is an α-club set $C \subseteq \alpha^+$ such that $C \subseteq \Gamma$.
(2) $\mathrm{Def}^P = \bigcap\{\mathrm{Hull}^P_\omega(\Gamma) \mid \Gamma \text{ is thick}\}$.
(3) P is *very sound* iff $P = \mathrm{Def}^P$.

It is easy to see that P is very sound iff $\alpha^P \subseteq \mathrm{Def}^P$.

The following is a uniqueness lemma for very sound hod pairs. We shall formulate it as a first order fact about least branch hod mice. One could abstract the first order properties of such mice that we shall use in its proof, but we are not going to do that.

LEMMA 11.4.3. (AD^+) *For any lbr hod pair* (W, Ψ) *with scope* HC, *the following is true in* W: *whenever* (P, Σ) *and* (Q, Λ) *are lbr hod pairs with scope* H_λ, *where* λ *is a limit of Woodin cardinals, and* P *and* Q *are* α^+*-universal and very sound, where* $\alpha < \lambda$, *then* $(P, \Sigma) = (Q, \Lambda)$.

PROOF. We work inside W. Let Σ_0 and Λ_0 be the restrictions of Σ and Λ to V_δ, where $\alpha < \delta < \lambda$ and δ is Woodin. We show that $(P, \Sigma_0) = (Q, \Lambda_0)$, and since δ is arbitrary, this is enough. Let w be the order of construction on V_δ, and $\mathcal{F}$ be the set of nice $E \restriction \eta$ such that E is on the $W|\delta$ sequence. Let Ψ_0 be the $\mathcal{F}$-iteration strategy determined by Ψ, and let $\mathbb{C}$ be the maximal $(w\mathcal{F}, \Psi_0)$-construction. $\mathbb{C}$ is good at all $\nu, k\rangle$ and δ is Woodin, so $\mathbb{C}$ reaches non-dropping iterates of (P, Σ_0) and (Q, Λ_0). Let (M, Ω) be the first pair in $\mathbb{C}$ that is a non-dropping iterate of one of these two, and assume without loss of generality that it is (P, Σ_0) that iterates to (M, Ω), while (Q, Λ_0) iterates past (M, Ω), perhaps not strictly.

Let $\mathcal{T}$ be the λ-separated tree on P with last model M, and

$$i \colon P \to M$$

the canonical embedding. i is given by an extender all of whose measures concentrate on bounded subsets of α, so i is continuous at points of cofinality α. It follows that $\mathrm{ran}(i)$ is α-club in $o(M)$. Let $\mathcal{U}$ be the λ-separated tree whereby Q iterates past M, and let

$$R = \mathcal{M}^{\mathcal{U}}_\theta$$

be the first model in $\mathcal{U}$ such that

$$M = R||o(M).$$

CLAIM 11.4.4. *Let* $[0,\eta]_U \cap D^{\mathcal{U}} =$, *and let* $\nu < o(\mathcal{M}_\eta^{\mathcal{U}})$ *be a successor cardinal of* $\mathcal{M}_\eta^{\mathcal{U}}$ *such that* $\text{lh}(E_\xi^{\mathcal{U}}) \leq \nu$ *for all* $\xi < \eta$*; then in* V, $\text{cof}(\nu) \leq \alpha$.

PROOF. Let $N = \mathcal{M}_\eta^{\mathcal{U}}$ and $j = i_{0,\eta}^{\mathcal{U}}$. j is continuous at α since α is regular in Q and not measurable in Q, and $\nu \leq j(\alpha)$. Let

$$\varepsilon = \sup\{\hat{\lambda}(E_\xi^{\mathcal{U}}) + 1 \mid \xi < \eta\};$$

then $\varepsilon < \nu$ by our hypothesis. Since the generators of j are contained in ε,

$$\nu \subseteq \{j(f)(a) \mid f \in Q|\alpha \wedge a \in [\varepsilon]^{<\omega}\}.$$

For each $f \in Q|\alpha$, let

$$\gamma_f = \sup\{j(f)(a) \mid a \in [\varepsilon]^{<\omega} \wedge j(f)(a) < \nu\}.$$

Since ν is regular in N, $\gamma_f < \nu$ for all $f \in Q|\alpha$. But ν is the sup over f of the γ_f, so the V cofinality of ν is $\leq \alpha$. ⊣

CLAIM 11.4.5. *Q-to-R does not drop.*

PROOF. Suppose otherwise. Since $i(\alpha)$ is the largest cardinal of M, $\rho(R) \leq i(\alpha)$. We would like to apply the condensation theorem of [81]. This will give us a club of collapsing structures $N \lhd M$ with limit R that we can pull back to P via i, and then fit together into a collapsing structure for P.[12] We just need to take a little care to avoid the protomouse case in the condensation proof. Let us say that R is *problematic* iff R is extender active, $k(R) = 0$, and $\text{crit}(\dot{F}^R) \leq i(\alpha)$.

Suppose that R is not problematic. Assume first that $k(R) = 0$ and R is passive, and let $p = p_1(R)$. It is not hard to see that $\text{cof}(o(R)) = \text{cof}(o(M)) = \alpha^+$. For $\xi < o(M)$ let γ_ξ be least such that $\xi \subseteq h_{R||\gamma_\xi}$"$(i(\alpha) \cup p)$. Let

$$N_\xi = \text{transitive collapse of } h_{R||\gamma_\xi}\text{“}(i(\alpha) \cup p),$$

and let

$$\pi_\xi : N_\xi \to R||\gamma_\xi$$

be the anticollapse. Letting $\tau_\xi = \text{crit}(\pi_\xi)$, it is easy to see that $\pi_\xi(\tau_\xi) = o(M)$, and $N_\xi \in M$. In fact, if τ_ξ is not an index on the M-sequence, then by [81],

$$N_\xi \trianglelefteq M.$$

By the non-subcompactness clause of universality, we have an α-club $C \subseteq \text{ran}(i)$ such that for all $\xi \in C$, $\xi = \tau_\xi$ and $N_\xi \trianglelefteq M$. For $\xi \in C$, $\rho(N_\xi) = \rho_1(N_\xi) = i(\alpha)$, and

$$p(N_\xi) = \pi_\xi^{-1}(p).$$

[12]This is just a variant of Solovay's proof that $\square_\kappa$ fails when κ is strongly compact.

For $\xi < \eta$ with both in C, we have a natural

$$\sigma_{\xi,\eta}: N_\xi \to N_\eta,$$

determined by $\sigma_{\xi,\eta} \restriction \tau_\xi = \text{id}$, and $\sigma_{\xi,\eta}(p(N_\xi)) = p(N_\eta)$. The full R is just the direct limit of the N_ξ, for $\xi \in C$, under the $\sigma_{\xi,\eta}$.

Now we pull back to P. Let D be an α-club in $o(P)$ such that $i``D \subseteq C$. For $\xi \in D$, let $M_\xi \trianglelefteq P$ be such that

$$i(J_\xi) = N_{i(\xi)},$$

and let

$$\varphi_{\xi,\eta}: J_\xi \to J_\eta$$

be given by $\varphi_{\xi,\eta} = i^{-1} \circ \sigma_{\xi,\eta} \circ i$. Note here that $N_{i(\xi)}$ is definable from $i(\xi)$ as the first level of Q collapsing $i(\xi)$ to $i(\alpha)$, so $N_{i(\xi)} \in \text{ran}(i)$, and J_ξ is the first level of P collapsing ξ to α. Note also that

$$\varphi_{\xi,\eta}(p(J_\xi)) = p(J_\eta).$$

Letting J be the direct limit of the J_ξ and r be the common value of $\varphi_{\xi,\infty}(p(J_\xi))$, for ξ in D, we see that $h_J``(\alpha \cup r)$ contains α^+, which is a contradiction.

If $k(R) = 0$ but R is active, the proof is similar. Letting $G = \dot{F}^R$, we define $\eta_\xi < \text{dom}(G)$ and $\gamma_\xi < o(R)$ cofinal in $\text{dom}(G)$ and $o(R)$ so that $i_G``\eta_\xi$ is cofinal in γ_ξ. Letting G_ξ be the fragment of G that represents $i_G \restriction R||\eta_\xi$, we replace $R||\gamma_\xi$ in the argument above by

$$S_\xi = (R||\gamma_\xi, G_\xi),$$

and let

$$N_\xi = \text{cHull}_1^{S_\xi}(i(\alpha) \cup p(R)).$$

Letting $\pi_\xi : N_\xi \to S_\xi$ be the anticollapse, there are club many ξ such that $\text{ran}(\pi_\xi) \cap \text{dom}(G)$ has supremum η_ξ. This is precisely where we use that $\text{crit}(G) > i(\alpha)$, and it is what guarantees that for such ξ, N_ξ is a premouse, rather than just a protomouse.[13] The rest of the argument goes as in the case that R is passive.

If $k(R) > 0$, then again the proof is similar, again using the condensation result of [81]. We use the fact that $r\Sigma_{k+1}$ over R is the same as Σ_1 over the mastercode structure R^k to find our approximations N_ξ to R. Since $k(R) > 0$, there is enough elementarity that protomice do not arise.

Finally, let us turn to the problematic case. Let $F = \dot{F}^R$. $o(M)$, $o(R) = \text{lh}(F)$, and $\text{dom}(F)$ have the same V-cofinality, namely α^+. Let

$$\mathcal{U}_1 = \mathcal{U} \restriction (\xi + 1)^\frown \langle F^+ \rangle$$

[13] That is, it guarantees that $\dot{F}^{N_\xi}$ measures all sets in N_ξ.

be the normal extension of $\mathcal{U} \restriction \xi + 1$ by F^+. Let $\beta = U_1\text{-pred}(\xi + 1)$ and

$$R_1 = \mathcal{M}^{\mathcal{U}_1}_{\xi+1}$$
$$= \text{Ult}(S_1, F),$$

where $S \trianglelefteq \mathcal{M}^{\mathcal{U}}_\beta$ is what F is applied to. It cannot be the case that $[0, \beta]_U \cap D^{\mathcal{U}} = \emptyset$ and $S = \mathcal{M}^{\mathcal{U}}_\beta$, for then by Claim 11.4.4, $\text{cof}^V(\text{dom}(F)) \leq \alpha$. It follows that

$$\rho(R_1) \leq \text{crit}(F).$$

If R_1 is not problematic, then we reach a contradiction just as above.[14] If R_1 is problematic, let F_1 be its last extender, and

$$\mathcal{U}_2 = \mathcal{U}_1^\frown \langle F_1^+ \rangle$$

be the corresponding normal extension, and R_2 its last model. Note that $F_1 = i_F(\dot{F}^S)$, so $\text{crit}(F_1) \leq o(M)$ implies $\text{crit}(F_1) < \text{crit}(F)$. If R_2 is not problematic, we are done, and otherwise its last extender F_2 is such that $\text{crit}(F_2) < \text{crit}(F_1)$. And so on. Eventually we reach an R_n that is not problematic, and a contradiction. $\dashv$

Now let $j = i^{\mathcal{U}}_{0,\xi}$ be the iteration map from Q to R. j is continuous at α, because α is regular but not measurable in Q.

We claim that $M = R$. For if not, $j(\alpha) \geq o(M)$. But then $o(M)$ is a successor cardinal in R that is $< o(R)$, and the generators of j are bounded in $o(M)$, so $\text{cof}(o(M)) \leq \alpha$ by Claim 11.4.4.

So $M = R$, and thus $i(\alpha) = j(\alpha)$. By the continuity of i and j at points of cofinality α, we have an α-club set $C \subseteq \alpha^+$ such that $i(\xi) = j(\xi)$ for all $\xi \in C$. This implies that $\text{Hull}^P(C)$ is isomorphic to $\text{Hull}^Q(C)$. By very soundness, $P = \text{Hull}^P(C)$ and $Q = \text{Hull}^Q(C)$. So $P = Q$, and then $i = j$ because the two agree on the generating set C. But then $\Sigma_0 = \Omega^i = \Omega^j = \Lambda_0$. $\dashv$

LEMMA 11.4.6. *(AD^+) Let (W, Ψ) be an lbr hod pair with scope* HC. *Working inside W, let α be regular but not subcompact, and $\alpha < \lambda$, where λ is a limit of Woodin cardinals. Let $P = W|\alpha^+$ and $\Sigma = \Psi_P$; then (P, Σ) is very sound.*

PROOF. We work in W. The proof of $\diamondsuit$ yields a sequence $\langle S_\gamma \mid \gamma < \alpha^+ \rangle$ that is definable over P (in fact, Σ_1 definable), and such that $S_\gamma \subseteq \gamma$ for all γ, and for all $A \subseteq o(P)$, there are stationarily many γ such that $\text{cof}(\gamma) = \alpha$ and $S_\gamma = A \cap \gamma$. Now let

$$\pi \colon \text{cHull}^P_\omega(\Gamma) \to P$$

[14] In the proof with R, we had that all of R was generated by $i(\alpha) \cup p(R)$. But all we really needed was that $o(M)$ is generated this way, and that still holds for R_1. This is because S is Σ_1 generated by $\text{crit}(F) \cup p(S)$, so R_1 is Σ_1 generated by $p(R_1) = i_F(p(S))$ together with the Dodd parameter and projectum of F. But the Dodd projectum of F is $\leq i(\alpha)$. So R_1 is Σ_1 generated by $i(\alpha) \cup q$, for some finite q. That is what we need.

be the anticollapse, where Γ is thick. There is an α-club C such that $\pi(\gamma) = \gamma$ for all $\gamma \in C$. But then for any $\beta < \alpha$ we can find $\gamma \in C$ such that $S_\gamma = \{\beta\}$. Since $\gamma \in \operatorname{ran}(\pi)$, $\beta \in \operatorname{ran}(\pi)$. Thus $\alpha \subseteq \operatorname{ran}(\pi)$, so π is the identity, as desired.[15] ⊣

The following definition is meant to be employed inside hod mice satisfying ZFC and having arbitrarily large Woodin cardinals and their set generic extensions.

DEFINITION 11.4.7. Let P be a least branch premouse and α be a cardinal; then we say P is *K-like at α* iff P is α^+-universal and very sound, and for δ the least Woodin cardinal $> \alpha$, there is a Σ such that (P, Σ) is an lbr hod pair with scope H_{δ^+}.

THEOREM 11.4.8 (AD^+). *Let (H, Ω) be an lbr hod pair with scope* HC, *and suppose $H \models \mathsf{ZFC} +$ "there are arbitrarily large Woodin cardinals". Let g be generic over H for a poset of size $< \nu$ in H, and let α be a successor cardinal of H above ν; then in $H[g]$, the following are equivalent:*

(1) P is K-like at α,
(2) $P = H|\alpha^+$.

PROOF. Lemmas 11.4.3 and 11.4.6 show that the equivalence is true in H itself. In $H[g]$ we must work above the size of the forcing.[16] We leave it to the reader to think through that case. ⊣

DEFINITION 11.4.9. We say that K^ν *exists* iff

(1) for every successor cardinal $\alpha > \nu$, there is a unique lpm $K^\nu(\alpha)$ such that $K^\nu(\alpha)$ is K-like at α, and
(2) if $\nu < \alpha < \beta$ and α, β are successor cardinals, then $K^\nu(\alpha) \trianglelefteq K^\nu(\beta)$.

If K^ν exists, then we set $K^\nu = \bigcup_\alpha K^\nu(\alpha)$.

COROLLARY 11.4.10. *Under the hypotheses of Theorem 11.4.8,*

(a) $H \models \forall \nu (V = K^\nu)$, and
(b) if g is H-generic for a poset of size $< \kappa$, then $H = (K^\nu)^{H[g]}$, for all $\nu \geq \kappa$.

So a hod mouse H as in the theorem satisfies $V =$ HOD, and in fact, it is the generic HOD, or gHOD, of its generic multiverse.[17] This should be compared with

THEOREM 11.4.11 (Woodin [85]). *Assume $\mathsf{AD}_\mathbb{R} + V = L(P(\mathbb{R}))$; then* HOD $\models$ $V =$ HOD, *and* HOD$|\theta$ *is the generic* HOD *of its own generic multiverse.*

[15]This proof is due to Gabriel Goldberg. We originally had a more complicated one that required the hypothesis that $W \models$ "there are no 1-extendible cardinals".

[16]This is provably necessary, because of the local nature of "K-like at α". If $\delta_0 < \delta_1$ are Woodin cardinals, and if $j \colon H \to M \subseteq H[g]$ comes from a $\mathbb{P}_{\delta_1}$-stationary tower ultrapower, it will be initial segments of M that are K-like at $\alpha < j(\delta_0)$ in the sense of $H[g]$.

[17]See [12].

This result is significantly more general than what we have proved, in that it applies to $\mathsf{AD}_{\mathbb{R}}$ models that have iteration strategies for mice with long extenders, and are therefore beyond the HOD analysis we have developed here. Our proof that HOD $\models V =$ HOD does have extra information in it, in the short-extender region to which it applies.

11.5. Further results

Our analysis of HOD in the derived model D of a HOD mouse was based on the fact that $D \models \mathsf{HPC}$. (This was the content of the first two claims in the proof of Theorem 11.3.2.) We used further facts about the way we had derived D, but with more work, one can avoid an appeal to them. Thus we get

THEOREM 11.5.1 ([73]). *Assume* $\mathsf{AD}_{\mathbb{R}}$ *and* HPC; *then* $V_\theta \cap \mathrm{HOD}$ *is the universe of a least branch premouse.*

Concerning the mouse capturing hypothesis of this theorem, we have

THEOREM 11.5.2 ([73]). *Assume* AD^+; *then*

(a) if HPC *holds, then for any* $\Gamma \subseteq P(\mathbb{R})$, $L(\Gamma, \mathbb{R}) \models \mathsf{HPC}$, *and*

(b) if LEC *holds, then for any* $\Gamma \subseteq P(\mathbb{R})$, $L(\Gamma, \mathbb{R}) \models \mathsf{LEC}$, *and*

(c) if there is an ω_1 *iteration strategy for a countable pure extender premouse with a long extender on its sequence, then for any* $\Gamma \subseteq P(\mathbb{R})$ *such that* $L(\Gamma, \mathbb{R}) \models \mathsf{NLE}$, *we have* $L(\Gamma, \mathbb{R}) \models \mathsf{LEC}$, *and hence* $L(\Gamma, \mathbb{R}) \models \mathsf{HPC}$.

Part (c) is pretty strong evidence that $\mathsf{AD}^+ + \mathsf{NLE}$ implies LEC, and hence HPC. Whether this is in fact true is perhaps the main open problem in the theory to which this book contributes. Parts (a) and (b) suggest that one ought to try to prove this via an induction on the Wadge hierarchy, and that is a natural thing to try on other counts, too. There are partial results in this direction, but the situation is in sufficient flux that it seems wisest not to attempt a discussion of them.

The proof of Theorem 11.5.1 gives a characterization of the Solovay sequence in terms of the Woodin cardinals in HOD.

DEFINITION 11.5.3. For any set X, $\theta(X)$ is the least ordinal α such that there is no ordinal definable surjection of X onto α.

If there is an ordinal definable map from X onto $X \times X$, then $\theta(X)$ is the supremum of the surjective images of X under maps that are ordinal definable from some parameter in X. This is our case of interest.

DEFINITION 11.5.4 (AD^+). The *Solovay sequence* $\langle \theta_\alpha \mid \alpha \leq \Omega \rangle$ is given by

$$\theta_0 = \theta(\mathbb{R}),$$

and if $\theta_\alpha < \theta$, then

$$\theta_{\alpha+1} = \theta(\mathbb{R} \cup \{A\}), \text{ for any (all) } A \text{ of Wadge rank } \theta_\alpha,$$
$$\theta_\lambda = \bigcup_{\alpha<\lambda} \theta_\alpha.$$

Ω is the least β such that $\theta_\beta = \theta$.

Assuming AD^+, if $\theta_\alpha < \theta$, then

$$\theta_{\alpha+1} = \theta(P(\theta_\alpha)).$$

This is easy to see, using the Coding Lemma and the fact that every set of reals of Wadge rank θ_α is θ_α-Suslin. The Solovay sequence is an important feature of any model AD^+, one that is tied to the pattern of scales in the model. It is definable, so it is in HOD. In fact, it has a natural identity within HOD.

Assume $\mathsf{AD}_\mathbb{R} + \mathsf{HPC}$. The proof of Theorem 11.5.1 then gives a canonical least branch premouse $\mathcal{H}$ whose universe is V_θ^{HOD}. We have shown in the last section that in fact $\mathcal{H}$ is definable over $(V_\theta^{\mathrm{HOD}}, \in)$, as the union of all universal, very sound premice. Let us say that δ is a *cutpoint of* HOD iff δ is a cutpoint of $\mathcal{H}$, in the sense that there is no extender E on the $\mathcal{H}$-sequence such that $\mathrm{crit}(E) < \delta \leq \mathrm{lh}(E)$.[18] It is easy to see that if δ is Woodin and a cutpoint of HOD, then there are no extenders on the $\mathcal{H}$-sequence with critical point δ.

THEOREM 11.5.5 ([73]). *Assume* $\mathsf{AD}_\mathbb{R} + V = L(P(\mathbb{R})) + \mathsf{HPC}$; *then the following are equivalent:*

(1) δ *is a cutpoint Woodin cardinal of* HOD,
(2) $\delta = \theta_0$, *or* $\delta = \theta_{\alpha+1}$ *for some* α.

In particular, θ_0 *is the least Woodin cardinal in* HOD.

That θ_0 and the $\theta_{\alpha+1}$ are Woodin in HOD is due to Woodin, cf. [26]. Woodin also proved an approximation to the statement that they are cutpoints of HOD (unpublished). The rest of (2) $\rightarrow$ (1), and all of (1) $\rightarrow$ (2), comes from [73].

One can characterize the next Woodin cardinal of HOD in terms of a modified Solovay sequence. The following definition is due to Grigor Sargsyan.[19]

DEFINITION 11.5.6. Assume AD^+. We set

$$\eta_0 = \theta({}^\omega\omega) = \theta_0,$$
$$\eta_{\alpha+1} = \theta({}^\omega\kappa), \text{ where } \kappa = (\eta_\alpha)^{+,\mathrm{HOD}},$$
$$\eta_\lambda = \bigcup_{\alpha<\lambda} \eta_\alpha.$$

One can show

[18]Presumably, every extender in HOD that coheres with the $\mathcal{H}$-sequence is actually on that sequence, but no one has actually proved this, so far as we know.

[19]One might call this the *Sargsyan sequence*.

THEOREM 11.5.7 ([73]). *Assume* $\mathsf{AD}_{\mathbb{R}} + \mathsf{HPC}$*; then for any* $\delta < \theta$*,* δ *is a successor Woodin cardinal of* HOD *iff* $\delta = \eta_{\alpha+1}$ *for some* α.

Of course, "successor Woodin" means "least Woodin above some ordinal". The Sargsyan sequence may grow more slowly than the Solovay sequence. Assuming $\mathsf{AD}_{\mathbb{R}} + \mathsf{HPC}$, Theorem 11.5.7 implies that this happens if and only if HOD has extenders overlapping Woodin cardinals.

It is also interesting to see what strong determinacy theories are true in the derived models of lbr hod pairs (P,Σ) such that P reaches reasonably large cardinals. There are some results in this direction in [73].

The key to the theorems above is an analysis of optimal Suslin representations for mouse pairs. That in turn rests on a strengthening of strong hull condensation that [64] calls *very strong hull condensation*. Roughly speaking, this property amounts to condensation under *weak tree embeddings*, a more general kind of tree embedding than the kind we have defined in Definition 6.4.1.[20] [64] shows

THEOREM 11.5.8 ([64]). *Assume* AD^+*, and let* (P,Σ) *be a mouse pair with scope* HC*; then* Σ *has very strong hull condensation.*

Given a stack $\langle \mathcal{T},\mathcal{U}\rangle$ on P with last model Q, there is a natural attempt $X(\mathcal{T},\mathcal{U})$ at a normal tree on P with last model Q. We say that Σ *fully normalizes well* iff whenever $\langle \mathcal{T},\mathcal{U}\rangle$ is by Σ, then $X(\mathcal{T},\mathcal{U})$ exists and is by Σ, and $\Sigma_{\langle \mathcal{T},\mathcal{U}\rangle} = \Sigma_{X(\mathcal{T},\mathcal{U})}$. (See [64].) The construction of $X(\mathcal{T},\mathcal{U})$ produces a weak tree embedding from $X(\mathcal{T},\mathcal{U})$ into $W(\mathcal{T},\mathcal{U})$. Thus Theorem 11.5.8 yields

COROLLARY 11.5.9 ([64]). *Assume* AD^+*, and let* (P,Σ) *be a mouse pair with scope* HC*; then*

(a) Σ *fully normalizes well, and*
(b) Σ *is positional.*

From the proof of Corollary 11.5.9 we obtain a λ-separated tree $\mathcal{U}(P,\Sigma)$ on P that has last model $M_\infty(P,\Sigma)$, and is such that all its countable weak hulls are by Σ. This then gives us a Suslin representation for the fragment of Σ that is actually used in forming $M_\infty(P,\Sigma)$: to justify a countable tree $\mathcal{T}$ on P, we search for a weak tree embedding of $\mathcal{T}$ into $\mathcal{U}(P,\Sigma)$.

Not all of Σ is actually used in forming $M_\infty(P,\Sigma)$. Let us call a λ-separated tree $\mathcal{T}$ *relevant* iff $\mathcal{T}$ is by Σ, and there is a λ-separated $\mathcal{S}$ by Σ such that $\mathcal{T} \subseteq \mathcal{S}$, and $\mathcal{S}$ has a last model Q, and the branch P-to-Q does not drop. Call a P-stack s relevant if for $i+1 < \mathrm{dom}(s)$, the branch of $\mathcal{T}_i(s)$ to $M_\infty(\mathcal{T}_i(s))$ does not drop, and for $i+1 = \mathrm{dom}(s)$, $\mathcal{T}_i(s)$ is relevant. Let Σ^{rel} be the restriction of Σ to relevant trees. The Σ-iterations that go into forming $M_\infty(P,\Sigma)$ are all relevant, so Σ^{rel} is

[20] In a weak tree embedding, the connection between exit extenders required by Definition 6.4.1(d) is loosened. Rather than require that $t_\alpha(E_\alpha^{\mathcal{T}}) = E_{u(\alpha)}^{\mathcal{U}}$, we require that $t_\alpha(E_\alpha^{\mathcal{T}})$ be connected to $E_{u(\alpha)}^{\mathcal{U}}$ inside $\mathcal{M}_{u(\alpha)}^{\mathcal{U}}$ via a sequence of fine structural hulls. This sequence of hulls is an abstract version of the sequence that occurred in Claim 6.1.8 of our proof of full normalizability of trees of length two.

what we need to construct $M_\infty(P,\Sigma)$ and $\mathcal{U}(P,\Sigma)$. Moreover, $\mathcal{U}(P,\Sigma)$ acts as a kind of universal tree by Σ^{rel}, in that all countable trees by Σ^{rel} can be weakly embedded into it. This leads to

THEOREM 11.5.10 ([73]). *Assume* AD^+, *and let* (P,Σ) *be a mouse pair with scope* HC. *Let* κ *be the cardinality of* $o(M_\infty(P,\Sigma))$, *and let* $\text{Code}(\Sigma^{\text{rel}})$ *be the set of reals coding stacks by* Σ^{rel}; *then*

(a) $\text{Code}(\Sigma^{\text{rel}})$ *and its complement are* κ*-Suslin, and*

(b) $\text{Code}(\Sigma^{\text{rel}})$ *is not* α*-Suslin, for any* $\alpha < \kappa$.

In particular, κ *is a Suslin cardinal.*

Part (b) of the Theorem 11.5.10 follows at once from the Kunen-Martin theorem, and the fact that there is a wellfounded relation W on $\mathbb{R}$ of rank at least $o(M_\infty(P,\Sigma))$ such that W is arithmetic in $\text{Code}(\Sigma^{\text{rel}})$. [Let $(t,b)W(s,a)$ iff s and t are stacks by Σ^{rel} with last models M and N, $s \subseteq t$, P-to-N does not drop, and $i^t_{M,N}(a) > b$.]

The one can show the irrelevant part of Σ is also Suslin, although it may not be $o(M_\infty(P,\Sigma))$-Suslin. (It is possible that $M_\infty(P,\Sigma) = P$, because there are no non-dropping iterations of P!) So one gets

THEOREM 11.5.11 ([73]). *Assume* AD^+, *and let* (P,Σ) *be a mouse pair with scope* HC. *and let* $\text{Code}(\Sigma)$ *be the set of reals coding stacks by* Σ; *then* $\text{Code}(\Sigma)$ *and its complement are Suslin.*

Note here that since Σ is total on stacks by Σ, if $\text{Code}(\Sigma)$ is β-Suslin, then so is its complement.

Theorem 11.5.10 implies that $|o(M_\infty(P,\Sigma))|$ is a Suslin cardinal.[21] With more work along the same lines, one can show that for any cutpoint τ of $M_\infty(P,\Sigma)$, $|\tau|$ is a Suslin cardinal. In recent unpublished work, S. Jackson and G. Sargsyan have shown that all Suslin cardinals below $o(M_\infty(P,\Sigma))$ arise this way.[22] So we have

THEOREM 11.5.12 (Jackson, Sargsyan, S.). *Assume* AD^+, *let* (P,Σ) *be a mouse pair, and let* $\kappa \leq o(M_\infty(P,\Sigma))$. *The following are equivalent:*

(a) κ *is a Suslin cardinal,*

(b) $\kappa = |\tau|$, *where* τ *is a cutpoint of* $M_\infty(P,\Sigma)$ *or* $\tau = o(M_\infty(P,\Sigma))$.

The proof that (a) implies (b) by Jackson and Sargsyan shows that if κ is a regular Suslin cardinal, then κ itself is a cutpoint of $M_\infty(P,\Sigma)$. It is open whether that is also true for the other Suslin cardinals, the problematic case being when κ is the next Suslin cardinal after some regular Suslin cardinal.

Assuming $\mathsf{AD}^+ + \mathsf{HPC}$, we get at once from Theorem 11.5.12 that for any κ, κ is a Suslin cardinal iff $\kappa = |\tau|$, for some cutpoint τ of the distinguished extender sequence of HOD.

---

[21] κ is a Suslin cardinal iff there is a set A of reals such that A is κ-Suslin, but not α-Suslin for any $\alpha < \kappa$.

[22] See [17].

The correspondence between iteration strategies and definable scales is central to descriptive inner model theory. Theorem 11.5.12 captures one aspect of it.

REFERENCES

[1] ALESSANDRO ANDRETTA, ITAY NEEMAN, and JOHN STEEL, *The domestic levels of K^c are iterable*, ***Israel Journal of Mathematics***, vol. 125 (2001), pp. 157–201.

[2] STEWART BALDWIN, *Generalizing the Mahlo hierarchy, with applications to the Mitchell models*, ***Annals of Pure and Applied Logic***, vol. 25 (1983), pp. 103–127.

[3] ———, *Between strong and superstrong*, ***The Journal of Symbolic Logic***, vol. 51 (1986), pp. 547–559.

[4] ANTHONY J. DODD, *Strong cardinals*, Unpublished notes, circa 1982.

[5] ANTHONY J. DODD and RONALD B. JENSEN, *The core model*, ***Annals of Mathematical Logic***, vol. 20 (1981), pp. 43–75.

[6] ———, *The covering lemma for K*, ***Annals of Mathematical Logic***, vol. 22 (1982), pp. 1–30.

[7] ———, *The covering lemma for L^U*, ***Annals of Mathematical Logic***, vol. 22 (1982), pp. 127–155.

[8] ILIJAS FARAH, *The extender algebra and Σ^2_1 absoluteness*, ***Large Cardinals, Determinacy, and Other Topics. The Cabal Seminar, Volume IV*** (A. S. Kechris et al., editors), Lecture Notes in Logic, vol. 43, Association for Symbolic Logic, Ithaca, New York, 2020, pp. 155–192.

[9] QI FENG, MENACHEM MAGIDOR, and W. HUGH WOODIN, *Universally Baire sets of reals*, ***Set Theory of the Continuum (Berkeley, CA, 1989)*** (New York), Mathematical Sciences Research Institute Publications, vol. 26, Springer, 1992, pp. 208–242.

[10] GUNTER FUCHS, *λ-structures and s-structures: translating the iteration strategies*, ***Annals of Pure and Applied Logic***, vol. 162 (2011), pp. 710–751.

[11] ———, *λ-structures and s-structures: translating the models*, ***Annals of Pure and Applied Logic***, vol. 162 (2011), pp. 257–317.

[12] GUNTER FUCHS, JOEL HAMKINS, and JONAS RIETZ, *Set-theoretic geology*, ***Annals of Pure and Applied Logic***, vol. 166 (2015), pp. 464–501.

[13] GUNTER FUCHS, ITAY NEEMAN, and RALF SCHINDLER, *A criterion for coarse iterability*, ***Archive for Mathematical Logic***, vol. 49 (2010), pp. 447–467.

[14] KURT F. GÖDEL, ***The Consistency of the Axiom of Choice and the Generalized Continuum Hypothesis with the Axioms of Set Theory***, Annals of Mathematical Studies, vol. 3, Princeton University Press, Princeton, New Jersey, 1940.

[15] LEO A. HARRINGTON and ALEXANDER S. KECHRIS, *On the determinacy of games on ordinals*, ***Annals of Mathematical Logic***, vol. 20 (1981), pp. 109–154.

[16] GREGORY HJORTH, *A boundedness lemma for iterations*, ***The Journal of Symbolic Logic***, vol. 66 (2001), pp. 1058–1072.

[17] STEPHEN JACKSON, GRIGOR SARGSYAN, and JOHN STEEL, *Suslin cardinals and cutpoints in mouse limits*, https://arxiv.org/abs/2207.04042, (2022).

[18] RONALD B. JENSEN, *The fine structure of the constructible hierarchy*, ***Annals of Mathematical Logic***, vol. 4 (1972), pp. 229–308.

[19] ———, *A new fine structure for higher core models*, (1997), manuscript available at https://www.mathematik.hu-berlin.de/~raesch/org/jensen.html.

[20] ———, *Addendum to "A new fine structure for higher core models"*, (1998), manuscript available at https://www.mathematik.hu-berlin.de/~raesch/org/jensen.html.

[21] ———, *Smooth iterations*, (2018), preprint available at https://www.mathematik.hu-berlin.de/~raesch/org/jensen.html.

[22] RONALD B. JENSEN and JOHN R. STEEL, *K without the measurable*, ***The Journal of Symbolic Logic***, vol. 73 (2013), pp. 708–733.

[23] ALEXANDER S. KECHRIS, *Countable ordinals and the analytical hierarchy*, ***Pacific Journal of Mathematics***, vol. 60 (1975), pp. 223–227.

[24] ———, *The theory of countable analytical sets*, ***Transactions of the American Mathematical Society***, vol. 202 (1975), pp. 259–297.

[25] RICHARD KETCHERSID, ***Toward*** $\mathsf{AD}_{\mathbb{R}}$ ***from the Continuum Hypothesis and an*** ω_1***-dense ideal***, Ph.D. thesis, Berkeley, 2000.

[26] PETER KOELLNER and W. HUGH WOODIN, *Large cardinals from determinacy*, ***Handbook of Set Theory, Vol. 3*** (M. Foreman and A. Kanamori, editors), Springer-Verlag, 2010, pp. 1951–2120.

[27] KENNETH KUNEN, *Some applications of iterated ultrapowers in set theory*, ***Annals of Mathematical Logic***, vol. 1 (1970), pp. 179–227.

[28] PAUL LARSON, ***The Stationary Tower: Notes on a Course by Hugh Woodin***, University Lecture Series, vol. 32, American Mathematical Society, Providence, RI, 2004.

[29] DONALD A. MARTIN and JOHN R. STEEL, *A proof of projective determinacy*, ***Journal of the American Mathematical Society***, vol. 2 (1989), pp. 71–126.

[30] ———, *Iteration trees*, ***Journal of the American Mathematical Society***, vol. 7 (1994), pp. 1–73.

[31] WILLIAM J. MITCHELL, *Sets constructible from sequences of ultrafilters*, ***The Journal of Symbolic Logic***, vol. 39 (1974), pp. 57–66.

[32] ———, *Hypermeasurable cardinals*, ***Logic Colloquium '78 (Mons 1978)*** (Amsterdam – New York) (M. Boffa et al., editors), Studies in Logic and the Foundations of Mathematics, North Holland, 1979, pp. 303–316.

[33] ———, *Coherent sequences of measures: revisited*, ***The Journal of Symbolic Logic***, vol. 48 (1983), pp. 600–609.

[34] WILLIAM J. MITCHELL and RALF D. SCHINDLER, *A universal extender model without large cardinals in V*, ***The Journal of Symbolic Logic***, vol. 69 (2004), pp. 371–386.

[35] WILLIAM J. MITCHELL and JOHN R. STEEL, ***Fine Structure and Iteration Trees***, Lecture Notes in Logic, vol. 3, Springer-Verlag, Berlin, 1994.

[36] JOHN MYHILL and DANA SCOTT, *Ordinal definability*, ***Axiomatic Set Theory, Proceedings of Symposia in Pure Mathematics, Vol. XIII, Part I, UCLA, (1967)*** (Providence, R.I.), American Mathematical Society, 1971, pp. 271–278.

[37] ITAY NEEMAN, *Inner models in the region of a Woodin limit of Woodin cardinals*, ***Annals of Pure and Applied Logic***, vol. 116 (2002), pp. 67–155.

[38] ———, *Determinacy in* $L(\mathbb{R})$, ***Handbook of Set Theory, Vols. 1, 2, 3*** (M. Foreman and A. Kanamori, editors), Springer, Dordrecht, 2010, pp. 1877–1950.

[39] ITAY NEEMAN and JOHN STEEL, *A weak Dodd-Jensen lemma*, ***The Journal of Symbolic Logic***, vol. 64 (1999), pp. 1285–1294.

[40] ———, *Counterexamples to the unique and cofinal branches hypotheses*, ***The Journal of Symbolic Logic***, vol. 71 (2006), pp. 977–988.

[41] ———, *Equiconsistencies at subcompact cardinals*, ***Archive for Mathematical Logic***, vol. 55 (2016), pp. 207–238.

[42] GRIGOR SARGSYAN, ***A Tale of Hybrid Mice***, Ph.D. thesis, UC Berkeley, Ann Arbor, MI, 2009.

[43] ———, ***Hod Mice and the Mouse Set Conjecture***, Memoirs of the American Mathematical Society, vol. 236, American Mathematical Society, Providence, RI, 2015.

[44] GRIGOR SARGSYAN and NAM TRANG, *The largest Suslin axiom*, (2016), preprint.

[45] KENNETH SCHILLING, *On absolutely* Δ^1_2 *operations*, ***Fundamenta Mathematicae***, vol. 121 (1984), pp. 239–250.

[46] ERNEST SCHIMMERLING, *Woodin cardinals, Shelah cardinals, and the Mitchell-Steel core model*, ***Proceedings of the American Mathematical Society***, vol. 130 (2002), pp. 3385–3391.

[47] ERNEST SCHIMMERLING and JOHN STEEL, *Fine structure for tame inner models*, ***The Journal of Symbolic Logic***, vol. 61 (1996), pp. 621–639.

[48] ———, *The maximality of the core model*, ***Transactions of the American Mathematical Society***, vol. 351 (1999), pp. 3119–3141.

[49] ERNEST SCHIMMERLING and MARTIN ZEMAN, *Characterization of* $\square_\kappa$ *in core models*, ***Journal of Mathematical Logic***, vol. 4 (2004), pp. 1–72.

[50] RALF SCHINDLER, *Core models in the presence of Woodin cardinals*, ***The Journal of Symbolic Logic***, vol. 71 (2006), pp. 1145–1154.

[51] RALF SCHINDLER and JOHN STEEL, *The self-iterability of $L[\vec{E}]$*, ***The Journal of Symbolic Logic***, vol. 74 (2009), pp. 751–779.

[52] ———, ***The Core Model Induction***, 2014, preliminary draft available at www.math.uni-muenster.de/logik/Personen/rds.

[53] RALF SCHINDLER, JOHN STEEL, and MARTIN ZEMAN, *Deconstructing inner model theory*, ***The Journal of Symbolic Logic***, vol. 67 (2002), pp. 721–736.

[54] RALF SCHINDLER and MARTIN ZEMAN, *Fine structure*, ***Handbook of Set Theory, Vols. 1, 2, 3*** (M. Foreman and A. Kanamori, editors), Springer, Dordrecht, 2010, pp. 605–656.

[55] FARMER SCHLUTZENBERG, ***Measures in Mice***, Ph.D. thesis, Berkeley, 2007.

[56] ———, *Reconstructing copying and condensation*, (2015), preprint available at https://sites.google.com/site/schlutzenberg/home.

[57] ———, *Iterability for (transfinite) stacks*, ***Journal of Mathematical Logic***, vol. 21 (2021).

[58] ———, *Definability of E in self-iterable mice*, (March 2019), preprint available at https://sites.google.com/site/schlutzenberg/home.

[59] ———, *Fine structure from normal iterability*, (November 2019), preprint at arXiv:2011.10037v2 [math.LO].

[60] FARMER SCHLUTZENBERG and JOHN STEEL, *Comparison of fine structural mice by coarse iteration*, ***Archive for Mathematical Logic***, vol. 53 (2014), pp. 539–559.

[61] FARMER SCHLUTZENBERG and NAM TRANG, *Scales in hybrid mice over $\mathbb{R}$*, (2016), preprint available at https://sites.google.com/site/schlutzenberg/home.

[62] JACK SILVER, *Measurable cardinals and Δ^1_3 well orderings*, ***Annals of Mathematics***, vol. 94 (1971), pp. 414–446.

[63] BENJAMIN SISKIND, ***Aspects of Martin's Conjecture and Inner Model Theory***, Ph.D. thesis, Berkeley, 2021.

[64] BENJAMIN SISKIND and JOHN STEEL, *Full normalization for mouse pairs*, https://arxiv.org/abs/2207.11065, (2022).

[65] JOHN R. STEEL, $\mathrm{HOD}^{L(\mathbb{R})}$ *is a core model below* θ, ***The Bulletin of Symbolic Logic***, vol. 1 (1995), pp. 75–84.

[66] ———, ***The Core Model Iterability Problem***, Lecture Notes in Logic, vol. 8, Springer-Verlag, Berlin, 1996.

[67] ———, *Local K^c constructions*, ***The Journal of Symbolic Logic***, vol. 72 (2007), pp. 721–737.

[68] ———, *Derived models associated to mice*, ***Computational Prospects of Infinity. Part I. Tutorials***, Lecture Notes Series, Institute of Mathematical Sciences, National University of Singapore, vol. 14, World Scientific Publishers, Hackensack, NJ, 2008, pp. 105–193.

[69] ———, *The derived model theorem*, ***Logic Colloquium 2006*** (Cooper et al., editors), Lecture Notes in Logic, vol. 32, Association for Symbolic Logic, Chicago, IL, 2009, pp. 280–327.

[70] ———, *An outline of inner model theory*, ***Handbook of Set Theory, Vols. 1, 2, 3*** (M. Foreman and A. Kanamori, editors), Springer, Dordrecht, 2010, pp. 580–645.

[71] ———, *A theorem of Woodin on mouse sets*, ***Ordinal Definability and Recursion Theory. The Cabal Seminar, Volume III*** (A. S. Kechris et al., editors), Lecture Notes in Logic, vol. 43, Association for Symbolic Logic, Ithaca, New York, 2016, pp. 243–256.

[72] ———, HOD *in models of determinacy*, ***Ordinal Definability and Recursion Theory. The Cabal Seminar, Volume III*** (A. S. Kechris et al., editors), Lecture Notes in Logic, vol. 43, Association for Symbolic Logic, Ithaca, New York, 2016, pp. 3–48.

[73] ———, *Mouse pairs and Suslin cardinals*, ***Proceedings of the 2019 workshop "Higher Recursion Theory and Set Theory"*** (C. T. Chong et. al., editor), Lecture Notes Series, Institute for Mathematical Sciences, National University of Singapore, World Scientific Publishing Co., 2022.

[74] ———, *The comparison lemma*, (2022), preprint available at www.math.berkeley.edu/~steel.

[75] ———, *Remarks on a paper by Sargsyan*, (April 2013), notes available at www.math.berkeley.edu/~steel.

[76] ———, *Remarks on a hod mouse construction*, (August 2014), notes available at www.math.berkeley.edu/~steel.

[77] ———, *Local HOD computation*, (July 2016), notes available at www.math.berkeley.edu/~steel.

[78] ———, LSA *from least branch hod pairs*, (July 2016), notes available at www.math.berkeley.edu/~steel.

[79] ———, *Hod mice below* LST^-, (October 2013), notes available at www.math.berkeley.edu/~steel.

[80] ———, *Hod pair capturing and short tree strategies*, (October 2016), notes available at www.math.berkeley.edu/~steel.

[81] JOHN R. STEEL and NAM TRANG, *Condensation for mouse pairs*, https://arxiv.org/abs/2207.03559, (2022).

[82] JOHN R. STEEL and W. HUGH WOODIN, HOD *as a core model*, ***Ordinal Definability and Recursion Theory. The Cabal Seminar, Volume III*** (A. S. Kechris et al., editors), Lecture Notes in Logic, vol. 43, Association for Symbolic Logic, Ithaca, New York, 2016, pp. 257–348.

[83] PHILIP D. WELCH, Σ^* *fine structure*, ***Handbook of Set Theory, Vols. 1, 2, 3*** (M. Foreman and A. Kanamori, editors), Springer, Dordrecht, 2010, pp. 657–736.

[84] W. HUGH WOODIN, *Suitable extender models I*, ***Journal of Mathematical Logic***, vol. 10 (2010), pp. 101–339.

[85] ———, *In search of Ultimate L: the 19th Midrasha Mathematicae lectures*, ***The Bulletin of Symbolic Logic***, vol. 23 (2017), pp. 1–109.

[86] Martin Zeman, ***Inner Models and Large Cardinals***, De Gruyter Series in Logic and its Applications, vol. 5, Walter De Gruyter, Berlin, 2001.

[87] ———, *Dodd parameters and the λ-indexing of extenders*, ***Journal of Mathematical Logic***, vol. 04 (2004), pp. 73–108.

INDEX